Digital

Saunders College Publishing Electrical Engineering, formerly The Holt, Rinehart and Winston Series in Electrical Engineering

M. E. Van Valkenburg, Senior Consulting Editor
Adel S. Sedra, Series Editor/Electrical Engineering
Michael R. Lightner, Series Editor/Computer Engineering

ALLEN AND HOLBERG
CMOS Analog Circuit Design

BELANGER, ADLER AND RUMIN
Introduction to Circuits with Electronics: An Integrated Approach

BOBROW
Elementary Linear Circuit Analysis, 2/e

BOBROW
Fundamentals of Electrical Engineering

BUCKMAN
Guided Wave Photonics

CHEN
Linear System Theory and Design

CHEN
System and Signal Analysis

COMER
Digital Logic and State Machine Design, 2/e

COMER
Microprocessor-Based System Design

COOPER AND McGILLEM
Probabilistic Methods of Signal and System Analysis, 2/e

GHAUSI
Electronic Devices and Circuits: Discrete and Integrated

GUNGOR
Power Systems

GURU/HIZROGLU
Electric Machinery and Transformers

HOSTETTER, SAVANT AND STEPHANI
Design of Feedback Control Systems, 2/e

HOSTETTER
Digital Control System Design

HOUTS
Signal Analysis in Linear Systems

ISHII
Microwave Engineering

JONES
Introduction to Optical Fiber Communication Systems

KARNI AND BYATT
Mathematical Methods in Continuous and Discrete Systems

KENNEDY
Operational Amplifier Circuits: Theory and Application

KUO
Digital Control Systems, 2/e

LASTMAN AND SINHA
Microcomputer-Based Numerical Methods for Science and Engineering

LATHI
Modern Digital and Analog Communication Systems, 2/e

LEVENTHAL
Microcomputer Experimentation with the IBM PC

LEVENTHAL
Microcomputer Experimentation with the Intel SDK-86

LEVENTHAL
Microcomputer Experimentation with the Motorola MC6800 ECB

McGILLEM AND COOPER
Continuous and Discrete Signal and System Analysis, 3/e

MIRON
Design of Feedback Control Systems

NAVON
Semiconductor Microdevices and Materials

PAPOULIS
Circuits and Systems: A Modern Approach

RAMSHAW AND VAN HEESWIJK
Energy Conversion

SADIKU
Elements of Electromagnetics

SCHWARZ
Electromagnetics for Engineers

SCHWARZ AND OLDHAM
Electrical Engineering: An Introduction

SEDRA AND SMITH
Microelectronic Circuits, 3/e

SINHA
Control Systems

VAN VALKENBURG
Analog Filter Design

VRANESIC AND ZAKY
Microcomputer Structures

WARNER AND GRUNG
Semiconductor Device Electronics

WASER AND FLYNN
Introduction to Arithmetic for Digital Systems Designers

WOLOVICH
Robotics: Basic Analysis and Design

YARIV
Optical Electronics, 4/e

Digital Control Systems

Second Edition

Benjamin C. Kuo
University of Illinois at Urbana-Champaign

Saunders College Publishing
A Harcourt Brace Jovanovich College Publisher

Ft. Worth Philadelphia San Diego New York Orlando Austin
San Antonio Toronto Montreal London Sydney Tokyo

THIS BOOK IS PRINTED ON ACID-FREE, RECYCLED PAPER 

Text Typeface: Times Roman
Compositor: Techset Composition Limited
Senior Acquisitions Editor: Barbara Gingery
Assistant Editor: Laura Shur
Project Editor: Spectrum Publisher Services, Inc.
Copy Editor: Ellen Pearce
Manager of Art and Design: Carol Bleistine
Art Director: Doris Bruey
Text Designer: Rebecca Lemna
Cover Designer: Lawrence R. Didona
Director of EDP: Tim Frelik
Production Manager: Bob Butler

Printed in the United States of America
Library of Congress Catalog Card Number: 91-41422
Discrete and Integrated Circuit Electronics
ISBN: 0-03-012884-6

1234 039 987654321

Preface

This text covers the theory and applications of digital control systems and is an extensive revision of the first edition published in 1980. The author has written several books since 1963 on the subject of digital and sampled-data control systems. The first text, *Analysis and Synthesis of Sampled-Data Control Systems,* was published in 1963 by Prentice-Hall. The second book, *Discrete-Data Control Systems* was published in 1970 by the same publisher. The book represents a completely new approach from the 1963 edition by concentrating on the state variable approach and modern control theory. The first edition of *Digital Control Systems* published by Holt, Reinhart and Winston in 1980 was a complete revision of the 1970 text.

TEXT APPROACH AND ORGANIZATION

In the 1960s, at the height of the expansion of the aerospace industries and space exploration, development of sampled-data and digital control systems theory enjoyed the most prolific period. A large number of papers and books were published during this period on discrete-data and sampled-data control systems. In recent years although research and development activities in the aerospace sector have matured, the advances made in microprocessors, microcomputers and digital signal processors (DSPs) have infused new life to the growth of digital control systems. The popularity of microprocessor and DSP control has put new emphasis on digital control systems theory. Because microprocessors are slow digital computers, and are usually equipped with a limited wordlength, it is necessary to point out these constraints and the effects of time delays and amplitude quantization when designing a digital control system.

This book is intended to be used as an introductory college text for a senior or graduate course on digital control systems, or a reference for practicing engineers. As a college text, the material presented should be more than adequate for a one-semester course. This will give the instructor some flexibility in selecting the subjects he or she would like to cover. All the material has been class-tested by the author teaching at the University of Illinois at Urbana-Champaign for many years. As a reference, the book is written with self-study in mind.

Although most contemporary control systems encountered in the industry employ digital control, the fact is that the physical world is still an analog one. For example, most of the physical processes to be controlled are still analog systems. Therefore, the first course in control systems would probably be on continuous-data systems theory before digital control theory can be taught. As a result, it is assumed that the reader already has a knowledge on the basic principles of continuous-data control systems. Background on the basics of linear systems theory such as, matrix algebra, differential equations, Laplace transform, is essential.

NEW TO THIS EDITION

The second edition contains many new features and material not found in the first edition. Many new subjects, especially on designs, have been added or enhanced. Important highlights of the new revision are:

1. Separate discussions on controllability, observability, and stability.
2. Expanded discussions on sampling period selection.
3. Emphasis on computer-aided solutions.
4. A new and simpler approach to Nyquist criterion of stability.
5. Expanded coverage on design of digital control systems to include disturbance rejection, sensitivity considerations, zero-ripple deadbeat-response design.

CONTENT OVERVIEW

Chapter 1 gives a general introduction to the subject of digital control systems. Some typical examples of digital control systems are presented. **Chapter 2** covers signal conversion, signal processing, and the sampling operation. The contents in this chapter gives the motivation on the mathematical and analytical treatment of digital data and signals in control systems. **Chapter 3** covers z-transforms, modified z-transform, including nonuniform- and multirate-sampled systems. **Chapter 4** treats the subjects of pulse transfer function, block diagrams and signal flow graphs. **Chapter 5** gives a thorough coverage on the state-variable technique applied to digital control systems. The subjects on controllability, observability, and stability are discussed in **Chapter 6**. The analytical method of testing stability is deemphasized due to the fact that most stability tests with unknown system parameters are extremely tedious to be carried out by hand, and the stability of linear systems can always be studied by solving the characteristic polynomial by root-solving program of a computer. **Chapter 7** is on time-domain and z-domain analyses. Comparisons between continuous-data and sampled-data system responses are made. The relationships between the pole-zero configurations in the z-plane and the time response are made. Emphasis is placed on the prediction of time response from pole-zero locations. Steady-state analysis and the root locus method completes the chapter. **Chapter 8** is on frequency-domain analysis of digital control systems. The conventional methods such as the Nyquist criterion, Bode plot, Nichol's chart, gain margin and phase margin, are all extended to digital control systems. In particular, the treatment of the Nyquist criterion is new. **Chapter 9** is devoted to digital simulation and digital redesign. Frequency warping and prewarping are discussed. **Chapter 10** represents a comprehensive coverage on the design of linear digital control systems. The subjects covered are: PID controller, phase-lead and phase-lag controllers, feedforward control, disturbance rejection, pole-zero cancellation, sensitivity considerations, deadbeat-response design, notch controllers, etc. The approach is to integrate the time-domain, frequency-domain, and z-plane design methods, so that a given design problem is investigated with all these optional design approaches. State feedback and dynamic output feedback are also treated. **Chapter 11** is devoted to optimal control; the topics include maximum principle, optimal linear regulator design, dynamic programming, sampling-period sensitivity, and digital observers. The last chapter, **Chapter 12**, is on microprocessor and DSP control. Since the text is not designed to teach assembly language and how microprocessors and DSPs are programmed, the chapter is devoted to the introduction of single-board controllers, single-chip microprocessor and DSP control, and development systems.

LEARNING AIDS

Each chapter begins with **keywords and topics** that give a overview of the key topics covered in the chapter. Whenever applicable a chapter begins with an introduction that gives the motivation and background on the subjects treated in the chapter. Illustrative examples, many of them derived from practical systems, are found throughout the text. An abundant number of exercise problems are found at the end of each chapter starting from Chapter 2.

The second edition is complemented by the software, **DCSP**, **Digital Control Systems Programs**. **DCSP** is designed to provide computer-aided learning to the second edition. Practically all the illustrative examples in the book were solved with DCSP. The reader should find DCSP helpful in learning the material in the text, and particularly in solving the problems at the end of the chapters. While there are an abundance of software available commercially for the analysis and design of control systems, it is important to keep in mind that computer solutions cannot substitute for the understanding of the fundamentals. More details on the contents of DCSP are found under **Description of the DCSP Software Package**.

INSTRUCTOR'S MANUAL

The Instructor's Manual contains detailed solutions to all the problems in the text.

ACKNOWLEDGMENTS

The author is grateful to many of his graduate students whose indirect assistance in the form of thesis work and classroom discussions has contributed in many ways to the preparation of this book.

Appreciation also goes to the reviewers: Jules Thibault, Laval University (CANADA); Yokendra Kakad, University of North Carolina, Charlotte; Frederic Ham, Florida Institute of Technology; and Can Isik, Syracuse University. Their comments and suggestions were very helpful in the preparation of the final manuscript.

To the professional staff of Saunders College Publishing: Barbara Gingery, Senior Acquisition Editor, Laura Shur, Assistant Editor, and Bob Butler, Production Manager, I extend a special note of appreciation. Special thanks also goes to Elise Oranges of the Spectrum Publisher Services who did an excellent job in handling the production of this book.

Finally, the author wishes to pay tribute to his family, especially his wife, Margaret, for her support and tolerance in many ways to this and many other of my professional projects for the past 37 years. Without her moral support, these projects could never have been accomplished.

B. C. Kuo
Champaign, Illinois

Descriptions of the DCSP Software Package

The DCSP (Digital Control Systems Programs) package is a collection of programs that are designed to accompany *Digital Control Systems*, second edition. The programs found in the DCSP package are capable of solving reasonably complex linear digital control systems rapidly. All the illustrative examples as well as the problems in the text were solved using the programs in the DCSP.

The DCSP software programs are prepared for use on the IBM PC, XT, AT, PS/2®, or any compatible computer with DOS 3.0 or higher. The programs are supported by high-resolution color display and graphics that require IBM EGA or VGA or equivalent display adapters. The programs utilize a self-contained WINDOW environment, and no external WINDOW program is needed to run the programs. The programs also support a mouse, so that data and command entries and menu selections can all be executed by using a mouse. Results of the computer analysis and design can be printed on an Epson/IBM or any compatible dot-matrix printer. The graphs on the monitor screen can be dumped on an Epson/IBM dot-matrix printer or a LaserJet printer via screen dump.

The programs contained in the DCSP are listed as follows:

LINSYSD—state space analysis of linear sampled-data systems.
POLYR—polynomial root solving program.
PFFZ—partial-fraction expansion of *z*-domain transfer functions.
RL—root locus diagram of *z*-domain equations.
FZ—frequency-domain analysis of linear discrete-data systems.

SD—analysis and design of sampled-data control systems. Design methods include PID control, deadbeat-response design, selection of numerical-integration method of implementation of integral control, arbitrary digital controller.

INVZ—time response of digital control systems in the z-domain.

RWT—r-transformation and w-transformation of z-domain transfer functions.

Please use the Order Form at the end of the book for ordering the DCSP. Complimentary copies are provided to instructors who have adopted the text. To order DCSP for classes, send order to

SRL, Inc.
P.O. Box 2772, Station A
Champaign, IL 61825-2772

Contents

Introduction

KEYWORDS AND TOPICS

discrete-data systems • digital control systems • sampled data • advantages of sampled-data and digital control • examples of discrete-data and digital control systems

1-1 INTRODUCTION

In recent years significant progress has been made in the analysis and design of discrete-data and digital control systems. These systems have gained popularity and importance in industry due in part to the advances made in digital computers for controls and, more recently, in microprocessors (MP) and digital signal processors (DSP).

Discrete-data and digital control systems differ from the conventional continuous-data or analog systems in that the signals in one or more parts of these systems are in the form of either pulse trains or numerical codes. The terms *sampled-data control systems*, *discrete-data control systems*, and *digital control systems* have all been used loosely and interchangeably in the control systems literature. Strictly speaking, sampled data are pulse-amplitude modulated signals and are obtained by some means of sampling an analog signal. A pulse-amplitude modulated signal is often presented in the form of a pulse train with signal information carried by the amplitudes of the pulses. Digital data usually are those signals generated by digital computers or digital transducers; they are often in some kind of digitally coded form. It will be shown later that practical systems found in industry often contain analog, sampled, as well as digital data. Therefore, in this text we shall use the term *discrete-data systems* in a broad sense to describe all systems having some form

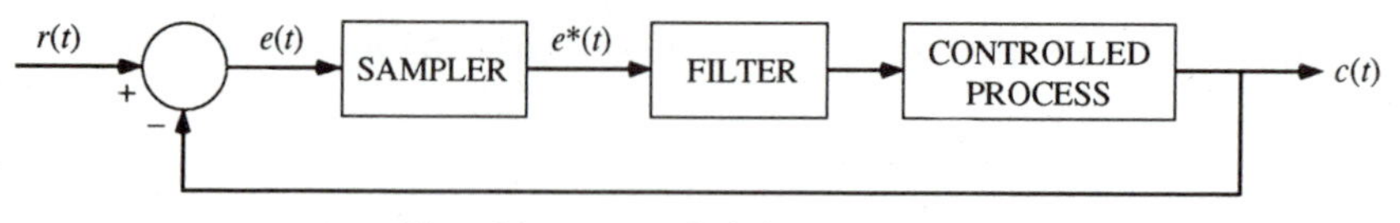

Figure 1-1. Closed-loop sampled-data control system.

of digital or sampled signals. Furthermore, the existing analytical and design methods are essentially the same whether the system contains sampled or digitally coded data.

1-1-1 Basic Elements of a Discrete-Data Control System

Figure 1-1 shows the basic elements of a typical closed-loop control system with sampled data. The sampler simply represents a device or operation that outputs a pulse train. No information is transmitted between two consecutive pulses. Figure 1-2 illustrates typical input and output of a sampler. A continuous input signal $e(t)$ is sampled by the sampler, and the output is a sequence of pulses. In the illustrated case, the sampler is assumed to have a uniform sampling rate. The magnitudes of the pulses at the sampling instants represent the values of the input signal $e(t)$ at the corresponding instants. In general, the sampling scheme has many variations; some of these are the *periodic*, *cyclic-rate*, *multirate*, *random*, and *pulsewidth-modulated* samplings. The most common type of sampling found in practical systems is the single-rate periodic sampling, such as that shown in Fig. 1-2.

The filter located between the sampler and the controlled process is used for the purpose of smoothing, since most controlled processes, such as the ones involving a conventional ac or dc motor, are naturally designed and constructed to receive analog signals.

The block diagram of a typical digital control system is shown in Fig. 1-3. The existence of digitally coded signals, such as binary-coded signals, in certain parts of the system requires the use of digital-to-analog (D/A) and analog-to-digital (A/D) converters. The digital computer block in Fig. 1-3 can be a

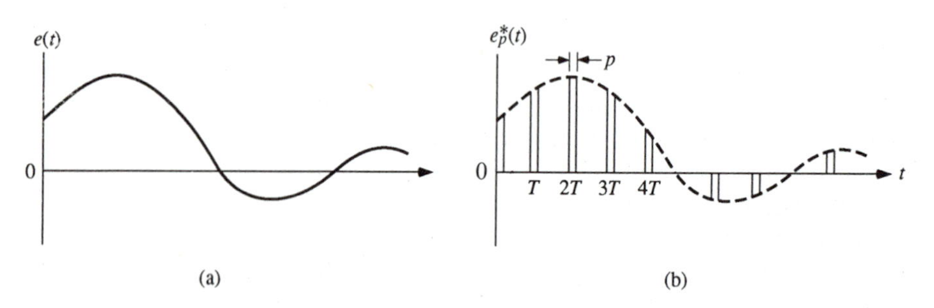

Figure 1-2. (a) Continuous-data input to sampler. (b) Discrete-data output of sampler.

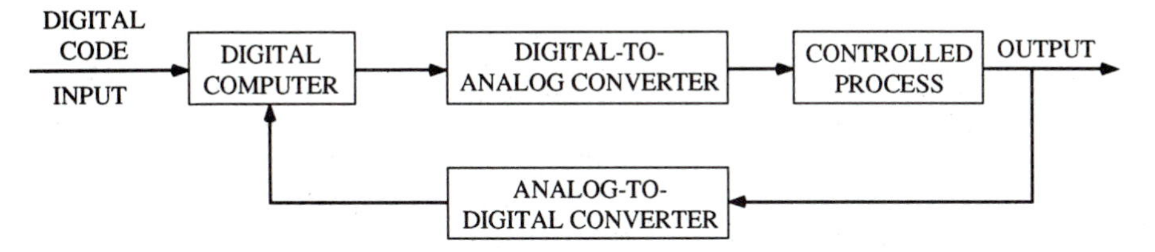

Figure 1-3. A typical digital control system.

special-purpose digital computer, a microprocessor, or a digital signal processor. Although there are basic differences between the hardware structures and components between sampled-data and digital control systems, we shall show that, from an analytical standpoint, both types of systems are treated by the same analytical tools.

1-1-2 Advantages of Discrete-Data Control Systems

The use of sampled data in control systems can be traced back at least 70 years. Some of these early applications of sampled data were for improving the performance of the control system in one form or another. For instance, in the chopper-bar galvanometer, described by Oldenbourg and Sartorious [1] and shown in Fig. 1-4, the sampling operation produces greater system sensitivity (gain) to a low-level input signal. With reference to Fig. 1-4, a small signal is normally applied to the galvanometer coil. The chopper bar is lowered periodically, and the projected pointer of the galvanometer causes the load to be driven in proportion to the signal strength. The torque applied to the load shaft is thus determined by the chopper-bar driver, rather than just by the torque developed in the galvanometer coil.

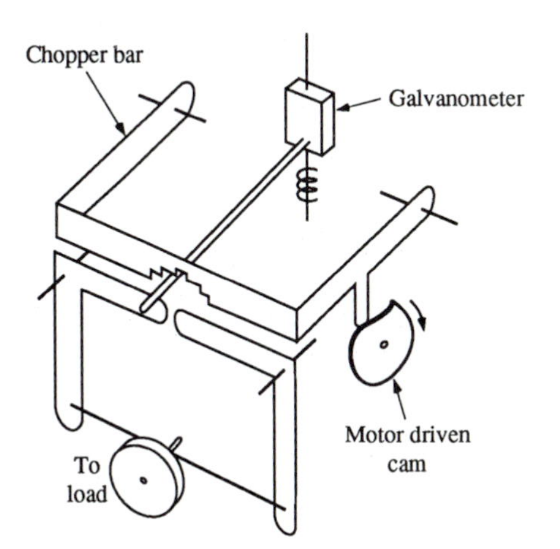

Figure 1-4. Chopper-bar galvanometer.

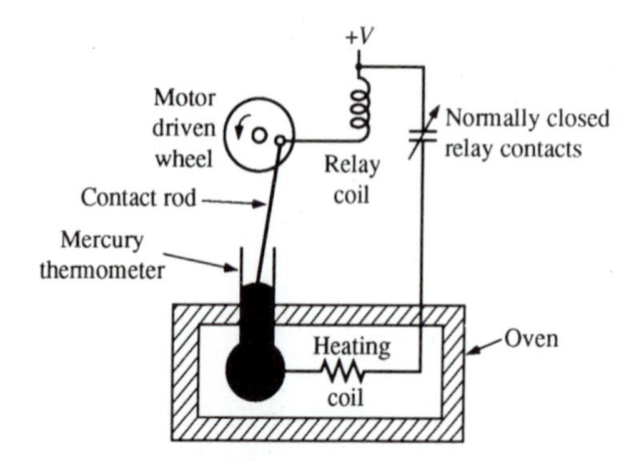

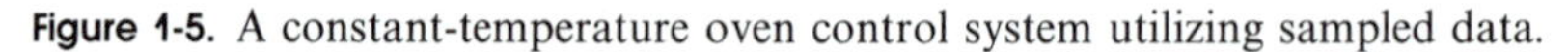
Figure 1-5. A constant-temperature oven control system utilizing sampled data.

Another early application of the sampling concept in control systems is the constant-temperature oven devised by Gouy [2]. The system consists of the elements shown in Fig. 1-5. The purpose of this system is to maintain a constant temperature in the oven at all times. Whenever the electrical contact rod is immersed in the mercury, current flows in the relay coil, causing the relay to open and thus interrupt the heating current. Since the contact is periodically dipped into the mercury, the heating current consists of a sequence of pulses. Furthermore, since the time that the contact is immersed in the mercury depends on the level of the mercury, and the level in turn depends on the temperature of the oven, the pulsewidths of the heating current are varied in proportion to the temperature of the oven, but the pulses all have equal amplitudes. A typical set of signals in the sampled-data temperature control system is shown in Fig. 1-6. In contrast to the conventional sampling scheme such as the one described in Fig. 1-2, in which the amplitude of the signal is sampled and transmitted by the amplitude of a pulse, the sampled data shown in Fig. 1-6 have a constant pulse amplitude, but with the signal information carried by the widths of the pulses. Therefore, we see that, in practice, there are many ways of sampling a signal, or perhaps more appropriately, of representing a signal in sampled form.

The two examples just given merely illustrate the early applications of sampling in control systems. To understand fully the merits and advantages of

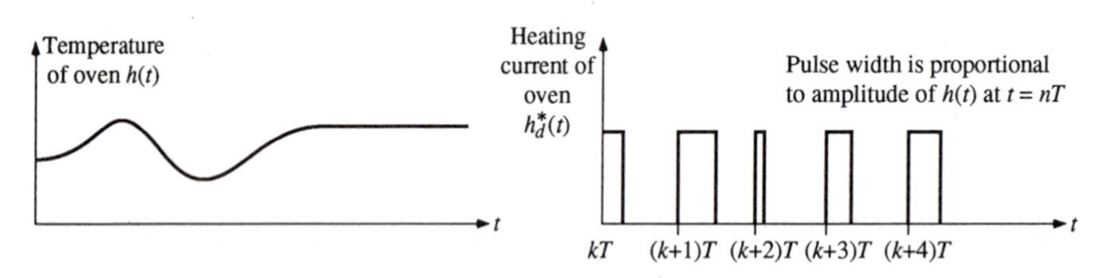

Figure 1-6. Input and output of pulsewidth sampler.

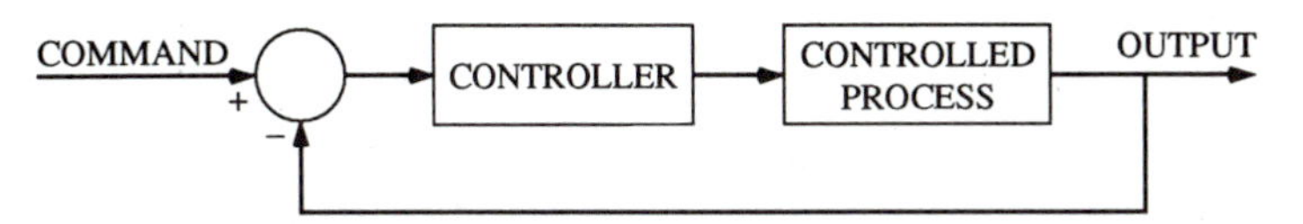

Figure 1-7. Basic elements of an analog control system.

using sampling and digital data in control systems, one should ask, *Why sampled data and digital control*? In other words, what are the advantages and characteristics of sampled data and digital control? We must first recognize that many physical systems have inherent sampling, or their behavior can be described by discrete-data or digital models. For instance, in a radar tracking system the signals transmitted and received by the system are in the form of pulse trains. The scanning operation of the radar performs the function of a sampler, converting both the azimuth and the elevation information to sampled data. Numerous phenomena, social, economic, or biological systems, exist whose dynamics can be modeled by discrete-data models.

To address the advantages and characteristics of discrete-data systems, we must first look at the characteristics of continuous-data control systems which have been in existence for many years. Figure 1-7 shows the block diagram containing the basic elements of a single-loop continuous-data control system. The controller is usually an electronic circuit that operates on an analog signal and outputs the same type of signal. The advantage with the analog controller is that the system operates in real time and is capable of a very high bandwidth. It is equivalent to having an infinite sampling frequency, so that the controller is effective at all times. The disadvantages of the analog controller are that its elements are usually hard-wired, so that their characteristics are fixed, making it more difficult to make design changes, and component aging and sensitivity to environmental changes can be quite severe. Analog components are also more susceptible to noise problems. Some of the advantages of digital control are listed as follows:

1. Digital components are less susceptible to aging and environmental variations.
2. They are less sensitive to noise and disturbance.
3. Digital processors are more compact and lightweight. Single-chip MPs and DSPs can be made very versatile and powerful for control applications.
4. The costs of MPs and DSPs are continuously going down.
5. MPs and DSPs allow more flexibility in programming: changing a design does not require an alteration in the hardware.
6. They are more reliable.
7. They provide improved sensitivity to parameter variations.

The chopper-bar galvanometer system illustrated in Fig. 1-4 is an example on how sampling can improve sensitivity. In this case, a small signal is amplified for control purposes through the sampling operation.

One important advantage of digital controllers, as summarized above, is that they are more versatile than analog controllers. The program that characterizes a digital controller can be modified easily to accommodate design changes, or adaptive performances, without any alteration to the hardware. Digital components in the form of chips, electronic parts, transducers, and encoders are often more reliable, more rugged in construction, and more compact in size than their analog counterparts. These and other significant distinctions are causing many analog control system applications to be converted to digital ones.

Digital control systems are not without disadvantages when compared with the analog systems, however. Some of the important disadvantages are as follows.

1. Limitations on computing speed and signal resolution due to the finite wordlength of the digital processor. In contrast, analog controllers operate in real time, and the resolution is theoretically infinite.
2. The finite wordlength of the digital processor often translates into system instability in the form of limit cycles in closed-loop systems.
3. The limitation on computing speed causes time delays in the control loop which may cause instability in closed-loop systems.

1-2 EXAMPLES OF DISCRETE-DATA AND DIGITAL CONTROL SYSTEMS

In this section we shall illustrate several examples of discrete-data and digital control systems. The objective is simply to show some of the essential components of discrete-data and digital control systems.

1-2-1 A Simplified Single-Axis Autopilot Control System

Figure 1-8(a) shows the block diagram of a simplified single-axis (pitch, yaw, or roll) analog autopilot control of an aircraft or missile. This is a typical analog or continuous-data control system in which the signals can all be represented as functions of the continuous-time variable t. The objective of the control is that the attitude of the airframe follow the command signal. The rate loop is incorporated here for the improvement of system stability. Instead of using the analog controller as shown in Fig. 1-8(a), a digital controller with the necessary A/D and D/A converters can be used for the same objective, as shown in Fig. 1-8(b). Since all the components of the system other than the digital controller are still analog, the A/D and D/A converters are necessary for signal conversions.

Figure 1-9 shows the digital autopilot control system in which the position and rate information are obtained by digital transducers, and the operations are represented on the block diagram by sample-and-hold devices. The sampler samples the analog signal at some uniform sampling rate, and the hold device

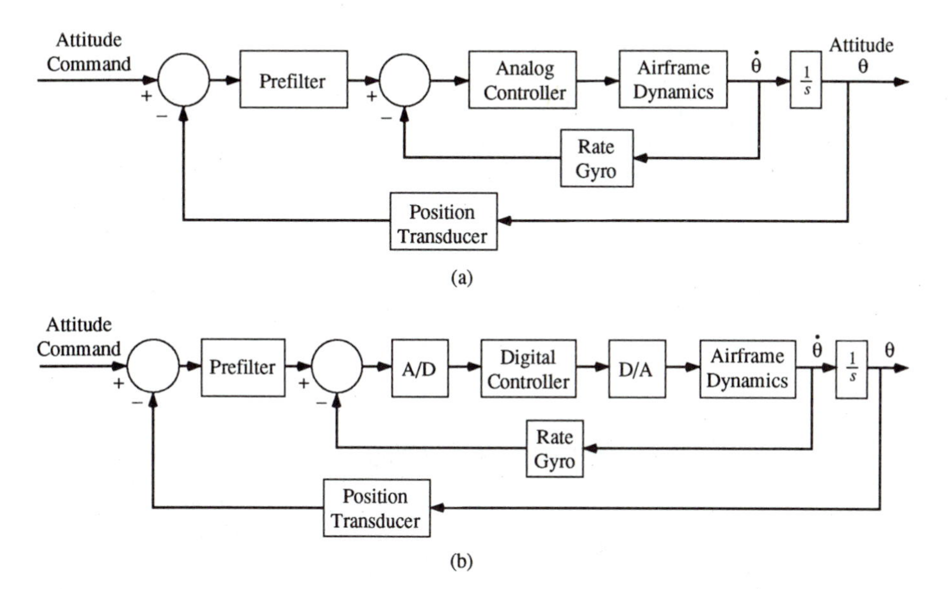

Figure 1-8. (a) A simplified single-axis autopilot control system with analog data. (b) A simplified single-axis autopilot control system with digital data.

simply holds the value of the pulse signal until the next sample comes along. Figure 1-9 illustrates the situation in which the two samplers have different sampling periods T_1 and T_2. In general, if the rate of variation of the signal in one loop is very much less than that of the other loop, the sampling period used for the slower loop can be longer. The system shown in Fig. 1-9 has samplers with different sampling periods and is called a *multirate sampled-data system.*

One of the advantages of using sampling and multirate sampling is that certain expensive components of the system can be used on a time-shared basis.

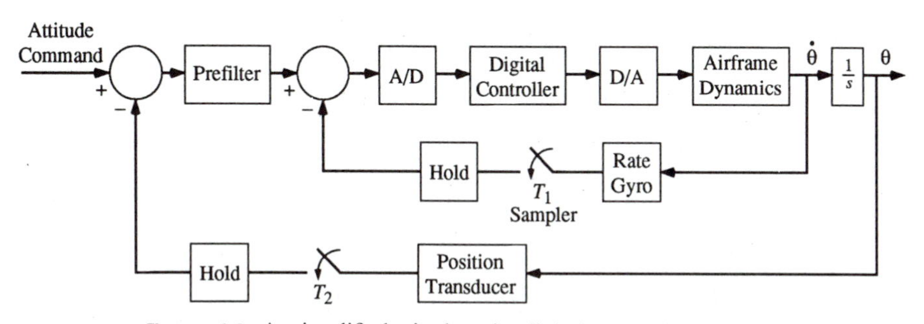

Figure 1-9. A simplified single-axis digital autopilot control system with multirate sampling.

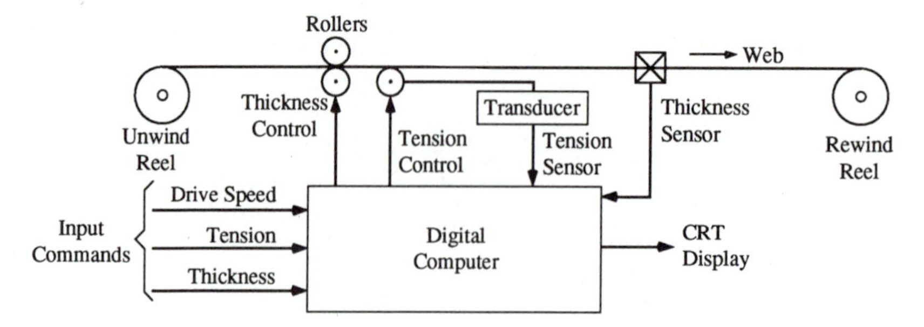

Figure 1-10. A digital computer-controlled rolling mill regulating system.

1-2-2 A Digital Computer-Controlled Rolling Mill Regulating System

Many industrial processes are controlled and monitored by digital computers and digital transducers. Practically all the modern steel rolling mills are regulated and controlled by digital computers. Figure 1-10 illustrates the basic elements of such a system. Figure 1-11 shows a block diagram of the thickness-control portion of the system.

1-2-3 A Digital Controller for a Turbine and Generator

Figure 1-12 shows the block diagram and the essential elements of a mini-computer system used for speed and voltage control as well as data acquisition of a turbine–generator unit. The D/A converter forms the interface between the digital computer and the speed and voltage controls. The data-acquisition system measures such variables as the generator speed, rotor angle, terminal voltage, field and armature current, and real and reactive power. Some of these variables may be measured by digital transducers whose outputs are then digitally multiplexed and sent to the computer, as shown in Fig. 1-13. The quantities measured by analog transducers are first sent through an analog multiplexer which performs a time-division multiplexing operation between a number of different input signals. Each input-signal channel is sequentially connected to the output of the multiplexer for a specific period of time. The

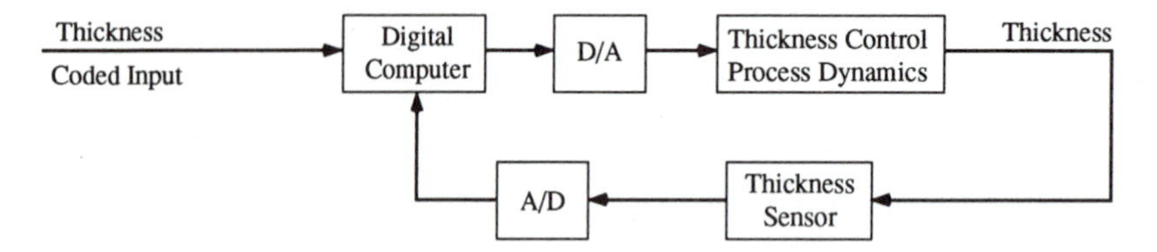

Figure 1-11. Thickness control in a rolling mill regulating system.

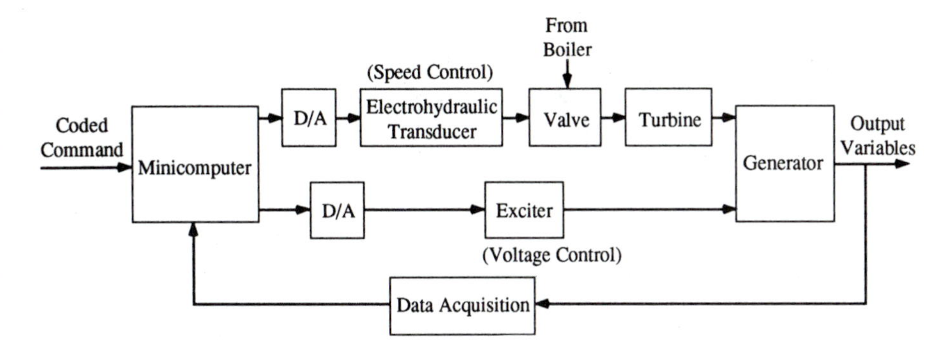

Figure 1-12. Computer control of a turbine and generator unit.

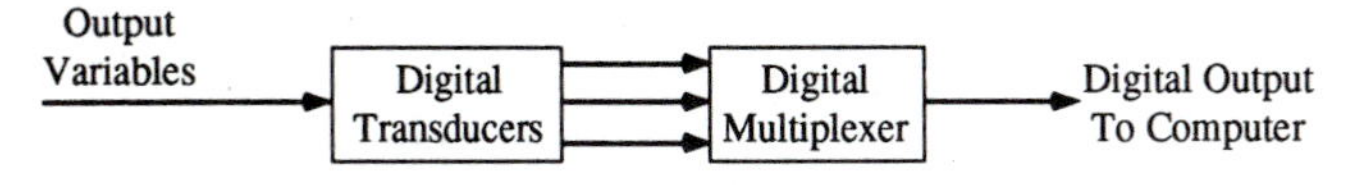

Figure 1-13. A digital data-acquisition system.

system that follows the multiplexer is thus time-shared between a number of signal channels. Figure 1-14 illustrates the data-acquisition system when the system variables are measured by analog transducers. The output of the analog multiplexer is connected to a sample-and-hold device which samples the output of the multiplexer at a fixed time interval and then holds the signal level at its output until the A/D converter completes its task.

1-2-4 A Step Motor Control System

Occasionally, we may come across a system which contains all-digital elements so that A/D and D/A converters for signal matching are unnecessary. Figure 1-15 shows such a system, which is used for the control of the read–write head of a computer memory disk. The prime mover used in the disk drive system is a step motor driven by pulse commands. The step motor moves one fixed displacement increment in response to each pulse input. Thus, the system may be considered an all-digital data system.

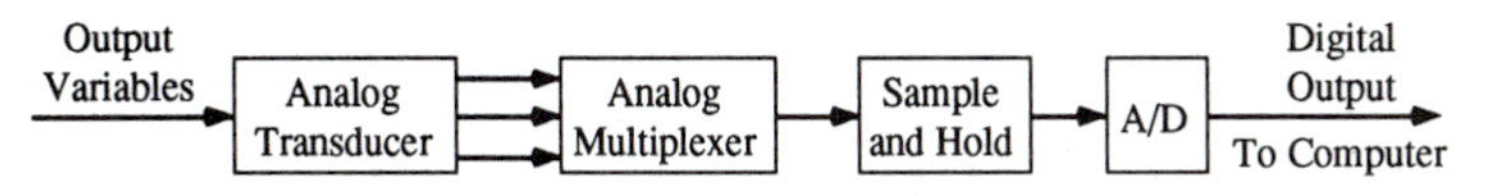

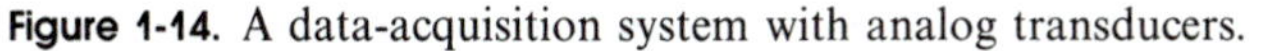
Figure 1-14. A data-acquisition system with analog transducers.

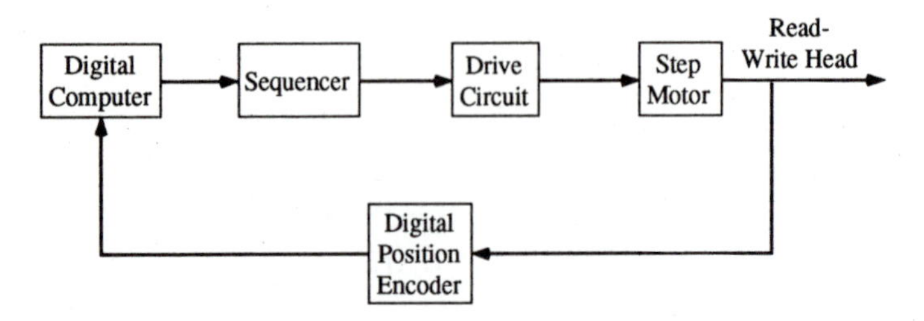

Figure 1-15. A step motor control system for read–write head positioning on a disk drive.

1-2-5 Microprocessor-Controlled Systems

Special-purpose microprocessors and DSPs for control system applications have become quite common in recent years. These compact and economical devices have improved the performance and feasibility of digital control systems dramatically. As an illustrative example, Fig. 1-16 shows the block diagram of a closed-loop dc motor control system that utilizes the GL-1200 controller manufactured by Galil Motion Control. The GL-1200 is a single-chip microprocessor that provides closed-loop position and velocity control of a dc motor. The controller accepts position feedback from a two-channel incremental encoder, and outputs a signal that is proportional to the position error and its derivative. Thus, the closed-loop dc motor control systems can be applied to position and velocity control applications with proportional and derivative (PD) control.

Figure 1-17 illustrates a similar dc motor control system to that of Fig. 1-16 using a National Semiconductor LM628 controller. The LM628 is a single-chip special-purpose microprocessor designed with proportional, integral, and derivative (PID) controls. The parameters of the PID controller can be programmed to meet design performance objectives of position and velocity controls of the dc motor system.

The microprocessors in Figs. 1-16 and 1-17 can all be replaced by a DSP, such as the TMS320C14 by Texas Instruments, which may result in greater

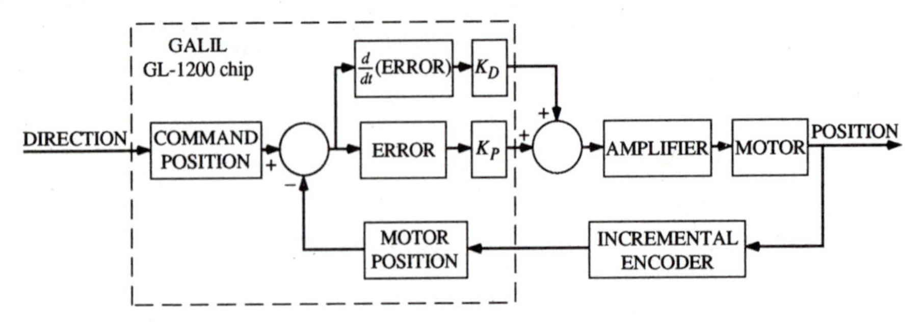

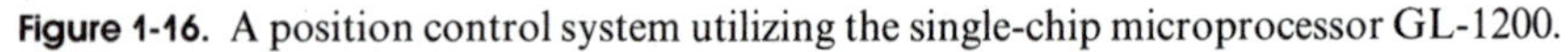
Figure 1-16. A position control system utilizing the single-chip microprocessor GL-1200.

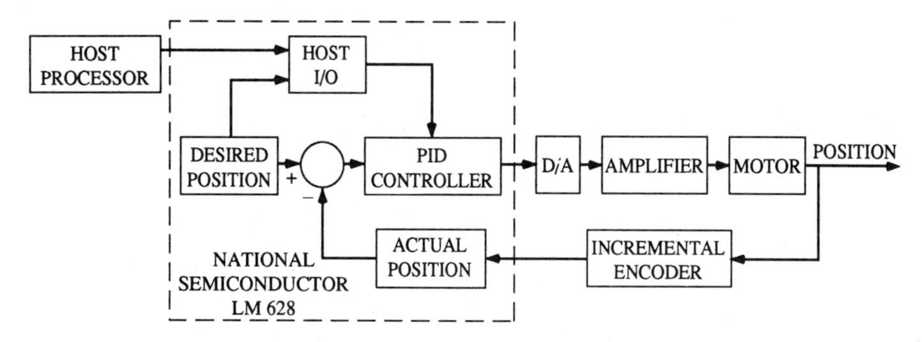

Figure 1-17. A position control system utilizing the single-chip microprocessor LM628.

improvement of system performance. The reason is that DSPs are capable of doing computations at a much faster rate than microprocessors. As a result, DSPs can be programmed to provide filter characteristics with sharp cutoff rates that are impossible to obtain with conventional microprocessors.

1-2-6 A Discrete-Data Model of an Interest-Payment Problem

Thus far we have illustrated digital control systems that are physical in nature. Numerous social and economic systems can also be modeled by a discrete-data system model. Specifically, we recognize that analog dynamic systems are described by difference equations with variables that are functions of the continuous variable t. For discrete-data systems, the system dynamics are described by difference equations. The variables for a discrete-data model are either functions of the discrete-time variable kT ($k = 1, 2, \dots$), where T is a positive real constant, or simply by functions of the discrete variable k.

Let us consider that the amount of capital $p(0)$ is borrowed initially. The interest rate on the unpaid balance is R percent per period. We assume that the principal and interest, U, are to be paid back in N equal payments. Let $p(k)$ denote the amount owed after the kth period. Then the following difference equation can be written for the problem:

$$p(k+1) = (1+R)p(k) - U \tag{1-1}$$

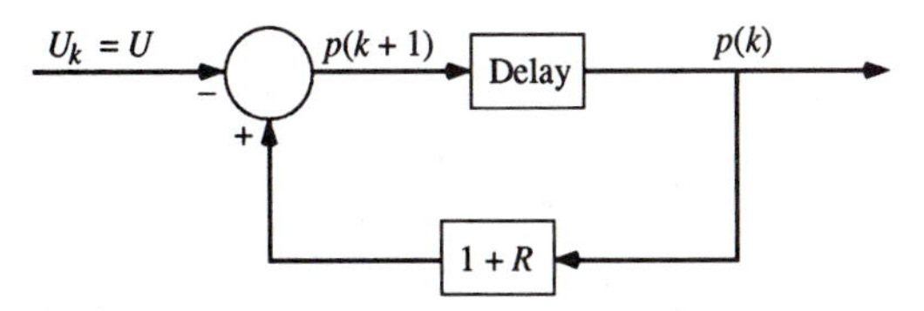

Figure 1-18. A discrete-data system model for an interest-payment problem.

where $p(k+1)$ denotes the amount still owed at the end of the $(k+1)$th period. The boundary conditions are the $p(0)$ given and $p(N) = 0$. The difference equation in Eq. (1-1) can be solved for U by a recursive method or the z-transform method (Chapter 3). The equation is also known as a first-order difference equation. A discrete-data system model is shown in Fig. 1-18 in the form of a block diagram.

References

1. Oldenbourg, R. C., and H. Sartorius, *The Dynamics of Automatic Control*, American Society of Mechanical Engineers, New York, 1948.
2. Gouy, G., "On a Constant Temperature Oven," *J. Physique*, vol. 6, series 3, pp. 479–483, 1897.
3. Bucella, T., and I. Ahmed, "Taking Control with DSPs," *Machine Design*, pp. 73–80, October 1989.

Signal Conversion and Processing

KEYWORDS AND TOPICS

signal processing • A/D and D/A converters • quantization • sample-and-hold • sampling process • sampling theorem • ideal sampler • data reconstruction, and filtering of sample signals

2-1 INTRODUCTION

Digital signal processing offers distinct advantages over analog signal processing. In general, digital circuits are more stable and predictable. They do not require adjustments and tuning in production or on-line operation. As the integrated circuit (IC) and large-scale integration (LSI) technology advances, digital signal processing provides advantages in cost, size, speed, and reliability over its analog counterpart. In 1979, Intel introduced the first commercially available real-time digital signal processor (DSP), the Intel 2920. The DSP is known to be an LSI signal processor that can compute and number crunch faster than a microprocessor. A microprocessor, on the other hand, is more versatile, and can provide more functions for system control.

Digital computers are used by control systems engineers for two primary purposes. The first is the simulation and computation of control systems dynamics. Control engineers often rely on digital computer simulations to conduct the analysis and design of complex control systems that would otherwise be too laborious to carry out by hand. Computer simulations are also used to check or present the results obtained by analytical means.

Another important application of the digital computer in control systems is as controllers or processors. In recent years, with the advancements made in microcomputers, a controller in most cases can be better implemented by a microcomputer or microprocessor. We have already pointed out the advantages of a digital controller against an analog controller in the last chapter. However, most of the processes to be controlled in the real world contain analog elements. Therefore, most of the so-called digital control systems usually contain analog as well as digital signals, and the process of signal conversion is essential so that the digital and analog components can be interfaced in the same system. For instance, the output signal of an analog device, such as an analog sensor, must first undergo an *analog-to-digital* (A/D) conversion before the signal can be processed by a digital controller. Sometimes the analog-to-digital conversion process may involve an *encoding* operation in that the signal is converted into some digital code. Similarly, the digitally coded signal from a digital controller or processor must first be decoded by a *digital-to-analog* (D/A) converter before it can be sent to an analog device for processing. In fact, many microprocessors designed for control purposes, such as the Intel MCS-96, have on-board A/D converters for data conversion. Since the signal from a digital processor occurs not only as code, but also in the form of a number sequence, the signal must also be smoothed out after decoding by a data-reconstruction device or low-pass filter before being sent to the analog process. It was illustrated in Chapter 1 that other signal-processing operations such as *time sharing* require the use of such interfacing devices as the *multiplexer*, *sample-and-hold* (S/H), etc. Because these components are important for signal processing and signal conditioning in digital control systems, they will be described and modeled in this chapter. However, since many texts are devoted to the subjects of signal processing and the hardware aspects, the reader should refer to the literature for a comprehensive coverage of the details of A/D and D/A converters and digital signal processing. The objective of the introductory treatment of A/D and D/A converters, multiplexers, and the sample-and-hold operation in this chapter is to establish the importance of signal processing in discrete-data and digital control systems and, most important, how these components are modeled and treated mathematically for analysis and design purposes.

The definitions of the signal-processing devices mentioned above are now given.

□ **D/A Converter.**

A digital-to-analog converter performs the task of decoding on a digitally coded input. The output of the D/A is an analog signal, usually in the form of a current or a voltage. ■

□ **A/D Converter.**

An analog-to-digital converter is a device which converts an analog signal to a digital-coded signal. ■

□ Sample-and-Hold Device.

A sample-and-hold device is used for many purposes in discrete-data and digital control systems. The S/H device makes a fast acquisition of an analog signal and then holds this signal at a constant value until the next acquisition (sample) is made. An S/H device is often an integral part of an A/D. ■

□ Multiplexer.

A multiplexer is used to couple signals from several sources so that they can be processed by the same processor or communication channel. The purpose is to time share the processor by all the incoming signals. ■

2-2 DIGITAL SIGNALS AND CODING

Signals in digital computers are represented by digital words or codes. The information carried by the digital code is generally in the form of discrete bits (logic pulses of "0" or "1") coded in a serial or parallel format. The numerical value of the digital word or code then represents the magnitude of the information in the variable the word represents.

Since it is simple to distinguish and implement just two states, "on" and "off," all modern digital computers are designed on the basis of the binary number system. A digital signal can be stored in a digital computer as a binary number of zeros and ones. Each of the binary digits (0 or 1) is referred to as a *bit*. The bit itself, however, is too small a number to be considered as the basic unit of information. Typically, bits are strung together to form larger, more useful units; 8 bits placed together form a *byte*, and several of the 8-bit bytes are grouped together to form a *word*. In general, a word may be of almost any bit length, from 4 bits to 128 bits or more. The distinction between a bit and a byte is somewhat like that between a letter of the alphabet and a word in the English language. While a letter may be said to be the smallest unit that can be written down, it usually has little value in itself until it is strung together with other letters to form a word. Figure 2-1 illustrates the relation between *word*, *byte*, and *bit*. In this case, the length of the word illustrated is 2 bytes or 16 bits.

The accuracy of a digital computer in its ability to store and manipulate digital signals is indicated by its *wordlength*. For example, a computer with an 8-bit word, such as the Intel 8080 microprocessor, can store only numbers with 8 bits of accuracy in its memory. Similarly, the registers and the accumulator are all 8 bits in length, so that temporary storage and computation can be carried out in only 8 bits of accuracy unless double-precision arithmetic is used. The Intel 80286 microprocessor, for instance, has a 16-bit word, and its accuracy and computational capability are much greater than the 8080.

Digital signals in a computer can be represented as *fixed-point numbers* or *floating-point numbers*. These number systems are described in Appendix A.

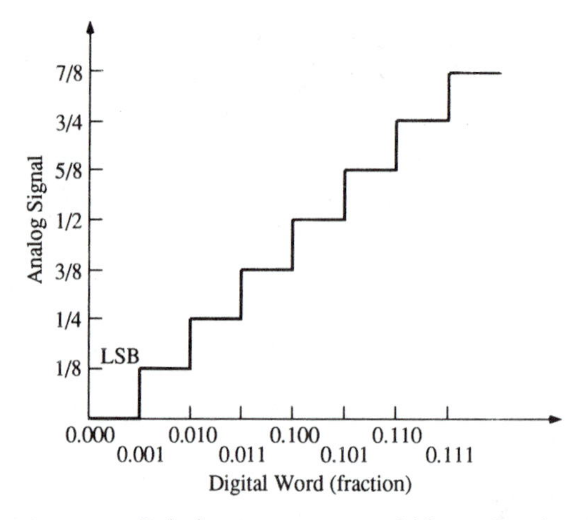

Figure 2-1. Relation between digital binary fractional code and decimal numbers.

2-3 DATA CONVERSION AND QUANTIZATION

In D/A and A/D conversions, the most-significant bit (MSB) and least-significant bit (LSB) and the weight of each in a digitally coded word are important in the understanding of the conversion process. The practical D/A and A/D converters based on the natural binary code make use of the fractional code. As shown in Table A-2 in Appendix A, for a 3-bit binary fractional code, the MSB has a weight of one-half of full scale (1/2 FS), the second bit has a weight of 1/4 FS, and the LSB has a weight of 1/8 FS. In general, for an n-bit binary fractional code, the MSB still has a weight of 1/2 FS, but the LSB has a weight of 2^{-n} FS. *Regardless of whether integer or fractional coding is utilized, an n-bit binary word defines 2^n distinct states, thus the word provides a resolution of one part in 2^n*. For example, Fig. 2-1 illustrates the resolution of $2^{-3} = 1/8$ of a 3-bit binary fractional code. The LSB in this case is 1/8, and FS = 1. Resolution may be improved by using more bits. For instance, a 4-bit binary fractional coded word still has a full-scale decimal equivalent of one, but the LSB is 2^{-4} or 1/16, and the resolution is improved to one part in 16. The diagram shown in Fig. 2-1 would then have 16 distinct levels. Note that the digital code used in general does not have to correspond to its analog signal, and vice versa. Therefore, the analog numbers represented on the vertical axis in Fig. 2-1 should be considered as a fractional part of full scale. The same principle also works for the integer code. Keep in mind that *to improve the resolution by increasing the number of bits, the same full-scale analog or digital signal value should be maintained.*

The operation of converting an analog signal to a digital number is called A/D conversion. If the number of bits in the digital word is finite, as it will be in a practical system, only a finite resolution can be attained by the A/D conversion; since the digital output can assume only a finite number of levels, the analog number is rounded off or truncated into a digital number. This approximation procedure is generally referred to as a *quantization* procedure. In the round-off operation, for example, the number 3.55 is rounded off to 3.6, and the number −3.55 is rounded off to −3.6, if the round-off is to one decimal place. For truncation, the numbers 3.56 and −3.56 are truncated to 3.5 and −3.5, respectively.

Figure 2-2(a) illustrates the round-off quantization relation between the analog and the digital binary integral codes for a 3-bit word for positive and

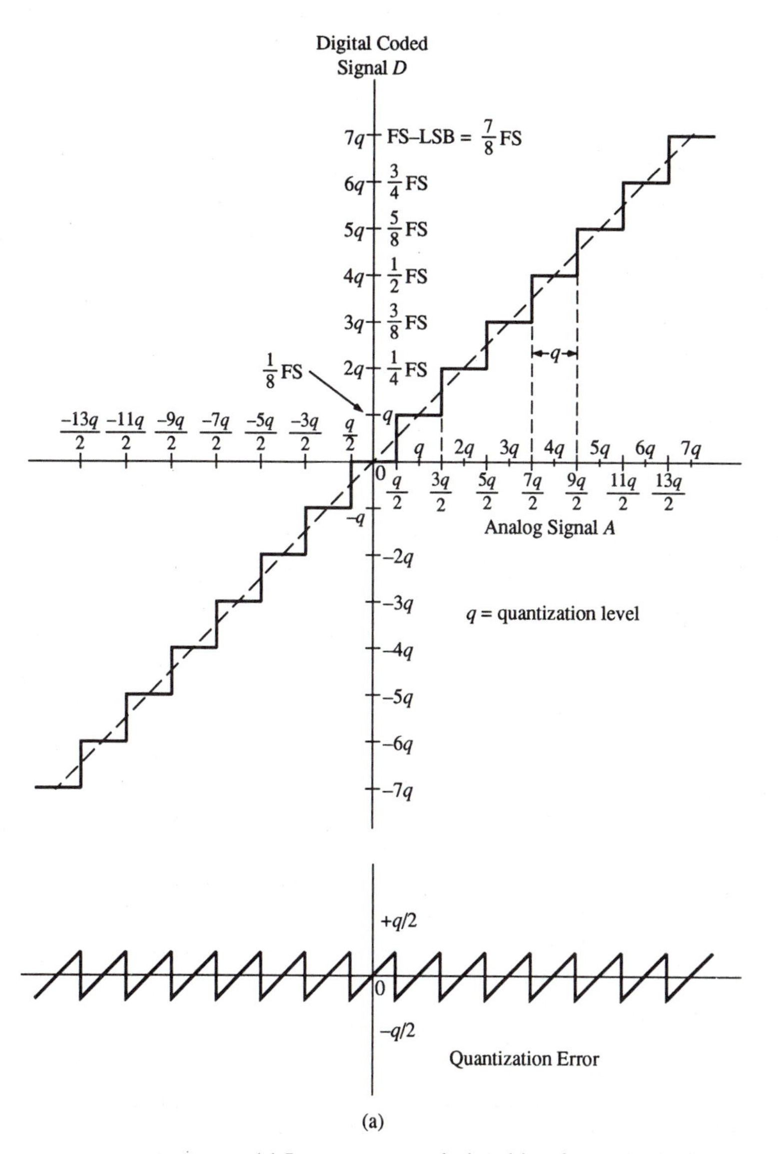

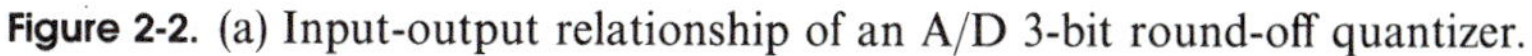

(a)

Figure 2-2. (a) Input-output relationship of an A/D 3-bit round-off quantizer.

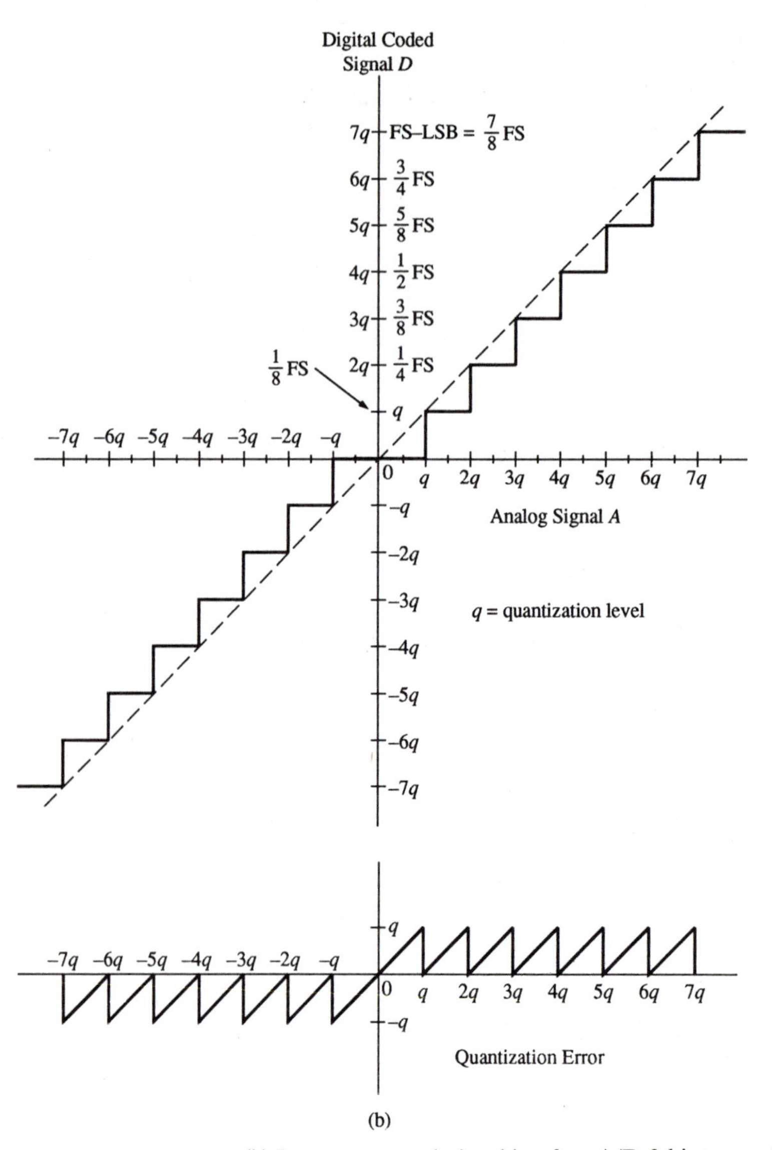

Figure 2-2. (b) Input-output relationship of an A/D 3-bit truncation quantizer.

negative numbers. Figure 2-2(b) illustrates the same situation except that the quantization is based on truncation. As shown in Fig. 2-2(a), the analog signal has decision levels at $\pm 0.5q$, $\pm 1.5q$, $\pm 2.5q, \ldots, \pm 6.5q$. The digital output D is related to the analog input A through the relationships tabulated in Table 2-1. Similarly, for the truncated quantization, the analog signal has decision levels at $\pm q$, $\pm 2q, \ldots, \pm 7q$, and the input-output relation of the quantization is given in Table 2-2.

Table 2-1 Relation Between Analog and Digital Numbers for a 3-Bit Analog-to-Digital Conversion (Round-Off)

Analog Number A	Binary-Coded Digital Number D			
	Binary Number			
	MSB ($\times 4q$)	($\times 2q$)	LSB ($\times q$)	
$\|A\| < 0.5q$	0	0	0	0
$0.5q \le \|A\| < 1.5q$	0	0	1	1/8 FS $= q =$ LSB
$1.5q \le \|A\| < 2.5q$	0	1	0	1/4 FS $= 2q$
$2.5q \le \|A\| < 3.5q$	0	1	1	3/8 FS $= 3q$
$3.5q \le \|A\| < 4.5q$	1	0	0	1/2 FS $= 4q$
$4.5q \le \|A\| < 5.5q$	1	0	1	5/8 FS $= 5q$
$5.5q \le \|A\| < 6.5q$	1	1	0	3/4 FS $= 6q$
$6.5q \le \|A\| < \infty$	1	1	1	7/8 FS = FS − LSB $= 7q$

Table 2-2 Relation Between Analog and Digital Numbers for a 3-Bit Analog-to-Digital Conversion (Truncation)

Analog Number A	Binary-Coded Digital Number D			
	Binary Number			
	MSB ($\times 4q$)	($\times 2q$)	LSB ($\times q$)	
$\|A\| \le q$	0	0	0	0
$q \le \|A\| < 2$	0	0	1	1/8 FS $= q =$ LSB
$2q \le \|A\| < 3q$	0	1	0	1/4 FS $= 2q$
$3q \le \|A\| < 4q$	0	1	1	3/8 FS $= 3q$
$4q \le \|A\| < 5q$	1	0	0	1/2 FS $= 4q$
$5q \le \|A\| < 6q$	1	0	1	5/8 FS $= 5q$
$6q \le \|A\| < 7q$	1	1	0	3/4 FS $= 6q$
$7q \le \|A\| < \infty$	1	1	1	7/8 FS = FS − LSB $= 7q$

Note that the quantizations shown in Figs. 2-2(a) and 2-2(b) are not one-to-one relations. The parameter q, which is equal to the least-significant bit, is known as the *quantization level.* For the 3-bit word illustrated, the LSB is 1/8 FS, as indicated in Tables 2-1 and 2-2. In Fig. 2-2(a) and (b), the dotted lines represent the ideal outputs if there is no quantization. The difference between the straight line and the quantization characteristic is the *quantization error.* Thus, for the truncated quantization, the maximum quantization error is $\pm q$, whereas for the round-off quantization, the maximum error is $\pm q/2$.

An important aspect of the saturation level of the quantization process, as shown in Fig. 2-2 and Tables 2-1 and 2-2, is that the maximum digital output, which corresponds to the binary number 111, is 7/8 FS, not FS. This would not have any effect on the accuracy of the A/D conversion as long as the magnitude of the analog input does not exceed 7/8 FS + $q/2$ for the round-off quantization, and 7/8 FS + q for the truncated quantization, since if the maximum level is not exceeded, the maximum quantization error will be bounded by $\pm q/2$ and $\pm q$, respectively. As an illustrative example, if the full-scale reference voltage of a 3-bit A/D conversion with round-off quantization is set at 10 V, referring to Fig. 2-2(a), FS = 10 V, and the quantization level is

$$q = \tfrac{1}{8}\,\text{FS} = 1.25 \qquad \text{V} \tag{2-1}$$

Then, the maximum value of the analog signal that the quantizer can convert without exceeding the quantization error of ± 0.625 is

$$\frac{7}{8}\,\text{FS} + \frac{q}{2} = 9.375 \qquad \text{V} \tag{2-2}$$

2-4 SAMPLE-AND-HOLD DEVICES

Sample-and-hold devices are used extensively in digital and sampled-data control systems. A sampler is a device that converts an analog signal into a train of amplitude-modulated pulses or a digital signal. A hold device simply maintains or "freezes" the value of the pulse or digital signal for a prescribed time duration. In a majority of the practical digital operations, sample-and-hold operations are performed by a single unit, and the device is commercially known as a sample-and-hold, or S/H, device.

One of the main sample-and-hold applications is to "freeze" fast-moving signals during all types of conversion operations. Another common usage of the S/H device is for the storage of multiplexer outputs while the signal is being converted. Peak detecting of a signal is another S/H application. We shall show in the following that the S/H operation is often used in conjunction with A/D and D/A converters.

The S/H operation is conceptually illustrated by the circuit shown in Fig. 2-3. The opening and closing of the switch or sampler are controlled by a *sample command.* When the switch is closed, the S/H device samples and *tracks* the input signal $e_s(t)$. When the switch is opened, the output is held at the voltage that the capacitor is charged to. Figure 2-4 illustrates typical input and output signals of the simple S/H device shown in Fig. 2-3 when the source resistance is zero. The time interval during which the sampler is closed is called the *sampling duration* p. In practice, the resistance R_s is not zero, and the capacitor

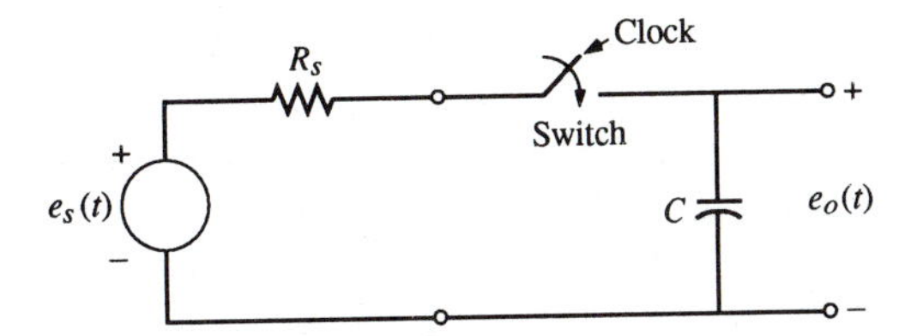

Figure 2-3. A simple circuit illustrating the sample-and-hold principle.

will charge toward the sampled input signal with a time constant R_sC. Furthermore, the operation of the sampler is not instantaneous, as it would take time for it to respond to the S/H commands.

A practical S/H operation may have many imperfections and errors, and the output of the device may deviate considerably from the ideal waveform illustrated in Fig. 2-4.

Figure 2-5 illustrates a typical input signal $e_s(t)$ and the corresponding output of a practical S/H device. The typical S/H output is characterized by several sources of time delays and imperfect holding during the hold mode. These characteristics are illustrated in Fig. 2-5 and are defined in the following.

□ Acquisition Time (T_a).

When the *sample command* is given to the S/H device, the unit does not begin to track the input signal instantaneously. The *acquisition time* is measured from the instant the sample command is given to the time when the S/H output enters and remains within a specified error band (say ± 1 percent) around theinput signal. Typical catalog information on acquisition time is given in terms of the percent of FS for a certain voltage step size; e.g., 0.1 percent FS for a 10-V step. ■

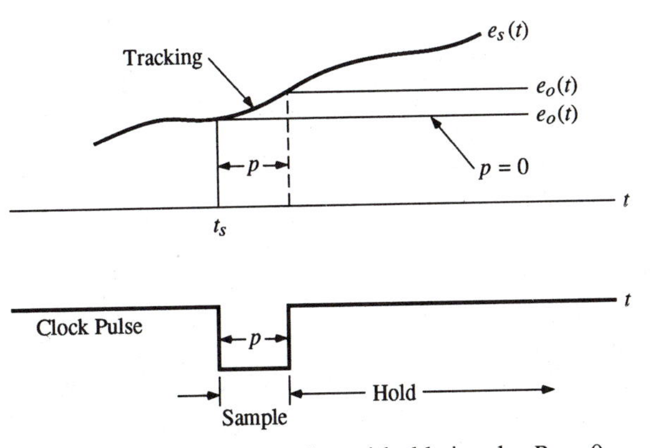

Figure 2-4. Simplified sample-and-hold signals; $R_s = 0$.

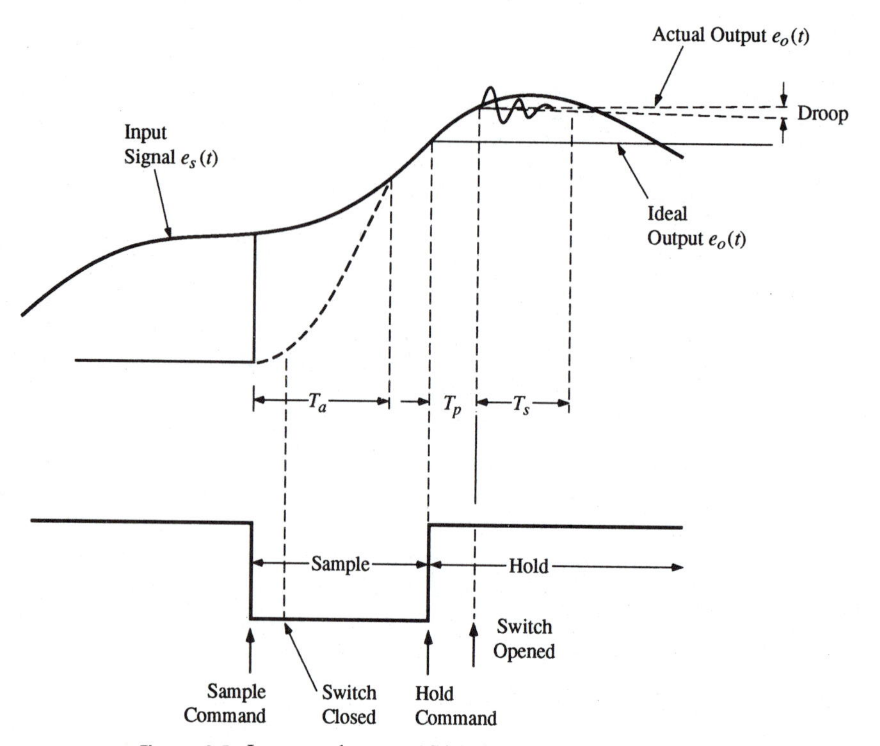

Figure 2-5. Input and output S/H signals with finite time delays.

□ Aperture Time (T_p).

When the *hold command* is given to the S/H device, while it is in the sample (track) mode, it will stay in this mode before reacting. The time between the issuance of the hold command and the time the sampler is opened is called the *aperture time*. This delay is usually caused by the switching circuit time delays within the S/H. For a given S/H device the aperture time is not constant, and the catalog information usually includes the worst case figure. For example, the aperture time of a typical commercial S/H unit may be on the order of 10 ns. ■

□ Settling Time (T_s).

In switching from the sample mode to the hold mode, transient caused by capacitance feedthrough from the digital logic circuitry through the electronic switch to the analog signal path can occur. The time required for the transient oscillation to settle to within a certain percent of FS is called the *settling time*. The typical settling time of a commercial S/H device is on

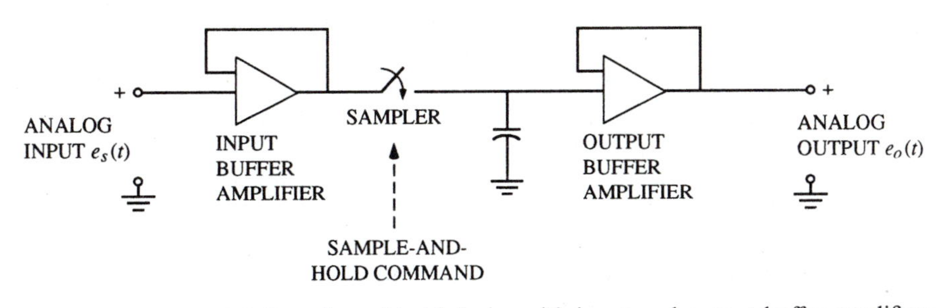

Figure 2-6. Sample-and-hold device with input and output buffer amplifiers.

the order of several nanoseconds to several microseconds, depending on the final accuracy required. ■

□ Hold-Mode Droop.

During the hold mode, the output voltage of the S/H device may decrease slightly, due to the leakage currents with the field-effect transistor (FET) switch and the buffer amplifier of the input circuit. The droop in the output of the S/H device can be greatly reduced by using a buffer amplifier with a very high input impedance at the S/H output. Similarly, an input buffer amplifier may be used so as to keep the input current of the S/H device relatively constant. An S/H device with input and output buffer amplifiers is illustrated by the block diagram of Fig. 2-6. ■

In digital systems the S/H operation is often controlled by a periodic clock. Figure 2-7 illustrates a uniform-rate S/H device. Both the actual and the ideal

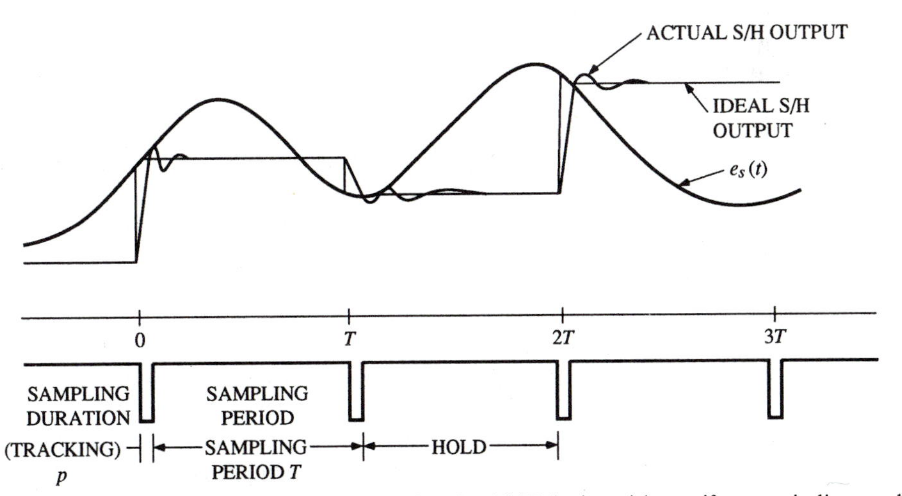

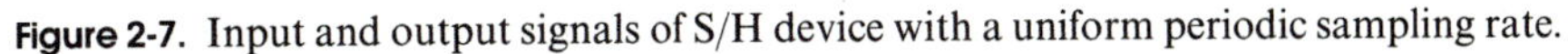

Figure 2-7. Input and output signals of S/H device with a uniform periodic sampling rate.

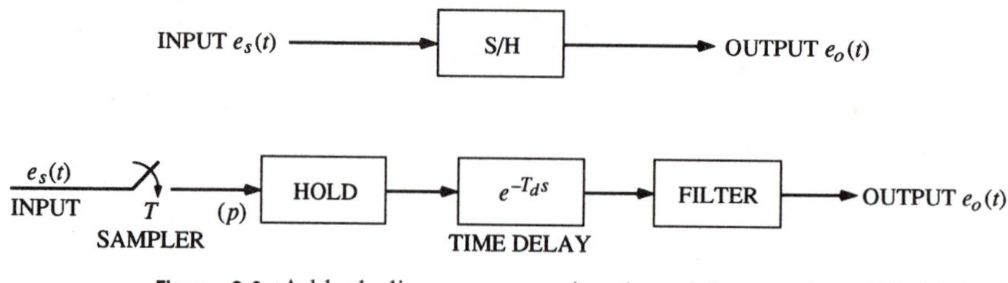

Figure 2-8. A block diagram approximation of the sample-and-hold device.

S/H outputs are shown for the given input analog signal. The time duration between the sample commands is called the *sampling period T*.

2-4-1 Block Diagram Representation of the S/H Device

Although the S/H device is available electronically as a single chip, for analytical purposes it is more convenient to treat the sampling and holding operations separately. Figure 2-8 illustrates an equivalent block diagram which isolates the S/H functions and the effects of all the delay times and transient oscillations. The sampler, which can be regarded as a pulse-amplitude modulator, has a sampling period T and a sampling duration p s. The hold device simply holds the sampled signal during the holding periods. The pure time delay T_d approximates the acquisition time and the aperture time delays, whereas the filter is for the representation of the finite time constant and dynamics of the buffer amplifiers. In general, the transfer function of the filter can be expressed as that of a second-order system,

$$G_f(s) = \frac{\omega_n^2}{s^2 + 2\zeta\omega_n s + \omega_n^2} \tag{2-3}$$

The sampler shown in Fig. 2-8 has a finite sampling duration p. In practice, an S/H device has a very small sampling duration p as compared with the sampling period T and the most significant time constant of the input analog signal. The delay times of the S/H device are also comparatively small, so that these can be neglected from the standpoint of the system dynamics. For example, for a given S/H device, the aperture time may be only 10 ns, and the acquisition time is 300 ns. The settling time may be another 100 ns; i.e., the transient oscillations will decay to a prescribed accuracy within 100 ns. Thus, the total time delay is only 410 ns, which may be neglected, since few control systems would be able to respond to signals faster than this time frame.

Therefore, for all practical purposes, if $p \ll T$ and the time delay due to sampling and holding is small, the S/H device can be modeled by the block diagram shown in Fig. 2-9. In this case, the sampler is called an *ideal sampler* since it is assumed to have a zero sampling duration; that is, $p = 0$.

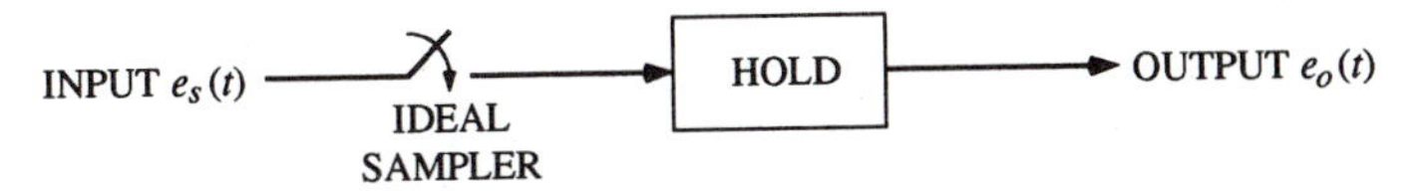

Figure 2-9. An ideal sample-and-hold device.

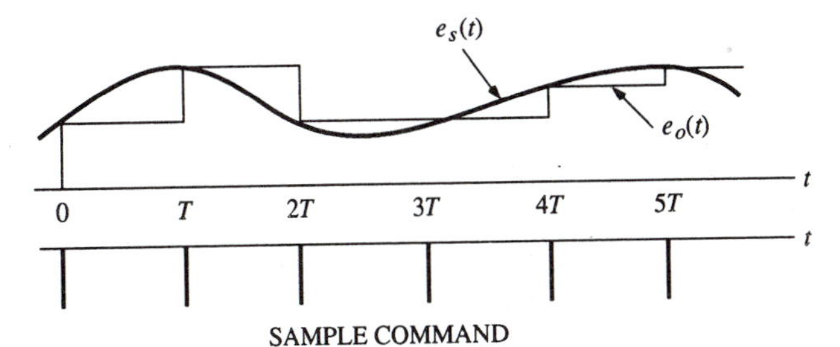

Figure 2-10. Input and output signals of an ideal S/H device.

Figure 2-10 shows typical input and output waveforms of an ideal S/H device. We shall show later that the ideal sampler model leads to convenient mathematical modeling of the S/H operation.

2-5 DIGITAL-TO-ANALOG (D/A) CONVERSION

Digital-to-analog conversion, or simply decoding, consists of transforming the numerical information contained in a digitally coded word into an equivalent analog signal. The basic elements of a D/A converter are portrayed by the block diagram in Fig. 2-11. The function of the logic circuit is to control the switching of the precision reference voltage or current source to the proper input terminals of the resistor network as a function of the digital value of each digital input

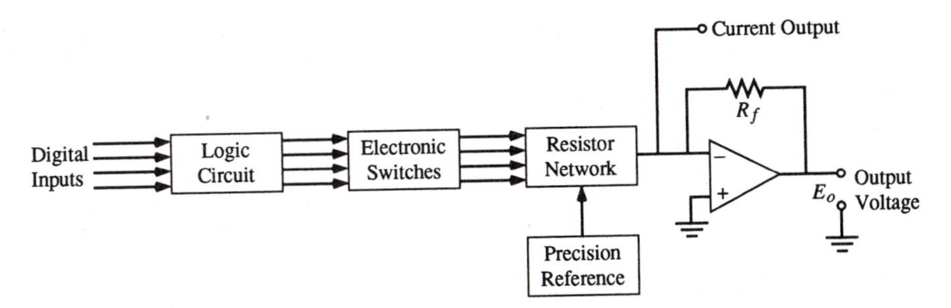

Figure 2-11. Basic elements of a D/A converter.

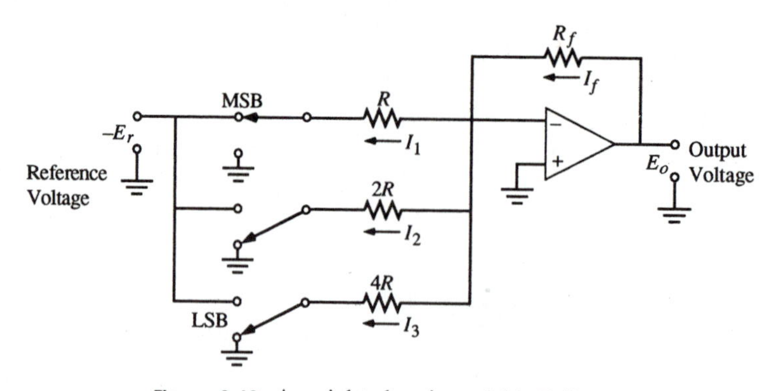

Figure 2-12. A weighted-resistor 3-bit D/A converter.

bit. Figure 2-12 illustrates a simple 3-bit binary D/A. The values of the summing resistors of the operational amplifier are weighted in a binary fashion. Each of these resistors is connected through an electronic switch to the reference voltage or to ground. When a binary 1 appears at the control logic circuit of a switch, it closes the switch and connects the resistor to the reference voltage. On the other hand, a binary 0 connects the resistor to ground. For the high-gain operational amplifier, the input impedance is very low so that the voltage at the summing point 0 is practically zero. Then, if the switch of the MSB branch is connected to the reference voltage $-E_r$, and the other two switches are connected to ground, thus corresponding to a digital word of 100, the output voltage E_o is

$$E_o = R_f I_f \tag{2-4}$$

Since $I_f = I_1$ and $I_1 = E_r/R$, we have

$$E_o = \frac{R_f E_r}{R} \tag{2-5}$$

If the digital word 110 is to be converted, the switches in the MSB and the next significant branches are connected to the reference. The output voltage becomes

$$E_o = \left[\frac{1}{R} + \frac{1}{2R}\right] R_f E_r = \frac{3}{2R} R_f E_r \tag{2-6}$$

The maximum value of the 3-bit word corresponds to the output voltage

$$E_o = \left[\frac{1}{R} + \frac{1}{2R} + \frac{1}{4R}\right] R_f E_r = \frac{7}{4R} R_f E_r \tag{2-7}$$

The LSB that corresponds to the digital word 001 is $R_f E_r/4R$. Thus, full scale is $8R_f E_r/4R = 2R_f E_r/R$. Table 2-3 gives the output voltages of the D/A conversion corresponding to all the 3-bit binary words for an FS of 10 V.

Table 2-3 Output Voltages of the D/A of Fig. 2-12 Corresponding to All Digital Inputs

Digital Word	Output Voltage E_o	Fraction of FS	Output Voltage for FS = 10 V
0 0 1	$\frac{R_f E_r}{4R}$	$\frac{FS}{8} = LSB$	1.25
0 1 0	$\frac{R_f E_r}{2R}$	$\frac{FS}{4}$	2.50
0 1 1	$\frac{3R_f E_r}{4R}$	$\frac{3}{8}$ FS	3.75
1 0 0	$\frac{R_f E_r}{R}$	$\frac{1}{2}$ FS	5.00
1 0 1	$\frac{5R_f E_r}{4R}$	$\frac{5}{8}$ FS	6.25
1 1 0	$\frac{6R_f E_r}{4R}$	$\frac{3}{4}$ FS	7.50
1 1 1	$\frac{7R_f E_r}{4R}$	$FS - LSB = \frac{7}{8}$ FS	8.75

Extending to an n-bit binary word, the network in Fig. 2-12 should contain n parallel resistor branches. The resistor in the LSB branch has a value of $2^{n-1}R$. The output voltage is written as

$$E_o = \left[\frac{a_0}{R} + \frac{a_1}{2R} + \cdots + \frac{a_{n-1}}{2^{n-1}R}\right] R_f E_r \tag{2-8}$$

where $a_0, a_1, \ldots, a_{n-1}$ are either 1 or 0, depending on the digital binary word which is to be converted.

Since a D/A converts a digital signal into an analog signal of corresponding magnitude, from the functional standpoint it may be regarded as a device consisting of a decoder and an S/H unit, as shown in Fig. 2-13. The decoder

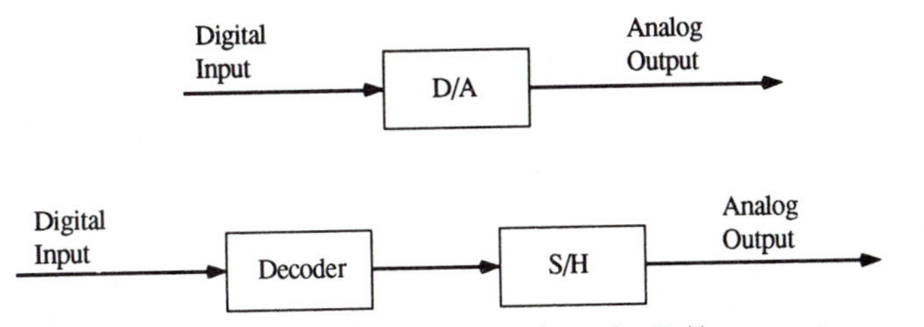

Figure 2-13. Block diagram representation of a D/A converter.

decodes the digital word into a number of an amplitude-modulated pulse. In reality, the sampler is redundant in the functional representation of Fig. 2-13. Since the S/H device is usually considered as one unit, the sampling operation is included even though it is not necessary. The transfer relation of the decoder is simply a constant gain, and in the ideal case, this gain is unity.

It is interesting to note that the operational amplifier at the output of the D/A converter is capable of generating oscillations and spikes in the transient output, and in practice, an S/H device may be used to eliminate these spikes. Thus, the use of the S/H unit in Fig. 2-13 as a functional representation actually has a realistic justification.

2-6 ANALOG-TO-DIGITAL CONVERSION

Analog-to-digital conversion, or simply encoding, consists of converting the numerical information contained in an analog signal into a digitally coded word. A/D conversion is a more complex process than D/A conversion and requires more elaborate circuitry. In comparison to the D/A converter, the A/D converter is generally more expensive and has slower response for the same conversion accuracy.

When a number is given as an input to an A/D, the converter performs the operations of *quantizing* and *encoding*. When a time-varying signal (voltage or current) is to be converted from analog to digital form, the A/D converter usually performs the following operations in succession: sample-and-hold, quantization, and encoding.

The sampling operation is needed to sample the analog signal at fixed periodic intervals. Theoretically, the holding operation is not needed; however, the A/D conversion time is not zero. To reduce the effect of signal variation during conversion, the sampled signal is held until the conversion is completed. Figure 2-14 gives the block diagram representation of an A/D converter.

The input to an A/D converter is usually in the form of a voltage or a current reading that is quantized during the conversion process. The ideal input-output relationship for a 3-bit natural binary A/D converter is essentially identical to the quantization characteristics shown in Fig. 2-2. While all the

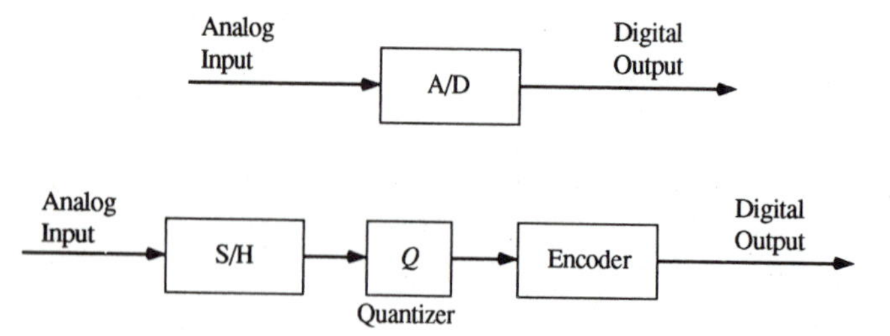

Figure 2-14. Block diagram representation of an A/D converter.

input values comprise a continuum, the output is partitioned into eight (2^3) discrete ranges. Thus, as Fig. 2-2(a) shows, in the case of the round-off quantization, an inherent $\pm 1/2$ LSB quantization uncertainty exists in the process in addition to the possible conversion errors. If greater resolution is needed, the number of bits in the output signal should be increased. However, this will also increase the circuit complexity and possibly the total time of conversion.

Although a large number of A/D circuits are available, only a few types are suitable for efficient and compact commercial units. The most commonly used A/D converters are of one of the following types:

1. Successive approximation.
2. Integration (single and dual slope).
3. Counter or servo type.
4. Parallel type.

Each of the A/D converter types listed has its own advantages and limitations. Each type is useful for a specific class of applications based on speed of conversion, cost, accuracy, and size. Commercial A/D units are available from 6 bits to 16 bits. Typical conversion accuracy is 0.01 percent FS $\pm 1/2$ LSB.

To illustrate the basic A/D conversion process, the successive-approximation type of A/D is briefly described in the following. Figure 2-15 shows the

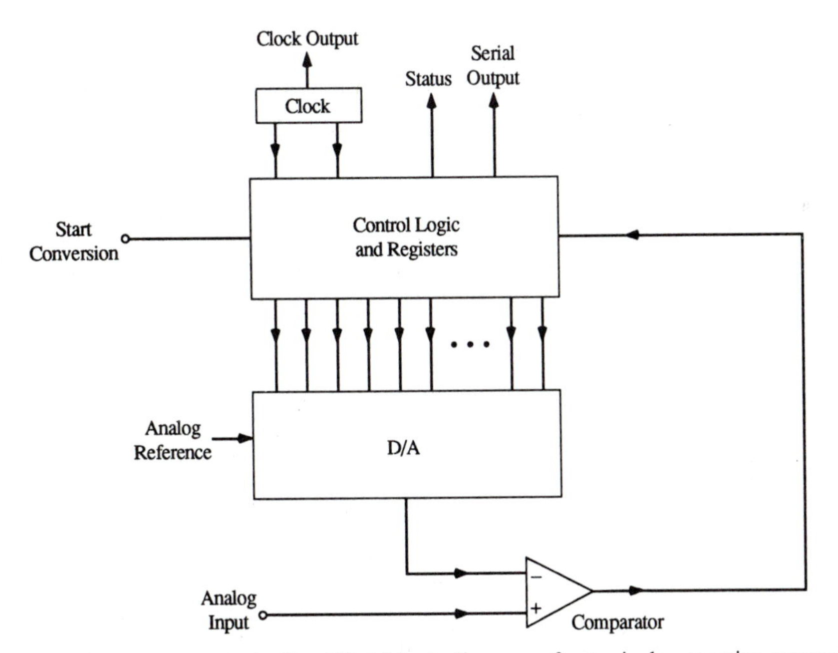

Figure 2-15. Simplified block diagram of a typical successive-approximation A/D converter.

simplified block diagram of a successive-approximation type of A/D converter. Basically, it consists of a comparator, a D/A converter, and some associated control logic. At the start of the conversion, all the bits of the output of the A/D converter are set to zero (clearing), and the MSB is then set to one. The MSB, representing one-half of full scale, is then converted by the D/A converter internally and compared with the analog input. If the input is greater than the converted MSB, then MSB = 1 is left on; otherwise, it is set to zero. The next significant bit is then turned on and compared and set, then a status line indicates that comparison is completed and the digital output is available for transmission. A typical timing diagram for an A/D converter of the successive-approximation type is shown in Fig. 2-16.

The conversion time of an A/D device acts as a time delay that is known to have adverse effects on the stability of closed-loop systems. In addition, the conversion time depends on the resolution of the A/D unit and the conversion method used. Typical conversion times of commercial A/Ds range from 100 ns to 200 μs.

In the simple case when the input analog signal is constant, the conversion time of the A/D unit is unimportant, since the input signal does not vary as it is compared with the various bits of the A/D converter. In practice, the input signal usually varies with time; as mentioned earlier, the analog signal is first

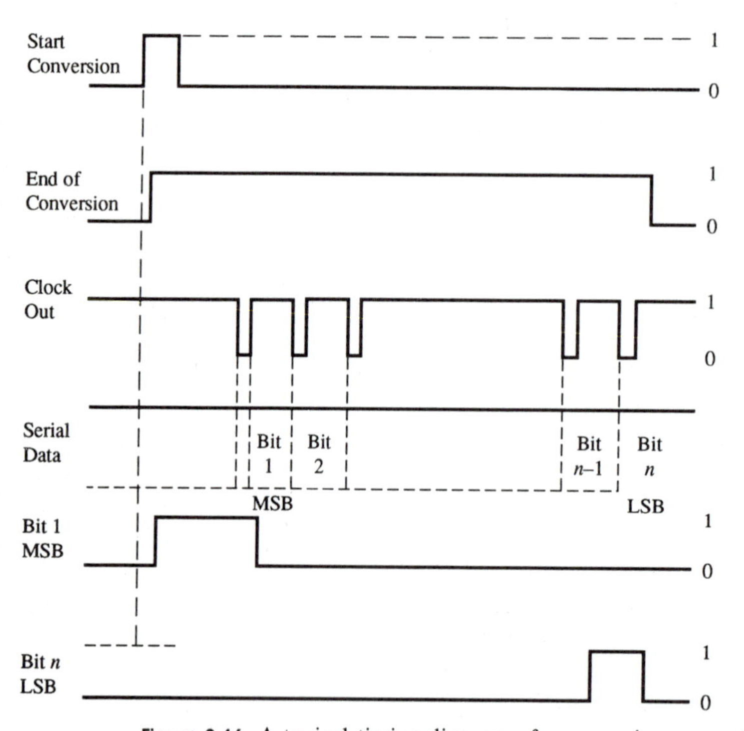

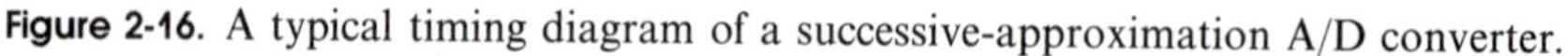

Figure 2-16. A typical timing diagram of a successive-approximation A/D converter.

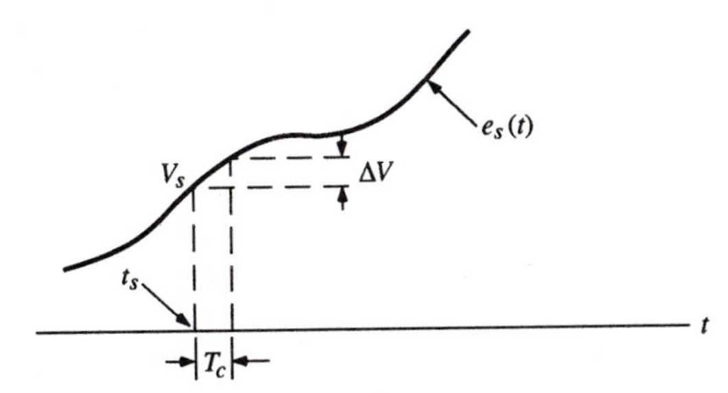

Figure 2-17. Conversion time and amplitude uncertainty in A/D conversion.

operated on by an S/H device, which samples and holds the input until the conversion is completed.

To illustrate the implications of the conversion time, let us consider the analog signal shown in Fig. 2-17. The objective of using the A/D converter is to convert the value V_s into a digital number. If the time required to make a measurement or conversion is T_c, a corresponding uncertainty exists in the measured amplitude if the signal varies over the interval T_c. Let us assume that the full-scale value of the input signal shown in Fig. 2-17 is denoted by V_{FS}, and the A/D device is an n-bit binary converter. This means that the signal is to be digitized to an n-bit resolution, or a resolution of one part in 2^n. The amplitude change of the signal over the conversion time T_c can be approximated by

$$\Delta V \cong \left.\frac{de_s(t)}{dt}\right|_{t=t_s} \times T_c \tag{2-9}$$

For a resolution of 2^{-n},

$$\frac{\Delta V}{V_{\text{FS}}} \leq 2^{-n} \tag{2-10}$$

Thus, using Eqs. (2-9) and (2-10), we have

$$\left.\frac{de_s(t)}{dt}\right|_{t=t_s} \leq 2^{-n}\frac{V_{\text{FS}}}{T_c} \tag{2-11}$$

The right-hand side of Eq. (2-11) represents the upper limit on the rate of change of the input signal over any conversion period, so that the resolution of the conversion can be kept at one part in 2^n. The reason for this is that the digital output can thus represent any point of the input signal during the conversion period.

In practice, we may measure an input signal by its highest frequency component of a sine wave. As an illustrative example, let us assume that the conversion time of a 10-bit binary A/D converter is 1 μs. Representing $e_s(t)$ as

$$e_s(t) = \frac{V_{\text{FS}}}{2}\sin\omega t \tag{2-12}$$

we have, from Eq. (2-11),

$$V_{FS}\omega \cos \omega t|_{t=0} \leq 2 \times 2^{-10} \frac{V_{FS}}{T_c} \tag{2-13}$$

In Eq. (2-13) we have chosen the sampling instant to be at $t = 0$, since this gives the largest value for $\cos \omega t$ (or the steepest slope for $\sin \omega t$). Thus,

$$\omega \leq \frac{2^{-9}}{10^{-6}} \cong 1953 \qquad \text{rad/s} \tag{2-14}$$

or the input signal should not vary at a rate in excess of 311 Hz. So, although the converter is capable of converting at a rate of 10^6 conversions per second, it cannot accurately encode signals whose frequency content is greater than 311 Hz. In general, the maximum frequency is given by

$$\omega_{max} = \frac{1}{2^{n-1}T_c} \tag{2-15}$$

where n is the number of bits of the conversion. Equation (2-15) shows that ω_{max} is inversely proportional to the conversion time T_c. As the number of bits (resolution) increases, the maximum allowable frequency decreases geometrically. Table 2-4 shows the frequency limitations for various combinations of wordlengths and conversion times T_c.

2-6-1 Sampling Period Considerations

From the preceding discussion we can see that the conversion time T_c plays a dominant role in the effect of an A/D conversion on system performance. While

Table 2-4 Frequency Limitation of A/D Due to Conversion Time T_c

Conversion Time T_c (s) \ Word Length n bits	4	6	8	10	12	14	16
	Maximum Frequency (rad/s)						
10^{-3} (1 ms)	125	31.25	7.8125	1.953	0.488	0.122	0.0305
10^{-4} (100 μs)	1250	312.5	78.125	19.53	4.88	1.22	0.305
10^{-5} (10 μs)	12,500	3125	781.25	195.3	48.8	12.2	3.05
10^{-6} (1 μs)	125,000	31,250	7812.5	1953	488	122	30.5
10^{-7} (100 ns)	1,250,000	312,500	78,125	19,530	4880	1220	305

the effect of the conversion time cannot be significantly reduced except through the use of a faster converter, the restriction on the maximum input frequency can be relaxed by using an S/H operation at the input of the A/D converter. The function of the S/H operation is to sample the input signal and then hold the value of the signal during the conversion period. In practice, the designer of a digital system is always confronted with the question, *How high should the sampling frequency be*? Let us designate the sampling frequency as f_s in Hz or ω_s in rad/s. The sampling frequency is related to the sampling period T (s) through

$$f_s = 1/T \qquad \text{Hz}$$

$$\omega_s = 2\pi f_s = 2\pi/T \qquad \text{rad/s}$$

In general, the sampling frequency of an S/H device depends on many signal-processing and system performance factors. In this section we can only address the factors related to signal processing.

The minimum sampling period of an S/H operation is bounded by the conversion time of the A/D unit and the delay times of the S/H device. For example, if the total conversion time of a 10-bit S/H and A/D combination is 1 μs, the minimum sampling period is also 1 μs. The corresponding maximum sampling frequency is 1 MHz. However, in practice, the limiting values of the sampling period and the sampling frequency are also governed by the characteristics of other system components, as well as by how fast the digital data can be processed. If multiplexing is involved or if the data are to be processed by a microprocessor, the S/H and A/D units are not the only limiting factors on the sampling frequency. Since the microprocessor is often a rather slow digital computer, data can be processed only at a certain maximum rate. Thus, in digital control systems the maximum sampling frequency is seldom limited by the S/H and A/D devices.

In the other extreme, we can easily see that a lower limit exists on how slow the sampling frequency can be. It is apparent that the S/H device should sample at a sufficiently fast rate so that the information contained in the input signal will not be lost through the sample-and-hold operation. We can use a sinusoidal signal to illustrate the point. Suppose we sample a 1-MHz sine wave only once every 1 μs. If the sampling instants always occur while the sine wave goes through zero, the resulting sampled data would erroneously represent a signal of zero magnitude. In fact, the sampled output would be zero if the sampling period is exactly 0.5 μs with the sampling instants at the zero crossings of the sine wave. Therefore, to represent the sine wave characteristics accurately, we have to sample at a frequency greater than twice the frequency of the signal. In general, given an input signal whose highest frequency component is ω_h rad/s, to retain vital information through sampling, the *minimum* sampling frequency is $2\omega_h$. This criterion of the minimum sampling frequency from the signal-processing standpoint is presented in the next section as the *sampling theorem*. For example, in the illustration given earlier on the 10-bit S/H and A/D conversion operations, the total delay time is 1 μs. If we use a sampling frequency of 1 MHz (the theoretical maximum sampling rate from the S/H and A/D standpoint), the input signal cannot contain frequency components in excess of 0.5 MHz.

In later chapters we will show that the stability of a closed-loop digital control system is closely related to the sampling period. In most cases, low sampling periods have a detrimental effect on the stability of closed-loop systems. Therefore, the selection of a proper sampling period for a closed-loop digital control system has to be made with stability as well as signal-processing considerations.

2-6-2 Simplified Block Diagram Representations of the A/D and D/A Converters

It is interesting to compare the block diagram representations of the D/A (Fig. 2-13) and the A/D (Fig. 2-14) converters. If the resolution of the A/D converter is very high, the nonlinear effect of the quantization can be neglected, and since the decoder and the encoder transfer relations can all be represented by constant gains, the two block diagrams essentially reduce to an S/H operation. This is indeed the case in the analytical studies of digital control systems. For instance, for analytical purposes, the digital autopilot system shown in Fig. 1-8 is represented by the block diagram of Fig. 2-18, where the A/D and D/A converters are replaced by S/H devices. In this case, the digital controller is represented by a discrete transfer function $D(z)$, and the analog elements are represented by their respective analog transfer functions. Perhaps another good reason for neglecting the quantization operation is that a quantizer is a nonlinear element. While the quantization nonlinearity can be treated analytically and the error produced estimated by quantization, virtually no general method exists that allows the design of a class of digital control systems with quantizers.

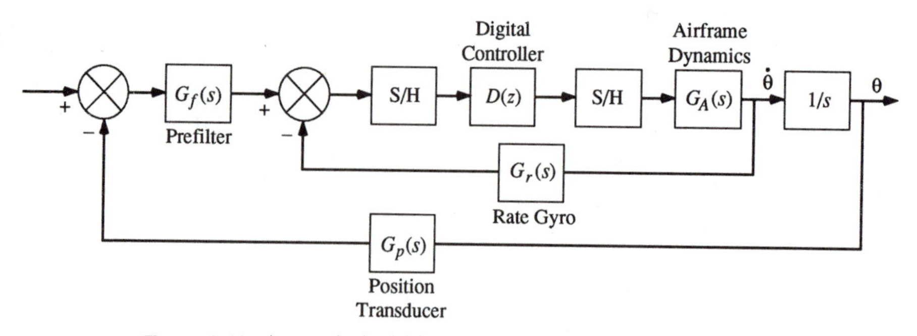

Figure 2-18. An analytical block diagram of the digital autopilot in Fig. 1-8.

2-7 MATHEMATICAL MODELING OF THE SAMPLING PROCESS

From the preceding sections we realize that the sampling operation in sampled-data and digital control systems is used to model either the sample-and-hold operation or the fact that the signal is digitally coded. Thus, the *sampler*, which is an essential element in a sampled-data or digital control system, can be either a realistic or fictitious component. If the sampler is used to represent S/H and A/D operations, it may involve delays, finite sampling duration, and quantization effects. If, on the other hand, the sampler is just used for the purpose of representing digital data, then the model would be much simpler. The sampling scheme of a digital or sampled-data system may assume a great variety of forms. For instance, a sampler may have a nonuniform or cyclic-variable sampling rate. It is also common to find digital control systems that contain multiple samplers with different sampling rates, as shown in the *multirate sampling system* example of Fig. 1-9. The sampling rates of the multirate samplers may or may not be integrally related, and they may not be synchronized; that is, the sampling starts at the same instant of time. In certain inherent sampling systems the sampling operation may be entirely random; i.e., the time interval between successive samples may be thought of as following some random scheme. We shall now examine some of the various sampling operations and develop mathematical models for the sampler so that the operations can be used for analytical purposes.

2-7-1 Finite-Pulsewidth Sampler

In general, the operation of a sampler may be regarded as one which converts a continuous-time signal into a pulse-modulated or digital signal. The most common type of modulation, as discussed in the sample-and-hold operation in the preceding sections, is the *pulse-amplitude modulation* (PAM). Figure 2-19 shows the block diagram representation of a periodic sampler with finite sampling duration. The pulse or sampling duration is p s, and the sampling period is T s. The finite sampling duration characteristic of a sampler should be considered if the value of p is not negligible when compared with the sampling period T. There are also physical systems that have natural properties which need to be modeled by a *finite-pulsewidth sampler*.

Consider that the input to the sampler is a continuous-time function $f(t)$. The output of the sampler, designated as $f_p^*(t)$, is a train of finite-width pulses whose amplitudes are modulated by the input $f(t)$. Figure 2-20 shows an

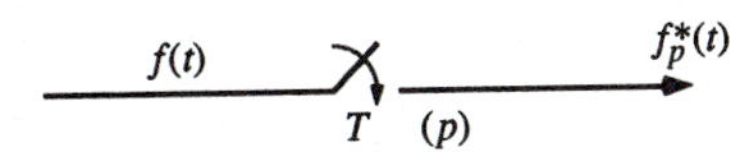

Figure 2-19. A uniform-rate sampler with finite sampling duration.

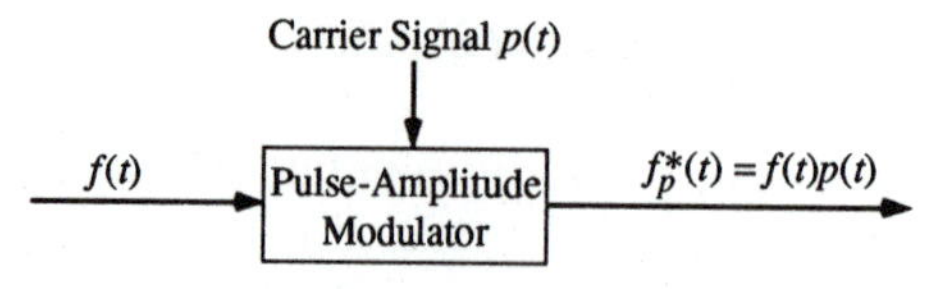

Figure 2-20. Pulse-amplitude modulator as a sampler.

equivalent block diagram representation of the sampler as a pulse-amplitude modulator. The input $f(t)$ is considered to be multiplied by a carrier signal $p(t)$ which is a train or periodic pulses each with unit height. Figure 2-21 illustrates typical waveforms of the input signal $f(t)$, the carrier $p(t)$, and the output $f_p^*(t)$.

Some sampling operations may be described by a *pulsewidth-modulation* (*PWM*) *sampler*. This type of sampler typically has outputs that are pulses with the pulsewidths varying as functions of the amplitudes of the inputs at the sampling instants but the amplitude of the output pulses is constant. Typical input and output signals of a pulsewidth modulator are illustrated in Fig. 2-22. A still more elaborate scheme is one in which the amplitudes as well as

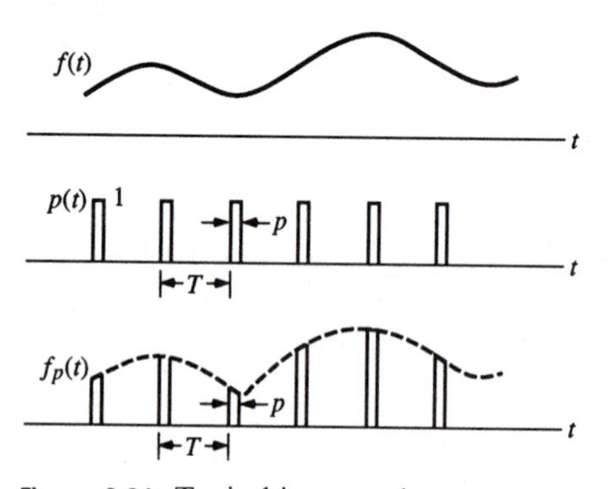

Figure 2-21. Typical input and output waveforms of a uniform-rate sampler.

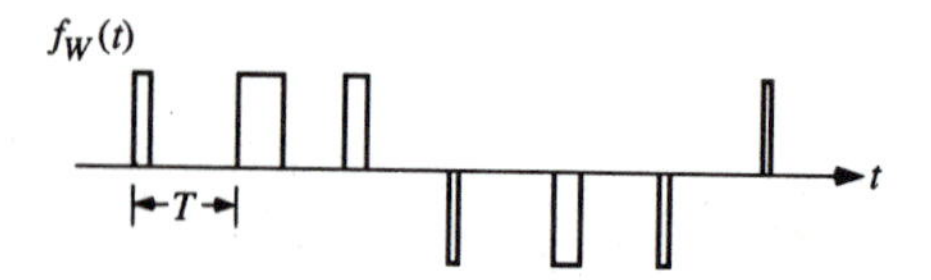

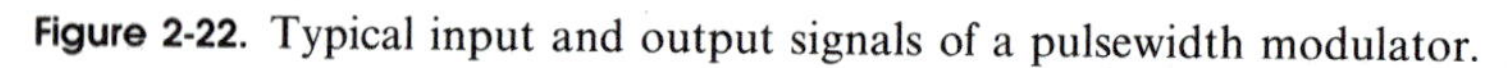

Figure 2-22. Typical input and output signals of a pulsewidth modulator.

the widths of the output pulses vary as some function of the input signal at the sampling instants. This type of sampler is known as a *pulsewidth-pulse-amplitude modulator*. In certain applications this more complex sampling may be used to improve the performance of digital or sampled-data control systems.

Only the mathematical modeling of the uniform-rate sampling operation is developed in this chapter. Once we have established the input-output relation of the uniform-rate sampler, the analysis can be easily extended to some of the other types of nonuniform-rate and multirate sampling.

It is important to point out that the uniform-rate sampler described in Figs. 2-19 and 2-20 is a *linear device*. The reason for this is that the sampler satisfies the *principle of superposition*. Unfortunately, the analog-to-digital conversion nature of the sampling operation precludes the description of the sampler by a conventional *transfer function*, as is characteristic for linear systems. Nevertheless, the linear property of the sampler allows us to use Fourier and Laplace transforms to arrive at the input-output transfer relation of the device.

For the uniform-rate finite-pulsewidth sampler described in Figs. 2-19 through 2-21, the output signal can be regarded as the product of the input $f(t)$ and the carrier $p(t)$ which is a unit pulse train with period T. The unit pulse train $p(t)$ contains pulses each with a unit magnitude, as shown in Fig. 2-21. Thus, $p(t)$ can be expressed as

$$p(t) = \sum_{k=-\infty}^{\infty} [u_s(t-kT) - u_s(t-kT-p)] \qquad (p < T) \tag{2-16}$$

where $u_s(t)$ is the unit-step function.

In Eq. (2-16) we have assumed that the sampling operation begins at $t = -\infty$, and the leading edge of the pulse at $t = 0$ coincides with $t = 0$, as shown in Fig. 2-23. The output of the sampler is written as

$$f_p^*(t) = f(t)p(t) \tag{2-17}$$

Substituting Eq. (2-16) into Eq. (2-17), we get

$$f_p^*(t) = f(t) \sum_{k=-\infty}^{\infty} [u_s(t-kT) - u_s(t-kT-p)] \qquad (p < T) \tag{2-18}$$

Equation (2-18) gives a time-domain description of the input-output relation of the uniform-rate finite-pulsewidth sampler.

It is of interest to investigate the frequency-domain characteristics of the sampler output. Since it is a pulse train, $f_p^*(t)$ generally contains much higher

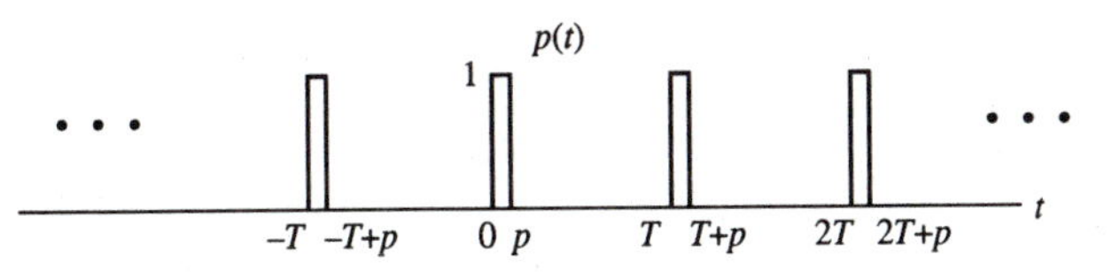

Figure 2-23. The unit pulse train.

frequency components than $f(t)$. Thus, the sampler may be regarded as a *harmonic generator.*

Since the unit pulse train $p(t)$ is a periodic function with period T, it can be represented by a Fourier series,

$$p(t) = \sum_{k=-\infty}^{\infty} C_n e^{jn\omega_s t} \tag{2-19}$$

where ω_s is the *sampling frequency* in rad/s and is related to the *sampling period* T by

$$\omega_s = 2\pi T \tag{2-20}$$

In Eq. (2-19) C_n denotes the complex Fourier series coefficients and is given by

$$C_n = \frac{1}{T}\int_0^T p(t)e^{-jn\omega_s t}\,dt \tag{2-21}$$

Since $p(t) = 1$ for $0 \leq t \leq p$, Eq. (2-21) becomes

$$C_n = \frac{1}{T}\int_0^p e^{-jn\omega_s t}\,dt = \frac{1 - e^{-jn\omega_s t}}{jn\omega_s T} \tag{2-22}$$

Using well-known trigonometric identities, C_n is written

$$C_n = \frac{p}{T}\frac{\sin(n\omega_s p/2)}{n\omega_s p/2}e^{-jn\omega_s p/2} \tag{2-23}$$

Substituting Eq. (2-23) into Eq. (2-19), we get

$$p(t) = \sum_{n=-\infty}^{\infty} \frac{\sin(n\omega_s p/2)}{n\omega_s p/2}e^{-jn\omega_s p/2}e^{jn\omega_s t} \tag{2-24}$$

Substitution of $p(t)$ from Eq. (2-24) into Eq. (2-17) yields

$$f_p^*(t) = \sum_{n=-\infty}^{\infty} C_n f(t)e^{jn\omega_s t} \tag{2-25}$$

where C_n is given by Eq. (2-23).

The Fourier transform of $f_p^*(t)$ is obtained as

$$F_p^*(j\omega) = \mathscr{F}[f_p^*(t)] = \int_{-\infty}^{\infty} f_p^*(t)e^{-j\omega t}\,dt \tag{2-26}$$

where $\mathscr{F}$ denotes the Fourier transform. Using the *complex shifting theorem* of the Fourier transform which states that

$$\mathscr{F}[e^{jn\omega_s t}f(t)] = F(j\omega - jn\omega_s) \tag{2-27}$$

the Fourier transform of Eq. (2-25) is written

$$F_p^*(j\omega) = \sum_{n=-\infty}^{\infty} C_n F(j\omega - jn\omega_s) \tag{2-28}$$

Since n extends from $-\infty$ to ∞, the last equation can also be written as

$$F_p^*(j\omega) = \sum_{n=-\infty}^{\infty} C_n F(j\omega + jn\omega_s) \tag{2-29}$$

where C_n is given by Eq. (2-22) or Eq. (2-23).

The significance of the finite-pulsewidth sampling operation as viewed from the frequency domain is explained by the following development using Eq. (2-29). First, taking the limit as $n \to 0$ of the Fourier coefficient in Eq. (2-23), we have

$$C_0 = \lim_{n \to 0} C_n = \frac{p}{T} \tag{2-30}$$

The $n = 0$ term in Eq. (2-29) is written

$$F_p^*(j\omega)|_{n=0} = C_0 F(j\omega) = \frac{p}{T} F(j\omega) \tag{2-31}$$

The last equation shows the important fact that the frequency components contained in the original continuous-time input, $f(t)$, are still present in the sampler output $f_p^*(t)$, except that the amplitude is multiplied by the factor p/T.

For $n \neq 0$, C_n is a complex quantity, but the magnitude of C_n may be written as

$$|C_n| = \frac{p}{T} \left| \frac{\sin(n\omega_s p/2)}{n\omega_s p/2} \right| \tag{2-32}$$

The magnitude of $F_p^*(j\omega)$ is written

$$|F_p^*(j\omega)| = \left| \sum_{n=-\infty}^{\infty} C_n F(j\omega + jn\omega_s) \right| \tag{2-33}$$

The frequency spectrum of the unit pulse train $p(t)$ is simply the plot of the Fourier coefficient C_n as a function of ω, when n assumes various integral values between $-\infty$ and ∞. The amplitude spectrum of C_n is simply the plot of the amplitude of the Fourier coefficient C_n as a function of ω, when n assumes various integral values between $-\infty$ and ∞. The amplitude spectrum of C_n is shown in Fig. 2-24(a). We see that the amplitude spectrum of C_n is not a continuous function but rather a line spectrum at discrete intervals of ω, with $\omega = n\omega_s$, for $n = \pm 1, \pm 2, \ldots$, as described by the right-hand side of Eq. (2-32).

Since both C_n and $F(j\omega + jn\omega_s)$ are complex, we cannot express the amplitude quantity in Eq. (2-33) as functions of the individual amplitude spectrum of $|C_n|$ and $|F(j\omega + jn\omega_s)|$. However, Eq. (2-33) leads to the following inequality:

$$|F_p^*(j\omega)| \leq \sum_{n=-\infty}^{\infty} |C_n||F(j\omega + jn\omega_s)| \tag{2-34}$$

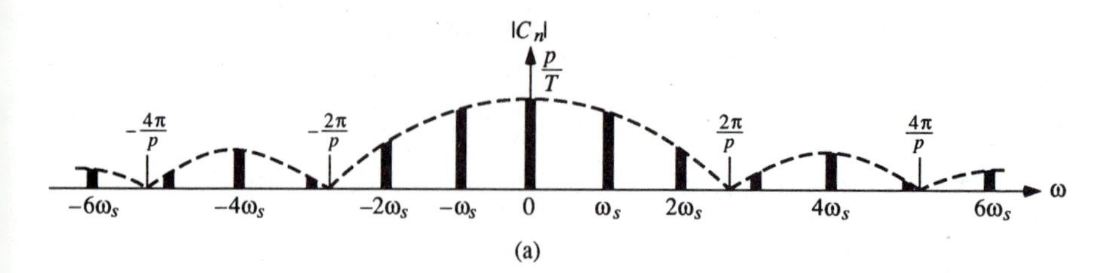

(a)

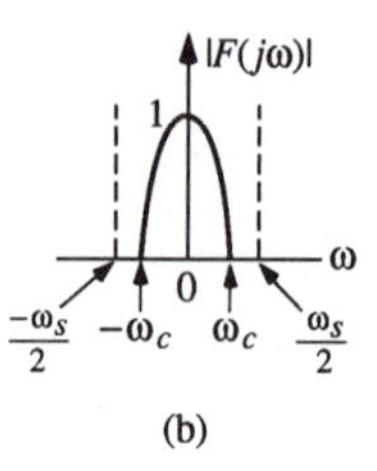

(b)

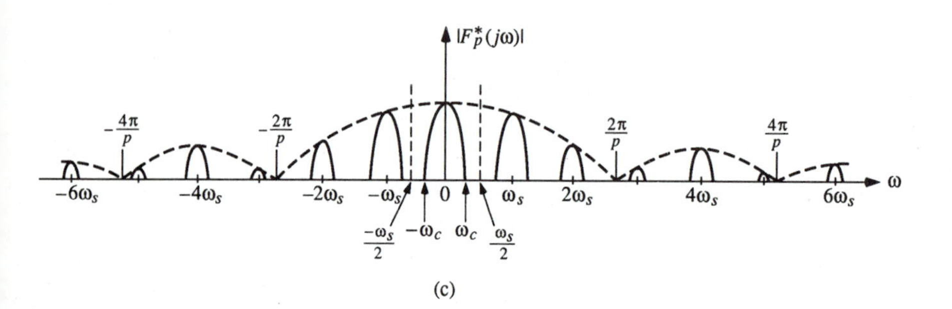

(c)

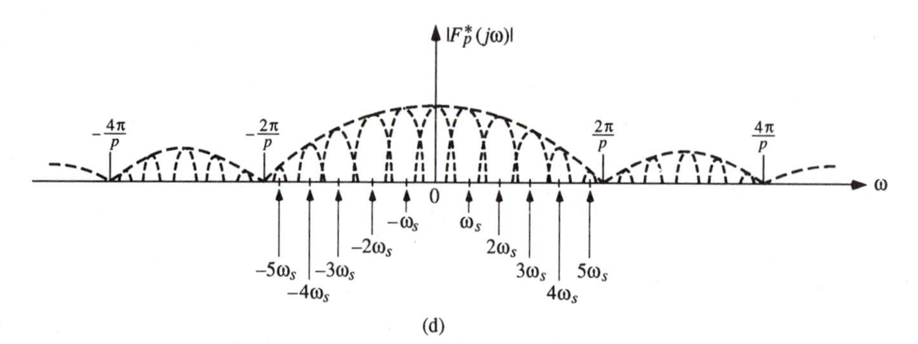

(d)

Figure 2-24. Amplitude spectra of input and output signals of a finite-pulsewidth sampler. (a) Amplitude spectrum of unit pulse train $p(t)$. (b) Amplitude spectrum of continuous input signal $f(t)$. (c) Amplitude spectrum of sampler output ($\omega_s > 2\omega_c$). (d) Amplitude spectrum of sampler output ($\omega_s < 2\omega_c$).

which may be used as a qualitative way of illustrating the amplitude spectrum of $F_p^*(j\omega)$. If the amplitude spectrum of the continuous-time input $f(t)$ is a band-limited signal, as shown in Fig. 2-24(b), the amplitude spectrum of $F_p^*(j\omega)$ will be of the form shown in Fig. 2-24(c). Note that the waveform shown in Fig. 2-24(c) is not exactly that of $|F_p^*(j\omega)|$, but it does indicate the general shape of the spectrum.

2-7-2 The Folding Frequency, Nyquist Frequency, and Alias Frequency

Examining the waveform conjectured as $|F_p^*(j\omega)|$ in Fig. 2-24(c), we see that the sampling operation retains the fundamental component of $F(j\omega)$, but in addition, the sampler output also contains the harmonic components, $F(j\omega + jn\omega_s)$, for $n = \pm 1, \pm 2, \ldots$. Thus, as mentioned before, the sampler may be considered as a harmonic generator whose output contains the weighed fundamental component, plus all the weighed complementary components at all frequencies separated by the sampling frequency ω_s. The band around zero frequency still carries all the information contained in the continuous input signal, but the same information is also repeated along the frequency axis; the amplitude of each component is weighed by the magnitude of its corresponding Fourier coefficient $|C_n|$. It should be pointed out that the frequency spectrum for $|F_p^*(j\omega)|$ shown in Fig. 2-24(c) is sketched with the assumption that the sampling frequency ω_s is greater than twice the highest frequency contained in $f(t)$, ω_c, that is, $\omega_s > 2\omega_c$. If $\omega_s < 2\omega_c$, distortion in the frequency spectrum of $|F_p^*(j\omega)|$ will appear because of the overlapping of the harmonic components. Figure 2-24(d) shows that if $\omega_s < 2\omega_c$, the spectrum of $|F_p^*(j\omega)|$ around zero frequency bears little resemblance to that of the original signal. Therefore, theoretically, the original signal can be recovered from the spectrum shown in Fig. 2-24(c) by means of an ideal low-pass filter with a bandwidth that lies between ω_c and $\omega_s - \omega_c$, whereas the input signal cannot be recovered from the spectrum shown in Fig. 2-24(d), due to the overlapping of the complementary components.

The phenomenon of the overlapping of the high-frequency components with the fundamental component in the frequency spectrum of the sampled signal is sometimes referred to as *folding*. The frequency $\omega_s/2$ is often known as the *folding frequency* (rad/s). The frequency ω_c is sometimes referred to as the *Nyquist frequency*.

Another effect due to the result of frequency folding is known as *aliasing*. Figure 2-25(a) shows a periodic signal that is sampled at a rate greater than twice per cycle (sampling period $T < T_c/2$). Apparently, the sampled output has the same period as the input. If the periodic signal is sampled at a rate less than twice the input frequency ($T > T_c/2$), as shown in Fig. 2-25(b), the output signal will have a different frequency from the input (the phenomenon is called *aliasing*), and the output frequency is referred to as the *alias frequency*. The period of the output is called the *alias period*.

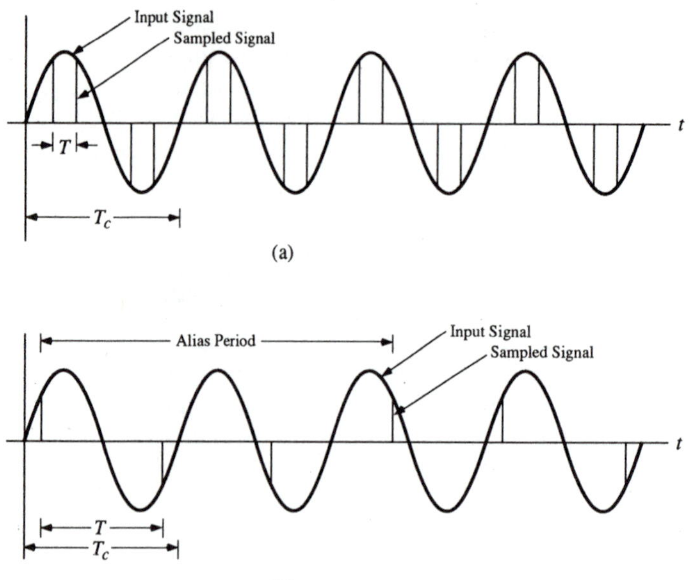

Figure 2-25. (a) Sampling frequency is greater than two times the frequency of the input signal. No aliasing. (b) Aliasing caused by inadequate sampling rate. T is the sampling period, T_c the period of input signal.

The requirement that ω_s be at least twice as large as the highest frequency (Nyquist frequency) contained in the signal $f(t)$ is formally known as the *sampling theorem*, which is stated more formally in Sec. 2-8.

In reality, of course, most signals encountered in systems do not have an abrupt cutoff in the amplitude spectrum as shown in Fig. 2-24(b), so that some folding effect will exist even though the sampling frequency is made greater than the folding frequency. More on the subject and the physical implications of the folding of frequency on the performance of digital control systems will be discussed later.

2-8 THE SAMPLING THEOREM

When sampling is used in a given system, the question most often encountered is, What should be the proper sampling rate, and what are the limits on the rate of sampling? Naturally, any sampler constructed with physical components must have a finite maximum sampling rate. From a theoretical standpoint, when the sampling frequency reaches infinity, the sampler is essentially closed completely, and the sampled-data or digital system should revert to a continuous-data system. We can easily show this in the finite-pulsewidth sampling, but the ideal sampling case must be treated carefully since impulses are involved in the interpretation of the zero-pulsewidth sampling operation.

As far as the lower limit of the sampling frequency is concerned, we know intuitively that if the continuous-time signal changes rapidly with respect to time, sampling the signal at too low a rate will lose vital information on the signal between the sampling instants. Consequently, it may not be able to reconstruct the original signal from the information contained in the sampled data. We conclude from the amplitude spectra shown in Fig. 2-24 of the preceding section that the lowest sampling frequency for possible signal reconstruction is $2\omega_c$, where ω_c is the highest frequency contained in $f(t)$. Formally, this is known as the *sampling theorem.* The theorem states that

> *If a signal contains no frequency higher than ω_s rad/s, it is completely characterized by the values of the signal measured at instants of time separated by $T = \pi/\omega_c$ s.* ■

The sampling theorem, originally due to Shannon, was devised for communication systems in which sampling and digital coding are mainly used for the purposes of transmission, multiplexing, noise rejection, etc., and data reconstruction is a very important element in these systems. While many similarities exist between sampling and digital coding in communication systems and control systems, the objectives are not always the same. For example, in digital control systems, data reconstruction may not be of primary importance. The objective of the data hold is for the smoothing and filtering of the sampled data, since the controlled processes are usually analog devices and are preferably driven by analog signals. Since most control systems are closed-loop systems, stability is often a problem. We shall show later that the stability of a sampled-data or digital control system is directly related to the sampling frequency. As a matter of fact, *a low sampling frequency normally has a detrimental effect on the stability of a closed-loop system.* Thus, often we may have to select a sampling rate that is much higher than the theoretical minimum governed by the sampling theorem.

It should also be pointed out that the waveforms shown in Fig. 2-24 are idealized in that the amplitude spectrum of $f(t)$ is assumed to be band-limited. In practical situations, the signals found in the real world do not exhibit a sharp cutoff, as illustrated, although in many cases the amplitude of the frequency spectrum does decrease as the frequency increases. In addition, an ideal low-pass filter exists only theoretically, which means that the sampling theorem is valuable only as a guideline for the purposes of data reconstruction.

2-8-1 An Addendum to the Sampling Theorem

An interesting note on the sampling theorem may be introduced here. A signal can still be defined completely by the sampled data at a rate less than $2\omega_c$ rad/s provided that the derivatives of the signal are known at the sampling instants as well as the amplitude information. L. J. Fogel [10] and others have proved that, if a signal $f(t)$ contains no frequency higher than ω_c rad/s, it is completely characterized by the values $f^{(n)}(kT)$, $f^{(n-1)}(kT)$, ..., $f^{(1)}(kT)$, and $f(kT)$, $k = 0, 1, 2, \ldots$, where

$$f^{(n)}(kT) = \left.\frac{d^n f(t)}{dt^n}\right|_{t=kT} \tag{2-35}$$

This means that when the values of the first derivative of $f(t)$ at $t = kT$, for $k = 0, 1, 2, \ldots$, are known in addition to the values of $f(kT)$, the minimum sampling rate is ω_c, which is one-half of that required when $f(kT)$ alone is measured. The addition of each succeeding higher order derivative allows the time between samples to become longer according to $(1/2)(n + 1)(2\pi/\omega_c)$, where n is the order of the highest derivative when all lower ordered derivatives are observed for each sample.

2-9 MATHEMATICAL MODELING OF SAMPLING BY CONVOLUTION INTEGRAL

Alternate descriptions of the sampled signal $f_p^*(t)$ in the transform domain can be obtained by applying the complex-convolution theorem of the Laplace transform to Eq. (2-17). Several alternate expressions for the input-output transfer relations of the finite-pulsewidth sampler, including the Fourier transform relation in Eq. (2-29), are given in the following. The details of the derivations by complex convolution are found in Appendix B.

Fourier Transform

$$F_p^*(j\omega) = \sum_{n=-\infty}^{\infty} \frac{p}{T} \frac{\sin(n\omega_s p/2)}{n\omega_s p/2} e^{-jn\omega_s p/2} F(j\omega + jn\omega_s) \tag{2-36}$$

Laplace Transform

$$\begin{aligned} F_p^*(s) &= \sum_{n=-\infty}^{\infty} \frac{p}{T} \frac{\sin(n\omega_s p/2)}{n\omega_s p/2} e^{-jn\omega_s p/2} F(s + jn\omega_s) \\ &= \sum_{n=-\infty}^{\infty} \frac{1 - e^{-jn\omega_s p}}{jn\omega_s T} F(s + jn\omega_s) \end{aligned} \tag{2-37}$$

Laplace Transform [$F(s)$ has k simple poles]

$$F_p^*(s) = \sum_{n=-\infty}^{\infty} \frac{N(\xi_n)}{D'(\xi_n)} \frac{1 - e^{-p(s-\xi_n)}}{(s - \xi_n)[1 - e^{-T(s-\xi_n)}]} \tag{2-38}$$

$$F(\xi) = N(\xi)/D(\xi) \tag{2-39}$$

$$D'(\xi_n) = \left.\frac{dD(\xi)}{d\xi}\right|_{\xi=\xi_n} \tag{2-40}$$

where ξ_n is the nth simple pole of $F(\xi)$, $n = 1, 2, \ldots, k$.

Laplace Transform [$F(s)$ has k poles with multiplicity $m_n \geq 1$]

$$F_p^*(s) = \sum_{n=1}^{k} \sum_{i=1}^{m_n} \frac{(-1)^{m_n - i} K_{ni}}{(m_n - i)!} \frac{\partial^{m_n - i}}{\partial s^{m_n - i}} \left[\frac{1 - e^{-ps}}{s(1 - e^{-Ts})} \right] \Bigg|_{s = s - s_n} \tag{2-41}$$

$$K_{ni} = \frac{1}{(i-1)!} \frac{\partial^{i-1}}{\partial s^{i-1}} (s - s_n)^{m_n} F(s) \Bigg|_{s = s_n} \tag{2-42}$$

2-10 FLAT-TOP APPROXIMATION OF FINITE-PULSEWIDTH SAMPLING

In the analysis of the finite-pulsewidth sampler, if the sampling duration p is very much smaller than the sampling period T and the smallest time constant of the signal $f(t)$, the sampler output $f_p^*(t)$ can be approximated by a sequence of flat-topped pulses, since the variation of $f(t)$ in the sampling duration will not be significant. Then, $f_p^*(t)$ can be written as

$$f_p^*(t) = \begin{cases} f(kT) & \text{for} \quad kT \leq t < kT + p \\ 0 & \text{for} \quad kT + p \leq t < (k+1)T \end{cases} \tag{2-43}$$

where $k = 0, 1, 2, \ldots$.

The function $f_p^*(t)$ can be expressed as an infinite series,

$$f_p^*(t) = \sum_{k=0}^{\infty} f(kT)[u_s(t - kT) - u_s(t - kT - p)] \tag{2-44}$$

where $u_s(t)$ is the unit-step function.

Taking the Laplace transform on both sides of Eq. (2-44), we get

$$F_p^*(s) = \sum_{k=0}^{\infty} f(kT) \left[\frac{1 - e^{-ps}}{s} \right] e^{-kTs} \tag{2-45}$$

Since it has been assumed that the sampling duration p is very small, the term e^{-ps} can be approximated by taking only the first two terms of its power-series expansion. Then,

$$1 - e^{-ps} = 1 - \left[1 - ps + \frac{(ps)^2}{2!} - \cdots \right] \cong ps \tag{2-46}$$

Thus, Eq. (2-45) is simplified to

$$F_p^*(s) \cong p \sum_{k=0}^{\infty} f(kT) e^{-kTs} \tag{2-47}$$

In the time domain, Eq. (2-47) is equivalent to

$$f_p^*(t) \cong p \sum_{k=0}^{\infty} f(kT) \delta(t - kT) \tag{2-48}$$

where $\delta(t)$ is the unit impulse function.

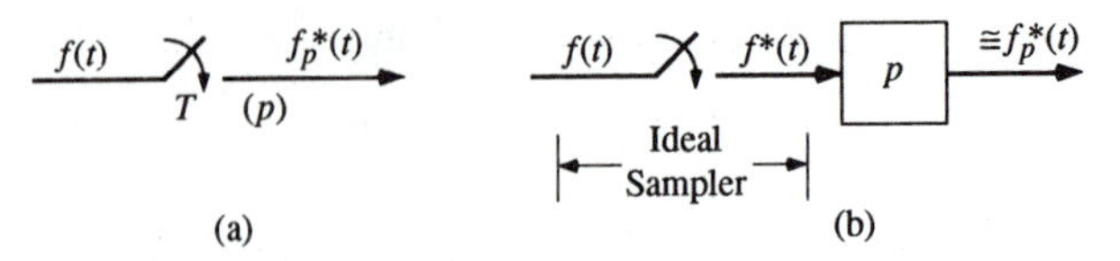

Figure 2-26. (a) Finite-pulsewidth sampler. (b) Ideal sampler connected in series with an attenuator p. (a) and (b) are equivalent if p is much smaller than T and the smallest time constant of $f(t)$.

2-10-1 The Ideal Sampler

The right-hand side of Eq. (2-48) is recognized as a train of impulses with the strength of the impulse at $t = kT$ equal to $pf(kT)$. This means that the finite-pulsewidth sampler can be approximated by an *impulse modulator* whose block diagram can still be represented by that in Fig. 2-20, except that the carrier signal is replaced by the unit impulse train,

$$\delta_T(t) = \sum_{k=0}^{\infty} \delta(t - kT) \tag{2-49}$$

Thus, the finite-pulsewidth sampler can be approximated by an *ideal sampler* connected in series with an attenuator with attenuation p. The equivalent situation is illustrated in Fig. 2-26(a) and (b).

An ideal sampler is defined as a sampler which closes and opens instantaneously, every T seconds, for a zero time duration.

The output of the ideal sampler is expressed as

$$f^*(t) = \sum_{k=0}^{\infty} f(kT)\delta(t - kT) = f(t)\delta_T(t) \tag{2-50}$$

where $f(t)$ is the input of the sampler, and sampling is assumed to begin at $t = 0$.

Remember that if the finite-pulsewidth sampler is followed by a hold device, then only the values of the function $f(t)$ at the sampling instants are of importance; the output of the hold device is the same whether the sampler is approximated by a flat-top sampler or an ideal sampler, and the attenuation p should be dropped.

Taking the Laplace transform on both sides of Eq. (2-50), we get

$$F^*(s) = \sum_{k=0}^{\infty} f(kT)e^{-kTs} \tag{2-51}$$

which is the Laplace transform of the output of the ideal sampler.

Typical input and output signals of an ideal sampler are illustrated in Fig. 2-27. The output of the ideal sampler is shown to be a train of impulses with the respective areas (strengths) of the impulses equal to the magnitudes of the input signal at the corresponding sampling instants. Since an impulse function

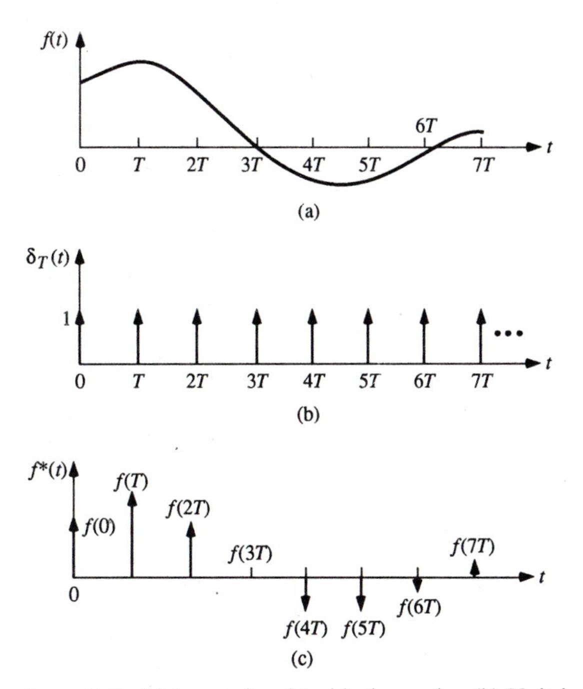

Figure 2-27. (a) Input signal to ideal sampler. (b) Unit impulse train. (c) Output of ideal sampler. $f^*(t) = f(t)\delta_T(t)$.

is defined to have zero pulsewidth and infinite pulse amplitude, in Fig. 2-27 the impulses are depicted as arrows with amplitudes representing the strengths of the impulses.

2-10-2 Alternate Expressions of $F^*(s)$

We can derive alternate expressions for $F^*(s)$ for the ideal sampler that are comparable to those of $F_p^*(s)$ in Eqs. (2-37) through (2-42). The time-domain and the transform-domain expressions for the ideal sampling process are summarized in the following. Details of the derivations are found in Appendix B.

Representation as an Impulse Train in the Time Domain

$$f^*(t) = \sum_{k=0}^{\infty} f(kT)\delta(t - kT) \tag{2-52}$$

Frequency Domain

$$F^*(s) = \frac{f(0+)}{2} + \frac{1}{T}\sum_{n=-\infty}^{\infty} F(s + jn\omega_s) \tag{2-53}$$

Laplace Transform (as an impulse train)

$$F^*(s) = \sum_{k=0}^{\infty} f(kT)e^{-kTs} \tag{2-54}$$

Laplace Transform [$F(s)$ has k simple poles]

$$F^*(s) = \sum_{n=1}^{k} \frac{N(\xi_n)}{D'(\xi_n)} \frac{1}{1 - e^{-T(s-\xi_n)}} \tag{2-55}$$

where

$$F(\xi) = N(\xi)/D(\xi) \tag{2-56}$$

$$D'(\xi_n) = \left.\frac{dD(\xi)}{d\xi}\right|_{\xi=\xi_n} \tag{2-57}$$

and ξ_n is the nth simple pole of $F(\xi)$, $n = 1, 2, \ldots, k$.

Laplace Transform [$F(s)$ has k poles with multiplicity $m_n \geq 1$]

$$F^*(s) = \sum_{n=1}^{k} \sum_{i=1}^{m_n} \frac{(-1)^{m_n-i}K_{ni}}{(m_n-1)!} \frac{\partial^{m_n-i}}{\partial s^{m_n-i}} \left[\frac{1}{1-e^{-Ts}}\right]\Bigg|_{s=s-s_n} \tag{2-58}$$

$$K_{ni} = \frac{1}{(i-1)!} \frac{\partial^{i-1}}{\partial s^{i-1}} [(s-s_n)^{m_n}F(s)]\Bigg|_{s=s_n} \tag{2-59}$$

or the residue at the pole $\xi = \xi_n$ which has a multiplicity of m_n (>1) is

$$\frac{1}{(m_n-1)!} \frac{\partial^{m_n-1}}{\partial \xi^{m_n-1}} \left[(\xi-\xi_n)^{m_n}F(\xi)\frac{1}{1-e^{-T(s-\xi)}}\right]_{\xi=\xi_n} \tag{2-60}$$

Now we present the following examples to illustrate the applications of Eqs. (2-52) through (2-59).

Example 2-1

Consider that the exponential function $f(t) = e^{-at}u_s(t)$, where a is a constant, is sampled by the ideal sampler at intervals of T s. From Eq. (2-52), the output of the sampler is written

$$f^*(t) = \sum_{k=0}^{\infty} f(kT)\delta(t-kT) = \sum_{k=0}^{\infty} e^{-akT}\delta(t-kT) \tag{2-61}$$

The Laplace transform of $f^*(t)$ is

$$F^*(s) = \sum_{k=0}^{\infty} e^{-akT}e^{-kTs} = 1 + e^{-(s+a)T} + e^{-2(s+a)T} + \cdots \tag{2-62}$$

The infinite series in the last equation can be written in a closed form by first multiplying both sides of the equation by $e^{-(s+a)T}$, giving

$$e^{-(s+a)T}F^*(s) = e^{-(s+a)T} + e^{-2(s+a)T} + e^{-3(s+a)T} + \cdots \tag{2-63}$$

Now subtracting the last equation from Eq. (2-62) and solving for $F^*(s)$, we get

$$F^*(s) = \frac{1}{1 - e^{-(s+a)T}} \tag{2-64}$$

provided that $|e^{-(s+a)T}| < 1$, which is the condition that the infinite series will converge.

Now let us solve the same problem using Eq. (2-55). The Laplace transform of $f(t)$ is

$$F(s) = \frac{N(s)}{D(s)} = \frac{1}{s + a} \tag{2-65}$$

which has a simple pole at $s = -a$. From Eq. (2-65), $N(s) = 1$, and $D(s) = s + a$. Thus, $D'(s) = 1$. Equation (2-55) gives

$$F^*(s) = \frac{N(\xi_1)}{D'(\xi_1)} \frac{1}{1 - e^{-T(s-\xi_1)}} \tag{2-66}$$

where $\xi_1 = -a$. Thus,

$$F^*(s) = \frac{1}{1 - e^{-(s+a)T}} \tag{2-67}$$

which is identical to the result in Eq. (2-64).

We can also arrive at the same answer in Eq. (2-67) using Eq. (2-58). In this case, $k = 1$ and the multiplicity of the pole is also unity. Thus, both n and i take on the value of 1 only, and Eq. (2-58) gives

$$F^*(s) = \left. \frac{1}{1 - e^{-Ts}} \right|_{s=s+a} = \frac{1}{1 - e^{-(s+a)T}} \tag{2-68}$$

Example 2-2

As an example of the application of Eqs. (2-58) and (2-60) to an $F(s)$ that has multiple-order poles, let us consider the sampling of a unit-ramp function by an ideal sampler. Let

$$f(t) = tu_s(t) \tag{2-69}$$

The Laplace transform of $f(t)$ is

$$F(s) = \frac{1}{s^2} \tag{2-70}$$

Since $F(s)$ has a second-order pole at $s = 0$, to find $F(s)$, Eq. (2-58) should be used. In this case $k = 1$, since there is only one distinct pole, and $m_n = m_1 = 2$ (second order). Equation (2-59) gives

$$K_{11} = s^2 F(s)|_{s=0} = 1 \tag{2-71}$$

$$K_{12} = \frac{\partial}{\partial s} s^2 F(s)\bigg|_{s=0} = 0 \tag{2-72}$$

Thus, from Eq. (2-58), we get

$$F^*(s) = (-1)K_{11}\left[\frac{\partial}{\partial s}\frac{1}{1 - e^{-Ts}}\right] = \frac{Te^{-Ts}}{(1 - e^{-Ts})^2} \tag{2-73}$$

Equation (2-60) can be applied directly to find $F^*(s)$. We have

$$F^*(s) = \frac{\partial}{\partial \xi}\left[\xi^2 \frac{1}{\xi^2}\frac{1}{1 - e^{-T(s-\xi)}}\right]_{\xi=0} = \frac{Te^{-Ts}}{(1 - e^{-Ts})^2} \tag{2-74}$$

In principle, Eq. (2-53) can be used to get the same results in the last two examples, but the mathematics involved are more complex. For this reason, Eq. (2-53) is seldom used for the determination of a closed-form solution of $F^*(s)$. If $f(t)$ is given, and the signal is sampled by an ideal sampler, then Eq.

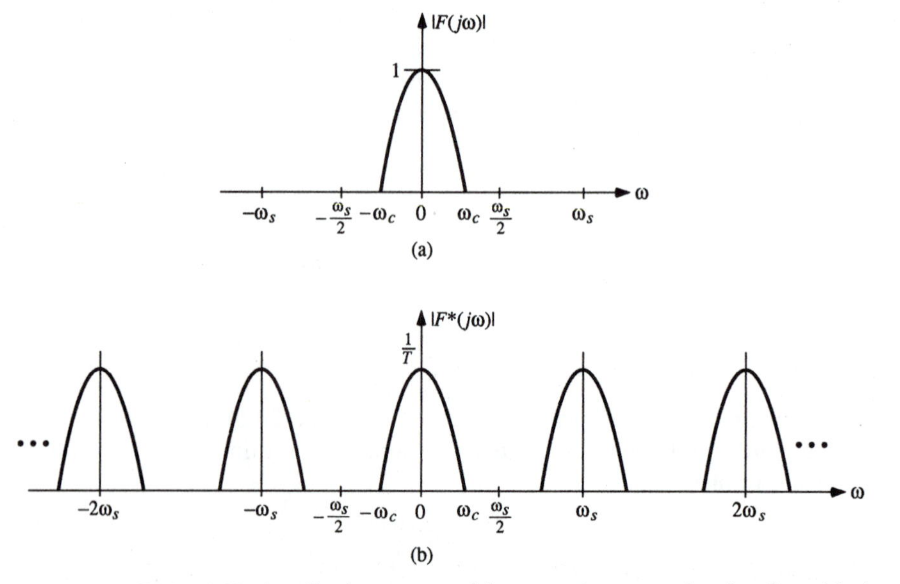

Figure 2-28. Amplitude spectra of input and output signals of an ideal sampler. (a) Amplitude spectrum of continuous input $f(t)$. (b) Amplitude spectrum of sampler output ($\omega_s > 2\omega_c$).

(2-54) can be used to find $F^*(s)$, although the disadvantage of this equation is that the answer appears in the infinite-series form, and we have to find the closed-form solution. Alternatively, we can take the Laplace transform of $f(t)$ and use either Eq. (2-55) or (2-58), depending on the pole properties of $F(s)$. If $F(s)$ is given, then we can go directly to Eq. (2-55) or (2-58); there is no need to take the inverse Laplace transform of $F(s)$ to first get $f(t)$ and then use Eq. (2-54).

Equation (2-53) is useful in interpreting the operation of the ideal sampler as a harmonic generator. As shown by Eq. (2-53), the ideal sampler reproduces in its output the spectrum of the continuous input $f(t)$ as well as the harmonic components at integral multiples of the sampling frequency ω_s, and all the components have the same amplitude of $1/T$. Let us assume that the amplitude spectrum of the continuous input $f(t)$ is band-limited, as shown in Fig. 2-28(a), where ω_s is the sampling frequency in rad/s and ω_c is the highest frequency contained in $f(t)$. Again, if the sampling frequency is less than $2\omega_c$, distortion will occur in the output spectrum because of the overlapping of the sidebands in $|F^*(j\omega)|$. It is also interesting to note that, for the finite-pulsewidth sampling, the amplitude of the spectrum of $|F_p^*(j\omega)|$ decreases as ω increases, whereas in the case of ideal sampling, the amplitude is constant for all ω.

2-11 SOME s-PLANE PROPERTIES OF $F^*(s)$

It is of interest to examine some of the properties of the ideal sampler exhibited by the Laplace transformed function $F^*(s)$. Understanding these properties will aid us in the analysis and design of control systems with digital and sampled data.

1. $F^*(s)$ is a periodic function with period ω_s.

The property of $F^*(s)$ as a periodic function is apparent in view of Eq. (2-53) and Fig. 2-28. Analytically, we can show this by substituting $s + jm\omega_s$ for s in Eq. (2-53), where m is an integer. We have

$$F^*(s + jm\omega_s) = \frac{f(0+)}{2} + \frac{1}{T}\sum_{n=-\infty}^{\infty} F(s + jn\omega_s + jm\omega_s) \tag{2-75}$$

Since the limits of the summation in the last equation are from $-\infty$ to ∞, the equation is still $F^*(s)$ whatever the integral value of m. Thus,

$$F^*(s + jm\omega_s) = F^*(s) \tag{2-76}$$

for any integer m. We can arrive at the same answer by using Eq. (2-54).

Equation (2-76) indicates that given any point $s = s_1$ in the s-plane, the function $F^*(s)$ has the *same* value at all periodic points $s = s_1 + jm\omega_s$ where m is an integer. In Fig. 2-29 the s-plane is divided into an infinite number of periodic strips, each with a width of ω_s, the sampling frequency. The strip between $\omega = -\omega_s/2$ and $\omega = \omega_s/2$ is called the *primary strip*, and all

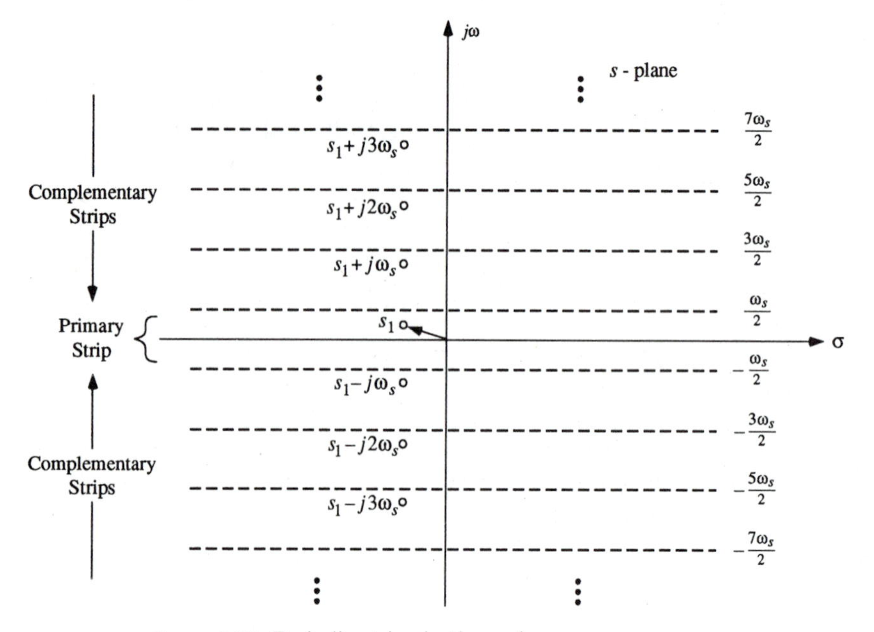

Figure 2-29. Periodic strips in the s-plane.

others occurring at higher frequencies, both positive and negative, are called the *complementary strips. The function $F^*(s)$ has the same value at all congruent points in the various periodic strips.*

2. If the function $F(s)$ has a pole at $s = s_1$, then $F^*(s)$ has poles at $s = s_1 + jm\omega_s$ where m is an integer from $-\infty$ to ∞.

   Equation (2-53) again shows this property. Figure 2-30 shows that if $F(s)$ has two real poles in the s-plane, $F^*(s)$ will have an infinite number of poles located at periodic locations that are spaced by integral multiples of ω_s in the complementary strips. This pole configuration of the sampled signal is useful in demonstrating the significance of the sampling theorem and the folding frequency $\omega_s/2$. Figure 2-30 shows that when the signal $f(t)$ is sampled, the poles of $F(s)$ are folded out about frequencies that are integral multiples of the folding frequency to form the poles of $F^*(s)$. As long as the poles of $F^*(s)$ lie inside the primary strip of $-\omega_s/2 < \omega < \omega_s/2$, which corresponds to the situation where the sampling frequency is at least twice the highest frequency contained in $f(t)$, these poles will be folded uniquely into the complementary strips by the sampling operation. Then, at least in principle, an ideal filter with a bandpass characteristic of $|\omega_c| < \omega_s/2$ would eliminate all the harmonic poles so that the end result would be the exact recovery of $F(s)$ or $f(t)$. On the other hand, if the sampling theorem is not satisfied by the sampler so that the relative position of the poles of a given $F(s)$ with respect to the folding frequency is shown in Fig. 2-31(a), the sampled function $F^*(s)$ will have poles that are folded

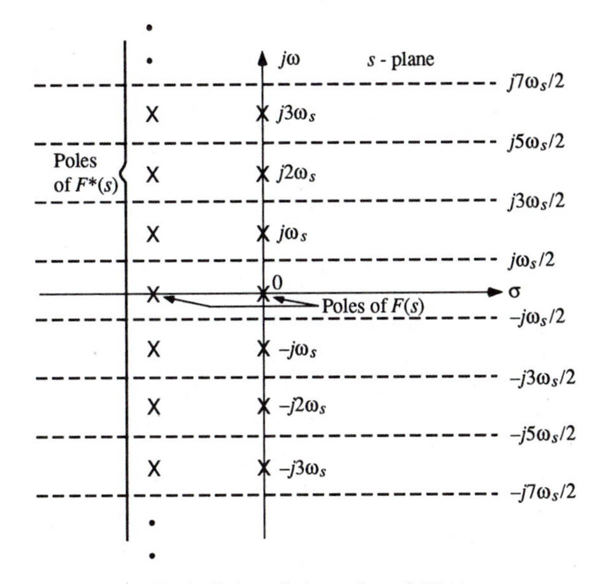

Figure 2-30. Periodicity of the poles of $F^*(s)$.

back into the primary strip, as shown in Fig. 2-31(b). The folding back of the high-frequency poles of $F^*(s)$ into the primary strip would prevent the recovery of the original signal from $f^*(t)$ even by means of an ideal filter. In control systems, the folding of the high-frequency poles back to the primary strip, due to an inadequate sampling rate, may cause design difficulties.

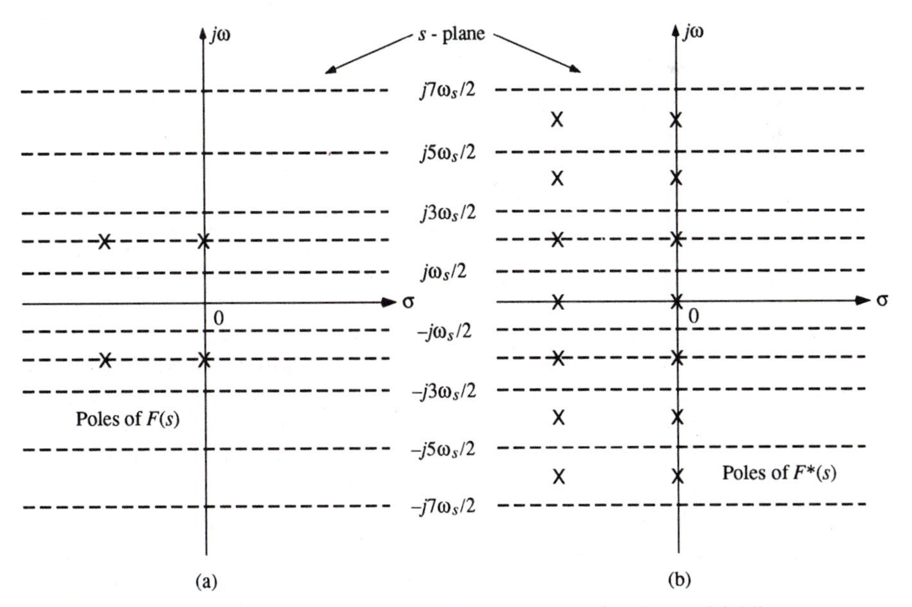

Figure 2-31. Poles of $F(s)$ and $F^*(s)$ to show the effects of folding.

2-12 DATA RECONSTRUCTION AND FILTERING OF SAMPLED SIGNALS

Most control systems have controlled processes that inherently contain analog devices. For instance, ac and dc motors are important prime movers often used in control systems. These devices are inherently driven by analog current or voltage inputs whether the control system is analog or digital. Thus, when sampled-data or digital signals appear in a control system, they should be first converted into analog signals before being applied to the controlled process. Another way of looking at the problem is that the high-frequency harmonic components in a signal $f^*(t)$ that resulted from the sampling operation must be removed before the signal is applied to the analog portion of the system. In general, a low-pass filter or a data-reconstruction device may be used for this purpose. In control systems, due to simplicity and low cost, the hold operation of the sample-and-hold device discussed in Sec. 2-4, which holds the value of a given sampled pulse until the next one arrives, turns out to be the most popular filtering device.

We shall show that the hold device is the simplest form of a general data-reconstruction problem. The problem of data reconstruction can be regarded as: *given a sequence of numbers*, $f(0)$, $f(T)$, $f(2T), \ldots, f(kT), \ldots$, *a continuous-data signal* $f(t)$, $t \geq 0$, *is to be reconstructed from information contained in the sequence.* This data-reconstruction process may be regarded as an *extrapolation process*, since the continuous-data signal is to be constructed based on information available only at past sampling instants. For instance, the original signal $f(t)$ between two consecutive sampling instants, kT and $(k+1)T$, is to be estimated based on the values of $f(t)$ at all preceding sampling instants of kT, $(k-1)T$, $(k-2)T, \ldots, 0$; that is, $f(kT)$, $f(k-1)T$, $f(k-2)T, \ldots, f(0)$.

A well-known method of generating this desired approximation is based on the power-series expansion of $f(t)$ in the interval between the sampling instants kT and $(k+1)T$. The approximation is written as

$$f_k(t) = f(kT) + f^{(1)}(kT)(t - kT) + \frac{f^{(2)}(kT)}{2!}(t - kT)^2 + \cdots \qquad \text{(2-77)}$$

where

$$f_k(t) = f(t) \qquad \text{for} \quad kT \leq t < (k+1)T \qquad \text{(2-78)}$$

$$f^{(n)}(kT) = \left. \frac{d^{(n)}f(t)}{dt^n} \right|_{t=kT} \qquad \text{(2-79)}$$

for $n = 1, 2, \ldots$.

To evaluate the coefficients of the power series in Eq. (2-77), the derivatives of the function $f(t)$ must be obtained at the sampling instants. Since the only available information on $f(t)$ is its magnitudes at the sampling instants, the derivatives of $f(t)$ must be estimated from the values of $f(kT)$. A simple expression involving only two data pulses gives an estimate of the first derivative of $f(t)$ at $t = kT$:

$$f^{(1)}(kT) = \frac{1}{T}\,[f(kT) - f[(k-1)T]] \tag{2-80}$$

Similarly, we can approximate the second derivative of $f(t)$ at $t = kT$ as

$$f^{(2)}(kT) = \frac{1}{T}\,[f^{(1)}(kT) - f^{(1)}[(k-1)T]] \tag{2-81}$$

where

$$f^{(1)}[(k-1)T] = \frac{1}{T}\,[f[(k-1)T] - f[(k-2)T]] \tag{2-82}$$

Substituting of Eqs. (2-81) and (2-82) into Eq. (2-80), we get

$$f^{(2)}(kT) = \frac{1}{T^2}\,[f(kT) - 2f[(k-1)T] + f[(k-2)T]] \tag{2-83}$$

Thus, we have expressed the second derivative of $f(t)$ at $t = kT$ in terms of $f(kT)$ and the sampled values at two preceding sampling instants. We can see that, if necessary, we can express any higher order derivatives in terms of more of the past sampled values of $f(kT)$. For instance, $f^{(3)}(kT)$ would involve $f(kT)$, $f[(k-1)T]$, $f[(k-2)T]$, and $f[(k-3)T]$. Thus, the higher the order of the derivative to be approximated, the larger will be the number of delayed pulses required. In general, the number of delayed pulse data required to approximate the value of $f^{(n)}(kT)$ is $n + 1$. Thus, the extrapolating device described above consists essentially of a series of time delays, and in theory the number of delays depends on the degree of accuracy of the estimate of the time function $f(t)$ during the time interval from kT to $(k + 1)T$.

An adverse effect of the time delay on the stability of closed-loop control systems is well known. Therefore, an attempt to utilize the higher order derivatives of $f(t)$ for the purpose of more accurate extrapolation is often met with serious stability problems in most control systems. Furthermore, a high-order extrapolation also requires complex circuitry and results in a higher cost in construction.

2-13 THE ZERO-ORDER HOLD

For the two reasons cited above, quite frequently only the first term of Eq. (2-77) is used for the approximation of $f(t)$ during the time interval $kT \leq t < (k + 1)T$. Alternatively,

$$f_k(t) = f(kT) \tag{2-84}$$

The device that performs this type of extrapolation is known as the *zero-order extrapolator*, since the polynomial used is of the zeroth order. It is also better

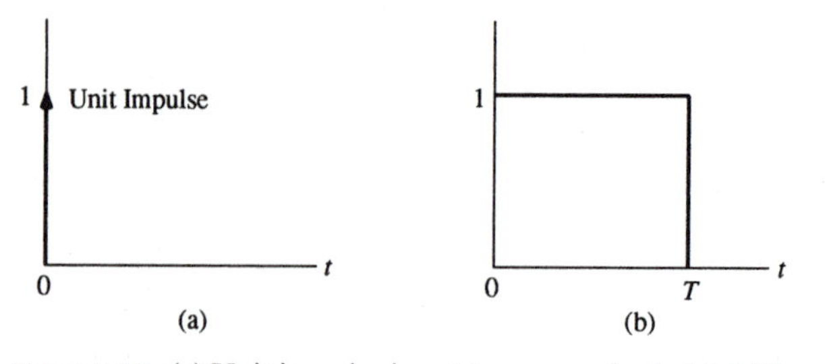

Figure 2-32. (a) Unit impulse input to zero-order hold. (b) Impulse response of zero-order hold.

known as the *zero-order hold* (zoh) since it holds the value of the sampled value $f(kT)$ for $kT \leq t < (k+1)T$ until the next sample $f[(k+1)T]$ arrives. The zero-order hold has the same characteristics as the hold portion of the S/H device discussed in Sec. 2-4.

We can easily see that the zoh is a linear device, since it satisfies the principle of superposition. To derive the transfer function of the zoh, we refer to the impulse response shown in Fig. 2-32(b), which is apparent in view of Eq. (2-84). The impulse response is expressed as

$$g_{h0}(t) = u_s(t) - u_s(t - T) \tag{2-85}$$

where $u_s(t)$ is the unit-step function. The transfer function of the zoh is then obtained by taking the Laplace transform of Eq. (2-85). The result is

$$G_{h0}(s) = \frac{1 - e^{-Ts}}{s} \tag{2-86}$$

Figure 2-33(b) illustrates the waveform of the output of the zoh $h(t)$ when its input is a typical pulse sequence $f(kT)$, $k = 0, 1, 2, \ldots$, or a sampled signal $f^*(t)$. The waveforms in Fig. 2-33 clearly indicate that the accuracy of the zero-order hold as an extrapolating device depends greatly on the magnitude of the sampling frequency ω_s. As the sampling frequency increases to infinity or the sampling period T approaches zero, the output of the zoh $h(t)$ approaches the continuous-time signal $f(t)$. We shall verify this analytically in the next chapter.

2-13-1 Frequency-Domain Characteristics of the Zero-Order Hold

Since the zoh is a data-reconstruction or filtering device, it is of interest to examine its frequency-domain characteristics. Replacing s by $j\omega$ in Eq. (2-86), we get

$$G_{h0}(j\omega) = \frac{1 - e^{-j\omega T}}{j\omega} \tag{2-87}$$

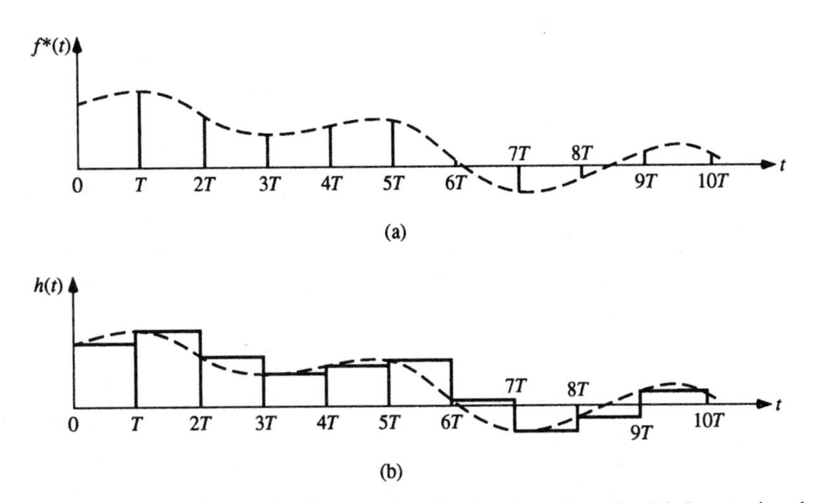

Figure 2-33. Zero-order hold operation in the time domain. (a) Input signal $f(t)$ and sampled signal $f^*(t)$. (b) Output waveform of zero-order hold.

The last equation can be conditioned as follows:

$$G_{h0}(j\omega) = \frac{2e^{-j\omega T/2}(e^{j\omega T/2} - e^{-j\omega T/2})}{j2\omega}$$

$$= \frac{2\sin(\omega T/2)}{\omega} e^{-j\omega T/2} \tag{2-88}$$

or

$$G_{h0}(j\omega) = T \frac{\sin(\omega T/2)}{\omega T/2} e^{-j\omega T/2} \tag{2-89}$$

Since T is the sampling period in seconds, and $T = 2\pi/\omega_s$, where ω_s is the sampling frequency in rad/s, Eq. (2-89) becomes

$$G_{h0}(j\omega) = \frac{2\pi}{\omega_s} \frac{\sin(\pi\omega/\omega_s)}{\pi\omega/\omega_s} e^{-j(\pi\omega/\omega_s)} \tag{2-90}$$

The magnitude of $G_{h0}(j\omega)$ is written as

$$|G(j\omega_s)| = \frac{2\pi}{\omega_s} \left| \frac{\sin(\pi\omega/\omega_s)}{\pi\omega/\omega_s} \right| \tag{2-91}$$

The phase of $G_{h0}(j\omega)$ is written as

$$\angle G(j\omega_s) = \angle \sin(\pi\omega/\omega_s) - \pi\omega/\omega_s \qquad \text{rad} \tag{2-92}$$

The sign of $\sin(\pi\omega/\omega_s)$ changes for every integral value of $\pi\omega/\omega_s$. We can regard the change of sign from $+$ to $-$ as a phase change of -180 degrees.

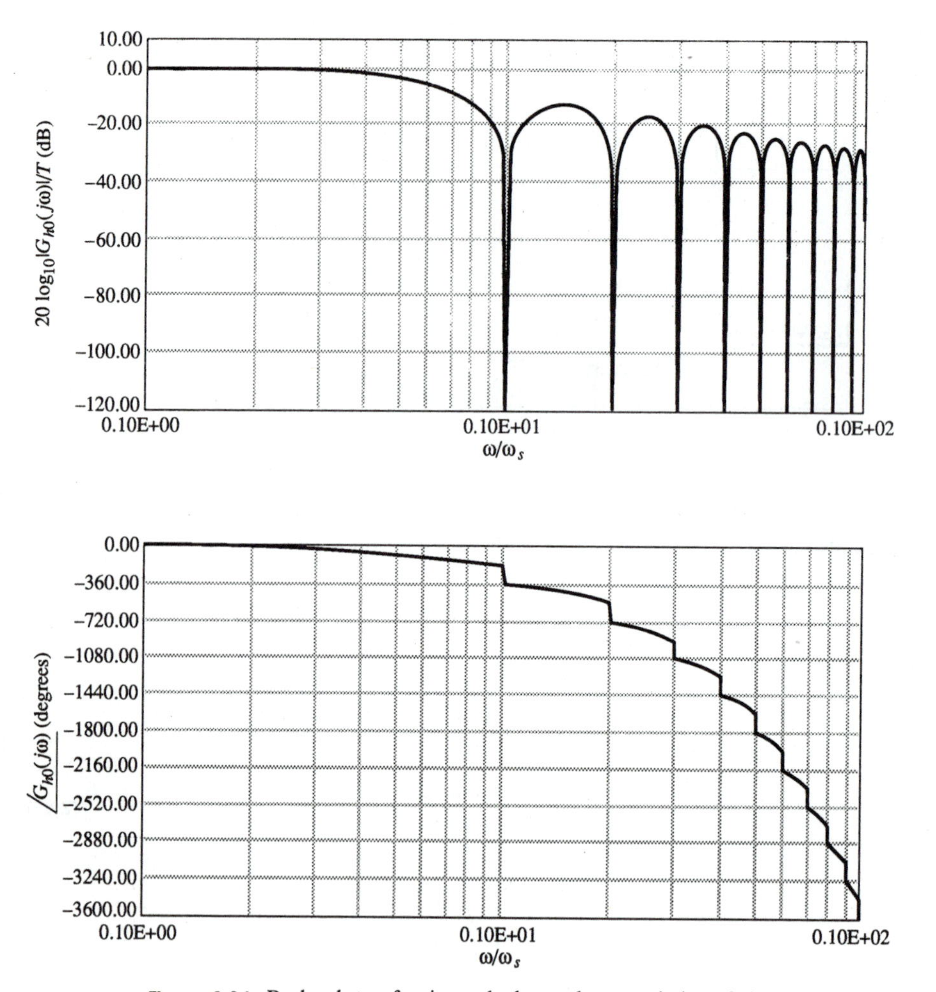

Figure 2-34. Bode plots of gain and phase characteristics of the zero-order hold.

The Bode plots of the gain and phase characteristics of $G_{h0}(j\omega)/T$ are shown in Fig. 2-34 as functions of ω/ω_s. A better illustration of the zoh as a low-pass filter is to plot the gain and phase characteristics of $G_{h0}(j\omega)$ in the regular coordinates, as shown in Fig. 2-35. However, when compared with the characteristics of the ideal low-pass filter that has a cutoff frequency at $\omega = \omega_s/2$, the amplitude characteristic of $G_{h0}(j\omega)$ is zero at $\omega = \omega_s$ (as well as integral multiples of ω_s) instead of cutting off sharply at $\omega_s/2$. At $\omega = \omega_s/2$, the magnitude of $G_{h0}(j\omega)/T$ is 0.636.

The phase characteristic of $G_{h0}(j\omega)$ is linear over the frequency intervals of $k\omega_s \leq \omega < (k+1)\omega_s$, $k = 0, 1, 2, \ldots$, with jump discontinuities of -180 degrees at integral multiples of ω_s.

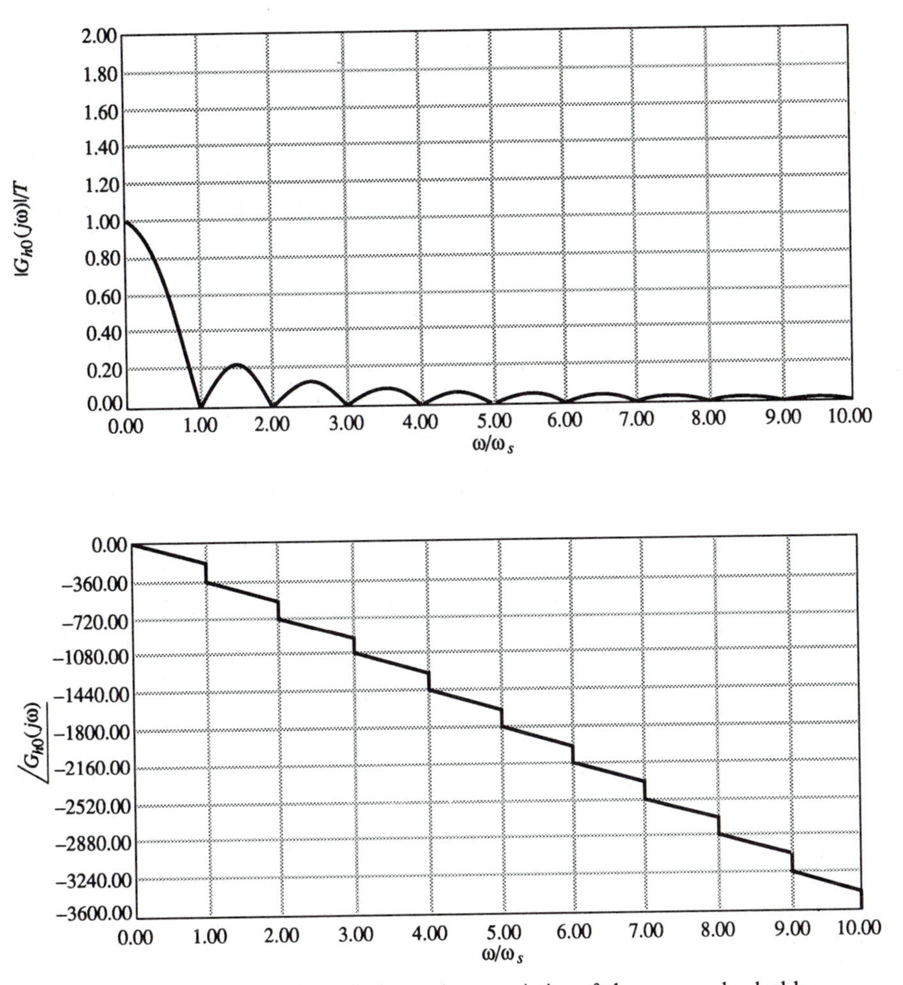

Figure 2-35. Gain and phase characteristics of the zero-order hold.

2-14 THE FIRST-ORDER HOLD

If the first two terms of the power series in Eq. (2-77) are used to extrapolate the time function $f(t)$ over the time interval $kT \le t < (k+1)T$, the device is called a *first-order hold* (foh). The equation for the foh is

$$f_k(t) = f(kT) + f^{(1)}(kT)(t - kT) \tag{2-93}$$

where the first-order derivative of $f(t)$ at $t = kT$ is approximated as

$$f^{(1)}(kT) = \frac{f(kT) - f[(k-1)T]}{T} \tag{2-94}$$

Substituting Eq. (2-94) in Eq. (2-93), we get

$$f_k(t) = f(kT) + \frac{f(kT) - f[(k-1)T]}{T}(t - kT) \tag{2-95}$$

where $kT \le t < (k+1)T$. Equation (2-95) shows that the output of the foh between two consecutive sampling instants is a ramp function. The slope of the ramp is equal to the difference of $f(kT)$ and $f[(k-1)T]$.

The impulse response of the foh is obtained by applying a unit impulse at $t = 0$ as input. The corresponding output, or the impulse response, is determined from Eq. (2-95) by setting $k = 0, 1, 2, \ldots,$ for the various time intervals. When $k = 0$, the impulse response for $0 \le t < T$ is described by

$$f_0(t) = f(0) + \frac{f(0) - f(-T)}{T}t \tag{2-96}$$

For a unit impulse input, $f(0) = 1$ and $f(-T) = 0$. Thus, the impulse response of the foh for $0 \le t < T$ is

$$g_{h1}(t) = 1 + \frac{1+t}{T} \tag{2-97}$$

Setting $k = 1$ in Eq. (2-95), we have

$$f_1(t) = f(T) + \frac{f(T) - f(0)}{T}(t - T) \tag{2-98}$$

Since $f(0) = 1$ and $f(T) = 0$, the impulse response of the foh over the time interval $0 \le t < T$ is

$$g_{h1}(t) = 1 - \frac{t}{T} \tag{2-99}$$

The impulse response of the foh for $t > T$ is zero, since $f(t) = 0$ for $t \ge 2T$. The impulse response of the foh is shown in Fig. 2-36(b). Functionally, the impulse response in Fig. 2-36(b) is written

$$\begin{aligned} g_{h1}(t) = u_s(t) + \frac{t}{T}u_s(t-T) - 2u_s(t-T) - \frac{2(t-T)}{T}u_s(t-T) \\ + \frac{(t-2T)}{T}u_s(t-2T)u_s(t-2T) + u_s(t-2T) \end{aligned} \tag{2-100}$$

The transfer function of the foh is obtained by taking the Laplace transform of the last equation, and after simplifying,

$$G_{h1}(s) = \frac{1+Ts}{T}\left[\frac{1 - e^{-Ts}}{s}\right]^2 \tag{2-101}$$

or

$$G_{h1}(s) = \frac{1+Ts}{T}[G_{h0}(s)]^2 \tag{2-102}$$

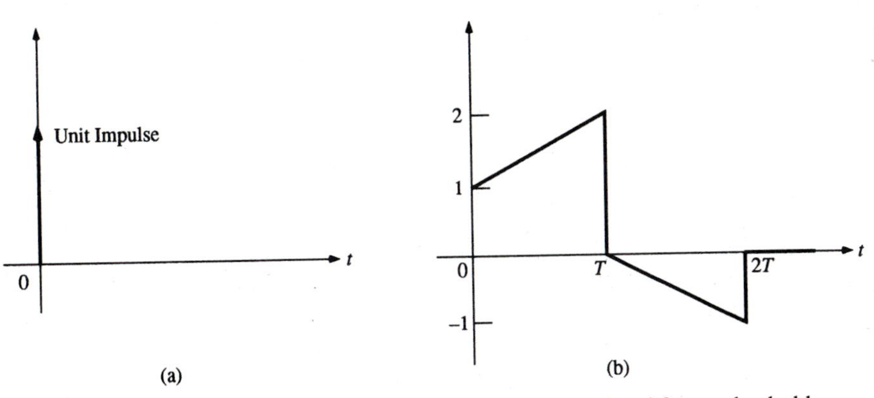

Figure 2-36. (a) Unit impulse. (b) Impulse response of first-order hold.

Figure 2-37 illustrates the reconstruction of a time function $f(t)$ from the sampled signal $f(kT)$ using an foh. In this case, the advantage of the foh over the zoh is not obvious. In fact, the zoh actually gives a better approximation in certain sampling intervals. It appears that the ramp functions contained in the impulse response of the foh may cause "overcorrections" in the approximation.

2-14-1 Frequency-Domain Characteristics of the First-Order Hold

The frequency response of the foh is obtained by substituting $s = j\omega$ in Eq. (2-101). We have

$$G_{h1}(j\omega) = \frac{1 + j\omega T}{T}\left[\frac{1 - e^{j\omega T}}{j\omega}\right]^2 \tag{2-103}$$

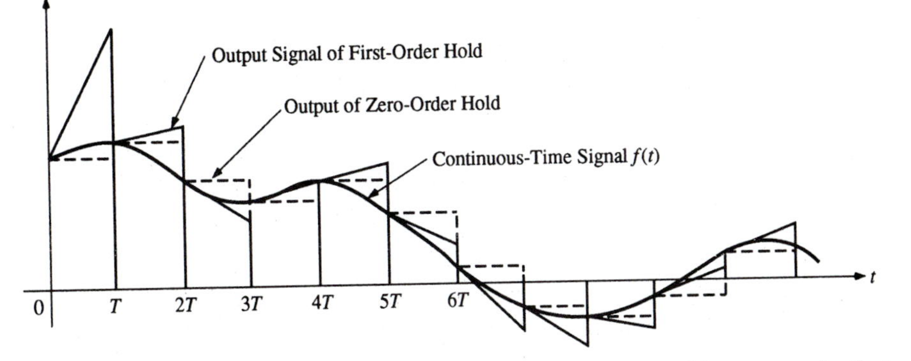

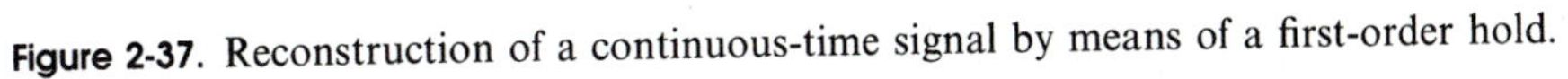

Figure 2-37. Reconstruction of a continuous-time signal by means of a first-order hold.

The magnitude and phase of $G_{h1}(j\omega)$ are obtained as

$$|G_{h0}(j\omega)| = \frac{2\pi}{\omega_s}\sqrt{1 + \frac{4\pi^2\omega^2}{\omega_s^2}}\left[\frac{\sin \pi\omega/\omega_s}{\pi\omega/\omega_s}\right]^2 \tag{2-104}$$

$$\angle G_{h1}(j\omega) = \tan^{-1}\left[\frac{2\pi\omega}{\omega_s}\right] - \frac{2\pi\omega}{\omega_s} \quad \text{rad} \tag{2-105}$$

Figure 2-38 illustrates the Bode plots of the frequency-response curves of the foh in dB magnitude versus ω/ω_s and phase versus ω/ω_s. Figure 2-39 shows the same curves in absolute coordinates. From the magnitude characteristic we see that the foh cuts off at $\omega = \omega_s$, just as in the case of the zoh, but

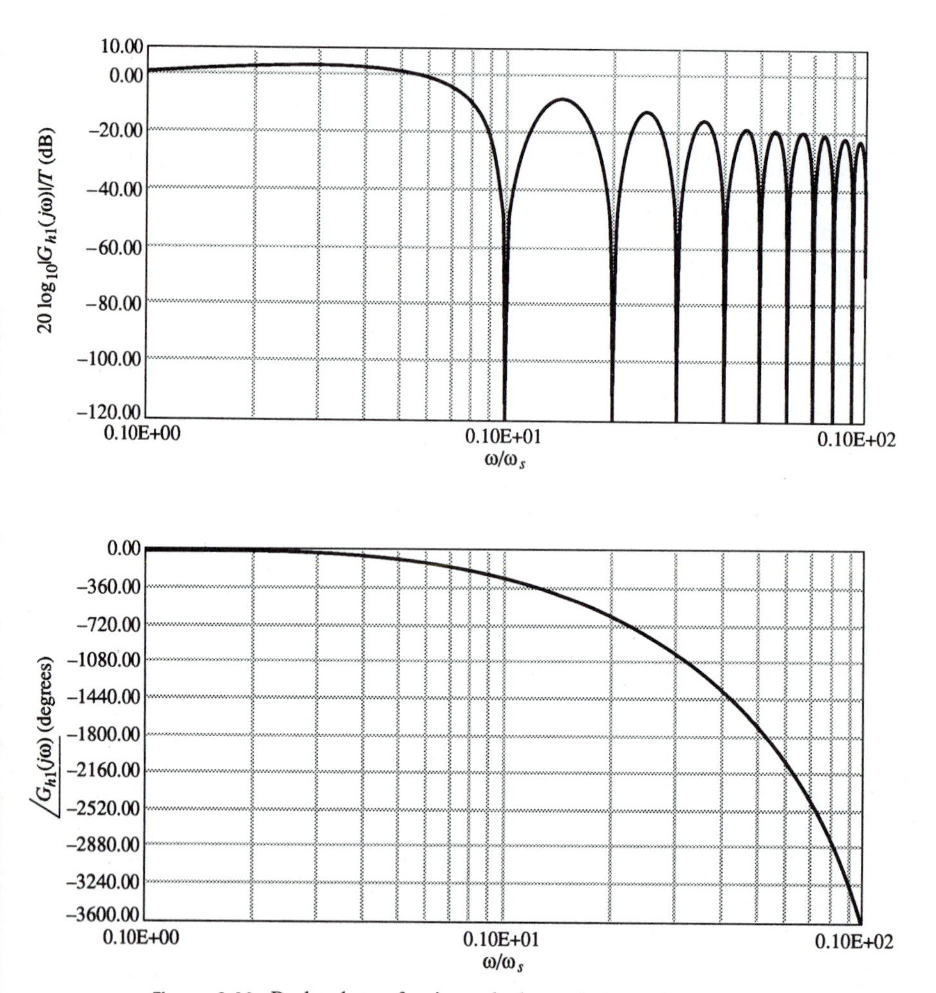

Figure 2-38. Bode plots of gain and phase characteristics of first-order hold.

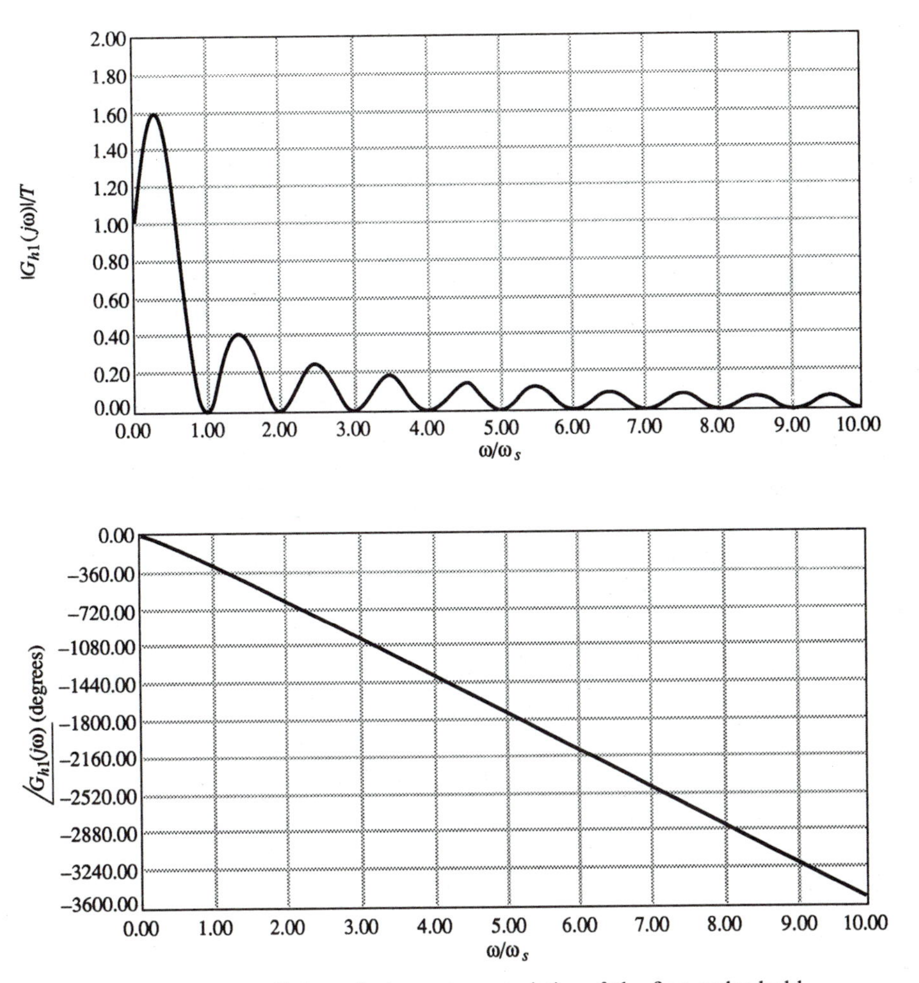

Figure 2-39. Gain and phase characteristics of the first-order hold.

at a sharper rate. However, the peak values of the foh in each ω/ω_s interval are higher than those of the zoh, which explains the ramp components in the output of the foh. The phase characteristic of the foh, as given by Eq. (2-105), is smooth for all values of ω.

Figure 2-40 shows a comparison between the frequency responses of the foh and zoh for $\omega/\omega_s = 0$ to $\omega/\omega_s = 5$. Note that at low frequencies over each ω/ω_s interval, the phase lag produced by the zoh exceeds that of the foh, but the opposite is true at high frequencies. In feedback control systems, the high-frequency characteristics of the open-loop transfer function normally govern the stability of the closed-loop system. Thus, application of the foh, in general, will create more stability problems than the zoh.

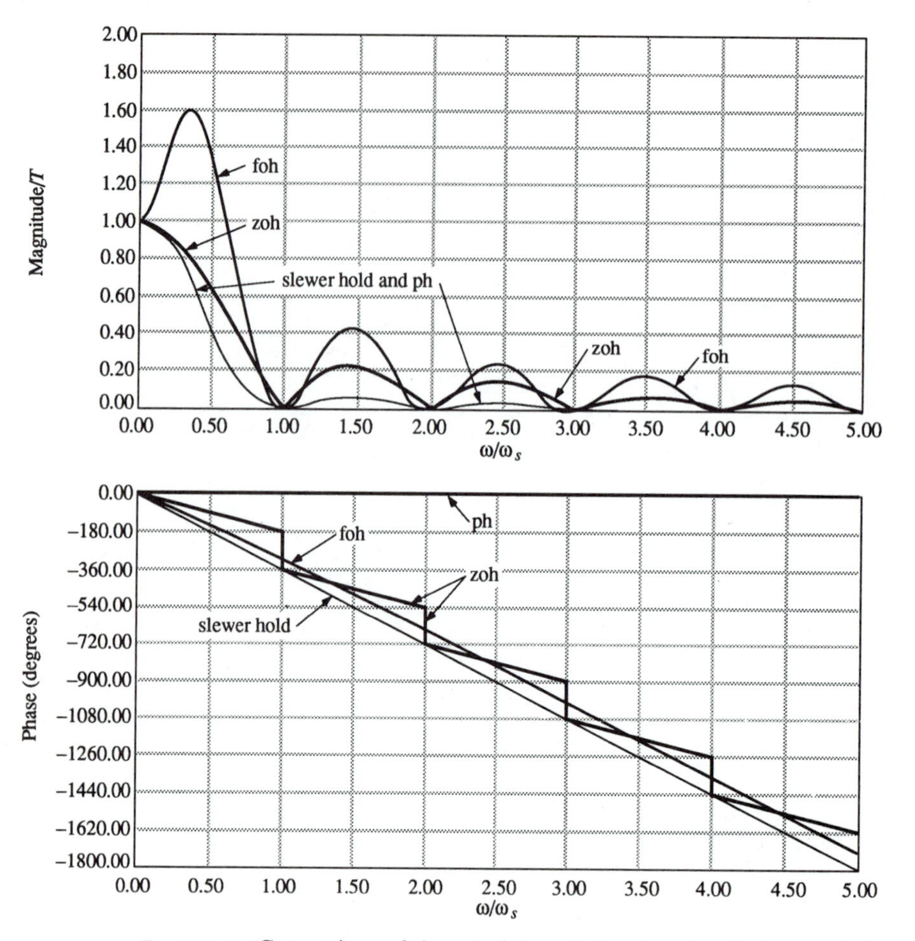

Figure 2-40. Comparison of the zoh, foh, and the slewer hold characteristics.

2-15 THE POLYGONAL HOLD AND THE SLEWER HOLD

From the standpoint of reconstructing a continuous-time signal from its sampled data, a smooth extrapolating scheme would seem to be simply to connect all the sample points by straight lines. Figure 2-41 illustrates such an approximation scheme. Since the output of such a data hold is in the form of a polygon between two consecutive sampling periods, it is often referred to as a *polygonal hold*, or ph. It is easy to see that the polygonal output $h(t)$ shown in Fig. 2-41 can be considered as the sum of the two triangular signals $f_1(t)$ and $f_2(t)$ shown in Fig. 2-42. Thus, the impulse response of the polygonal hold is the triangle response shown in Fig. 2-43. Obviously, the polygonal hold described is not *causal*, and cannot be physically realized, since the output of the device precedes the input. As shown in Fig. 2-41, to connect all the sampled

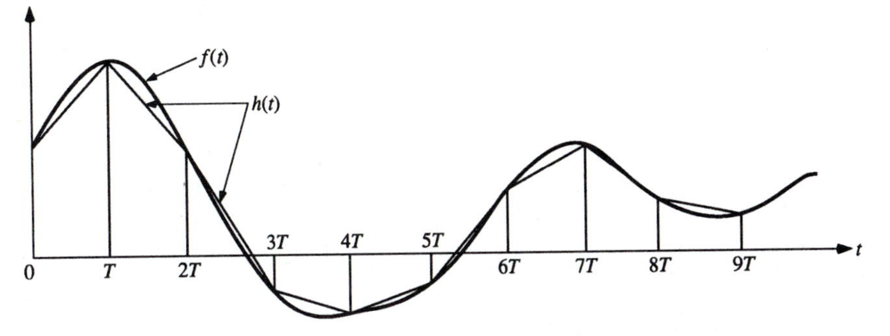

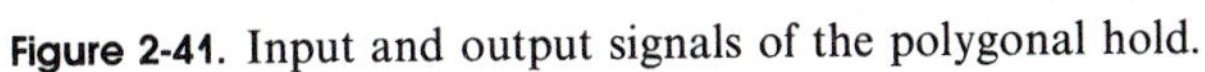
Figure 2-41. Input and output signals of the polygonal hold.

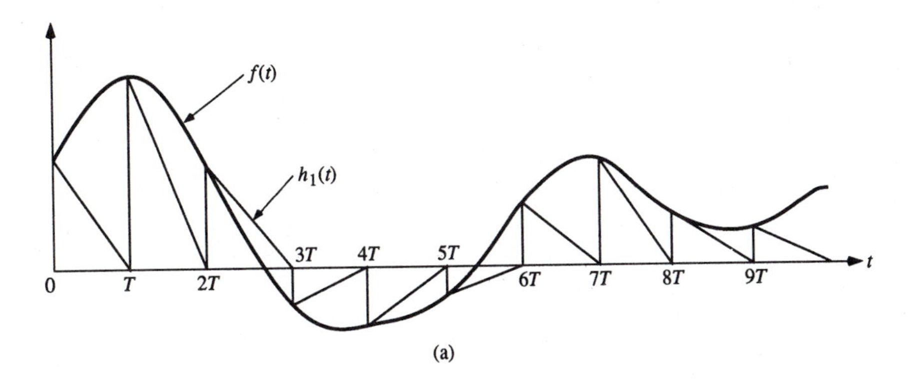

(a)

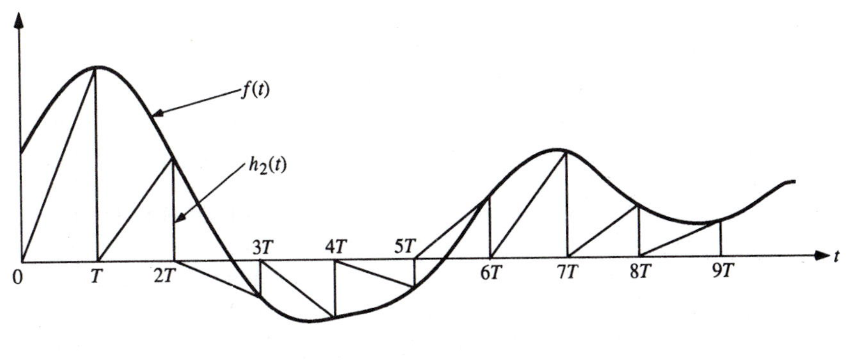

(b)

Figure 2-42. The polygonal hold output interpreted as the sum of two triangular hold outputs $h_1(t)$ and $h_2(t)$.

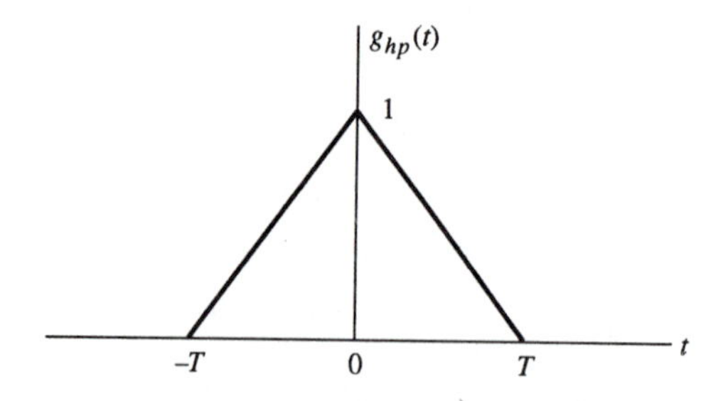

Figure 2-43. Impulse response of the polygonal hold.

points by straight lines, we have to know the sampled value $f[(k+1)T]$ at $t = kT$.

For reasons that will be apparent, let us proceed with the investigation of the properties of the ph even though it is physically unrealizable. The impulse response in Fig. 2-43 is written as

$$g_{hp}(t) = -\frac{(t+T)}{T} u_s(t+T) - \frac{2t}{T} u_s(t) + \frac{(t-T)}{T} u_s(t-T) \tag{2-106}$$

The transfer function of the ph is obtained by taking the Laplace transform of $g_{hp}(t)$.

$$G_{hp}(s) = \frac{e^{Ts} - 2 + e^{-Ts}}{Ts^2} \tag{2-107}$$

Equation (2-107) can be written as

$$G_{hp}(s) = \frac{(1 - e^{-Ts})^2}{Ts^2} e^{Ts} \tag{2-108}$$

or

$$G_{hp}(s) = \frac{[G_{ho}(s)]^2 e^{Ts}}{T} \tag{2-109}$$

2-15-1 Frequency-Domain Characteristics of the Polygonal Hold

To get the frequency-response characteristics of the ph, we substitute $s = j\omega$ in $G_{hp}(s)$ of Eq. (2-108). The result is

$$G_{hp}(j\omega) = \frac{2(1 - \cos \omega T)}{\omega^2 T} \tag{2-110}$$

Thus, the phase of the ph is zero for all frequencies. The gain and phase versus ω/ω_s characteristics of the ph are shown in Fig. 2-40 for up to $\omega/\omega_s = 5$. The amplitude characteristic of the ph shows that it is a better low-pass filter

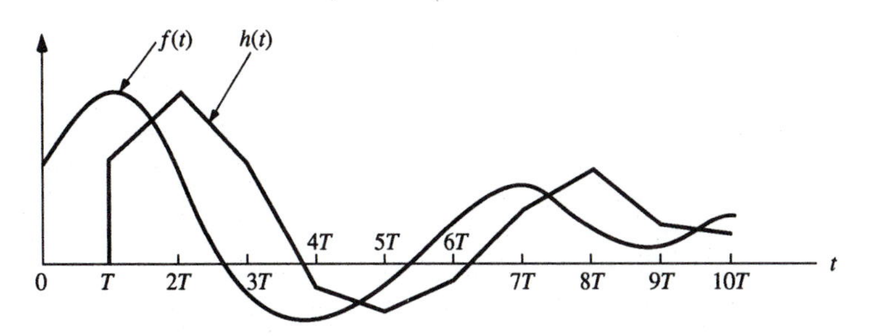

Figure 2-44. Input and output signals of the slewer hold.

than the zoh and the foh. The amplitude of $G_{hp}(j\omega)$ drops off very rapidly beyond $\omega/\omega_s = 1$.

2-15-2 The Slewer Hold

The e^{Ts} term in Eq. (2-109) is the reason that the ph is physically unrealizable. We can devise a data hold that retains only the part in Eq. (2-109) without the time-advance term. The result is called the *slewer hold*, or sh [21], [22]. Figure 2-44 shows the time-domain interpretation of the sh which is the output of the ph delayed by one sampling period T. The transfer function of the sh is written as

$$G_{hs}(s) = \frac{[G_{h0}(s)]^2}{T} = \frac{(1 - e^{-Ts})^2}{Ts^2} \tag{2-111}$$

The frequency-domain characteristics of the sh are described as

$$|G_{hs}(j\omega)| = \frac{2(1 - \cos \omega T)}{\omega^2 T} \tag{2-112}$$

$$\angle G_{hs}(j\omega) = -\frac{2\pi\omega}{\omega_s} \quad \text{rad} \tag{2-113}$$

Thus, the amplitude characteristic of the sh is identical to that of the ph, but the phase is a straight line, as shown in Fig. 2-40.

PROBLEMS

2-1 Consider that a new car is purchased with a loan of $p(0)$ dollars over a period of N months, at a monthly interest rate of R percent. The principal and interest are to be paid back in N equal payments. Determine the amount of each of the monthly payments U. The difference equation for this problem is given by Eq. (1-1). Solve the problem when $p(0) = \$15{,}000$, $N = 36$ months, and $R = 1$ percent $= 0.01$.

2-2 Consider that a new car is to be purchased on an installment loan of $p(0)$ over a period of N months. The current interest rate is R percent per month. The principal and interest are to be paid in N equal monthly payments U. Assume that the purchaser can afford to outlay only \$300 per month for 48 months for the purchase. Determine the maximum principal that can be borrowed for $R = 1.2$ percent $= 0.012$.

2-3 Consider that $p(0)$ dollars are borrowed over a period of N months at a monthly interest rate of R percent on the unpaid balance. The principal and interest are to be paid back at the end of each period with each payment doubling the preceding one. Write a difference equation that expresses the amount owed at the end of the $(k+1)$th period in terms of that of the previous period and the initial payment $u(0)$. Express the solution for $p(N)$ in terms of $p(0)$ and $u(0)$. For $N = 5$, $R = 0.01$, $p(0) = \$50{,}000$, find $u(0)$, $p(k)$, $k = 1, 2, 3, 4$, and the total amount (principal and interest) paid over the entire period.

2-4 It is desired to determine a recursive expression (difference equation) for finding the rth root of a real number N. We start by writing the Taylor series expansion of function $f(x)$ about a point x_n:

$$f(x) = f(x_n) + (x - x_n)f^{(1)}(x_n) + \frac{(x - x_n)^2}{2!} f^{(2)}(x_n) + \cdots \qquad \text{(P2-4-1)}$$

where

$$f^{(k)}(x_n) = \left. \frac{d^k f(x)}{dt^k} \right|_{x = x_n}$$

If the series is truncated after two terms, we have

$$f(x) = f(x_n) + (x - x_n)f^{(1)}(x_n) \qquad \text{(P2-4-2)}$$

Let x represent the next iterate x_{n+1}, and let x also represent a solution of the equation $f(x) = 0$. Then Eq. (P2-4-2) becomes

$$0 = f(x_n) + (x_{n+1} - x_n)f^{(1)}(x_n) \qquad \text{(P2-4-3)}$$

or

$$x_{n+1} = x_n - \frac{f(x_0)}{f^{(1)}(x_n)} \qquad \text{(P2-4-4)}$$

Equation (P2-4-4) is a difference equation useful for solving $f(x) = 0$.

a. Choose $f(x)$ appropriately and determine the corresponding difference equation for finding the rth root of a number N. Do five iterations to calculate the cube root of six using $x_0 = 1$ as the initial guess.

b. Repeat part **a** to find the cube root of seven.

2-5 Find the decimal equivalent of the following fixed-point binary numbers. The first bit is a sign bit.

a. 11011011

b. 0101.1111

c. 001101011.011

2-6 Find the decimal equivalent of the following floating-point binary numbers.

a. $+.1011101 + 011$

b. $+.1101 - 11$

2-7

a. Find the maximum conversion time required to digitize a 1-kHz sinusoidal signal $v(t) = V \sin \omega t$ to 10-bit resolution.

b. Show that, for a fixed resolution, the relation between conversion time and frequency for the sinusoidal signal described in part **a** is a straight line in the log-log coordinates. Illustrate the cases for 10-bit and 16-bit resolution for the conversion time scale of 10^{-9} to 10^{-3} s on the vertical axis and frequency from 1 Hz to 100 kHz on the horizontal axis, on log-log coordinates.

2-8 Find the output voltages of an 8-bit D/A converter with 10 V full scale for the digital input words listed below. Also fill in the weighting factor for each bit in terms of a fraction of full scale.

Binary word	Fraction of FS	Output voltage for FS = 10 V
00000001		
00000010		
00000100		
00001000		
00100000		
10000000		
11111111		

2-9 A machine gun mounted on a helicopter gunship may be modeled by the mass-spring-friction combination shown in Fig. P2-9. The force acting on the mass is in the form of blows of impulses of strength F at time intervals of $t = kT$, $k = 0, 1, 2, \ldots$. Assume that, at $t = 0$, M is at rest in its equilibrium position of $x(0) = 0$. Write the equation of motion for the system, and solve for the displacement $x(t)$ of M for $t > 0$ by means of the Laplace transform method: $K =$ linear spring constant $= M(u^2 + \omega_s^2)$, $u =$ constant, $\omega_s = 2\pi/T$, and $f =$ viscous frictional coefficient $= 2Mu$.

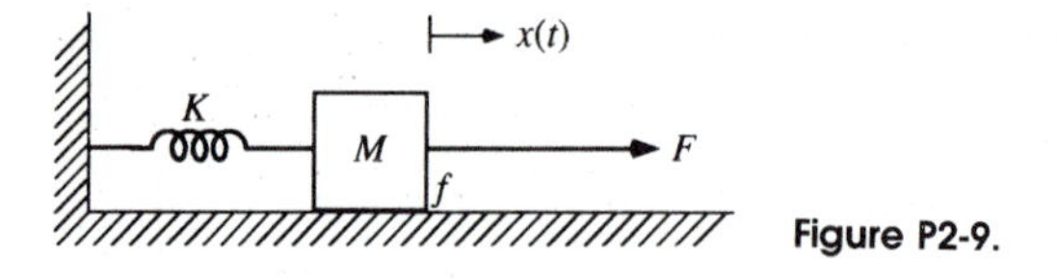

Figure P2-9.

2-10 The following signals are sampled by an ideal sampler with sampling period T. Determine the sampling output $f^*(t)$, and evaluate the pulse transform $F^*(s)$ by the Laplace transform method and the complex convolution method.

a. $f(t) = 3te^{-at}$, where a is a real constant
b. $f(t) = \sin 2t$
c. $f(t) = t \sin \omega t$
d. $f(t) = e^{-at} \sin 2t$
e. $f(t) = e^{-2(t-T)}u_s(t - T)$, where $u_s(t)$ is a unit-step function

2-11 Prove that if a signal $f(t)$ contains no frequency components higher than ω_c rad/s, it is completely characterized by the values $f^{(k)}(nT)$, $f^{(k-1)}(nT), \ldots,$ $f^{(1)}(nT)$, and $f(nT)$, $n = 0, 1, 2, \ldots,$ of the signal measured at instants of time separated by $T = (1/2)(k + 1)(2\pi/\omega_c)$ s, where

$$f^{(k)}(nT) = \left. \frac{d^k f(t)}{dt^k} \right|_{t=nT}$$

2-12

a. Find $f^*(t)$ for $f(t) = e^{-a|t|}$ $(-\infty < t < \infty)$. Find the two-sided Laplace transform of $f^*(t)$, $F^*(s)$. Determine the region of convergence of $F^*(s)$.
b. Find $f^*(t)$ of $f(t) = e^{-at} - e^{at}$, $t \geq 0$, and find $F^*(s)$. Compare the two $F^*(s)$ functions obtained in parts **a** and **b**. Discuss your findings.

2-13 A "delayed" sampler is considered to close for a short duration p at instants $t = \Delta, \Delta + T, \Delta + 2T, \ldots, \Delta + nT, \ldots,$ where $T > \Delta > 0$ and T is the sampling period in seconds. The input to the sampler, $f(t)$, is continuous. The output of the sampler denoted by $f_p^*(t)_\Delta$ is assumed to be a flat-topped pulse train, since $\Delta \ll T$. Derive the pulse transform of $f_p^*(t)_\Delta$, $F_p^*(s, \Delta)$. [Note that this can also be referred to as the "delayed pulse transform" of $f^*(t)$.] For small p, assume that $1 - e^{-ps} \cong ps$. Determine the delayed pulse transform of $f(t) = e^{-2t}$.

2-14 The delayed pulse transform described in Problem 2-13 can be applied to finite-pulsewidth consideration in sampled-data systems. As shown in Fig. P2-14(a), a pulse of width p can be considered as the resultant of N elementary pulses of width Δ, such that $N\Delta = p$ and Δ is very small. Therefore, a practical sampler can be represented by N samplers $S_0, S_\Delta, S_{2\Delta}, \ldots, S_{(N-1)\Delta}$, connected in parallel as shown in Fig. P2-14(b). This implies that sampler S_0 samples at instants $0, T, 2T, \ldots, nT, \ldots$; sampler S_Δ is actuated at instants Δ, $\Delta + T$, $\Delta + 2T, \ldots, \Delta + nT$, etc. Derive the pulse transform of $f_p^*(t)$.

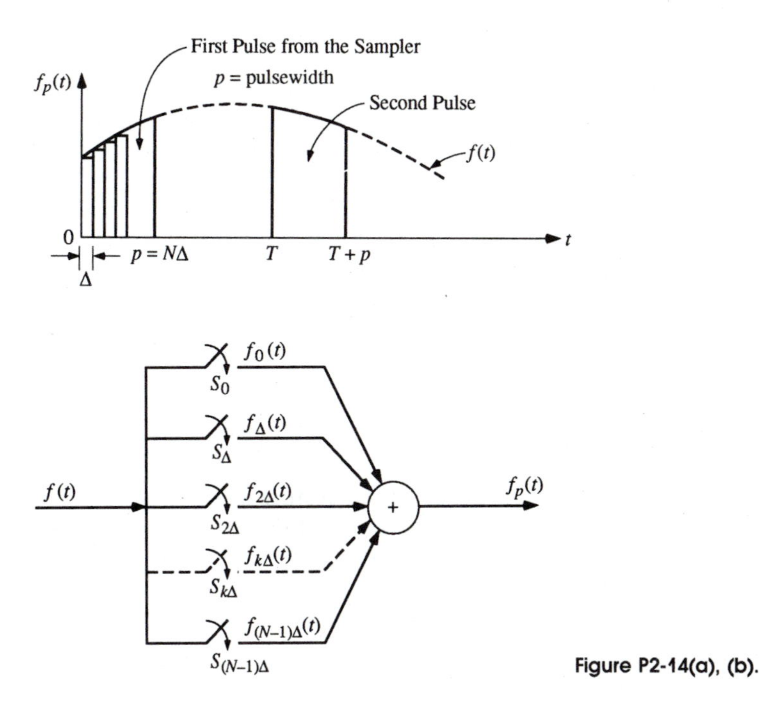

Figure P2-14(a), (b).

2-15 It is pointed out in this chapter that the frequency characteristic of a zero-order hold device has a rapid attenuation for low-frequency signals, while that of a first-order hold has an overshoot at low frequencies. It is suggested that a "fractional-order hold" can be devised by extrapolating the function in any given sampling interval with only a fraction of, rather than the full, first difference. The process is illustrated in Fig. P2-15. In the first sampling period

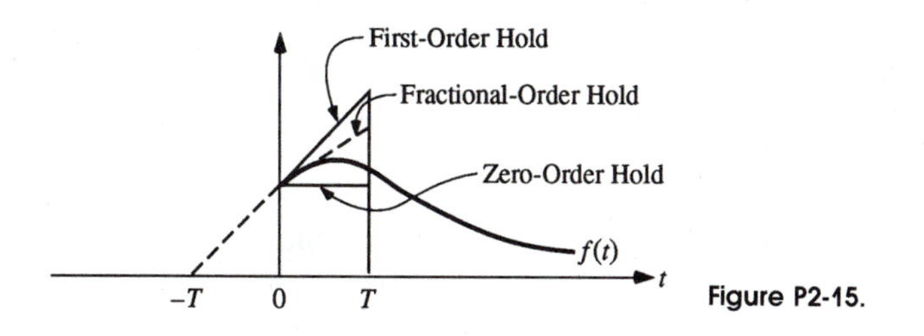

Figure P2-15.

from 0 to T s, the extrapolated function is a straight line whose slope is a ($a < 1$).

a. Derive the transfer function of the fractional-order hold.

b. Plot the frequency characteristics (amplitude and phase) of the fractional-order hold for $a = 0.2$, 0.3, and 0.5. Discuss your results.

2-16 Two different schemes of triangular hold are shown in Fig. P2-16(a) and (b). Write the impulse responses and the transfer functions of the triangular holds. Which one of these is physically unrealizable?

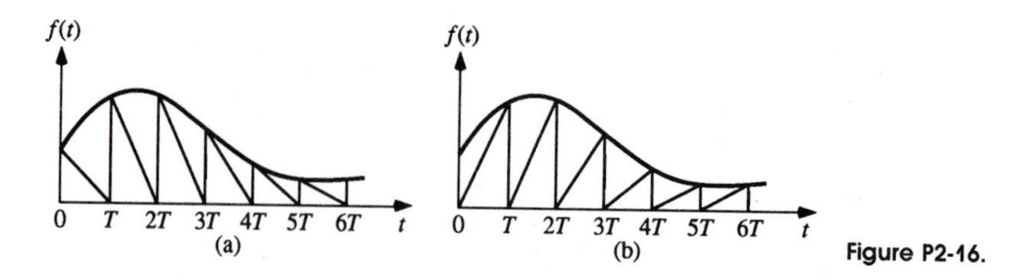

Figure P2-16.

2-17 Show that the polygonal hold discussed in Sec. 2-15 can be made up by adding the two triangular holds of Problem 2-16. Show that the sum of the transfer functions of the two triangular holds is equal to that of the polygonal hold given in Eq. (2-107).

2-18 This problem deals with the digital speed measurement of a control system with an incremental encoder. Consider an incremental encoder that generates two sinusoidal signals in quadrature as the encoder disk rotates. The outputs of the two channels are shown in Fig. P2-18. Note that the two output signals generate four zero crossings per cycle. These zero crossings may be used for position indication or speed calculations in control systems. Consider that the encoder is mounted on the output shaft of a motor which directly drives a printwheel of an electronic typewriter or word processor. The printwheel has 96 character positions and the encoder disk has 480 cycles.

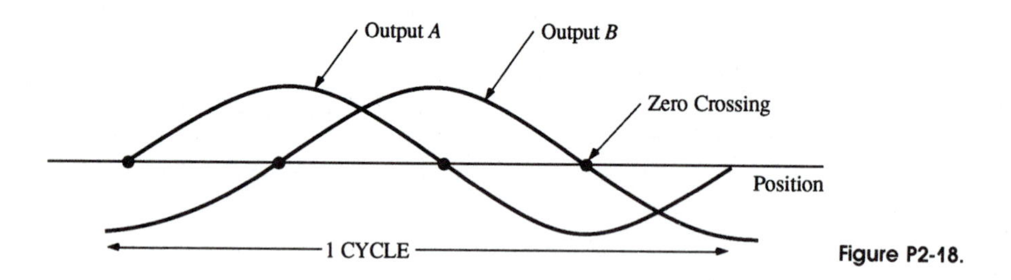

Figure P2-18.

a. Determine how many zero crossings of the encoder outputs correspond to one printwheel character.

b. Consider that a 500-kHz clock is used to measure the speed of the printwheel in conjunction with the encoder signals. The speed is measured by counting the number of pulses between consecutive zero crossings of the encoder outputs. What is the minimum velocity in revolutions per minute (RPM) that can be measured by this setup without overflow if the counter has 12 bits?

c. The maximum excursion of the printwheel is 180 degrees in either direction, or 48 characters. This must be accomplished in 30 ms or less. Assuming that the printwheel represents a pure inertial load, and that the 48-character motion is accomplished with a constant-acceleration profile for half the distance and constant deceleration for the remaining distance, find the maximum velocity in RPM the encoder setup must measure, in number of zero crossings per second.

d. If the encoder-counter system is implemented in a control system for the control of the position of the printwheel, what is the range of the sampling periods in seconds to which the system will subjected based on the minimum velocity constraint found in part **b**, and the maximum velocity requirement found in part **c**? Each zero crossing of the encoder corresponds to one sampling operation.

2-19 A digital speed measurement system makes use of an incremental shaft encoder and counts the number of pulses generated by the encoder over a given time interval. Let the resolution of the shaft encoder be P pulses/revolution. The encoder output is processed by a microprocessor that takes up to T_s s to count and register one pulse. The shaft speed is denoted by N RPM or ω rad/s.

a. Find the highest shaft speed that can be measured in RPM and in rad/s in terms of the parameters P and T_s. Find the highest measurable shaft speed if $P = 1000$ pulses/revolution and $T_s = 50\ \mu$s.

b. The shaft speed is computed by the microprocessor by counting the number of pulses during a sampling period of T s. For an encoder with P pulses/revolution, find the resolution of this speed measuring system in RPM/pulse and in rad/s/pulse in terms of P and T. Find the resolution with $P = 1000$ pulses/revolution and $T = 0.1$ s.

c. For analytical purposes it is desirable to devise a model for the speed measurement system described above. The sampled-data model of the speed measurement system is shown in Fig. P2-19. The input variable is the shaft speed $\omega(t)$ in rad/s. The output is sampled data which represents the digital measurement of the shaft speed in pulses per sampling period T. Find the constant G_s in terms of P and T. Note that T_s does not enter the analysis in parts **b** and **c**.

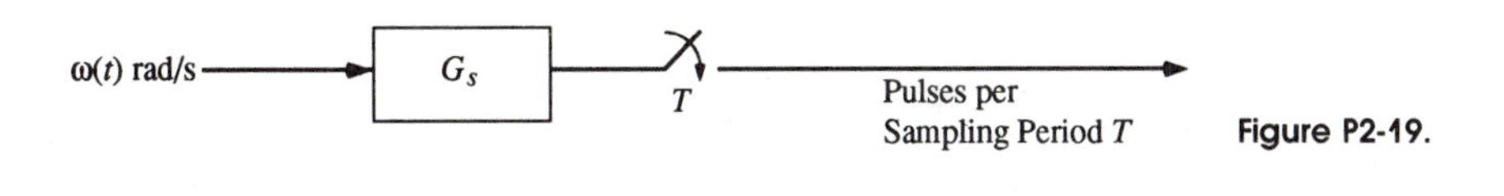

Figure P2-19.

References

Digital Signal Processing

1. Rabiner, L. R., and C. M. Rader, Eds., *Digital Signal Processing*, IEEE Press, New York, 1972.
2. Helms, H. D., and L. R. Rabiner, Eds., *Literature in Digital Signal Processing: Terminology and Permuted Title Index*, IEEE Press, New York, 1972.
3. Openheim, A. V., and R. W. Schafer, *Digital Signal Processing*, Prentice-Hall, Englewood Cliffs, N. J., 1975.

A/D and D/A Converters

4. Moeschele, D. F., Jr., *Analog-to-Digital/Digital-to-Analog Conversion Techniques*, John Wiley & Sons, New York, 1968.

5. Sheingold, D. M., Ed., *Analog-Digital Conversion Handbook*, Analog Devices, Inc., Norwood, Mass., 1972.
6. Schmid, M., *Electronic Analog/Digital Conversion*, Van Nostrand Rheinhold, New York, 1970.

Sampling and Data Reconstruction

7. Balakrishnan, A. V., "A Note of the Sampling Principle for Continuous Signals," *IRE Trans. Information Theory*, vol. IT-3, pp. 143–146, June 1957.
8. Shannon, C. E., "Communication in the Presence of Noise," *Proc. IRE*, vol. 37, pp. 10–21, January 1949.
9. Shannon, C. E., B. M. Oliver, and J. R. Pierce, "The Philosophy of Pulse Code Modulation," *Proc. IRE*, vol. 36, pp. 1324–1331, November 1948.
10. Fogel, L. J., "A Note of the Sampling Theorem," *IRE Trans. Information Theory*, vol. 1, pp. 47–48, March 1955.
11. Jegerman and L. J. Fogel, "Some General Aspects of Sampling Theorem," *IRE Trans. Information Theory*, vol. 2, pp. 139–146, December 1956.
12. Linden, D. A., "A Discussion of Sampling Theorems," *Proc. IRE*, vol. 47, pp. 1219–1226, July 1959.
13. Oliver, R. M., "On the Functions Which Are Represented by the Expansions of the Interpolation-Theory," *Proc. Royal Society* (Edinburgh), vol. 35, pp. 181–194, 1914–1915.
14. Barker, R. H., "The Reconstruction of Sampled-Data," *Proc. Conf. Data Processing and Automatic Computing Machines*, Salisbury, Australia, June 1957.
15. Porter, A., and F. Stoneman, "A New Approach to the Design of Pulse Monitored Servo Systems," *JIEE*, London, vol. 97, part II, pp. 597–610, 1950.
16. Ragazzini, J. R., and L. H. Zadeh, "The Analysis of Sampled-Data Systems," *Trans. AIEE*, vol. 71, part II, pp. 225–234, 1952.
17. Tsypkin, Y. Z., "Sampled-Data Systems with Extrapolating Devices," *Avtomat. Telemekh*, vol. 19, no. 5, pp. 389–400, 1958.
18. Linden, D. A., and N. M. Abramson, "A Generalization of the Sampling Theorem," Tech. Rep. 1551-2, Solid-State Elec. Lab., Stanford University, August 1959.
19. Jury, E. I., "Sampling Schemes in Sampled-Data Control Systems," *IRE Trans. Automatic Control*, vol. AC-6, pp. 86–88, February 1961.
20. Beutler, F. J., "Sampling Theorems and Bases in a Hilbert Space," *Information and Control*, vol. 4, pp. 97–117, 1961.
21. Goff, K. W., "Dynamics in Direct Digital Control, Part I," *ISA Journal*, vol. 13, no. 11, pp. 45–49, November 1966; part II, vol. 13, no. 12, pp. 44–54, December 1966.
22. Whitbeck, R. F., and L. G. Hofmann, "Analysis of Digital Flight Control Systems with Flying Qualities Applications," Technical Report, Systems Technology, Inc., Hawthorne, Calif., vol. II, September 1978.

The z-transform

KEYWORDS AND TOPICS

z-transform • z-plane properties • z-transform theorems • delayed z-transform • modified z-transform • difference equations • applications of z-transform

3-1 MOTIVATION FOR USING THE z-TRANSFORM

In general, the methods for analyzing and designing control systems can be classified as either *conventional* or *modern.* The classical methods are characterized by the transform techniques and transfer functions, whereas modern control theory is based on the modeling of systems by state variables and state equations. The Laplace transform is the basic tool in the conventional analysis and design of analog control systems. In principle, the Laplace transform can also be used to model digital control systems. However, from the preliminary development presented in Chapter 2, typical Laplace transform expressions of systems involving digital or sampled signals all contain exponential terms in the form of e^{Ts}. This inevitably would make the manipulation of the transform expressions in the Laplace domain unduly difficult. From a heuristic standpoint, we may regard this as a motivation for the introduction of the z-transformation.

3-1-1 Definition of the z-Transform

Let the output of an ideal sampler be designated by $f^*(t)$, which is given by Eq. (2-50). The Laplace transform of $f^*(t)$ is given by Eq. (2-51) as

$$\mathscr{L}[f^*(t)] = F^*(s) = \sum_{k=0}^{\infty} f(kT)e^{-kTs} \tag{3-1}$$

Since the expression of $F^*(s)$ contains the term e^{-Ts}, unlike the majority of transfer functions of continuous-data systems, it is not a rational function of s. When terms involving e^{-Ts} appear in a transfer function other than as a multiplying factor, difficulties in taking the inverse Laplace transform may arise. Therefore, it is desirable first to transform the irrational function $F^*(s)$ into a rational function, say $F(z)$, through a transformation from the complex variable s to another complex variable z. An obvious choice for this transformation is

$$z = e^{Ts} \tag{3-2}$$

although $z = e^{-Ts}$ would be just as acceptable.

Solving for s in Eq. (3-2), we obtain

$$s = \frac{1}{T} \ln z \tag{3-3}$$

In these last two equations, T is the sampling period in seconds, and z is a complex variable whose real and imaginary parts are related to those of s through

$$\text{Re } z = e^{T\sigma} \cos \omega T \tag{3-4}$$

and

$$\text{Im } z = e^{T\sigma} \sin \omega T \tag{3-5}$$

with

$$s = \sigma + j\omega \tag{3-6}$$

The relationship between s and z in Eq. (3-2) may be defined as the *z-transformation*. When Eq. (3-2) is substituted in Eq. (3-1), we have

$$F^*\left[s = \frac{1}{T} \ln z\right] = F(z) = \sum_{k=0}^{\infty} f(kT)z^{-k} \tag{3-7}$$

which, when written in closed form, will be a rational function of z. Therefore, we can define $F(z)$ as the *z-transform* of $f(t)$, that is,

$$F(z) = z\text{-transform of } f(t) = \mathscr{Z}[f(t)] \tag{3-8}$$

where $\mathscr{Z}$ denotes "the z-transform of."

In view of the way the z-transform operation is defined in Eq. (3-7), and Eq. (3-1), we can also write the z-transform of $f(t)$ as

$$F(z) = \begin{cases} [\text{Laplace transform of } f^*(t)]|_{s=(\ln z)/T} \\ [F^*(s)]|_{s=(\ln z)/T} \end{cases} \tag{3-9}$$

Since the z-transform of $f(t)$ is obtained from the Laplace transform of $f^*(t)$ by performing the transformation $z = e^{Ts}$, we can say that, in general, *any function $f(t)$ that is Laplace transformable also has a z-transform.*

In summary, the operation of taking the z-transform of a continuous-data function $f(t)$ involves the following three steps.

1. $f(t)$ is sampled by an ideal sampler to give $f^*(t)$. Alternatively, the values of $f(t)$ at $t = kT$ are defined as the sequence $f(kT)$, $k = 0, 1, 2$.
2. The Laplace transform of $f^*(t)$ is taken to give $F^*(s)$:

$$F^*(s) = \mathscr{L}[f^*(t)] = \sum_{k=0}^{\infty} f(kT)e^{-kTs} \tag{3-10}$$

3. Replace e^{Ts} by z in $F^*(s)$ to get $F(z)$:

$$F(z) = \sum_{k=0}^{\infty} f(kT)z^{-k} \tag{3-11}$$

Equation (3-11) represents a useful expression for evaluating the z-transform of the function $f(t)$. The only disadvantage of Eq. (3-11) is that the expression is an infinite series of z^{-k}, so that additional effort is needed to obtain a closed-form expression for $F(z)$.

Depending on the starting point, the z-transform can also be defined under the following two scenarios.

1. The values of $f(t)$ at $t = kT$ are defined as the sequence $f(kT)$, $k = 0, 1, 2, \ldots$. Then, $F^*(s)$ and $F(z)$ are redefined in Eqs. (3-10) and (3-11), respectively.
2. The sequence of numbers of events $f(k)$ for $k = 0, 1, 2, \ldots$ is given. The z-transform of $f(k)$ is defined as

$$z\text{-transform of } f(k) = \mathscr{Z}[f(k)] = \sum_{k=0}^{\infty} f(k)z^{-k} \tag{3-12}$$

Notice that in this case the sampling operation and the sampling period are not mentioned at all.

3-1-2 Relationship Between the Laplace Transform and the z-Transform

It is of interest to compare the defining equations of the Laplace transform and the z-transform. Given $f(t)$ as a function that is Laplace transformable, the Laplace transform and the z-transform of $f(t)$ are

$$F(s) = \mathscr{L}[f(t)] = \int_0^{\infty} f(t)e^{-st}\,dt \tag{3-13}$$

$$F(z) = \mathscr{Z}[f(t)] = \sum_{k=0}^{\infty} f(kT)z^{-k} \tag{3-14}$$

respectively. Since $F(z)$ is obtained by first sampling $f(t)$ with an ideal sampler at a sampling period T, there is a *general misconception* that when the sampling period T approaches zero, $F(z)$ will approach $F(s)$. The fact is that, in general,

$$\lim_{T \to 0} F(z) \neq F(s) \tag{3-15}$$

Since $f^*(t)$ represents a train of impulses, spaced T s apart, as T becomes infinitesimally small, the impulse train simply collapses into a "bunch" of impulses at $t = 0$, and the result in no way resembles $f(t)$. The time equivalent of Eq. (3-15) is

$$\lim_{T \to 0} f^*(t) \neq f(t) \tag{3-16}$$

How do we explain the fact that, in reality, a sampled signal should revert to the continuous-data signal as the sampling period approaches zero? We shall address this problem in Chapter 4.

3-1-3 An Alternate Expression for $F(z)$

Often in the analysis of linear systems, the transfer function $F(s)$ is given, and the z-transform $F(z)$ is to be determined. In this case, it is not necessary to find $f(t)$ first and then go through the three steps outlined in the previous section. The following development shows how $F(z)$ can be determined directly from $F(s)$.

$F(s)$ Has Simple Poles

The z-transform of the function $f(t)$ can be determined directly from the Laplace transform of $f(t)$, $F(s)$. Replacing e^{-Ts} by z^{-1} in Eq. (2-55), we have

$$F(z) = \sum_{n=1}^{k} \frac{N(\xi_n)}{D'(\xi_n)} \frac{1}{1 - e^{\xi_n T} z^{-1}} \tag{3-17}$$

where

$$F(\xi) = \frac{N(\xi)}{D(\xi)} \tag{3-18}$$

has k simple poles at $\xi = \xi_n$, $n = 1, 2, \ldots, k$. $D'(\xi_n)$ is defined as

$$D'(\xi_n) = \left. \frac{dD(\xi)}{d\xi} \right|_{\xi = \xi_n} \tag{3-19}$$

$F(s)$ with Multiple-Order Poles

If $F(s)$ has multiple-order poles, $s_1, s_2, \ldots, s_k$, with multiplicity, $m_1, m_2, \ldots, m_k$, respectively, the z-transform of $F(s)$ is written by using Eq. (2-58). We get

$$F(z) = \sum_{n=1}^{k} \sum_{i=1}^{m_n} \frac{(-1)^{m_n - i} K_{ni}}{(m_n - i)!} \left[\frac{\partial^{m_n - 1}}{\partial s^{m_n - i}} \frac{1}{1 - e^{-Ts}} \right] \Bigg|_{s = s - s_n,\ z = e^{Ts}} \tag{3-20}$$

where

$$K_{ni} = \frac{1}{(i-1)!}\left[\frac{\partial^{i-1}}{\partial s^{i-1}}(s-s_n)^{m_n}F(s)\right]\Bigg|_{s=s_n} \tag{3-21}$$

For a pole at $s = s_n$ with multiplicity m_n (>1), the residue can also be written as

$$\frac{1}{(m_n-1)!}\frac{\partial^{m_n-1}}{\partial s^{m_n-1}}\left[(s-s_n)^{m_n}F(s)\frac{1}{1-e^{Ts}z^{-1}}\right]\Bigg|_{s=s_n} \tag{3-22}$$

3-2 EXAMPLES OF THE EVALUATION OF z-TRANSFORMS

The defining equation of the z-transform in Eq. (3-11) is useful for the evaluation of $F(z)$ when either $f(t)$ or $f(kT)$ is given. This means that when $f(t)$ is given, we replace t by kT in $f(t)$ and substitute it in Eq. (3-11) to get $F(z)$. Sometimes a sequence of numbers or a time series is given in the form of $f(0)$, $f(T)$, $f(2T)$, $\ldots$, $f(kT)$, $\ldots$; then $F(z)$ is given by the sum of $f(kT)z^{-k}$ from $k = 0$ to $k = \infty$. In general, the time function or time series can be of any form without restriction if we take Eq. (3-11) at its face value. However, as mentioned earlier, since $F(z)$ is defined using the Laplace transform, strictly, $f(t)$ must be Laplace transformable. Another consideration is that, to be able to express $F(z)$ in a closed form for system analysis purposes, the infinite series of Eq. (3-11) must be convergent.

Equations (3-17) and (3-20) are useful in generating the z-transforms when $F(s)$ is known. Equation (3-17) is for $F(s)$ with only simple poles, real or complex, whereas Eq. (3-20) is for $F(s)$ that has at least one multiple-order pole.

The following examples illustrate how the defining z-transform equations are utilized, as well as some of the fine points of evaluating z-transforms. For day-to-day engineering applications, a z-transform table such as the one given in Appendix C may be utilized, and one does not have to resort to the chores of deriving $F(z)$ from $F(s)$ or $f(t)$ using the defining equations.

Example 3-1

Find the z-transform of the unit-step function $u_s(t)$ which is defined as

$$\begin{aligned} u_s(t) &= 1 \qquad t > 0 \\ u_s(t) &= 0 \qquad t < 0 \end{aligned} \tag{3-23}$$

We proceed according to the steps outlined in the previous section.

1. The unit-step function is sampled by an ideal sampler. This yields a train of unit impulses (impulses with strength of each impulse equal to one) starting

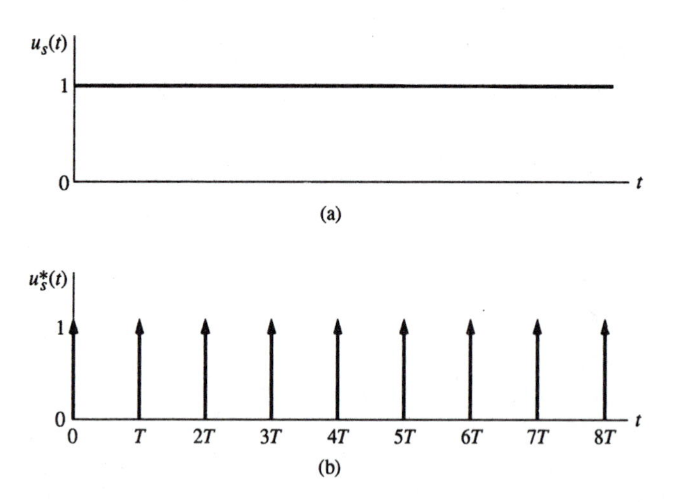

Figure 3-1. (a) Unit-step function. (b) Sampled unit-step function.

at $t = 0$ and occurring at $t = kT$, $k = 0, 1, 2, \ldots$. The input and the output of the ideal sampler are shown in Fig. 3-1. The sampled signal is described by

$$u_s^*(t) = \delta_T(t) = \sum_{k=0}^{\infty} \delta(t - kT) \tag{3-24}$$

2. Taking the Laplace transform on both sides of Eq. (3-24), we get

$$U_s^*(s) = \Delta_T(s) = \sum_{k=0}^{\infty} e^{-kTs} \tag{3-25}$$

To express $U_s^*(s)$ in a closed form, we multiply both sides of Eq. (3-25) by e^{-Ts} and then subtract the result from Eq. (3-25). The final result is

$$U_s^*(s) = \Delta_T(s) = \frac{1}{1 - e^{-Ts}} \tag{3-26}$$

for $|e^{-Ts}| < 1$.

3. Replacing e^{Ts} by z in Eq. (3-26), we have the z-transform of the unit-step function,

$$U_s(z) = \mathscr{Z}[u_s(t)] = \frac{1}{1 - z^{-1}} = \frac{z}{z - 1} \tag{3-27}$$

for $|z^{-1}| < 1$, or $|z| > 1$.

It should be pointed out that in the above development we purposefully demonstrated the complete three steps of evaluating $F(z)$ from $f(t)$. In practice, we can simply substitute $f(kT)$ into Eq. (3-11) without going through the Laplace transform expressions.

The same result can be obtained by using Eq. (3-17). The Laplace transform of $u_s(t)$ is $U_s(s) = 1/s$ which has a simple pole at $s = 0$. Thus, from Eq. (3-18), $N(\xi) = 1$ and $D(\xi) = \xi$.

$$D'(\xi) = \frac{dD(\xi)}{d\xi} = 1 \tag{3-28}$$

The z-transform of the unit-step function is written from Eq. (3-17) as

$$U_s(z) = \left.\frac{1}{1 - e^{T\xi}z^{-1}}\right|_{\xi=0} = \frac{z}{z-1} \tag{3-29}$$

Example 3-2

Find the z-transform of the exponential function $f(t) = e^{-at}$, where a is a real constant.

Without going through the detailed steps as in Example 3-1, we substitute $f(t)$ into Eq. (3-11). We have

$$F(z) = \sum_{k=0}^{\infty} f(kT)z^{-k} = \sum_{k=0}^{\infty} e^{-akT}z^{-k} \tag{3-30}$$

This infinite series converges for all values of z that satisfy $|e^{-aT}z^{-1}| < 1$. To obtain the closed-form expression of Eq. (3-30), we multiply both sides of the equation by $e^{-aT}z^{-1}$, and subtract the result from Eq. (3-30). After rearranging, the final result is

$$F(z) = \frac{1}{1 - e^{-aT}z^{-1}} = \frac{z}{z - e^{-aT}} \tag{3-31}$$

for $|e^{-aT}z^{-1}| < 1$, or $|z^{-1}| < e^{aT}$.

We can demonstrate that Eq. (3-17) leads to the same result. The Laplace transform of e^{-at} is

$$F(s) = \frac{1}{s+a} \tag{3-32}$$

which has a simple pole at $s = -a$. In Eq. (3-17), $N(\xi) = 1$, $D(\xi) = \xi + a$, and $D'(\xi) = 1$. Thus, Eq. (3-17) gives

$$F(z) = \left.\frac{N(\xi_1)}{D'(\xi_1)}\frac{1}{1 - e^{\xi_1 T}z^{-1}}\right|_{\xi_1 = a}$$

$$= \frac{1}{1 - e^{-aT}z^{-1}} = \frac{z}{z - e^{-aT}} \tag{3-33}$$

Example 3-3

Find the z-transform of $f(t) = \sin \omega t$.

Substituting $f(t)$ into Eq. (3-11), we get

$$F(z) = \sum_{k=0}^{\infty} \sin \omega kTz^{-k} \tag{3-34}$$

In this case it is more convenient to express $\sin \omega kT$ in the exponential form. Then Eq. (3-34) becomes

$$F(z) = \sum_{k=0}^{\infty} \frac{e^{j\omega kT} - e^{-j\omega kT}}{2j} z^{-k} \tag{3-35}$$

This infinite series is convergent for $|z^{-1}| < 1$ and can be written in a closed form by using the same method as described in Examples 3-1 and 3-2. The result is

$$F(z) = \frac{1}{2j}\left[\frac{1}{1 - e^{j\omega kT}z^{-1}} - \frac{1}{1 - e^{-j\omega kT}z^{-1}}\right] \tag{3-36}$$

After simplification, the last equation is written

$$F(z) = \frac{z \sin \omega T}{z^2 - 2z \cos \omega T + 1} \tag{3-37}$$

Now to get the same result with Eq. (3-17), we recognize that the Laplace transform of $f(t)$ is

$$F(s) = \mathscr{L}[\sin \omega t] = \frac{\omega}{s^2 + \omega^2} \tag{3-38}$$

From Eq. (3-17), we have

$$N(\xi) = \omega$$

$$D(\xi) = \xi^2 + \omega^2$$

$$D'(\xi) = 2\xi$$

The poles of $F(s)$ are at $\xi = \xi_1 = j\omega$ and $\xi = \xi_2 = -j\omega$. Thus,

$$N(\xi_1) = N(\xi_2) = \omega$$

since $N(\xi)$ is not a function of ξ.

$$D'(\xi_1) = 2j\omega$$

$$D'(\xi_2) = -2j\omega$$

Equation (3-17) now gives

$$F(z) = \sum_{n=1}^{2} \frac{N(\xi_n)}{D'(\xi_n)} \frac{1}{1 - e^{-T\xi_n} z^{-1}}$$

$$= \frac{1}{2j}\left[\frac{1}{1 - e^{j\omega T} z^{-1}} - \frac{1}{1 - e^{-j\omega T} z^{-1}}\right] \tag{3-39}$$

which apparently will lead to the same result as in Eq. (3-37).

Example 3-4

Find the z-transform of the ramp function $f(t) = tu_s(t)$.

Using Eq. (3-11) first, we have

$$F(z) = \sum_{k=0}^{\infty} kTz^{-k} = Tz^{-1} + 2Tz^{-2} + \cdots \tag{3-40}$$

To express $F(z)$ in closed form, we first multiply both sides of Eq. (3-40) by z^{-1}, resulting in

$$z^{-1}F(z) = Tz^{-2} + 2Tz^{-3} + \cdots \tag{3-41}$$

Subtracting the last equation from Eq. (3-40), we get

$$(1 - z^{-1})F(z) = Tz^{-1} + Tz^{-2} + Tz^{-3} + \cdots \tag{3-42}$$

Apparently, we cannot reduce the right-hand side of Eq. (3-40) into a finite number of terms in one operation. Now multiplying both sides of Eq. (3-42) again by z^{-1} and subtracting the result from Eq. (3-42), we have

$$(1 - z^{-1})^2 F(z) = Tz^{-1} \tag{3-43}$$

Thus the closed-form solution of $F(z)$ is

$$F(z) = \frac{Tz^{-1}}{(1 - z^{-1})^2} = \frac{Tz}{(z - 1)^2} \tag{3-44}$$

The Laplace transform of $f(t)$ is

$$F(s) = \mathscr{L}[f(t)] = \mathscr{L}[tu_s(t)] = 1/s^2 \tag{3-45}$$

which has a double pole at $s = 0$. To obtain the closed-form solution of $F(z)$ directly from $F(s)$, we have to use Eq. (3-20), since $F(s)$ in this case has a second-order pole. Thus, from Eqs. (3-20) and (3-21) we identify that $k = 1$, $s_1 = 0$, and $m_1 = 2$. Equation (3-20) gives

$$K_{1i} = \frac{1}{(i - 1)!}\left[\frac{\partial^{i-1}}{\partial s^{i-1}} s^2 \frac{1}{s^2}\right]_{s=0} \tag{3-46}$$

Thus, $K_{11} = 1$ and $K_{12} = 0$. Equation (3-20) gives

$$F(z) = \sum_{i=1}^{2} \frac{(-1)^{2-i} K_{1i}}{(2-i)!} \left[\frac{\partial^{2-i}}{\partial s^{2-i}} \frac{1}{1 - e^{-Ts} s} \right]_{z = e^{Ts}} = \frac{Tz}{(z-1)^2} \tag{3-47}$$

which is the same result as in Eq. (3-44).

We can also use Eq. (3-22) to arrive at the same result as in Eq. (3-47). We have

$$F(z) = \frac{\partial}{\partial s} \left[s^2 \frac{1}{s^2} \frac{1}{1 - e^{Ts} z^{-1}} \right]_{s=0} = \frac{Tz}{(z-1)^2} \tag{3-48}$$

Example 3-5

Find the z-transform of the sequence $f(k) = (1/2)^k$ for $k = 0, 1, 2, \ldots$.

Multiplying $f(k)$ by z^{-k} and taking the sum from $k = 0$ to $k = \infty$, or substituting $f(k)$ into Eq. (3-12), we have

$$F(z) = \sum_{k=0}^{\infty} f(k) z^{-k} = \sum_{k=0}^{\infty} (1/2)^k z^{-k} \tag{3-49}$$

$$= 1 + \frac{1}{2} z^{-1} + \frac{1}{4} z^{-2} + \frac{1}{8} z^{-3} + \cdots \tag{3-50}$$

Notice that the sampling period T does not appear in this case, and it can be regarded as equal to unity.

To express $F(z)$ in a closed form, we multiply both sides of Eq. (3-50) by $(1/2)z^{-1}$ and subtract the result from Eq. (3-50). After solving for $F(z)$, we get

$$F(z) = \frac{1}{(1 - \frac{1}{2} z^{-1})} = \frac{z}{(z - \frac{1}{2})} \tag{3-51}$$

3-3 RELATIONSHIP BETWEEN THE s-PLANE AND THE z-PLANE

At this point it is necessary to study the relationship between the s-plane and z-plane. In the analysis and design of continuous-data control systems, often the pole-zero configuration of the transfer function in the s-plane is referred to. The notions that the left half of the s-plane is the stable region and the right half of the s-plane corresponds to an unstable region are well established. For relative stability, the left half of the s-plane is further divided into regions in which the transfer function poles must preferably be located. In a similar way, the poles and zeros of a transfer function in the z-transform govern the performance of a digital system. In addition to establishing the significance of pole-zero locations in the z-plane, it would be useful to study the relation between the s-plane and the z-plane, so that some of the s-plane techniques can be extended to the study of digital control systems.

In Chapter 2 we have pointed out that when a system is subject to sampled or digital data, one of the properties of the sampled function $F^*(s)$ is that it has an infinite number of poles, located periodically along a vertical axis with intervals of $\pm m\omega_s$, with $m = 0, 1, 2, \ldots$, in the s-plane, where ω_s is the sampling frequency in rad/s. Since it would be difficult to deal with transfer functions with infinite numbers of poles, the z-transform essentially *folds* these into a finite number of poles in the z-plane.

As shown in Fig. 2-29 the s-plane is divided into an infinite number of *periodic strips* along the $j\omega$ axis. The span of these strips is one sampling frequency ω_s rad/s. The *primary strip* extends from $\omega = -\omega_s/2$ to $+\omega_s/2$, and the *complementary strips* extend from $-\omega_s/2$ to $-3\omega_s/2$, $-3\omega_s/2$ to $-5\omega_s/2, \ldots$ for negative frequencies, and from $\omega_s/2$ to $3\omega_s/2$, $3\omega_s/2$ to $5\omega_s/2, \ldots$ for positive frequencies.

If we consider only the primary strip shown in Fig. 3-2(a), the path described by (1)-(2)-(3)-(4)-(5)-(1) in the left half of the s-plane is mapped into a *unit circle*

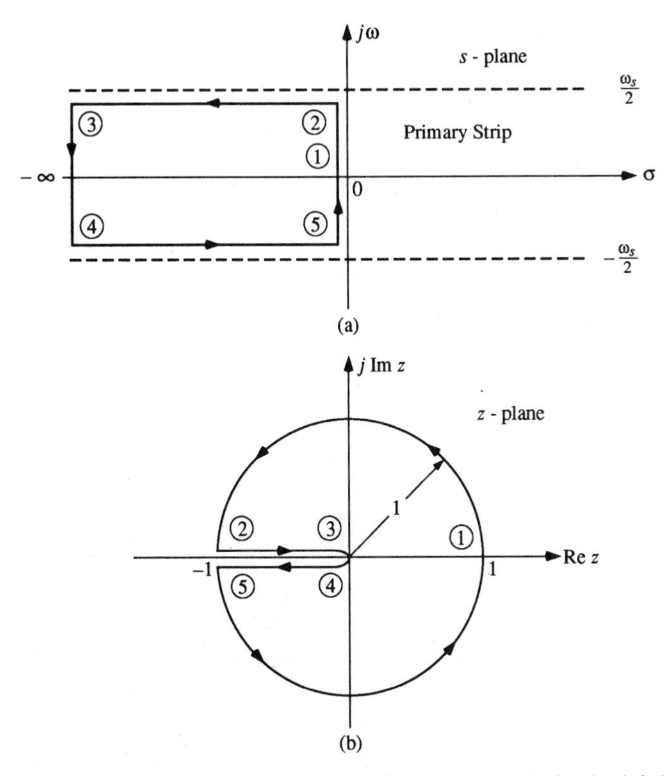

Figure 3-2. Mapping of the primary strip in the left half of the s-plane into the z-plane by the z-transform.

(circle with unit radius), centered at the origin in the z-plane, by the transformation $z = e^{Ts}$, as shown in Fig. 3-2(b). Since

$$e^{(s+jm\omega_s)T} = e^{Ts}e^{j2\pi m} = e^{Ts} = z \quad (3\text{-}52)$$

when m is an integer, all the complementary strips in the left half of the s-plane are also mapped into the same unit circle in the z-plane. Thus, we can draw the following conclusions from the general mapping between the s- and z-planes by the z-transform.

All the points in the left half of the s-plane correspond to points inside the unit circle in the z-plane.

All the points in the right half of the s-plane correspond to points outside the unit circle in the z-plane.

Points on the $j\omega$ axis in the s-plane correspond to points on the unit circle $|z| = 1$ in the z-plane.

We shall now investigate the mapping of the *constant-damping loci*, the *constant-frequency loci*, and the *constant-damping-ratio loci* from the s-plane to the z-plane.

3-3-1 The Constant-Damping Loci

The damping factor σ_1 is the real part of a pole of the transfer function in the s-domain. It represents the rate of rise or decay of the time response of the system, since the inverse of σ_1 is the time constant in seconds. A large σ_1 would correspond to a small time constant, and thus a faster decay or rise in the time response. Thus, the constant-damping loci in the s-plane are vertical lines parallel to the imaginary axis. The loci in the left half s-plane denote *positive damping*, since the system is stable, and the loci in the right half s-plane correspond to *negative damping*.

The *constant-damping loci* in the z-plane are described by

$$z = e^{Ts} = e^{T\sigma_1} \quad (3\text{-}53)$$

Since σ_1 is a real constant, the constant-damping loci in the z-plane are concentric circles with the center at $z = 0$ and the radii given by Eq. (3-53), as shown in Fig. 3-3. We should develop the notion that poles on the constant-damping loci closer to the unit circle should correspond to time responses that have slower decay or rise time.

3-3-2 The Constant-Frequency Loci

For any given frequency $\omega = \omega_1$ in the s-domain, the constant-frequency locus in the s-plane is a horizontal line, $s = j\omega_1$. The corresponding z-transform is

$$z = e^{Ts} = e^{j\omega_1 T} \quad (3\text{-}54)$$

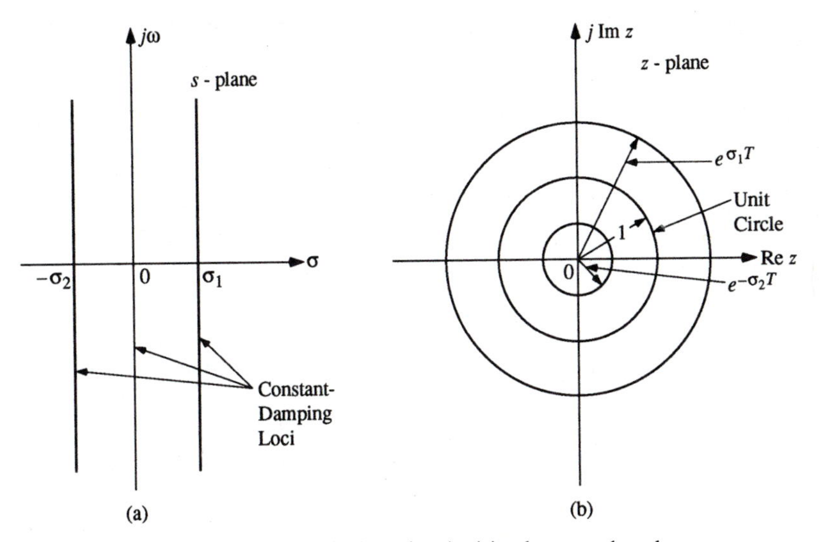

Figure 3-3. Constant-damping loci in the s- and z-planes.

which represents a straight line emanating from the origin at an angle of $\theta = \omega_1 T$ rad, measured from the positive real axis as shown in Fig. 3-4. Figure 3-4 shows several other typical constant-frequency loci in the s- and z-planes. Thus, poles located in the first and fourth quadrants in the z-plane would have frequencies that are lower than those in the second and the third quadrants.

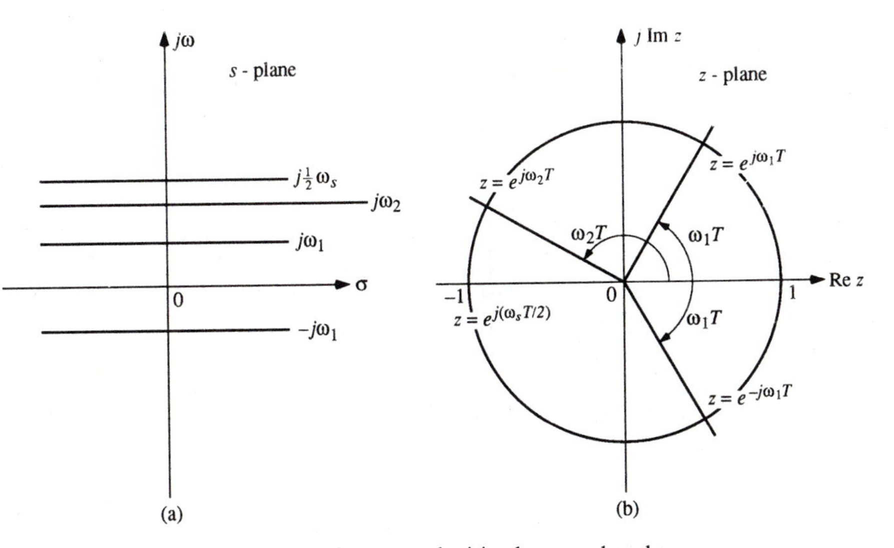

Figure 3-4. Constant-frequency loci in the s- and z-planes.

3-3-3 The Constant-Damping-Ratio Loci

In the design of control systems, we often specify a certain desired damping ratio ζ for the dominant poles of the system transfer function, such as $\zeta = 0.707$. The constant-damping-ratio locus in the s-plane is described by

$$s = -\omega \tan \beta + j\omega \tag{3-55}$$

where

$$\beta = \sin^{-1} \zeta \tag{3-56}$$

and β in degrees is measured from the $j\omega$ axis, as shown in Fig. 3-5(a). The z-transform relation is

$$z = e^{Ts} = e^{T(-\omega \tan \beta + j\omega)} = e^{-2\pi\omega \tan \beta/\omega_s} \angle 2\pi\omega/\omega_s \tag{3-57}$$

where we have substituted $2\pi/\omega_s$ for T.

For a given damping ratio ζ, which gives a β described by Eq. (3-56), the constant-damping-ratio locus described by Eq. (3-57) is a *logarithmic spiral* in the z-plane, except for $\zeta = 0$ ($\beta = 0°$) and $\zeta = 1$ ($\beta = 90°$).

Figure 3-5(b) shows the constant-damping-ratio locus for $\zeta = 0.5$ or $\beta = 30°$. Notice that each one-half revolution of the logarithmic spiral corresponds to the passage of the constant-ζ locus in the s-plane through a change of $\omega_s/2$ rad/s along the $j\omega$ axis.

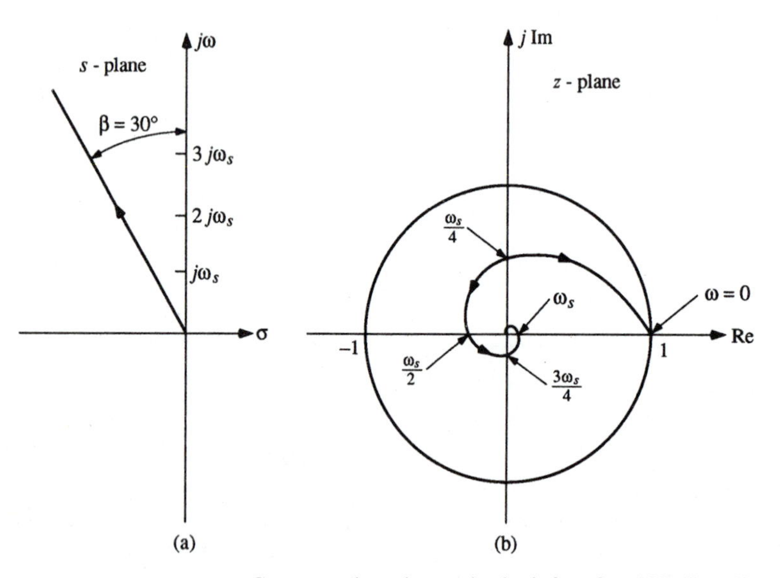

Figure 3-5. Constant-damping-ratio loci for $\beta = 30°$ ($\zeta = 50$ percent) in the s- and z-planes.

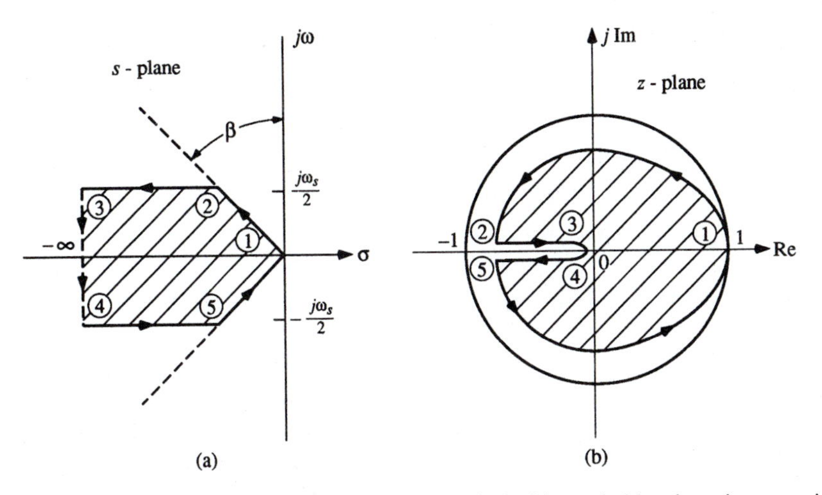

Figure 3-6. Constant-damping-ratio loci bounded by the primary strip in the s-plane and the corresponding loci in the z-plane.

Since most control systems have low-pass characteristics, usually only the poles located in the vicinity of the primary strip are of importance. This means that for practical purposes only the portion of the constant-damping-ratio locus that corresponds to the frequency range from $\omega = 0$ to $\omega = \omega_s/2$ is of importance. For example, if it is desired to realize a damping ratio of 70.7 percent or better ($\beta \geq 45°$), the corresponding regions covered by the primary strip in the s-plane and the z-plane are shown in Fig. 3-6.

3-4 THE INVERSE z-TRANSFORM

Just as in the Laplace transform analysis, the z-transform is used to facilitate the algebraic manipulation of the transfer functions. Eventually, the inverse z-transform must be conducted to get from the transform to the real domain.

3-4-1 Nonuniqueness of the z-Transform

It is well known that the single-sided Laplace transform and its inverse transform are unique, so that if $F(s)$ is the Laplace transform of $f(t)$, then $f(t)$ is the inverse Laplace transform of $F(s)$. However, in the z-transform, given the function $F(z)$, its inverse z-transform is generally not unique. Starting from the function $f(t)$, its z-transform is $F(z)$, but the inverse z-transform is not necessarily, and generally not, equal to $f(t)$. The correct result of the inverse z-transform of $F(z)$ is $f(kT)$ which is equal to $f(t)$ only at the sampling instants. This result

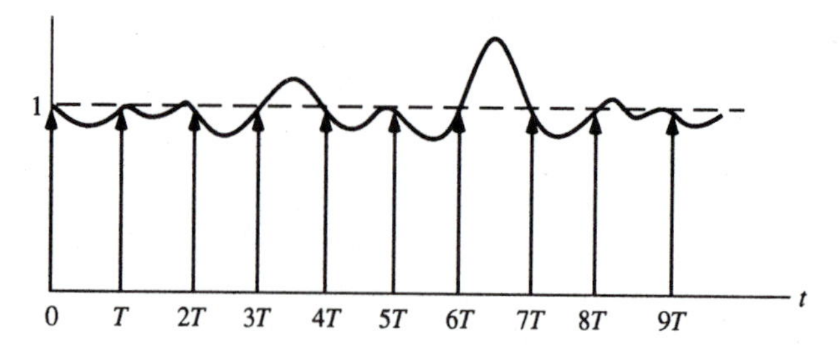

Figure 3-7. Illustration of the nonuniqueness of the inverse z-transform.

should not be surprising, since once $f(t)$ is sampled by the ideal sampler, the information between the sampling instants is totally lost, and we cannot expect to recover $f(t)$ from $F(z)$. Figure 3-7 illustrates the simple case of taking the z-transform of a unit-step function, and the associated sampling operation yields a train of unit impulses with the sampling period of Ts. The inverse z-transform of $F(z)$ is not unique, since $f(t)$ can be any function of t that has a value of unity at the sampling instants.

The problem of nonuniqueness in the inverse z-transform is one of the limitations of the z-transform method. However, the simplicity and the advantages of the z-transform far outweigh this disadvantage. Furthermore, if the sampling period T is carefully chosen, relative to the variation of the time function $f(t)$, the sequence $f(kT)$, and thus the z-transform $F(z)$, can still be accurate representations of the continuous function $f(t)$.

3-4-2 Defining Equation of the Inverse z-Transform

Given the z-transform function $F(z)$, the inverse z-transform of $F(z)$ is denoted by

$$f(kT) = \mathscr{Z}^{-1}[F(z)] \tag{3-58}$$

In general, the inverse z-transform can be carried out with one of the following four methods.

1. Transform Table

For simple functions, the inverse z-transform can be looked up from the z-transform table such as the one in Appendix C.

2. Partial-Fraction Expansion Method

This method is parallel to the partial-fraction expansion method used in the Laplace transform, with a minor modification. In the analysis of continuous-data systems, for example, the inverse Laplace transform of a function $F(s)$ with simple poles at $-a$, $-b$, and $-c$ is obtained by expanding $F(s)$ as

$$F(s) = \frac{K_1}{s+a} + \frac{K_2}{s+b} + \frac{K_3}{s+c} \tag{3-59}$$

where K_1, K_2, and K_3 are constant coefficients. The inverse Laplace transform of $F(s)$ is written as

$$f(t) = K_1 e^{-at} + K_2 e^{-bt} + K_3 e^{-ct} \tag{3-60}$$

In the case of the z-transform, we do not ordinarily expand $F(z)$ into a form similar to Eq. (3-59). The reason is that if we examine the z-transform table, we find that most of the common z-transform functions have at least a zero at $z = 0$. For example, the z-transforms of step, ramp, and parabolic time functions are tabulated below, and these z-transforms all have at least one z in the numerator. If we perform the partial-fraction expansion of $F(z)$ in the same fashion as in Eq. (3-59), then the z in the numerator would not be carried over to the individual fractions, and we would not be able to find corresponding real functions for $f(kT)$.

Step Function

$$F(z) = \frac{z}{z-1} \tag{3-61}$$

Ramp Function

$$F(z) = \frac{Tz^2}{(z-1)^2} \tag{3-62}$$

Parabolic Function

$$F(z) = \frac{T^2 z(z+1)}{(z-1)^3} \tag{3-63}$$

A simple remedy to this situation is to "store" away the z in the numerator of $F(z)$ (if there is one) and carry out the partial-fraction expansion of $F(z)/z$. After the expansion is performed, the z in the denominator of $F(z)$ is then multiplied across to each term of the fractions. Of course, if $F(z)$ does not have any zeros at $z = 0$, such as when the system has a pure time delay that is an integral multiple of the sampling period T, then the partial-fraction expansion of $F(z)$ is carried out in the same way as in Eq. (3-59). For instance,

$$\mathscr{Z}^{-1}\left[\frac{1}{z-1}\right] = u_s(kT) \tag{3-64}$$

where $k = 1, 2, 3, \ldots$. Alternatively, Eq. (3-64) denotes the sequence 1, 1, 1, ..., starting at $k = 1$.

Example 3-6

Given the z-transform

$$F(z) = \frac{(1 - e^{-aT})z}{(z - 1)(z - e^{-aT})} \tag{3-65}$$

where a is a constant and T is the sampling period. The inverse z-transform of $F(z)$, $f(kT)$, is determined by using the partial-fraction method. Since $F(z)$ has a zero at $z = 0$, we perform the partial fraction of $F(z)/z$ as

$$\frac{F(z)}{z} = \frac{1}{z - 1} - \frac{1}{z - e^{-aT}} \tag{3-66}$$

Then,

$$F(z) = \frac{z}{z - 1} - \frac{z}{z - e^{-aT}} \tag{3-67}$$

From the z-transform table in Appendix C, the inverse z-transform of $F(z)$ is found to be

$$f(kT) = 1 - e^{-akT} \tag{3-68}$$

for $k = 0, 1, 2, \ldots$. The inverse z-transform can also be interpreted as a sampled time function $f^*(t)$, given by

$$f^*(t) = \sum_{k=0}^{\infty} f(kT)\delta(t - kT) = \sum_{k=0}^{\infty} (1 - e^{-akT})\delta(t - kT) \tag{3-69}$$

Notice that the time function $f(t)$ cannot be determined from the inverse z-transform, since no information on the behavior of $f(t)$ is available between the sampling instants.

A computer program such as the PFEZ of the DCSP software package can be quite useful for the partial-fraction expansion of z-transform functions.

3. The Power-Series Method

The defining equation of the z-transform in Eq. (3-7) suggests that the inverse z-transform of $F(z)$ can be determined simply by expanding the function into an infinite series in powers of z^{-1}, as follows:

$$F(z) = f(0) + f(T)z^{-1} + f(2T)z^{-2} + \cdots + f(kT)z^{-k} + \cdots \tag{3-70}$$

Thus, apparently, the coefficient of the z^{-k} term ($k = 0, 1, 2, \ldots$) represents the value of $f(kT)$.

Given the function $F(z)$, the best way to express it into a power series as in Eq. (3-70) is to divide its numerator by its denominator by long division.

For high-order systems, a computer program such as the INVZ of the DCSP software package is devised for this purpose when $F(z)$ is a rational function with real coefficients.

The main difference between the partial-fraction expansion method and the series expansion method is that the former gives a closed-form solution to $f(kT)$, whereas the latter gives a sequence of numbers. Since the two methods are equivalent, for a given rational function $F(z)$, the infinite series in Eq. (3-70) can always be written in a closed form if necessary, although, in general, the determination of the closed-form expression is not trivial except for simple cases.

Example 3-7

For the function $F(z)$ given in Eq. (3-65), by dividing the numerator polynomial by the denominator, we get

$$F(z) = (1 - e^{-aT})z^{-1} + (1 - e^{-2aT})z^{-2} + (1 - e^{-3aT})z^{-3} + \cdots \tag{3-71}$$

Thus, it is not difficult to see that the closed-form solution of $f(kT)$ is

$$f(kT) = (1 - e^{-kaT}) \tag{3-72}$$

where, in this case, k can start from zero or one and the result in Eq. (3-72) would be the same. Equation (3-71) clearly indicates that $f(0) = 0$.

4. The Inversion Formula Method

The inverse Laplace transform of $F(s)$ is

$$f(t) = \frac{1}{2\pi j}\int_{c-j\infty}^{c+j\infty} F(s)e^{st}\,dt \tag{3-73}$$

where c denotes the abscissa of convergence and is chosen so that it is greater than the real parts of all the singularities of the integrand, $F(s)e^{st}$. We shall show that the inverse z-transform is given by an analogous expression,

$$f(kT) = \frac{1}{2\pi j}\oint_{\Gamma} F(z)z^{k-1}\,dz \tag{3-74}$$

where Γ is a closed path in the z-plane that encloses all the singularities of $F(z)z^{k-1}$.

To derive Eq. (3-74), we substitute $t = kT$ in Eq. (3-73), and we have

$$f(kT) = \frac{1}{2\pi j}\int_{c-j\infty}^{c+j\infty} F(s)e^{kTs}\,ds \tag{3-75}$$

The path of integration of the line integral in Eq. (3-75) is the straight line which extends from $-j\infty$ to $+j\infty$ along $\sigma = \text{Re}\,(s) = c$, as shown in Fig. 3-8(a).

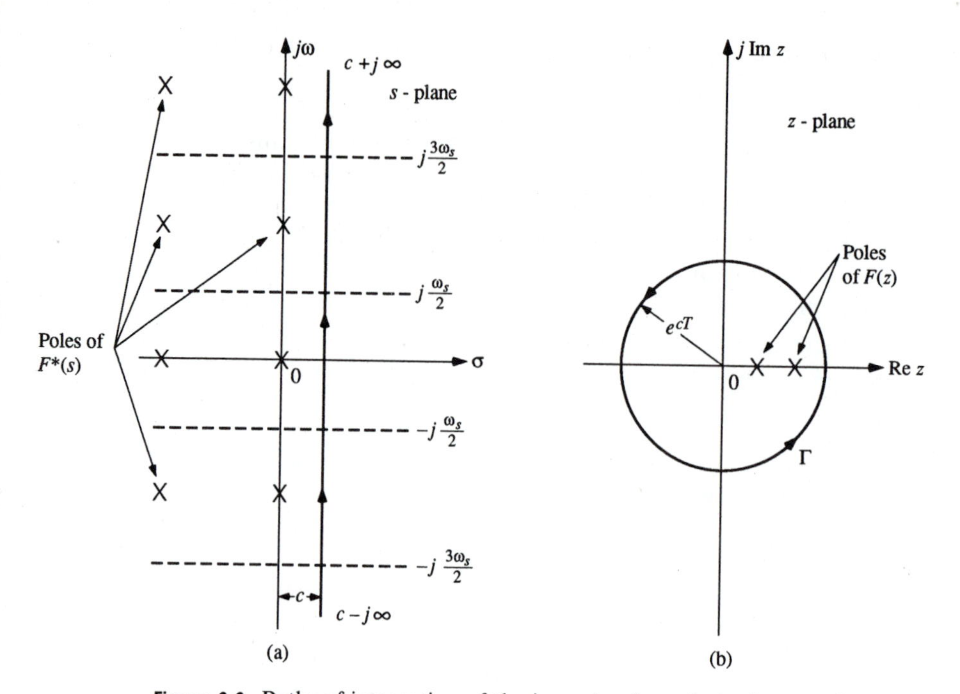

Figure 3-8. Paths of integration of the inversion formula in the s- and z-planes.

This path of integration is seen to pass through the periodic strips in the s-plane vertically, and thus the integral of Eq. (3-75) may be broken up into a sum of integrals, each performed within one of the periodic strips. Equation (3-75) is now written as

$$f(kT) = \frac{1}{2\pi j} \sum_{i=-\infty}^{\infty} \int_{c+j\omega_s(i-1/2)}^{c+j\omega_s(i+1/2)} F(s)e^{kTs}\, ds \tag{3-76}$$

where $\omega_s = 2\pi/T$. Replacing s by $s + ji\omega_s$, where i is an integer, Eq. (3-76) becomes

$$f(kT) = \frac{1}{2\pi j} \sum_{i=-\infty}^{\infty} \int_{c-j\omega_s/2}^{c+j\omega_s/2} F(s + ji\omega_s)e^{kT(s+ji\omega_s)}\, d(s + ji\omega_s) \tag{3-77}$$

Since $\exp[kT(s + ji\omega_s)] = \exp(kTs)$ for integral values of j and i, and $d(s + ji\omega_s) = ds$, Eq. (3-77) can be written

$$f(kT) = \frac{1}{2\pi j} \sum_{i=-\infty}^{\infty} \int_{c-j\omega_s/2}^{c+j\omega_s/2} F(s + ji\omega_s)e^{kTs}\, ds \tag{3-78}$$

Interchanging the summation and integration operations, Eq. (3-78) becomes

$$f(kT) = \frac{1}{2\pi j} \int_{c-j\omega_s/2}^{c+j\omega_s/2} \sum_{i=-\infty}^{\infty} F(s + ji\omega_s)e^{kTs}\, ds \tag{3-79}$$

Since, from Eqs. (B-26) and (2-53),

$$F^*(s) = \frac{1}{T} \sum_{k=-\infty}^{\infty} F(s + jk\omega_s) \tag{3-80}$$

Eq. (3-79) becomes

$$f(kT) = \frac{T}{2\pi j} \int_{c-j\omega_s/2}^{c+j\omega_s/2} F^*(s) e^{kTs} \, ds \tag{3-81}$$

Now, substituting $z = e^{Ts}$ in Eq. (3-81), and since

$$F^*(s)|_{s=\ln z/T} = F(z) \tag{3-82}$$

$$e^{kTs} = z^k \tag{3-83}$$

and

$$ds = d\left[\frac{1}{T} \ln z\right] = \frac{1}{T} z^{-1} \, dz \tag{3-84}$$

Eq. (3-81) becomes

$$f(kT) = \frac{1}{2\pi j} \oint_\Gamma F(z) z^{k-1} \, dz \tag{3-85}$$

which is the inversion formula for z-transform we set out to prove in Eq. (3-74). The path of integration in the s-plane from $s = -c - j(\omega_s/2)$ to $s = c + j(\omega_s/2)$ is mapped onto the circle $|z| = e^{cT}$ in the z-plane, as shown in Fig. 3-8(b). Since $F^*(s)$ does not have singularities on or to the right of the path of integration $s = c - j\infty$ to $s = c + j\infty$ in the s-plane, all the singularities of $F(z)z^{k-1}$ must lie inside the circle Γ which is described by $|z| = e^{cT}$ in the z-plane.

Example 3-8

The inverse z-transform of $F(z)$ in Eq. (3-65) is now determined by use of the inversion formula of Eq. (3-74).

Substituting Eq. (3-65) into Eq. (3-74), we have

$$f(kT) = \frac{1}{2\pi j} \oint_\Gamma \frac{z(1 - e^{-aT})}{(z-1)(z - e^{-aT})} z^{k-1} \, dz \tag{3-86}$$

where Γ is a circle with its center at $z = 0$ and is large enough to enclose the poles of $F(z)$ at $z = 1$ and $z = e^{-aT}$. The contour integral of Eq. (3-86) is usually evaluated by means of the residues theorem of complex variable theory. If the integrand of the integral contains a rational function—that is, with a finite number of poles and zeros—the residues method is essentially the same as the partial-fraction expansion method. Thus, Eq. (3-86) is written

$$f(kT) = \sum \text{residues of } F(z)z^{k-1} \text{ at the poles of } F(z) \tag{3-87}$$

where the residue of $F(z)z^{k-1}$ at a given pole of $F(z)$ is essentially the same as the coefficient of the partial-fraction expansion of $F(z)z^{k-1}$ at that pole. Thus, Eq. (3-87) gives

$$f(kT) = \sum \text{residues of } \frac{z(1 - e^{-aT})}{(z-1)(z - e^{-aT})} z^{k-1} \text{ at } z = 1 \text{ and } z = e^{-aT}$$

$$= 1 - e^{-akT} \tag{3-88}$$

for $k = 0, 1, 2, \ldots$. This result agrees with that obtained in Eqs. (3-68) and (3-72) by the two preceding methods.

3-5 THEOREMS OF THE z-TRANSFORM

Some of the problems dealing with the evaluation and manipulations of the z-transform can be simplified by identifying some of the properties of the z-transform. These properties are given in the form of theorems in the following.

□ 1. Addition and Subtraction

If $f_1(t)$ and $f_2(t)$ have z-transforms $F_1(z)$ and $F_2(z)$, respectively, and

$$F_1(z) = \mathscr{Z}[f_1(t)] = \sum_{k=0}^{\infty} f_1(kT)z^{-k} \tag{3-89}$$

and

$$F_2(z) = \mathscr{Z}[f_2(t)] = \sum_{k=0}^{\infty} f_2(kT)z^{-k} \tag{3-90}$$

then, the z-transform of $f_1(t) \pm f_2(t)$ is

$$\mathscr{Z}[f_1(t) \pm f_2(t)] = F_1(z) \pm F_2(z) \quad \blacksquare \tag{3-91}$$

Proof: By definition,

$$\begin{aligned}\mathscr{Z}[f_1(t) \pm f_2(t)] &= \sum_{k=0}^{\infty} [f_1(kT) \pm f_2(kT)]z^{-k} \\ &= \sum_{k=0}^{\infty} f_1(kT)z^{-k} \pm \sum_{k=0}^{\infty} f_1(kT)z^{-k} \\ &= F_1(z) \pm F_2(z)\end{aligned} \tag{3-92}$$

●

□ 2. Multiplication by a Constant

If $F(z)$ is the z-transform of $f(t)$, then

$$\mathscr{Z}[af(t)] = a\mathscr{Z}[f(t)] = aF(z) \tag{3-93}$$

where a is a constant. ■

Proof: By the z-transform definition,

$$\mathscr{Z}[af(t)] = \sum_{k=0}^{\infty} af(kT)z^{-k} = a\sum_{k=0}^{\infty} f(kT)z^{-k} = aF(z) \tag{3-94}$$

●

□ 3. Real Translation (Shifting) Theorem

Right Shift (Time Delay)

If $f(t)$ has the z-transform $F(z)$, then the z-transform of $f(t)$ shifted to the right by nT, where n is a positive integer, is

$$\mathscr{Z}[f(t - nT)u_s(t - nT)] = z^{-n}F(z) \quad ■ \tag{3-95}$$

Proof: By the z-transform definition,

$$\mathscr{Z}[f(t - nT)u_s(t - nT)] = \sum_{k=0}^{\infty} f(kT - nT)z^{-k} \tag{3-96}$$

which can be written as

$$\mathscr{Z}[f(t - nT)u_s(t - nT)] = z^{-n}\sum_{k=0}^{\infty} f(kT - nT)z^{-(k-n)} \tag{3-97}$$

Since $f(t) = 0$ for $t < 0$, Eq. (3-97) becomes

$$\mathscr{Z}[f(t - nT)u_s(t - nT)] = z^{-n}\sum_{k=n}^{\infty} f(kT - nT)z^{-(k-n)} \tag{3-98}$$

Now, letting $p = k - n$ in Eq. (3-98) yields

$$\begin{aligned}\mathscr{Z}[f(t - nT)u_s(t - nT)] &= z^{-n}\sum_{p=0}^{\infty} f(pT)z^{-p} \\ &= z^{-n}F(z)\end{aligned} \tag{3-99}$$

●

From the shifting theorem, we realize that the transform variable z^{-1} corresponds to a delay of one sampling period in the time domain. Thus, z^{-1} in digital control systems is usually considered as a *delay operator.*

It should be pointed out that the above shifting theorem is applicable only to a real function that has a delay that is an integral multiple of the sampling period. In practice, we often encounter situations where the time delay is not an integral multiple of the sampling period, and these should be treated in a different manner.

Left Shift (Time Advance)

The z-transform of a time function $f(t)$ which is shifted to the left, or advanced in time by nT, where n is a positive integer, is written

$$\mathscr{Z}[f(t+nT)u_s(t+nT)] = z^n\left[F(z) - \sum_{k=0}^{n-1} f(kT)z^{-k}\right] \tag{3-100}$$

Notice that the result is not simply $z^nF(z)$, as one would expect. ■

Proof: The z-transform of $f(t+nT)u_s(t+nT)$ is written as

$$\mathscr{Z}[f(t+nT)u_s(t+nT)] = \sum_{k=0}^{\infty} f(kT+nT)z^{-k} \tag{3-101}$$

since, by definition of the z-transform, the sequence starts at $k = 0$.

Equation (3-101) is conditioned as

$$\mathscr{Z}[f(t+nT)u_s(t+nT)] = z^n \sum_{k=0}^{\infty} f(kT+nT)z^{-(k+n)} \tag{3-102}$$

or

$$\mathscr{Z}[f(t+nT)u_s(t+nT)] = z^n\left[\sum_{k=n}^{\infty} f(kT)z^{-k}\right]$$

$$= z^n\left[F(z) - \sum_{k=0}^{n-1} f(kT)z^{-k}\right] \tag{3-103}$$

which is the desired result. Since the z-transform defined here is one-sided—that is, for $k \geq 0$, when $f(t)$ is shifted to the left by nT and then sampled by the ideal sampler starting at $t = 0$ the impulses that correspond to the sampling of the shifted function for $t < 0$ are lost. This accounts for the last term on the right side of Eq. (3-103). ●

Example 3-9

The z-transform of a unit-step function which is delayed by one sampling period T is determined by using the right-shift theorem of Eq. (3-95).

$$\mathscr{Z}[u_s(t-T)] = z^{-1}\mathscr{Z}[u_s(t)] = z^{-1}\frac{z}{z-1} = \frac{1}{z-1} \tag{3-104}$$

The z-transform of a unit-step function that is advanced by one sampling period T is determined by using the left-shift theorem of Eq. (3-100).

$$\mathscr{Z}[u_s(t+T)] = z[F(z) - f(0)]$$

$$= z\left[\frac{z}{z-1} - 1\right] = \frac{z}{z-1} \tag{3-105}$$

which is the same as the z-transform of the unshifted unit-step function. The result is expected, since when a left-shifted unit-step function is sampled, the result is still a train of unit impulses starting at $t = 0$.

□ 4. Complex Translation Theorem

If $f(t)$ has the z-transform $F(z)$, then

$$\mathscr{Z}[e^{\mp at}f(t)] = [F(s \pm a)]|_{z=e^{Ts}} = F(ze^{\pm aT}) \tag{3-106}$$

where a is a constant. ■

Proof: From the z-transform definition we can write the z-transform of $e^{\mp at}f(t)$ as

$$\mathscr{Z}[e^{\mp at}f(t)] = \sum_{k=0}^{\infty} f(kT)e^{\mp akT}z^{-k} \tag{3-107}$$

Let $z_1 = ze^{\pm aT}$; then Eq. (3-107) becomes

$$\mathscr{Z}[e^{\mp at}f(t)] = \sum_{k=0}^{\infty} f(kT)z_1^{-k} = F(z_1)$$

$$= F(ze^{\pm aT}) \tag{3-108}$$

●

The complex transition theorem is useful for the derivation of the z-transforms of functions that are multiplied by $e^{\mp at}$.

Example 3-10

The z-transform of $g(t) = e^{-at}\sin \omega t$ is derived in the following using the complex translation equation of Eq. (3-106).

From the z-transform table in Appendix C, the z-transform of $f(t) = \sin \omega t$ is

$$F(z) = \mathscr{Z}[\sin \omega t] = \frac{z \sin \omega T}{z^2 - 2z\cos \omega T + 1} \tag{3-109}$$

Thus, from Eq. (3-106),

$$G(z) = \mathscr{Z}[e^{-at}\sin\omega t] = F(ze^{aT}) = \frac{ze^{-aT}\sin\omega T}{z^2 - 2ze^{-aT}\cos\omega T + e^{-2aT}} \tag{3-110}$$

This agrees with the z-transform of $e^{-at}\sin\omega t$ found in Appendix C.

□ 5. Initial Value Theorem

If the function $f(t)$ has the z-transform $F(z)$, then

$$\lim_{k\to 0} f(kT) = \lim_{z\to\infty} F(z) \tag{3-111}$$

if the limit exists.

The theorem simply states that the value of $f(0)$ is equal to the value of $F(z)$ at $z = \infty$. ■

Proof: From the defining equation of the z-transform, we write

$$F(z) = \sum_{k=0}^{\infty} f(kT)z^{-k} = f(0) + f(T)z^{-1} + f(2T)z^{-2} + \cdots \tag{3-112}$$

Taking the limit on both sides of Eq. (3-112) as z approaches infinity, we get

$$\lim_{z\to\infty} F(z) = f(0) = \lim_{k\to 0} f(kT) \tag{3-113}$$

□ 6. Final Value Theorem

If the function $f(t)$ has the z-transform $F(z)$ and if the function $(1 - z^{-1})F(z)$ does not have poles on or outside the unit circle $|z| = 1$ in the z-plane, then

$$\lim_{k\to\infty} f(kT) = \lim_{z\to 1} (1 - z^{-1})F(z) \tag{3-114}$$

The final value theorem of the z-transform is very useful for the determination of the final value of the time response of a digital control system from the z-transform expression of the output, without carrying out the inverse z-transform. ■

Proof: Let us consider two finite sequences,

$$\sum_{k=0}^{n} f(kT)z^{-k} = f(0) + f(T)z^{-1} + \cdots + f(nT)z^{-n} \tag{3-115}$$

and

$$\sum_{k=0}^{n} f[(k-1)T]z^{-k} = f(0)z^{-1} + f(T)z^{-2} + \cdots + f[(n-1)T]z^{-n} \tag{3-116}$$

where n is a positive integer. Since $f(t) = 0$ for all $t < 0$, the term $f(-T)$ that would have appeared in Eq. (3-116) is zero. Now, comparing Eqs. (3-115) and (3-116), we see that the latter can be written as

$$\sum_{k=0}^{n} f[(k-1)T]z^{-k} = z^{-1}\sum_{k=0}^{n-1} f(kT)z^{-k} \tag{3-117}$$

Taking the difference between Eqs. (3-115) and (3-117), and letting z approach 1, we get

$$\lim_{z\to 1}\left[\sum_{k=0}^{n} f(kT)z^{-k} - z^{-1}\sum_{k=0}^{n-1} f(kT)z^{-k}\right] = \sum_{k=0}^{n} f(kT) - \sum_{k=0}^{n-1} f(kT) = f(nT) \tag{3-118}$$

Taking the limit as n approaches infinity on both sides of Eq. (3-118), we get

$$\lim_{n\to\infty} f(nT) = \lim_{n\to\infty}\lim_{z\to 1}\left[\sum_{k=0}^{n} f(kT)z^{-k} - z^{-1}\sum_{k=0}^{n-1} f(kT)z^{-k}\right] \tag{3-119}$$

Interchanging the limits on the right-hand side of Eq. (3-119), and since

$$\lim_{n\to\infty}\sum_{k=0}^{n} f(kT)z^{-k} = \lim_{n\to\infty}\sum_{k=0}^{n-1} f(kT)z^{-k} = F(z) \tag{3-120}$$

we have the final value theorem

$$\lim_{n\to\infty} f(nT) = \lim_{z\to 1}(1 - z^{-1})F(z) \tag{3-121}$$

●

Example 3-11

Given the z-transform

$$F(z) = \frac{0.792z^2}{(z-1)(z^2 - 0.416z + 0.208)} \tag{3-122}$$

the final value of $f(kT)$ is determined from $F(z)$ by use of the final-value theorem, Eq. (3-114). Since

$$(1 - z^{-1})F(z) = \frac{0.792z}{z^2 - 0.416z + 0.208} \tag{3-123}$$

does not have poles or zeros on or outside the unit circle $|z| = 1$, the final value theorem can be applied. Thus, from Eq. (3-114),

$$\lim_{k\to\infty} f(kT) = \lim_{z\to 1}\frac{0.792z}{z^2 - 0.416z + 0.208} = 1 \tag{3-124}$$

This result is checked readily by expanding $F(z)$ of Eq. (3-122) into a power series in z^{-1}.

$$F(z) = 0.792z^{-1} + 1.12z^{-2} + 1.091z^{-3} + 1.010z^{-4} + 0.983z^{-5} + 0.989z^{-6} + 0.990z^{-7} + \cdots \tag{3-125}$$

When a sufficient number of terms are carried out in the power-series expansion of $F(z)$, we can easily see that the coefficients of the series converge rapidly to the final steady-state value of unity.

□ 7. Partial Differentiation Theorem

Let the z-transform of the function $f(t, a)$ be represented by $F(z, a)$, where a is an independent variable or constant. The z-transform of the partial derivative of $f(t, a)$ with respect to a is given by

$$\mathscr{Z}\left[\frac{\partial}{\partial a}[f(t, a)]\right] = \frac{\partial}{\partial a} F(z, a) \quad \blacksquare \tag{3-126}$$

Proof: From the z-transform definition,

$$\mathscr{Z}\left[\frac{\partial}{\partial a}[f(t, a)]\right] = \sum_{k=0}^{\infty} \frac{\partial}{\partial a} f(kT, a)z^{-k}$$

$$= \frac{\partial}{\partial a} \sum_{k=0}^{\infty} f(kT, a)z^{-k} = \frac{\partial}{\partial a} F(z, a) \tag{3-127}$$

●

The following example illustrates the usefulness of the partial differentiation theorem in the derivation of a certain class of functions.

Example 3-12

The z-transform of $f(t) = te^{-at}$ is determined by the use of the partial differentiation theorem:

$$\mathscr{Z}[te^{-at}] = \mathscr{Z}\left[-\frac{\partial}{\partial a} e^{-at}\right] \tag{3-128}$$

The z-transform of e^{-at} is $z/(z - e^{-aT})$. Thus, from Eq. (3-126), the z-transform of te^{-at} is written

$$\mathscr{Z}[te^{-at}] = \mathscr{Z}\left[-\frac{\partial}{\partial a} e^{-at}\right] = -\frac{\partial}{\partial a} \mathscr{Z}[e^{-at}]$$

$$= -\frac{\partial}{\partial a}\left[\frac{z}{z - e^{-aT}}\right] = \frac{Tze^{-aT}}{(z - e^{-aT})^2} \tag{3-129}$$

□ 8. Real Convolution Theorem

If the functions $f_1(t)$ and $f_2(t)$ have the z-transforms $F_1(z)$ and $F_2(z)$, respectively, and $f_1(t) = 0$ and $f_2(t) = 0$ for $t < 0$, then

$$F_1(z)F_2(z) = \mathscr{Z}\left[\sum_{n=0}^{k} f_1(nT)f_2(kT - nT)\right] \quad \blacksquare \tag{3-130}$$

Proof: By definition of the z-transform, the right-hand side of Eq. (3-130) is written as

$$\begin{aligned}\mathscr{Z}\left[\sum_{n=0}^{k} f_1(nT)f_2(kT - nT)\right] &= \sum_{k=0}^{\infty}\sum_{n=0}^{k} f_1(nT)f_2(kT - nT)z^{-k} \\ &= \sum_{k=0}^{\infty}\sum_{n=0}^{\infty} f_1(nT)f_2(kT - nT)z^{-k}\end{aligned} \tag{3-131}$$

The upper limit of the second summation in the last equation is changed to ∞ since $f_1(kT)$ and $f_2(kT)$ equal 0 for $k < 0$. Letting $m = k - n$ and interchanging the order of the summation in Eq. (3-131), we get

$$\mathscr{Z}\left[\sum_{n=0}^{k} f_1(nT)f_2(kT - nT)\right] = \sum_{n=0}^{\infty} f_1(nT)z^{-k} \sum_{m=-n}^{\infty} f_2(mT)z^{-m} \tag{3-132}$$

Since $f_2(kT) = 0$ for $k < 0$, the last equation becomes

$$\mathscr{Z}\left[\sum_{n=0}^{\infty} f_1(nT)z^{-n} \sum_{n=0}^{\infty} f_1(nT)z^{-n} \sum_{m=0}^{\infty} f_2(mT)z^{-m}\right] = F_1(z)F_2(z) \tag{3-133}$$

which is the desired result. ●

The real convolution theorem of the z-transform is analogous to that of the Laplace transform which states that

$$F_1(s)F_2(s) = \mathscr{L}\left[\int_0^t f_1(\tau)f_2(t - \tau)\, d\tau\right] \tag{3-134}$$

where $F_1(s)$ and $F_2(s)$ are the Laplace transforms of $f_1(t)$ and $f_2(t)$, respectively.

An important fact to remember is that the inverse transform (Laplace or z) of the product of two transform functions in general is not equal to the product of the two functions in the real domain; that is,

$$\mathscr{Z}^{-1}[F_1(z)F_2(z)] \neq f_1(kT)f_2(kT) \tag{3-135}$$

and

$$\mathscr{L}^{-1}[F_1(s)F_2(s)] \neq f_1(t)f_2(t) \tag{3-136}$$

Another way of expressing the real convolution of two real sequences $f_1(kT)$ and $f_2(kT)$ is

$$\mathscr{Z}^{-1}[F_1(z)F_2(z)] = f_1(kT) * f_2(kT) \tag{3-137}$$

where the asterisk denotes real convolution as defined in Eq. (3-130).

□ 9. Complex Convolution Theorem

If the z-transforms of $f_1(t)$ and $f_2(t)$ are $F_1(z)$ and $F_2(z)$, respectively, then the z-transform of the product of the two functions is

$$\mathscr{Z}[f_1(t)f_2(t)] = \frac{1}{2\pi j}\oint_\Gamma \frac{F_1(\zeta)F_2(z\zeta^{-1})}{\zeta}\,d\zeta \tag{3-138}$$

where Γ is a circle in the z-plane which lies in the region (annulus) described by

$$\sigma_1 < |\zeta| < \frac{|z|}{\sigma_2} \tag{3-139}$$

and

$$|z| > \max(\sigma_1, \sigma_2, \sigma_1\sigma_2) \tag{3-140}$$

where

$$\sigma_1 = \text{radius of convergence of } F_1(\zeta)$$
$$\sigma_2 = \text{radius of convergence of } F_2(\zeta) \quad \blacksquare$$

Proof: By definition, the z-transform of the product of $f_1(t)$ and $f_2(t)$ is written

$$\mathscr{Z}[f_1(t)f_2(t)] = \sum_{k=0}^{\infty} f_1(kT)f_2(kT)z^{-k} \tag{3-141}$$

For this z-transform series to converge absolutely, the magnitude of z must be greater than the largest value of σ_1, σ_2, and $\sigma_1\sigma_2$; that is, Eq. (3-140) must be satisfied. We may express $f_1(kT)$ as an inverse z-transform relation:

$$f_1(kT) = \frac{1}{2\pi j}\oint_\Gamma F_1(\zeta)\zeta^{k-1}\,d\zeta \tag{3-142}$$

where Γ denotes a circle that encloses all the singularities of $F_1(\zeta)\zeta^{k-1}$ in the complex ζ-plane. This is equivalent to requiring that ζ satisfy the left-hand inequality condition of Eq. (3-139), or $|\zeta| > \sigma_1$.

Substituting Eq. (3-142) into Eq. (3-141), we obtain

$$\mathscr{Z}[f_1(t)f_2(t)] = \frac{1}{2\pi j}\oint_\Gamma \frac{F_1(\zeta)}{\zeta} F_2(z\zeta^{-1})\,d\zeta \tag{3-143}$$

Since

$$F_2(z\zeta^{-1}) = \sum_{k=0}^{\infty} f_2(kT)(z\zeta^{-1})^{-k} \tag{3-144}$$

which is absolutely convergent for $|z\zeta^{-1}| > \sigma_2$, or $|\zeta| < |z|/\sigma_2$, Eq. (3-141) becomes

$$\mathscr{Z}[f_1(t)f_2(t)] = \frac{1}{2\pi j}\oint_\Gamma \frac{F_1(\zeta)F_2(z\zeta^{-1})}{\zeta}\,d\zeta \tag{3-145}$$

with

$$\sigma_1 < |\zeta| < |z|/\sigma_2 \tag{3-146}$$

The analogous complex translation theorem of the Laplace transform states that

$$\mathscr{L}[f_1(t)f_2(t)] = \frac{1}{2\pi j}\int_{c-j\infty}^{c+j\infty} F_1(\zeta)F_2(s-\zeta)\,d\zeta \tag{3-147}$$

with $\sigma = \text{Re}(s) > \max(\sigma_1, \sigma_2, \sigma_1 + \sigma_2)$, and $\sigma_1 < \text{Re}(\zeta) = c < \sigma - \sigma_2$.

Thus, for the Laplace transform and the z-transform, remember that in general the transform of the product of $f_1(t)f_2(t)$ is not equal to the product of the transformed functions; i.e.,

$$\mathscr{L}[f_1(t)f_2(t)] \neq F_1(s)F_2(s) \tag{3-148}$$

and

$$\mathscr{Z}[f_1(t)f_2(t)] \neq F_1(z)F_2(z) \tag{3-149}$$

The complex convolution integral in Eq. (3-138) can also be expressed symbolically as

$$\mathscr{Z}[f_1(t)f_2(t)] = F_1(z)*F_2(z) \tag{3-150}$$

where the asterisk denotes the complex convolution integral operation.

Example 3-13

The z-transform of the time function $f(t) = te^{-at}$ considered in Example 3-12 is now determined by use of the complex convolution theorem.

Let $f_1(t) = t$ and $f_2(t) = e^{-at}$. Then,

$$F_1(z) = \mathscr{Z}[t] = \frac{Tz}{(z-1)^2} \qquad |z| > 1 = \sigma_1 \tag{3-151}$$

$$F_2(z) = \mathscr{Z}[e^{-at}] = \frac{z}{z - e^{-aT}} \qquad |z| > e^{-aT} = \sigma_2 \tag{3-152}$$

Substituting Eqs. (3-151) and (3-152) into Eq. (3-138), we get

$$\mathscr{Z}[f_1(t)f_2(t)] = \frac{1}{2\pi j}\oint_\Gamma \frac{T\zeta}{\zeta(\zeta-1)^2}\,\frac{z\zeta^{-1}}{z\zeta^{-1} - e^{-aT}}\,d\zeta \tag{3-153}$$

where Γ is a circular path that lies in the annular ring described by

$$1 < |\zeta| < \frac{|z|}{e^{-aT}} = |z|e^{aT} \tag{3-154}$$

and $|z| > 1$.

Therefore, the integration path of Eq. (3-153) encloses only the poles of the integrand at $\zeta = 1$. The usual way of evaluating the contour integral of Eq. (3-153) is to apply the residues theorem of complex variables. In the present case, since the integrand of the integral in Eq. (3-153) is a rational function, we can use the partial-fraction expansion method. The coefficient of the expansion that corresponds to the $(\zeta - 1)^2$ term is written

$$\mathscr{Z}[f_1(t)f_2(t)] = \frac{\partial}{\partial \zeta}\left[\frac{Tz\zeta^{-1}}{(z\zeta^{-1} - e^{-aT})}\right]_{\zeta=1} = \frac{Tze^{-aT}}{(z - e^{-aT})^2} \tag{3-155}$$

which agrees with the result obtained in Eq. (3-129) of Example 3-12.

3-6 LIMITATIONS OF THE z-TRANSFORM METHOD

From the discussions of the z-transform in the preceding sections, we see that the z-transform can be a very convenient tool for representing linear digital and sampled-data control systems. However, in applying the z-transform method, one must realize the limitations and the conditions of the method. The following considerations should be kept in mind when applying the z-transform.

1. *Ideal Sampler Assumption.* The derivation of the z-transform of a continuous-data function $f(t)$ is based first on the sampling of the function by an ideal sampler. The result of this is that the z-transform $F(z)$ represents the function $f(t)$ only at the sampling instants.
2. *Nonuniqueness of the Inverse z-Transform.* Given $F(z)$, the inverse z-transform of $F(z)$ gives only a unique solution to $f(kT)$. Strictly, the solution to $f(t)$ is unknown.
3. *Accuracy of the z-Transform Method.* The accuracy of the z-transform method depends on the magnitude of the sampling frequency ω_s or the sampling period T, relative to the highest frequency component contained in the function $f(t)$. If the sampling period is too large or the sampling frequency is too low, relative to the variation of $f(t)$, the z-transform solution may be erroneous, since $f^*(t)$ would not be a good representation of $f(t)$ at the sampling instants. In reality, although $f(t)$ cannot be uniquely determined from $f(kT)$, the inverse z-transform of $F(z)$, if the sampling period is adequately small, a good approximation of $f(t)$ can be obtained by connecting the sequence of $f(kT)$ by a smooth curve.

3-7 APPLICATIONS OF THE z-TRANSFORM

One of the most important applications of the z-transform is in the solution of linear difference equations. The solution of linear difference equations with the z-transform is parallel to the solution of linear differential equations by the Laplace transform.

Let us represent an nth-order linear difference equation with constant coefficients as

$$\begin{aligned} c(k+n) &+ a_1 c(k+n-1) + a_2 c(k+n-2) + \cdots + a_{n-1} c(k+1) + a_n c(k) \\ &= r(k+m) + b_1 r(k+m-1) + \cdots + b_{m-1} r(k+1) + b_m r(k) \end{aligned} \quad \text{(3-156)}$$

where n and m ($m < n$) are positive integers. The problem is this: given the n initial conditions $c(0), c(1), \ldots, c(n-1)$, and the input $r(k)$ for $k \geq 0$, find $c(k)$ for $k \geq 0$.

The z-transform solution of Eq. (3-156) involves the use of the real translation theorems given by Eqs. (3-95) and (3-100). In terms of the notation given in Eq. (3-156), the important relations are summarized as follows:

$$\begin{aligned} \mathscr{Z}[c(k)] &= C(z) \\ \mathscr{Z}[c(k+1)] &= z[C(z) - c(0)] \\ &\vdots \\ \mathscr{Z}[c(k+n)] &= z^n C(z) - z^n c(0) - z^{n-1} c(1) - \cdots - zc(n-1) \end{aligned} \quad \text{(3-157)}$$

The following examples illustrate the z-transform solutions of linear difference equations.

3-7-1 Interest-Payment Problem

The difference equation of the car-loan problem stated in Problem 2-1 is written

$$p(k+1) = (1+R)p(k) - U \quad \text{(3-158)}$$

where

$p(k)$ = amount owed at the end of the kth period
R = monthly interest rate (percent)
U = monthly payment (constant).

The amount borrowed initially is $p(0)$. The total time duration for the loan is N months. Equation (3-158) is a linear first-order difference equation with constant coefficients that can be solved by means of the z-transform method. However, before embarking on the z-transform method, let us examine the real-domain solution using the recursive method.

Substituting $k = 0$ in Eq. (3-158), we have

$$p(1) = (1+R)p(0) - U \quad \text{(3-159)}$$

Setting $k = 1$ in Eq. (3-158) and making use of Eq. (3-159), we have

$$\begin{aligned} p(2) &= (1 + R)p(1) - U \\ &= (1 + R)^2 p(0) - (1 + R)U - U \end{aligned} \tag{3-160}$$

Continuing the process, for $k = N - 1$, we have

$$p(N) = (1 + R)^N p(0) - X \tag{3-161}$$

where

$$X = (1 + R)^{N-1}U - (1 + R)^{N-2}U - \cdots - (1 + R)U - U \tag{3-162}$$

The sum of the last series can be obtained by multiplying both sides of the equation by $(1 + R)$, and then subtracting Eq. (3-162) from the resulting equation. The result is

$$RX = (1 + R)^N U - U \tag{3-163}$$

Solving for X from the last equation, we have

$$X = \frac{(1 + R)^N}{R} - \frac{1}{R} \tag{3-164}$$

Substitution of X from Eq. (3-164) into Eq. (3-161) gives the solution for $p(N)$.

$$p(N) = \frac{U}{R} + \left(p(0) - \frac{U}{R}\right)(1 + R)^N \tag{3-165}$$

Thus, we see that, while the real-domain recursive method of solving the difference equation is conceptually straightforward, the procedure usually involves the determination of the sum of a series, which can be tedious.

A systematic method of solving linear difference equations in the real domain using characteristic equations may be found in reference [15].

The z-transform solution of Eq. (3-158) is carried out by first multiplying both sides of the equation by z^{-k} and then applying the summation from $k = 0$ to ∞. The result is

$$\sum_{k=0}^{\infty} p(k + 1)z^{-k} = (1 + R)\sum_{k=0}^{\infty} p(k)z^{-k} - \sum_{k=0}^{\infty} Uz^{-k} \tag{3-166}$$

Applying the shifting theorem of Eq. (3-100) or the relations in Eq. (3-157) and the fact that U is a constant, Eq. (3-166) becomes

$$zP(z) - zp(0) = (1 + R)P(z) - \frac{Uz}{z - 1} \tag{3-167}$$

where $P(z)$ denotes the z-transform of $p(k)$.

Solving for $P(z)$ from Eq. (3-167), we have

$$P(z) = \frac{zp(0)}{z - R - 1} - \frac{Uz}{(z - 1)(z - R - 1)} \tag{3-168}$$

Performing partial-fraction expansion on the last term of Eq. (3-168), as described in Sec. 3-4-2, Eq. (3-168) becomes

$$P(z) = \frac{Uz}{R(z-1)} + \frac{[p(0) - U/R]z}{z - R - 1} \tag{3-169}$$

Taking the inverse z-transform on both sides of the last equation, and setting $k = N$, we obtain

$$p(N) = \frac{U}{R} + \left(p(0) - \frac{U}{R}\right)(1 + R)^N \tag{3-170}$$

which is identical to the result in Eq. (3-165).

Since the amount owed at the end of N months is zero, $p(N) = 0$. Solving for U in Eq. (3-170) gives the monthly loan payment as

$$U = \frac{R(1+R)^N}{(1+R)^N - 1} p(0) \tag{3-171}$$

where N is any positive integer greater than zero.

3-7-2 A Ladder Network Problem

The z-transform can be used to solve certain network problems that are often difficult to solve by conventional methods. Figure 3-9 shows an all-resistive ladder network that can be described by a linear difference equation. Let the loop current of the kth stage of the ladder network be designated as $i(k)$. The loop equation of the $(k + 1)$st stage is written

$$i(k+2) - 3i(k+1) + i(k) = 0 \tag{3-172}$$

The boundary conditions, as observed from Fig. 3-9, are

$$V = 2Ri(0) - Ri(1) \tag{3-173}$$

$$i(N+1) = 0 \tag{3-174}$$

Due to the form of the boundary conditions, a closed-form solution of Eq. (3-172) is difficult to obtain by using the recursive method, although by substituting $k = 0, 1, 2, \ldots$ into Eq. (3-172) continuously, we can obtain $i(N)$ in terms of V, R, and $i(0)$ for any positive integral N.

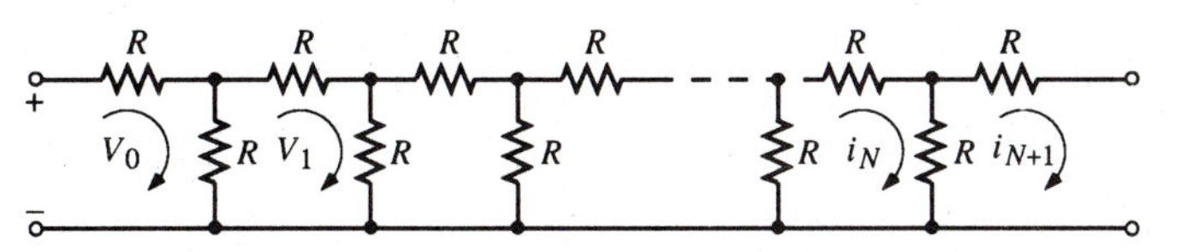

Figure 3-9. Ladder network.

Taking the z-transform on both sides of Eq. (3-172), we have

$$z^2 I(z) - z^2 i(0) - zi(1) - 3zI(z) + 3zi(0) + I(z) = 0 \tag{3-175}$$

Substituting $i(1)$ in terms of V and $i(0)$ from Eq. (3-173) into Eq. (3-175) and solving for $I(z)$, we get

$$I(z) = \frac{z[(z-1)i(0) - V/R]}{z^2 - 3z + 1} = \frac{z[(z-1)i(0) - V/R]}{(z - 0.382)(z - 2.618)} \tag{3-176}$$

Performing partial-fraction expansion on the last equation, we get

$$I(z) = \frac{[0.2764i(0) + 0.4472V/R]z}{z - 0.382} + \frac{[0.7236i(0) - 0.4472V/R]z}{z - 2.618} \tag{3-177}$$

Taking the inverse z-transform of Eq. (3-177), we have

$$\begin{aligned} i(k) &= [0.2764(0.382)^k + 0.7236(2.618)^k]i(0) \\ &\quad + 0.4472[(0.382)^k - (2.618)^k]V/R \end{aligned} \tag{3-178}$$

The initial condition $i(0)$ is determined by applying the end condition of Eq. (3-174) to Eq. (3-178). Thus,

$$i(0) = \frac{0.4472[(2.618)^{N+1} - (0.382)^{N+1}]}{0.2764(0.382)^{N+1} + 0.7236(2.618)^{N+1}} \tag{3-179}$$

In this example we have demonstrated that the independent variable k in a difference equation is not related to time.

3-7-3 Solution of Difference Equations with Complex Poles

The characteristic equations of the difference equations in the last two examples all have real roots. In this section we shall show the z-transform solution of a difference equation that has a characteristic equation with complex roots. Consider the linear nonhomogeneous difference equation,

$$y(k+2) + 0.4y(k+1) + 0.1y(k) = -(0.5)^{k+1} \tag{3-180}$$

The initial conditions are $y(0) = 0$ and $y(1) = 0$. Taking the z-transform on both sides of Eq. (3-180), applying the initial conditions, and solving for $Y(z)$, we have

$$\begin{aligned} Y(z) &= \frac{-0.5z}{(z - 0.5)(z^2 + 0.4z + 0.1)} \\ &= \frac{-0.5z}{(z - 0.5)(z + 0.2 + j0.245)(z + 0.2 - j0.245)} \end{aligned} \tag{3-181}$$

The partial-fraction expansion of Eq. (3-181) is

$$Y(z) = \frac{-0.909z}{z - 0.5} + \frac{1.377e^{j1.234}z}{z + 0.2 + j0.245} + \frac{1.377e^{-j1.234}z}{z + 0.2 - j0.245} \tag{3-182}$$

The inverse z-transform of $Y(z)$ is

$$\begin{aligned} y(k) &= -0.909(0.5)^k + 1.377e^{j1.234}(-0.2 - j0.245)^k \\ &\quad + 1.377e^{-j1.234}(-0.2 + j0.245)^k \\ &= -0.909(0.5)^k + 1.377(0.316)^k[e^{j(2.255k-1.234)} + e^{-j(2.255k-1.234)}] \end{aligned} \quad (3\text{-}183)$$

Thus,

$$y(k) = -0.909(0.5)^k + 2.754(0.316)^k \cos(2.255k - 1.234) \qquad (k \geq 0) \quad (3\text{-}184)$$

3-8 SIGNALS BETWEEN THE SAMPLING INSTANTS

In the preceding sections we pointed out that the z-transform method is effective only for systems in which the signals can be adequately represented by their values at the sampling instants. When the inherent sampling rate of a given system is too low relative to the frequency contained in the signals, additional effort may be needed in gaining knowledge of the signals between the sampling instants. The *submultiple sampling method*, the *delayed z-transform*, and the *modified z-transform* are devised for the recovery of signal information between the sampling instants. These methods are also useful as analytical tools for the study of digital control systems with nonuniform or multirate sampling.

3-8-1 The Submultiple Sampling Method—Multirate z-Transform

Consider that the inherent sampling period of a digital control system is T s. To describe the signal between the sampling instants that are separated by T s, we introduce a *fictitious sampler* with a sampling period T/N, where N is a positive integer greater than one. Figure 3-10 shows the block diagram of a sampled-data system with a sampler S_1 that has a sampling period T at the input. We shall show in the next chapter that, to analyze the system with the z-transform method, we must introduce a fictitious sampler S_1 with sampling

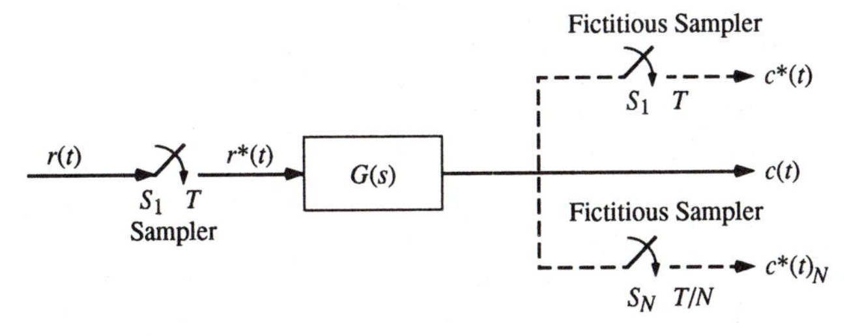

Figure 3-10. A sampled-data system with a multirate sampler at the output.

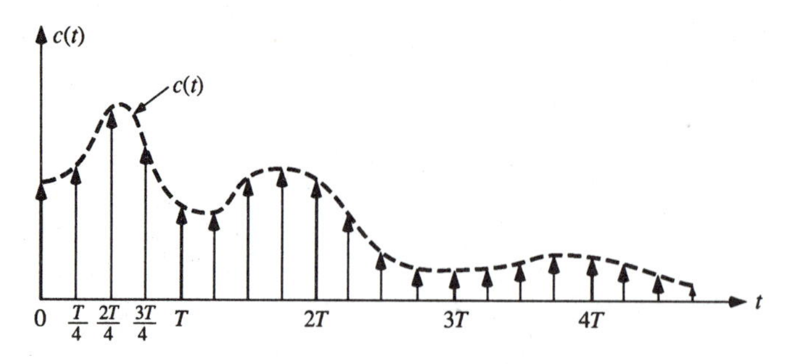

Figure 3-11. Output response of a multirate sampler with a sampling period of $T/4$.

period T at the output, so that $c(t)$ is characterized by the sampled signal $c^*(t)$. However, if $c(t)$ contains high-frequency components that are too fast for the sampling period T, we can introduce another fictitious sampler S_N at the output which has a sampling period of T/N, where N is a positive integer greater than one. The output of the fictitious sampler S_N is written as

$$c^*(t)_N = \sum_{k=0}^{\infty} c(kT/N)\delta(t - kT/N) \tag{3-185}$$

which contains $N - 1$ additional pulses between kT and $(k+1)T$ of the sampler S_1. Figure 3-11 shows a typical signal sampled by the multirate sampler with a sampling period of $T/4$.

Taking the z-transform on both sides of Eq. (3-185) we get

$$C(z)_N = \mathscr{Z}[c^*(t)_N] = \sum_{k=0}^{\infty} c(kT/N)z^{-k/N} \tag{3-186}$$

which is defined as the *multirate z-transform*. Comparison of the last equation with the defining equation of the ordinary z-transform in Eq. (3-11) reveals that $C(z)_N$ can be obtained directly from $C(z)$ by replacing z by $z^{1/N}$ and T by T/N; that is,

$$C(z)_N = C(z)|_{z=z^{1/N},\ T=T/N} \tag{3-187}$$

We should not be concerned with $C(z)_N$ having fractional powers of z, $z^{k/N}$. By letting $z_N = z^{1/N}$, $z^{k/N}$ becomes z_N^k. Thus, $C(z)_N$ will contain all integral powers of the variable z_N. The expansion of $C(z)_N$ in powers of z_N^{-1} gives the values of $c(t)$ at the submultiple sampling instants of $t = kT/N$.

Note that the submultiple sampling and the multirate z-transform are used only when real or fictitious samplers with sampling rates that are integral multiples of the basic sampling period T are found in a system. If the samplers found in one system all have a sampling period of T/N, then we can simply redefine the sampling period with a new value of the same.

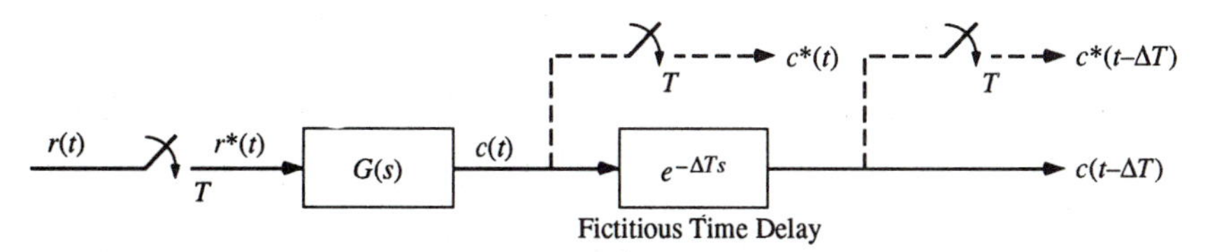

Figure 3-12. A sampled-data system with fictitious time delay and sampler.

3-8-2 The Delayed z-Transform and the Modified z-Transform

Another way of representing the details of a signal between the sampling instants with sampling period T is to delay the analog signal by ΔT, where $0 < \Delta < 1$; then the delayed signal is sampled by a conventional sampler at $t = kT$, $k = 0, 1, 2, \ldots$. By varying the amount of the time delay ΔT, we can recover as much information about the signal between the sampling instants as we wish. Figure 3-12 shows a system with a fictitious time delay ΔT at the output, and the delayed signal is sampled by a sampler with period T.

The delayed output of the system shown in Fig. 3-12 is $c(t - \Delta T)u_s(t - \Delta T)$. The sampled output of the fictitious time delay is expressed as

$$c^*(t - \Delta T) = \sum_{k=0}^{\infty} c(kT - \Delta T)\delta(t - kT) \tag{3-188}$$

The z-transform of the last equation is defined as the *delayed z-transform* and is written

$$C(z, \Delta) = \sum_{k=0}^{\infty} c(kT - \Delta T)z^{-k} \tag{3-189}$$

Figure 3-13 illustrates the steps of first shifting the signal $c(t)$ by ΔT and then sampling the shifted signal $c(t - \Delta T)$ by the ideal sampler starting from $t = 0$. Note that, since Δ is less than one, we cannot use the shifting theorem of Eq. (3-95).

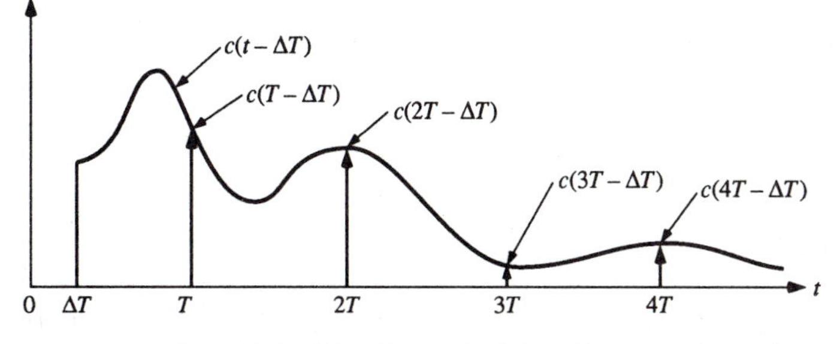

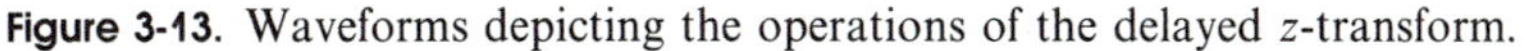

Figure 3-13. Waveforms depicting the operations of the delayed z-transform.

Comparing the operation of the delayed z-transform with that of the multirate sampling of Fig. 3-11, we see that, in the multirate sampling, the samples in between the regular sampling instants are obtained in one operation by the selection of N, whereas in the delayed z-transform, only one sample between two adjacent sampling instants is recovered for each value of ΔT.

Although the application of the delayed z-transform given in Eq. (3-189) seems straightforward, the fact that, when $\Delta \neq 0$, the first sample is always zero may cause some analytical problems. The following example will illustrate this difficulty.

Example 3-14

The delayed z-transform of $f(t) = e^{-at}u_s(t)$ is determined from Eq. (3-189) as follows:

$$F(z, \Delta) = \sum_{k=0}^{\infty} f(kT - \Delta T)z^{-k} = \sum_{k=0}^{\infty} e^{-a(kT-\Delta T)}z^{-k}$$

$$= e^{a\Delta T} \sum_{k=0}^{\infty} e^{-akT}z^{-k} \tag{3-190}$$

The summation term on the right-hand side of Eq. (3-190) is recognized as the z-transform of $f(t)$. Thus, Eq. (3-190) becomes

$$F(z, \Delta) = e^{a\Delta T} \frac{z}{z - e^{-aT}}$$

$$= e^{a\Delta T} + e^{-a(T-\Delta T)}z^{-1} + e^{-a(2T-\Delta T)}z^{-2} + \cdots \tag{3-191}$$

Thus, the first term in the series expansion of $F(z, \Delta)$, $e^{a\Delta T}$, is misleading and should be interpreted as zero, since it came from $f(-\Delta T)$.

To overcome the difficulty with the delayed z-transform in Example 3-14, we introduce a factor m, such that

$$m = 1 - \Delta \tag{3-192}$$

Since Δ lies between zero and one, m also lies in the same range.

Substituting Eq. (3-192) into Eq. (3-189) yields

$$C(z, m) = C(z, \Delta)|_{\Delta=1-m} = \sum_{k=0}^{\infty} c(kT - T + mT)z^{-k} \tag{3-193}$$

Now, using the shifting theorem in Eq. (3-95), the last equation is written as

$$C(z, m) = z^{-1} \sum_{k=0}^{\infty} c(kT + mT)z^{-k} \tag{3-194}$$

where $0 < m < 1$.

Equation (3-194) is defined as the *modified z-transform* of $c(t)$ and is denoted as

$$\mathscr{Z}_m[c(t)] = C(z, m) \tag{3-195}$$

or

$$\mathscr{Z}_m[c(t)] = C(z, \Delta)|_{\Delta=1-m} \tag{3-196}$$

Note that, when $\Delta = 0$, the delayed z-transform reverts to the z-transform. However, when $\Delta = 0$, $m = 1$, and Eq. (3-194) gives

$$C(z, m)|_{m=1} = z^{-1} \sum_{k=0}^{\infty} c[(k+1)T]z^{-k} = C(z) - c(0) \tag{3-197}$$

Thus, when $m = 1$, $\Delta = 0$, the modified z-transform is not equal to $C(z)$ unless the function $c(t)$ is zero at $t = 0$.

When $m = 0$ $(\Delta = 1)$, the function $c(t)$ is first delayed by one sampling period before taking the z-transform. Thus,

$$C(z, m)|_{m=0} = C(z, 0) = z^{-1}C(z) \tag{3-198}$$

Figure 3-14 illustrates the following steps of taking the modified z-transform.

1. The time function $c(t)$ is first shifted to the left (time advance) by mT, where $0 < m < 1$. This gives $c(t + mT)$.
2. The shifted time function $c(t + mT)$ is sampled by an ideal sampler starting from $t = 0$.
3. The sampled sequence is shifted to the right by one sampling instant T.

Now applying the modified z-transform to the function $f(t)$ in Example 3-14, we get

$$F(z, m) = z^{-1} \sum_{k=0}^{\infty} e^{-a(kT+mT)} z^{-k}$$

$$= z^{-1} e^{-amT} \frac{z}{z - e^{-aT}} = \frac{e^{-amT}}{z - e^{-aT}} \tag{3-199}$$

Thus, the expansion of $F(z, m)$ of Eq. (3-199) will show that the first term is $e^{-amT}z^{-1}$, and the z^{-k} term will carry the coefficient $e^{-a(kT+mT)}$.

Alternate Expressions for the Modified z-Transform

Two alternate expressions for the modified z-transform of Eq. (3-194) are summarized in the following. These expressions are derived using the complex convolution of the Laplace transform, and detailed derivations are given in Appendix B.

$$C(z, m) = z^{-1} \sum \text{residues of } C(\xi) \frac{e^{mT\xi}}{1 - e^{T\xi}z^{-1}} \text{ at the poles of } C(\xi) \tag{3-200}$$

$$C(z, m) = \frac{1}{T} \sum_{n=-\infty}^{\infty} C(s + jn\omega_s) e^{-(1-m)(s+jn\omega_s)T} \Big|_{z=e^{Ts}} \tag{3-201}$$

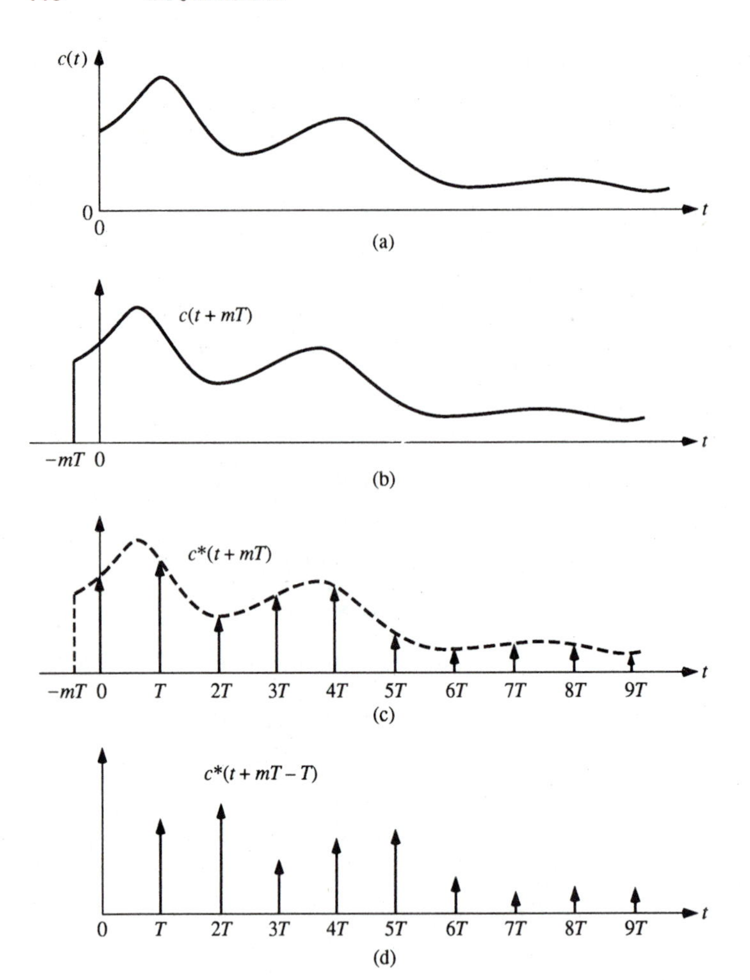

Figure 3-14. Steps illustrating the modified z-transform. (a) Time function $c(t)$. (b) $c(t)$ is shifted left by mT, $0 < m < 1$. (c) $c(t + mT)$ is sampled starting at $t = 0$. (d) The sampled sequence is shifted to the right by T.

Setting $m = 1$ in Eq. (3-201), we get

$$C(z, m)\Big|_{m=1} = \frac{1}{T}\sum_{n=-\infty}^{\infty} C(s + jn\omega_s)\Big|_{z=e^{Ts}} = C(z) \tag{3-202}$$

In this case, since $c(0) = 0$ is implied in Eq. (3-201), $C(z, m) = C(z)$ when $m = 1$.

We have presented three alternate expressions for the modified z-transform of a real function in Eqs. (3-194), (3-200), and (3-201). These equations are subject to different conditions of validity and are useful for various purposes. Equation (3-194) is the most general in that it is valid for any function $c(t)$. Equation

(3-200) is valid for any $c(t)$ that has a Laplace transform $C(s)$. Equation (3-201) is valid only if $c(0) = 0$, and thus, it is not valid for functions that have a jump discontinuity at $t = 0$.

We shall illustrate the use of Eq. (3-200) for the evaluation of the modified z-transform. A table of modified z-transforms is given in Appendix C, along with the regular z-transforms.

Example 3-15

Consider the function $f(t) = e^{-at}u_s(t)$ used in Example 3-14. Using Eq. (3-200), we have

$$F(z, m) = z^{-1}\left[\text{residue of } \frac{1}{\xi + a}\frac{e^{mT\xi}}{1 - e^{T\xi}z^{-1}}\right] \text{at } \xi = -a$$

$$= z^{-1}\left[\frac{e^{mT\xi}}{1 - e^{T\xi}z^{-1}}\right]_{\xi = -a} = \frac{e^{-mTa}}{z - e^{-aT}} \tag{3-203}$$

The Inverse Modified z-Transform

The major advantage of the modified z-transform is that it gives information on a time function in between the sampling instants. The inverse z-transform of $F(z, m)$ gives the values of $f(t)$ in between the sampling instants for a given value of m. The inverse modified z-transform operation is denoted by $\mathscr{Z}_m^{-1}$ and may be carried out by either the *power-series method* or the *inversion formula.*

The Power-Series Method

Just as in the ordinary z-transform, the function $F(z, m)$ is expanded into a power series in z^{-1} by long division. Thus, $F(z, m)$ is written

$$F(z, m) = f(mT)z^{-1} + f(T + mT)z^{-2} + \cdots + f(kT - T + mT)z^{-k} + \cdots \tag{3-204}$$

The coefficient $f(kT - T + mT)$ corresponds to a value of $f(t)$ between $t = (k - 1)T$ and $t = kT$ for any value of m between 0 and 1, $k = 1, 2, \ldots$.

Example 3-16

We can expand the function $F(z, m)$ in Eq. (3-203) into a power series of z^{-1},

$$F(z, m) = e^{-amT}z^{-1} + e^{-(m+1)aT}z^{-2} + \cdots + e^{-(m+k)aT}z^{-(k+1)} + \cdots \tag{3-205}$$

The coefficient of the z^{-k} term in the infinite series represents the value of $f(t)$ between the sampling instants of $t = (k - 1)T$ and $t = kT$ where $k = 1, 2,$

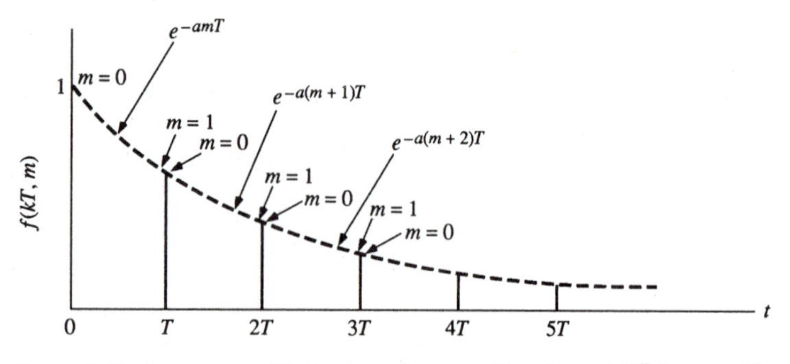

Figure 3-15. Inverse modified z-transform of $F(z, m) = e^{-amT}/(z - e^{-aT})$.

$\ldots$, and $0 < m < 1$. The time function $f(t)$ during the first sampling period is described by the coefficient e^{-amT} of the z^{-1} term in Eq. (3-205). When $m = 0$, the value of $f(t)$ at $t = 0^+$ is obtained; and when $m = 1$, e^{-amT} gives the value of $c(t)$ at $t = T^-$. Similarly, for the kth sampling period, $k = 1, 2, \ldots,$

$\boldsymbol{m = 0}$

$$f[(k-1)T^+] = f(kT, 0) = e^{-a(k-1)T}$$

$\boldsymbol{m = 1}$

$$f[(kT)^-] = f(kT, 1) = e^{-akT}$$

In general, the response between any two consecutive sampling instants is determined by assigning values to m continuously from zero to unity. The total time response $f(kT, m)$ which is the inverse modified z-transform of $F(z, m)$ is shown in Fig. 3-15.

The Inversion Formula

Just as in the z-transform method, the inverse modified z-transform can be carried out by means of the inversion integral

$$\mathscr{Z}_m^{-1}[F(z, m)] = f(kT, m) = \frac{1}{2\pi j}\oint_\Gamma F(z, m)z^{k-1}\,dz \qquad \text{(3-206)}$$

where Γ is a closed path in the z-plane that encloses all the singularities of $F(z, m)z^{k-1}$. We can show that when $m = 1$, Eq. (3-206) reverts to the inversion formula of the z-transform in Eq. (3-74). During the sampling period between $t = (k-1)T$ and $t = kT$, the value of $f[(n-1)T^+]$ is obtained from $f(kT, m)$ by setting $m = 0$; the value of $f(kT^+)$ is obtained by substituting $m = 1$ in $c(kT, m)$.

Example 3-17

The same function in Example 3-15 is now evaluated by use of the inversion formula. Substituting the result of $F(z, m)$ in Eq. (3-203) into Eq. (3-206), we get

$$f(kT, m) = \frac{1}{2\pi j} \oint_{\Gamma} \frac{e^{-mT}}{z - e^{-aT}} z^{k-1}\, dz \tag{3-207}$$

where Γ may be a circle in the z-plane, centered at the origin, and is large enough to enclose the pole of the integrand of Eq. (3-207) which is at $z = e^{-aT}$. The evaluation of the inversion integral of Eq. (3-207) is essentially that of finding the residue of the integrand at the pole. Thus,

$$f(kT, m) = \sum \text{residue of } \frac{e^{-mTa}}{z - e^{-aT}} z^{k-1} \text{ at } z = e^{-aT}$$

$$= e^{-mTa} z^{k-1} \big|_{z=e^{-aT}} = e^{-(k-1+m)aT} \tag{3-208}$$

which can be shown to be a closed-form solution to the problem in Example 3-16.

PROBLEMS

3-1 Determine the z-transforms of the following functions. You may perform partial-fraction expansion, and then use the z-transform table.

a. $F(s) = \dfrac{5}{s(s^2 + 4)}$

b. $F(s) = \dfrac{4}{s^2(s + 2)}$

c. $F(s) = \dfrac{2}{s^2 + s + 2}$

d. $F(s) = \dfrac{2(s + 1)}{s(s + 5)}$

e. $F(s) = \dfrac{10}{s(s^2 + s + 2)}$

3-2 Given that the z-transforms of $g(t)$ for $T = 1$ s is

$$G(z) = \frac{z(z - 0.2)}{4(z - 0.8)(z - 1)}$$

find the sequence $g(kT)$ for $k = 0, 1, 2, \ldots, 40$. Use a computer program such as the INVZ of the DCSP software package if one is available. What is the final value of $g(kT)$ when $k \to \infty$?

3-3 Given $f_1(t) = t^2u_s(t)$ and $f_2(t) = e^{-2t}u_s(t)$, find the z-transform of $f_1(t)f_2(t)$ by means of the partial differentiation theorem of z-transform.

3-4 Given $f_1(t) = t^2u_s(t)$ and $f_2(t) = e^{-2t}u_s(t)$, find the z-transform of $f_1(t)f_2(t)$ by means of the complex convolution theorem of z-transform.

3-5 Determine the z-transform of the following sequences. Express the results in closed form.

a. $f(k) = 0$ for $k = 0$ and even integers.
   $f(k) = 1$ for $k =$ odd integers.

b. $f(k) = 1$ for $k = 0$ and even integers.
   $f(k) = -1$ for $k =$ odd integers.

c. $f(k) = e^{-k}\sin 2k$.

d. $f(kT) = kTe^{-2kT}$.

3-6 The difference equation for Problem 2-3 is written

$$p(k+1) = (R+1)p(k) - 2^k u(0)$$

Solve $u(0)$ using the z-transform method. The boundary condition is $p(N) = 0$, where N is the number of months of the loan.

3-7 Find the inverse z-transform $f(k)$ of the following function.

$$F(z) = \frac{2z+1}{(z-0.1)^2}$$

3-8 Find the inverse z-transform $f(k)$ of the following function.

$$F(z) = \frac{2z}{z^2 - 1.2z + 0.5}$$

3-9 Given the function

$$f(k) = (0.1)^k u_s(k) + 0.5k(0.1)^k u_s(k)$$

find the z-transform of $f(k)$, $F(z)$.

3-10 Given the function

$$f(k) = (0.1)^k u_s(k) + 0.5k(0.1)^{k-1} u_s(k-1)$$

find the z-transform of $f(k)$, $F(z)$.

3-11 Find the inverse z-transform $f(kT)$ of the following functions.

a. $F(z) = \dfrac{z}{z^2+1}$

b. $F(z) = \dfrac{10z}{z^2-1}$

c. $F(z) = \dfrac{1}{z(z-0.2)}$

3-12 Find the inverse z-transform of

$$F(z) = \frac{z(z+1)}{(z-1)(z^2-z+1)}$$

by means of the following methods.

a. Real inversion formula.

b. Power-series expansion.

c. Partial-fraction expansion/z-transform table.

3-13 Solve the following difference equation using the z-transform method:

$$c(k+2) - 0.1c(k+1) - 0.2c(k) = r(k+1) + r(k)$$

where $r(k) = u_s(k)$ for $k = 0, 1, 2, \ldots$; $c(0) = 0$ and $c(1) = 0$.

3-14 Solve the following difference equation using the z-transform method:

$$c(k+2) - 1.5c(k+1) + c(k) = 2u_s(k)$$

where $c(0) = 0$ and $c(1) = 1$.

3-15

a. Given the transfer function of a digital control system as

$$\frac{C(z)}{R(z)} = \frac{\sum_{k=0}^{M} b_k z^k}{\sum_{k=0}^{N} a_k z^k}$$

where M and N are positive integers, and $N \geq M$. First, multiply the numerator and the denominator of the right-hand side of the last equation by z^{-N}/a_N, and then cross multiply both sides of the equation. Express $C(z)$ in terms of the rest of the terms. Now show that the value of the output $c(nT)$ for $n = 0, 1, 2, \ldots$ can be expressed as

$$c(nT) = \sum_{k=0}^{M} \frac{b_k}{a_N} r[(n-N+k)T] - \sum_{k=0}^{N-1} \frac{a_k}{a_N} c[(n-N+k)T]$$

where $r(nT) = 0$ and $c(nT) = 0$ for $n < 0$.

Note that the expression for $c(nT)$ given above represents an alternate method of finding the inverse z-transform of $C(z)$ given the values of the input sequence $r(kT)$, $k = 0, 1, 2, \ldots$. The advantage of this method is that the z-transform of the input, $R(z)$, need not be known or even exist, although the definition of $R(z)$ must be used in the above proof. (The material used for this problem was contributed by Dr. S. M. Seltzer.)

b. Given the transfer function

$$\frac{C(z)}{R(z)} = \frac{z+1}{z^2-z+1}$$

with the input sequence $r(kT) = 1$ for $k \geq 0$. Find $c(nT)$ for $n = 0, 1, 2, \ldots,$ using the expression given in part **a**.

3-16 If $F_1(s)$ is band limited, that is, $|F_1(j\omega)| = 0$ for $|\omega| \geq \omega_1$, and the sampling frequency ω_s is greater than or equal to $2\omega_1$, show that

$$\mathscr{Z}[F_1(s)F_2(s)] = TF_1(z)\bar{F}_2(z)$$

where

$$F_1(z) = \mathscr{Z}[F_1(s)]$$
$$\bar{F}_2(z) = \mathscr{Z}[\bar{F}_2(s)]$$

and

$$\bar{F}_2(s) = \begin{cases} F_2(s) & \text{for } \omega < \omega_1 \\ 0 & \text{for } \omega \geq \omega_1 \end{cases}$$

If $F_1(s)$ and $F_2(s)$ are both band limited with $|F_1(j\omega)| = 0$ for $|\omega| \geq \omega_1$ and $|F_2(j\omega)| = 0$ for $|\omega| \geq \omega_2$, and the sampling frequency $\omega_s \geq 2\max(\omega_1, \omega_2)$, show that

$$\mathscr{Z}[F_1(s)F_2(s)] = TF_1(z)F_2(z)$$

3-17 Consider the sampled-data system with nonintegral time delay shown in Fig. P3-17. Show that the z-transform of the output is

$$C_m(z) = z^{-1}\sum_{k=0}^{\infty} c[(k+m)T]z^{-k}$$

Assume that $c(t) = 0$ for $t < 0$.

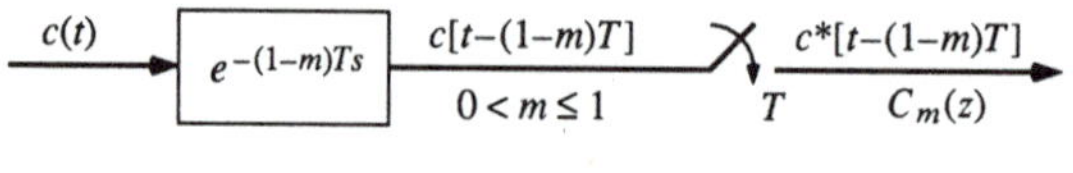

Figure P3-17.

References

Sampling and z-Transforms

1. Lago, G. V., "Additions to Sampled-Data Theory," *Proc. National Electronics Conf.*, vol. 10, pp. 758–766, 1954.
2. Helm, H. A., "The z-Transformation," *Bell System Technical Journal*, vol. 38, pp. 177–196, January 1959.
3. Ragazzini, J. R., and L. H. Zadeh, "The Analysis of Sampled-Data Systems," *Trans. AIEE*, vol. 71, part 2, pp. 225–234, 1952.
4. Kuo, B. C., *Analysis and Synthesis of Sampled-Data Control Systems*, Prentice-Hall, Englewood Cliffs, N.J., 1963.
5. Jury, E. I., "A General z-Transform Formula for Sampled-Data Systems," *IEEE Trans. Automatic Control*, vol. AC-12, pp. 606–608, October 1967.

6. Jury, E. I., *Theory and Application of the z-Transform Method,* John Wiley & Sons, New York, 1964.
7. Kliger, I., and W. C. Lipinski, "The z-Transform of a Product of Two Functions," *IEEE Trans. Automatic Control,* vol. AC-9, pp. 582–583, October 1964.
8. Will, P. M., "Variable Frequency Sampling," *IRE Trans. Automatic Control,* vol. AC-7, p. 126, October 1962.
9. Dorf, R. C., M. C. Farren, and C. A. Phillips, "Adaptive Sampling Frequency for Sampled-Data Control Systems," *IRE Trans. Automatic Control,* vol. AC-7, pp. 38–47, January 1962.
10. Cavin, R. K., III, D. L. Chenoweth, and C. L. Phillips, "The z-Transform of an Impulse Function," *IEEE Trans. Automatic Control,* vol. AC-12, p. 113, February 1967.
11. Phillips, C. L., "A Note on Sampled-Data Control Systems," *IEEE Trans. Automatic Control,* vol. AC-10, pp. 489–490, October 1965.
12. Lago, G. V., "Additions to the z-Transformation Theory for Sampled-Data Systems," *Trans. AIEE,* vol. 74, part 2, pp. 403–407, January 1955.
13. Hufnagel, R. E., "Analysis of Cyclic-Rate Sampled-Data Feedback-Control Systems," *Trans. AIEE* (Applications and Industry), vol. 77, pp. 421–425, November 1958.
14. Jury, E. I., and F. J. Mullin, "The Analysis of Sampled-Data Control Systems with a Periodically Time-Varying Sampling Rate," *IRE Trans. Automatic Control,* vol. AC-4, pp. 15–21, May 1959.
15. Gabel, R. A., and R. A. Roberts, *Signals and Linear Systems,* 3rd ed., John Wiley & Sons, New York, 1987.

Modified z-Transform

16. Jury, E. I., "Additions to the Modified z-Transform Method," *IRE WESCON Convention Record,* part 4, pp. 136–156, 1957.
17. Jury, E. I., and Farmanfarma, "Tables of z-Transforms and Modified z-Transforms of Various Sampled-Data Systems Configurations," Univ. of California, Berkeley, Electronics Research Lab., Report 136A, Ser. 60, 1955.
18. Mesa, W., and C. L. Phillips, "A Theorem on the Modified z-Transform," *IEEE Trans. Automatic Control,* vol. AC-10, p. 489, October 1965.

Transfer Functions, Block Diagrams, and Signal Flow Graphs

KEYWORDS AND TOPICS

pulse transfer function • z-transfer function • characteristic equation • block diagrams • signal flow graphs • multirate sampled systems

4-1 INTRODUCTION

Thus far our investigations of discrete-data and digital control systems have focused on the mathematical treatment of the sampled signals and digital data. Now we must consider the modeling and analysis of systems that are subject to discrete and digital data.

Just as in the analysis and design of linear analog-signal control systems, the concepts of transfer function, block diagram, and signal flow graph can all be applied to linear discrete and digital control systems.

4-2 THE PULSE TRANSFER FUNCTION AND THE z-TRANSFER FUNCTION

It is well known that the transfer function of the linear system with continuous-data input $r(t)$ shown by the block diagram in Fig. 4-1(a) is defined as

$$G(s) = \frac{C(s)}{R(s)} \tag{4-1}$$

where $R(s)$ and $C(s)$ are the Laplace transforms of the input $r(t)$ and the output $c(t)$, respectively. Note that, in defining the transfer function $G(s)$, we must

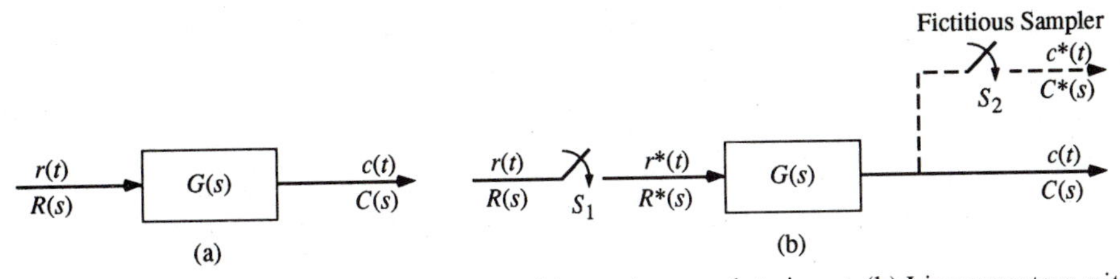

Figure 4-1. (a) Linear system with continuous-data input. (b) Linear system with sampled-data input.

assume that the *initial conditions of the systems are all set to zero.* If the same system is now subject to a sampled or digital signal $r^*(t)$, where the latter is simulated by the ideal sampler S_1, as shown in Fig. 4-1(b), the objective now is to find input-output relations of the discrete-data system and, if possible, an input-output transfer function in the Laplace and the z-domains. The Laplace transform of the output of the system in Fig. 4-1(b) is written as

$$C(s) = R^*(s)G(s) \tag{4-2}$$

where $R^*(s)$ is the Laplace transform of $r^*(t)$ and $G(s)$ is as defined in Eq. (4-1). Although this transfer relation is easy to achieve, it contains a mixture of discrete and analog signals and is difficult to manipulate analytically. Thus, it is desirable to express the system characteristics by a transfer relation that relates $r^*(t)$ to $c^*(t)$, the sampled form of the output $c(t)$. For this purpose we introduce a fictitious sampler S_2 at the output of the system, as shown in Fig. 4-1(b). The sampler S_2 is synchronized with the sampler S_1; that is, they have the same sampling period T and sample at the same instants of time. Now we have both $c(t)$ and $c^*(t)$ at our disposal. Two useful expressions for $C^*(s)$ follow directly from Eqs. (2-53) and (2-54), and are written below.

$$C^*(s) = \frac{1}{T}\sum_{n=-\infty}^{\infty} C(s + jn\omega_s) \qquad c(0) = 0 \tag{4-3}$$

$$C^*(s) = \sum_{k=0}^{\infty} c(kT)e^{-kTs} \tag{4-4}$$

Similar expressions can be written for $R^*(s)$.

From Fig. 4-1(b), $C^*(s)$ is written

$$C^*(s) = \frac{1}{T}\sum_{n=-\infty}^{\infty} C(s + jn\omega_s) = \frac{1}{T}\sum_{n=-\infty}^{\infty} R^*(s + jn\omega_s)G(s + jn\omega_s) \tag{4-5}$$

Using the periodic property of $R^*(s)$ of Eq. (2-76), i.e., $R^*(s + jn\omega_s) = R^*(s)$, the last equation becomes

$$C^*(s) = R^*(s)\frac{1}{T}\sum_{n=-\infty}^{\infty} G(s + jn\omega_s) \tag{4-6}$$

Now define $G^*(s)$ as

$$G^*(s) = \frac{1}{T} \sum_{n=-\infty}^{\infty} G(s + jn\omega_s) \tag{4-7}$$

Substituting Eq. (4-7) into Eq. (4-6), we get

$$C^*(s) = R^*(s)G^*(s) \tag{4-8}$$

The Pulse Transfer Function $G^*(s)$

We can define the *pulse transfer function* of the system in Fig. 4-1(b) as

$$G^*(s) = \frac{C^*(s)}{R^*(s)} \tag{4-9}$$

The z-Transfer Function $G(z)$

Substituting $z = e^{Ts}$ in Eq. (4-9), we have the *z-transfer function* of the system in Fig. 4-1(b):

$$G(z) = \frac{C(z)}{R(z)} \tag{4-10}$$

where $G(z)$ is also defined as

$$G(z) = \sum_{k=0}^{\infty} g(kT)z^{-k} \tag{4-11}$$

In the last equation $g(kT)$ denotes the sequence of the impulse response $g(t)$ of the system described by the transfer function $G(s)$. In other words, $g(t)$ is the inverse Laplace transform of $G(s)$. The sequence $g(kT)$, $k = 0, 1, 2, \ldots$, is also defined as the *impulse sequence* or the *weighting sequence* of the system G.

The conclusion is that the transformed expressions of a linear system with a transfer function in the sampled environment, whether it is in the s- or the z-domain, are defined in the same way as sampled signals. Based on the discussions given above, we can make the following observations concerning the transfer relations of discrete-data systems.

1. *To express the input-output transform relations in a uniformly sampled form, fictitious samplers must be used to sample the output of the system.*
2. *The z-transfer function characterizes the discrete-data system responses only at the sampling instants.* The information on the output of the system $c(t)$ between the sampling instants is lost.
3. *Since the input of the discrete-data system in Fig. 4-1(b) is actually described by the output of the sampler, for all practical purposes, we can simply ignore the sampler and regard the input to the system as $r^*(t)$.*

An alternative way of arriving at Eq. (4-10) is to use the impulse response method. Consider that at $t = 0$ a unit impulse is applied as input to the system in Fig. 4-1(b). The output of the system is simply the impulse response $g(t)$. The output of the fictitious sampler S_2 is described by

$$c^*(t) = g^*(t) = \sum_{k=0}^{\infty} g(kT)\delta(t - kT) \tag{4-12}$$

where $g(kT)$, for $k = 0, 1, 2, \ldots$, is the weighting sequence of the system. When the sampled signal $r^*(t)$ is applied as input to the linear system, the output of the system is written as a series of impulse responses, each weighted by the individual values of $r(kT)$. Thus, the output $c(t)$ is written

$$c(t) = r(0)g(t) + r(T)g(t - T) + r(2T)g(t - 2T) + \cdots \tag{4-13}$$

At $t = kT$, where k is a positive integer, the last equation gives

$$c(kT) = r(0)g(kT) + r(T)g[(k - 1)T] + \cdots + r(kT)g(0) \tag{4-14}$$

The last series is truncated at the $(k + 1)$st term since $g(t)$ is zero for $t < 0$. Equation (4-14) is written

$$c(kT) = \sum_{n=0}^{k} r(nT)g(kT - nT) \tag{4-15}$$

Taking the z-transform on both sides of the last equation, which involves multiplying both sides by z^{-k} and summing from $k = 0$ to $k = \infty$, we have

$$C(z) = \sum_{k=-\infty}^{\infty} \sum_{n=0}^{k} r(nT)g(kT - nT)z^{-k} \tag{4-16}$$

Now, using the real convolution theorem of the z-transform of Eq. (3-130), the last equation leads to

$$C(z) = R(z)G(z) \tag{4-17}$$

where $G(z)$ is as defined in Eq. (4-11).

4-2-1 Discrete-Data System with Cascaded Elements Separated by a Sampler

When a discrete-data or digital control system contains cascaded elements, care must be taken in deriving the transfer relations for the overall system. Figure 4-2 illustrates a discrete-data system with cascaded elements G_1 and G_2.

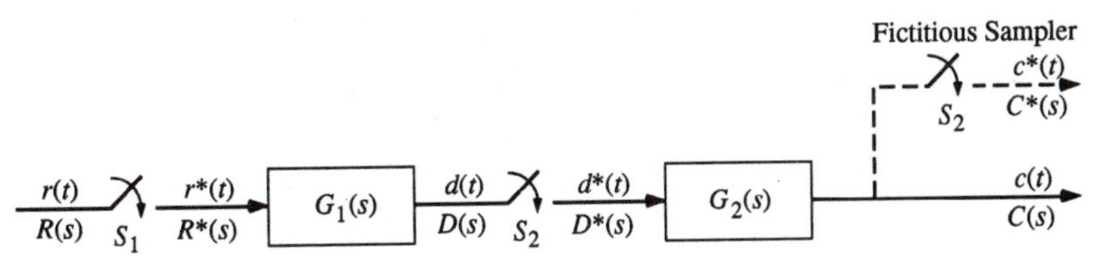

Figure 4-2. Digital system with sampler-separated elements.

The two elements are separated by a second sampler S_2 which is identical and synchronized to the sampler at the input, S_1. To facilitate the z-transform method, a fictitious sampler S_2 is introduced at the system output, as shown in Fig. 4-2. The z-transform of the output of the overall system is derived in the following.

The transfer relations of the two systems G_1 and G_2 are

$$D(z) = G_1(z)R(z) \tag{4-18}$$

and

$$C(z) = G_2(z)D(z) \tag{4-19}$$

where $D(z)$ is the z-transform of $d(t)$, the output of G_1. Substitution of Eq. (4-18) in Eq. (4-19) yields

$$C(z) = G_1(z)G_2(z)R(z) \tag{4-20}$$

Therefore, we can state that *the z-transfer function of two linear systems separated by a sampler is equal to the product of the z-transfer functions of the two systems.*

4-2-2 Discrete-Data System with Cascaded Elements not Separated by a Sampler

Figure 4-3 shows a discrete-data system that contains two elements not separated by a sampler. The z-transform of the output of the overall system should be written as

$$C(z) = \mathscr{Z}[G_1(s)G_2(s)]R(z) \tag{4-21}$$

Apparently, since $G_1(s)$ and $G_2(s)$ are not separated by a sampler, they should be treated as one function, and the z-transform operation must be taken on the product of the two functions as a whole.

We can introduce the following notation for this purpose:

$$\mathscr{Z}[G_1(s)G_2(s)] = G_1G_2(z) = G_2G_1(z) \tag{4-22}$$

which reads as "the z-transform of $G_1(s)G_2(s)$." Notice that, in general,

$$G_1G_2(z) \neq G_1(z)G_2(z) \tag{4-23}$$

except in some special situations.

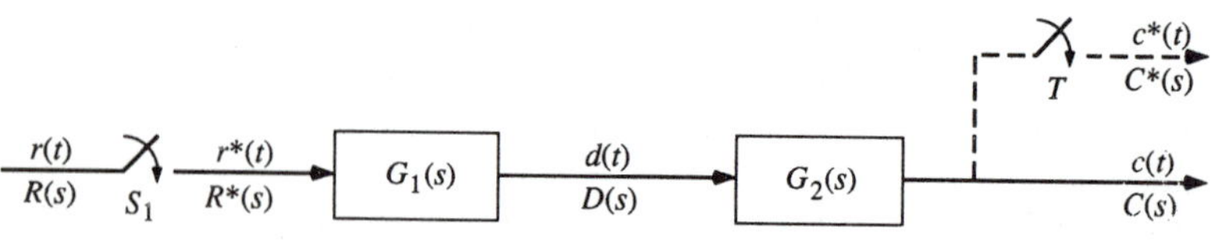

Figure 4-3. Digital system with cascaded elements.

Equation (4-21) is written

$$C(z) = G_1G_2(z)R(z) \tag{4-24}$$

The above analysis can be extended to systems with more than two elements, and the extension is straightforward.

For complex system configurations and closed-loop systems, a block diagram algebra similar to that of analog systems, using the fundamental transfer relations defined above, can be used to find the input-output transfer relationships. However, the sampled signal flow graph method presented in Sec. 4-4 is generally more convenient to apply.

4-3 PULSE TRANSFER FUNCTION OF THE ZERO-ORDER HOLD AND THE RELATION BETWEEN $G(s)$ AND $G(z)$

Since the zero-order hold (zoh) is an important element in discrete-data control systems, its transfer function should be treated with special interest.

The transfer function of the zoh is given in Eq. (2-86) and is repeated here:

$$G_{h0}(s) = \frac{1 - e^{-Ts}}{s} \tag{4-25}$$

Taking the z-transform on both sides of the last equation, we get

$$G_{h0}(z) = \mathscr{Z}\left[\frac{1 - e^{-Ts}}{s}\right] \tag{4-26}$$

Since the term e^{-Ts} in the last equation represents a delay of one sampling period, from the right-shift theorem of the z-transform given in Eq. (3-95), $G_{h0}(z)$ is written

$$G_{h0}(z) = (1 - z^{-1})\mathscr{Z}\left[\frac{1}{s}\right]$$

$$= (1 - z^{-1})\frac{z}{z - 1} = 1 \tag{4-27}$$

This result is expected since the zoh simply holds the discrete signal for one sampling period, and taking the z-transform of the zoh would revert it to the original sampled signal again. However, the above exercise simply does not have any physical usefulness, since a zoh is almost never found just situated by itself. A situation that is commonly found in discrete-data control systems is shown in Fig. 4-4, that is, a sample-and-hold (S/H) unit is followed by a linear system with transfer function $G(s)$. The transfer function of the linear system block is $G(s)$, and we are interested in finding the transform relation between the input $r^*(t)$ and the sampled output $c^*(t)$. Again, we introduce a fictitious

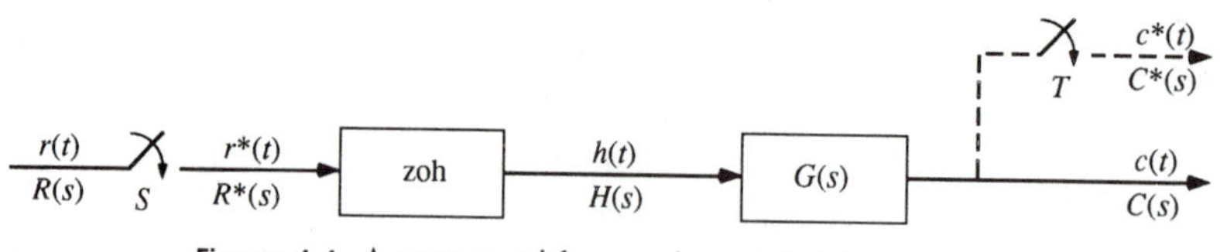

Figure 4-4. A system with sample-and-hold.

sampler at the system output. The z-transform of the output $c(t)$ is written as

$$C(z) = \mathscr{Z}[G_{h0}(s)G(s)]R(z) \tag{4-28}$$

where

$$\begin{aligned}\mathscr{Z}[G_{h0}(s)G(s)] &= \mathscr{Z}\left[\frac{1 - e^{-Ts}}{s} G(s)\right] \\ &= (1 - z^{-1})\mathscr{Z}\left[\frac{G(s)}{s}\right]\end{aligned} \tag{4-29}$$

The last equation represents the important z-transfer function of the combination of an S/H device and a linear system with the transfer function $G(s)$. Notice that the $(1 - z^{-1})$ is factored out as a result of the delay theorem, and the z-transform of $G(s)/s$ must be taken as one function.

It was mentioned in Chapter 2 that, in theory, when the sampling frequency reaches infinity, a discrete-data system reverts to a corresponding continuous-data system, i.e., the S/H can be eliminated. However, analytically, this *does not* mean that, given

$$\mathscr{Z}[G(s)] = G(z) \tag{4-30}$$

then

$$\lim_{T \to 0} G(z) = G(z) \tag{4-31}$$

Since the z-transform theory is based on the impulse-amplitude modulation of a continuous-data signal with the sampling period of T s, setting T to zero simply "bunched" all the impulses together, and so it does not make much physical sense. In other words, if the signal $r(t)$ is sampled by an ideal sampler to give $r^*(t)$, then setting the sampling period T to zero or, equivalently, ω_s to infinity *does not* revert $r^*(t)$ to $r(t)$. This explains why Eq. (4-31) is not true in general. However, if we first send the sampled signal $f^*(t)$ through a zoh with the output designated as $h(t)$, and then take the limit of $T \to 0$, then

$$\lim_{T \to 0} h(t) = r(t) \tag{4-32}$$

as well as

$$\lim_{T \to 0} H(s) = R(s) \tag{4-33}$$

Thus, the significance of the results in the last two equations is that *if a continuous-data signal $r(t)$ is sent through an S/H device with a sampling period T, the output of the latter can be reverted to $r(t)$ by setting the sampling period T to zero.*

Example 4-1

As an illustrative example of the limiting process of the S/H unit, consider that the input $r(t)$ in the system of Fig. 4-4 is

$$r(t) = e^{-at}u_s(t) \tag{4-34}$$

The Laplace transform of $r(t)$ is

$$R(s) = \frac{1}{s + a} \tag{4-35}$$

The Laplace transform of the sampled signal $r^*(t)$ is [using Eq. (2-64)]

$$R^*(s) = \frac{e^{Ts}}{e^{Ts} - e^{-aT}} \tag{4-36}$$

The Laplace transform of the output of the zoh is

$$H(s) = G_{h0}(s)R^*(s) = \frac{1 - e^{-Ts}}{s} \frac{e^{Ts}}{e^{Ts} - e^{-aT}} \tag{4-37}$$

Taking the limit as T approaches zero in $H(s)$, we get

$$\lim_{T \to 0} H(s) = \lim_{T \to 0} \frac{1 - e^{-Ts}}{s} \frac{e^{Ts}}{e^{Ts} - e^{-aT}} \tag{4-38}$$

The evaluation of the above limit requires the use of L'Hopital's theorem of calculus. The result is

$$\lim_{T \to 0} H(s) = \frac{1}{s + a} = R(s) \tag{4-39}$$

This shows that the limit of the output of an S/H operation as T approaches zero is equal to the input of the S/H unit.

A more useful and important property of the z-transform is to show that

$$\lim_{T \to 0} \mathscr{Z}[G_{h0}(s)G(s)] = G(s) \tag{4-40}$$

We can prove the above relation by substituting Eq. (4-25) for $G_{h0}(s)$. We have

$$\lim_{T \to 0} \mathscr{Z}[G_{h0}(s)G(s)] = \lim_{T \to 0} \mathscr{Z}\left[\frac{1 - e^{-Ts}}{s} G(s)\right] \tag{4-41}$$

Expanding e^{-Ts} into a power series and taking only the first two terms, we have

$$e^{-Ts} \cong 1 - Ts \tag{4-42}$$

Equation (4-41) is simplified to

$$\lim_{T \to 0} \mathscr{Z}[G_{h0}(s)G(s)] = \lim_{T \to 0} TG(z) \tag{4-43}$$

The last equation implies that

$$\lim_{T \to 0} [G_{h0}(s)G(s)]^* = \lim_{T \to 0} TG^*(s) \tag{4-44}$$

where $[G_{h0}(s)G(s)]^*$ denotes the pulse transform of $G_{h0}(s)G(s)$. Now, substituting the frequency-domain expression for $G^*(s)$ in Eq. (4-44), we get

$$\lim_{T \to 0} [G_{h0}(s)G(s)]^* = \lim_{T \to 0} \sum_{n=-\infty}^{\infty} G(s + jn\omega_s) \tag{4-45}$$

The infinite series on the right-hand side of the last equation becomes $G(s)$ as T approaches zero, since, as shown in Fig. 2-34, all the harmonic components are moved out to infinity as ω_s approaches infinity. Thus, Eq. (4-45) becomes

$$\lim_{T \to 0} [G_{h0}(s)G(s)]^* = G(s) \tag{4-46}$$

Since $[G_{h0}(s)G(s)]^*$ is essentially the same as $\mathscr{Z}[G_{h0}(s)G(s)]$ except for a change of variable, $z = e^{Ts}$, the identity in Eq. (4-40) is proven.

Example 4-2

Consider that the transfer function of the system in Fig. 4-4 is

$$G(s) = \frac{K}{s(s + a)} \tag{4-47}$$

where K and a are constants. The z-transform of the combination of the zero-order hold and $G(s)$ is

$$\begin{aligned} \mathscr{Z}[G_{h0}(s)G(s)] &= (1 - z^{-1})\mathscr{Z}\left[\frac{K}{s^2(s + a)}\right] \\ &= \frac{KT}{a(z - 1)} - \frac{K(1 - e^{-aT})}{a^2(z - e^{-aT})} \end{aligned} \tag{4-48}$$

Taking the limit as $T \to 0$, and approximating e^{Ts} by $1 + Ts$ and e^{-aT} by $1 - aT$, we get

$$\lim_{T\to 0} \mathscr{Z}[G_{h0}(s)G(s)] = \lim_{T\to 0} \frac{KT}{a(Ts + 1 - 1)} - \lim_{T\to 0} \frac{K(1 - 1 + aT)}{a^2(1 + Ts - 1 + aT)}$$

$$= \frac{K}{as} - \frac{K}{a(s+a)} = \frac{K}{s(s+a)} = G(s) \quad (4\text{-}49)$$

which is the desired result.

4-4 CLOSED-LOOP SYSTEMS

Due to the advantages of feedback, most control systems are of the closed-loop configuration, and the same holds true for discrete-data control systems. However, since most discrete-data control systems have a mixture of analog and discrete data, and the samplers can be located at any point in the system, the analysis of these systems naturally will take on more variations over that of their analog counterparts. Figure 4-5 shows a simple single-loop system with a sampler located in the forward path. The time-domain variables and their Laplace transforms are designated on the diagram, and a fictitious sampler is added at the output to generate the sampled output $c^*(t)$. The applied input is $r(t)$, but due to the sampling operation, the system sees only $r^*(t)$. We shall derive the transfer functions between $R^*(s)$ and $C^*(s)$ and between $R^*(s)$ and $C(s)$.

For the present analysis, we shall rely on the algebraic method of first writing down a set of linearly independent equations and then eliminating the unwanted variables from these equations, leaving only the input and output variables. For systems with more complex configurations, the algebraic method can cause confusion unless the equations are written in a systematic way. The following guidelines should be helpful.

1. *Establish the input and the output variables of the system.* The input of a sampler is regarded as an output of the system, and the output of a sampler is regarded as an input to the system.
2. *Write the transfer relations between the inputs and outputs following the principle of cause and effect.*

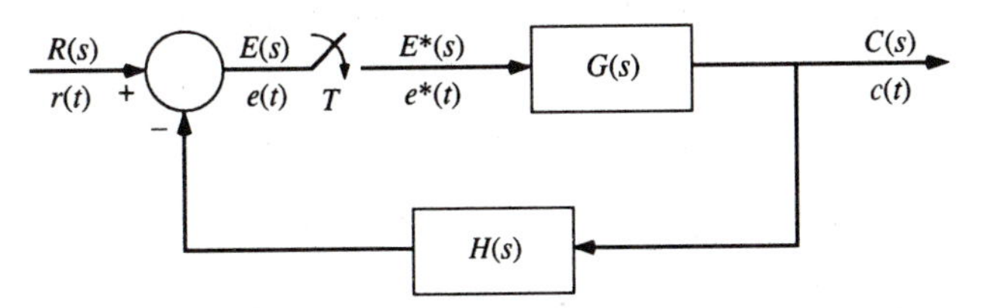

Figure 4-5. Discrete-data control system with sampler in the forward path.

For the system in Fig. 4-5, the output of the sampler $e^*(t)$ is regarded as an input to the system, and the input of the sampler $e(t)$ can be regarded as an output, in addition to the true output $c(t)$. Once the roles of these variables are defined, the sampler can essentially be eliminated, since its input $e(t)$ and output $e^*(t)$ are well defined. To summarize, for the system in Fig. 4-5, the input and output variables are defined as follows:

input variables: $r(t)$, $e^*(t)$
output variables: $e(t)$, $c(t)$, $c^*(t)$.

Now we write the input-output relations of the system using the cause-and-effect principle; i.e.,

$$\text{output} = \text{gain} \times \text{input} \tag{4-50}$$

From Fig. 4-5, the following cause-and-effect equations are written:

$$E(s) = R(s) - G(s)H(s)E^*(s) \tag{4-51}$$

$$C(s) = G(s)E^*(s) \tag{4-52}$$

Notice that only the output variables are found on the left-hand side of the equations, and only input variables are found on the right-hand side. Now, taking the pulse transform on both sides of Eq. (4-51), we get

$$E^*(s) = R^*(s) - GH^*(s)E^*(s) \tag{4-53}$$

where

$$GH^*(s) = [G(s)H(s)]^*$$
$$= \frac{1}{T}\sum_{n=-\infty}^{\infty} G(s + jn\omega_s)H(s + jn\omega_s) \tag{4-54}$$

Solving for $E^*(s)$ from Eq. (4-53), we have

$$E^*(s) = \frac{1}{1 + GH^*(s)} R^*(s) \tag{4-55}$$

Substituting the last equation in Eq. (4-52) gives the transfer relation between $R^*(s)$ and $C(s)$:

$$C(s) = \frac{G(s)}{1 + GH^*(s)} R^*(s) \tag{4-56}$$

Taking the pulse transform on both sides of Eq. (4-52), we get

$$C^*(s) = [G(s)E^*(s)]^*$$
$$= G^*(s)E^*(s) \tag{4-57}$$

where it has been recognized that the pulse transform of $E^*(s)$ is still $E^*(s)$, since

$$[E^*(s)]^* = \frac{1}{T}\sum_{n=-\infty}^{\infty} E^*(s + jn\omega_s) = E^*(s) \tag{4-58}$$

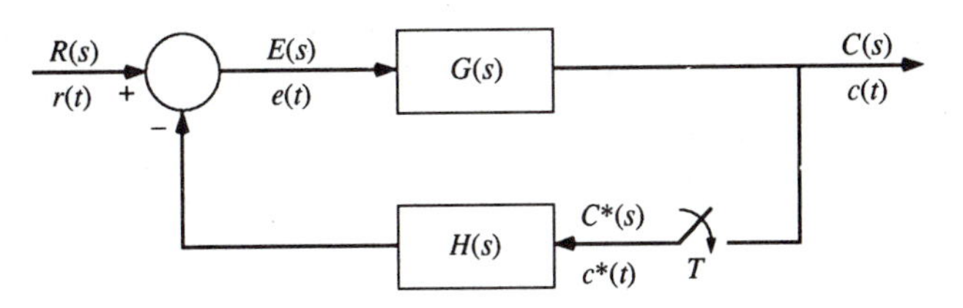

Figure 4-6. Discrete-data control system with sampler in the feedback path.

Now, substituting Eq. (4-55) into Eq. (4-57), the transfer function between $R^*(s)$ and $C^*(s)$ is written as

$$\frac{C^*(s)}{R^*(s)} = \frac{G^*(s)}{1 + GH^*(s)} \tag{4-59}$$

The last equation can readily be converted to the z-transform form.

$$\frac{C(z)}{R(z)} = \frac{G(z)}{1 + GH(z)} \tag{4-60}$$

where $GH(z) = \mathscr{Z}[G(s)H(s)]$.

Although Eq. (4-56) gives the transfer function relation between the continuous-data output $c(t)$ and the input $r^*(t)$, the expression has little practical value since the transfer function has mixed analog and discrete quantities, making it difficult to utilize analytically.

Figure 4-6 illustrates a discrete-data system in which the sampler is located in the feedback path. In this case, the signal $r(t)$ and the output of the sampler, $c^*(t)$, are the input of the system; $e(t)$ and $c(t)$ are regarded as the outputs of the system. Writing the Laplace transformed quantities of the output variables as functions of the inputs, we have the following equations:

$$E(s) = R(s) - H(s)C^*(s) \tag{4-61}$$

$$C(s) = G(s)R(s) - G(s)H(s)C^*(s) \tag{4-62}$$

The reader should examine these equations closely and notice that, in addition to the system's transfer functions, the right-hand sides of the two equations contain only the input signal, whereas the left-hand sides contain only the output variables. Taking the pulse transform on both sides of Eqs. (4-61) and (4-62), we have

$$E^*(s) = R^*(s) - H^*(s)C^*(s) \tag{4-63}$$

$$C^*(s) = GR^*(s) - GH^*(s)C^*(s) \tag{4-64}$$

where

$$GR^*(s) = [G(s)R(s)]^* \tag{4-65}$$

and

$$GH^*(s) = [G(s)H(s)]^* \tag{4-66}$$

Solving for $C^*(s)$ from Eq. (4-64), we get

$$C^*(s) = \frac{1}{1 + GH^*(s)} GR^*(s) \tag{4-67}$$

The z-transform of the output $c^*(t)$ is written directly from the last equation:

$$C(z) = \frac{1}{1 + GH(z)} GR(z) \tag{4-68}$$

Of interest in this case is the fact that we can no longer define the input-output transfer function of this system by either $C^*(s)/R^*(s)$ or $C(z)/R(z)$. Since the input $r(t)$ is not directly sampled by a sampler, the sampled signal $r^*(t)$ simply does not exist.

To express the continuous-data output $C(s)$ in terms of the input, we substitute Eq. (4-67) into Eq. (4-62); the result is

$$C(s) = G(s)R(s) - \frac{G(s)H(s)}{1 + GH^*(s)} GR^*(s) \tag{4-69}$$

4-4-1 The Characteristic Equation

The characteristic equation plays an important role in the study of linear systems. It can be defined from the basis of the difference equation or the transfer function.

Consider that a linear time-invariant discrete-data system is described by the nth-order difference equation,

$$\begin{aligned} c(k+n) + a_{n-1}c(k+n-1) + a_{n-2}c(k+n-2) + \cdots + a_1c(k+1) + a_0c(k) \\ = b_mr(k+m) + b_{m-1}r(k+m-1) + \cdots + b_1r(k+1) + b_0r(k) \end{aligned} \tag{4-70}$$

where $r(k)$ and $c(k)$ denote the kth input and output, respectively. The difference equation can be solved by means of the z-transform. The input-output transfer function $C(z)/R(z)$ is obtained by taking the z-transform on both sides of Eq. (4-70) and assuming zero initial conditions. Thus,

$$G(z) = \frac{C(z)}{R(z)} = \frac{b_mz^m + b_{m-1}z^{m-1} + \cdots + b_1z + b_0}{z^n + a_{n-1}z^{n-1} + \cdots + a_1z + a_0} \tag{4-71}$$

The characteristic equation of the system is defined as the equation obtained by equating the denominator of $G(z)$ *to zero; that is,*

$$z^n + a_{n-1}z^{n-1} + \cdots + a_1z + a_0 = 0 \tag{4-72}$$

We shall show later that the roots of the characteristic equation determine the absolute stability of the system.

Example 4-3

Consider that the forward-path transfer function $G(s)$ of the system shown in Fig. 4-5 is

$$G(s) = \frac{10}{s(s+5)} \tag{4-73}$$

and the feedback-path transfer function is $H(s) = 1$. The sampling period is $T = 0.1$ s. Applying Eq. (4-60), we have

$$G(z) = GH(z) = \mathscr{Z}[G(s)H(s)] = \mathscr{Z}\left(\frac{10}{s(s+5)}\right) = \frac{2(1-e^{-5T})z}{(z-1)(z-e^{-5T})}$$

$$= \frac{0.7869z}{(z-1)(z-0.6065)} \tag{4-74}$$

Now substituting Eq. (4-74) into Eq. (4-60), we have the closed-loop transfer function as

$$\frac{C(z)}{R(z)} = \frac{0.7869z}{z^2 - 0.8196z + 0.6065} \tag{4-75}$$

Thus, the characteristic equation of the system is

$$z^2 - 0.8196z + 0.6065 = 0 \tag{4-76}$$

4-4-2 Causality and Physical Realizability

A *causal system* is one in which its output does not precede the input. In other words, a causal system produces an output at any time kT depending only on input values that have occurred for times up to and including kT. The transfer function of a linear causal system is also said to be *physically realizable*, since the system must be composed of physical components.

It is shown in Eq. (4-11) that, by expanding the transfer function $G(z)$ into a power series in z^{-1}, the coefficients of the series represent the values of the weighting sequence of the system. The coefficients of the z^{-k} term correspond to the value of the weighting sequence $g(kT)$ at $t = kT$. Clearly, *for the discrete-data system to be causal, or physically realizable, the power-series expansion of $G(z)$ must not contain any positive power in z.* Any positive power in z in the series simply indicates "prediction," or that the output precedes the input. Therefore, *for the $G(z)$ in Eq. (4-71) to be a physically realizable transfer function, n must be greater than or equal to m.* When $m = n$, $G(z)$ is referred to as a *proper transfer function.* This condition exists rarely in physical systems as the output responds to a jump discontinuity in the input. When $n > m$, the transfer function is referred to as *strictly proper.*

Quite often, in the analysis and design of discrete-data systems, the transfer function of a digital process is expressed in terms of negative powers of z as

$$G(z) = \frac{b_m + b_{m-1}z^{-1} + \cdots + b_0 z^{-m}}{a_n + a_{n-1}z^{-1} + \cdots + a_0 z^{-n}} \tag{4-77}$$

where n and m are positive integers. Notice that no matter what the values of m and n are, $G(z)$ has the same number of poles and zeros. In this case, $G(z)$ is physically realizable if it does not contain any factor z^{-k} $(k > 0)$ in its denominator, or $a_n \neq 0$.

4-5 THE SAMPLED SIGNAL FLOW GRAPH

In the preceding section we demonstrated the algebraic method of analyzing closed-loop discrete-data control systems. For systems with multiple samplers and loops, block diagram algebra may become unwieldy. It is well known that transfer functions of linear continuous-data systems can be determined from signal flow graphs using Mason's gain formula [1], [2]. In this section we extend the signal flow graph method to the analysis of discrete-data systems. For reference purposes, a review of the Mason's gain formula for signal flow graphs is given in Appendix D.

Since most discrete-data control systems contain analog as well as discrete signals, Mason's gain formula cannot be applied directly to the system's original signal flow graph or block diagram. The bottom line is that Mason's gain formula can be applied to a signal flow graph with node variables that are either all analog or all discrete but not a mixture of the two. Therefore, the first step in applying the signal flow graph to discrete-data systems is to express the system's equations in terms of discrete-data variables only. The steps for the sampled signal flow graph method for discrete-data systems are outlined as follows.

1. *With the system's block diagram as the starting point, construct an equivalent signal flow graph for the system.* This is also known as the signal flow graph of the system, since it is entirely equivalent to the block diagram representation. As an illustration, Fig. 4-7 shows the block diagram of the system in Fig. 4-5 and its equivalent signal flow graph.
2. *Following the definitions of the input and output variables of the system given in the block diagram algebra in Sec. 4-4, write the cause-and-effect equations of the system from the equivalent signal flow graph.* For instance, for the system in Fig. 4-7 these equations are already written in Eqs. (4-51) and (4-52). *Note that Mason's gain formula should be applied to the original signal flow graph in accordance with all the defined input and output nodes.* The signal flow graph shown in Fig. 4-7(b) does not have a loop in the sense of the conventional definition, so that Eqs. (4-51) and (4-52) are obtained with the trivial application of the gain formula. In complex situations, the

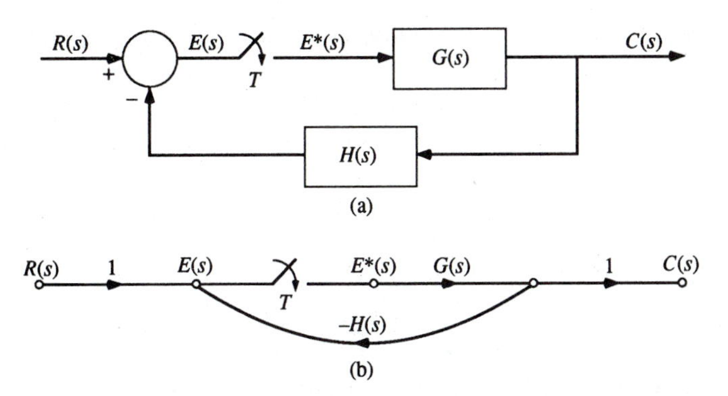

Figure 4-7. (a) Block diagram of a closed-loop digital system. (b) Equivalent signal flow graph of the system.

signal flow graph will have loops that are not broken by the samplers, and the determinant Δ of the signal flow graph would not be equal to unity.

3. *Take the pulse transform on both sides of the cause-and-effect equations obtained in Step 2.* The purpose of this step is to convert all the system variables into discrete-data form. For example, the pulse-transformed form of Eqs. (4-51) and (4-52) are Eqs. (4-53) and (4-57), respectively.
4. *Draw a sampled signal flow graph using the equations with only discrete-data variables obtained in Step 3.* For Eqs. (4-53) and (4-57), the sampled signal flow graph is drawn as shown in Fig. 4-8.
5. *Once the sampled signal flow graph is drawn, the transfer function relation between any pair of input and output nodes, defined according to the signal flow graph terminology, can be determined by use of Mason's gain formula.* In the present case, applying the gain formula to the sampled signal flow graph of Fig. 4-8 with $C^*(s)$ and $E^*(s)$ as output nodes, we get

$$C^*(s) = \frac{G^*(s)}{1 + GH^*(s)} R^*(s) \tag{4-78}$$

and

$$E^*(s) = \frac{1}{1 + GH^*(s)} R^*(s) \tag{4-79}$$

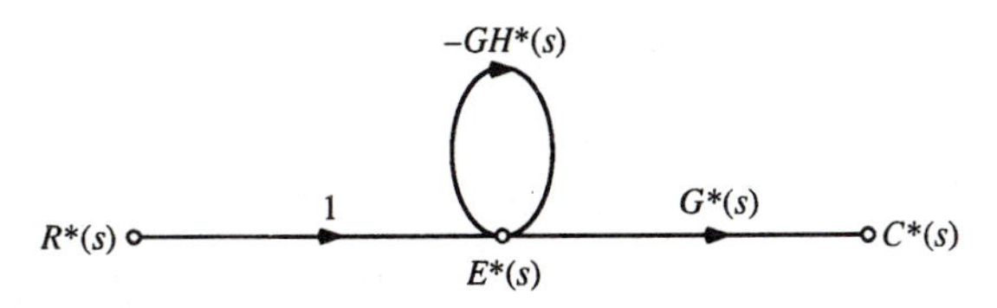

Figure 4-8. Sampled signal flow graph of the system shown in Fig. 4-7.

6. *The composite signal flow graph* [5] *is obtained by combining the equivalent and the sampled signal flow graphs, according to the relation between these flow graphs. The transfer function relationships between the inputs and the continuous-data outputs are determined from the composite signal flow graph by use of Mason's gain formula.* In fact, all the output variables of the system, discrete and analog, can be determined from the composite signal flow graph.

More specifically, the composite signal flow graph is formed by connecting the output nodes of samplers on the sampled signal flow graph with unity-gain branches to the same nodes on the equivalent signal flow graph. For example, the composite signal flow graph of the system in Fig. 4-7 is shown in Fig. 4-9.

Application of Mason's gain formula to the composite signal flow graph in Fig. 4-9 yields the input-output transfer relations for all the discrete- and the continuous-data outputs. Thus, from Fig. 4-9 we get

$$C(s) = \frac{G(s)}{1 + GH^*(s)} R^*(s) \tag{4-80}$$

and

$$E(s) = R(s) - \frac{G(s)H(s)}{1 + GH^*(s)} R^*(s) \tag{4-81}$$

The discrete-data output relations in Eqs. (4-78) and (4-79) can also be determined from the composite signal flow graph in Fig. 4-9 by using the gain formula.

The sampled signal flow graph method outlined above can be applied to linear multiloop multisampler systems, provided that the samplers are synchronized and of the same sampling frequency. The method can also be applied to the modified z-transform relations so that responses between the sampling instants can be recovered.

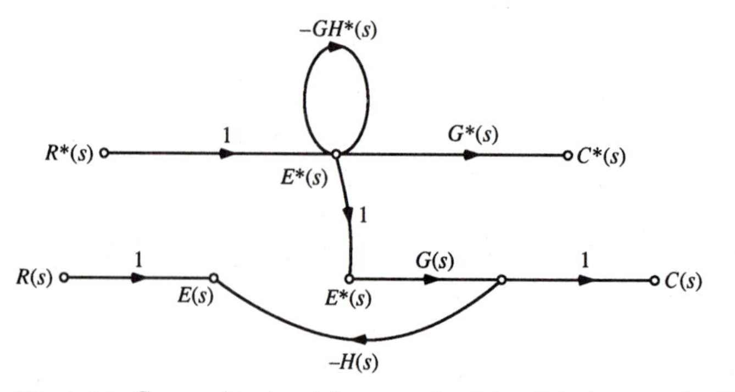

Figure 4-9. Composite signal flow graph of the digital system in Fig. 4-7.

When one is more proficient with the application of the sampled signal flow graph method, Step 2 can be bypassed, and the writing of the input-output equations and the taking of the pulse transform in Step 3 can be combined.

Example 4-4

The block diagram of a multiloop discrete-data control system is shown in Fig. 4-10(a). The equivalent signal flow graph of the system is shown in Fig. 4-10(b). By regarding the output of the samplers as inputs to the system, the input and the output variables of the system are identified as follows:

input variables: $r^*(t)$, $e^*(t)$, $c^*(t)$
output variables: $e(t)$, $c(t)$.

The reader should not be puzzled by the fact that $c^*(t)$ is treated as an input of the system. The above definitions are based on the signal flow graph terminology, and the roles of $c(t)$ and $c^*(t)$ should be distinguished. Also, in any signal flow graph, any noninput node variable can be regarded as an output. Thus, the two outputs defined above are only the ones that are essential for the present system. If desired, we can also designate the output node variable of $G_1(s)$ as an output.

The cause-and-effect equations between the inputs and the outputs designated above are written by applying Mason's gain formula to the original signal flow graph of Fig. 4-10(b). We can take the pulse transform of these equations at the same time. The results are

$$E^*(s) = R^*(s) - G_1G_2^*(s)E^*(s) + G_2H^*(s)C^*(s) \tag{4-82}$$

$$C^*(s) = G_1G_2^*(s)E^*(s) - G_2H^*(s)C^*(s) \tag{4-83}$$

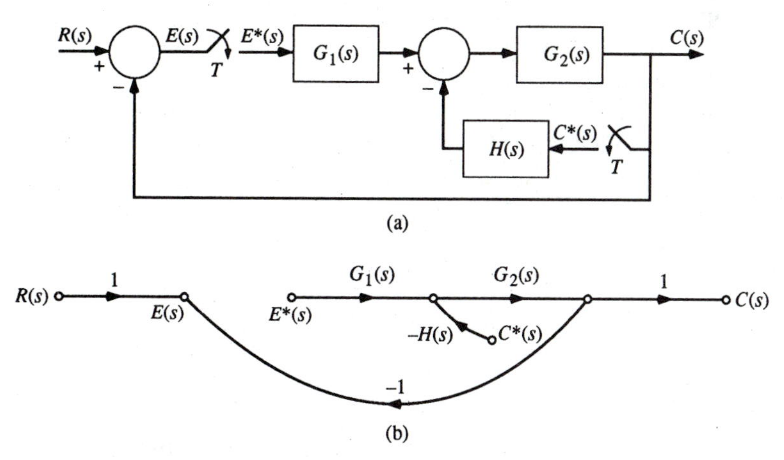

Figure 4-10. (a) Block diagram of a multiloop discrete-data control system. (b) Original signal flow graph.

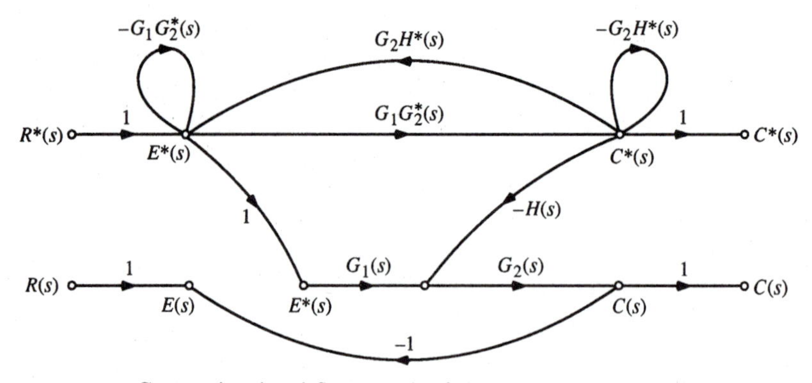

Figure 4-11. Composite signal flow graph of the system in Fig. 4-10.

The sampled signal flow graph is drawn as the top part of Fig. 4-11 using the last two equations. The composite signal flow graph is obtained by directing branches with unity gains from $E^*(s)$ of the sampled signal flow graph to the same node on the original signal flow graph and the same between $C^*(s)$ and $C^*(s)$.

Applying Mason's gain formula to the composite signal flow graph of Fig. 4-11, we get the following input-output transfer function relations:

$$E^*(s) = \frac{1 + G_2H^*(s)}{1 + G_1G_2^*(s) + G_2H^*(s)} R^*(s) \tag{4-84}$$

$$C^*(s) = \frac{G_1G_2^*(s)}{1 + G_1G_2^*(s) + G_2H^*(s)} R^*(s) \tag{4-85}$$

$$C(s) = \frac{G_1(s)G_2(s)[1 + G_2H^*(s)] - G_2(s)H(s)G_1G_2^*(s)}{1 + G_1G_2^*(s) + G_2H^*(s)} R^*(s) \tag{4-86}$$

In determining $E^*(s)$ and $C^*(s)$ in Eqs. (4-84) and (4-85), respectively, only the sampled signal flow graph portion of Fig. 4-11 is of concern, since no forward paths exist between $R(s)$ and these two variables. In obtaining the analog output $C(s)$ in Eq. (4-86), there are two forward paths between $R^*(s)$ and $C(s)$, and one of these has a nontouching loop with a loop gain of $-G_2H^*(s)$. Since no forward paths lie between $R(s)$ and $C(s)$, the analog input does not appear in Eq. (4-86).

If necessary, the z-transform equivalents of Eqs. (4-84) and (4-85) are written as

$$E(z) = \frac{1 + G_2H(z)}{1 + G_1G_2(z) + G_2H(z)} R(z) \tag{4-87}$$

$$C(z) = \frac{G_1G_2(z)}{1 + G_1G_2(z) + G_2H(z)} R(z) \tag{4-88}$$

Example 4-5

In this example we illustrate a discrete-data system that has a loop not broken by the samplers. The block diagram of the system is shown in Fig. 4-12(a). The following steps are carried out.

1. The original signal flow graph of the system is drawn in Fig. 4-12(b). The input and the output variables of the signal flow graph are

   input variables: $r(t)$, $e^*(t)$
   output variables: $e(t)$, $c(t)$.

   Notice that the signal flow graph has an analog loop with a loop gain of $-G_2(s)$. The $D(s)$ block with samplers at its input and output is a block diagram representation of a digital controller that has discrete-data input and output. The transfer function $D(s)$ and the sampler at the output can be replaced by the pulse transfer function $D^*(s)$, as shown in Fig. 4-12.
2. The following equations are written from Fig. 4-12(a) using Mason's gain formula, with $r(t)$ and $e^*(t)$ as inputs and $e(t)$ and $c(t)$ as outputs, and taking the pulse transform at the same time:

$$E^*(s) = \left(\frac{R(s)}{1 + G_2(s)}\right)^* - D^*(s)\left(\frac{G_1(s)G_2(s)}{1 + G_2(s)}\right)^* E^*(s) \tag{4-89}$$

$$C^*(s) = \left(\frac{R(s)G_2(s)}{1 + G_2(s)}\right)^* + D^*(s)\left(\frac{G_1(s)G_2(s)}{1 + G_2(s)}\right)^* E^*(s) \tag{4-90}$$

Notice that in this case, similar to the system in Fig. 4-6, the input signal does not appear as an independent entity, since it is not sampled directly.

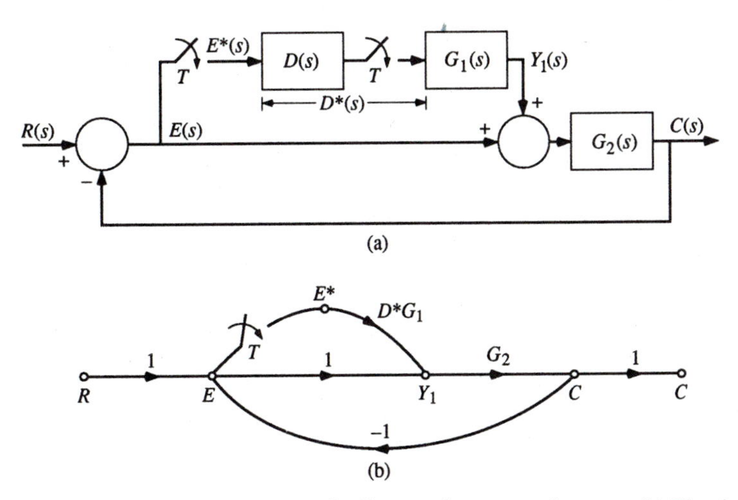

Figure 4-12. (a) Block diagram of a discrete-data control system. (b) The signal flow graph.

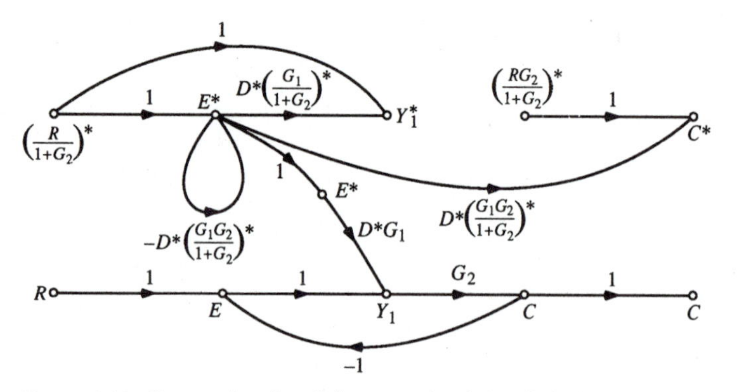

Figure 4-13. Composite signal flow graph of the digital system in Fig. 4-12.

3. The composite signal flow graph of the system is drawn as shown in Fig. 4-13.
4. Applying Mason's gain formula to the composite signal flow graph of Fig. 4-13, the transfer relation between the input and the analog output $C(s)$ and the discrete output $C^*(s)$ are determined as follows.

   The composite signal flow graph has a total of three inputs. These are

$$\left(\frac{R(s)}{1+G_2(s)}\right)^* \qquad \left(\frac{R(s)G_2(s)}{1+G_2(s)}\right)^* \qquad R(s)$$

The analog input $R(s)$ does not affect the output $C^*(s)$, and in fact, although the original signal flow graph has an analog loop that is not affected by the samplers, that part of the flow graph will not contribute to the sampled signal flow graph. Thus, for the output $C^*(s)$, we have

$$C^*(s) = \left(\frac{R(s)G_2(s)}{1+G_2(s)}\right)^* + \frac{D^*(s)\left(\dfrac{G_1(s)G_2(s)}{1+G_2(s)}\right)^*}{1+D^*(s)\left(\dfrac{G_1(s)G_2(s)}{1+G_2(s)}\right)^*}\left(\frac{R(s)}{1+G_2(s)}\right)^* \tag{4-91}$$

or, in the z-transform form,

$$C(z) = \left(\frac{RG_2(s)}{1+G_2(s)}\right)(z) + \frac{D(z)\left(\dfrac{G_1(s)G_2(s)}{1+G_2(s)}\right)(z)}{1+D(z)\left(\dfrac{G_1(s)G_2(s)}{1+G_2(s)}\right)(z)}\left(\frac{R(s)}{1+G_2(s)}\right)(z) \tag{4-92}$$

where

$$\left(\frac{R(s)G_2(s)}{1+G_2(s)}\right)(z) = \mathscr{Z}\left(\frac{R(s)G_2(s)}{1+G_2(s)}\right) \tag{4-93}$$

and the other z-transform quantities are defined similarly.

Applying the gain formula to the composite signal flow graph in Fig. 4-13 using $C(s)$ as the output node, we have

$$C(s) = \frac{R(s)G_2(s)}{1 + G_2(s)} + \frac{\left(\dfrac{G_1(s)G_2(s)}{1 + G_2(s)}\right)D^*(s)\left(\dfrac{R(s)}{1 + G_2(s)}\right)^*}{1 + D^*(s)\left(\dfrac{G_1(s)G_2(s)}{1 + G_2(s)}\right)^*} \tag{4-94}$$

The composite signal flow graph in Fig. 4-13 also includes information on the node variables Y_1 and Y_1^*, in case these are of interest.

The sampled signal flow graph is not the only signal flow graph method available for discrete-data systems. The *direct signal flow graph* [3] is an alternate method which allows the evaluation of the input-output transfer relations of discrete-data systems by direct inspection. The method depends on an entirely different set of terminologies and definitions than those of Mason's signal flow graph.

4-6 THE MODIFIED z-TRANSFER FUNCTION

In Sec. 3-8-2 we introduced the modified z-transform method for the description of the response between the sampling instants of a sampled signal. In this section we apply the method to discrete-data systems and define the *modified z-transfer function.*

Consider that we are interested in determining the response between the sampling instants of the system shown in Fig. 4-14(a). First, we insert a fictitious time delay ΔT, $0 < \Delta < 1$, at the output of the system, and the delayed output

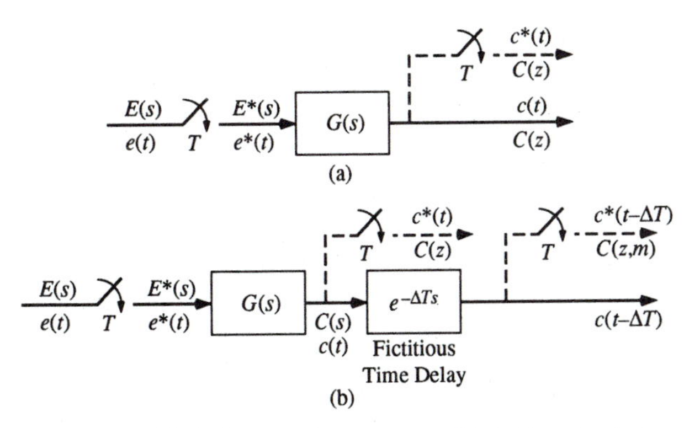

Figure 4-14. (a) A discrete-data system. (b) A discrete-data system with a fictitious time delay.

$c(t - \Delta T)$ is sampled by a fictitious sampler, as shown in Fig. 4-14(b). As shown in Eq. (3-194), the modified z-transform of the output $c(t)$ is defined as

$$\mathscr{Z}_m[c(t)] = C(z, m) = z^{-1} \sum_{k=0}^{\infty} c(kT + mT)z^k \tag{4-95}$$

The modified z-transform in the last equation can be used for the determination of the input-output transfer relation of the system shown in Fig. 4-14. Substituting $C(s) = G(s)E^*(s)$ in Eq. (3-201), we get

$$C(z, m) = \frac{1}{T} \sum_{n=-\infty}^{\infty} G(s + jn\omega_s)E^*(s + jn\omega_s)e^{-(1-m)(s+jn\omega_s)T}\Big|_{z=e^{Ts}} \tag{4-96}$$

Since $E^*(s + jn\omega_s) = E^*(s)$, the last equation becomes

$$\begin{aligned} C(z, m) &= E(z)\frac{1}{T} \sum_{n=-\infty}^{\infty} G(s + jn\omega_s)e^{-(1-m)(s+jn\omega_s)T}\Big|_{z=e^{Ts}} \\ &= E(z)G(z, m) \end{aligned} \tag{4-97}$$

In the last equation $G(z, m)$ denotes the modified z-transform of $G(s)$ and is expressed as

$$G(z, m) = \frac{1}{T} \sum_{n=-\infty}^{\infty} G(s + jn\omega_s)e^{-(1-m)(s+jn\omega_s)T}\Big|_{z=e^{Ts}} \tag{4-98}$$

Therefore, the modified z-transform of a system with the transfer function $G(s)$ is defined in exactly the same way as that of a signal.

The conclusion of the result in Eq. (4-97) is that for the system in Fig. 4-14, which has a sampler at the input only, the modified z-transform of the output $c(t)$ is the product of the modified z-transform of the system transfer function $G(s)$ and the z-transform of the input $E(s)$. Alternatively,

$$\mathscr{Z}_m[C(s)] = C(z, m) = \mathscr{Z}_m[G(s)E^*(s)] = G(z, m)E(z) \tag{4-99}$$

Further, *the modified z-transform of a sampled signal $e^*(t)$ is just the z-transform of the signal*, or

$$\mathscr{Z}_m[E^*(s)] = E(z) \tag{4-100}$$

Example 4-6

Consider that the system shown in Fig. 4-14 has the transfer function

$$G(s) = \frac{1}{s + a} \tag{4-101}$$

where a is a constant. The input to the system is a unit-step function, $e(t) = u_s(t)$. The output of the system is to be evaluated by the modified z-transform method.

The modified z-transform of $c(t)$ is written

$$C(z, m) = G(z, m)E(z) \tag{4-102}$$

The z-transform of $e^*(t)$ is

$$E(z) = \frac{z}{z - 1} \tag{4-103}$$

and the modified z-transform of $G(s)$ is already determined in Eq. (3-203),

$$G(z, m) = \frac{e^{-maT}}{z - e^{-aT}} \tag{4-104}$$

Thus,

$$C(z, m) = \frac{e^{-maT}}{z - e^{-aT}} \frac{z}{z - 1} \tag{4-105}$$

which can be expanded into an infinite series of z^{-1},

$$C(z, m) = e^{-maT}z^{-1} + e^{-maT}(1 + e^{-aT})z^{-2} + e^{-maT}(1 + e^{-aT} + e^{-2aT})z^{-3} + \cdots \tag{4-106}$$

It can be shown that the coefficient of z^{-k} is

$$\begin{aligned} c[(k + m - 1)T] &= e^{-maT}(1 + e^{-aT} + e^{-2aT} + \cdots + e^{-kaT}) \\ &= \frac{e^{-maT}(1 - e^{-kaT})}{1 - e^{-aT}} \end{aligned} \tag{4-107}$$

which gives the output response for the time duration $kT \leq t < (k + 1)T$ for $k \geq 0$ and when m is varied between zero and one.

The z-transform of the output $c(t)$ is written

$$\begin{aligned} C(z) = G(z)E(z) &= \frac{z}{z - e^{-aT}} \frac{z}{z - 1} \\ &= 1 + (1 + e^{-aT})z^{-1} + (1 + e^{-aT} + e^{-2aT})z^{-2} + \cdots \end{aligned} \tag{4-108}$$

Comparing Eq. (4-108) with Eq. (4-105) we see that

$$C(z, m)|_{m=0} = z^{-1}C(z) \tag{4-109}$$

which is derived earlier in Eq. (3-198). Since the input $e(t)$ and the impulse response of $G(s)$ both have jump discontinuities at $t = 0$, $C(z, 1) \neq C(z)$.

Figure 4-15 shows the response of $c(t)$ and the sampled values at $t = kT$. Notice that the z-transform result is misleading, since the output has jump discontinuities at the sampling instants kT. By varying the value of m between zero and one, all the data between the sampling instants can be specified by $C(z, m)$.

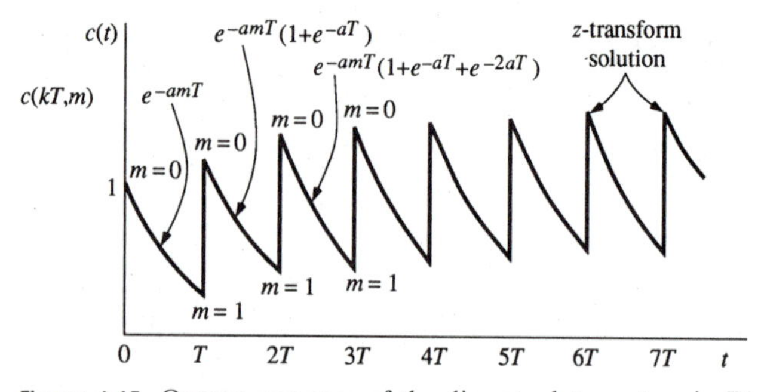

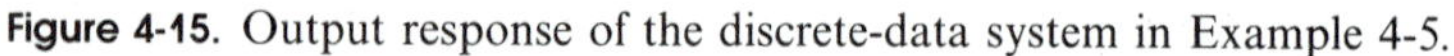
Figure 4-15. Output response of the discrete-data system in Example 4-5.

Closed-Loop Systems

The modified z-transform method can be applied to closed-loop discrete-data systems without complications. Figure 4-16 shows a closed-loop system with a fictitious time delay positioned at the system output. The following cause-and-effect equations are written from the block diagram by treating the output of the sampler, $e^*(t)$, as an input, in addition to the input $r(t)$, and the output signals of the system are $e(t)$ and $c(t - \Delta T)$:

$$C(z, m) = G(z, m)E(z) \tag{4-110}$$

$$E(z) = R(z) - GH(z)E(z) \tag{4-111}$$

Solving for $E(z)$ in terms of $R(z)$ and substituting the result in Eq. (4-110), we get

$$C(z, m) = \frac{G(z, m)}{1 + GH(z)} R(z) \tag{4-112}$$

which is the modified z-transform of the output $c(t)$.

Although the modified z-transform was originally devised for the purpose of describing signals between sampling instants, we shall show later that the method is also useful for the analysis of multirate sampled-data systems.

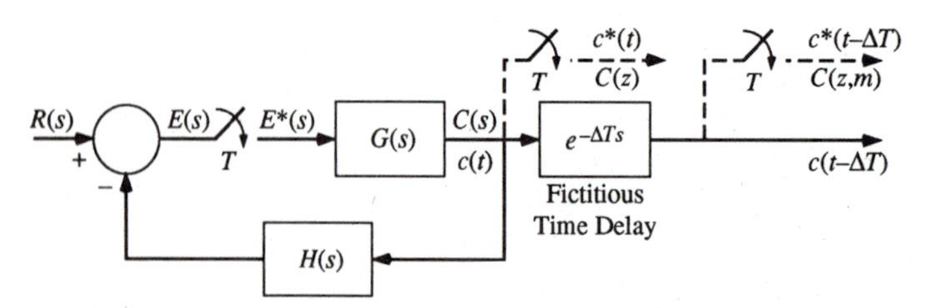

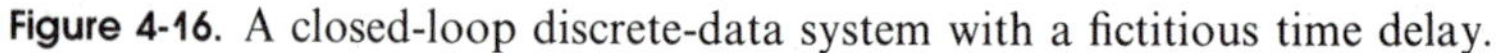
Figure 4-16. A closed-loop discrete-data system with a fictitious time delay.

4-7 MULTIRATE DISCRETE-DATA SYSTEMS

The discrete-data systems considered in the preceding sections all have a uniform sampling rate. When the data are digital, the rates of the input and the output data of digital processors and controllers throughout the system are all identical and uniform. For sampled-data systems, the samplers throughout the system all have the same uniform sampling rate.

In practice, the bandwidths of various portions of a discrete-data system may be far apart. Thus, it is more appropriate to sample a slowly varying signal at a low sampling rate, while a signal with high-frequency content should be sampled at a relatively high rate. When a discrete-data system has different sampling rates at various locations, the system is referred to as a *multirate discrete-data system.* Figure 4-17 illustrates the block diagram of a typical digital missile autopilot control system. Due to the difference between the dynamics in the position and the rate feedback loops, the samplers in the system have two sampling periods, T_1 and T_2. Note that the two data holds in the system have different characteristics, since they hold the sampled data for periods of T_1 and T_2, respectively.

Due to the variety of sampling schemes, the analysis and design of multirate discrete-data systems are usually quite complex. It would be helpful first to establish the basic types of multirate sampling. Figure 4-18 shows three basic types of open-loop discrete-data systems with multirate sampling. The system shown in Fig. 4-18(a) has a slow-rate sampler at the input and a fast-rate sampler at the output. This type of system is referred to as the *slow-fast multirate-sampled system.* Figure 4-18(b) shows a system in which the sampler at the input has a faster rate than that of the output sampler, and the system is referred to as a *fast-slow multirate-sampled system.* A digital control system whose controllers have different input and output data rates can be modeled by the block diagram in Fig. 4-18(c). In general, the sampling periods T_1 and T_2 may not be related by an integral factor. For analytical purposes, the mathematics are greatly simplified if we assume that T_1 and T_2 are integrally related; that is, with reference to Fig. 4-18(a), $T_1 = NT_2$, where N is a positive integer greater than one.

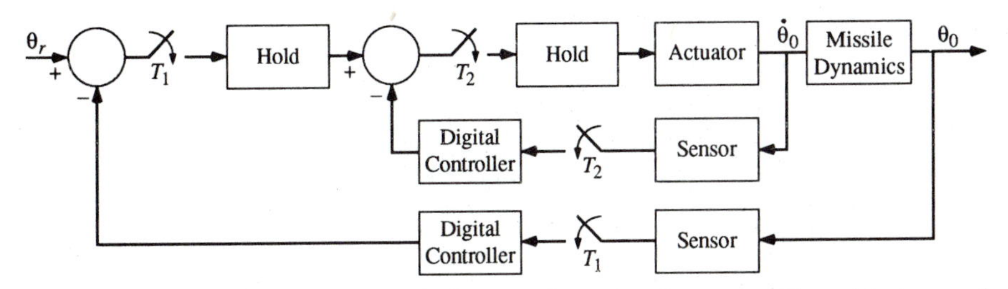

Figure 4-17. A digital missile autopilot control system with multirate samplers.

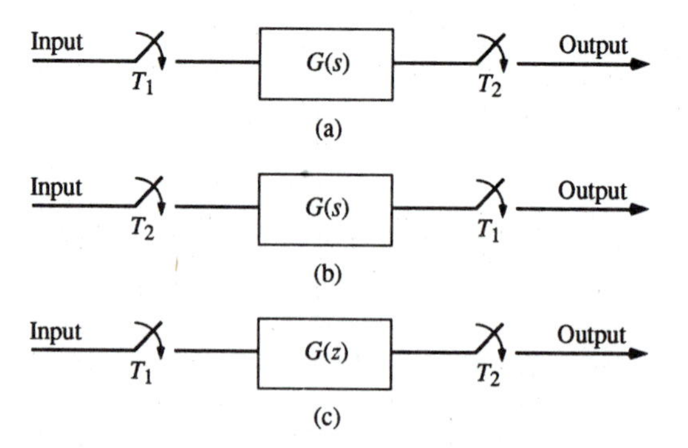

Figure 4-18. Open-loop multirate discrete-data systems. (a) Slow-fast sampled system ($T_1 > T_2$). (b) Fast-slow sampled system ($T_1 > T_2$). (c) All-digital system.

4-7-1 Slow-Fast Multirate-Sampled Systems

Consider the slow-fast sampled system shown in Fig. 4-18(a), with $T_1 = NT_2$, where N is a positive integer greater than one. Several methods are available for the analytical study of this type of system. Each of these methods is useful for a particular situation, and the methods are described as follows.

1. Fictitious-Sampler Method

Since the input sampler samples at a lower rate, or a larger sampling period T_1, we can insert a fictitious fast-rate sampler with the sampling period $T_2 = T_1/N$ at the input, as shown in Fig. 4-19. Since the input to the fictitious sampler, $r^*(t)$, is a pulse train with sampling period T_1, the output of the fictitious fast sampler is still $r^*(t)$. The situation is very similar to the submultiple sampling method described in Sec. 3-8-1. The input-output transfer relation of the system shown in Fig. 4-19 is written

$$C(z)_N = G(z)_N R(z) \tag{4-113}$$

where $G(z)_N$ is as defined in Eq. (3-186),

$$G(z)_N = \sum_{k=0}^{\infty} g(kT/N)z^{-k/N} \tag{4-114}$$

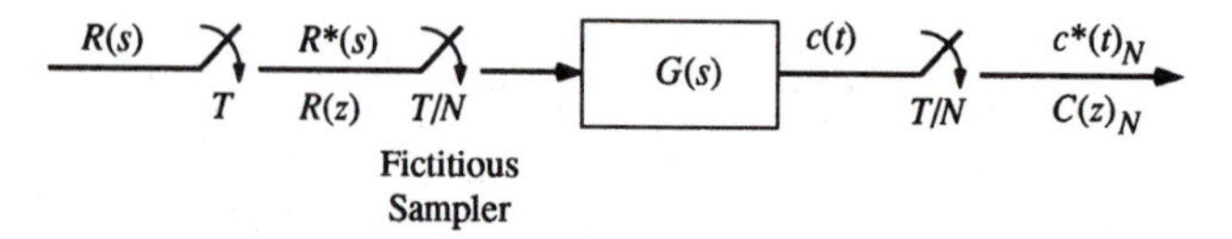

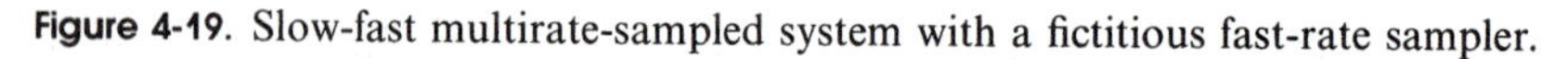
Figure 4-19. Slow-fast multirate-sampled system with a fictitious fast-rate sampler.

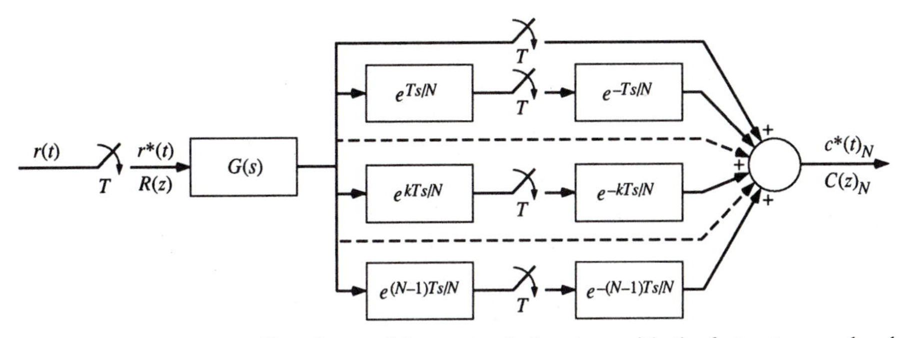

Figure 4-20. Slow-fast multirate-sampled system with the fast-rate sampler decomposed into N slow-rate samplers with time-advance and time-delay units.

and $C(z)_N$ is defined in the same way. As in Eq. (3-187), $G(z)_N$ can be found from $G(z)$ through the following transformation:

$$G(z)_N = G(z)|_{z = z^{1/N},\, T = T/N} \tag{4-115}$$

2. Sampler-Decomposition Method

A versatile method of analyzing the slow-fast multirate system of Fig. 4-18(a) is to decompose the fast-rate sampler into N parallel-connected slow-rate samplers with time-delay and time-advance units, as shown in Fig. 4-20. Since now the samplers are all at the same sampling rate, the ordinary z-transform method of analysis may be applied. The z-transform of the output of the system $c^*(t)_N$ is written

$$C(z)_N = \sum_{k=0}^{N-1} z^{-k/N} \mathscr{Z}[e^{kTs/N} G(s)] R(z) \tag{4-116}$$

where the z-transforms of the time delays at the output are written as

$$\mathscr{Z}[e^{-kTs/N}] = z^{-k/N} \tag{4-117}$$

for $k = 1, 2, \ldots, N-1$.

Comparing Eq. (4-116) with Eq. (4-113), we see that $G(z)_N$ can be written as

$$G(z)_N = \sum_{k=0}^{N-1} z^{-k/N} \mathscr{Z}[e^{kTs/N} G(s)] \tag{4-118}$$

The z-transform of $e^{kTs/N}G(s)$ can be determined from the modified z-transform of $G(s)$. By definition of the z-transform,

$$\mathscr{Z}[e^{kTs/N} G(s)] = \sum_{n=0}^{\infty} g\left[\left(n + \frac{k}{N}\right)T\right] z^{-n} \tag{4-119}$$

Since k/N is always less than unity in the present application, comparing the right-hand side of the last equation with the defining equation of the modified z-transform in Eq. (3-194), we have the following relationship:

$$\mathscr{Z}[e^{kTs/N}G(s)] = [zG(z, m)]|_{m=k/N} \tag{4-120}$$

or

$$\mathscr{Z}[e^{kTs/N}G(s)] = zG(z, k/N) \tag{4-121}$$

Substituting Eq. (4-121) into Eq. (4-118), we get

$$G(z)_N = \sum_{k=0}^{N-1} z^{1-k/N}G(z, k/N) \tag{4-122}$$

Therefore, the output of the system in Fig. 4-20 is written

$$C(z)_N = R(z) \sum_{k=0}^{N-1} z^{1-k/N}G(z, k/N) \tag{4-123}$$

Example 4-7

Consider that the transfer function of the slow-fast multirate-sampled system in Fig. 4-18(a) is

$$G(s) = \frac{1}{s(s+1)} \tag{4-124}$$

The input of the system is a unit-step function $u_s(t)$. The sampling periods are $T_1 = 1$ s and $T_2 = 1/2$ s.

The modified z-transform of $G(s)$ for the sampling period of 1 s is

$$G(z, m) = \frac{1}{z-1} - \frac{e^{-m}}{z - e^{-1}} \tag{4-125}$$

Using Eq. (4-123), the z-transform of the output with the fast-rate sampling is

$$C(z)_2 = \sum_{k=0}^{1} z^{1-k/2}\left[\frac{1}{z-1} - \frac{e^{-k/2}}{z-0.368}\right]\frac{z}{z-1} \tag{4-126}$$

The last equation is expanded to read

$$C(z)_2 = \left[\frac{z + z^{1/2}}{z-1} - \frac{z + e^{-1/2}z^{1/2}}{z-0.368}\right]\frac{z}{z-1} \tag{4-127}$$

Letting $z_2 = z^{1/2}$, then $z_2^2 = z$; the last equation is written

$$\begin{aligned} C(z)_2 &= \left[\frac{z_2^2 + z_2}{z_2^2 - 1} - \frac{z_2^2 + 0.607z_2}{z_2^2 - 0.368}\right]\frac{z_2^2}{z_2^2 - 1} \\ &= \left[\frac{0.393z_2^3 + 0.632z_2^2 + 0.239z_2}{z_2^4 - 1.368z_2^2 + 0.368}\right]\frac{z_2^2}{z_2^2 - 1} \end{aligned} \tag{4-128}$$

Table 4-1 $c^*(t)_2$ and $c^*(t)$ of the Multirate-Sampled System in Example 4-7

$c^*(t)_2$		$c^*(t)$	
Sample Periods	Values	Sample Periods	Values
0	0.000000	0	0.000000
1	0.393000	1	0.632000
2	0.631551	2	1.496576
3	1.169351	3	2.446740
4	1.495796	4	3.428401
5	2.086948	5	4.421651
6	2.445778	6	5.419168
7	3.056587	7	6.418253
8	3.427348	8	7.417916
9	4.045400	9	8.417791
10	4.420558	10	9.417746
11	5.041279	11	10.417728
12	5.418056	12	11.417722
13	6.039761	13	12.417719
14	6.417135	14	13.417717
15	7.039202	15	14.417716
16	7.416796	16	15.417716
17	8.038996	17	16.417715
18	8.416671	18	17.417713
19	9.038919	19	18.417711
20	9.416624		
21	10.038890		
22	10.416605		
23	11.038877		
24	11.416596		
25	12.038872		
26	12.416593		
27	13.038869		
28	13.416591		
29	14.038868		
30	14.416591		
31	15.038868		
32	15.416590		
33	16.038866		
34	16.416588		
35	17.038864		
36	17.416586		
37	18.038862		
38	18.416584		
39	19.038860		

Canceling the common factor in the last equation, $C(z)_2$ is reduced to

$$C(z)_2 = \left[\frac{0.393z_2}{(z_2 - 1)(z_2 - 0.607)}\right]\frac{z_2^2}{z_2^2 - 1} \tag{4-129}$$

The expression inside the brackets, in the last equation is recognized to be $G(z)_2$ which can be obtained from $G(z)$ using Eq. (4-115).

The last equation is expanded into a power series in z_2^{-1} by long division, using the computer program INVZ of the DCSP software package, and the response of $c(t)$ at the sampling instants of kT_2 is tabulated in Table 4-1 for $k = 0$ to 39. The reader can check that if the output sampler in the system of Fig. 4-18(a) were with the sampling period $T_1 = 1$ s, then the z-transform of the system output is

$$C(z) = \left[\frac{0.632z}{z^2 - 1.368z + 0.368}\right]\frac{z}{z - 1} \tag{4-130}$$

The response of $c(t)$ at the sampling instants of kT_1 is obtained by expanding $C(z)$ into a power series of z^{-1}, and the results are tabulated in Table 4-1 for $k = 0$ to 19. Notice that the values of $c(kT_1)$ agree with every other value of $c(kT_2)$, starting from $k = 0$.

3. Infinite-Series Representation

The input-output transfer function relation in Eq. (4-113) for the slow-fast sampled system can be written as a pulse transform relation in the Laplace domain; that is,

$$C^*(s)_N = G^*(s)_N R^*(s) \tag{4-131}$$

Substituting T/N for T in the infinite-series representation of $G^*(s)$, as in Eq. (2-53) and assuming that $g(0) = 0$, $G^*(s)_N$ is written as

$$G^*(s)_N = \frac{N}{T}\sum_{n=-\infty}^{\infty} G\left[s + j\frac{2\pi nN}{T}\right] \tag{4-132}$$

Thus, $C^*(s)_N$ can be expressed in the same way. These pulse transform relations are useful for frequency-domain analysis. Another application of Eq. (4-132) is demonstrated in the following development.

4-7-2 Fast-Slow Multirate-Sampled Systems

The discrete-data system with the fast-slow sampling shown in Fig. 4-18(b) can be analyzed by several alternate methods. These methods are described in the following.

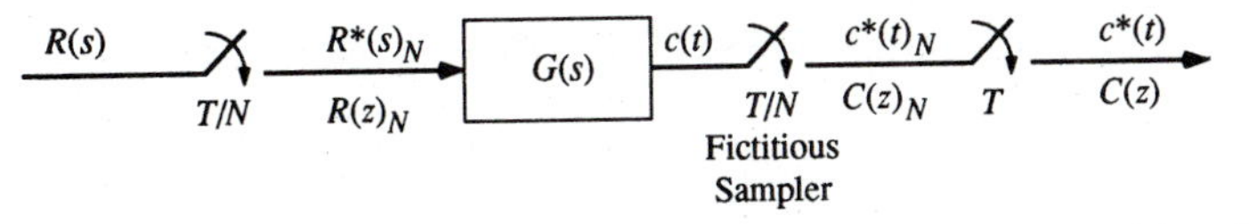

Figure 4-21. Fast-slow multirate-sampled system with a fictitious sampler.

1. Fictitious-Sampler Method

One method of analyzing the fast-slow multirate-sampled system shown in Fig. 4-18(b) is to insert a fictitious fast-rate sampler with sampling period T/N in front of the slow-rate sampler at the output. The result is shown in Fig. 4-21. Since the output of the system is not affected by the fictitious fast-rate sampler, the two systems in Figs. 4-18(b) and 4-21 are the same from the analytical standpoint. From Fig. 4-21 the z-transform of the output of the fictitious sampler is

$$C(z)_N = G(z)_N R(z)_N \tag{4-133}$$

where $G(z)_N$ is defined in Eqs. (4-114) and (4-115), and $R(z)_N$ is similarly defined. Once $C(z)_N$ is determined, which corresponds to the z-transform of the fast-rate sampled output $c^*(t)_N$, the output of the fictitious slow-rate sampler $c^*(t)$ is obtained by "extracting" the samples at $t = kT$, $k = 0, 1, 2, \ldots,$ from $c^*(t)_N$, or obtaining $c(kT)$ from the set $c(kT/N)$. Then, $C(z)$ is simply the z-transform of $c^*(t)$.

We present three different methods of extracting $C(z)$ from $C(z)_N$.

(a) *The Frequency-Domain Method*

First, we write

$$C^*(s) = \frac{1}{T}\sum_{n=-\infty}^{\infty} C\left[s + j\frac{2\pi n}{T}\right] \tag{4-134}$$

Letting $n = pN + k$, where $-\infty \le p \le \infty$, and $k = 0, 1, 2, \ldots, N-1$, the last equation becomes

$$C^*(s) = \frac{1}{T}\sum_{pN+k=-\infty}^{\infty} C\left[s + j\frac{2\pi pN}{T} + j\frac{2\pi k}{T}\right] \tag{4-135}$$

which is then written as

$$C^*(s) = \frac{1}{T}\sum_{k=0}^{N-1}\frac{T}{N}\frac{N}{T}\sum_{p=-\infty}^{\infty} C\left[s + j\frac{2\pi pN}{T} + j\frac{2\pi k}{T}\right] \tag{4-136}$$

In view of the relation for $G^*(s)_N$ in Eq. (4-132), the last equation is written

$$C^*(s) = \frac{1}{N}\sum_{k=0}^{N-1} C^*\left[s + j\frac{2\pi k}{T}\right]_N \tag{4-137}$$

Now substituting Eq. (4-131) in the last equation, we get

$$C^*(s) = \frac{1}{N} \sum_{k=0}^{N-1} G^*\left[s + j\frac{2\pi k}{T}\right]_N R^*\left[s + j\frac{2\pi k}{T}\right] \tag{4-138}$$

Since k is an integer, $R^*(s + j2\pi k/T) = R^*(s)$; Eq. (4-138) becomes

$$C^*(s) = G^*(s)R^*(s) \tag{4-139}$$

or

$$C(z) = G(z)R(z) \tag{4-140}$$

Equation (4-137) gives the slow-sampled signal in terms of the fast-sampled signal. Using the complex translation theorem of the z-transform in Eq. (3-106), the z-domain equivalent of Eq. (4-137) is

$$C(z) = \frac{1}{N} \sum_{k=0}^{N-1} C[z_N e^{j2\pi k/N}]_N \tag{4-141}$$

Thus, given $C(z)_N$, the z-transform of the fast-sampled function, $C(z)$ can be obtained by replacing z_N by $z_N e^{2\pi k/N}$ in $C(z)_N$ and substituting it in the summation of Eq. (4-141).

(b) *The Time-Domain Method*

Another more comprehensible method of finding $C(z)$ from $C(z)_N$ is to make use of the equivalent time response of $c(kT)$ and $c(kT/N)$. The method is outlined as follows.

1. Take the inverse z-transform of $C(z)_N$ to give $c(kT/N)$.
2. Replace T/N by T in $c(kT/N)$ to give $c(kT)$.
3. Take the z-transform of $c(kT)$ to get $C(z)$.

(c) *The Residues Method*

The z-transform of the output of the fast-rate sampler in Fig. 4-21 is written

$$C(z)_N = \sum_{k=0}^{\infty} c(kT/N)z^{-k/N} \tag{4-142}$$

Using the inversion formula, the inverse z-transform of $C(z)_N$ is

$$c(kT/N) = \frac{1}{2\pi j} \oint_{\Gamma} C(z_N)z_N^{k-1}\, dz_N \tag{4-143}$$

where

$$C(z_N) = C(z)_N \tag{4-144}$$

with $z^{1/N}$ replaced by z_N.

The z-transform of the output of the system in Fig. 4-21 is

$$C(z) = \sum_{n=0}^{\infty} c(nT)z^{-n} \tag{4-145}$$

Letting $n = k/N$ and substituting Eq. (4-143) into Eq. (4-145), we get

$$\begin{aligned} C(z) &= \sum_{n=0}^{\infty} \frac{1}{2\pi j} \oint_{\Gamma} C(z_N) z_N^{nN-1}\, dz_N z^{-n} \\ &= \frac{1}{2\pi j} \oint_{\Gamma} C(z_N) \sum_{n=0}^{\infty} z_N^{nN} z^{-n} \frac{dz_N}{z_N} \end{aligned} \tag{4-146}$$

The last equation is simplified by writing the infinite series in a closed form. We have the result

$$C(z) = \frac{1}{2\pi j} \oint_{\Gamma} C(z_N) \frac{1}{1 - z_N^N z^{-1}} \frac{dz_N}{z_N} \tag{4-147}$$

The contour integral of Eq. (4-147) is carried out along the path Γ in the z_N-plane. Assuming that the contour Γ encloses all the poles of $C(z_N)z_N^{-1}$, using the residues theorem, Eq. (4-147) is written

$$C(z) = \sum \text{residues of } C(z_N) \frac{z_N^{-1}}{1 - z_N^N z^{-1}} \text{ at the poles of } C(z_N)z_N^{-1} \tag{4-148}$$

It should be pointed out that in the last equation, although $z_N^N = z$, the terms z_N^N and z^{-1} in the denominator should not cancel each other.

Example 4-8

Consider that the transfer function $G(s)$ of the system in Fig. 4-21 is as given in Eq. (4-124), and $T = 1$ s, $N = 2$. The input $r(t)$ is a unit-step function. The z-transform of the output of the fast sampler is written as $C(z)_N = G(z)_N R(z)_N$. Thus,

$$C(z)_2 = \left[\frac{0.393 z_2}{(z_2 - 1)(z_2 - 0.607)}\right] \frac{z_2}{z_2 - 1} \tag{4-149}$$

We shall use the three methods described above to find the z-transform of the output $C(z)$. The z-transform of the output of the slow sampler $c^*(t)$ is obtained by substituting the last equation in Eq. (4-141). We have

$$\begin{aligned} C(z) &= \frac{1}{2} \sum_{k=0}^{1} C(z_2 e^{j2\pi k/2})_2 \\ &= \frac{1}{2}\left[\frac{0.393 z_2^2}{(z_2 - 1)^2 (z_2 - 0.607)} + \frac{0.393 z_2^2}{(-z_2 - 1)^2 (-z_2 - 0.607)}\right] \end{aligned} \tag{4-150}$$

After simplification and substituting $z = z_2^2$, the last equation leads to

$$C(z) = \left[\frac{1.0246z + 0.2386}{z^2 - 1.368z + 0.368}\right]\frac{z}{(z-1)^2} \tag{4-151}$$

Using the time-domain method, we start again with Eq. (4-149). Carrying out the partial-fraction expansion, $C(z)_2$ is written

$$C(z)_2 = \frac{z_2}{(z_2-1)^2} - \frac{1.5445z_2}{z_2 - 1} + \frac{1.5445z_2}{z_2 - 0.607} \tag{4-152}$$

The inverse z-transform of the last equation is

$$c(kT/2) = 2(kT/2) - 1.5445 + 1.5445e^{-kT/2} \tag{4-153}$$

for $k = 0, 1, 2, \ldots$. The equivalent $c(kT)$ is obtained from the last equation by replacing $T/2$ by T. Thus,

$$c(kT) = 2kT - 1.5445 + 1.5445e^{-kT} \tag{4-154}$$

for $k = 0, 1, 2, \ldots$. The z-transform of $c(kT)$ is simply

$$C(z) = \frac{2z}{(z-1)^2} - \frac{1.5445z}{z-1} + \frac{1.5445z}{z - 0.368} = \frac{(1.0246z + 0.2386)}{(z-1)^2(z-0.368)} \tag{4-155}$$

which is the correct result.

The last method to be illustrated is the residues method of Eq. (4-148). Substituting Eq. (4-149) into Eq. (4-148), we get

$$C(z) = \sum \text{residues of} \frac{0.393z_2^2}{(z_2^2-1)^2(z_2-0.607)}\frac{z_2^{-1}}{1 - z_2^2z^{-1}} \tag{4-156}$$

at the poles $z_2 = 1$ and $z_2 = 0.607$. These residues are evaluated as follows.

Residue at $z_2 = 0.607$

$$\left.\frac{0.393z_2}{(z_2-1)^2(1-z_2^2z^{-1})}\right|_{z_2=0.607} = \frac{1.5445z}{z-0.368} \tag{4-157}$$

Residue at $z_2 = 1$

$$\frac{d}{dz_2}\left[\frac{0.393z_2}{(z_2-0.607)(1-z_2^2z^{-1})}\right]_{z_2=1} = \frac{-1.5445z^2 + 3.5445z}{(z-1)^2} \tag{4-158}$$

Thus,

$$C(z) = \frac{1.5445z}{z-0.368} - \frac{1.5445z^2 - 3.5445z}{(z-1)^2} = \frac{(1.0246z + 0.2386)z}{(z-1)^2(z-0.368)} \tag{4-159}$$

which is the expected result.

The responses of $c^*(t)_2$ and $c^*(t)$ are determined by expanding $C(z)_2$ and $C(z)$ into power series of z_2^{-1} and z^{-1}, respectively, and are tabulated in Table 4-2.

Table 4-2 $c^*(t)_2$ and $c^*(t)$ of the Multirate-Sampled System in Example 4-8

$c^*(t)_2$		$c^*(t)$	
Sample Periods	Values	Sample Periods	Values
0	0.000000	0	0.000000
1	0.393000	1	1.024600
2	1.024551	2	2.664853
3	1.800902	3	4.531666
4	2.665147	4	6.481853
5	3.582742	5	8.462722
6	4.532722	6	10.454881
7	5.502357	7	12.451195
8	6.483922	8	14.449038
9	7.472728	9	16.447441
10	8.465928	10	18.446051
11	9.461794	11	20.444735
12	10.459277	12	22.443445
13	11.457741	13	24.442163
14	12.456799	14	26.440884
15	13.456215	15	28.439604
16	14.455849	16	30.438324
17	15.455612	17	32.437046
18	16.455454	18	34.435768
19	17.455341	19	36.434490
20	18.455256		
21	19.455183		
22	20.455116		
23	21.455050		
24	22.454979		
25	23.454903		
26	24.454821		
27	25.454733		
28	26.454639		
29	27.454538		
30	28.454430		
31	29.454315		
32	30.454193		
33	31.454063		
34	32.453926		
35	33.453777		
36	34.453617		
37	35.453445		
38	36.453262		
39	37.453068		

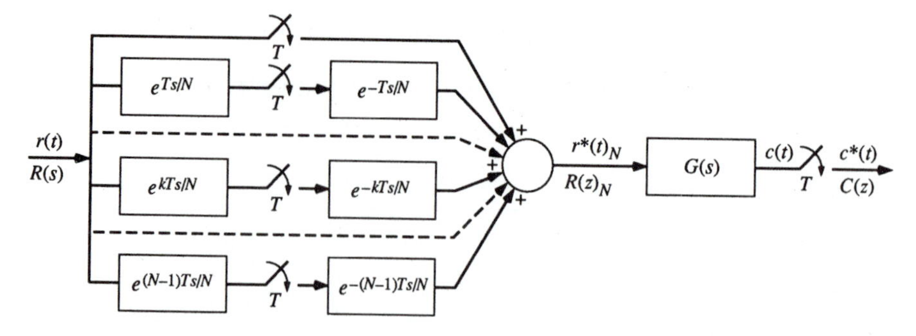

Figure 4-22. Fast-slow multirate-sampled system with the fast-rate sampler decomposed into N slow-rate samplers with time-advance and time-delay units.

2. Sampler-Decomposition Method

The fast-rate sampler is decomposed into N slow-rate samplers with time-advance and time-delay units, as shown in Fig. 4-22. The input-output relation of the system is written in the z-domain as

$$C(z) = \sum_{k=0}^{N-1} \mathscr{Z}[R(s)e^{kTs/N}]\mathscr{Z}[G(s)e^{-kTs/N}] \tag{4-160}$$

The first term on the right side of the last equation, when $k = 0$, can be evaluated by the modified z-transform method described by Eq. (4-121); that is,

$$\mathscr{Z}[R(s)e^{kTs/N}] = zR(z, k/N) \tag{4-161}$$

The second term in Eq. (4-160) is also related to the modified z-transform. From the definition of the modified z-transform, we have

$$G(z, m)|_{m=1-\Delta} = G(z, \Delta) \tag{4-162}$$

Thus, we have

$$\mathscr{Z}[G(s)e^{-kTs/N}] = G(z, m)|_{m=1-k/N} \tag{4-163}$$

where $k/N < 1$. This means that the second term in Eq. (4-160) is determined by setting $m = 1 - k/N$ in the modified z-transform of $G(s)$.

Equation (4-160) is now written as

$$C(z) = R(z)G(z) + \sum_{k=0}^{N-1} zR\left(z, \frac{k}{N}\right)G\left(z, 1 - \frac{k}{N}\right) \tag{4-164}$$

Example 4-9

Consider that the transfer function $G(s)$ of the system in Fig. 4-21 is as in Eq. (4-124), and $T = 1$ s, $N = 2$. The input $r(t)$ is a unit-step function. Using Eq.

(4-164), the z-transform of the output of the system is written

$$C(z) = R(z)G(z) + zR(z, 1/2)G(z, 1/2) \tag{4-165}$$

where $R(z, 1/2)$ and $G(z, 1/2)$ represent $R(z, m)$ and $G(z, m)$ with $m = 1/2$, respectively. The modified z-transforms of $R(s)$ and $G(s)$ are, respectively,

$$R(z, m) = \frac{1}{z-1} \tag{4-166}$$

$$G(z, m) = \frac{1}{z-1} - \frac{e^{-mT}}{z-e^{-T}} \tag{4-167}$$

Substitution of $R(z)$, $G(z)$, and the last two equations into Eq. (4-165) yields

$$C(z) = \frac{z}{z-1}\left[\frac{z}{z-1} - \frac{z}{z-e^{-T}}\right] + \frac{z}{z-1}\left[\frac{1}{z-1} - \frac{e^{-T/2}}{z-e^{-T}}\right]$$

$$= \frac{(1.0246z + 0.2386)z}{(z-1)^2(z-0.368)} \tag{4-168}$$

which agrees with the result in Eq. (4-159).

4-7-3 Multirate Systems with All-Digital Elements

It is common to have multirate digital systems in which some components are entirely digital so that the inputs and outputs of these digital components are digitally coded, and no physical samplings are involved. We shall show that the mathematical relations derived in the preceding sections for the multirate sampled systems are still applicable for this case.

Figure 4-23 shows the block diagram representation of an all-digital multirate system with a slow-rate input and a fast-rate output. In this case, the input of the system consists of digital data that are represented by a pulse train with a period of T s. The output data are represented by a pulse train with a period of T/N. The input-output z-transfer function relation is written as

$$C(z)_N = G(z)_N R(z) \tag{4-169}$$

where $C(z)_N$ and $G(z)_N$ are defined only in the form of

$$G(z)_N = \sum_{k=0}^{\infty} g(kT/N)z^{-k/N} \tag{4-170}$$

Figure 4-23. A slow-fast all-digital multirate system.

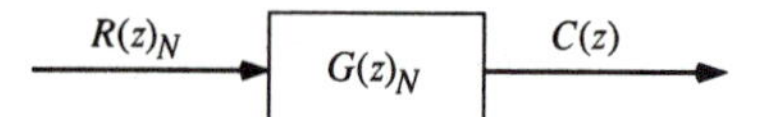

Figure 4-24. A fast-slow all-digital multirate system.

The slow-rate output is written as

$$C(z) = G(z)R(z) \tag{4-171}$$

or $C(z)$ may be determined from $C(z)_N$ using Eq. (4-141).

Figure 4-24 shows the block diagram representation of a fast-slow all-digital multirate system. In this case, we can first write

$$C(z)_N = G(z)_N R(z)_N \tag{4-172}$$

and then the slow-rate output is given by

$$C(z) = \frac{1}{N} \sum_{m=0}^{N-1} G(z_N e^{j2\pi m/N})_N R(z_N e^{j2\pi m/N})_N \tag{4-173}$$

4-7-4 Closed-Loop Multirate-Sampled Systems

In this section we apply some of the methods of analysis of multirate-sampled systems introduced in the preceding sections to closed-loop control systems.

Example 4-10

Consider the closed-loop digital control system with multirate sampling shown in Fig. 4-25(a). The fast-rate sampler is decomposed into synchronized slow-rate samplers with time-advance and time-delay units, as shown in Fig. 4-25(b). Notice that the zero-order hold of the fast-rate sampler has a different transfer function than that of the slow-rate sampler. In any case the hold period of the zoh must be identical with the sampling period.

The transfer function of the zoh of the sampler with the sampling period T is known to be

$$G_{h0}(s) = \frac{1 - e^{-Ts}}{s} \tag{4-174}$$

The transfer function of the zoh associated with the sampler with the sampling period $T/2$ is

$$G_{h02}(s) = \frac{1 - e^{-Ts/2}}{s} \tag{4-175}$$

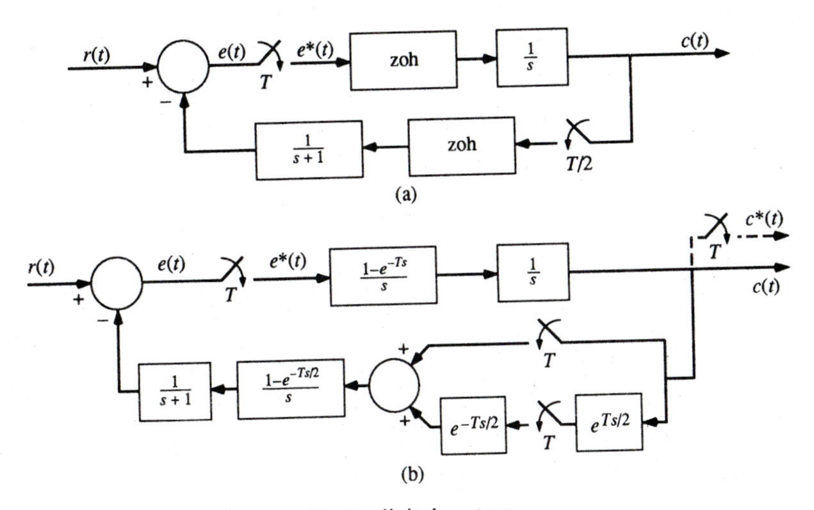

Figure 4-25. Closed-loop multirate digital system.

The z-transfer function relation of the system is written from Fig. 4-25(b).

$$\frac{C(z)}{R(z)} = \frac{\mathscr{Z}[G_{h0}(s)G(s)]}{1+\mathscr{Z}[G_{h02}(s)H(s)]\mathscr{Z}[G_{h0}(s)G(s)]+\mathscr{Z}[e^{Ts/2}G_{h0}(s)G(s)]\mathscr{Z}[e^{-Ts/2}G_{h02}(s)H(s)]} \tag{4-176}$$

where

$$\mathscr{Z}[G_{h0}(s)G(s)] = (1-z^{-1})\mathscr{Z}\left[\frac{1}{s^2}\right] = \frac{T}{z-1} \tag{4-177}$$

$$\begin{aligned}\mathscr{Z}[G_{h02}(s)H(s)] &= \mathscr{Z}\left[\frac{1-e^{-Ts/2}}{s}\frac{1}{s+1}\right]\\ &= \mathscr{Z}\left[\frac{1}{s(s+1)}\right] - \mathscr{Z}\left[\frac{e^{-Ts/2}}{s(s+1)}\right]\end{aligned} \tag{4-178}$$

The last term in the last equation is evaluated by use of Eq. (4-161). Thus,

$$\mathscr{Z}\left[\frac{e^{-Ts/2}}{s(s+1)}\right] = \mathscr{Z}_m\left[\frac{1}{s(s+1)}\right]_{m=1/2} = \frac{1}{z-1} - \frac{e^{-T/2}}{z-e^{-T}} \tag{4-179}$$

After substituting Eq. (4-179) into Eq. (4-178) and simplifying, we have

$$\mathscr{Z}[e^{Ts/2}G_{h0}(s)G(s)] = \mathscr{Z}\left[e^{Ts/2}\frac{1-e^{-Ts}}{s}\frac{1}{s}\right] = (1-z^{-1})\mathscr{Z}\left[\frac{e^{Ts/2}}{s^2}\right] \tag{4-180}$$

and using Eq. (4-121),

$$\mathscr{Z}\left[\frac{e^{Ts/2}}{s^2}\right] = z\mathscr{Z}_m\left[\frac{1}{s^2}\right]_{m=1/2} = \frac{zT/2}{z-1} + \frac{Tz}{(z-1)^2} \tag{4-181}$$

$$\mathscr{Z}[e^{-Ts/2}G_{h1}(s)H(s)] = \mathscr{Z}\left[e^{-Ts/2}\frac{1-e^{-Ts/2}}{s}\frac{1}{s+1}\right]$$

$$= \mathscr{Z}\left[\frac{e^{-Ts/2}}{s(s+1)}\right] - \mathscr{Z}\left[\frac{e^{-Ts}}{s(s+1)}\right]$$

$$= \frac{1-e^{-T/2}}{z-e^{-T}} \tag{4-182}$$

Now, substituting all these components in Eq. (4-176) and simplifying, the transfer function of the system is

$$\frac{C(z)}{R(z)} = \frac{T(z-e^{-T})}{z^2 - \left[1 + e^{-T} - \dfrac{T}{2} + \dfrac{T}{2}e^{-T/2}\right]z + e^{-T} - Te^{-T} + \dfrac{T}{2}e^{-T/2} + \dfrac{T}{2}} \tag{4-183}$$

Example 4-11

Figure 4-26(a) shows the block diagram of a closed-loop control system with multirate sampling. The sampler S_1 has a sampling period of $T/2$ s, and the sampler S_2 samples with a sampling period of $T/3$ s, as illustrated by the sampling schemes in Fig. 4-26(b). A fictitious sampler S_1 is applied at the output, so that $c(t)$ is sampled at the sampling instants of S_1. Figure 4-26(c) shows the block diagram of the system with the samplers decomposed into parallel paths with advances and delays. Notice that the zero-order holds associated with the samplers have different hold periods.

The transfer functions of the two zero-order holds are

$$G_{h02}(s) = \frac{1-e^{-Ts/2}}{s} \tag{4-184}$$

$$G_{h03}(s) = \frac{1-e^{-Ts/3}}{s} \tag{4-185}$$

The outputs of the decomposed samplers are designated as X_0^*, X_1^*, X_2^*, Y_0^*, Y_1^*, and Y_2^*, as shown in Fig. 4-26(c). It is important to regard these as "inputs," and the samplers can all be disregarded. The output $c^*(t)_2$ and its z-transform, $C(z)_2$, describe $c(t)$ at $t = 0, T/2, T, 3T/2, \ldots$.

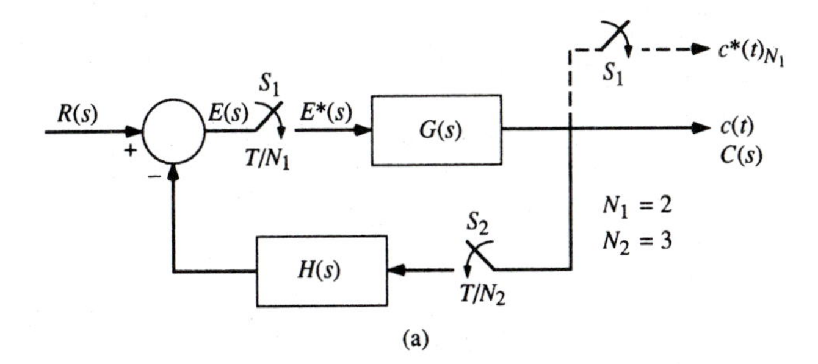

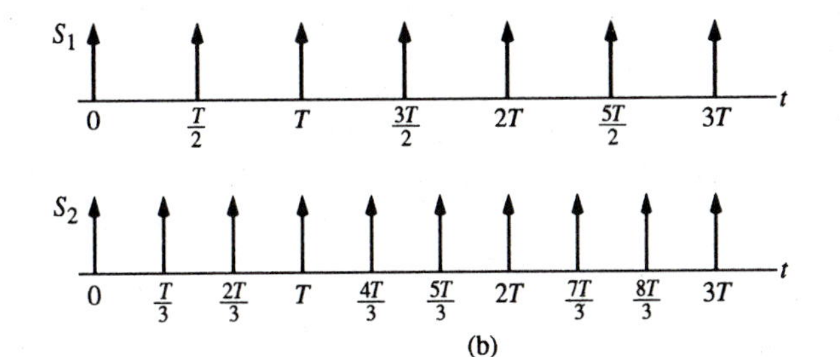

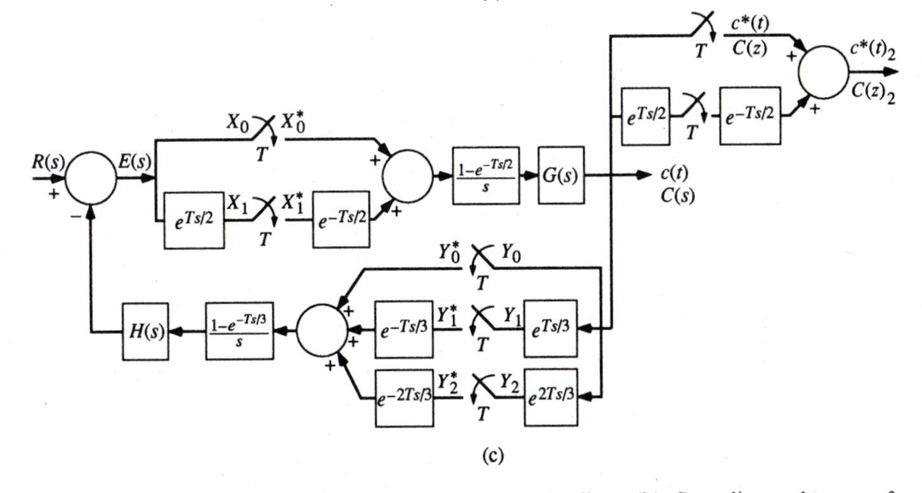

Figure 4-26. (a) Control system with multirate sampling. (b) Sampling schemes of multirate samplers. (c) System with equivalent decomposed samplers.

Writing $C(s)$ as a function of X_0^* and X_1^*, we have

$$C(s) = [G_{h02}(s)G(s)]X_0^* + [e^{-Ts/2}G_{h02}(s)G(s)]X_1^* \tag{4-186}$$

The Laplace transform of the output $c^*(t)_2$ is

$$\begin{aligned} C^*(s)_2 = {} & \{[G_{h02}(s)G(s)]^* + (e^{-Ts/2})^*[e^{Ts/2}G_{h02}(s)G(s)]\}X_0^* \\ & + \{[e^{-Ts/2}G_{h02}(s)G(s)]^*X_0^* + (e^{-Ts/2})^*[G_{h02}(s)G(s)]^*\}X_1^* \end{aligned} \tag{4-187}$$

Next, we write the pulsed transform of X_0, X_1, Y_0, Y_1, and Y_2 in terms of all the inputs, including R,

$$X_0^* = R^*(s) - [G_{h03}(s)H(s)]^*Y_0^* - [e^{-Ts/3}G_{h03}(s)H(s)]^*Y_1^* - [e^{-2Ts/3}G_{h03}(s)H(s)]^*Y_2^* \tag{4-188}$$

$$X_1^* = [e^{Ts/2}R(s)]^* - [G_{h03}(s)H(s)]^*Y_0^* - [e^{-Ts/3}G_{h03}(s)H(s)]^*Y_1^* - [e^{-2Ts/3}G_{h03}(s)H(s)]^*Y_2^* \tag{4-189}$$

$$Y_0^* = [G_{h02}(s)G(s)]^*X_0^* + [e^{-Ts/2}G_{h02}(s)G(s)]^*X_1^* \tag{4-190}$$

$$Y_1^* = [e^{Ts/3}G_{h02}(s)G(s)]^*X_0^* + [e^{-Ts/6}G_{h02}(s)G(s)]^*X_1^* \tag{4-191}$$

$$Y_2^* = [e^{2Ts/3}G_{h02}(s)G(s)]^*X^* + [e^{Ts/6}G_{h02}(s)G(s)]^*X_1^* \tag{4-192}$$

Solving for X_0^* and X_1^* from the last five equations, we get

$$X_0^* = \frac{R^*(s)[1 + F_1^*(s)] - F_1^*(s)[e^{Ts/2}R(s)]}{1 + F_1^*(s) + F_2^*(s)} \tag{4-193}$$

$$X_1^* = \frac{[e^{Ts/2}R(s)]^*[1 + F_2^*(s)] - R^*(s)F_2^*(s)}{1 + F_1^*(s) + F_2^*(s)} \tag{4-194}$$

where

$$F_1^*(s) = [e^{-Ts/2}G_{h02}(s)G(s)]^*[G_{h03}H(s)]^* + [e^{-Ts/6}G_{h02}(s)G(s)]^*[e^{-Ts/3}G_{h03}(s)H(s)]^* + [e^{Ts/6}G_{h02}(s)G(s)]^*[e^{-2Ts/3}G_{h03}(s)H(s)]^* \tag{4-195}$$

$$F_2^*(s) = [G_{h02}(s)G(s)]^*[G_{h03}(s)H(s)]^* + [e^{Ts/3}G_{h02}(s)G(s)]^*[e^{-Ts/3}G_{h03}(s)H(s)]^* + [e^{2Ts/3}G_{h02}(s)G(s)]^*[e^{-2Ts/3}G_{h03}(s)H(s)]^* \tag{4-196}$$

Substituting Eqs. (4-193) and (4-194) into Eq. (4-186), and taking the z-transform, we get

$$C(z) = \frac{\mathscr{Z}(G_{h02}G)[1 + F_1(z)] - \mathscr{Z}(e^{-Ts/2}G_{h02}G)F_2(z)}{1 + F_1(z) + F_2(z)}R(z) + \frac{\mathscr{Z}(e^{-Ts/2}G_{h02}G)[1 + F_2(z)] - \mathscr{Z}(G_{h02}G)F_1(z)}{1 + F_1(z) + F_2(z)}\mathscr{Z}(e^{Ts/2}R) \tag{4-197}$$

where the function of s notation (s) has been omitted.

Making use of Eqs. (4-161) and (4-163), the following z-transforms are identified:

$$\mathscr{Z}(e^{Ts/2}R) = zR(z, 1/2) \tag{4-198}$$

$$\mathscr{Z}(G_{h02}G) = \mathscr{Z}[G(s)/s] - \mathscr{Z}_m[G(s)/s]_{m=1/2} \tag{4-199}$$

$$\begin{aligned}\mathscr{Z}(e^{-Ts/2}G_{h02}G) &= \mathscr{Z}[e^{-Ts/2}G(s)/s] - \mathscr{Z}[e^{-Ts}G(s)/s] \\ &= \mathscr{Z}_m[G(s)/s]_{m=1/2} - z^{-1}\mathscr{Z}[G(s)/s]\end{aligned} \tag{4-200}$$

$$\begin{aligned}\mathscr{Z}(e^{-Ts/6}G_{h02}G) &= \mathscr{Z}[e^{-Ts/6}G(s)/s] - \mathscr{Z}[e^{-2Ts/3}G(s)/s] \\ &= \mathscr{Z}_m[G(s)/s]_{m=5/6} - \mathscr{Z}_m[G(s)/s]_{m=1/3}\end{aligned} \tag{4-201}$$

$$\begin{aligned}\mathscr{Z}(e^{Ts/3}G_{h02}G) &= \mathscr{Z}[e^{Ts/3}G(s)/s] - \mathscr{Z}[e^{-Ts/6}G(s)/s] \\ &= z\mathscr{Z}_m[G(s)/s]_{m=1/3} - \mathscr{Z}_m[G(s)/s]_{m=5/6}\end{aligned} \tag{4-202}$$

$$\begin{aligned}\mathscr{Z}(e^{2Ts/3}G_{h02}G) &= \mathscr{Z}[e^{2Ts/3}G(s)/s] - \mathscr{Z}[e^{Ts/6}G(s)/s] \\ &= z\mathscr{Z}_m[G(s)/s]_{m=2/3} - z\mathscr{Z}_m[G(s)/s]_{m=1/6}\end{aligned} \tag{4-203}$$

$$\begin{aligned}\mathscr{Z}(e^{Ts/2}G_{h02}G) &= \mathscr{Z}[e^{Ts/2}G(s)/s] - \mathscr{Z}[G(s)/s] \\ &= z\mathscr{Z}_m[G(s)/s]_{m=1/2} - \mathscr{Z}[G(s)/s]\end{aligned} \tag{4-204}$$

$$\begin{aligned}\mathscr{Z}(e^{Ts/6}G_{h02}G) &= \mathscr{Z}[e^{Ts/6}G(s)/s] - \mathscr{Z}[e^{-Ts/3}G(s)/s] \\ &= z\mathscr{Z}_m[G(s)/s]_{m=1/6} - \mathscr{Z}_m[G(s)/s]_{m=2/3}\end{aligned} \tag{4-205}$$

$$\begin{aligned}\mathscr{Z}(G_{h03}H) &= \mathscr{Z}[H(s)/s] - \mathscr{Z}[e^{-Ts/3}H(s)/s] \\ &= \mathscr{Z}[H(s)/s] - \mathscr{Z}_m[H(s)/s]_{m=2/3}\end{aligned} \tag{4-206}$$

$$\begin{aligned}\mathscr{Z}(e^{-Ts/3}G_{h03}H) &= \mathscr{Z}[e^{-Ts/3}H(s)/s] - \mathscr{Z}[e^{-Ts}H(s)/s] \\ &= \mathscr{Z}_m[H(s)/s]_{m=2/3} - \mathscr{Z}_m[H(s)/s]_{m=1/3}\end{aligned} \tag{4-207}$$

$$\begin{aligned}\mathscr{Z}(e^{-2Ts/3}G_{h03}H) &= \mathscr{Z}[e^{-2Ts/3}H(s)/s] - \mathscr{Z}[e^{-Ts}H(s)/s] \\ &= \mathscr{Z}_m[H(s)/s]_{m=1/3} - z^{-1}\mathscr{Z}[H(s)/s]\end{aligned} \tag{4-208}$$

The $C(z)$ in Eq. (4-197) gives information on $c(t)$ only at the sampling instants, $t = kT$, $k = 0, 1, 2, 3, \ldots$. To describe $c(t)$ at the sampling instants of S_1, we substitute X_0^* and X_1^* from Eqs. (4-193) and (4-194), respectively, into Eq. (4-187), and taking the z-transform, we get

$$\begin{aligned}C(z)_2 = {} & \frac{\begin{array}{c}[\mathscr{Z}(G_{h02}G) + z^{-1/2}\mathscr{Z}(e^{Ts/2}G_{h02}G)][1 + F_1(z)] - [\mathscr{Z}(e^{-Ts/2}G_{h02}G) \\ + z^{-1}\mathscr{Z}(e^{Ts/2}G_{h02}G)]F_2(z)\end{array}}{1 + F_1(z) + F_2(z)} R(z) \\ & + \frac{\begin{array}{c}[\mathscr{Z}(e^{-Ts/2}G_{h02}G) + z^{-1}\mathscr{Z}(e^{Ts/2}G_{h02}G)][1 + F_2(z)] \\ - [\mathscr{Z}(G_{h02}G) + z^{-1/2}\mathscr{Z}(e^{Ts/2}G_{h02}G)]F_1(z)\end{array}}{1 + F_1(z) + F_2(z)} \mathscr{Z}(e^{Ts/2}R)\end{aligned} \tag{4-209}$$

where, for simplicity, the function of s notation (s) has been deleted.

Since $C(z)_2$ contains $z^{1/2}$, for it to be useful, we should define $z_2 = z^{1/2}$, and thus $z = z_2^2$, so that $C(z)_2$ will contain only integral powers of z_2.

4-7-5 Cyclic-Rate Sampled Systems

There is a class of discrete-data systems in which the samplers operate with a periodically variable rate, or cyclic rate. The block diagram of a cyclic-rate discrete-data control system is shown in Fig. 4-27(a). The sampler S_1 operates with a constant sampling period T, whereas the cyclic-rate sampler S_c operates with a variable period at instants $0, T/N_1, T/N_2, \ldots, T/N_n; T, (1 + 1/N_1)T, (1 + 1/N_2)T, \ldots, (1 + 1/N_n)T; 2T, (2 + 1/N_1)T, (2 + 1/N_2)T, \ldots, (2 + 1/N_n)T; 3T, \ldots,$ where $N_1 > N_2 > \cdots > N_n > 1$. Figure 4-27(b) reveals that S_c can again be replaced by a parallel combination of the decomposed samplers with time advances and delays, as shown in Fig. 4-27(c). Figure 4-27(d) shows the block diagram of the system with the equivalent decomposed samplers. Notice that the zero-order hold that follows S_c must hold the individual samples at varying time durations.

The Laplace transform of the output $C(s)$ is written

$$C(s) = \frac{D_0^*(s)G_0(s) + \sum_{k=1}^{n} D_k^*(s)G_k(s)}{1 + D_0^*(s)G_0^*(s) + \sum_{k=1}^{n} D_k^*(s)G_k^*(s)} \tag{4-210}$$

where

$$D_0^*(s) = \left(\frac{1 - e^{-Ts}}{s} D(s)\right)^* \tag{4-211}$$

$$D_k^*(s) = \left(e^{Ts/N_k} \frac{1 - e^{-Ts}}{s} D(s)\right) \qquad k = 1, 2, \ldots, n \tag{4-212}$$

$$G_0(s) = \frac{1 - e^{-Ts/N_1}}{s} G(s) \tag{4-213}$$

$$G_k(s) = e^{-Ts/N_k} \frac{1 - e^{-Ts[(1/N_{k+1}) - (1/N_k)]}}{s} G(s) \tag{4-214}$$

$$G_n(s) = e^{-Ts/N_n} \frac{1 - e^{-Ts(1 - 1/N_n)}}{s} G(s) \tag{4-215}$$

Starting with $C(s)$ in Eq. (4-210), we can obtain $C(z)$ by taking the z-transform on both sides of the equation. However, $C(z)$ gives information on $c(t)$ only at $t = kT, k = 0, 1, 2, \ldots$. To get additional information on $c(t)$ between the basic sampling instants, we can apply a fictitious sampler S_c at the output, or apply the modified z-transform to $C(s)$ to get $C(z, m)$.

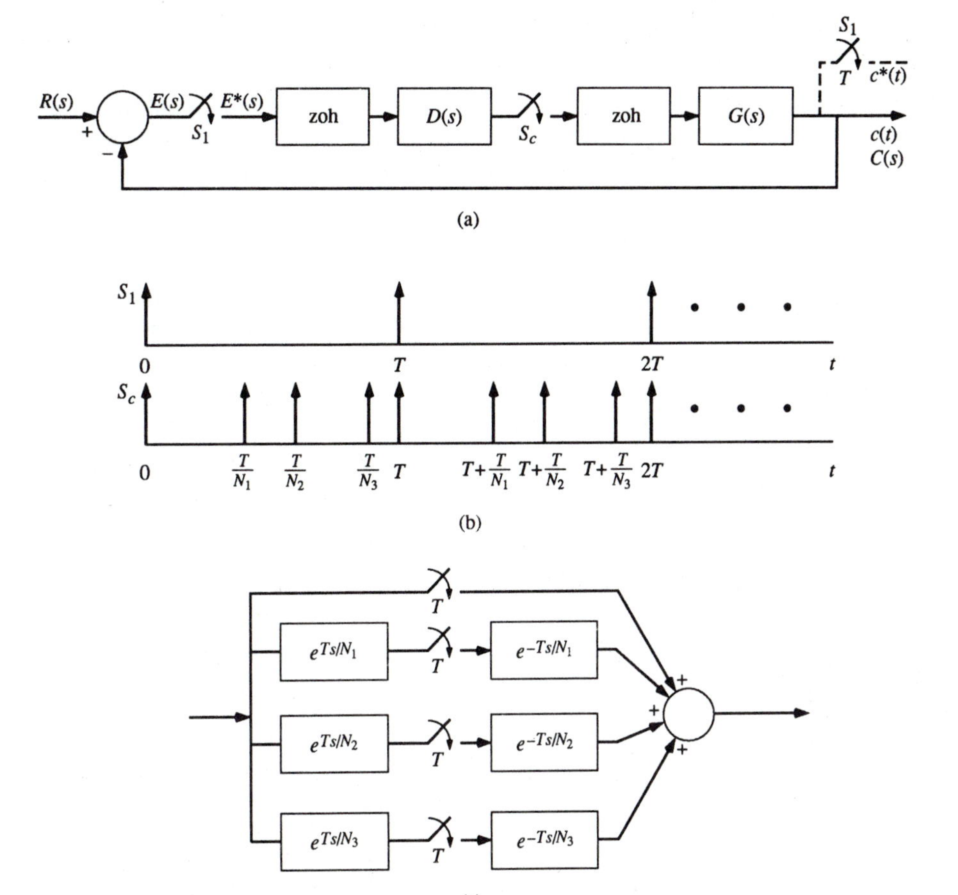

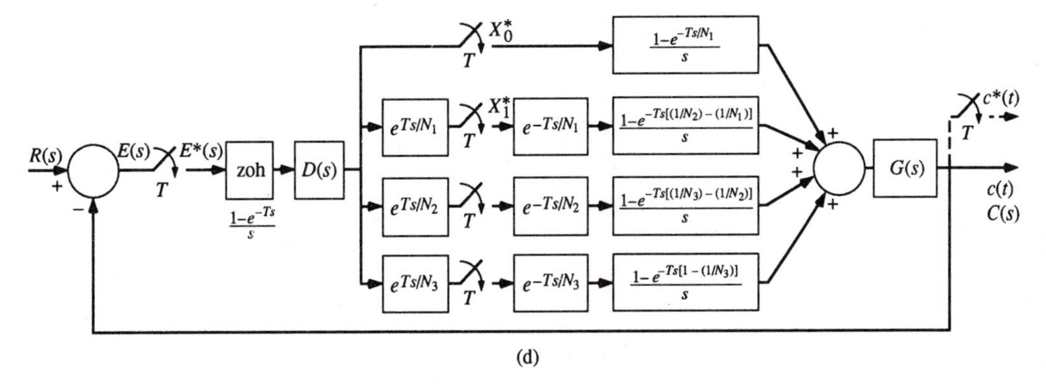

Figure 4-27. (a) Control system with cyclic-rate sampling. (b) Sampling scheme of samplers. (c) Decomposed samplers with time advances and delays. (d) Equivalent system with decomposed samplers.

Example 4-12

Figure 4-28(a) shows the block diagram of a discrete-data control system with cyclic-rate sampling. The sampler operates at instants of $t = 0,\ T/N,\ T,\ (1 + 1/N)T,\ 2T,\ (2 + 1/N)T,\ 3T, \ldots$, where $N = 3$ and $T = 0.1$ s. The sampling scheme is portrayed in Fig. 4-28(b). Decomposing the cyclic-rate sampler into samplers with period T and time-advance and -delay units, we have the equivalent block diagram shown in Fig. 4-28(c). A fictitious sampler S_c is applied at the output, so that we have the flexibility of getting the output data at the sampling instants $t = kT$, $k = 0, 1, 2, \ldots$, or at the cyclic-rate sampling instants.

Let us define the following transfer functions:

$$G_0(s) = \frac{1 - e^{-Ts/N}}{s} G(s) \tag{4-216}$$

$$G_1(s) = \frac{1 - e^{-Ts(1-1/N)}}{s} G(s) \tag{4-217}$$

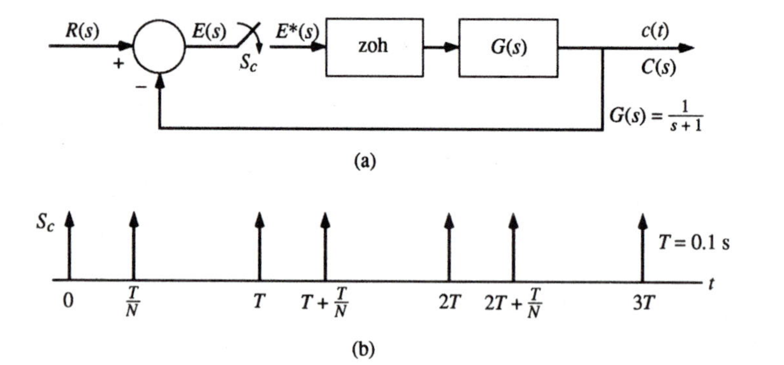

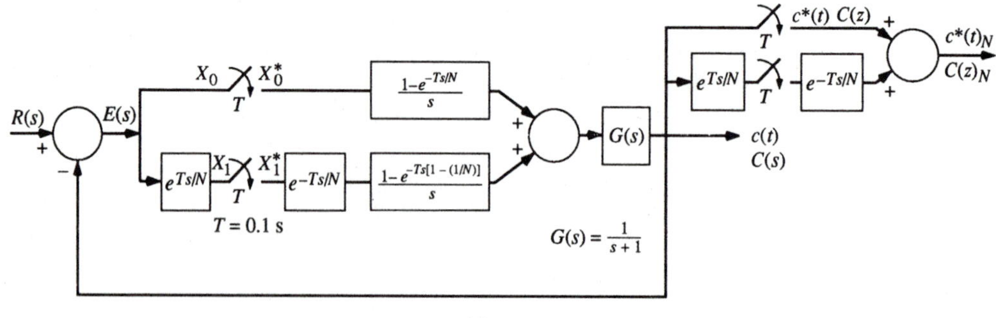

Figure 4-28. (a) Control system with cyclic-rate sampling. (b) Sampling scheme. (c) Equivalent system with decomposed samplers.

Writing the pulsed transform of the signals at the inputs of the samplers in Fig. 4-28(c) as functions of the inputs, we have

$$X_0^*(s) = R^*(s) - G_0^*(s)X_0^*(s) - [e^{-Ts/N}G_1(s)]^*X_1^*(s) \tag{4-218}$$

$$X_1^*(s) = [e^{Ts/N}R(s)]^* - [e^{Ts/N}G_0(s)]^*X_0^*(s) - G_1^*(s)X_1^*(s) \tag{4-219}$$

The pulsed transforms of the outputs $c(t)$ and $c^*(t)_N$ are written in terms of $X_0^*(s)$ and $X_1^*(s)$:

$$C^*(s) = G_0^*(s)X_0^*(s) + [e^{-Ts/N}G_1(s)]^*X_1^*(s) \tag{4-220}$$

$$\begin{aligned} C^*(s)_N &= \{G_0^*(s) + (e^{-Ts/N})^*[e^{Ts/N}G_0(s)]^*\}X_0^*(s) \\ &\quad + \{[e^{-Ts/N}G_1(s)]^* + (e^{-Ts/N})^*G_1^*(s)\}X_1^*(s) \\ &= C^*(s) + (e^{-Ts/N})^*\{[e^{Ts/N}G_0(s)]^*X_0^*(s) + G_1^*(s)X_1^*(s)\} \end{aligned} \tag{4-221}$$

The last four equations can be solved using the pulsed signal flow graph method. Figure 4-29 shows the pulsed signal flow graph of the system in Fig. 4-28(c). Applying the gain formula to Fig. 4-29 and taking the z-transform at the same time, we have

$$C(z) = \frac{G_0(z)[1 + G_1(z)] - \mathscr{Z}[e^{Ts/N}G_0(s)]\mathscr{Z}[e^{-Ts/N}G_1(s)]R(z)}{\Delta(z)} + \frac{\mathscr{Z}[e^{-Ts/N}G_1(s)]}{\Delta(z)}\mathscr{Z}[e^{Ts/N}R(s)] \tag{4-222}$$

$$C(z)_2 = \frac{G_0(z)[1 + G_1(z)] + \mathscr{Z}[e^{Ts/N}G_0(s)]\{z^{-1/N} - \mathscr{Z}[e^{-Ts/N}G_1(s)]}{\Delta(z)}R(z) + \frac{\mathscr{Z}[e^{-Ts/N}G_1(s)](1 + z^{-1/N}\{1 + G_0(z) + \mathscr{Z}[e^{Ts/N}G_0(s)]\})}{\Delta(z)}\mathscr{Z}[e^{Ts/N}R(s)] \tag{4-223}$$

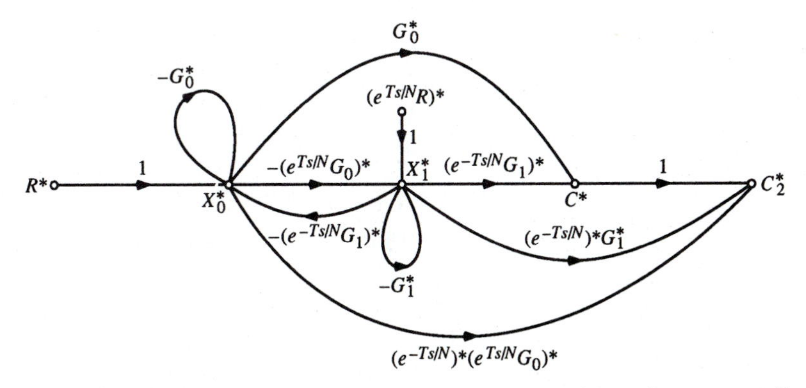

Figure 4-29. Sampled signal flow graph of the system with cyclic-rate sampling in Fig. 4-28.

where

$$\Delta(z) = 1 + G_0(z) + G_1(z) + G_0(z)G_1(z) - \mathscr{Z}[e^{Ts/N}G_0(s)]\mathscr{Z}[e^{-Ts/N}G_1(s)] \tag{4-224}$$

The functions in $C(z)$ and $C(z)_2$ are determined by using Eqs. (4-161) and (4-163).

$$G_0(z) = \mathscr{Z}\left(\frac{1 - e^{-Ts/N}}{s(s+1)}\right) = \frac{e^{-T(1-1/N)} - e^{-T}}{z - e^{-T}} = \frac{0.0307}{z - 0.9048} \tag{4-225}$$

$$G_1(z) = \mathscr{Z}\left(\frac{1 - e^{-Ts(1-1/N)}}{s(s+1)}\right) = \frac{e^{-T/N} - e^{-T}}{z - e^{-T}} = \frac{0.0624}{z - 0.9048} \tag{4-226}$$

$$\mathscr{Z}[e^{Ts/N}G_0(s)] = \frac{(1 - e^{-T/N})z}{z - e^{-T}} = \frac{0.0328z}{z - 0.9048} \tag{4-227}$$

$$\mathscr{Z}[e^{-Ts/N}G_1(s)] = \frac{1 - e^{-T(1-1/N)}}{z - e^{-T}} = \frac{0.0645}{z - 0.9048} \tag{4-228}$$

$$R(z) = \mathscr{Z}[e^{Ts/N}R(s)] = \mathscr{Z}[e^{Ts/N}/s] = \frac{z}{z-1} \tag{4-229}$$

Substituting the last five equations in Eqs. (4-222) and (4-223) and simplifying, we obtain the output transforms

$$C(z) = \frac{0.0931z^2 - 0.0842z}{z^3 - 2.7186z^2 + 2.455z - 0.7364} \tag{4-230}$$

$$C(z)_2 = \frac{0.0931z - 0.0842 + 0.0328z^{-1/3}z(z - 0.9048)}{z^2 - 1.7186z + 0.7364}\,\frac{z}{z-1} \tag{4-231}$$

Letting $z_3 = z^{1/3}$, and thus $z_3^3 = z$, Eq. (4-231) becomes

$$C(z)_2 = \frac{0.0328z_3^8 + 0.0931z_3^6 + 0.0327z_3^5 - 0.0842z_3^3 - 0.0564z_3^2}{z_3^9 - 2.7186z_3^6 + 2.455z_3^3 - 0.7364} \tag{4-232}$$

The unit-step responses $c^*(t)$ and $c^*(t)_2$ are obtained by performing long division on $C(z)$ in Eq. (4-230) and $C(z)_2$ in Eq. (4-232), respectively. Table 4-3 shows the values of the responses at the sampling instants. Notice that the output of the fictitious sampler S_c at the system output also contains the samples at the basic sampling period kT, $k = 0, 1, 2, \ldots$.

Table 4-3 $c^*(t)_2$ of the Multirate-Sampled System in Example 4-12

Sample Periods	Values	Sample Periods	Values
0	0.000000	30	0.436737
1	0.032800	31	0.450663
2	0.000000	32	0.000000
3	0.093100	33	0.448615
4	0.121870	34	0.462030
5	0.000000	35	0.000000
6	0.168902	36	0.458277
7	0.194392	37	0.471278
8	0.000000	38	0.000000
9	0.230616	39	0.466135
10	0.253437	40	0.478799
11	0.000000	41	0.000000
12	0.280857	42	0.472526
13	0.301507	43	0.484916
14	0.000000	44	0.000000
15	0.321755	45	0.477721
16	0.340638	46	0.489889
17	0.000000	47	0.000000
18	0.355046	48	0.481943
19	0.372492	49	0.493932
20	0.000000	50	0.000000
21	0.382141	51	0.485375
22	0.398418	52	0.497218
23	0.000000	53	0.000000
24	0.404192	54	0.488163
25	0.419519	55	0.499888
26	0.000000	56	0.000000
27	0.422137	57	0.490428
28	0.436690	58	0.502057
29	0.000000	59	0.000000

PROBLEMS

4-1 The weighting sequence of a linear discrete-data system is

$$g(kT) = \begin{cases} 5e^{-(k-1)T} & \text{for } k \geq 1 \\ 0 & \text{for } k = 0 \end{cases}$$

Let the input to the system be $r(kT) = kT$ for $k \geq 0$. The sampling period is $T = 0.5$ s.

a. Find the z-transform of the system output.

b. Find the output sequence $c(kT)$ in closed form.

4-2 The weighting sequence of a linear discrete-data system is

$$g(k) = \begin{cases} 0.15(0.6)^k - 0.15(0.4)^k & k \geq 0 \\ 0 & k < 0 \end{cases}$$

a. Find the transfer function $G(z)$ of the system.

b. Let the input be $r(k) = u_s(k)$, the unit-step sequence. Find the output $c(k)$ in closed form. Find the final value of $c(k)$ as $k \to \infty$.

4-3 For the system shown in Fig. P4-3, find the output at the sampling instants $c(kT)$. The input is a unit impulse, and the sampling period is 0.1 s. Find the final value of $c(kT)$ as $k \to \infty$.

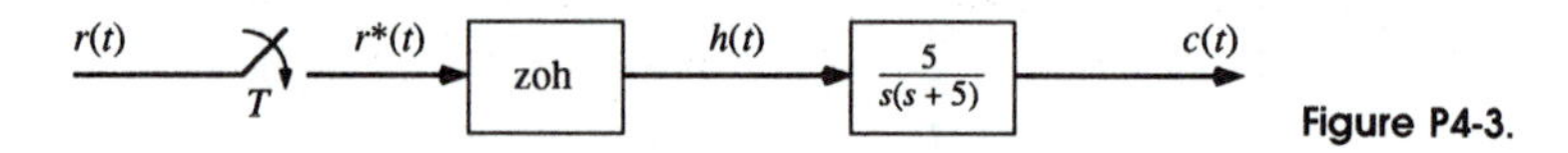

Figure P4-3.

4-4 Find the output at the sampling instants $c(kT)$ of the system shown in Fig. P4-4.

r(t) — T — r*(t) — First-Order Hold — h(t) — $\frac{1}{s+1}$ — c(t)

Figure P4-4.

a. The input is a unit-step function, and the sampling period is 0.1 s. Find the final value of $c(kT)$ as $k \to \infty$.

b. The input is a unit-ramp function, and the sampling period is 0.1 s.

4-5 The digital computer block diagrams of digital controllers are shown in Fig. P4-5. Find the input-output transfer functions $C(z)/R(z)$ of the systems.

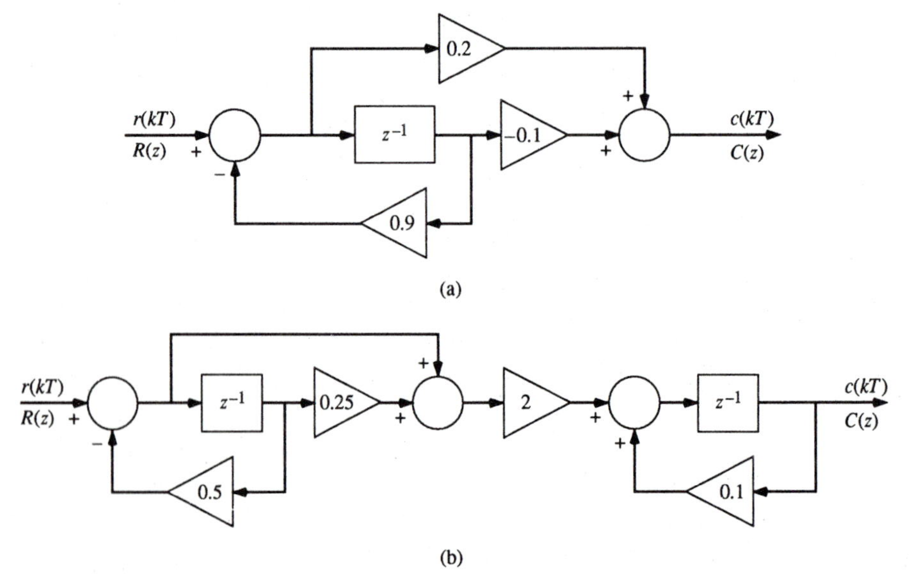

Figure P4-5(a), (b).

4-6 The block diagram of a flight control system utilizing a dc motor is shown in Fig. P4-6. Derive the open-loop transfer function $G(z) = \Theta_o(z)/\Theta_e(z)$ and the closed-loop transfer function $\Theta_o(z)/\Theta_i(z)$ with the following sampling periods.

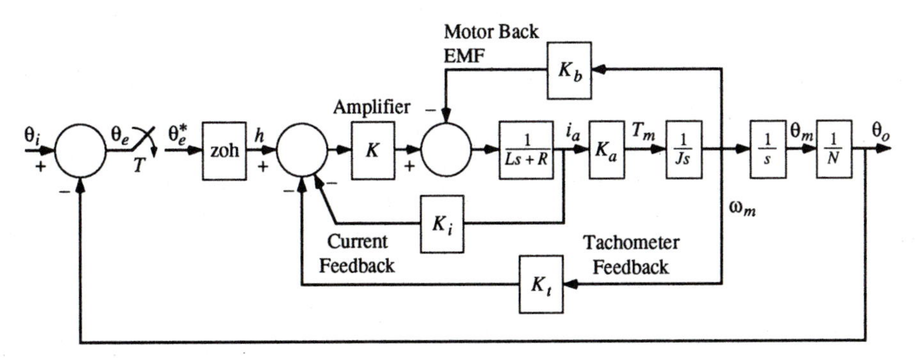

Figure P4-6.

a. $T = 0.01$ s.
b. $T = 0.1$ s.

The system parameters are given as:

Motor resistance	$R = 5\ \Omega$
Motor inductance	$L \cong 0$ H
Motor and load inertia	$J = 0.001$ oz-in-s^2
Gear ratio	$N = 100$
Motor torque constant	$K_a = 10$ oz-in/A
Motor back-EMF constant	$K_b = 0.07$ V/rad/s
Tachometer feedback gain	$K_t = 0.0005$ V/rad/s
Current feedback gain	$K_i = 1$ V/A
Amplifier gain	$K = 20$

4-7 Repeat Problem 4-6 with $L = 0.05$ H. All other system parameters are as given in Problem 4-6.

4-8 A variation of the flight control system described in Problem 4-6 is shown in Fig. P4-8. The system parameters are identical to those given in Problem

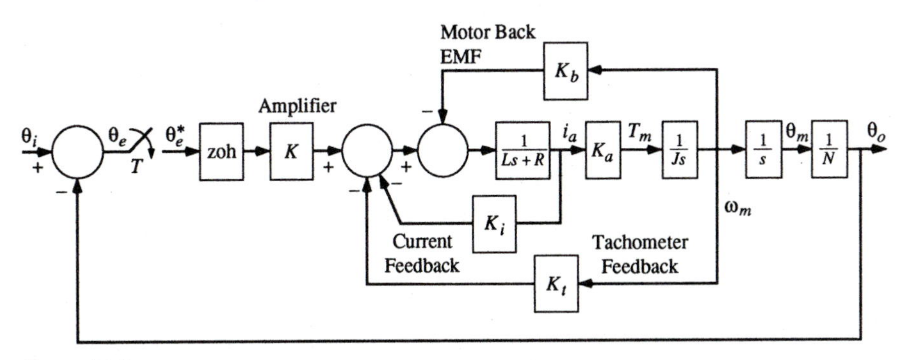

Figure P4-8.

4-6. Derive the open-loop transfer function $G(z) = \Theta_o(z)/\Theta_e(z)$ and the closed-loop transfer function $\Theta_o(z)/\Theta_i(z)$ with the following sampling periods.

a. $T = 0.01$ s.

b. $T = 0.1$ s.

4-9 Repeat Problem 4-8 with $L = 0.05$ H. All other system parameters are as given in Problem 4-6.

4-10 The block diagram of a digital control system of the Large Space Telescope is shown in Fig. P4-10. The system incorporates proportional plus integral control and rate feedback.

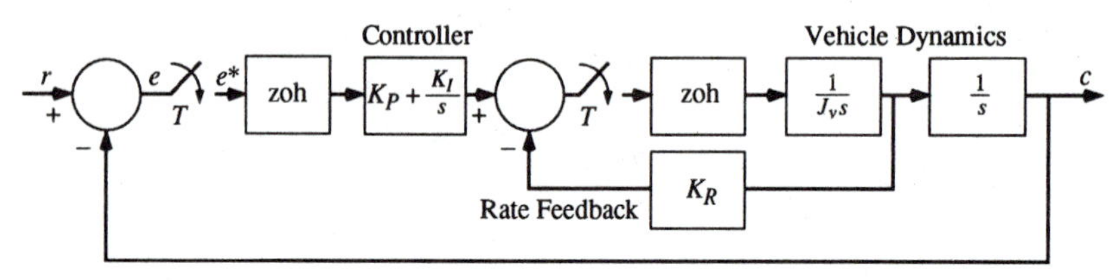

Figure P4-10.

a. Find the open-loop transfer function $G(z) = C(z)/E(z)$ and the closed-loop transfer function $C(z)/R(z)$.

b. A continuous-data counterpart of the digital control system is obtained by deleting the sample-and-hold units. Find the open-loop and closed-loop transfer functions of the continuous-data system by taking the limit of $G(z)$ and $C(z)/R(z)$ as T approaches zero.

4-11 The block diagram of the digital control of the idle speed on an automobile engine is shown in Fig. P4-11. The function of each block of the diagram is indicated on the figure. The system variables are described as follows:

α = throttle position (in)
v = input voltage (V)
ω = engine speed (rad/s)
T_e = engine torque (ft-lb)
T_d = disturbance torque (ft-lb)
T = sampling period = 0.05 s

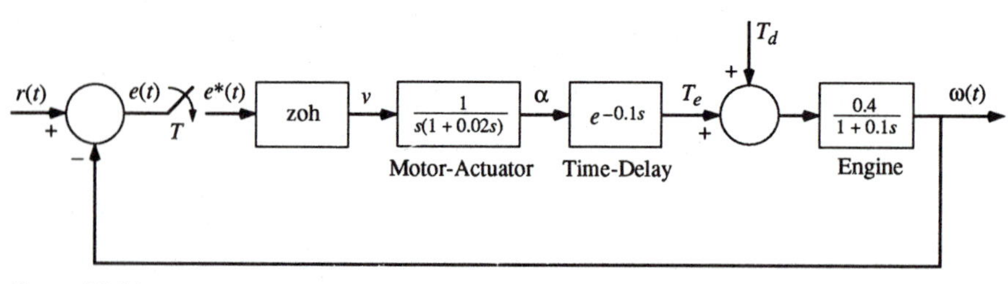

Figure P4-11.

a. Find the open-loop transfer function $G(z) = \Omega(z)/E(z)$ and the closed-loop transfer function $\Omega(z)/R(z)$.
b. Consider that the disturbance torque is a step function with magnitude T_d. Find the z-transform of the output $\Omega(z)$ due to T_d acting alone ($r = 0$). Find the steady-state value of $\omega(kT)$ as $k \to \infty$ in terms of T_d.

4-12 The open-loop transfer function of the digital control system shown in Fig. P4-12 is

$$G(z) = \frac{27.8(s + 1)}{s(s^2 + 6s + 48.6)}$$

The sampling period is 1 s.

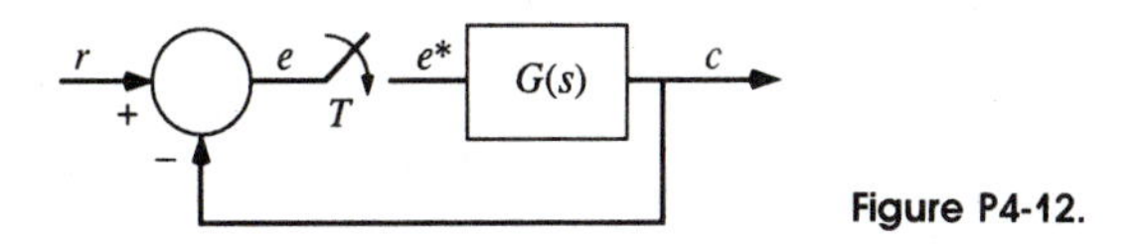

Figure P4-12.

a. Determine the modified z-transform of the output $C(z, m)$ for a unit-step input.
b. Find $C(z)$ by setting $m = 1$ in $C(z, m)$.
c. Find $c(kT, m)$ by means of the inversion formula. Sketch $c(kT, m)$ with several values of m ($0 < m < 1$) so that an accurate estimate of $c(t)$ can be obtained.
d. Find $c(kT)$ from $C(z)$ and compare with $c(kT, m)$ for $m \neq 1$.

4-13 Figure P4-13 shows a discrete-data system with a pulsewidth sampler. The input to the pulsewidth sampler is designated as $r(t)$. The sampled output $r_d(t)$ is written

$$r_d(t) = E \sum_{n=0}^{\infty} SGN[r(nT)][u_s(t - nT) - u_s(t - nT - p_n)]$$

where p_n, the variable pulse duration, is given as

$$p_n = \begin{cases} a|r(nT)| & \text{for } a|r(nT)| \leq T \\ T & \text{for } |r(nT)| > T \end{cases}$$

a and E are constants, and T is the sampling period. R. E. Andeen suggested in his paper that the pulsewidth sampler may be replaced by an equivalent arrangement with a pulse-amplitude sampler and a special hold device as shown

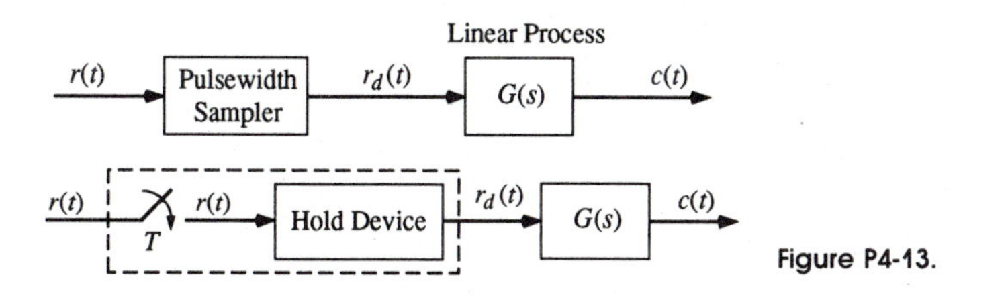

Figure P4-13.

in Fig. P4-13, if the principle of equivalent area is applied. In other words, it can be shown that two signals $r_d(t)$ and $r'_d(t)$ are equivalent for suitably small T if

$$\int_{(n-1)T}^{nT} r_d(t)\,dt = \int_{(n-1)T}^{nT} r'_d(t)\,dt \qquad \text{(P4-13-1)}$$

where $r'_d(t)$ is now the output of the equivalent sample-and-hold device. The pulse width of $r'_d(t)$ is chosen to be $aR_{\max}$, where $R_{\max}$ is the largest value of $r(t)$ that is measurable.

a. Is the pulsewidth sampler a linear or nonlinear device? Give reasons for your answer.
b. Find the transfer function $H(s)$ of the hold device so that Eq. (P4-13-1) is satisfied.
c. Find $C(z)$, if $G(s) = b/(s + b)$ and $r(t) = u_s(t)$.

4-14 For the digital control systems shown in Fig. P4-14, determine $C(s)$ and $C(z)$ using a sampled signal flow graph. Do not attempt to solve the problems algebraically.

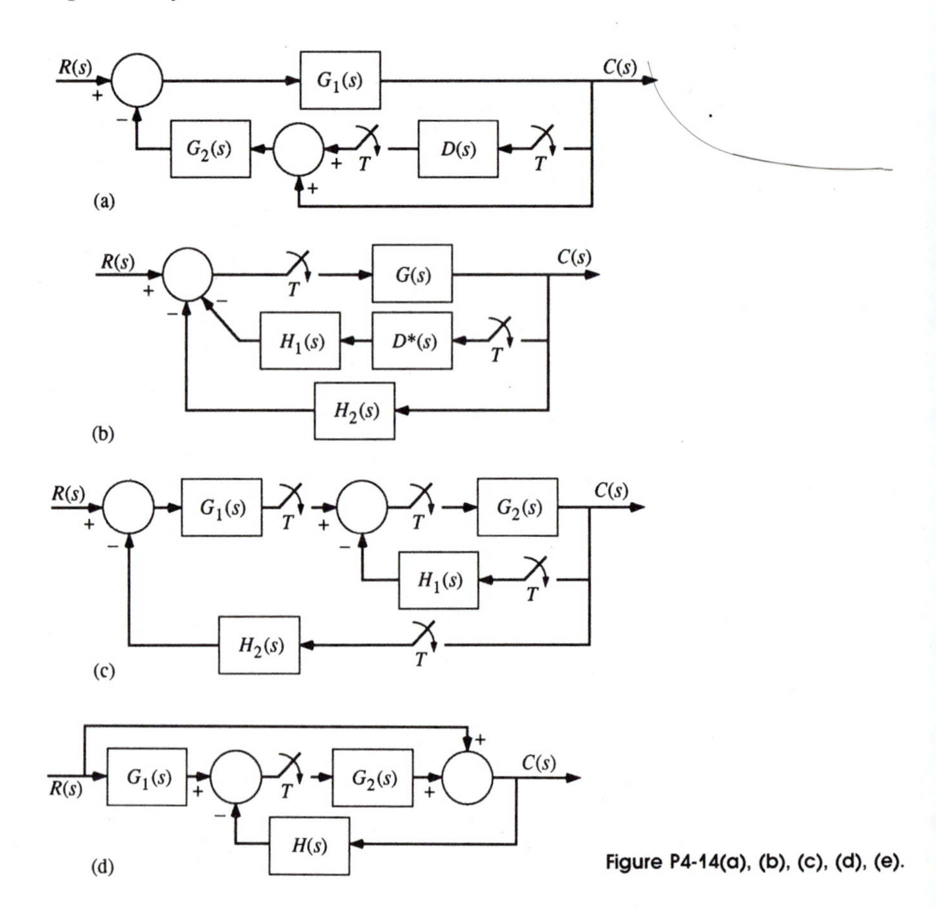

Figure P4-14(a), (b), (c), (d), (e).

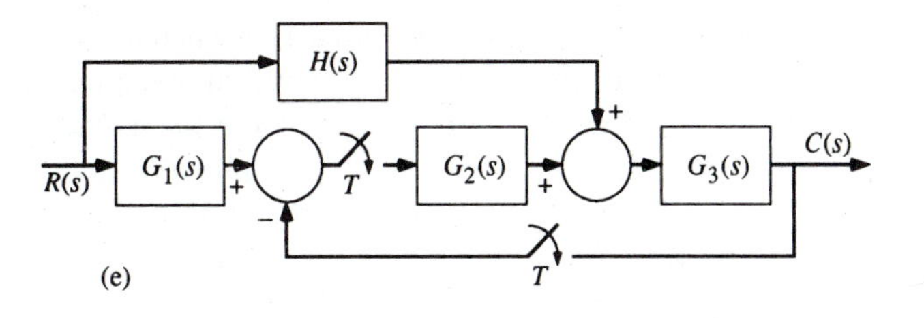

4-15 The block diagram of a multirate discrete-data control system is shown in Fig. P4-15. Find the closed-loop transfer function $C(z)/R(z)$ of the system. The sampling period is 1 s; N is an unspecified integer ≥ 1. The transfer functions are

$$D(s) = \frac{1}{s+1} \qquad G(s) = \frac{K}{s}$$

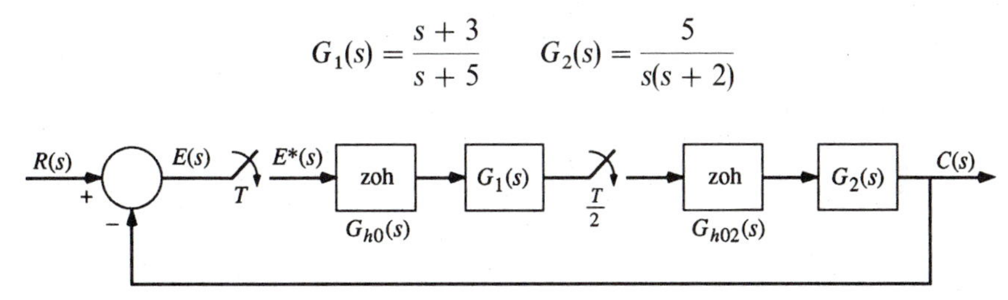

Figure P4-15.

4-16 The block diagram of a multirate discrete-data control system is shown in Fig. P4-16. Find the open-loop transfer function $G(z) = C(z)/E(z)$ and the closed-loop transfer function $C(z)/R(z)$. The sampling period is $T = 0.1$ s, and the transfer functions are

$$G_1(s) = \frac{s+3}{s+5} \qquad G_2(s) = \frac{5}{s(s+2)}$$

Figure P4-16.

4-17 The block diagram of a multirate discrete-data control system is shown in Fig. P4-17. Find the z-transform of the output $C(z)$ when the input is a unit-step function, that is, $r(t) = u_s(t)$. The sampling period T and the transfer functions $G_1(s)$ and $G_2(s)$ are as given in Problem 4-16.

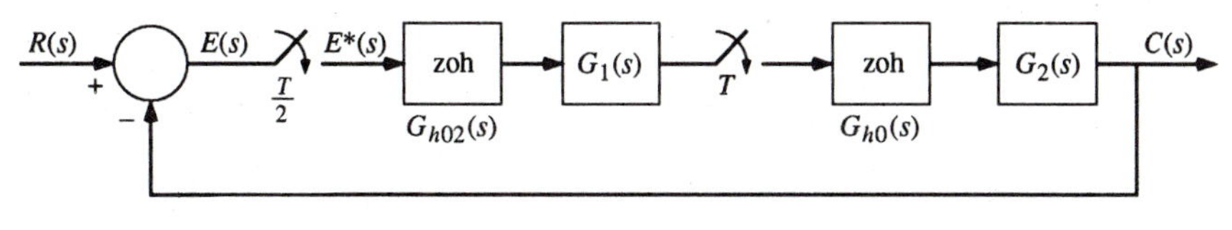

Figure P4-17.

4-18 A bilinear transformation is sometimes applied to the analysis of discrete-data control systems to transform from the z-plane to a w-plane so that the unit circle $|z| = 1$ is mapped onto the imaginary axis in the w-plane. Equations (4-147) and (4-148) give the z-transform of the slow-sampler output $C(z)$ in terms of the output of the fast sampler, $C(z_N)$. Let the bilinear transformation be $z = (w + 1)/(w - 1)$, such that

$$z_N = z^{1/N} = \frac{w_N + 1}{w_N - 1}$$

$$C(w_N) = C(z_N)|_{z_N = (w_N + 1)/(w_N - 1)}$$

Show that $C(w)$ and $C(w_N)$ are related through

$$C(w) = -\sum \text{residues of} \frac{2C(w_N)}{1 - \left(\dfrac{w_N + 1}{w_N - 1}\right)^N \dfrac{w - 1}{w + 1}} \frac{1}{w_N^2 - 1}$$

at the poles of $C(w_N)/(w_N + 1)$. Given $C(z_N) = z_N/(z_N - e^{-T/N})$, find $C(w)$ using the transformation method described above.

4-19 The block diagram of a multirate discrete-data control system is shown in Fig. P4-19. Write the z-transform of the output $C(z)$ in terms of the z-transforms of the system transfer functions. Use the sampler-decomposition method.

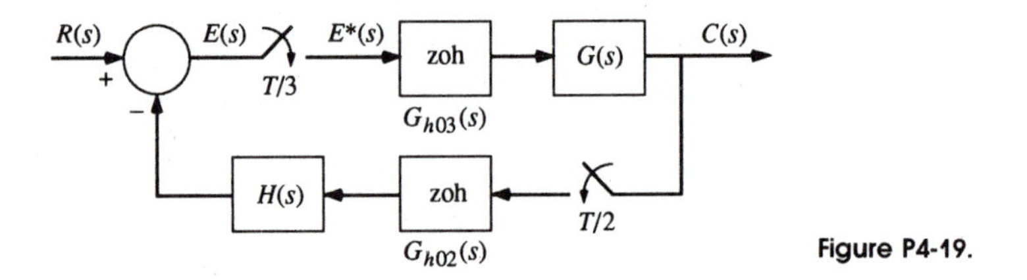

Figure P4-19.

4-20 The block diagram of a digital control system with a nonuniform sampling rate is shown in Fig. P4-20. The sampler operates at time instants of

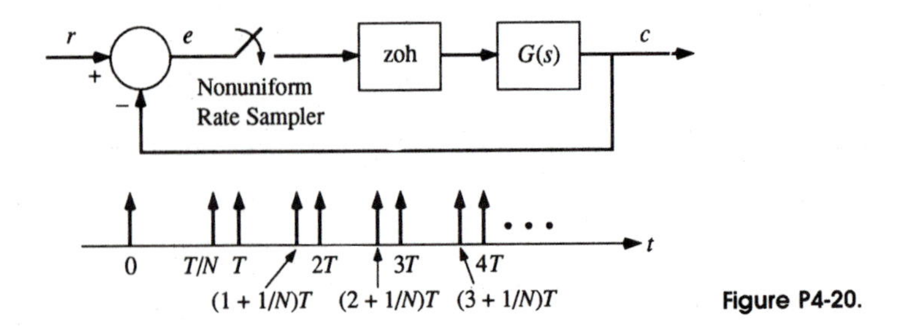

Figure P4-20.

$t = 0, T/N, T, (1 + 1/N)T, 2T, (2 + 1/N)T, 3T, \ldots$, where N is a positive integer greater than one. Show that the z-transform of the output can be written as

$$C(z) = \frac{1}{\Delta(z)}\{[G_0(z)[1 + G_1(z)] - \mathscr{Z}[e^{Ts/N}G_0(s)]\mathscr{Z}[e^{-Ts/N}G_1(s)]]R(z)$$
$$+ \mathscr{Z}[e^{Ts/N}R(s)]\mathscr{Z}[1 + G_0(z)]\mathscr{Z}[e^{-Ts/N}G_1(s)] - \mathscr{Z}[e^{Ts/N}G_1(s)]G_0(z)\}$$

where

$$\Delta(z) = 1 + G_0(z) + G_1(z) - \mathscr{Z}[e^{Ts/N}G_0(s)]\mathscr{Z}[e^{-Ts/N}G_1(s)] + G_0(z)G_1(z)$$

$$G_0(z) = \mathscr{Z}\left(\frac{1 - e^{-Ts/N}}{s}\right)G(s)$$

$$G_1(z) = \mathscr{Z}\left(\frac{1 - e^{-Ts(1-1/N)}}{s}\right)G(s)$$

Find $C(z)$ for $G(s) = K/(s + 1)$, $r(t) = u_s(t)$, and $T = 1$ s. The values of K and N are not specified.

4-21 A special type of multirate sampling scheme assumes that the input and output of a system are pulse trains with pulses occurring at $t = kT/N$, $k = 0, 1, 2, \ldots$, except that the input pulses change values only at $t = kT$, $k = 0, 1, 2, \ldots$. Figure P4-21 shows the block diagram of the system with S_1 representing the special multirate sampler at the input. The sampler at the output is a regular multirate sampler with sampling period T/N. Show that the sampler S_1 can be decomposed into a combination of sampler with period T and N parallel delays. Draw a block diagram for the equivalent system. Derive the z-transform of the output $c^*(t)_N$, $C(z_N)$, in terms of the transfer function of the system and the input transform $R(z)$.

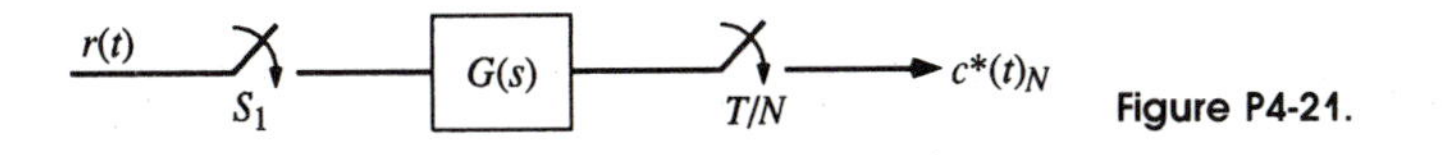

Figure P4-21.

4-22 Figure P4-22 shows the block diagram of an all-digital data system, with the input and output data rates as described by the sampling operations of Problem 4-21. Draw a block diagram with samplers and delays as an equivalent analytical model for the system. Derive the output transform, $C(z_N)$, in terms of the system transform $G(z_N)$ and the input transform $R(z)$.

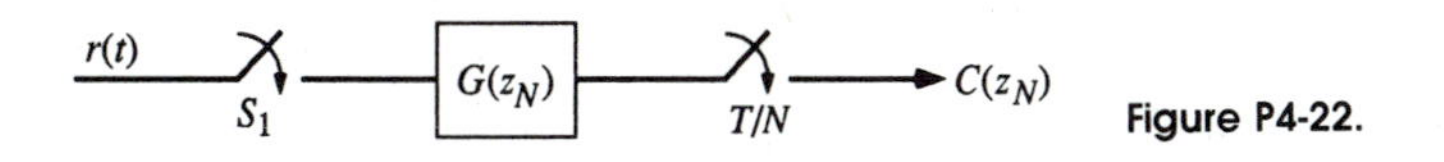

Figure P4-22.

4-23 The block diagram of a multirate digital control system is shown in Fig. P4-23. Derive the transform relations $E(z)/R(z)$ and $C(z_N)/R(z)$ and show that the denominators of these two functions are different. Comment on this phenomenon. Add fictitious samplers in the system whenever necessary.

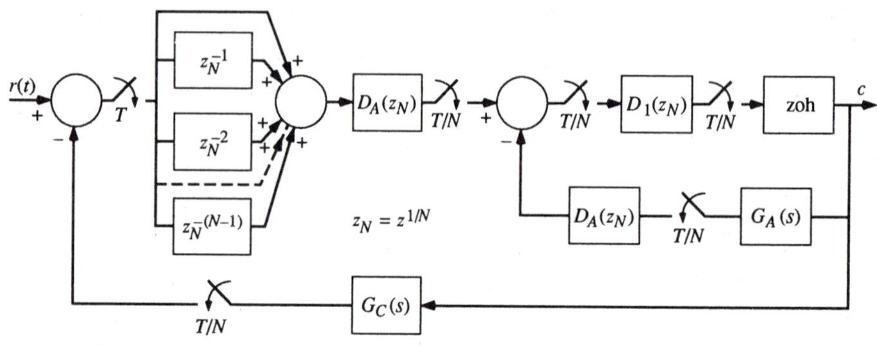

Figure P4-23.

4-24 The block diagram of a digital control system with nonuniform-rate sampling is shown in Fig. P4-24. The sampler S_v samples at time instants of $t = 0, T/N, T, (1 + 1/N)T, 2T, (2 + 1/N)T, 3T, \ldots$, where N is a positive number greater than one. For $T = 1$ s,

$$D(s) = \frac{1}{s+1} \qquad G(s) = \frac{K}{s}$$

find the closed-loop transfer function $C(z)/R(z)$ with K and N as parameters.

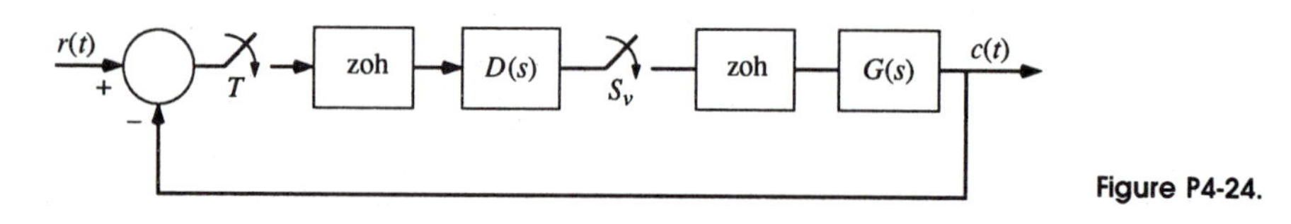

Figure P4-24.

References

1. Mason, S. J., "Feedback Theory—Some Properties of Flow Graphs," *Proc. IRE*, vol. 41, pp. 1144–1156, September 1953.
2. Mason, S. J., "Feedback Theory—Further Properties of Signal Flow Graphs," *Proc. IRE*, vol. 44, pp. 960–966, July 1956.
3. Sedlar, M., and G. A. Bekey, "Signal Flow Graphs of Sampled-Data Systems: A New Formulation," *IEEE Trans. Automatic Control*, vol. AC-12, no. 2, pp. 154–161, April 1967.
4. Ash, R., W. E. Kim, and G. M. Kranc, "A General Flow Graph Technique for the Solution of Multiloop Sampled Systems," *Trans. ASME Journal Basic Engineering*, pp. 360–370, June 1960.

5. Kuo, B. C., "Composite Flow Graph Technique for the Solution of Multiloop, Multisampler, Sampled Systems," *IRE Trans. Automatic Control*, vol. AC-6, pp. 343–344, 1961.
6. Salzer, J. M., "Signal Flow Reduction in Sampled-Data Systems," *IRE WESCON Convention Record*, part 4, pp. 166–170, 1957.
7. Kamen, E., *Introduction to Signals and Systems*, MacMillan Publishing Co., New York, 1987.
8. Gabel, R. A., and R. A. Roberts, *Signals and Linear Systems*, 3rd ed., John Wiley & Sons, New York, 1987.

Pulse Transfer Function

9. Franklin, G. F., J. D. Powell, and M. L. Workman, *Digital Control of Dynamic Systems*, 2nd ed., Addison-Wesley, Reading, Mass., 1990.
10. Phillips, C. L., and H. T. Nagle, Jr., *Digital Control System Analysis and Design*, Prentice-Hall, Englewood Cliffs, N.J., 1984.
11. Ogata, K., *Discrete-Time Control Systems*, Prentice-Hall, Englewood Cliffs, N.J., 1987.
12. Barker, R. H., "The Pulse Transfer Function and Its Application to Servo Systems," *Proc. IEE*, London, vol. 99, part 4, pp. 302–317, December 1952.
13. Lendaris, G. G., and E. I. Jury, "Input-Output Relationships for Multiple Sampled-Loop Systems," *Trans. AIEE*, vol. 79, part 2, pp. 375–385, January 1960.

Multirate-Sampled Systems

14. DuPlessis, R. M., "Two Digital Computer Programs with Multirate Sampled-Data System Analysis," *IRE Trans. Automatic Control*, vol. AC-6, pp. 85–86, February 1961.
15. Jury, E. I., "A Note on Multirate Sampled-Data Systems," *IEEE Trans. Automatic Control*, vol. AC-12, pp. 319–320, June 1967.
16. Bekey, G. A., "A Survey of Techniques for the Analysis of Sampled-Data Systems with a Variable Sampling Rate," Technical Documentary Report No. ASD-TDR-62-35, Flight Control Laboratory, Wright-Patterson Air Force Base, Ohio, May 1962.
17. Whitbeck, R. F., and D. G. J. Didaleusky, "Multi-Rate Digital Control Systems with Simulation Applications," vol. I: Technical Report, AFWAL-TR-80-3101, Wright-Patterson Air Force Base, Ohio, September 1980.

The State-Variable Technique

KEYWORDS AND TOPICS

state variables • state equations • state transition equations • state transition matrix • digital simulation • characteristic equation • eigenvalues • eigenvectors • similarity transformation • Jordan canonical form • phase-variable canonical form • state diagram • decomposition • multirate sampling

5-1 INTRODUCTION

In the preceding chapter we introduced the conventional methods of analyzing discrete-data control systems using z-transforms, transfer functions, block diagrams, or signal flow graphs. As an alternative, we can use the so-called *modern* approach which utilizes the *state-variable* formulation. Most practicing engineers feel more comfortable with the conventional methods since they are simpler to comprehend and are supported by a wealth of graphical techniques. The state-variable technique, on the other hand, has the following advantages, at least in discrete-data control systems studies, over the conventional methods.

1. The state-variable formulation is natural and convenient for computer solutions.
2. The state-variable approach allows a unified representation of digital systems with various types of sampling schemes.
3. The state-variable method allows a unified representation of single-variable and multivariable systems.
4. The state-variable method can be applied to nonlinear and time-varying systems.

In the state-variable formulation, a continuous-data system is represented by a set of first-order differential equations, called *state equations*. For a discrete-data system, when all the variables are defined in discrete time kT, the state equations are in the form of first-order *difference equations*. As mentioned earlier, a discrete-data control system often contains continuous-data as well as digital-data components, and the state equations of the system will generally consist of both first-order difference as well as first-order differential equations. For this reason, we shall begin by reviewing the state equations and their solutions of linear continuous-data systems.

5-2 STATE EQUATIONS AND STATE TRANSITION EQUATIONS OF CONTINUOUS-DATA SYSTEMS

Consider that an nth-order continuous-data system with p inputs and q outputs, as shown in Fig. 5-1, is characterized by the following set of n first-order differential equations (state equations):

$$\frac{d\mathbf{x}_i(t)}{dt} = f_i[x_1(t), x_2(t), \ldots, x_n(t), u_1(t), u_2(t), \ldots, u_p(t), t] \tag{5-1}$$

$$i = 1, 2, \ldots, n$$

where $x_1(t), x_2(t), \ldots, x_n(t)$ are the state variables, $u_1(t), u_2(t), \ldots, u_p(t)$ are the input variables, and f_i denotes the ith functional relationship. In general, f_i can be a linear or a nonlinear function.

The q outputs of the system are related to the state variables and the inputs through the *output equations* which are of the form

$$c_k(t) = g_k[x_1(t), x_2(t), \ldots, x_n(t), u_1(t), u_2(t), \ldots, u_p(t), t] \tag{5-2}$$

$$k = 1, 2, \ldots, q$$

Similar remarks can be made for g_k as for f_i.

The state equations and the output equations as a set are called the *dynamic equations* of the system.

It is customary to write the dynamic equations in vector-matrix form. In fact, the compact matrix notation is considered one of the advantages of the state-variable method. In matrix form,

State Equation

$$\frac{d\mathbf{x}(t)}{dt} = \mathbf{f}[\mathbf{x}(t), \mathbf{u}(t), t] \tag{5-3}$$

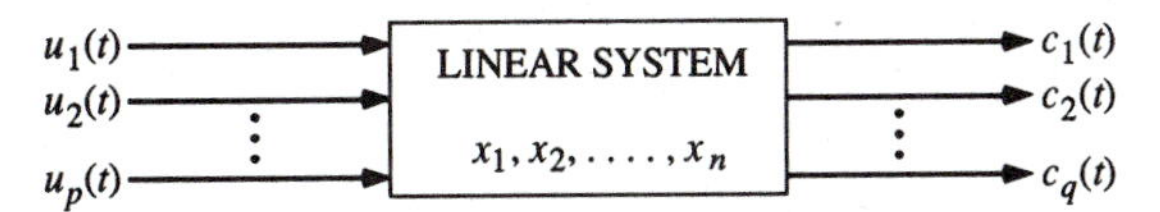

Figure 5-1. A linear system with p inputs, q outputs, and n state variables.

Output Equation

$$\mathbf{c}(t) = \mathbf{g}[\mathbf{x}(t), \mathbf{u}(t), t] \tag{5-4}$$

where $\mathbf{x}(t)$ is an $n \times 1$ column matrix and is called the *state vector*; that is,

$$\mathbf{x}(t) = \begin{bmatrix} x_1(t) \\ x_2(t) \\ \vdots \\ x_n(t) \end{bmatrix} \tag{5-5}$$

The input vector $\mathbf{u}(t)$ is a $p \times 1$ column matrix and is denoted as

$$\mathbf{u}(t) = \begin{bmatrix} u_1(t) \\ u_2(t) \\ \vdots \\ u_p(t) \end{bmatrix} \tag{5-6}$$

The output vector $\mathbf{c}(t)$ is defined as

$$\mathbf{c}(t) = \begin{bmatrix} c_1(t) \\ c_2(t) \\ \vdots \\ c_q(t) \end{bmatrix} \tag{5-7}$$

which is a $q \times 1$ column matrix or vector.

If the system is linear and time-invariant, the dynamic equations are written as

$$\frac{d\mathbf{x}(t)}{dt} = \mathbf{A}\mathbf{x}(t) + \mathbf{B}\mathbf{u}(t) \tag{5-8}$$

and

$$\mathbf{c}(t) = \mathbf{D}\mathbf{x}(t) + \mathbf{E}\mathbf{u}(t) \tag{5-9}$$

where $\mathbf{A}$ is an $n \times n$ matrix, $\mathbf{B}$ is an $n \times p$ matrix, $\mathbf{D}$ is a $q \times n$ matrix, and $\mathbf{E}$ is a $q \times p$ matrix, all with constant coefficients.

5-2-1 The State Transition Matrix—Solution of the Homogeneous State Equations

The *state transition matrix*, $\boldsymbol{\phi}(t)$, is defined as an $n \times n$ matrix that satisfies the homogeneous state equation,

$$\frac{d\mathbf{x}(t)}{dt} = \mathbf{A}\mathbf{x}(t) \tag{5-10}$$

Thus, $\boldsymbol{\phi}(t)$ satisfies the equation

$$\frac{d\boldsymbol{\phi}(t)}{dt} = \mathbf{A}\boldsymbol{\phi}(t) \tag{5-11}$$

Let $\mathbf{x}(0)$ denote the initial state at $t = 0$. Then, $\boldsymbol{\phi}(t)$ is also defined by the matrix equation

$$\mathbf{x}(t) = \boldsymbol{\phi}(t)\mathbf{x}(0) \tag{5-12}$$

which is the solution of the homogeneous state equation for $t \geq 0$.

To determine the state transition matrix $\boldsymbol{\phi}(t)$, we take the Laplace transform on both sides of Eq. (5-10); we get

$$s\mathbf{X}(s) - \mathbf{x}(0) = \mathbf{A}\mathbf{X}(s) \tag{5-13}$$

Solving for $\mathbf{X}(s)$ from the last equation, we get

$$\mathbf{X}(s) = (s\mathbf{I} - \mathbf{A})^{-1}\mathbf{x}(0) \tag{5-14}$$

where we assume that the matrix $(s\mathbf{I} - \mathbf{A})$ is nonsingular. Taking the inverse Laplace transform on both sides of the last equation yields

$$\mathbf{x}(t) = \mathscr{L}^{-1}[(s\mathbf{I} - \mathbf{A})^{-1}]\mathbf{x}(0) \qquad t \geq 0 \tag{5-15}$$

Comparing Eqs. (5-15) and (5-12), the state transition matrix is written

$$\boldsymbol{\phi}(t) = \mathscr{L}^{-1}[(s\mathbf{I} - \mathbf{A})^{-1}] \tag{5-16}$$

An alternative method of solving the homogeneous state equation is to assume a solution, as in the classical method of solving linear ordinary differential equations, and then substitute it in the state equation to show that it is indeed a correct solution. Let the solution of Eq. (5-10) be

$$\mathbf{x}(t) = e^{\mathbf{A}t}\mathbf{x}(0) \tag{5-17}$$

for $t \geq 0$, where $e^{\mathbf{A}t}$ represents a power series of the matrix $\mathbf{A}t$,

$$e^{\mathbf{A}t} = \mathbf{I} + \mathbf{A}t + \frac{1}{2!}\mathbf{A}^2t^2 + \frac{1}{3!}\mathbf{A}^3t^3 + \cdots \tag{5-18}$$

Taking the derivative of the last equation with respect to t, we get

$$\frac{d(e^{\mathbf{A}t})}{dt} = \mathbf{A}e^{\mathbf{A}t} \tag{5-19}$$

which has the same form as Eq. (5-11).

In addition to Eq. (5-16), we have obtained another useful expression for the state transition matrix,

$$\boldsymbol{\phi}(t) = e^{\mathbf{A}t} = \mathbf{I} + \mathbf{A}t + \frac{1}{2!}\mathbf{A}^2t^2 + \frac{1}{3!}\mathbf{A}^3t^3 + \cdots \tag{5-20}$$

For nonzero initial time t_0, the solution of the homogeneous state equation is written

$$\mathbf{x}(t) = \boldsymbol{\phi}(t - t_0)\mathbf{x}(t_0) \tag{5-21}$$

5-2-2 Properties of the State Transition Matrix

Several useful properties of the state transition matrix $\boldsymbol{\phi}(t)$ are stated as follows:

$$\boldsymbol{\phi}(0) = \mathbf{I} \qquad \text{(identity matrix)} \tag{5-22}$$

$$\boldsymbol{\phi}(t) \text{ is nonsingular for finite elements in } \mathbf{A} \tag{5-23}$$

$$\boldsymbol{\phi}^{-1}(t) = \boldsymbol{\phi}(-t) \tag{5-24}$$

$$\boldsymbol{\phi}(t_1 - t_2)\boldsymbol{\phi}(t_2 - t_3) = \boldsymbol{\phi}(t_1 - t_3) \qquad \text{for any } t_1, t_2, t_3 \tag{5-25}$$

5-2-3 Solution of the Nonhomogeneous State Equations—The State Transition Equation

The nonhomogeneous state equation

$$\frac{d\mathbf{x}(t)}{dt} = \mathbf{A}\mathbf{x}(t) + \mathbf{B}\mathbf{u}(t) \tag{5-26}$$

is now solved using the Laplace transform method. Taking the Laplace transform on both sides of the last equation, we have

$$s\mathbf{X}(s) - \mathbf{x}(0) = \mathbf{A}\mathbf{X}(s) + \mathbf{B}\mathbf{U}(s) \tag{5-27}$$

Solving for $\mathbf{X}(s)$ in the last equation, we get

$$\mathbf{X}(s) = (s\mathbf{I} - \mathbf{A})^{-1}\mathbf{x}(0) + (s\mathbf{I} - \mathbf{A})^{-1}\mathbf{B}\mathbf{U}(s) \tag{5-28}$$

The solution of Eq. (5-26) is obtained by taking the inverse Laplace transform of Eq. (5-28). The result is

$$\mathbf{x}(t) = \mathscr{L}^{-1}[(s\mathbf{I} - \mathbf{A})^{-1}]\mathbf{x}(0) + \mathscr{L}^{-1}[(s\mathbf{I} - \mathbf{A})^{-1}\mathbf{B}\mathbf{U}(s)] \tag{5-29}$$

Using the definition of the state transition matrix in Eq. (5-16) and the convolution integral, the last equation is written as

$$\mathbf{x}(t) = \boldsymbol{\phi}(t)\mathbf{x}(0) + \int_0^t \boldsymbol{\phi}(t - \tau)\mathbf{B}\mathbf{u}(\tau)\, d\tau \qquad t \geq 0 \tag{5-30}$$

This solution of the state equation is also called the *state transition equation.*

By setting the initial time at $t = t_0$ and using $\mathbf{x}(t_0)$ as the initial state, we can show by using Eq. (5-30) and some of the properties of $\boldsymbol{\phi}(t)$ that the state transition equation is written generally as

$$\mathbf{x}(t) = \boldsymbol{\phi}(t - t_0)\mathbf{x}(t_0) + \int_{t_0}^t \boldsymbol{\phi}(t - \tau)\mathbf{B}\mathbf{u}(\tau)\, d\tau \qquad t \geq t_0 \tag{5-31}$$

The state transition equation for $t \geq t_0$ is particularly useful for the analysis of discrete-data systems, since the transition of state is carried out from one discrete time instant to the next, and the initial time t_0 is advanced each time to the next time instant.

Once the state transition equation is determined, the output vector $\mathbf{c}(t)$ is written

$$\mathbf{c}(t) = \mathbf{D}\boldsymbol{\phi}(t - t_0)\mathbf{x}(t_0) + \int_{t_0}^{t} \mathbf{D}\boldsymbol{\phi}(t - \tau)\mathbf{B}\mathbf{u}(\tau)\, d\tau + \mathbf{E}\mathbf{u}(t) \tag{5-32}$$

5-3 STATE EQUATIONS OF DISCRETE-DATA SYSTEMS WITH SAMPLE-AND-HOLD DEVICES

A typical multichannel discrete-data system with sample-and-hold (S/H) devices or zero-order holds (zoh) is shown in Fig. 5-2. The outputs of the zoh are described by

$$u_i(t) = u_i(kT) = e_i(kT) \qquad kT \le t < (k+1)T \tag{5-33}$$

where $k = 0, 1, 2, \ldots$, and $i = 1, 2, \ldots, p$.

Let the dynamics of the linear system be represented by the state equations in Eq. (5-8). The state transition equation is given by Eq. (5-31) for all t and t_0. Since the inputs are constant between any two consecutive sampling periods, the input vector $\mathbf{u}(\tau)$ in Eq. (5-31) has the property

$$\mathbf{u}(\tau) = \mathbf{u}(kT) \qquad \text{for } kT \le \tau < (k+1)T \tag{5-34}$$

Thus, in Eq. (5-31) $\mathbf{u}(\tau)$ can be placed outside the integral, and Eq. (5-31) becomes

$$\mathbf{x}(t) = \boldsymbol{\phi}(t - t_0)\mathbf{x}(t_0) + \int_{t_0}^{t} \boldsymbol{\phi}(t - \tau)\mathbf{B}\, d\tau\mathbf{u}(kT) \tag{5-35}$$

which is valid for the time interval, $kT \le t \le (k+1)T$.

Setting $t_0 = kT$, Eq. (5-35) becomes

$$\mathbf{x}(t) = \boldsymbol{\phi}(t - kT)\mathbf{x}(kT) + \int_{kT}^{t} \boldsymbol{\phi}(t - \tau)\mathbf{B}\, d\tau\mathbf{u}(kT) \tag{5-36}$$

for $kT \le t \le (k+1)T$. The last equation describes the state vector $\mathbf{x}(t)$ at all times between the sampling instants kT and $(k+1)T$, $k = 0, 1, 2, \ldots$.

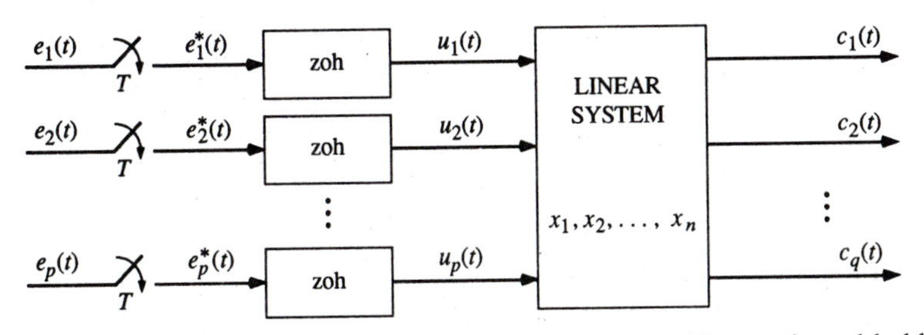

Figure 5-2. A multichannel digital system with sample-and-hold devices.

Equation (5-36) can be simplified by setting

$$\boldsymbol{\theta}(t - kT) = \int_{kT}^{t} \boldsymbol{\phi}(t - \tau)\mathbf{B}\, d\tau \tag{5-37}$$

Thus, Eq. (5-36) is written as

$$\mathbf{x}(t) = \boldsymbol{\phi}(t - kT)\mathbf{x}(kT) + \boldsymbol{\theta}(t - kT)\mathbf{u}(kT) \tag{5-38}$$

Note that although the ith component of $\mathbf{u}(kT)$, $u_i(kT)$, is held constant only for the time interval from $t = kT$ to $t = (k + 1)T^-$, the solution in Eq. (5-38) is valid for the entire interval including $t = (k + 1)T$, since $\mathbf{x}(t)$ is a continuous function of t.

For numerical iteration, it is more convenient to describe $\mathbf{x}(t)$ only at the sampling instants. We let $t = (k + 1)T$, and Eq. (5-38) becomes

$$\mathbf{x}[(k + 1)T] = \boldsymbol{\phi}(T)\mathbf{x}(kT) + \boldsymbol{\theta}(T)\mathbf{u}(kT) \tag{5-39}$$

where, from Eq. (5-20),

$$\boldsymbol{\phi}(T) = e^{\mathbf{A}T} = \mathbf{I} + \mathbf{A}T + \frac{1}{2!}\mathbf{A}^2T^2 + \frac{1}{3!}\mathbf{A}^3T^3 + \cdots \tag{5-40}$$

From Eq. (5-37),

$$\boldsymbol{\theta}(T) = \int_{kT}^{(k+1)T} \boldsymbol{\phi}[(k + 1)T - \tau]\mathbf{B}\, d\tau \tag{5-41}$$

By setting $m = -kT + \tau$, we get $dm = d\tau$, $(k + 1)T - \tau = T - m$; the last equation is written as

$$\boldsymbol{\theta}(T) = \int_{0}^{T} \boldsymbol{\phi}(T - \tau)\mathbf{B}\, d\tau \tag{5-42}$$

where we have replaced m by τ. Equation (5-39) represents a set of first-order difference equations, referred to as the *discrete state equations* of the sampled-data system shown in Fig. 5-2. These state equations, however, describe the dynamics of the system only at the sampling instants. In other words, by setting $t_0 = kT$ and $t = (k + 1)T$ in Eq. (5-38), all information on the system between the sampling instants is lost.

In a similar manner, the output equation in Eq. (5-9) is discretized by setting $t = kT$. Then,

$$\mathbf{c}(kT) = \mathbf{D}\mathbf{x}(kT) + \mathbf{E}\mathbf{u}(kT) \tag{5-43}$$

The discrete state equations in Eq. (5-39) along with the output equation in Eq. (5-43) are defined to be the *dynamic equations* of the system in Fig. 5-2.

In certain situations, the sampling operation is not done with respect to time, and the latter is not the independent variable. Under these circumstances the discrete dynamic equations describe the discrete "events" and are expressed as

$$\mathbf{x}(k+1) = \boldsymbol{\phi}(1)\mathbf{x}(k) + \boldsymbol{\theta}(1)\mathbf{u}(k) \tag{5-44}$$

$$\mathbf{c}(k) = \mathbf{D}\mathbf{x}(k) + \mathbf{E}\mathbf{u}(k) \tag{5-45}$$

which are obtained from Eqs. (5-39) and (5-43), respectively, by setting $T = 1$. Another way of looking at the notation is that in Eqs. (5-44) and (5-45) the sampling period T has been normalized to be one.

5-4 STATE EQUATIONS OF DIGITAL SYSTEMS WITH ALL-DIGITAL ELEMENTS

When a digital system is composed of all-digital elements and the inputs and outputs of the system are all digital, the system may be described by the following discrete dynamic equations:

$$\mathbf{x}(k+1) = \mathbf{A}\mathbf{x}(k) + \mathbf{B}\mathbf{u}(k) \tag{5-46}$$

$$\mathbf{c}(k) = \mathbf{D}\mathbf{x}(k) + \mathbf{E}\mathbf{u}(k) \tag{5-47}$$

The difference between Eqs. (5-39) and (5-46) is that, in the former, the coefficient matrix $\boldsymbol{\phi}(k)$ is the state transition matrix of the $\mathbf{A}$ matrix of the continuous-data process, and $\boldsymbol{\theta}(k)$ is a function of $\boldsymbol{\phi}(t)$ and $\mathbf{B}$; thus they must all conform with the properties of the state transition matrix. In Eq. (5-46), however, the matrices $\mathbf{A}$ and $\mathbf{B}$ can be arbitrary and depend entirely on the characteristics of the digital system.

5-5 DIGITAL SIMULATION AND APPROXIMATION

Discrete state equations result when an analog system is approximated by a discrete-data model. Let us consider the following dynamic equations of an analog system.

$$\dot{\mathbf{x}}(t) = \mathbf{A}\mathbf{x}(t) + \mathbf{B}\mathbf{u}(t) \tag{5-48}$$

$$\mathbf{c}(t) = \mathbf{D}\mathbf{x}(t) + \mathbf{E}\mathbf{u}(t) \tag{5-49}$$

We may perform a digital approximation of the system at $t = kT$ by approximating the derivative of $\mathbf{x}(t)$ at $t = kT$ as

$$\dot{\mathbf{x}}(t)\big|_{t=kT} = \frac{1}{T}\,[\mathbf{x}[(k+1)T - \mathbf{x}(kT)]] \tag{5-50}$$

Then, Eq. (5-48) is approximated by

$$\frac{1}{T}\,[[\mathbf{x}(k+1)T - \mathbf{x}(kT)]] = \mathbf{A}\mathbf{x}(kT) + \mathbf{B}\mathbf{u}(kT) \tag{5-51}$$

Similarly, Eq. (5-49) is discretized as

$$\mathbf{c}(kT) = \mathbf{D}\mathbf{x}(kT) + \mathbf{E}\mathbf{u}(kT) \tag{5-52}$$

In the form of a discrete state equation, Eq. (5-51) is rearranged as

$$\mathbf{x}[(k+1)T] = (\mathbf{I} + T\mathbf{A})\mathbf{x}(kT) + T\mathbf{B}\mathbf{u}(kT) \tag{5-53}$$

5-6 THE STATE TRANSITION EQUATIONS

The solutions of the discrete state equations in Eq. (5-39), (5-44), or (5-46) are the state transition equations of the discrete-data system. Since these equations are all of the same form, their solutions will also be very similar. However, Eqs. (5-39) and (5-44) contain the state transition matrix $\boldsymbol{\phi}(T)$ of the $\mathbf{A}$ matrix of Eq. (5-8), which is nonsingular. Thus, the transition of state for the systems described by Eqs. (5-39) and (5-44) can take place in the forward or backward direction. For the discrete state equations in Eq. (5-46), generally, the matrix $\mathbf{A}$ has no restrictions. However, if the state transition is to progress in the negative direction, $\mathbf{A}$ must be nonsingular.

We shall derive the state transition equations of the state equations in Eq. (5-39), and then extend these to Eqs. (5-44) and (5-46).

5-6-1 The Recursive Method

The most straightforward method of solving Eq. (5-39) is by recursion. The following equations are written by substituting $k = 0, 1, 2, \ldots$ in Eq. (5-39) successively.

$k = 0$

$$\mathbf{x}(T) = \boldsymbol{\phi}(T)\mathbf{x}(0) + \boldsymbol{\theta}(T)\mathbf{u}(0) \tag{5-54}$$

$k = 1$

$$\begin{aligned}\mathbf{x}(2T) &= \boldsymbol{\phi}(T)\mathbf{x}(T) + \boldsymbol{\theta}(T)\mathbf{u}(T)\\ &= \boldsymbol{\phi}(T)[\boldsymbol{\phi}(T)\mathbf{x}(0) + \boldsymbol{\theta}(T)\mathbf{u}(0)] + \boldsymbol{\theta}(T)\mathbf{u}(T)\\ &= \boldsymbol{\phi}(T)\boldsymbol{\phi}(T)\mathbf{x}(0) + \boldsymbol{\phi}(T)\boldsymbol{\theta}(T)\mathbf{u}(0) + \boldsymbol{\theta}(T)\mathbf{u}(T)\end{aligned} \tag{5-55}$$

Since Eq. (5-25) implies

$$\boldsymbol{\phi}(T)\boldsymbol{\phi}(T) = \boldsymbol{\phi}(2T) \tag{5-56}$$

Eq. (5-55) is written

$$\mathbf{x}(2T) = \boldsymbol{\phi}(2T)\mathbf{x}(0) + \boldsymbol{\phi}(T)\boldsymbol{\theta}(T)\mathbf{u}(0) + \boldsymbol{\theta}(T)\mathbf{u}(T) \tag{5-57}$$

Continuing the process, with $k = N - 1$,

$$\mathbf{x}(NT) = \boldsymbol{\phi}[(N-1)T]\mathbf{x}[(N-1)T] + \boldsymbol{\theta}[(N-1)T]\mathbf{u}[(N-1)T] \tag{5-58}$$

Substituting the first $N - 1$ equations for $\mathbf{x}[(N-1)T], \mathbf{x}[(N-2)T], \ldots,$ $\mathbf{x}(T)$, into the last equation successively, we have

$$\mathbf{x}(NT) = \boldsymbol{\phi}(NT)\mathbf{x}(0) + \sum_{i=0}^{N-1} \boldsymbol{\phi}[(N-i-1)T]\boldsymbol{\theta}(T)\mathbf{u}(iT) \tag{5-59}$$

which is the solution of Eq. (5-39), given the initial state $\mathbf{x}(0)$ and the input $\mathbf{u}(iT)$, for $i = 0, 1, \ldots, N - 1$. In deriving Eq. (5-59) we have used the property

$$\underbrace{\boldsymbol{\phi}(T)\boldsymbol{\phi}(T)\cdots\boldsymbol{\phi}(T)}_{\longleftarrow N \longrightarrow} = \boldsymbol{\phi}(NT) \tag{5-60}$$

which follows directly from the property of the state transition matrix $\boldsymbol{\phi}(t)$ in Eq. (5-25). Equation (5-60) implies that if $\boldsymbol{\phi}(T)$ is a state transition matrix that satisfies all the properties in Eqs. (5-22) to (5-25), then multiplying $\boldsymbol{\phi}(T)$ by itself N times is equivalent to replacing all the T in $\boldsymbol{\phi}(T)$ by NT.

The time frame of Eq. (5-59) can be shifted forward by any positive integer M, so that the solution of $\mathbf{x}[(N + M)T]$ is given in terms of the initial state $\mathbf{x}(MT)$ and inputs $\mathbf{u}(iT)$ for $i = M, M + 1, \ldots, N + M - 1$. This is accomplished by solving for $\mathbf{x}(0)$ from Eq. (5-59),

$$\mathbf{x}(0) = \boldsymbol{\phi}(-NT)\mathbf{x}(NT) - \sum_{i=0}^{N-1} \boldsymbol{\phi}[(-i-1)T]\boldsymbol{\theta}(T)\mathbf{u}(iT) \tag{5-61}$$

where we have utilized the properties of

$$\begin{aligned} [\boldsymbol{\phi}(NT)]^{-1} &= \boldsymbol{\phi}(-NT) \\ \boldsymbol{\phi}(-NT)\boldsymbol{\phi}[(N-i-1)T] &= \boldsymbol{\phi}[(-i-1)T] \end{aligned} \tag{5-62}$$

Replacing N with $N + M$ in Eq. (5-59), we have

$$\mathbf{x}[(N+M)T] = \boldsymbol{\phi}[(N+M)T]\mathbf{x}(0) + \sum_{i=0}^{N+M-1} \boldsymbol{\phi}[(N+M-i-1)T]\boldsymbol{\theta}(T)\mathbf{u}(iT) \tag{5-63}$$

Substituting Eq. (5-61) into the last equation and simplifying, the state transition equation for any initial state $\mathbf{x}(MT)$ is written as

$$\mathbf{x}[(N+M)T] = \boldsymbol{\phi}[(NT)T]\mathbf{x}(MT) + \sum_{i=0}^{N-1} \boldsymbol{\phi}[(N-i-1)T]\boldsymbol{\theta}(T)\mathbf{u}[(M+i)T] \tag{5-64}$$

The solutions of Eqs. (5-44) and (5-46) are easily obtained and are presented as follows.

State Equation

$$\mathbf{x}(k+1) = \boldsymbol{\phi}(1)\mathbf{x}(k) + \boldsymbol{\theta}(1)\mathbf{u}(k) \tag{5-44}$$

State Transition Equation

$$\mathbf{x}(N) = \boldsymbol{\phi}(N)\mathbf{x}(0) + \sum_{i=0}^{N-1} \boldsymbol{\phi}(N - i - 1)\boldsymbol{\theta}(1)\mathbf{u}(i) \tag{5-65}$$

State Equation

$$\mathbf{x}(k + 1) = \mathbf{A}\mathbf{x}(k) + \mathbf{B}\mathbf{u}(k) \tag{5-66}$$

State Transition Equation

$$\mathbf{x}(N) = \mathbf{A}^N\mathbf{x}(0) + \sum_{i=0}^{N-1} \mathbf{A}^{N-i-1}\mathbf{B}\mathbf{u}(i) \tag{5-67}$$

where

$$\mathbf{A}^N = \underbrace{\mathbf{A} \cdot \mathbf{A} \cdot \mathbf{A} \cdots \mathbf{A}}_{\longleftarrow N \longrightarrow} \tag{5-68}$$

5-6-2 The z-Transform Method

The z-transform method can be applied to solve the linear time-invariant discrete state equations in Eqs. (5-39), (5-44), and (5-46). This is carried out in the following for Eq. (5-39).

The state equations in Eq. (5-39) are repeated as

$$\mathbf{x}[(k + 1)T] = \boldsymbol{\phi}(T)\mathbf{x}(kT) + \boldsymbol{\theta}(T)\mathbf{u}(kT) \tag{5-39}$$

Taking the Laplace transform on both sides of the last equation, we have

$$z\mathbf{X}(z) - z\mathbf{x}(0) = \boldsymbol{\phi}(T)\mathbf{X}(z) + \boldsymbol{\theta}(T)\mathbf{U}(z) \tag{5-69}$$

where $\mathbf{X}(z)$ is defined as the z-transform of the state vector $\mathbf{x}(kT)$ and is given as

$$\mathbf{X}(z) = \sum_{k=0}^{\infty} \mathbf{x}(kT)z^{-k} \tag{5-70}$$

and the same applies to $\mathbf{U}(z)$. Note that taking the z-transform of a matrix is equivalent to taking the z-transform of every element of the matrix. Solving for $\mathbf{X}(z)$ from Eq. (5-70), we have

$$\mathbf{X}(z) = [z\mathbf{I} - \boldsymbol{\phi}(T)]^{-1}z\mathbf{x}(0) + [z\mathbf{I} - \boldsymbol{\phi}(T)]^{-1}\boldsymbol{\theta}(T)\mathbf{U}(z) \tag{5-71}$$

The inverse z-transform of the last equation is

$$\mathbf{x}(kT) = \mathscr{Z}^{-1}[[z\mathbf{I} - \boldsymbol{\phi}(T)]^{-1}z]^{-1}\mathbf{x}(0) + \mathscr{Z}^{-1}[[z\mathbf{I} - \boldsymbol{\phi}(T)]^{-1}\boldsymbol{\theta}(T)\mathbf{U}(z)] \tag{5-72}$$

We shall show that the inverse z-transform of $[z\mathbf{I} - \boldsymbol{\phi}(T)]^{-1}z$ is the discrete state transition matrix $\boldsymbol{\phi}(kT)$.

The z-transform of $\boldsymbol{\phi}(kT)$ is defined in the usual manner as

$$\boldsymbol{\Phi}(z) = \sum_{k=0}^{\infty} \boldsymbol{\phi}(kT)z^{-k} \tag{5-73}$$

Premultiplying both sides of the last equation by $\boldsymbol{\phi}(T)z^{-1}$ and subtracting the result from Eq. (5-73), we have

$$[\mathbf{I} - \boldsymbol{\phi}(T)z^{-1}]\boldsymbol{\Phi}(z) = \mathbf{I} \tag{5-74}$$

Thus,

$$\boldsymbol{\Phi}(z) = [\mathbf{I} - \boldsymbol{\phi}(T)z^{-1}]^{-1} = [z\mathbf{I} - \boldsymbol{\phi}(T)]^{-1}z \tag{5-75}$$

Taking the inverse z-transform on both sides of the last equation, we have

$$\boldsymbol{\phi}(kT) = \mathscr{Z}^{-1}[[z\mathbf{I} - \boldsymbol{\phi}(T)]^{-1}z] \tag{5-76}$$

Thus, Eq. (5-70) represents the z-transform method of determining the state transition matrix of a discrete state equation.

The last term of Eq. (5-72) is evaluated by the use of the real convolution theorem of Eqs. (3-130) and (5-76). Therefore, it can be shown that

$$\mathscr{Z}^{-1}[[z\mathbf{I} - \boldsymbol{\phi}(T)]^{-1}\boldsymbol{\theta}(T)\mathbf{U}(z)] = \sum_{i=0}^{k-1} \boldsymbol{\phi}[(k-i-1)T]\boldsymbol{\theta}(T)\mathbf{u}(iT) \tag{5-77}$$

The entire state transition equation is

$$\mathbf{x}(kT) = \boldsymbol{\phi}(kT)\mathbf{x}(0) + \sum_{i=0}^{k-1} \boldsymbol{\phi}[(k-i-1)T]\boldsymbol{\theta}(T)\mathbf{u}(iT) \tag{5-78}$$

which is of the same form as Eq. (5-59). In a similar manner, the state equations of Eqs. (5-44) and (5-46) can be solved by means of the z-transform method illustrated above. Note that taking the z-transform of $\mathbf{A}^k$ implies

$$\mathscr{Z}(\mathbf{A}^k) = \sum_{k=0}^{\infty} \mathbf{A}^k z^{-k} \tag{5-79}$$

Example 5-1

In this example we shall illustrate the analysis of an open-loop discrete-data system by the state-variable method presented above. The block diagram of the system under consideration is shown in Fig. 5-3. The dynamic equations that describe the linear process are

$$\begin{bmatrix} \dfrac{dx_1(t)}{dt} \\ \dfrac{dx_2(t)}{dt} \end{bmatrix} = \begin{bmatrix} 0 & 1 \\ -2 & -3 \end{bmatrix} \begin{bmatrix} x_1(t) \\ x_2(t) \end{bmatrix} + \begin{bmatrix} 0 \\ 1 \end{bmatrix} u(t) \tag{5-80}$$

$$c(t) = x_1(t) \tag{5-81}$$

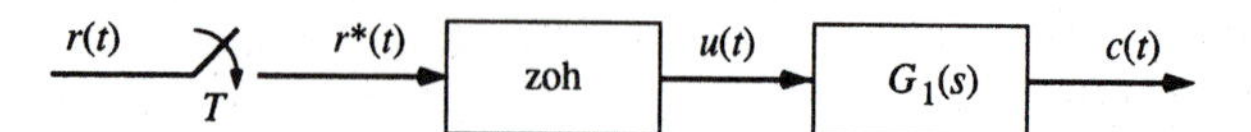

Figure 5-3. An open-loop digital system.

where $x_1(t)$ and $x_2(t)$ are the state variables, $c(t)$ is the scalar output, and $u(t)$ is the scalar input. Also, since $u(t)$ is the output of the zero-order hold,

$$u(t) = u(kT) = r(kT) \tag{5-82}$$

for $kT \le t < (k+1)T$.

Comparing Eq. (5-80) with the standard state equation form of Eq. (5-10), we have

$$\mathbf{A} = \begin{bmatrix} 0 & 1 \\ -2 & -3 \end{bmatrix} \qquad \mathbf{B} = \begin{bmatrix} 0 \\ 1 \end{bmatrix} \tag{5-83}$$

The following matrix is formed:

$$(s\mathbf{I} - \mathbf{A}) = \begin{bmatrix} s & -1 \\ 2 & s+3 \end{bmatrix} \tag{5-84}$$

Therefore,

$$(s\mathbf{I} - \mathbf{A})^{-1} = \frac{1}{s^2 + 3s + 2} \begin{bmatrix} s+3 & 1 \\ -2 & s \end{bmatrix} \tag{5-85}$$

The state transition matrix of **A** is obtained by taking the inverse Laplace transform of $(s\mathbf{I} - \mathbf{A})^{-1}$. Therefore, from Eq. (5-16),

$$\boldsymbol{\phi}(t) = \mathscr{L}^{-1}[(s\mathbf{I} - \mathbf{A})^{-1}] = \begin{bmatrix} 2e^{-t} - e^{-2t} & e^{-t} - e^{-2t} \\ -2e^{-t} + 2e^{-2t} & -e^{-t} + 2e^{-2t} \end{bmatrix} \tag{5-86}$$

Substitution of **B** in Eq. (5-83) and $\boldsymbol{\phi}(t)$ of Eq. (5-86) into Eq. (5-42) yields

$$\begin{aligned} \boldsymbol{\theta}(T) &= \int_0^T \boldsymbol{\phi}(T - \tau)\mathbf{B}\, d\tau \\ &= \int_0^T \begin{bmatrix} e^{-(T-\tau)} - e^{-2(T-\tau)} \\ -e^{-(T-\tau)} + 2e^{-2(T-\tau)} \end{bmatrix} d\tau = \begin{bmatrix} 0.5 - e^{-T} + 0.5e^{-2T} \\ e^{-T} - e^{-2T} \end{bmatrix} \end{aligned} \tag{5-87}$$

Now substituting Eq. (5-86) with $t = T$ and Eq. (5-87) into Eq. (5-39), the discrete state equation of the system is written

$$\begin{aligned} \begin{bmatrix} x_1[(k+1)T] \\ x_2[(k+1)T] \end{bmatrix} &= \begin{bmatrix} 2e^{-T} - e^{-2T} & e^{-T} - e^{-2T} \\ -2e^{-T} + 2e^{-2T} & -e^{-T} + 2e^{-2T} \end{bmatrix} \begin{bmatrix} x_1(kT) \\ x_2(kT) \end{bmatrix} \\ &\quad + \begin{bmatrix} 0.5 - e^{-T} + 0.5e^{-2T} \\ e^{-T} - e^{-2T} \end{bmatrix} u(kT) \end{aligned} \tag{5-88}$$

Using Eq. (5-59), the solution of Eq. (5-88) is

$$\begin{bmatrix} x_1(NT) \\ x_2(NT) \end{bmatrix} = \begin{bmatrix} 2e^{-NT} - e^{2NT} & e^{-NT} - e^{-2NT} \\ -2e^{-NT} + 2e^{-2NT} & -e^{-NT} + 2e^{-2NT} \end{bmatrix} \begin{bmatrix} x_1(0) \\ x_2(0) \end{bmatrix}$$
$$+ \sum_{k=0}^{N-1} \begin{bmatrix} (1 - e^{-T})e^{-(N-k-1)T} & -0.5(1 - e^{-2T})e^{-2(N-k-1)T} \\ -(1 - e^{-T})e^{-(N-k-1)T} & (1 - e^{-2T})e^{-2(N-k-1)T} \end{bmatrix} u(kT) \tag{5-89}$$

where N is any positive integer.

5-7 RELATIONSHIP BETWEEN STATE EQUATIONS AND TRANSFER FUNCTIONS

It is of interest to investigate the relationship between the transfer function and the state-variable methods.

Consider that a discrete-data system with multiple inputs and outputs is described by the transfer function

$$\mathbf{C}(z) = \mathbf{G}(z)\mathbf{U}(z) \tag{5-90}$$

where

$$\mathbf{U}(z) = \begin{bmatrix} U_1(z) \\ U_2(z) \\ \vdots \\ U_p(z) \end{bmatrix} \tag{5-91}$$

is the $p \times 1$ input transform vector;

$$\mathbf{C}(z) = \begin{bmatrix} C_1(z) \\ C_2(z) \\ \vdots \\ C_q(z) \end{bmatrix} \tag{5-92}$$

is the $q \times 1$ output transform vector;

$$\mathbf{G}(z) = \begin{bmatrix} G_{11}(z) & G_{12}(z) & \cdots & G_{1p}(z) \\ G_{21}(z) & G_{22}(z) & \cdots & G_{2p}(z) \\ \vdots & & & \\ G_{q1}(z) & G_{q2}(z) & \cdots & G_{qp}(z) \end{bmatrix} \tag{5-93}$$

is the $q \times p$ z-transfer function matrix.

The elements of $\mathbf{G}(z)$ may be of the form

$$\mathbf{G}_{ij}(z) = \frac{b_m + b_{m-1}z^{-1} + \cdots + b_0 z^{-m}}{a_n + a_{n-1}z^{-1} + \cdots + a_0 z^{-n}} \tag{5-94}$$

where $1 \le i \le q$ and $1 \le j \le p$.

In the state-variable formulation, the discrete-data system is described by the following dynamic equations:

$$\mathbf{x}[(k+1)T] = \mathbf{A}\mathbf{x}(kT) + \mathbf{B}\mathbf{u}(kT) \tag{5-95}$$

$$\mathbf{c}(kT) = \mathbf{D}\mathbf{x}(kT) + \mathbf{E}\mathbf{u}(kT) \tag{5-96}$$

Taking the z-transform on both sides of Eq. (5-95) and solving for $\mathbf{X}(z)$, we get

$$\mathbf{X}(z) = (z\mathbf{I} - \mathbf{A})^{-1}z\mathbf{x}(0) + (z\mathbf{I} - \mathbf{A})^{-1}\mathbf{B}\mathbf{U}(z) \tag{5-97}$$

Substituting Eq. (5-97) into the z-transform of Eq. (5-96) yields

$$\mathbf{C}(z) = \mathbf{D}(z\mathbf{I} - \mathbf{A})^{-1}z\mathbf{x}(0) + \mathbf{D}(z\mathbf{I} - \mathbf{A})^{-1}\mathbf{B}\mathbf{U}(z) + \mathbf{E}\mathbf{U}(z) \tag{5-98}$$

To get the transfer function, we assume that the initial state $\mathbf{x}(0)$ is a null matrix; thus, Eq. (5-98) becomes

$$\mathbf{C}(z) = [\mathbf{D}(z\mathbf{I} - \mathbf{A})^{-1}\mathbf{B} + \mathbf{E}]\mathbf{U}(z) \tag{5-99}$$

Comparing Eq. (5-99) with Eq. (5-90), we see that the z-transfer function matrix of the system may be written as

$$\mathbf{G}(z) = \mathbf{D}(z\mathbf{I} - \mathbf{A})^{-1}\mathbf{B} + \mathbf{E} \tag{5-100}$$

If the discrete-data system has sample-and-hold operations and is described by the dynamic equations in Eqs. (5-8) and (5-9), we only have to replace $\mathbf{A}$ and $\mathbf{B}$ in Eq. (5-100) by $\boldsymbol{\phi}(T)$ and $\boldsymbol{\theta}(T)$, respectively. Of course, this is performed with the assumption that a z-transfer function can be written for the system with sample-and-hold operations. As was illustrated in Sec. 4-4, sometimes z-transfer functions in the form of Eq. (5-90) cannot be defined for a discrete-data system if one or more of the inputs is not sampled. Under this condition, only input-output transfer relations exist.

The inverse transform of the transfer function matrix is called the *impulse response matrix* or the *weighting sequence matrix* $\mathbf{g}(kT)$. Taking the inverse z-transform on both sides of Eq. (5-100), we get

$$\mathbf{g}(kT) = \mathbf{D}\boldsymbol{\phi}[(k-1)T]\mathbf{B} + \mathbf{E}\delta(0) \tag{5-101}$$

where $\delta(0)$ is the impulse function occurring at $t = 0$ and has a magnitude of one.

Since $\boldsymbol{\phi}[(k-1)T] = \mathbf{0}$ for $k < 1$, $\mathbf{g}(kT)$ in Eq. (5-101) can be written as

$$\mathbf{g}(kT) = \mathbf{E} \qquad k = 0 \tag{5-102}$$

$$\mathbf{g}(kT) = \mathbf{D}\boldsymbol{\phi}[(k-1)T]\mathbf{B} \qquad k \geq 1 \tag{5-103}$$

Example 5-2

As an illustrative example, let us derive the transfer function of the open-loop system shown in Fig. 5-3, with the data given in Example 5-1.

First, we use the state-variable method. From Example 5-1, we have

$$\boldsymbol{\phi}(T) = \begin{bmatrix} 2e^{-T} - e^{-2T} & e^{-T} - e^{-2T} \\ -2e^{-T} + 2e^{-2T} & -e^{-T} + 2e^{-2T} \end{bmatrix} \quad (5\text{-}104)$$

$$\boldsymbol{\theta}(T) = \begin{bmatrix} 0.5 - e^{-T} + 0.5e^{-2T} \\ e^{-T} - e^{-2T} \end{bmatrix} \quad (5\text{-}105)$$

$$\mathbf{D} = [1 \quad 0] \qquad \text{and} \qquad E = 0 \quad (5\text{-}106)$$

The transfer function matrix is written

$$G(z) = \mathbf{D}(z\mathbf{I} - \mathbf{A})^{-1}\mathbf{B} + E \quad (5\text{-}107)$$

Substituting Eqs. (5-104), (5-105), and (5-106) into Eq. (5-107), we have

$$G(z) = \frac{(0.5 - e^{-T} + 0.5e^{-2T})z + 0.5e^{-T}(1 - e^{-T})^2}{z^2 - (e^{-T} + e^{-2T})z + e^{-3T}} \quad (5\text{-}108)$$

The characteristic equation of the system is

$$z^2 - (e^{-T} + e^{-2T})z + e^{-3T} = 0 \quad (5\text{-}109)$$

To use the z-transform method, we must first determine the transfer function of the linear process $G_1(s)$ of Fig. 5-3, which is given by

$$G_1(s) = \mathbf{D}(s\mathbf{I} - \mathbf{A})^{-1}\mathbf{B} \quad (5\text{-}110)$$

where $\mathbf{A}$ and $\mathbf{B}$ are given in Eq. (5-83). Thus,

$$G_1(s) = \frac{C(s)}{U(s)} = \frac{1}{s^2 + 3s + 2} \quad (5\text{-}111)$$

The z-transfer function of the overall system is

$$G(z) = \frac{C(z)}{R(z)} = \mathscr{Z}\left(\frac{1 - e^{-Ts}}{s} G_1(s)\right)$$

$$= (1 - z^{-1})\mathscr{Z}\left(\frac{1}{s(s+1)(s+2)}\right) \quad (5\text{-}112)$$

Evaluating the z-transform in Eq. (5-112), we have the same result as in Eq. (5-108) for $G(z)$.

5-8 CHARACTERISTIC EQUATION, EIGENVALUES, AND EIGENVECTORS

The characteristic equation of a linear discrete-data system was defined with respect to the transfer function in Sec. 4-4-1. In this section we shall define the characteristic equation starting from the matrix $\mathbf{A}$ [or from $\boldsymbol{\phi}(T)$] of the state equation. Consider that an nth-order discrete-data system is represented by the vector-matrix state equation in Eq. (5-39). The characteristic equation of the system, often referred to as the *characteristic equation of* $\boldsymbol{\phi}(T)$, is defined as the determinant of $z\mathbf{I} - \boldsymbol{\phi}(T)$ equated to zero; that is,

$$|z\mathbf{I} - \boldsymbol{\phi}(T)| = 0 \tag{5-113}$$

Notice that the matrix $\boldsymbol{\theta}(T)$ of the nonhomogeneous part of the state equation is not related to the characteristic equation at all.

The characteristic equation in Eq. (5-113) can also be derived according to the definition of Sec. 4-4-1 as equating the denominator of the transfer function matrix to zero. With reference to Eq. (5-100), the transfer function matrix is written as

$$\mathbf{G}(z) = \mathbf{D}\frac{[\Delta_{ij}]'}{|z\mathbf{I} - \boldsymbol{\phi}(T)|}\boldsymbol{\theta}(T) + \mathbf{E} \tag{5-114}$$

where Δ_{ij} represents the cofactor of the ijth element of the matrix $z\mathbf{I} - \boldsymbol{\phi}(T)$ and $[\Delta_{ij}]'$ is the matrix transpose of Δ_{ij}. Apparently, setting the denominator polynomial of Eq. (5-114) to zero yields the same results as in Eq. (5-113).

If the system is modeled by the dynamic equations of Eqs. (5-46) and (5-47), as in the case of the all-digital system, the characteristic equation is given as

$$|z\mathbf{I} - \mathbf{A}| = 0 \tag{5-115}$$

5-8-1 Eigenvalues

The roots of the characteristic equation are defined as the *eigenvalues* of the matrix $\mathbf{A}$ [or of $\boldsymbol{\phi}(T)$].

Example 5-3

For the system described in Example 5-2, the characteristic equation is given in Eq. (5-109). The roots of the characteristic equation, or the eigenvalues of $\boldsymbol{\phi}(T)$, are

$$z = e^{-T} \qquad \text{and} \qquad z = e^{-2T} \tag{5-116}$$

Some of the important properties of the eigenvalues are given as follows [The state equation in Eq. (5-46) is used with the understanding that $\mathbf{A}$ is replaced by $\boldsymbol{\phi}(T)$ if the state equation is of the form of Eq. (5-39)]:

1. If the coefficients of the characteristic equation are scalar quantities, the eigenvalues are either real or in complex conjugate pairs.
2. The trace of $\mathbf{A}$, which is defined as the sum of the elements on the main diagonal of $\mathbf{A}$, is given by

$$\text{trace of } \mathbf{A} = \text{tr}\,(\mathbf{A}) = z_1 + z_2 + \cdots + z_n \tag{5-117}$$

where z_i, $i = 1, 2, \ldots, n$, are the eigenvalues of $\mathbf{A}$.

3. The determinant of **A** is related to the eigenvalues through

$$|\mathbf{A}| = z_1 z_2 \cdots z_n \tag{5-118}$$

4. If **A** is nonsingular with eigenvalues z_i, $i = 1, 2, \ldots, n$, then $1/z_i$, $i = 1, 2, \ldots, n$, are the eigenvalues of $\mathbf{A}^{-1}$.
5. If z_i is an eigenvalue of **A**, then it is an eigenvalue of $\mathbf{A}'$.
6. If **A** is a real symmetric matrix, then its eigenvalues are all real.
7. For square matrices **A** and **B**,

$$|z\mathbf{I} - \mathbf{AB}| = |z\mathbf{I} - \mathbf{BA}| \tag{5-119}$$

Then the eigenvalues of **AB** are the same as that of **BA**.

5-8-2 Eigenvectors

The $n \times 1$ vector $\mathbf{p}_i$ which satisfies the matrix equation

$$(z_i\mathbf{I} - \mathbf{A})\mathbf{p}_i = \mathbf{0} \tag{5-120}$$

where z_i, $i = 1, 2, \ldots, n$, denotes the ith eigenvalue of **A**, is called the *eigenvector of* **A** *associated with the eigenvalue* z_i.

For distinct eigenvalues the eigenvectors can be solved directly from Eq. (5-120).

Example 5-4

Consider the discrete-data system in Example 5-2. The eigenvectors of the eigenvalues $z_1 = e^{-2T}$ and $z_2 = e^{-T}$ are designated as $\mathbf{p}_1$ and $\mathbf{p}_2$, respectively. By definition, the eigenvectors must satisfy the following equation:

$$[z_i\mathbf{I} - \boldsymbol{\phi}(T)]\mathbf{p}_i = \mathbf{0} \tag{5-121}$$

for $i = 1$ and 2. Substituting $\boldsymbol{\phi}(T)$ from Eq. (5-104) into Eq. (5-121), we have

$$\begin{bmatrix} z_i - 2e^{-T} + e^{-2T} & -e^{-T} + e^{-2T} \\ 2e^{-T} - 2e^{-2T} & z_i + e^{-T} - 2e^{-2T} \end{bmatrix} \mathbf{p}_i = \mathbf{0} \tag{5-122}$$

for $i = 1$ and 2. Letting

$$\mathbf{p}_1 = \begin{bmatrix} p_{11} \\ p_{21} \end{bmatrix} \qquad \mathbf{p}_2 = \begin{bmatrix} p_{12} \\ p_{22} \end{bmatrix} \tag{5-123}$$

from Eq. (5-122) we first arrive at, for $z_1 = e^{-2T}$, the following two scalar equations:

$$2(-e^{-T} + e^{-2T})p_{11} + (-e^{-T} + e^{-2T})p_{21} = 0 \tag{5-124}$$

$$2(e^{-T} - e^{-2T})p_{11} + (e^{-T} - e^{-2T})p_{21} = 0 \tag{5-125}$$

It is easy to show that these two equations are linearly dependent, so we may assign an arbitrary value to either p_{11} or p_{21} and solve for the other. Let $p_{11} = 1$, then $p_{21} = -2$. Note that we cannot assign p_{11} to be zero, since then p_{21} would be zero also. Keep in mind that an eigenvector cannot be a null vector.

Similarly, for $z_2 = e^{-T}$, Eq. (5-122) leads to

$$(-e^{-T} + e^{-2T})p_{12} + (-e^{-T} + e^{-2T})p_{22} = 0 \tag{5-126}$$

$$2(e^{-T} - e^{-2T})p_{12} + 2(e^{-T} - e^{-2T})p_{22} = 0 \tag{5-127}$$

Again, for these dependent equations, if we let $p_{12} = 1$, we get $p_{22} = -1$. Thus, the eigenvectors are

$$\mathbf{p}_1 = \begin{bmatrix} 1 \\ -2 \end{bmatrix} \quad \text{for } z_1 = e^{-2T} \tag{5-128}$$

$$\mathbf{p}_2 = \begin{bmatrix} 1 \\ -1 \end{bmatrix} \quad \text{for } z_2 = e^{-T} \tag{5-129}$$

It is important to keep in mind that the eigenvectors are *not unique*, in general. In the present case, the elements of the eigenvectors are related through $p_{21} = -2p_{11}$, and $p_{22} = -p_{12}$.

For distinct eigenvalues, the eigenvectors of **A** can also be determined by using any nonzero columns of the matrix, adj $(z_i\mathbf{I} - \mathbf{A})$, $i = 1, 2, \ldots, n$.

Example 5-5

Using the same system in Example 5-4, the adjoint of the matrix $[z_i\mathbf{I} - \boldsymbol{\phi}(T)]$ is written

$$\text{adj}\,[z_i\mathbf{I} - \boldsymbol{\phi}(T)] = \text{adj}\begin{bmatrix} z_i - 2e^{-T} + e^{-2T} & -e^{-T} + e^{-2T} \\ 2e^{-T} - 2e^{-2T} & z_i + e^{-T} - 2e^{-2T} \end{bmatrix}$$

$$= \begin{bmatrix} z_i + e^{-T} - 2e^{-2T} & e^{-T} - e^{-2T} \\ -2e^{-T} + 2e^{-2T} & z_i - 2e^{-T} + e^{-2T} \end{bmatrix} \tag{5-130}$$

for $i = 1, 2$. Then, for $z_1 = e^{-2T}$ and $z_2 = e^{-T}$,

$$\text{adj}\,[z_1\mathbf{I} - \boldsymbol{\phi}(T)] = \begin{bmatrix} e^{-T} - e^{-2T} & e^{-T} - e^{-2T} \\ -2e^{-T} + 2e^{-2T} & -2e^{-T} + 2e^{-2T} \end{bmatrix} \tag{5-131}$$

$$\text{adj}\,[z_2\mathbf{I} - \boldsymbol{\phi}(T)] = \begin{bmatrix} 2e^{-T} - 2e^{-2T} & e^{-T} - e^{-2T} \\ -2e^{-T} + 2e^{-2T} & -e^{-T} + e^{-2T} \end{bmatrix} \tag{5-132}$$

From Eq. (5-131), we set $\mathbf{p}_1$ equal to one column of the adjoint matrix, and after dividing by the common factor, $e^{-T} - e^{-2T}$, we have

$$\mathbf{p}_1 = \begin{bmatrix} 1 \\ -2 \end{bmatrix} \qquad \text{for } z_1 = e^{-2T} \tag{5-133}$$

Similarly, from Eq. (5-132), the eigenvector $\mathbf{p}_2$ is obtained:

$$\mathbf{p}_2 = \begin{bmatrix} 1 \\ -1 \end{bmatrix} \qquad \text{for } z_2 = e^{-T} \tag{5-134}$$

Properties of Eigenvectors

Some of the important properties of eigenvectors are summarized as follows.

1. The eigenvector $\mathbf{p}_i$ cannot be a null vector.
2. The rank of $(z_i\mathbf{I} - \mathbf{A})$, where z_i, $i = 1, 2, \ldots, n$, denotes the distinct eigenvalues of $\mathbf{A}$, is $n - 1$.
3. The eigenvector $\mathbf{p}_i$ is given by any nonzero column of the matrix adj $(z_i\mathbf{I} - \mathbf{A})$, where z_i denotes the ith distinct eigenvalue of $\mathbf{A}$.
4. If the matrix $\mathbf{A}$ has n distinct eigenvalues, then the set of n eigenvectors $\mathbf{p}_i$, $i = 1, 2, \ldots, n$, is linearly independent.
5. If $\mathbf{p}_i$ is an eigenvector of $\mathbf{A}$, then $k\mathbf{p}_i$ is also, where k is a scalar quantity.

Eigenvectors of Multiple-Order Eigenvalues

When one or more eigenvalues of $\mathbf{A}$ is of multiple order, a full set of n linearly independent eigenvectors may or may not exist. The number of linearly independent eigenvectors associated with a given eigenvalue z_i of multiplicity m_i is equal to the *degeneracy* d_i of $z_i\mathbf{I} - \mathbf{A}$. The degeneracy d_i of $z_i\mathbf{I} - \mathbf{A}$ is defined as

$$d_i = n - r_i \tag{5-135}$$

where n is the dimension of $\mathbf{A}$ and r_i is the rank of $z_i\mathbf{I} - \mathbf{A}$. *There are always d_i linearly independent eigenvectors associated with* $\mathbf{p}_i$. Furthermore,

$$1 \le d_i \le m_i \tag{5-136}$$

The eigenvectors of a matrix $\mathbf{A}$ with multiple-order eigenvalues are determined according to the methods described below.

Full Degeneracy ($d_i = m_i$)

For the eigenvalue z_i which has multiplicity m_i, the fully degenerated case has a full set of m_i eigenvectors associated with z_i. These eigenvectors are found from the nonzero columns of

$$\frac{1}{(m_i - 1)!}\left(\frac{d^{m_i - 1}}{dz^{m_i - 1}}[\text{adj}\,(z\mathbf{I} - \mathbf{A})]\right)\Bigg|_{z = z_i} \tag{5-137}$$

Example 5-6

Consider the matrix

$$\mathbf{A} = \begin{bmatrix} 3 & 0 & 0 \\ 2 & 4 & 1 \\ 2 & 1 & 4 \end{bmatrix} \tag{5-138}$$

The characteristic equation of $\mathbf{A}$ is

$$|z\mathbf{I} - \mathbf{A}| = (z-3)^2(z-5) = 0 \tag{5-139}$$

The eigenvalues of $\mathbf{A}$ are $z_1 = z_2 = 3$, and $z_3 = 5$. Thus, the eigenvalue $z_1 = 3$ has a multiplicity of two, and $z_3 = 5$ is of simple order. To check the degeneracy of $z\mathbf{I} - \mathbf{A}$ for $z_1 = 3$, we form the matrix $z_1\mathbf{I} - \mathbf{A}$ as follows.

$$z_1\mathbf{I} - \mathbf{A} = \begin{bmatrix} z-3 & 0 & 0 \\ -2 & z-4 & -1 \\ -2 & -1 & z-4 \end{bmatrix}\Bigg|_{z=3} = \begin{bmatrix} 0 & 0 & 0 \\ -2 & -1 & -1 \\ -2 & -1 & -1 \end{bmatrix} \tag{5-140}$$

which has a rank of one. Thus, the degeneracy of $z_1\mathbf{I} - \mathbf{A}$ is

$$d_1 = n - r_1 = 3 - 1 = 2 \tag{5-141}$$

Since z_1 is of multiplicity two, we say that $z_1\mathbf{I} - \mathbf{A}$ is of *full degeneracy*.

Now, using Eq. (5-137), we get

$$\left(\frac{d}{dz}\operatorname{adj}(z\mathbf{I} - \mathbf{A})\right)\Bigg|_{z=z_1}$$

$$= \frac{d}{dz}\begin{bmatrix} (z-3)(z-5) & 0 & 0 \\ 2(z-3) & (z-3)(z-4) & z-3 \\ 2(z-3) & z-3 & (z-3)(z-4) \end{bmatrix}\Bigg|_{z=z_1}$$

$$= \begin{bmatrix} 2z-8 & 0 & 0 \\ 2 & 2z-7 & 1 \\ 2 & 1 & 2z-7 \end{bmatrix}\Bigg|_{z=3} = \begin{bmatrix} -2 & 0 & 0 \\ 2 & -1 & 1 \\ 2 & 1 & -1 \end{bmatrix} \tag{5-142}$$

Thus, the two independent columns of the last matrix are selected as the eigenvectors.

$$\mathbf{p}_1 = \begin{bmatrix} -1 \\ 1 \\ 1 \end{bmatrix} \qquad \mathbf{p}_2 = \begin{bmatrix} 0 \\ -1 \\ 1 \end{bmatrix} \tag{5-143}$$

For the eigenvalue $z_3 = 5$, the eigenvector is found in the usual manner by setting $(z_3\mathbf{I} - \mathbf{A})\mathbf{p}_3 = \mathbf{0}$ and solving for $\mathbf{p}_3$, or by using any nonzero column of $\operatorname{adj}(z_3\mathbf{I} - \mathbf{A})$. The result is

$$\mathbf{p}_3 = \begin{bmatrix} 0 \\ 1 \\ 1 \end{bmatrix} \tag{5-144}$$

Simple Degeneracy ($d_i = 1$)

When the degeneracy of $z_i\mathbf{I} - \mathbf{A}$ is equal to one (simple degeneracy), only *one* eigenvector is associated with z_i regardless of the multiplicity of z_i. The eigenvector associated with z_i can be determined using the same method as for the case of the distinct eigenvalues. However, for the m_ith-order eigenvalue, there are $m_i - 1$ additional vectors called the *generalized eigenvectors*. These $m_i - 1$ generalized eigenvectors $\mathbf{p}_{i2}, \mathbf{p}_{i3}, \ldots, \mathbf{p}_{im_{i-1}}$ are found from the following $m_i - 1$ equations:

$$\begin{aligned}(z_i\mathbf{I} - \mathbf{A})\mathbf{p}_{i2} &= -\mathbf{p}_{i1} \\ (z_i\mathbf{I} - \mathbf{A})\mathbf{p}_{i3} &= -\mathbf{p}_{i2} \\ &\vdots \\ (z_i\mathbf{I} - \mathbf{A})\mathbf{p}_{im_i} &= -\mathbf{p}_{im_{i-1}}\end{aligned} \tag{5-145}$$

where $\mathbf{p}_{i1}$ is the eigenvector of z_i determined by solving the following set of matrix equations:

$$(z_i\mathbf{I} - \mathbf{A})\mathbf{p}_{i1} = \mathbf{0} \tag{5-146}$$

Example 5-7

Consider the matrix

$$\mathbf{A} = \begin{bmatrix} 1 & 2 \\ -2 & -3 \end{bmatrix} \tag{5-147}$$

which has the characteristic equation

$$|z\mathbf{I} - \mathbf{A}| = z^2 + 2z + 1 = 0 \tag{5-148}$$

The eigenvalues of $\mathbf{A}$ are $z_1 = z_2 = -1$. Thus, the eigenvalue $z_1 = -1$ has a multiplicity of two. To check the degeneracy of the matrix $z_1\mathbf{I} - \mathbf{A}$, we form

$$z_1\mathbf{I} - \mathbf{A} = \begin{bmatrix} z_1 - 1 & -2 \\ 2 & z_1 + 3 \end{bmatrix} = \begin{bmatrix} -2 & -2 \\ 2 & 2 \end{bmatrix} \tag{5-149}$$

which has a rank of one. Thus, the degeneracy of $z_1\mathbf{I} - \mathbf{A}$ is one. This means that we can find only one independent eigenvector for the eigenvalue z_1 from

$$\text{adj}\,(z_1\mathbf{I} - \mathbf{A}) = \begin{bmatrix} 2 & 2 \\ -2 & -2 \end{bmatrix} \tag{5-150}$$

Thus, the eigenvector of $z_1 = -1$ is

$$\mathbf{p}_1 = \begin{bmatrix} 1 \\ -1 \end{bmatrix} \tag{5-151}$$

To find the generalized eigenvector, we set

$$(z_1\mathbf{I} - \mathbf{A})\mathbf{p}_2 = -\mathbf{p}_1 = \begin{bmatrix} -1 \\ 1 \end{bmatrix} \tag{5-152}$$

or

$$\begin{bmatrix} -2 & -2 \\ 2 & 2 \end{bmatrix}\mathbf{p}_2 = \begin{bmatrix} -1 \\ 1 \end{bmatrix} \tag{5-153}$$

Solving the last equation, we get the solution for $\mathbf{p}_2$ as

$$\mathbf{p}_2 = \begin{bmatrix} 0 \\ 0.5 \end{bmatrix} \tag{5-154}$$

As a summary of the discussions on the state-variable analysis of discrete-data and digital systems given in the preceding sections, a tabulation and comparison of the results are given in Table 5-1.

5-9 DIAGONALIZATION OF THE A MATRIX

Solving the state equations of a linear time-invariant digital system would be a simple matter if these equations were decoupled from each other, that is, if $\mathbf{A}$ were a diagonal matrix. For instance, if the state equations of an nth-order digital system are of the form

$$x_i(k+1) = z_i x_i(k) + \sum_{j=1}^{r} \gamma_j u_j(k) \tag{5-155}$$

$i = 1, 2, \ldots, n$, the solutions of these state equations, given $x_i(0)$ and $u_j(k)$ for $k \geq 0$, are simply

$$x_i(k) = z_i^k x_i(0) + \sum_{j=1}^{r} \sum_{m=0}^{k} \gamma_j z_i^{m-1} u_j(k-m) \tag{5-156}$$

Therefore, the state transition matrix $\boldsymbol{\phi}(k)$ is also a diagonal matrix with elements z_i^k, $i = 1, 2, \ldots, n$, on the main diagonal.

In general, if $\mathbf{A}$ has distinct eigenvalues, it can be diagonalized by a *similarity transformation*. Let us consider the discrete state equation,

$$\mathbf{x}(k+1) = \mathbf{A}\mathbf{x}(k) + \mathbf{B}\mathbf{u}(k) \tag{5-157}$$

where $\mathbf{x}(k)$ is an n-vector, $\mathbf{u}(k)$ an r-vector, and $\mathbf{A}$ has distinct eigenvalues, $z_1, z_2, \ldots, z_n$. Let $\mathbf{P}$ be a nonsingular matrix that transforms the state vector $\mathbf{x}(k)$ into $\mathbf{y}(k)$, i.e.,

$$\mathbf{x}(k) = \mathbf{P}\mathbf{y}(k) \tag{5-158}$$

Table 5-1 Tabulation of Results of State-Variable Analysis of Linear Systems

	Continuous-Data System	Digital System with Sample-and-Hold Devices	Digital System
State Variables	$\mathbf{x}(t)$	$\mathbf{x}(kT)$	$\mathbf{x}(k)$
State equations	$\dot{\mathbf{x}}(t) = \mathbf{A}\mathbf{x}(t) + \mathbf{B}\mathbf{u}(t)$	$\mathbf{x}[(k+1)T] = \boldsymbol{\phi}(T)\mathbf{x}(kT) + \boldsymbol{\theta}(T)\mathbf{u}(kT)$	$\mathbf{x}(k+1) = \mathbf{A}\mathbf{x}(k) + \mathbf{B}\mathbf{u}(k)$
State transition matrix	$\boldsymbol{\phi}(t) = e^{\mathbf{A}t}$	$\boldsymbol{\phi}(kT) = [\boldsymbol{\phi}(T)]^k$	$\boldsymbol{\phi}(k) = \mathbf{A}^k$
Transforms of state transition matrix	$\boldsymbol{\Phi}(s) = (s\mathbf{I} - \mathbf{A})^{-1}$	$\boldsymbol{\Phi}(z) = [z\mathbf{I} - \boldsymbol{\phi}(T)]^{-1}z$	$\boldsymbol{\Phi}(z) = (z\mathbf{I} - \mathbf{A})^{-1}z$
Impulse response matrix	$\mathbf{g}(t) = \mathbf{D}\boldsymbol{\phi}(t)\mathbf{B} + \mathbf{E}\delta(t)$	$\mathbf{g}(kT) = \mathbf{D}\boldsymbol{\phi}[(k-1)T]\mathbf{B} \quad k \geq 1$ $= \mathbf{E} \quad k = 0$	$\mathbf{g}(k) = \mathbf{D}\boldsymbol{\phi}(k-1)\mathbf{B} \quad k \geq 1$ $= \mathbf{E} \quad k = 0$
Transfer matrix	$\mathbf{G}(s) = \mathbf{D}(s\mathbf{I} - \mathbf{A})^{-1}\mathbf{B} + \mathbf{E}$	$\mathbf{G}(z) = \mathbf{D}[z\mathbf{I} - \boldsymbol{\phi}(T)]^{-1}\mathbf{B} + \mathbf{E}$	$\mathbf{G}(z) = \mathbf{D}(z\mathbf{I} - \mathbf{A})^{-1}\mathbf{B} + \mathbf{E}$
State transition equation	$\mathbf{x}(t) = \boldsymbol{\phi}(t - t_0)\mathbf{x}(t_0) + \int_{t_0}^{t} \boldsymbol{\phi}(t - \tau)\mathbf{B}\mathbf{u}(\tau)\,d\tau$	$\mathbf{x}(NT) = \boldsymbol{\phi}(NT)\mathbf{x}(0) + \sum_{k=0}^{N-1} \boldsymbol{\phi}[(N-k-1)T]\boldsymbol{\theta}(T)\mathbf{u}(kT)$	$\mathbf{x}(N) = \mathbf{A}^N\mathbf{x}(0) + \sum_{k=0}^{N-1} \mathbf{A}^{N-k-1}\mathbf{B}\mathbf{u}(k)$
Characteristic equation	$\lvert s\mathbf{I} - \mathbf{A}\rvert = 0$	$\lvert z\mathbf{I} - \boldsymbol{\phi}(T)\rvert = 0$	$\lvert z\mathbf{I} - \mathbf{A}\rvert = 0$

and

$$\mathbf{y}(k) = \mathbf{P}^{-1}\mathbf{x}(k) \tag{5-159}$$

The transformed state equations are represented as

$$\mathbf{y}(k+1) = \mathbf{\Lambda y}(k) + \mathbf{\Gamma u}(k) \tag{5-160}$$

where

$$\mathbf{\Lambda} = \begin{bmatrix} z_1 & 0 & 0 & \cdots & 0 \\ 0 & z_2 & 0 & \cdots & \\ 0 & 0 & z_3 & \cdots & 0 \\ \cdots & \cdots & \cdots & \cdots & \cdots \\ 0 & 0 & 0 & \cdots & z_n \end{bmatrix} \qquad (n \times n) \tag{5-161}$$

The decoupled state equations in Eq. (5-160) are known as the *canonical form*. To find the matrix **P**, we substitute Eq. (5-158) into Eq. (5-157), and using the identity of Eq. (5-159), we have

$$\mathbf{\Lambda} = \mathbf{P}^{-1}\mathbf{AP} \tag{5-162}$$

and

$$\mathbf{\Gamma} = \mathbf{P}^{-1}\mathbf{B} \qquad (n \times r) \tag{5-163}$$

Several methods exist for determining the matrix **P**, given the matrix **A** and its eigenvalues. However, we shall show that *the columns of* **P** *are always the eigenvectors of* **A**. Let $\mathbf{p}_i$ represent the eigenvectors of **A** that correspond to z_i. Then,

$$\mathbf{P} = [\mathbf{p}_1 \quad \mathbf{p}_2 \quad \cdots \quad \mathbf{p}_n] \tag{5-164}$$

The proof of this relationship is carried out using the definition of the eigenvector, Eq. (5-120), which is written as

$$z_i\mathbf{p}_i = \mathbf{A}\mathbf{p}_i \qquad i = 1, 2, \ldots, n \tag{5-165}$$

Forming the $n \times n$ matrix,

$$\begin{aligned} [z_1\mathbf{p}_1 \quad z_2\mathbf{p}_2 \quad \cdots \quad z_n\mathbf{p}_n] &= [\mathbf{Ap}_1 \quad \mathbf{Ap}_2 \quad \cdots \quad \mathbf{Ap}_n] \\ &= \mathbf{AP} \end{aligned} \tag{5-166}$$

which is also written as

$$[\mathbf{p}_1 \quad \mathbf{p}_2 \quad \cdots \quad \mathbf{p}_n]\mathbf{\Lambda} = \mathbf{P\Lambda} = \mathbf{AP} \tag{5-167}$$

This leads to Eq. (5-162).

5-10 JORDAN CANONICAL FORM

The condition that the **A** matrix cannot be diagonalized does not mean that the system cannot be transformed into a form for which the solution of the state equations can be written by inspection. When **A** cannot be diagonalized, a similarity transformation $\Lambda = \mathbf{P}^{-1}\mathbf{AP}$ exists such that Λ is the *Jordan canonical form*, which is nearly a diagonal matrix. As an illustrative example of the Jordan canonical form, consider that $\mathbf{A}$ has eigenvalues z_1, z_2, z_3, z_3, and z_3, the last three being identical. The Jordan canonical form matrix Λ is given by

$$\Lambda = \begin{bmatrix} z_1 & 0 & 0 & 0 & 0 \\ 0 & z_2 & 0 & 0 & 0 \\ 0 & 0 & z_3 & 1 & 0 \\ 0 & 0 & 0 & z_3 & 1 \\ 0 & 0 & 0 & 0 & z_3 \end{bmatrix} \tag{5-168}$$

A Jordan canonical form matrix generally has the following properties.

1. The diagonal elements of the matrix Λ are the eigenvalues of $\mathbf{A}$.
2. All the elements below the principal diagonal are zeros.
3. Some of the elements immediately above the principal diagonal are ones.
4. The submatrices formed by each eigenvalue of Λ as shown by the dotted sections in Eq. (5-168) are called the *Jordan blocks*.

As an example, the Jordan block associated with an eigenvalue z_1 that has a multiplicity of four is

$$\Lambda = \begin{bmatrix} z_1 & 1 & 0 & 0 \\ 0 & z_1 & 1 & 0 \\ 0 & 0 & z_1 & 1 \\ 0 & 0 & 0 & z_1 \end{bmatrix} \tag{5-169}$$

There is a good reason for using the Jordan canonical form even though it is not a diagonal matrix. Consider the following state equation

$$\mathbf{y}(k+1) = \Lambda\mathbf{y}(k) \tag{5-170}$$

where Λ is as given in Eq. (5-169). It is possible to find the state transition matrix $\boldsymbol{\phi}(k)$ in a systematic manner with almost the same ease as in the case of a diagonal matrix. Notice that the last state equation in Eq. (5-170) is entirely decoupled from the other equations. Thus, the solution to $y_4(k)$ is

$$y_4(k) = z_1^k y_4(0) \tag{5-171}$$

The third state equation reads

$$y_3(k+1) = z_1 y_3(k) + y_4(k) \tag{5-172}$$

Since $y_4(k)$ is already given in Eq. (5-171), Eq. (5-172) is easily solved to yield

$$y_3(k) = z_1^k y_3(0) + kz_1^{k-1} y_4(0) \tag{5-173}$$

Similarly, the second state equation is

$$y_2(k+1) = z_1 y_2(k) + y_3(k) \tag{5-174}$$

Again, substituting $y_3(k)$ from Eq. (5-173) into Eq. (5-174) and solving, we have

$$y_2(k) = z_1^k y_2(0) + kz_1^{k-1} y_3(0) + \frac{k(k-1)}{2!} z_1^{k-2} y_4(0) \tag{5-175}$$

Continuing with the same process, the solution of $y_1(k)$ is

$$y_1(k) = z_1^k y_1(0) + kz_1^{k-1} y_2(0) + \frac{k(k-1)}{2!} z_1^{k-2} y_3(0) + \frac{k(k-1)(k-2)}{3!} z_1^{k-3} y_4(0) \tag{5-176}$$

In matrix form, the state transition equation is written as

$$\mathbf{y}(k) = \boldsymbol{\phi}(k)\mathbf{y}(0) \tag{5-177}$$

where

$$\boldsymbol{\phi}(k) = z_1^k \begin{bmatrix} 1 & kz_1^{-1} & \frac{k(k-1)}{2!} z_1^{-2} & \frac{k(k-1)(k-2)}{3!} z_1^{-3} \\ 0 & 1 & kz_1^{-1} & \frac{k(k-1)}{2!} z_1^{-2} \\ 0 & 0 & 1 & kz_1^{-1} \\ 0 & 0 & 0 & 1 \end{bmatrix} \tag{5-178}$$

In general, if z_1 is of mth-order multiplicity, the state transition matrix is of the form

$$\boldsymbol{\phi}(k) = z_1^k \begin{bmatrix} 1 & kz_1^{-1} & \frac{k(k-1)}{2!} z_1^{-2} & \cdots & \frac{k(k-1)\cdots(k-m+2)}{(m-1)!} z_1^{-m+1} \\ 0 & 1 & kz_1^{-1} & & \cdot \\ 0 & 0 & 1 & & \cdot \\ \cdot & \cdot & \cdot & & \frac{k(k-1)}{2!} z_1^{-2} \\ \cdot & \cdot & \cdot & & kz_1^{-1} \\ 0 & \cdot & \cdot & & 1 \end{bmatrix} \tag{5-179}$$

Now we shall determine the matrix **P** that will transform **A** with multiple-order eigenvalues into a Jordan canonical form. The matrix **P** is again formed using the eigenvectors of **A**, as in Eq. (5-164). The eigenvectors associated with the distinct eigenvalues of **A** are determined in the usual manner. Consider that the jth eigenvalue z_j is of mth order. The eigenvectors associated with an mth-order Jordan canonical form are found by referring to the Jordan block of the form in Eq. (5-169). Then, using $\mathbf{\Lambda} = \mathbf{P}^{-1}\mathbf{AP}$, the following relationship must hold:

$$[\mathbf{p}_1 \quad \mathbf{p}_2 \quad \cdots \quad \mathbf{p}_m]\mathbf{\Lambda} = \mathbf{A}[\mathbf{p}_1 \quad \mathbf{p}_2 \quad \cdots \quad \mathbf{p}_m] \tag{5-180}$$

The last matrix equation is expanded to

$$\begin{aligned} z_j\mathbf{p}_1 &= \mathbf{Ap}_1 \\ \mathbf{p}_1 + z_j\mathbf{p}_2 &= \mathbf{Ap}_2 \\ \mathbf{p}_2 + z_j\mathbf{p}_3 &= \mathbf{Ap}_3 \\ &\vdots \\ \mathbf{p}_{m-1} + z_j\mathbf{p}_m &= \mathbf{Ap}_m \end{aligned} \tag{5-181}$$

After rearranging, these equations become

$$\begin{aligned} (z_j\mathbf{I} - \mathbf{A})\mathbf{p}_1 &= \mathbf{0} \\ (z_j\mathbf{I} - \mathbf{A})\mathbf{p}_2 &= -\mathbf{p}_1 \\ (z_j\mathbf{I} - \mathbf{A})\mathbf{p}_3 &= -\mathbf{p}_2 \\ &\vdots \\ (z_j\mathbf{I} - \mathbf{A})\mathbf{p}_m &= -\mathbf{p}_{m-1} \end{aligned} \tag{5-182}$$

The generalized eigenvectors, $\mathbf{p}_1, \mathbf{p}_2, \ldots, \mathbf{p}_m$, are determined from these equations above.

5-11 METHODS OF COMPUTING THE STATE TRANSITION MATRIX

In the early sections of this chapter the state transition matrix of a digital system was defined. It is worthwhile to emphasize the difference between the formulations of the state transition matrix under two different conditions. For the sampled-data system,

$$\dot{\mathbf{x}}(t) = \mathbf{Ax}(t) + \mathbf{Bu}(t) \tag{5-183}$$

where

$$\mathbf{u}(t) = \mathbf{u}(kT) \qquad kT \le t < (k+1)T \tag{5-184}$$

The discretized state equations are expressed as

$$\mathbf{x}[(k+1)T] = \boldsymbol{\phi}(T)\mathbf{x}(kT) + \boldsymbol{\theta}(T)\mathbf{u}(kT) \tag{5-185}$$

where $\boldsymbol{\phi}(T)$ is the state transition matrix of $\mathbf{A}$ with t replaced by T; that is,

$$\boldsymbol{\phi}(T) = e^{\mathbf{A}T} = \boldsymbol{\phi}(t)|_{t=T} \tag{5-186}$$

The state transition matrix of $\boldsymbol{\phi}(T)$ is given by

$$\boldsymbol{\phi}(NT) = \underbrace{\boldsymbol{\phi}(T)\boldsymbol{\phi}(T)\cdots\boldsymbol{\phi}(T)}_{\longleftarrow N \longrightarrow} \tag{5-187}$$

Since $\boldsymbol{\phi}(t)$ is the state transition matrix of $\mathbf{A}$, $\boldsymbol{\phi}(NT)$ is determined from $\boldsymbol{\phi}(t)$ by substituting NT for t.

The state equations of a digital control system are expressed as

$$\mathbf{x}(k+1) = \mathbf{A}\mathbf{x}(k) + \mathbf{B}\mathbf{u}(k) \tag{5-188}$$

The state transition matrix is defined as

$$\boldsymbol{\phi}(N) = \mathbf{A}^N = \underbrace{\mathbf{A}\cdot\mathbf{A}\cdot\mathbf{A}\cdot\cdots\cdot\mathbf{A}}_{\longleftarrow N \longrightarrow} \tag{5-189}$$

In this case, $\boldsymbol{\phi}(N)$ is obtained by multiplying $\mathbf{A}$ by itself N times.

We shall first present two methods of computing $\boldsymbol{\phi}(NT)$ given $\boldsymbol{\phi}(T)$ in the sampled-data case, or $\boldsymbol{\phi}(NT)$ given $\mathbf{A}$ in the digital case.

5-11-1 The Cayley-Hamilton Theorem Method

Given the matrix $\boldsymbol{\phi}(T)$ or $\mathbf{A}$, as the case may be, Eq. (5-187) or Eq. (5-189) can be computed using the Cayley-Hamilton theorem.

The theorem states that *every square matrix must satisfy its own characteristic equation.* For example, consider that the characteristic equation of $\mathbf{A}$ $(n \times n)$ is written as

$$z^n + a_{n-1}z^{n-1} + a_{n-2}z^{n-2} + \cdots + a_1 z + a_0 = 0 \tag{5-190}$$

then,

$$\mathbf{A}^n + a_{n-1}\mathbf{A}^{n-1} + a_{n-2}\mathbf{A}^{n-2} + \cdots + a_1\mathbf{A} + a_0\mathbf{I} = \mathbf{0} \tag{5-191}$$

Thus,

$$\mathbf{A}^n = -(a_{n-1}\mathbf{A}^{n-1} + a_{n-2}\mathbf{A}^{n-2} + \cdots + a_1\mathbf{A} + a_0\mathbf{I}) \tag{5-192}$$

Similarly, if the state equation is given in the form of Eq. (5-185), the state transition matrix $\boldsymbol{\phi}(nT)$ must satisfy the characteristic equation of $\boldsymbol{\phi}(T)$ $(n \times n)$; that is,

$$\boldsymbol{\phi}(nT) = -\{a_{n-1}\boldsymbol{\phi}[(n-1)T] + a_{n-2}\boldsymbol{\phi}[(n-2)T] + \cdots + a_1\boldsymbol{\phi}(T) + a_0\mathbf{I})\} \tag{5-193}$$

where a_i, $i = 0, 1, 2, \ldots, n-1$, denotes the coefficients of the characteristic equation of $\boldsymbol{\phi}(T)$.

Example 5-8

To illustrate the use of the Cayley-Hamilton theorem, consider that the matrix $\mathbf{A}$ in Eq. (5-188) is

$$\mathbf{A} = \begin{bmatrix} 3 & 2 \\ 2 & 3 \end{bmatrix} \tag{5-194}$$

The characteristic equation of $\mathbf{A}$ is

$$|z\mathbf{I} - \mathbf{A}| = z^2 - 6z + 5 = 0 \tag{5-195}$$

Applying the Cayley-Hamilton theorem, we have the matrix equation

$$\mathbf{A}^2 = 6\mathbf{A} - 5\mathbf{I} \tag{5-196}$$

Thus, $\mathbf{A}^2$ is expressed in terms of $\mathbf{A}$. The crux of the Cayley-Hamilton theorem is that $\mathbf{A}^N$ can be expressed as an algebraic sum of $\mathbf{A}^{N-1}$, $\mathbf{A}^{N-2}, \ldots$, for any $N \geq n$. By repeatedly applying the theorem, $\mathbf{A}^N$ can eventually be expressed in terms of $\mathbf{A}$. For example, to find $\mathbf{A}^3$, we simply multiply both sides of Eq. (5-196) by $\mathbf{A}$. Thus,

$$\mathbf{A}^3 = 6\mathbf{A}^2 - 5\mathbf{A} = 31\mathbf{A} - 30\mathbf{I} \tag{5-197}$$

Similarly,

$$\mathbf{A}^4 = 6\mathbf{A}^3 - 5\mathbf{A}^2 = 156\mathbf{A} - 155\mathbf{I} \tag{5-198}$$

and so on.

5-11-2 The z-Transform Method

For the sampled-data system of Eq. (5-185), the state transition matrix of $\phi(T)$ is expressed in terms of the z-transform in Eq. (5-76) as

$$\phi(NT) = \mathscr{Z}^{-1}\{[z\mathbf{I} - \phi(T)]^{-1}z\} \tag{5-199}$$

For the digital control system of Eq. (5-188), the state transition matrix of $\mathbf{A}$ is expressed as

$$\phi(N) = \mathscr{Z}^{-1}[(z\mathbf{I} - \mathbf{A})^{-1}z] \tag{5-200}$$

In these last two equations the evaluation of the state transition matrix involves the matrix inverse and then the inverse z-transform. For second-order systems, these analytical steps can be carried out with ease. However, for higher order systems, the amount of work involved can be quite tedious if carried out by hand.

The task of finding $(z\mathbf{I} - \mathbf{A})^{-1}z$ in Eq. (5-200), given $\mathbf{A}$, can be simplified by the following method. Let

$$\mathbf{F} = (z\mathbf{I} - \mathbf{A})^{-1}z \tag{5-201}$$

Premultiplying both sides of the last equation by $z\mathbf{I} - \mathbf{A}$, and rearranging, we get

$$z\mathbf{F} = z\mathbf{I} + \mathbf{A}\mathbf{F} \tag{5-202}$$

Premultiplying both sides of the last equation by $z\mathbf{I} + \mathbf{A}$, and rearranging, we get

$$z^2\mathbf{F} = A^2\mathbf{F} + z\mathbf{A} + z^2\mathbf{I} \tag{5-203}$$

By repeatedly multiplying by $z\mathbf{I} + \mathbf{A}$, we have

$$z^3\mathbf{F} = \mathbf{A}^3\mathbf{F} + z\mathbf{A}^2 + z^2\mathbf{A} + z^3\mathbf{I} \tag{5-204}$$

$$z^4\mathbf{F} = \mathbf{A}^4\mathbf{F} + z\mathbf{A}^3 + z^2\mathbf{A}^2 + z^3\mathbf{A} + z^4\mathbf{I} \tag{5-205}$$

$$\vdots$$

$$z^n\mathbf{F} = \mathbf{A}^n\mathbf{F} + z\mathbf{A}^{n-1} + z^2\mathbf{A}^{n-2} + \cdots + z^{n-1}\mathbf{A} + z^n\mathbf{I} \tag{5-206}$$

Let the characteristic equation of $\mathbf{A}$ be as in Eq. (5-190). We form the following equations by using the coefficients of the characteristic equation and Eqs. (5-202) through (5-206):

$$\begin{aligned}
a_0\mathbf{F} &= a_0\mathbf{F} \\
a_1 z\mathbf{F} &= a_1\mathbf{A}\mathbf{F} + a_1 z\mathbf{I} \\
a_2 z^2\mathbf{F} &= a_2\mathbf{A}^2\mathbf{F} + a_2 z\mathbf{A} + a_2 z^2\mathbf{I} \\
a_3 z^3\mathbf{F} &= a_3\mathbf{A}^3\mathbf{F} + a_3 z\mathbf{A}^2 + a_3 z^2\mathbf{A} + a_3 z^3\mathbf{I} \\
&\vdots \\
a_{n-1}z^{n-1}\mathbf{F} &= a_{n-1}\mathbf{A}^{n-1}\mathbf{F} + a_{n-1}z\mathbf{A}^{n-2} + a_{n-1}z^2\mathbf{A}^{n-3} + \cdots \\
&\quad + a_{n-1}z^{n-2}\mathbf{A} + a_{n-1}z^{n-1}\mathbf{I} \\
z^n\mathbf{F} &= \mathbf{A}^n\mathbf{F} + z\mathbf{A}^{n-1} + z^2\mathbf{A}^{n-2} + \cdots + z^{n-1}\mathbf{A} + z^n\mathbf{I}
\end{aligned} \tag{5-207}$$

The equations in Eq. (5-207) are summed on both sides to give

$$\begin{aligned}
\sum_{i=0}^{n} a_i z^i\mathbf{F} &= \sum_{i=0}^{n} a_i\mathbf{A}^i\mathbf{F} + \sum_{i=1}^{n} a_i\mathbf{A}^{i-1}z + \sum_{i=2}^{n} a_i\mathbf{A}^{i-2}z^2 + \cdots \\
&\quad + \sum_{i=n-1}^{n} a_i\mathbf{A}^{i-n+1}z^{n-1} + z^n\mathbf{I}
\end{aligned} \tag{5-208}$$

where $a_n = 1$. Due to the Cayley-Hamilton theorem, the first term on the right-hand side of the last equation is a null matrix. Thus, Eq. (5-208) leads to

$$\mathbf{F} = \frac{\sum_{j=1}^{n} z^j \sum_{i=j}^{n} a_i\mathbf{A}^{i-j}}{\sum_{i=0}^{n} a_i z^i} = \frac{\sum_{j=1}^{n} z^j \sum_{i=j}^{n} a_i\mathbf{A}^{i-j}}{|z\mathbf{I} - \mathbf{A}|} \tag{5-209}$$

The numerator of the last equation is also known as the matrix,

$$[\text{adj}\,(z\mathbf{I} - \mathbf{A})]z$$

Once $\mathbf{F}$ is evaluated from Eq. (5-209), $\boldsymbol{\phi}(N)$ is given by

$$\boldsymbol{\phi}(N) = \mathscr{Z}^{-1}(\mathbf{F}) \tag{5-210}$$

For the case in Eq. (5-199), simply replace $\mathbf{A}$ by $\boldsymbol{\phi}(T)$ in Eq. (5-209).

Example 5-9

Consider that the digital control system described by Eq. (5-188) has the **A** matrix

$$\mathbf{A} = \begin{bmatrix} 0 & 1 & 0 \\ 0 & 0 & 1 \\ -6 & -11 & -6 \end{bmatrix} \tag{5-211}$$

The characteristic equation of **A** is

$$\begin{aligned} |z\mathbf{I} - \mathbf{A}| &= z^3 + 6z^2 + 11z + 6 \\ &= (z+1)(z+2)(z+3) = 0 \end{aligned} \tag{5-212}$$

The coefficients of the characteristic equation are $a_3 = 1$, $a_2 = 6$, $a_1 = 11$, and $a_0 = 6$.

Using Eq. (5-209), the matrix **F** is written as

$$\mathbf{F} = (z\mathbf{I} - \mathbf{A})^{-1}z = \frac{\sum_{j=1}^{3} z^j \sum_{i=j}^{3} a_i \mathbf{A}^{i-j}}{|z\mathbf{I} - \mathbf{A}|}$$

$$= \frac{a_3 z^3 \mathbf{I} + (a_2\mathbf{I} + a_3\mathbf{A})z^2 + (a_1\mathbf{I} + a_2\mathbf{A} + a_3\mathbf{A}^2)z}{|z\mathbf{I} - \mathbf{A}|} \tag{5-213}$$

Substituting the coefficients a_3, a_2, a_1, and **A** into the last equation, we have

$$\mathbf{F} = \frac{z^3\mathbf{I} + \begin{bmatrix} 6 & 1 & 0 \\ 0 & 6 & 1 \\ -6 & -11 & 0 \end{bmatrix} z^2 + \begin{bmatrix} 11 & 6 & 1 \\ -6 & 0 & 0 \\ 0 & -6 & 0 \end{bmatrix} z}{(z-1)(z-2)(z-3)} \tag{5-214}$$

Performing the partial-fraction expansion, Eq. (5-214) gives

$$\mathbf{F} = \frac{z}{2(z-1)} \begin{bmatrix} 18 & 7 & 1 \\ -6 & 7 & 1 \\ -6 & -17 & 1 \end{bmatrix} - \frac{z}{z-2} \begin{bmatrix} 27 & 8 & 1 \\ -6 & 16 & 2 \\ -12 & -28 & 4 \end{bmatrix}$$

$$+ \frac{z}{2(z-3)} \begin{bmatrix} 38 & 9 & 1 \\ -6 & 27 & 3 \\ -18 & -39 & 9 \end{bmatrix} \tag{5-215}$$

Now taking the inverse z-transform on both sides of the last equation, we get

$$\boldsymbol{\phi}(k) = \frac{1}{2}\begin{bmatrix} 18 & 7 & 1 \\ -6 & 7 & 1 \\ -6 & -17 & 1 \end{bmatrix} - \begin{bmatrix} 27 & 8 & 1 \\ -6 & 16 & 2 \\ -12 & -28 & 4 \end{bmatrix} e^{-0.693k} + \frac{1}{2}\begin{bmatrix} 38 & 9 & 1 \\ -6 & 27 & 3 \\ -18 & -39 & 9 \end{bmatrix} e^{-1.1k} \tag{5-216}$$

We can check this result by verifying that $\boldsymbol{\phi}(0) = \mathbf{I}$, the identity matrix.

5-11-3 Computing the State Transition Matrix $\boldsymbol{\phi}(T)$

The discussions conducted thus far have been concentrated on the computation of the state transition matrix $\boldsymbol{\phi}(NT)$ or $\boldsymbol{\phi}(N)$ of the sampled-data or digital control system. When a sampled-data system is encountered, the starting point is Eq. (5-183) in which the matrices $\mathbf{A}$ and $\mathbf{B}$ are given. In order to find $\boldsymbol{\phi}(NT)$ using Eq. (5-187), we must first find $\boldsymbol{\phi}(T)$ which is given by Eq. (5-186). Earlier in the chapter the state transition matrix of $\mathbf{A}$, $\boldsymbol{\phi}(t)$, is given by Eqs. (5-16) and (5-20). Equation (5-16) is the Laplace transform solution, whereas Eq. (5-20) is the power-series representation. We shall investigate various methods of computing $\boldsymbol{\phi}(t)$, given the matrix $\mathbf{A}$.

The Laplace Transform Method

Repeating Eq. (5-16),

$$\boldsymbol{\phi}(t) = \mathscr{L}^{-1}[(s\mathbf{I} - \mathbf{A})^{-1}] \tag{5-217}$$

The matrix inverse of $s\mathbf{I} - \mathbf{A}$ can be conducted using essentially the same procedure as described in Eqs. (5-201) through (5-209). Thus,

$$(s\mathbf{I} - \mathbf{A})^{-1} = \frac{\sum_{j=1}^{n} s^j \sum_{i=j}^{n} a_i \mathbf{A}^{i-j}}{s|s\mathbf{I} - \mathbf{A}|} \tag{5-218}$$

where n is the dimension of $\mathbf{A}$, and

$$|s\mathbf{I} - \mathbf{A}| = s^n + a_{n-1}s^{n-1} + \cdots + a_1 s + a_0 \tag{5-219}$$

Direct Power-Series Expansion Method

The power-series representation of $\boldsymbol{\phi}(T)$ is

$$\boldsymbol{\phi}(T) = e^{\mathbf{A}T} = \mathbf{I} + \mathbf{A}T + \frac{\mathbf{A}^2T^2}{2!} + \cdots \tag{5-220}$$

This expression can be programmed recursively. For example, the kth term of the series is $\mathbf{A}^k T^k/k!$, and the $(k+1)$st term is $\mathbf{A}^{k+1}T^{k+1}/(k+1)!$. Thus, we can write

$$(k+1)\text{st term} = \frac{\mathbf{A}T}{k+1} \times k\text{th term} \tag{5-221}$$

$k = 0, 1, 2, \ldots$. Normally, when computing the series, a check is made on the convergence, and the recursive iteration can be stopped after N terms.

Truncated Power-Series Expansion Method (Cayley-Hamilton Theorem)

The power series of $\boldsymbol{\phi}(T)$ can be truncated after N terms, where N is some positive integer. Then,

$$\boldsymbol{\phi}(T) \cong \boldsymbol{\phi}_N(T) = \sum_{k=0}^{N} \frac{\mathbf{A}^k T^k}{k!} \tag{5-222}$$

Applying the Cayley-Hamilton theorem to the last equation, we have

$$\boldsymbol{\phi}_N(T) = \sum_{j=0}^{n-1} \alpha_j \mathbf{A}^j \tag{5-223}$$

where n is the dimension of $\mathbf{A}$, and α_j are constants that depend on the coefficients of the characteristic equation of $\mathbf{A}$ and can be computed recursively. In general, N and n are not related, so N can be greater than n.

The Eigenvalue Method (Sylvester's Expansion Theorem: A Has Distinct Eigenvalues Only)

For systems with distinct eigenvalues, Sylvester's expansion theorem states that if

$$f(\mathbf{A}) = \sum_{k=1}^{\infty} c_k \mathbf{A}^k \tag{5-224}$$

then

$$f(\mathbf{A}) = \sum_{i=1}^{n} f(s_i)\mathbf{F}(s_i) \tag{5-225}$$

where s_i, $i = 1, 2, \ldots, n$, are the eigenvalues (all distinct) of $\mathbf{A}$.

$$\mathbf{F}(s_i) = \sum_{\substack{j=1 \\ i \neq j}}^{n} \frac{\mathbf{A} - s_j\mathbf{I}}{s_i - s_j} \qquad i = 1, 2, \ldots, n \tag{5-226}$$

For the state transition matrix problem, we have

$$f(\mathbf{A}) = \boldsymbol{\phi}(T) = \sum_{k=0}^{\infty} \frac{\mathbf{A}^k T^k}{k!} = e^{\mathbf{A}T} \tag{5-227}$$

Thus,

$$f(s_i) = e^{s_i T} \tag{5-228}$$

and from Eq. (5-225),

$$\phi(T) = \sum_{i=1}^{n} e^{s_i T}\mathbf{F}(s_i) \tag{5-229}$$

where $\mathbf{F}(s_i)$ is given by Eq. (5-226).

Example 5-10

To illustrate the eigenvalue method described above, consider the matrix **A** given in Eq. (5-194) but for the state equations in Eq. (5-8). The eigenvalues of **A** are $s_1 = 5$ and $s_2 = 1$, which are distinct. From Eq. (5-228),

$$f(s_1) = e^{5T} \tag{5-230}$$

$$f(s_2) = e^{T} \tag{5-231}$$

Equation (5-226) gives

$$\mathbf{F}(s_1) = \frac{\mathbf{A} - s_2\mathbf{I}}{s_1 - s_2} = \begin{bmatrix} 0.5 & 0.5 \\ 0.5 & 0.5 \end{bmatrix} \tag{5-232}$$

$$\mathbf{F}(s_2) = \frac{\mathbf{A} - s_1\mathbf{I}}{s_2 - s_1} = \begin{bmatrix} 0.5 & -0.5 \\ -0.5 & 0.5 \end{bmatrix} \tag{5-233}$$

Thus,

$$\phi(T) = f(s_1)\mathbf{F}(s_1) + f(s_2)\mathbf{F}(s_2)$$

$$= 0.5\begin{bmatrix} e^{5T} + e^{T} & e^{5T} - e^{T} \\ e^{5T} - e^{T} & e^{5T} + e^{T} \end{bmatrix} \tag{5-234}$$

5-12 RELATIONSHIP BETWEEN STATE EQUATIONS AND HIGH-ORDER DIFFERENCE EQUATIONS

In the preceding sections we discussed state equations and their solutions for linear discrete-data and digital systems. Although it is usually possible to write the state equations directly from a system, in practice, the digital system may already be described by a high-order difference equation or a transfer function. Therefore, it is useful to investigate how state equations can be written directly from the difference equations or the transfer functions.

The procedure of arriving at the state equations from the transfer function is referred to as *decomposition*, and the subject is treated in Sec. 5-15. It will be shown that it is generally simpler first to transform the high-order difference equation into a transfer function and then apply a decomposition scheme to get the state equations. In this section we shall first establish some basic relationships on the formulation of the state equations of a high-order system.

Let us consider that a single-variable linear digital or discrete-data system is described by the following nth-order difference equation:

$$c(k+n) + a_{n-1}c(k+n-1) + a_{n-2}c(k+n-2) + \cdots + a_1 c(k+1) + a_0 c(k) = r(k) \qquad \text{(5-235)}$$

where $c(k)$ is the output variable and $r(k)$ is the input. The coefficients, $a_j, j = 0, 1, 2, \ldots, n-1$, are real constants.

The problem is to represent Eq. (5-235) by n state equations and q ($q \leq n$) output equations. The first step involves the defining of the state variables as functions of $c(k)$. Although the state variables of a given system are not unique, the most convenient way to define the state variables in the case of Eq. (5-235) is

$$\begin{aligned} x_1(k) &= c(k) \\ x_2(k) &= c(k+1) = x_1(k+1) \\ &\vdots \\ x_n(k) &= c(k+n-1) \end{aligned} \qquad \text{(5-236)}$$

After substitution of the relations in Eq. (5-236) into Eq. (5-235), and rearranging, the state equations are written as

$$\begin{aligned} x_1(k+1) &= x_2(k) \\ x_2(k+1) &= x_3(k) \\ &\vdots \\ x_n(k+1) &= -a_0 x_1(k) - a_1 x_2(k) - \cdots - a_{n-1} x_n(k) + r(k) \end{aligned} \qquad \text{(5-237)}$$

If we designate only $c(k)$ as the output variable, the output equation is simply

$$c(k) = x_1(k) \qquad \text{(5-238)}$$

The state equations in Eq. (5-237) are written in vector-matrix form:

$$\mathbf{x}(k+1) = \mathbf{A}\mathbf{x}(k) + \mathbf{B}r(k) \qquad \text{(5-239)}$$

where $\mathbf{x}(k)$ is the $n \times 1$ state vector, and $r(k)$ is the scalar input. The coefficient matrices are

$$\mathbf{A} = \begin{bmatrix} 0 & 1 & 0 & 0 & 0 & \cdots & 0 \\ 0 & 0 & 1 & 0 & 0 & \cdots & 0 \\ 0 & 0 & 0 & 1 & 0 & \cdots & 0 \\ \cdots & \cdots & \cdots & \cdots & \cdots & \cdots & \cdots \\ 0 & 0 & 0 & 0 & 0 & \cdots & 1 \\ -a_0 & -a_1 & -a_2 & -a_3 & -a_4 & \cdots & -a_{n-1} \end{bmatrix} \quad (n \times n) \qquad \text{(5-240)}$$

$$\mathbf{B} = \begin{bmatrix} 0 \\ 0 \\ \vdots \\ 0 \\ 1 \end{bmatrix} \quad (n \times n) \qquad \text{(5-241)}$$

The output equation in vector-matrix form is

$$c(k) = \mathbf{D}\mathbf{x}(k) \tag{5-242}$$

where

$$\mathbf{D} = [1 \quad 0 \quad 0 \quad \cdots \quad 0] \qquad (n \times 1) \tag{5-243}$$

The state equations in Eq. (5-239) with the coefficient matrices **A** and **B** in the basic forms of Eqs. (5-240) and (5-241) are called the *phase-variable canonical form* (PVCF).

In the sections to follow we shall show that a system represented in the phase-variable canonical form has certain unique and useful characteristics that are helpful in the analysis and design of linear discrete-data control systems.

Note that the high-order difference equation in Eq. (5-235) is not general, since there are no high-order terms on the side of the input. In general, when terms such as $r(k+1), r(k+2), \ldots$ are included, the procedure of assigning state variables would not be as straightforward as in Eq. (5-236). Under such a situation, it is more convenient first to transform the high-order difference equation into a transfer function and then conduct a decomposition to get the dynamic equations.

5-13 TRANSFORMATION TO PHASE-VARIABLE CANONICAL FORM —POLE-PLACEMENT DESIGN

One of the useful characteristics of the phase-variable canonical form is that *a system in this form can always have its eigenvalues assigned arbitrarily by state feedback*. In modern control theory, this type of design principle is often referred to as the *pole-placement design*, where the word "pole" here refers to the poles of the closed-loop transfer function which are the same as the eigenvalues of **A**.

□ Theorem 5-1

Let the state equations of a linear digital or discrete-data control system be represented by

$$\mathbf{x}(k+1) = \mathbf{A}\mathbf{x}(k) + \mathbf{B}u(k) \tag{5-244}$$

where **A** and **B** are in PVCF, as in Eqs. (5-240) and (5-241).

The eigenvalues of the system can be arbitrarily assigned by the state feedback

$$u(k) = -\mathbf{G}\mathbf{x}(k) \tag{5-245}$$

where **G** is the feedback matrix and is given by

$$\mathbf{G} = [g_1 \quad g_2 \quad \cdots \quad g_n] \tag{5-246}$$

and $g_1, g_2, \ldots, g_n$ are real constants.

A block diagram portraying the state feedback is shown in Fig. 5-4. ■

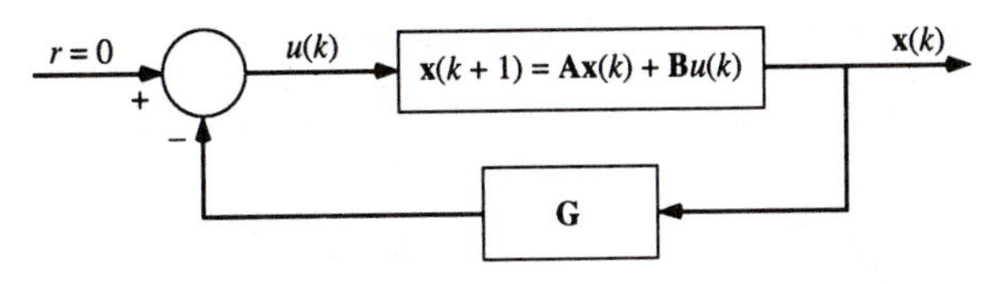

Figure 5-4. A linear digital system with state feedback.

Proof: Substituting the input $u(k)$ from Eq. (5-245) into Eq. (5-244), we have the state equations of the closed-loop system,

$$\mathbf{x}(k+1) = (\mathbf{A} - \mathbf{BG})\mathbf{x}(k) \tag{5-247}$$

The characteristic equation of the closed-loop system is

$$|z\mathbf{I} - \mathbf{A} + \mathbf{BG}| = 0 \tag{5-248}$$

Substitution of Eqs. (5-240), (5-241), and (5-246) into Eq. (5-248) gives

$$\begin{vmatrix} z & -1 & 0 & 0 & 0 & \cdots & 0 \\ 0 & z & -1 & 0 & 0 & \cdots & 0 \\ 0 & 0 & z & -1 & 0 & \cdots & 0 \\ \cdots & \cdots & \cdots & \cdots & \cdots & \cdots & \cdots \\ 0 & 0 & 0 & 0 & 0 & \cdots & -1 \\ a_0 + g_1 & a_1 + g_2 & a_2 + g_3 & a_3 + g_4 & a_4 + g_5 & \cdots & z + a_{n-1} + g_n \end{vmatrix} = 0 \tag{5-249}$$

This characteristic equation is evaluated to be

$$z^n + (a_{n-1} + g_n)z^{n-1} + (a_{n-2} + g_{n-1})z_{n-2} + \cdots + (a_1 + g_2)z + (a_0 + g_1) = 0 \tag{5-250}$$

Since each of the feedback gains is associated with only one of the coefficients of the last equation, it is apparent that if the eigenvalues of the characteristic equation of the closed-loop system are arbitrarily assigned, which corresponds to the coefficients of the equation being arbitrary, then equating these coefficients to the corresponding ones in Eq. (5-249) would yield n linearly independent equations for solving the feedback gains $g_1, g_2, \ldots, g_n$.

When a system is not in the phase-variable canonical form, a similarity transformation exists that transforms the matrices **A** and **B** into the proper forms of Eqs. (5-240) and (5-241), respectively. The following theorem describes a method of achieving this. ●

□ Theorem 5-2

Let the state equations of a linear digital or discrete-data system be represented by

$$\mathbf{x}(k+1) = \mathbf{Ax}(k) + \mathbf{B}u(k) \tag{5-251}$$

where $\mathbf{x}(k)$ is an n-vector, $u(t)$ is a scalar input, $\mathbf{A}$ is any $n \times n$ coefficient matrix, and $\mathbf{B}$ is any $n \times 1$ matrix, such that

$$\mathbf{S} = [\mathbf{B} \quad \mathbf{AB} \quad \mathbf{A}^2\mathbf{B} \quad \cdots \quad \mathbf{A}^{n-1}\mathbf{B}] \tag{5-252}$$

is nonsingular. Then there exists a nonsingular transformation

$$\mathbf{y}(k) = \mathbf{M}\mathbf{x}(k) \tag{5-253}$$

or

$$\mathbf{x}(k) = \mathbf{M}^{-1}\mathbf{y}(k) \tag{5-254}$$

which transforms Eq. (5-251) to the phase-variable canonical form

$$\mathbf{y}(k+1) = \mathbf{A}_1\mathbf{y}(k) + \mathbf{B}_1 u(k) \tag{5-255}$$

where

$$\mathbf{A}_1 = \begin{bmatrix} 0 & 1 & 0 & 0 & \cdots & 0 \\ 0 & 0 & 1 & 0 & \cdots & 0 \\ 0 & 0 & 0 & 1 & \cdots & 0 \\ \cdots & \cdots & \cdots & \cdots & \cdots & \cdots \\ 0 & 0 & 0 & 0 & & 1 \\ -a_0 & -a_1 & -a_2 & -a_3 & \cdots & -a_{n-1} \end{bmatrix} \tag{5-256}$$

$$\mathbf{B}_1 = \begin{bmatrix} 0 \\ 0 \\ \vdots \\ 0 \\ 1 \end{bmatrix} \tag{5-257}$$

The matrix $\mathbf{M}$ is given by

$$\mathbf{M} = \begin{bmatrix} \mathbf{M}_1 \\ \mathbf{M}_1\mathbf{A} \\ \vdots \\ \mathbf{M}_1\mathbf{A}^{n-2} \\ \mathbf{M}_2\mathbf{A}^{n-1} \end{bmatrix} \qquad (n \times n) \tag{5-258}$$

where

$$\mathbf{M}_1 = [0 \quad 0 \quad \cdots \quad 1][\mathbf{B} \quad \mathbf{AB} \quad \mathbf{A}^2\mathbf{B} \quad \cdots \quad \mathbf{A}^{n-1}\mathbf{B}]^{-1} \tag{5-259}$$

■

Proof: Let the matrix $\mathbf{M}$ be written as

$$\mathbf{M} = \begin{bmatrix} m_{11} & m_{12} & \cdots & m_{1n} \\ m_{21} & m_{22} & \cdots & m_{2n} \\ \cdots & \cdots & \cdots & \cdots \\ m_{n1} & m_{n2} & \cdots & m_{nn} \end{bmatrix} = \begin{bmatrix} \mathbf{M}_1 \\ \mathbf{M}_2 \\ \vdots \\ \mathbf{M}_n \end{bmatrix} \qquad (n \times n) \tag{5-260}$$

where

$$\mathbf{M}_i = [m_{i1} \quad m_{i2} \quad \cdots \quad m_{in}] \qquad i = 1, 2, \ldots, n \tag{5-261}$$

Equating the first row on both sides of Eq. (5-253) gives

$$y_1(k) = \mathbf{M}_1\mathbf{x}(k) \tag{5-262}$$

Advancing both sides of the last equation in time or stage by one, we get

$$y_1(k+1) = \mathbf{M}_1\mathbf{x}(k+1) \tag{5-263}$$

Substitution of the state equations of Eq. (5-251) in the last equation, and in view of Eq. (5-256), gives

$$y_1(k+1) = \mathbf{M}_1\mathbf{A}\mathbf{x}(k) + \mathbf{M}_1\mathbf{B}u(k) = y_2(k) \tag{5-264}$$

Since Eq. (5-253) stipulates that $\mathbf{y}(k)$ is a function of $\mathbf{x}(k)$ only, in Eq. (5-264) $\mathbf{M}_1\mathbf{B} = 0$. Thus, Eq. (5-264) becomes

$$y_1(k+1) = y_2(k) = \mathbf{M}_1\mathbf{A}\mathbf{x}(k) \tag{5-265}$$

Advancing the time or stage by one again in the last equation gives

$$y_1(k+2) = y_2(k+1) = y_3(k) = \mathbf{M}_1\mathbf{A}\mathbf{x}(k+1) = \mathbf{M}_1\mathbf{A}^2\mathbf{x}(k) \tag{5-266}$$

where for the same reason as in Eq. (5-264), $\mathbf{M}_1\mathbf{AB} = 0$.

Repeating the above procedure a total of $n - 1$ times leads to

$$y_{n-1}(k+1) = y_n(k) = \mathbf{M}_1\mathbf{A}^{n-1}\mathbf{x}(k) \tag{5-267}$$

with $\mathbf{M}_1\mathbf{A}^{n-2}\mathbf{B} = 0$. Therefore, collecting the above results, we have

$$y(k) = \begin{bmatrix} y_1(k) \\ y_2(k) \\ \vdots \\ y_n(k) \end{bmatrix} = \mathbf{M}\mathbf{x}(k) = \begin{bmatrix} \mathbf{M}_1 \\ \mathbf{M}_1\mathbf{A} \\ \vdots \\ \mathbf{M}_1\mathbf{A}^{n-1} \end{bmatrix} \mathbf{x}(k) \tag{5-268}$$

Thus, for any $\mathbf{x}(k)$,

$$\mathbf{M} = \begin{bmatrix} \mathbf{M}_1 \\ \mathbf{M}_1\mathbf{A} \\ \vdots \\ \mathbf{M}_1\mathbf{A}^{n-1} \end{bmatrix} \tag{5-269}$$

and, in addition, $\mathbf{M}_1$ should satisfy the condition

$$\mathbf{M}_1\mathbf{B} = \mathbf{M}_1\mathbf{AB} = \cdots = \mathbf{M}_1\mathbf{A}^{n-2}\mathbf{B} = 0 \tag{5-270}$$

From Eq. (5-253) we write

$$\begin{aligned} \mathbf{y}(k+1) = \mathbf{M}\mathbf{x}(k+1) &= \mathbf{MA}\mathbf{x}(k) + \mathbf{MB}u(k) \\ &= \mathbf{MAM}^{-1}\mathbf{y}(k) + \mathbf{MB}u(k) \end{aligned} \tag{5-271}$$

Comparing Eq. (5-271) with Eq. (5-255), we have

$$\mathbf{A}_1 = \mathbf{MAM}^{-1} \tag{5-272}$$

and

$$\mathbf{B}_1 = \mathbf{MB} \tag{5-273}$$

Then, from Eq. (5-269),

$$\mathbf{B}_1 = \begin{bmatrix} \mathbf{M}_1\mathbf{B} \\ \mathbf{M}_1\mathbf{AB} \\ \vdots \\ \mathbf{M}_1\mathbf{A}^{n-1}\mathbf{B} \end{bmatrix} = \begin{bmatrix} 0 \\ 0 \\ \vdots \\ 1 \end{bmatrix} \tag{5-274}$$

Since $\mathbf{M}_1$ is a $1 \times n$ matrix, Eq. (5-274) is written

$$\mathbf{M}_1[\mathbf{B} \quad \mathbf{AB} \quad \mathbf{A}^2\mathbf{B} \quad \cdots \quad \mathbf{A}^{n-1}\mathbf{B}] = [0 \quad 0 \quad \cdots \quad 1] \tag{5-275}$$

Thus, $\mathbf{M}_1$ is written as

$$\begin{aligned} \mathbf{M}_1 &= [0 \quad 0 \quad \cdots \quad 1][\mathbf{B} \quad \mathbf{AB} \quad \mathbf{A}^2\mathbf{B} \quad \cdots \quad \mathbf{A}^{n-1}\mathbf{B}]^{-1} \\ &= [0 \quad 0 \quad \cdots \quad 1]\mathbf{S}^{-1} \end{aligned} \tag{5-276}$$

where $\mathbf{S}$ is as defined in Eq. (5-252) and is assumed to be nonsingular. It will be shown later that the nonsingular condition of $\mathbf{S}$ is equivalent to the system being completely *state controllable.* Once $\mathbf{M}_1$ is determined from Eq. (5-276), the matrix $\mathbf{M}$ is given by Eq. (5-269). ●

□ Theorem 5-3

The two systems represented by the state equations of Eqs. (5-251) and (5-255), which are related through the similarity transformation of Eqs. (5-272) and (5-273), have the same eigenvalues. ■

Proof: The eigenvalues of the system in Eq. (5-251) are the roots of the characteristic equation:

$$|z\mathbf{I} - \mathbf{A}| = 0 \tag{5-277}$$

For the system of Eq. (5-255), the characteristic equation is

$$|z\mathbf{I} - \mathbf{A}_1| = |z\mathbf{I} - \mathbf{MAM}^{-1}| = 0 \tag{5-278}$$

Equation (5-278) is written as

$$|z\mathbf{MM}^{-1} - \mathbf{MAM}^{-1}| = 0 \tag{5-279}$$

or

$$|\mathbf{M}(z\mathbf{I} - \mathbf{A})\mathbf{M}^{-1}| = 0 \tag{5-280}$$

Since the determinant of the product of matrices is equal to the product of the determinants, we have

$$|\mathbf{M}(z\mathbf{I} - \mathbf{A})\mathbf{M}^{-1}| = |\mathbf{M}||z\mathbf{I} - \mathbf{A}||\mathbf{M}^{-1}| = |z\mathbf{I} - \mathbf{A}| \tag{5-281}$$

This proves that the characteristic equations of the two systems are the same, and their eigenvalues must be identical. ●

The significance of Theorem 5-3 is that, *given any digital or discrete-data system with a single input and its state equations given by Eq. (5-251), if the matrix* **S** *of Eq. (5-252) is nonsingular, the eigenvalues of the system can be arbitrarily assigned using the state feedback* $u(k) = -\mathbf{G}\mathbf{x}(k)$. Furthermore, we can transform the system into a PVCF (if it is not already one) so that the constant feedback gains of $u(t) = -\mathbf{G}_1\mathbf{y}(k)$ of the transformed system can be easily determined once the eigenvalues are assigned. The feedback matrix **G** of the original system is determined from

$$\mathbf{G} = \mathbf{G}_1\mathbf{M} \tag{5-282}$$

For the general multiple-input case, the necessary and sufficient conditions for eigenvalue assignment of a system with state feedback are still that the matrix **S** ($n \times nr$) must be of rank n. The proof is mathematically more involved, and it is not given here.

The following example illustrates the pole-placement design using the PVCF transformation.

Example 5-11

Consider that the state equations of a second-order digital control system are represented by

$$\mathbf{x}(k+1) = \mathbf{A}\mathbf{x}(k) + \mathbf{B}u(k) \tag{5-283}$$

where

$$\mathbf{A} = \begin{bmatrix} 1 & -1 \\ 0 & 1 \end{bmatrix} \qquad \mathbf{B} = \begin{bmatrix} 1 \\ 1 \end{bmatrix} \tag{5-284}$$

Find the feedback gain matrix **G** with the state feedback $u(k) = -\mathbf{G}\mathbf{x}(k)$ such that the eigenvalues of the closed-loop system are $z_1 = 0.4$ and $z_2 = 0.6$.

The eigenvalues of **A** (of the open-loop system) are $z = 1, 1$, so that the state feedback will move these eigenvalues inside the unit circle in the z-plane, thus stabilizing the system. Although it is possible to solve for the elements of **G** by equating the coefficients of the desired characteristic equation,

$$(z - 0.4)(z - 0.6) = z^2 - z + 0.24 = 0 \tag{5-285}$$

to those of the closed-loop system,

$$|z\mathbf{I} - \mathbf{A} + \mathbf{B}\mathbf{G}| = 0 \tag{5-286}$$

let us first transform the system into the PVCF and, at the same time, find out if the eigenvalues can be arbitrarily assigned. From Eq. (5-252),

$$\mathbf{S} = [\mathbf{B} \quad \mathbf{AB}] = \begin{bmatrix} 1 & 0 \\ 1 & 1 \end{bmatrix} \tag{5-287}$$

which is nonsingular. This shows that the eigenvalues of the open-loop system can be arbitrarily assigned. From Eq. (5-276),

$$\mathbf{M}_1 = [0 \quad 1] \quad \mathbf{S}^{-1} = [-1 \quad 1] \tag{5-288}$$

From Eq. (5-269) we get

$$\mathbf{M} = \begin{bmatrix} \mathbf{M}_1 \\ \mathbf{M}_1\mathbf{A} \end{bmatrix} = \begin{bmatrix} -1 & 1 \\ -1 & 2 \end{bmatrix} \tag{5-289}$$

The system in the PVCF is

$$\mathbf{y}(k+1) = \mathbf{A}_1\mathbf{y}(k) + \mathbf{B}_1 u(k) \tag{5-290}$$

where

$$\mathbf{A}_1 = \mathbf{MAM}^{-1} = \begin{bmatrix} 0 & 1 \\ -1 & 2 \end{bmatrix} \tag{5-291}$$

$$\mathbf{B}_1 = \mathbf{MB} = \begin{bmatrix} 0 \\ 1 \end{bmatrix} \tag{5-292}$$

The characteristic equation of the closed-loop system in PVCF is

$$|z\mathbf{I} - \mathbf{A}_1 + \mathbf{B}_1\mathbf{G}_1| = 0 \tag{5-293}$$

where

$$\mathbf{G}_1 = [g_1^* \quad g_2^*] \tag{5-294}$$

is the feedback matrix of the transformed system. Expanding Eq. (5-293), we get

$$z^2 + (g_2^* - 2)z + (g_1^* + 1) = 0 \tag{5-295}$$

Equating the corresponding coefficients of Eqs. (5-285) and (5-295), we have

$$\mathbf{G}_1 = [g_1^* \quad g_2^*] = [-0.76 \quad 1] \tag{5-296}$$

The feedback gain matrix of the original system is determined from Eq. (5-282).

$$\mathbf{G} = \mathbf{G}_1\mathbf{M} = [-0.76 \quad 1]\begin{bmatrix} -1 & 1 \\ -1 & 2 \end{bmatrix} = [-0.24 \quad 1.24] \tag{5-297}$$

It is simple to show that the eigenvalues of $\mathbf{A} - \mathbf{BG}$ are at 0.4 and 0.6. Thus, the state feedback control that gives the desired eigenvalues is

$$u(k) = [0.24 \quad -1.24]\mathbf{x}(k) \tag{5-298}$$

5-14 THE STATE DIAGRAM

The conventional signal flow graph method and the sampled signal flow graphs introduced in Chapter 4 apply only to algebraic equations. Therefore, these conventional signal flow graphs can be used only for the derivation of the input-output relations in the transform domain. In this section we shall apply the method of the *state transition signal flow graph* or the *state diagram* to represent difference state equations.

For a continuous-data system, the state diagram [1] resembles the block diagram of the analog computer program. Therefore, once the state diagram is drawn, the problem can be solved either by an analog computer or by pencil and paper. Although analog computers have almost become obsolete as a tool for solving system problems, we utilize the notation and relation here only as a comparison with their digital counterparts. For a digital or discrete-data system, the state diagram describes the operations of a hybrid or digital computer, so that the problem can again be solved either by machine or analytical methods.

5-14-1 State Diagrams of Continuous-Data Systems

In this section we review the fundamentals of the state diagram of continuous-data systems, especially since most of the discrete-data systems in practice also contain continuous-data components.

The fundamental linear operations that can be performed on an analog computer are *multiplication by a constant*, *addition*, *sign change*, and *integration* of variables. We now show how the state diagram elements are defined for each of these operations.

1. Multiplication by a Constant

Multiplication of a machine variable by a real constant on an analog computer is done by potentiometers and amplifiers. Consider the operation

$$x_2(t) = ax_1(t) \tag{5-299}$$

where a is a real constant. If a lies between zero and unity, a potentiometer is used for the operation. If the magnitude of a is greater than one, an operational amplifier (inverting or noninverting, depending on the sign of a) is used to realize Eq. (5-299). The analog computer block diagram symbols of the potentiometer and the operational amplifier are shown in Fig. 5-5(a) and (b), respectively. Since Eq. (5-299) is an algebraic equation, it can be represented by a signal flow graph as shown in Fig. 5-5(c). This signal flow graph can be regarded as a state diagram or an element of a state diagram if the variables $x_1(t)$ and $x_2(t)$ are state variables or linear combinations of state variables. Since the Laplace transform of Eq. (5-299) is

$$X_2(s) = aX_1(s) \tag{5-300}$$

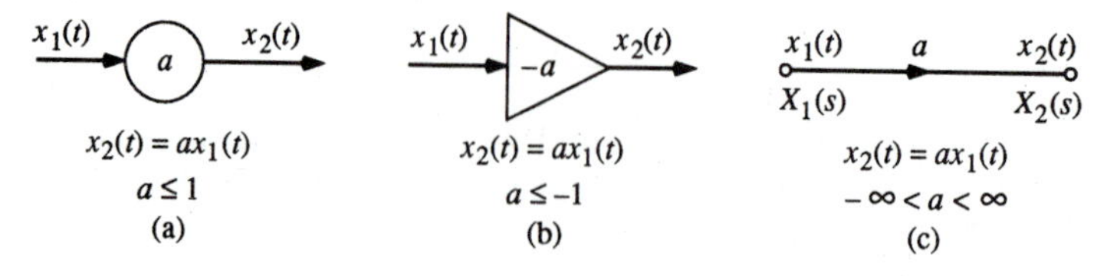

Figure 5-5. (a) Block diagram symbol of a potentiometer. (b) Block diagram symbol of an operational amplifier. (c) State diagram of the operation of multiplying a variable by a constant.

the state diagram symbol of Fig. 5-5(c) also portrays the relation between the transformed variables.

2. Algebraic Sum of Two or More Machine Variables

The algebraic sum of two or more machine variables of an analog computer is obtained by means of an operational amplifier, as shown in Fig. 5-6(a). The illustrated case portrays the algebraic equation

$$x_4(t) = a_1 x_1(t) + a_2 x_2(t) + a_3 x_3(t) \tag{5-301}$$

where the magnitudes of all the coefficients are assumed to be greater than one. The amplifier is assumed to be noninverting, so the gains are all designated as a_1, a_2, and a_3. The state diagram representation of Eq. (5-301) is shown in Fig. 5-6(b).

3. Integration

The integration of a machine variable on an analog computer is carried out by means of a computer element called the *integrator*. If $x_1(t)$ is the output of the integrator with the initial condition $x_1(t_0)$ given at $t = t_0$, and $x_2(t)$ is the input, the integrator performs the following operation:

$$x_1(t) = \int_{t_0}^{t} a x_2(\tau)\, d\tau + x_1(t_0) \tag{5-302}$$

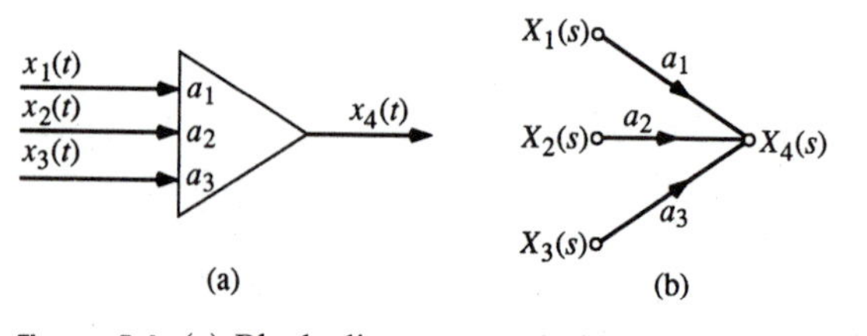

Figure 5-6. (a) Block diagram symbol of the operational amplifier used as a summing device. (b) State diagram representation of the summing operation.

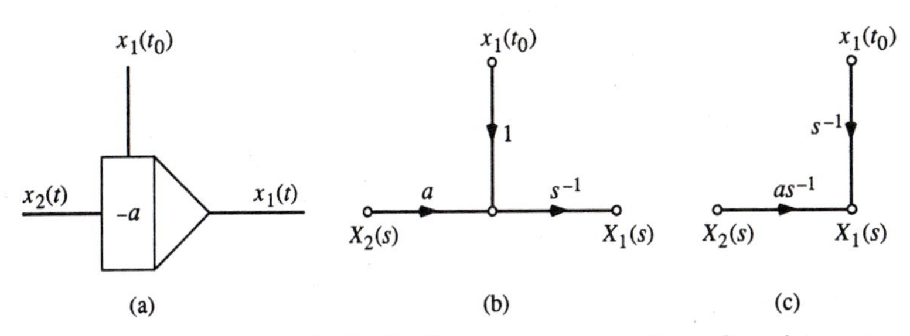

Figure 5-7. (a) Block diagram representation of an integrator. (b), (c) State diagram symbols of an integrator.

where a is a real constant. The block diagram symbol of the integrator is shown in Fig. 5-7(a). In general, the integrator can be used simultaneously as a summing and amplifying device. To determine the state diagram symbol of the integrating operation, we take the Laplace transform on both sides of Eq. (5-302), and we have

$$X_1(s) = a\,\frac{X_2(s)}{s} + a\int_{t_0}^{0} x_2(\tau)\,d\tau + \frac{x_1(t_0)}{s} \tag{5-303}$$

Since the past history of the integrator prior to t_0 is presented by $x_1(t_0)$ and the state transition is considered to begin at $t = t_0$, $x_2(\tau) = 0$ for $0 < \tau < t_0$. Thus, Eq. (5-303) becomes

$$X_1(s) = \frac{aX_2(s)}{s} + \frac{x_1(t_0)}{s} \qquad t \geq t_0 \tag{5-304}$$

It should be pointed out that the transformed equation of Eq. (5-304) is defined only for the period $t \geq t_0$. Therefore, the inverse Laplace transform of $X_1(s)$ should read

$$x_1(t) = \mathscr{L}^{-1}[X_1(s)] = a\int_{t_0}^{t} x_2(\tau)\,d\tau + x_1(t_0) \tag{5-305}$$

A state diagram representation of the integration operation is obtained from the signal flow graph description of Eq. (5-304). Two equivalent versions of the state diagram symbols of an integrator are shown in Fig. 5-7(b) and (c).

Besides leading to computer solutions, a state diagram can also provide the following information on a linear dynamic system:

1. Dynamic equations; that is, state equations and output equations.
2. State transition equations.
3. Transfer functions.
4. Definitions of state variables.

The following example is used to illustrate the applications of the state diagram for analyzing linear continuous-data systems.

Example 5-12

Consider that a linear dynamic system is described by the differential equation

$$\frac{d^2c(t)}{dt^2} + 3\frac{dc(t)}{dt} + 2c(t) = r(t) \tag{5-306}$$

where $c(t)$ is the output and the input $r(t)$ is a unit-step function. The initial conditions of the system are represented by $c(t_0)$ and $dc(t_0)/dt$, all specified at $t = t_0$.

A state diagram of the system is drawn by first equating the highest order derivative term to the rest of the terms in Eq. (5-306), as shown in Fig. 5-8. The state variables of the system, $x_1(t)$ and $x_2(t)$, are defined as the outputs of the integrators. The transformed state transition equations of the system are written directly from the state diagram using Mason's gain formula:

$$\begin{bmatrix} X_1(s) \\ X_2(s) \end{bmatrix} = \frac{1}{(s+1)(s+2)}\begin{bmatrix} s+3 & 1 \\ -2 & s \end{bmatrix}\begin{bmatrix} x_1(t_0) \\ x_2(t_0) \end{bmatrix} + \begin{bmatrix} \dfrac{1}{(s+1)(s+2)} \\ \dfrac{s}{(s+1)(s+2)} \end{bmatrix} R(s) \tag{5-307}$$

For a unit-step input, $R(s) = 1/s$, the inverse Laplace transform of Eq. (5-307) is

$$\begin{bmatrix} x_1(t_0) \\ x_2(t_0) \end{bmatrix} = \begin{bmatrix} 2e^{-(t-t_0)} - e^{-2(t-t_0)} & e^{-(t-t_0)} - e^{-2(t-t_0)} \\ -2e^{-(t-t_0)} + e^{-2(t-t_0)} & -e^{-(t-t_0)} + e^{-2(t-t_0)} \end{bmatrix}\begin{bmatrix} x_1(t_0) \\ x_2(t_0) \end{bmatrix}$$
$$+ \begin{bmatrix} \frac{1}{2} - e^{-(t-t_0)} + \frac{1}{2}e^{-2(t-t_0)} \\ e^{-(t-t_0)} - e^{-2(t-t_0)} \end{bmatrix} \qquad t \geq t_0 \tag{5-308}$$

which is the state transition equation of the system. Notice that, in deriving the last equation, the initial time is taken to be $t = t_0$, so that the following inverse Laplace transform relation has been used:

$$\mathscr{L}^{-1}\left[\frac{1}{s+a}\right] = e^{-a(t-t_0)} \qquad t \geq t_0 \tag{5-309}$$

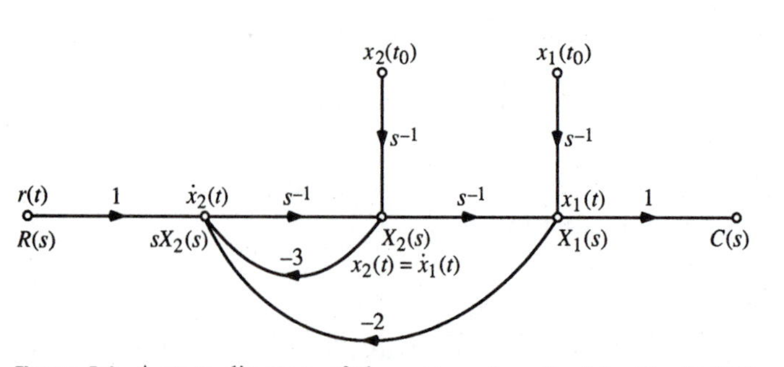

Figure 5-8. A state diagram of the system described by Eq. (5-306).

The reader may derive the state transition equation using the expression in Eq. (5-31) for comparison.

The state equations of the system are obtained from the state diagram by applying Mason's gain formula to the nodes $dx_1(t)/dt$ and $dx_2(t)/dt$ as outputs, with $r(t)$, $x_1(t)$, and $x_2(t)$ as inputs. The branches with the gain $1/s$ are deleted from the state diagram in this operation. Thus, the dynamic equations of the system are written as

$$\begin{bmatrix} \dfrac{dx_1(t)}{dt} \\ \dfrac{dx_2(t)}{dt} \end{bmatrix} = \begin{bmatrix} 0 & 1 \\ -2 & -3 \end{bmatrix} \begin{bmatrix} x_1(t) \\ x_2(t) \end{bmatrix} + \begin{bmatrix} 0 \\ 1 \end{bmatrix} r(t) \tag{5-310}$$

$$c(t) = \begin{bmatrix} 1 & 0 \end{bmatrix} \begin{bmatrix} x_1(t) \\ x_2(t) \end{bmatrix} \tag{5-311}$$

The transfer function of the system is ordinarily obtained by taking the Laplace transform of Eq. (5-306) and assuming zero initial conditions. However, by applying Mason's gain formula to the state diagram of Fig. 5-8 between $R(s)$ and $C(s)$ and setting the initial states to zero, we get

$$\frac{C(s)}{R(s)} = \frac{1}{s^2 + 3s + 2} \tag{5-312}$$

5-14-2 State Diagrams of Digital Systems

When a digital system is described by difference equations or discrete state equations, a state diagram may be drawn to represent the relationships between the discrete state variables. In contrast to the similarity between an analog computer diagram and a continuous-data state diagram, the digital state diagram portrays the operations on a digital computer.

Some of the basic linear operations of a digital computer are *multiplication by a constant*, *addition of variables*, and *time delay or storage*. The mathematical descriptions of these basic digital operations and their corresponding z-transform expressions are given below.

1. Multiplication by a Constant

$$x_2(kT) = ax_1(kT) \tag{5-313}$$

$$X_2(z) = aX_1(z) \tag{5-314}$$

2. Summing

$$x_3(kT) = x_1(kT) + x_2(kT) \tag{5-315}$$

$$X_3(z) = X_1(z) + X_2(z) \tag{5-316}$$

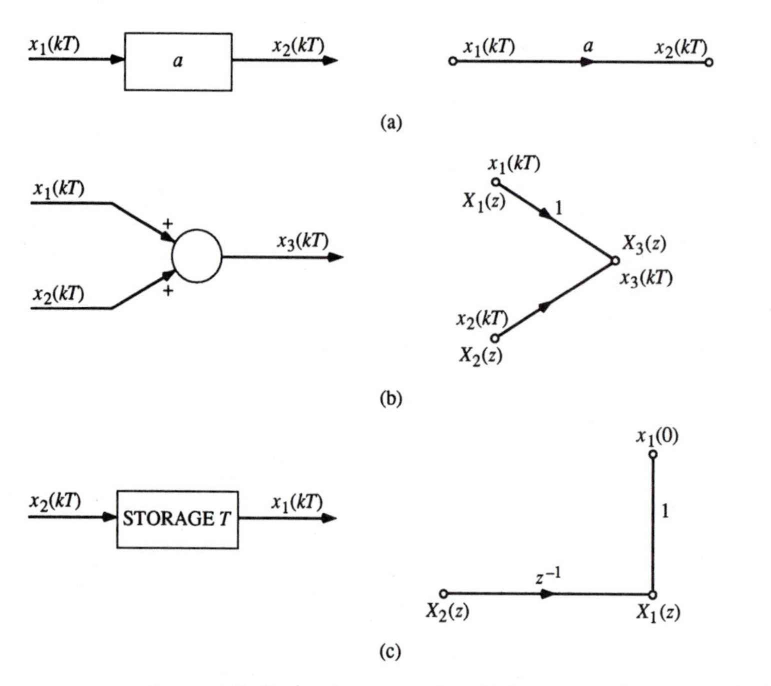

Figure 5-9. Basic elements of a digital state diagram and the corresponding digital computer operations.

3. Time Delay or Storage

$$x_2(kT) = x_1[(k+1)T] \tag{5-317}$$

$$X_2(z) = zX_1(z) - zx_1(0) \tag{5-318}$$

or

$$X_1(z) = z^{-1}X_2(z) + x_1(0) \tag{5-319}$$

The state diagram representations and the corresponding digital computer diagrams of these operations are shown in Fig. 5-9. The following example serves to illustrate the construction and applications of the state diagram of a digital system.

Example 5-13

Consider that a digital system is described by the difference equation,

$$c(k+2) + 2c(k+1) + 3c(k) = r(k) \tag{5-320}$$

The state diagram of the system is drawn as shown in Fig. 5-10 by first equating the highest-order term to the rest of the terms in Eq. (5-320).

The state variables of the system are designated as the outputs of the time-delay units of the state diagram. By setting the initial states to zero and

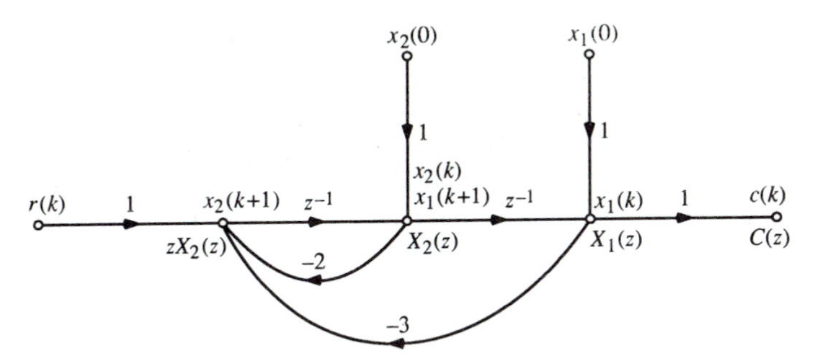

Figure 5-10. A state diagram of the digital system described by Eq. (5-320).

deleting the time-delay units on the state diagram, the state equations of the system are written as

$$\begin{bmatrix} x_1(k+1) \\ x_2(k+1) \end{bmatrix} = \begin{bmatrix} 0 & 1 \\ -3 & -2 \end{bmatrix} \begin{bmatrix} x_1(k) \\ x_2(k) \end{bmatrix} + \begin{bmatrix} 0 \\ 1 \end{bmatrix} r(k) \tag{5-321}$$

and the output equation is

$$c(k) = [1 \quad 0] \begin{bmatrix} x_1(k) \\ x_2(k) \end{bmatrix} \tag{5-322}$$

The state transition equations of the system, which are the solutions of the state equation in Eq. (5-321), are obtained from the state diagram using the gain formula, when $X_1(z)$ and $X_2(z)$ are regarded as output nodes and $R(z)$, $x_1(0)$, and $x_2(0)$ are input nodes. We have

$$\begin{bmatrix} X_1(z) \\ X_2(z) \end{bmatrix} = \frac{1}{\Delta} \begin{bmatrix} 1 + 2z^{-1} & z^{-1} \\ -3z^{-1} & 1 \end{bmatrix} \begin{bmatrix} x_1(0) \\ x_2(0) \end{bmatrix} + \frac{1}{\Delta} \begin{bmatrix} z^{-2} \\ z^{-1} \end{bmatrix} R(z) \tag{5-323}$$

where

$$\Delta = 1 + 2z^{-1} + 3z^{-2} \tag{5-324}$$

Equation (5-323) is the state transition equation in the z-transform domain. The general form of the equation is given by Eq. (5-97). One advantage of using the state diagram is that Eq. (5-323) is obtained simply by using the gain formula. This saves the effort of performing the matrix inverse of $(z\mathbf{I} - \mathbf{A})$ when Eq. (5-97) is used.

The state transition equation in the time domain can be obtained by performing the inverse z-transform on both sides of Eq. (5-323).

The transfer function between the output and input is determined from the state diagram by applying the gain formula between $R(z)$ and $C(z)$, with zero initial states. We have

$$\frac{C(z)}{R(z)} = \frac{X_1(z)}{R(z)} = \frac{z^{-2}}{1 + 2z^{-1} + 3z^{-1}} = \frac{1}{z^2 + 2z + 3} \tag{5-325}$$

5-15 DECOMPOSITION OF DISCRETE-DATA TRANSFER FUNCTIONS

Before embarking on the state diagrams of discrete-data systems with sample-and-hold devices, it is appropriate first to discuss the state representation of z-transfer functions. In general, the transfer function of a digital controller or system $D(z)$ may be realized by a digital processor, digital circuit, microprocessor, or a general-purpose digital computer. The steps involved in realizing the state diagram and dynamic equations of a z-transfer function are termed *decomposition*.

There are three basic types of decomposition: *direct decomposition*, *cascade decomposition*, and *parallel decomposition*. These three methods of decomposition are described in the following.

5-15-1 Direct Decomposition

Consider that the transfer function of a digital controller or process is

$$D(z) = \frac{C(z)}{R(z)} = \frac{b_m + b_{m-1}z^{-1} + b_{m-2}z^{-2} + \cdots + b_1 z^{-m+1} + b_0 z^{-m}}{a_n + a_{n-1}z^{-1} + a_{n-2}z^{-2} + \cdots + a_1 z^{-n+1} + a_0 z^{-n}} \tag{5-326}$$

where n and m are positive integers. We established in Sec. 4-4-2 that, for physical realizability, $a_n \neq 0$.

To find the state diagram realization of Eq. (5-326) by direct decomposition, we first multiply the numerator and the denominator of the right-hand side of Eq. (5-326) by a variable $X(z)$. We have

$$\frac{C(z)}{R(z)} = \frac{(b_m + b_{m-1}z^{-1} + b_{m-2}z^{-2} + \cdots + b_1 z^{-m+1} + b_0 z^{-m})X(z)}{(a_n + a_{n-1}z^{-1} + a_{n-2}z^{-2} + \cdots + a_1 z^{-n+1} + a_0 z^{-n})X(z)} \tag{5-327}$$

Next we equate the numerators on both sides of the last equation, and then do the same with the denominators. The resulting equations are

$$C(z) = (b_m + b_{m-1}z^{-1} + b_{m-2}z^{-2} + \cdots + b_1 z^{-m+1} + b_0 z^{-m})X(z) \tag{5-328}$$

$$R(z) = (a_n + a_{n-1}z^{-1} + a_{n-2}z^{-2} + \cdots + a_1 z^{-n+1} + a_0 z^{-n})X(z) \tag{5-329}$$

To construct a state diagram, Eq. (5-329) must first be written in a cause-and-effect relation. Therefore, dividing both sides of Eq. (5-329) by a_n, and then equating $X(z)$ in terms of the other terms, we get

$$X(z) = \frac{1}{a_n}R(z) - \frac{a_{n-1}}{a_n}z^{-1}X(z) - \frac{a_{n-2}}{a_n}z^{-2}X(z) - \cdots - \frac{a_1}{a_n}z^{-n+1}X(z) - \frac{a_0}{a_n}z^{-n}X(z) \tag{5-330}$$

The state diagram portraying Eqs. (5-328) and (5-330) is drawn as shown in Fig. 5-11 for $n = m = 3$. For simplicity, the initial states are excluded from the diagram. A digital computer program can be derived from the state diagram

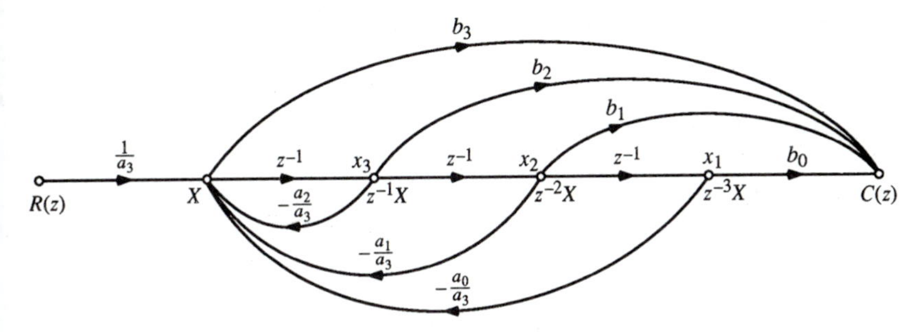

Figure 5-11. State diagram representation of the transfer function of Eq. (5-326) with $n = m = 3$ by direct decomposition.

of Fig. 5-11; the branches with the gain z^{-1} are realized by a time delay of T s which is the sampling period.

The state diagram of Fig. 5-11 can also be used for analytical purposes. By defining the variables at the output nodes of all the time-delay units as state variables, the dynamic equations and the state transition equations can all be determined from the state diagram by applying the gain formula.

The discrete state equations are obtained from Fig. 5-11 by applying the gain formula with $x_1(k+1)$, $x_2(k+1)$, and $x_3(k+1)$ as outputs. After deleting the time-delay branches with gains z^{-1}, the inputs are $r(k)$, $x_1(k)$, $x_2(k)$, and $x_3(k)$. The state equations are

$$x_1(k+1) = x_2(k)$$

$$x_2(k+1) = x_3(k)$$

$$x_3(k+1) = -\frac{a_2}{a_3}x_1(k) - \frac{a_1}{a_3}x_2(k) - \frac{a_0}{a_3}x_3(k) + \frac{1}{a_3}r(k) \tag{5-331}$$

From these state equations, we can see clearly that *direct decomposition will always yield a state-variable model in the phase-variable canonical form.* This is apparently one advantage of using direct decomposition.

5-15-2 Cascade Decomposition

If the numerator and the denominator of a transfer function $D(z)$ are expressed as products of first-order factors, each of these factors can be realized by a simple digital program or state diagram. The entire $D(z)$ is then represented by the cascade connection of the individual programs.

Consider that the transfer function of a digital process is written as

$$D(z) = \frac{C(z)}{R(z)} = \frac{K(z+z_1)(z+z_2)\cdots(z+z_m)}{(z+p_1)(z+p_2)\cdots(z+p_n)} \tag{5-332}$$

where K, z_i, $i = 1, 2, \ldots, m$, and p_j, $j = 1, 2, \ldots, n$, are real constants.

It has been established in Sec. 4-4-2 that for $D(z)$ to be physically realizable, $n \geq m$.

Writing $D(z)$ as a product of the constant K and n first-order transfer functions yields

$$D(z) = KD_1(z)D_2(z) \cdots D_n(z) \tag{5-333}$$

where

$$D_i(z) = \frac{z + z_i}{z + p_i} = \frac{1 + z_i z^{-1}}{1 + p_i z^{-1}} \qquad i = 1, 2, \ldots, m \tag{5-334}$$

and

$$D_j(k) = \frac{1}{z + p_j} = \frac{z^{-1}}{1 + p_j z^{-1}} \qquad j = m + 1, \ldots, n \tag{5-335}$$

The state diagram representation of Eq. (5-334) is constructed in Fig. 5-12(a) and that of Eq. (5-335) is in Fig. 5-12(b), all by direct decomposition. The overall state diagram for $D(z)$ is obtained by connecting the m diagrams of Fig. 5-12(a) and the $n - m$ diagrams of Fig. 5-12(b) with the gain branch of K, all in series.

The advantage with the cascade decomposition is that the gain K and the constants z_i and p_j are all preserved in the state diagram. In design situations these parameters may be varied individually, and the effects can be observed independently.

5-15-3 Parallel Decomposition

A transfer function $D(z)$ with real poles can be realized by parallel decomposition, in which case only the denominator of $D(z)$ must be factored. Let the

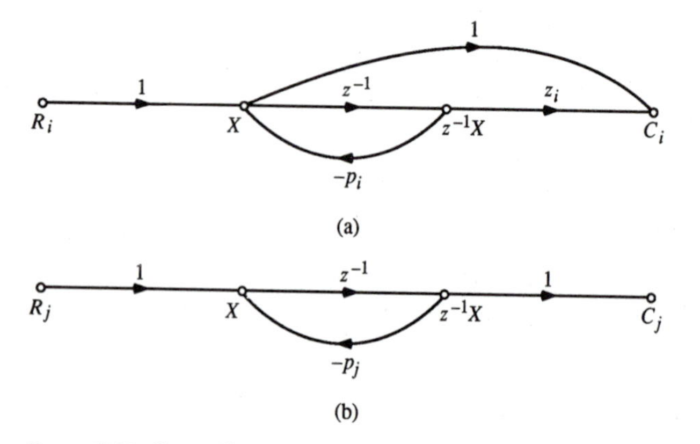

Figure 5-12. State diagram representations of Eqs. (5-334) and (5-335).

transfer function of a digital system be written as

$$D(z) = \frac{C(z)}{R(z)} = K \frac{z^m + b_{m-1}z^{m-1} + \cdots + b_1 z + b_0}{z^n + a_{n-1}z^{n-1} + \cdots + a_1 z + a_0} \tag{5-336}$$

where $n > m$. Assuming for illustration purposes that there are no pole-zero cancellations in $D(z)$ and that, among the n *real* eigenvalues, i are distinct and the rest are of multiplicity $n - i$, then $D(z)$ is expanded by partial-fraction expansion as

$$D(z) = \sum_{k=1}^{i} \frac{K_k}{z + p_k} + \sum_{k=i+1}^{n} \frac{K_k}{(z + p_k)^{k-i}} \tag{5-337}$$

If $D(z)$ has at least one zero at $z = 0$, then the partial-fraction expansion rule of z-transfer functions should be followed. Equation (5-337) implies that the individual state diagram components of $D(z)$ are all of the form of Fig. 5-12(b), and the overall state diagram is obtained by connecting these in parallel. Since the state-variable analysis is hinged on minimal order, *the state diagram of $D(z)$ should consist of a minimum number of time-delay units.*

The parallel decomposition of a transfer function will lead to a set of state equations that is in the canonical form for distinct eigenvalues or Jordan canonical form for systems with multiple-order eigenvalues. The following example illustrates the important features of parallel decomposition.

Example 5-14

Consider the following transfer function of a digital process,

$$D(z) = \frac{C(z)}{R(z)} = \frac{10(z^2 + z + 1)}{z^2(z - 0.5)(z - 0.8)} \tag{5-338}$$

which has eigenvalues at $z = 0, 0, 0.5$, and 0.8. We want to decompose this transfer function by parallel decomposition, and then write the dynamic equations of the system.

Performing partial-fraction expansion on $D(z)$, we have

$$D(z) = \frac{-233.33}{z - 0.5} + \frac{127.08}{z - 0.8} + \frac{25}{z^2} + \frac{106.25}{z} \tag{5-339}$$

It should be emphasized that since $D(z)$ is of the fourth order, the state diagram should have only four time-delay units. Notice that Eq. (5-339) has a total of five first-order components, and realizing the expanded $D(z)$ by first-order state diagram components would require five basic components of the form of Fig. 5-12(b). Figure 5-13 shows a minimum-order realization of $D(z)$ using only four time-delay units. Notice that the "trick" is to share one of the time-delay units for the double pole at $z = 0$.

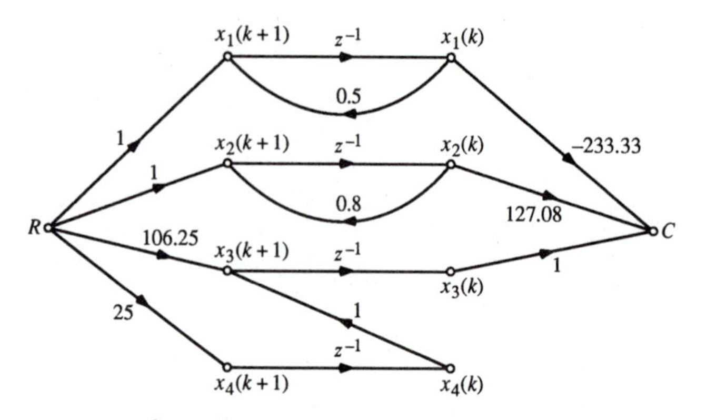

Figure 5-13. State diagram of Eq. (5-338) by parallel decomposition.

The state equations of the system are now written from the state diagram in a routine fashion. We have

$$\begin{bmatrix} x_1(k+1) \\ x_2(k+1) \\ x_3(k+1) \\ x_4(k+1) \end{bmatrix} = \begin{bmatrix} 0.5 & 0 & 0 & 0 \\ 0 & 0.8 & 0 & 0 \\ 0 & 0 & 0 & 1 \\ 0 & 0 & 0 & 0 \end{bmatrix} \begin{bmatrix} x_1(k) \\ x_2(k) \\ x_3(k) \\ x_4(k) \end{bmatrix} + \begin{bmatrix} 1 \\ 1 \\ 106.25 \\ 25 \end{bmatrix} \tag{5-340}$$

which is recognized as being in Jordan canonical form. The output equation is written as

$$c(k) = [-233.33 \quad 127.08 \quad 1 \quad 0]\mathbf{x}(k) \tag{5-341}$$

Since the state diagram of any system is not unique, in the present case, the gains of the individual partial-fractioned components of $D(z)$ can be allocated to either the input or the output branches of the state diagram. This would change the components of the **B** and **D** matrices.

5-16 STATE DIAGRAMS OF DISCRETE-DATA SYSTEMS—SYSTEMS WITH ZERO-ORDER HOLDS

A discrete-data control system usually contains digital as well as analog elements, and the two types of elements are coupled by sample-and-hold devices. A typical discrete-data control system is illustrated by the block diagram in Fig. 5-14. The system consists of a digital controller, a zero-order hold (or D/A

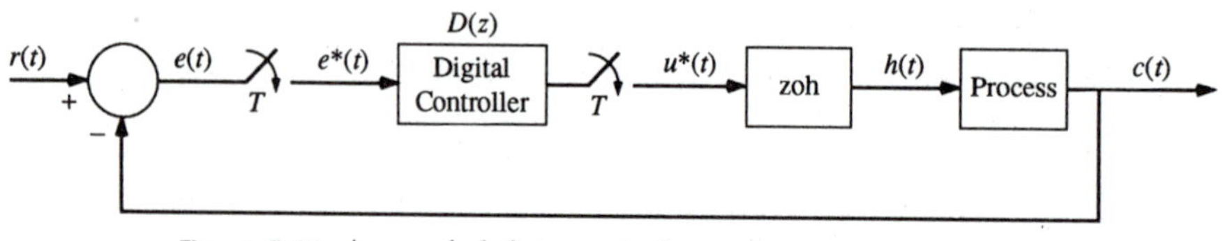

Figure 5-14. A sampled-data control system.

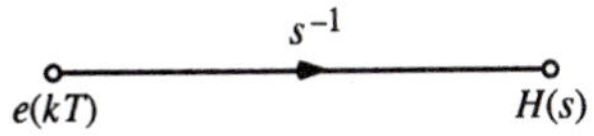

s^{-1}

$e(kT)$ $H(s)$

Figure 5-15. State diagram of the zero-order hold for $kT \le t < (k+1)T$.

converter), and a continuous-data process which is to be controlled. We show in this section how state diagrams and the state-variable analysis of this type of system are conducted.

5-16-1 State Diagram of the Zero-Order Hold

Before the state diagram for the discrete-data system in Fig. 5-14 can be drawn, we must establish the state diagram representation of the zero-order hold. Let the input and the output of the zero-order hold be denoted by $e^*(t)$ and $h(t)$, respectively. Then, for the time interval $kT \le t < (k+1)T$,

$$h(t) = e(kT) \tag{5-342}$$

Taking the Laplace transform on both sides of Eq. (5-342) we have

$$H(s) = \frac{e(kT)}{s} \tag{5-343}$$

for $kT \le t < (k+1)T$. Therefore, the state diagram of the zero-order hold consists of a single branch with a gain of s^{-1}, connected between the nodes $e(kT)$ and $H(s)$, as shown in Fig. 5-15. It should be pointed out that while the gain s^{-1} of the zero-order hold is analytically an integrator, it is not counted into the order of the system.

The following example illustrates how the state diagram and state-variable analysis are conducted for a discrete-data control system with zoh.

Example 5-15

The block diagram of a discrete-data control system is shown in Fig. 5-16. The state diagram, the state equations, and the state transition equations are to be obtained.

Using the method of direct decomposition, the state diagram of the controlled process $G(s)$ is drawn as in Fig. 5-17. The state diagram of the overall

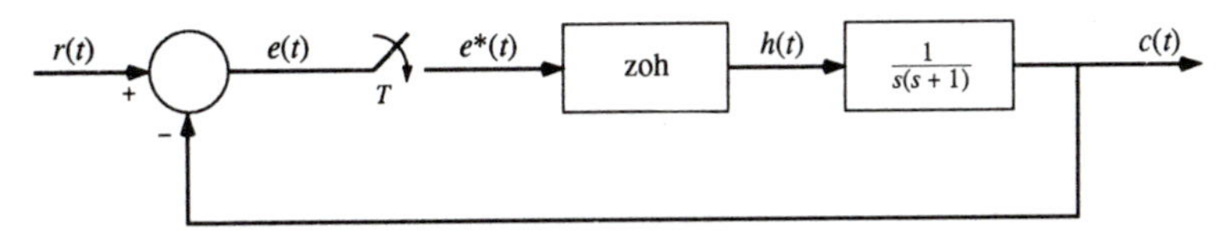

Figure 5-16. A sampled-data control system.

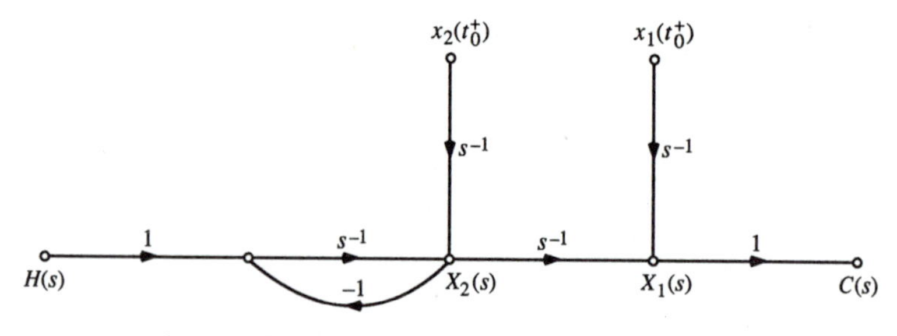

Figure 5-17. A state diagram of $G(s) = 1/s(s+1)$.

system is constructed by connecting the state diagram of Fig. 5-17 to that of the zero-order hold and using the relationship

$$\begin{aligned} e(kT) &= r(kT) - c(kT) \\ &= r(kT) - x_1(kT) \end{aligned} \tag{5-344}$$

Also, letting $t_0 = kT$ and

$$h(kT^+) = h(t) = e(kT) \qquad kT \le t < (k+1)T \tag{5-345}$$

the complete state diagram of the system is shown in Fig. 5-18.

The state transition equations in transformed vector-matrix form are written directly from the state diagram using $X_1(s)$ and $X_2(s)$ as outputs and $x_1(kT)$, $x_2(kT)$, and $r(kT)$ as inputs:

$$\begin{bmatrix} X_1(s) \\ X_2(s) \end{bmatrix} = \begin{bmatrix} \dfrac{1}{s} - \dfrac{1}{s^2(s+1)} & \dfrac{1}{s(s+1)} \\ -\dfrac{1}{s(s+1)} & \dfrac{1}{s+1} \end{bmatrix} \begin{bmatrix} x_1(kT) \\ x_2(kT) \end{bmatrix} + \begin{bmatrix} \dfrac{1}{s^2(s+1)} \\ \dfrac{1}{s(s+1)} \end{bmatrix} r(kT) \tag{5-346}$$

for $kT \le t \le (k+1)T$.

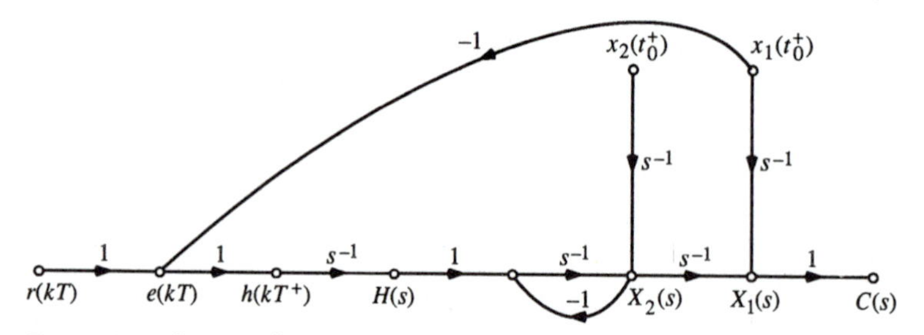

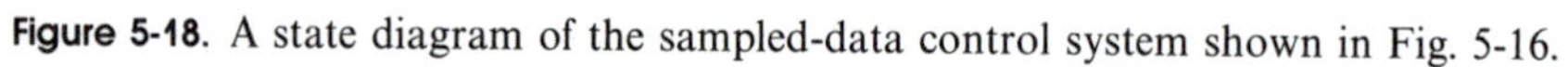

Figure 5-18. A state diagram of the sampled-data control system shown in Fig. 5-16.

Taking the inverse Laplace transform on both sides of the last equation, we have

$$\begin{bmatrix} x_1(t) \\ x_2(t) \end{bmatrix} = \begin{bmatrix} 2-(t-kT)-e^{-(t-kT)} & 1-e^{-(t-kT)} \\ -1+e^{-(t-kT)} & e^{-(t-kT)} \end{bmatrix} \begin{bmatrix} x_1(kT) \\ x_2(kT) \end{bmatrix} + \begin{bmatrix} (t-kT)-1+e^{-(t-kT)} \\ 1-e^{-(t-kT)} \end{bmatrix} r(kT) \quad (5\text{-}347)$$

for $kT \leq t \leq kT$.

If only the responses at the sampling instants are of interest, we let $t = (k+1)T$; then Eq. (5-347) becomes

$$\begin{bmatrix} x_1(k+1)T \\ x_2(k+1)T \end{bmatrix} = \begin{bmatrix} 2-T-e^{-T} & 1-e^{-T} \\ -1+e^{-T} & e^{-T} \end{bmatrix} \begin{bmatrix} x_1(kT) \\ x_2(kT) \end{bmatrix} + \begin{bmatrix} T-1+e^{-T} \\ 1-e^{-T} \end{bmatrix} \quad (5\text{-}348)$$

Notice that Eq. (5-348) is of the form of Eq. (5-39):

$$\mathbf{x}[(k+1)T] = \boldsymbol{\phi}(T)\mathbf{x}(kT) + \boldsymbol{\theta}(T)r(kT) \quad (5\text{-}349)$$

Let the sampling period T be 1 s and $r(t)$ be a unit-step function. We have

$$\boldsymbol{\phi}(T) = \begin{bmatrix} 0.632 & 0.632 \\ -0.632 & 0.368 \end{bmatrix} \qquad \boldsymbol{\theta}(T) = \begin{bmatrix} 0.368 \\ 0.632 \end{bmatrix} \quad (5\text{-}350)$$

Then,

$$[z\mathbf{I} - \boldsymbol{\phi}(1)]^{-1}z = \begin{bmatrix} z-0.632 & -0.632 \\ 0.632 & z-0.368 \end{bmatrix}^{-1} z = \frac{1}{z^2 - z + 0.632} \begin{bmatrix} z-0.368 & 0.632 \\ -0.632 & z-0.632 \end{bmatrix} \quad (5\text{-}351)$$

and

$$\boldsymbol{\phi}(k) = \mathscr{Z}^{-1}[[z\mathbf{I} - \boldsymbol{\phi}(1)]^{-1}z] = \begin{bmatrix} e^{-0.23k}(-0.378\sin 0.88k + \cos 0.88k) & e^{-0.23k}\sin 0.88k \\ -e^{-0.23k}\sin 0.88k & e^{-0.23k}(-0.786\sin 0.88k + \cos 0.88k) \end{bmatrix} \quad (5\text{-}352)$$

Also,

$$\boldsymbol{\phi}[(N-k-1)]\boldsymbol{\theta}(1) = \begin{bmatrix} e^{-0.23(N-k-1)}[0.493\sin 0.88(N-k-1) + 0.368\cos 0.88(N-k-1)] \\ e^{-0.23(N-k-1)}[-0.865\sin 0.88(N-k-1) + 0.632\cos 0.88(N-k-1)] \end{bmatrix} \quad (5\text{-}353)$$

Therefore, the discrete state transition equations of the system are written as

$$\mathbf{x}(N) = \begin{bmatrix} e^{-0.23N}(-0.378 \sin 0.88N + \cos 0.88N) & e^{-0.23N} \sin 0.88N \\ -e^{-0.23N} \sin 0.88N & -e^{-0.23N}(-0.786 \sin 0.88N + \cos 0.88N) \end{bmatrix} \begin{bmatrix} x_1(0) \\ x_2(0) \end{bmatrix} + \sum_{k=0}^{N-1} \begin{bmatrix} e^{-0.23(N-k-1)}[0.493 \sin 0.88(N-k-1) + 0.368 \cos 0.88(N-k-1)] \\ e^{-0.23(N-k-1)}[-0.865 \sin 0.88(N-k-1) + 0.632 \cos 0.88(N-k-1)] \end{bmatrix} \quad \text{(5-354)}$$

where $N = 1, 2, 3, \ldots$.

Example 5-16

In this example we conduct a state-variable analysis of a discrete-data control system that has a digital controller. Consider the block diagram shown in Fig. 5-14. The digital controller, which may be a digital processor or a microprocessor, is described by

$$D(z) = \frac{b_1 + b_0 z^{-1}}{1 + a_0 z^{-1}} \quad \text{(5-355)}$$

where a_0, b_0, and b_1 are real constants. The transfer function of the analog controlled process is as given in the system of Example 5-15.

The state diagram, the state equations, and the state transition equations of the system are to be determined.

Applying the direct decomposition scheme to $D(z)$ in Eq. (5-355), we get

$$D(z) = \frac{U(z)}{E(z)} = \frac{(b_1 + b_0 z^{-1})X(z)}{(1 + a_0 z^{-1})X(z)} \quad \text{(5-356)}$$

Equating the numerator terms on both sides of Eq. (5-356), we have

$$U(z) = (b_1 + b_0 z^{-1})X(z) \quad \text{(5-357)}$$

Equating the denominator terms on both sides of Eq. (5-356) and solving for $X(z)$, we have

$$X(z) = E(z) - a_0 z^{-1} X(z) \quad \text{(5-358)}$$

The state diagram of the digital controller is constructed as shown in Fig. 5-19, using Eqs. (5-357) and (5-358).

From Fig. 5-19, the dynamic equations of $D(z)$ are written

State Equation

$$x_3(k + 1)T = e(kT) - a_0 x_3(kT) \quad \text{(5-359)}$$

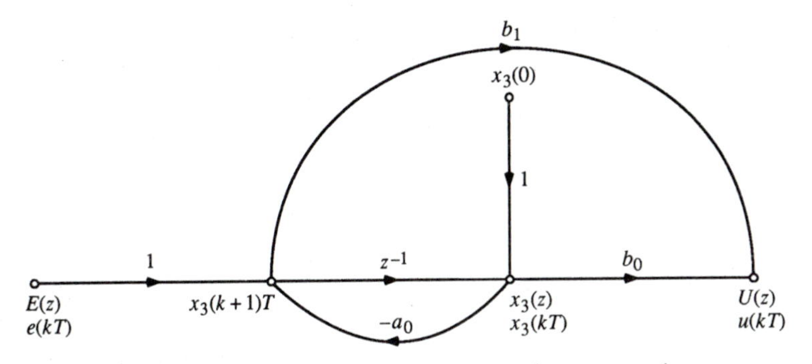

Figure 5-19. State diagram for $D(z) = (b_1 + b_0 z^{-1})/(1 + a_0 z^{-1})$.

Output Equation

$$u(kT) = b_1 e(kT) + (b_0 - b_1 a_0) x_3(kT) \tag{5-360}$$

The state diagram of $G(s)$ and the zero-order hold were established earlier in Example 5-15, Fig. 5-18. Utilizing the portion for $G(s)$ and the zoh in Fig. 5-18, the state diagram of the complete system in Fig. 5-14 is drawn in Fig. 5-20. Notice that the time-delay branch with the gain z^{-1} for the digital controller is shown as the dotted line in Fig. 5-20, since the state diagram is now in the continuous-time or Laplace domain. The inputs of the state diagram are $r(kT)$, $x_1(kT)$, $x_2(kT)$, and $x_3(kT)$. These nodes satisfy the condition of an output node from the signal flow graph standpoint.

Applying the gain formula to Fig. 5-20 with $X_1(s)$ and $X_2(s)$ as output nodes, we get

$$\begin{bmatrix} X_1(s) \\ X_2(s) \end{bmatrix} = \begin{bmatrix} \dfrac{1}{s} - \dfrac{1}{s^2(s+1)} & \dfrac{1}{s(s+1)} & \dfrac{b_0 - b_1 a_0}{s^2(s+1)} \\ \dfrac{-b_1}{s(s+1)} & \dfrac{1}{s+1} & \dfrac{b_0 - b_1 a_0}{s(s+1)} \end{bmatrix} \begin{bmatrix} x_1(kT) \\ x_2(kT) \\ x_3(kT) \end{bmatrix} + \begin{bmatrix} \dfrac{b_1}{s^2(s+1)} \\ \dfrac{b_1}{s(s+1)} \end{bmatrix} r(kT) \tag{5-361}$$

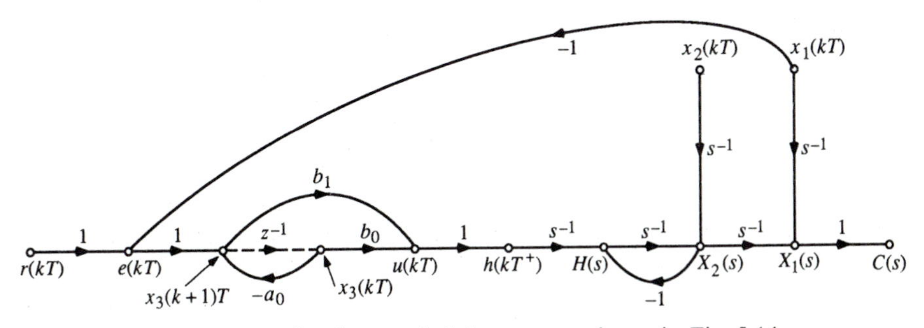

Figure 5-20. State diagram for the sampled-data system shown in Fig. 5-14.

The output of the state variable $x_3(kT)$ appears only as $x_3(k+1)T$. The equation relating x_3 from Fig. 5-20 is in the form of a state equation,

$$x_3(k+1)T = -x_1(kT) - a_0 x_3(kT) + r(kT) \tag{5-362}$$

Applying inverse Laplace transform to Eq. (5-361) and substituting $t = (k+1)T$, together with Eq. (5-362), we have the discrete state equation of the system,

$$\begin{bmatrix} x_1(k+1)T \\ x_2(k+1)T \\ x_3(k+1)T \end{bmatrix} = \begin{bmatrix} 1 - b_1(T - 1 + e^{-T}) & 1 - e^{-T} & (b_0 - b_1 a_0)(T - 1 + e^{-T}) \\ -b_1(1 - e^{-T}) & e^{-T} & (b_0 - b_1 a_0)(1 - e^{-T}) \\ -1 & 0 & -a_0 \end{bmatrix}$$

$$\times \begin{bmatrix} x_1(kT) \\ x_2(kT) \\ x_3(kT) \end{bmatrix} + \begin{bmatrix} b_1(T - 1 + e^{-T}) \\ b_1(1 - e^{-T}) \\ 1 \end{bmatrix} r(kT) \tag{5-363}$$

The output equation is simply $c(kT) = x_1(kT)$.

Equation (5-363) can now be solved by means of the same method described in Example 5-15.

5-17 STATE-VARIABLE ANALYSIS OF RESPONSE BETWEEN SAMPLING INSTANTS

The state-variable method is very convenient for the evaluation of system response between the sampling instants of discrete-data control systems. The method represents a modern alternative to the modified z-transform method.

Equation (5-31) gives the state transition equation which expresses the state vector $\mathbf{x}(t)$ in terms of the initial state $\mathbf{x}(t_0)$ and input vector $\mathbf{u}(t)$ for $t \geq t_0$. When $\mathbf{u}(t)$ is a constant vector for $t_0 \leq t < t$, Eq. (5-31) is written as

$$\mathbf{x}(t) = \boldsymbol{\phi}(t - t_0)\mathbf{x}(t_0) + \boldsymbol{\theta}(t - t_0)\mathbf{u}(t_0) \tag{5-364}$$

where

$$\boldsymbol{\theta}(t - t_0) = \int_{t_0}^{t} \boldsymbol{\phi}(t - \tau)\mathbf{B}\, d\tau \tag{5-365}$$

If the response between the sampling instants is desired, we let $t_0 = kT$, and

$$t = (k + \Delta)T \tag{5-366}$$

where $k = 0, 1, 2, \ldots$, and $0 \leq \Delta \leq 1$. Then, Eq. (5-364) becomes

$$\mathbf{x}(k + \Delta)T = \boldsymbol{\phi}(\Delta T)\mathbf{x}(kT) + \boldsymbol{\theta}(\Delta T)\mathbf{u}(kT) \tag{5-367}$$

By varying the value of Δ between 0 and 1, all information on $\mathbf{x}(t)$ for all t can be obtained.

Example 5-17

For the system described in Example 5-15, the state transition equation in Eq. (5-348) gives values of the state variables at the sampling instants only. Letting $t = (k + \Delta)T$ in Eq. (5-347), we have

$$\begin{bmatrix} x_1(kT + \Delta T) \\ x_2(kT + \Delta T) \end{bmatrix} = \begin{bmatrix} 2 - \Delta T - e^{-\Delta T} & 1 - e^{-\Delta T} \\ -1 + e^{-\Delta T} & e^{-\Delta T} \end{bmatrix} \begin{bmatrix} x_1(kT) \\ x_2(kT) \end{bmatrix} + \begin{bmatrix} \Delta T - 1 + e^{-\Delta T} \\ 1 - e^{-\Delta T} \end{bmatrix} r(kT) \tag{5-368}$$

By assigning values for Δ between zero and one, we can obtain all the values of $x_1(t)$ and $x_2(t)$ between the sampling instants.

Let the sampling period T be 1 s, and $\Delta = 0.5$. The input $r(t)$ is a unit-step function. The two sets of state equations are

$$\begin{bmatrix} x_1(k+1) \\ x_2(k+1) \end{bmatrix} = \begin{bmatrix} 0.632 & 0.632 \\ -0.632 & 0.368 \end{bmatrix} \begin{bmatrix} x_1(k) \\ x_2(k) \end{bmatrix} + \begin{bmatrix} 0.368 \\ 0.632 \end{bmatrix} \tag{5-369}$$

$$\begin{bmatrix} x_1(k+0.5) \\ x_2(k+0.5) \end{bmatrix} = \begin{bmatrix} 0.894 & 0.394 \\ -0.394 & 0.607 \end{bmatrix} \begin{bmatrix} x_1(k) \\ x_2(k) \end{bmatrix} + \begin{bmatrix} 0.107 \\ 0.394 \end{bmatrix} \tag{5-370}$$

The responses of $x_1(t)$ and $x_2(t)$ for $t = kT$ and $t = (k + 0.5)T$ are tabulated below for $k = 0$ to $k = 20$. The initial states are set to zero.

t	0	0.5	1.0	1.5	2.0	2.5	3.0	3.5	4.0	4.5	5.0	5.5
$x_1(t)$	0	0.107	0.368	0.684	1.000	1.249	1.399	1.448	1.399	1.291	1.147	1.008
$x_2(t)$	0	0.394	0.632	0.632	0.632	0.383	0.233	−0.016	−0.167	−0.258	0.314	−0.248
t	6.0	6.5	7.0	7.5	8.0	8.5	9.0	9.5	10.0	10.5	11.0	11.5
$x_1(t)$	0.895	0.824	0.802	0.819	0.868	0.930	0.994	1.045	1.077	1.088	1.081	1.061
$x_2(t)$	−0.208	−0.085	−0.010	0.072	0.122	0.086	0.128	0.080	0.051	0.001	−0.030	−0.050
t	12.0	12.5	13.0	13.5	14.0	14.5	15.0	15.5	16.0	16.5	17.0	17.5
$x_1(t)$	1.033	1.005	0.981	0.966	0.961	0.949	0.949	0.959	0.974	0.991	1.006	1.017
$x_2(t)$	−0.062	−0.051	−0.044	−0.019	−0.041	−0.009	0.010	0.026	0.036	0.032	0.029	0.016
t	18.0	18.5	19.0	19.5	20.0							
$x_1(t)$	1.023	1.023	1.019	1.012	1.004							
$x_2(t)$	0.007	−0.005	−0.012	−0.014	−0.016							

Figure 5-21 shows the time response of $x_1(t)$ at $t = k$ and $k - 0.5$ for $k = 0$ to 20. Notice that although the response is not too oscillatory, the additional information at the midpoints of the sampling periods does indicate that the maximum overshoot of the system is closer to 1.448, rather than 1.399, as presented by the results only at the sampling instants.

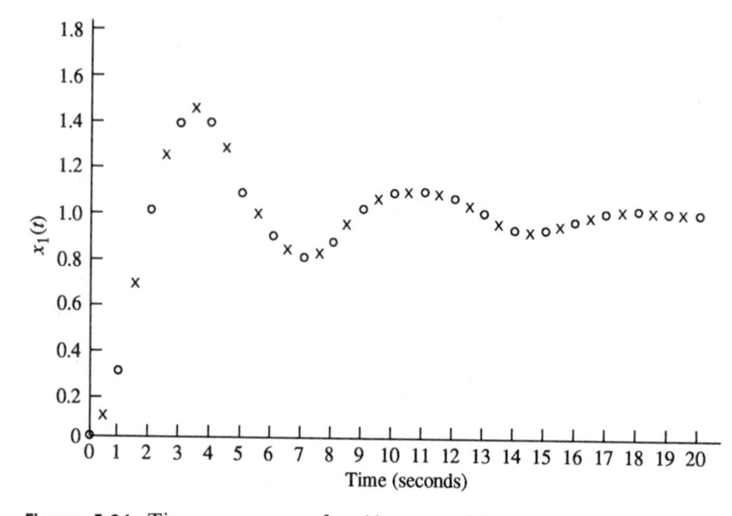

Figure 5-21. Time response of $x_1(t)$ at $t = kT$ and $t = (k + 0.5)T$ for the system in Example 5-17.

5-18 STATE-VARIABLE ANALYSIS OF SYSTEMS WITH MULTIRATE, SKIP-RATE, AND NONSYNCHRONOUS SAMPLINGS

We learned in Chapter 4 that, with the z-transform method, different types of sampling schemes generally require different equivalent models of the samplers. In this section we shall show that the state-variable method offers a unified approach to the analysis of discrete-data systems with a wide class of sampling schemes.

5-18-1 Cyclic-Rate Sampled System

Figure 4-27(a) of Sec. 4-7-5 describes a discrete-data system with cyclic-rate sampling. Figure 5-22 illustrates a similar system with the sampling taking place at kT and $kT + T_1$, for $k = 0, 1, 2, \ldots$, where $T_1 < T$. The state equations of the linear process at the sampling instants are described by

At $t = kT + T_1$

$$\mathbf{x}(kT + T_1) = \boldsymbol{\phi}(T_1)\mathbf{x}(kT) + \boldsymbol{\theta}(T_1)r(kT) \tag{5-371}$$

At $t = kT + T$

$$\mathbf{x}(kT + T) = \boldsymbol{\phi}(T - T_1)\mathbf{x}(kT + T_1) + \boldsymbol{\theta}(T - T_1)r(kT + T_1) \tag{5-372}$$

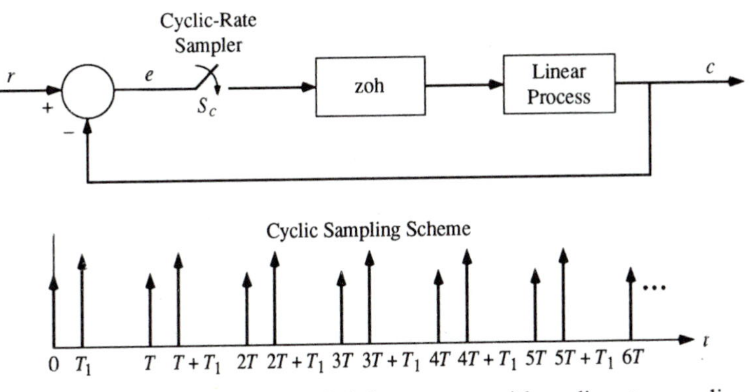

Figure 5-22. A sampled-data system with cyclic-rate sampling.

Substituting Eq. (5-371) into Eq. (5-372) yields

$$\mathbf{x}[(k+1)T] = \boldsymbol{\phi}(T - T_1)\boldsymbol{\phi}(T_1)\mathbf{x}(kT) + \boldsymbol{\phi}(T - T_1)\boldsymbol{\theta}(T_1)r(kT) + \boldsymbol{\theta}(T - T_1)r(kT + T_1) \tag{5-373}$$

In general, if the cyclic-rate sampler samples at $t = kT, kT + T_1, kT + T_1 + T_2, \ldots, (k+1)T, (k+1)T + T_1, (k+1)T + T_1 + T_2, \ldots, (k+1)T + T_1 + T_2 + \ldots + T_q, \ldots$ so that $T = T_1 + T_2 + \cdots + T_q$, then

$$\mathbf{x}(kT + T_1) = \boldsymbol{\phi}(T_1)\mathbf{x}(kT) + \boldsymbol{\theta}(T_1)r(kT) \tag{5-374}$$

$$\mathbf{x}(kT + T_1 + T_2) = \boldsymbol{\phi}(T_2)\mathbf{x}(kT + T_1) + \boldsymbol{\theta}(T_2)r(kT + T_1) \tag{5-375}$$

$$\vdots$$

$$\mathbf{x}(kT + T_1 + T_2 + \cdots + T_{q-1}) = \boldsymbol{\phi}(T_{q-1})\mathbf{x}(kT + T_1 + T_2 + \cdots + T_{q-2}) + \boldsymbol{\theta}(T_{q-1})r(kT + T_1 + T_2 + \cdots + T_{q-2}) \tag{5-376}$$

At $t = (k+1)T$

$$\mathbf{x}(k+1)T = \boldsymbol{\phi}(T_q)\mathbf{x}(kT + T_1 + T_2 + \cdots + T_{q-1}) + \boldsymbol{\theta}(T_q)r(kT + T_1 + T_2 + \cdots + T_{q-1}) \tag{5-377}$$

Substituting Eq. (5-374) into Eq. (5-375), and so on, Eq. (5-377) becomes

$$\begin{aligned}\mathbf{x}(k+1)T = {} & \boldsymbol{\phi}(T_q)\boldsymbol{\phi}(T_{q-1}) \cdots \boldsymbol{\phi}(T_2)\boldsymbol{\phi}(T_1)\mathbf{x}(kT) \\ & + \boldsymbol{\phi}(T_q)\boldsymbol{\phi}(T_{q-1}) \cdots \boldsymbol{\phi}(T_2)\boldsymbol{\theta}(T_1)r(kT) \\ & + \boldsymbol{\phi}(T_q)\boldsymbol{\phi}(T_{q-1}) \cdots \boldsymbol{\phi}(T_3)\boldsymbol{\theta}(T_2)r(kT + T_1) \\ & + \cdots + \boldsymbol{\phi}(T_q)\boldsymbol{\theta}(T_{q-1})r(kT + T_1 + T_2 + \cdots + T_{q-2}) \\ & + \boldsymbol{\theta}(T_q)r(kT + T_1 + T_2 + \cdots + T_{q-1})\end{aligned} \tag{5-378}$$

Example 5-18

Consider that for the system in Fig. 5-22, the linear process is described by the transfer function

$$G(s) = \frac{4}{s + 1} \tag{5-379}$$

The sampling period is $T = 1$ s, and $T_1 = 0.25$ s. The input signal is a unit-step function applied at $t = 0$. The system is initially at rest, and $c(0) = 0$. The output response of the system for $t > 0$ is desired. The periods of state transition in this case are $kT \le t \le kT + T_1$ and $kT + T_1 \le t \le (k + 1)T$, for $k = 0, 1, 2, \ldots$. The state diagram of the system for $t \ge t_0$, where t_0 can be kT or $kT + T_1$, is drawn as shown in Fig. 5-23. The state transition equation in the Laplace domain is written as

$$X_1(s) = \frac{1}{s + 1} x_1(t_0) - \frac{4}{s(s + 1)} x_1(t_0) + \frac{4}{s(s + 1)} r(t_0) \tag{5-380}$$

The inverse Laplace transform of $X_1(s)$ is

$$x_1(t) = e^{-(t - t_0)} x_1(t_0) - 4[1 - e^{-(t - t_0)}] x_1(t_0) + 4[1 - e^{-(t - t_0)}] r(t_0) \tag{5-381}$$

Thus,

$$\phi(t - t_0) = -4 + 5e^{-(t - t_0)} \tag{5-382}$$

$$\theta(t - t_0) = 4[1 - e^{-(t - t_0)}] \tag{5-383}$$

Setting $t_0 = kT$ and $t = kT + T_1$ in Eq. (5-381), we have

$$\begin{aligned} x_1(kT + T_1) &= \phi(T_1) x_1(kT) + \theta(T_1) r(kT) \\ &= (-4 + 5e^{-T_1}) x_1(kT) + 4(1 - e^{-T_1}) r(kT) \end{aligned} \tag{5-384}$$

Substituting $t_0 = kT + T_1$ and $t = (k + 1)T$ into Eq. (5-381), we get

$$x_1(k + 1)T = [-4 + 5e^{-(T - T_1)}] x_1(kT + T_1) + 4[1 - e^{-(T - T_1)}] r(kT + T_1) \tag{5-385}$$

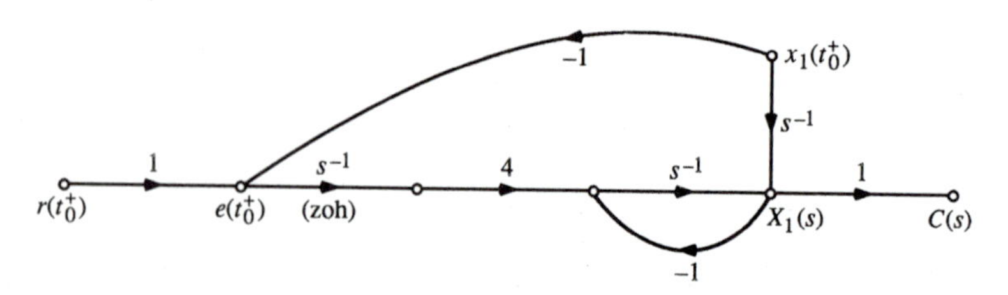

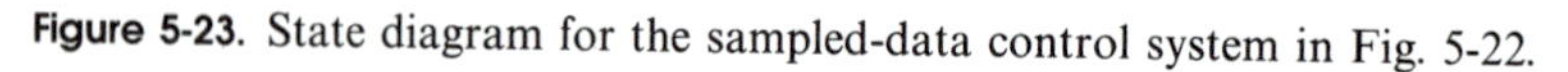

Figure 5-23. State diagram for the sampled-data control system in Fig. 5-22.

For $T_1 = 0.25$ s and $T = 1$ s, the last two equations become

$$\begin{aligned} x_1(k+0.25) &= \phi(0.25)x_1(kT) + \theta(0.25)r(kT) \\ &= -0.1x_1(kT) + 0.884r(kT) \end{aligned} \tag{5-386}$$

and

$$\begin{aligned} x_1(k+1) &= \phi(0.75)x_1(k+0.25) + \theta(0.75)r(kT+0.25) \\ &= -1.64x_1(k+0.25) + 2.1r(k+0.25) \end{aligned} \tag{5-387}$$

Equations (5-386) and (5-387) are two recursion relations from which the response of the system at $t = kT + T_1$ and $(k+1)T$ for $k = 0, 1, 2, \ldots$ can be computed. Note that the solutions are valid for any input $r(t)$ whose values at $t = kT$ and $kT + T_1$ are defined. For $r(t) = u_s(t)$, the following results are obtained.

t	0	0.25	1.00	1.25	2.00	2.25	3.00	3.25	4.00	4.25
$x_1(t)$	0	0.88	0.65	0.82	0.75	0.81	0.77	0.81	0.78	0.81

The unit-step response is shown in Fig. 5-24. In this case, it is shown that the response has a steady-state error of 0.2 or 20 percent.

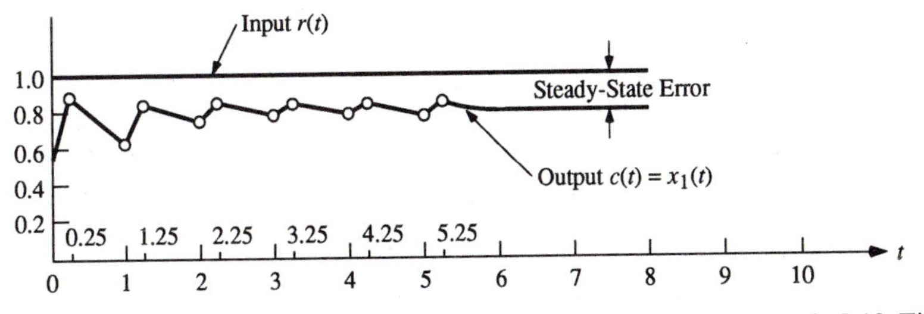

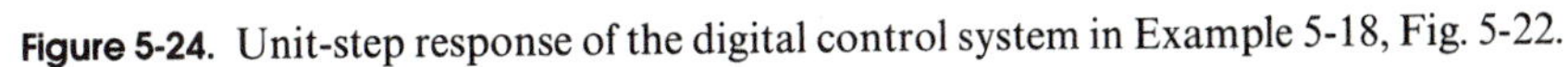

Figure 5-24. Unit-step response of the digital control system in Example 5-18, Fig. 5-22.

Example 5-19

In this example we illustrate that the state-variable method is conveniently applied to systems with nonsynchronous sampling. A discrete-data system with nonsynchronous sampling is shown in Fig. 5-25. The sampling instants of the sampler S_2 lag behind that of S_1 by T_1 s. We shall write the state transition

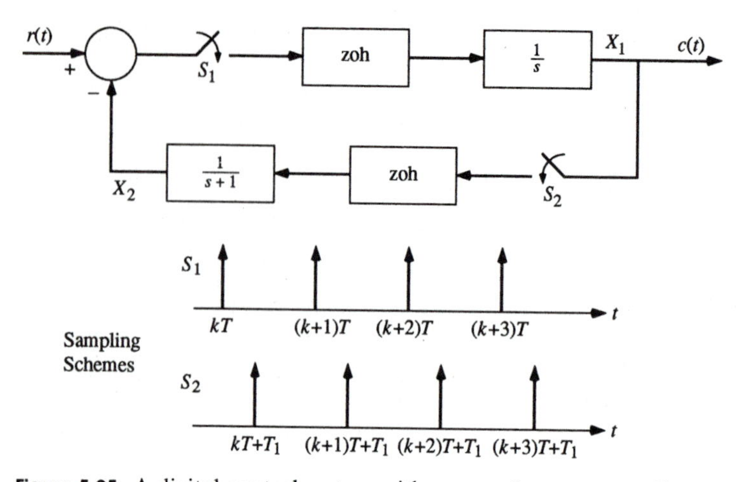

Figure 5-25. A digital control system with nonsynchronous sampling.

equations for the system using the same principle as discussed in this section. Since S_1 and S_2 close at different sampling instants, the forward path of the system has a transition interval from $t = kT$ to $t = (k + 1)T$, and the feedback path has a transition interval from $t = kT + T_1$ to $t = (k + 1)T + T_1$. The state diagrams for the two paths with the corresponding transition intervals are shown in Fig. 5-26.

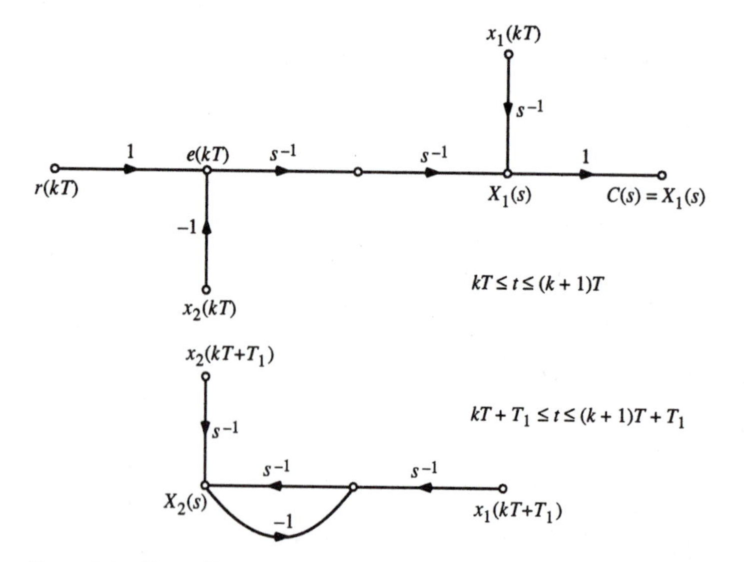

Figure 5-26. State diagrams for the digital control system with nonsynchronous sampling shown in Fig. 5-25.

The state transition equations in the Laplace transform domain for the forward path and the feedback path are written directly from Fig. 5-25:

$kT \le t \le (k+1)T$

$$X_1(s) = \frac{1}{s} x_1(kT) - \frac{1}{s^2} x_2(kT) + \frac{1}{s^2} r(kT) \tag{5-388}$$

$kT + T_1 \le t \le kT + T + T_1$

$$X_2(s) = \frac{1}{s(s+1)} x_1(kT + T_1) + \frac{1}{s+1} x_2(kT + T_2) \tag{5-389}$$

Taking the inverse Laplace transform on both sides of the last two equations, we get

$kT \le t \le (k+1)T$

$$x_1(t) = x_1(kT) - (t - kT)x_2(kT) + (t - kT)r(kT) \tag{5-390}$$

$kT + T_1 \le t \le kT + T + T_1$

$$x_2(t) = [1 - e^{-(t-kT-T_1)}]x_1(kT + T_1) + e^{-(t-kT-T_1)}x_2(kT + T_1) \tag{5-391}$$

Since the initial states are $x_1(kT)$, $x_1(kT + T_1)$, $x_2(kT)$, and $x_2(kT + T_1)$, it is necessary to compute these quantities from the two state transition equations. Thus, setting $t = kT + T_1$ and $t = (k+1)T$ in Eq. (5-390), we get

$$x_1(kT + T_1) = x_1(kT) - T_1 x_2(kT) + T_1 r(kT) \tag{5-392}$$

$$x_1(k+1)T = x_1(kT) - T x_2(kT) + T r(kT) \tag{5-393}$$

Setting $t = (k+1)T$ and $t = (k+1)T + T_1$ in Eq. (5-391), we get

$$x_2(k+1)T = [1 - e^{-(T-T_1)}]x_1(kT + T_1) + e^{-(T-T_1)}x_2(kT + T_1) \tag{5-394}$$

$$x_2[(k+1)T + T_1] = (1 - e^{-T})x_1(kT + T_1) + e^{-T}x_2(kT + T_1) \tag{5-395}$$

Now by setting $k = 0, 1, 2, 3, \ldots$ into these four state transition equations, the states and the output of the system can be computed at the sampling instants of both samplers.

PROBLEMS

5-1 Write the state equations and the output equations of the following difference equations.

a. $c(k+3) + 5c(k+2) + 3c(k+1) + 2c(k) = u(k)$

b. $c(k+3) + 5c(k+2) + 3c(k+1) + 2c(k) = u(k+1) + u(k)$

c. $c(k+4) + 2c(k+2) - c(k+1) + c(k) = 5u(k)$

5-2 Given the state equation

$$\mathbf{x}(k+1) = \mathbf{A}\mathbf{x}(k)$$

find the state transition matrix $\boldsymbol{\phi}(k)$ for the following cases.

a. $\mathbf{A} = \begin{bmatrix} 0 & 2 \\ 0 & -0.1 \end{bmatrix}$

b. $\mathbf{A} = \begin{bmatrix} 0 & 1 \\ 0.5 & 1 \end{bmatrix}$

c. $\mathbf{A} = \begin{bmatrix} 0 & 1 & 0 \\ 0 & 0 & 1 \\ 0 & -0.5 & 1.5 \end{bmatrix}$

d. $\mathbf{A} = \begin{bmatrix} 0 & 1 \\ -0.5 & -1 \end{bmatrix}$

5-3 Determine which matrices given below are candidates as state transition matrices of linear discrete-data systems obtained by discretizing the linear system

$$\frac{d\mathbf{x}(t)}{dt} = \mathbf{A}\mathbf{x}(t) + \mathbf{B}u(t)$$

a. $\boldsymbol{\phi}(T) = \begin{bmatrix} e^{-1.5T} & 0 \\ 0 & e^{-2T} \end{bmatrix}$

b. $\boldsymbol{\phi}(T) = \begin{bmatrix} e^{-1.5T} & 1 \\ 0 & e^{-2T} \end{bmatrix}$

c. $\boldsymbol{\phi}(T) = \begin{bmatrix} e^{-1.5T} & 1 - e^{-T} \\ 0 & e^{-2T} \end{bmatrix}$

d. $\boldsymbol{\phi}(T) = \begin{bmatrix} -e^{-T} + 2e^{-2T} & e^{-T} - e^{-2T} \\ -2e^{-T} + 2e^{-2T} & 2e^{-T} - e^{-2T} \end{bmatrix}$

5-4 Given the state equations of a linear system as

$$\frac{d\mathbf{x}(t)}{dt} = \mathbf{A}\mathbf{x}(t) + \mathbf{B}u(t)$$

with $u(t) = u(kT) = \text{constant}$ for $kT \le t < (k+1)T$. The system is discretized, resulting in the following discrete-data state equations:

$$\mathbf{x}[(k+1)T] = \boldsymbol{\phi}(T)\mathbf{x}(kT) + \boldsymbol{\theta}(T)u(kT)$$

Find the matrices $\boldsymbol{\phi}(T)$ and $\boldsymbol{\theta}(T)$.

a. $\mathbf{A} = \begin{bmatrix} 0 & 1 \\ -1 & -2 \end{bmatrix} \quad \mathbf{B} = \begin{bmatrix} 0 \\ 1 \end{bmatrix}$

b. $\mathbf{A} = \begin{bmatrix} -1 & 2 \\ -1 & 0 \end{bmatrix} \qquad \mathbf{B} = \begin{bmatrix} 0 \\ 1 \end{bmatrix}$

5-5 The block diagram of an open-loop discrete-data control system is shown in Fig. P5-5. Find $\mathbf{x}(N)$ as functions of $\mathbf{x}(0)$ and $u(k)$ for $k = 0, 1, 2, \ldots, N$.

a. $\mathbf{A} = \begin{bmatrix} 0 & 1 \\ -1 & 0 \end{bmatrix} \qquad \mathbf{B} = \begin{bmatrix} 1 \\ 1 \end{bmatrix} \qquad T = \pi/2 \text{ s}$

b. $\mathbf{A} = \begin{bmatrix} 0 & 1 \\ -4 & -5 \end{bmatrix} \qquad \mathbf{B} = \begin{bmatrix} 0 \\ 1 \end{bmatrix} \qquad T = 0.5 \text{ s}$

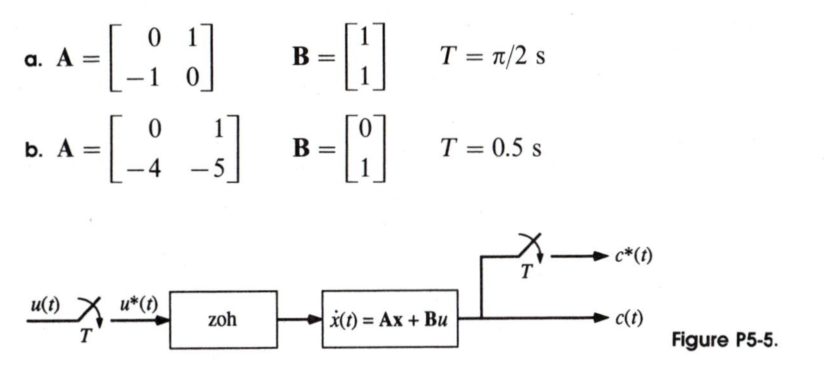

Figure P5-5.

5-6

a. Find the transfer function $\mathbf{X}(z)/U(z)$ for the system given in Problem 5-5-**a**.

b. Find the characteristic equation of the system described in Problem 5-5-**a**. Find the eigenvalues of **A**.

5-7

a. Find the transfer function $\mathbf{X}(z)/U(z)$ for the system given in Problem 5-5-**b**.

b. Find the characteristic equation of the system described in Problem 5-5-**b**. Find the eigenvalues of **A**.

5-8 Prove that if **A** is an $n \times n$ matrix with n distinct eigenvalues, then the set of n eigenvectors, p_i, $i = 1, 2, \ldots, n$, is linearly independent. [Hint: Let

$$a_1 p_1 + a_2 p_2 + \cdots + a_n p_n = 0 \tag{P5-1}$$

where $a_1, a_2, \ldots, p_n$ are constants. If it can be shown that $a_1 = a_2 = \cdots = a_n = 0$ is the only condition that can be satisfied by Eq. (P5-1), then the eigenvectors are linearly independent.]

5-9 Control theory can be applied to the control of economic systems. A well-known dynamic model of supply-and-demand of economic systems is due to Leontief [6] and is given as follows:

$$\frac{d\mathbf{x}(t)}{dt} = -\mathbf{F}(\mathbf{I} - \mathbf{A})\mathbf{x}(t) + \mathbf{F}\mathbf{d}(t)$$

$$\mathbf{s}(t) = (\mathbf{I} - \mathbf{A})\mathbf{x}(t)$$

where $\mathbf{F}$ and $\mathbf{A}$ are $n \times n$ coefficient matrices, $\mathbf{I}$ is an $n \times n$ identity matrix, $\mathbf{x}(t)$ is the $n \times 1$ state vector, $\mathbf{d}(t)$ is the $n \times 1$ demand vector, and $\mathbf{s}(t)$ is the $n \times 1$ supply vector. The elements of these above matrices are defined as

x_i = rate of production of the ith commodity.
a_{ij} = rate at which the ith commodity is used to produce one unit of the jth commodity per unit time.
s_i = supply rate of commodity i available for external consumption.
d_i = external demand rate for commodity i.
f_{ij} = matrix element giving the dynamics of the process.

a. Assuming that the demand $\mathbf{d}_i$, $i = 1, 2, \ldots, n$, is monitored periodically once every T weeks and that it is constant over two consecutive measures, devise a discrete-time model for the system in the form of

$$\mathbf{x}[(k+1)T] = \boldsymbol{\phi}(T)\mathbf{x}(kT) + \boldsymbol{\theta}(T)\mathbf{d}(kT)$$
$$\mathbf{s}(kT) = \boldsymbol{\phi}\mathbf{x}(kT)$$

given that

$$\mathbf{A} = \begin{bmatrix} 0 & 1 \\ 1 & 0 \end{bmatrix} \quad \mathbf{F} = \begin{bmatrix} 1 & 0 \\ 0 & 2 \end{bmatrix} \quad T = 1/7 \text{ week}$$

b. Find $\boldsymbol{\phi}(T)$ and $\boldsymbol{\theta}(T)$.
c. Find the transfer function matrix $\mathbf{G}(z)$ $(n \times n)$, where

$$\mathbf{S}(z) = \mathbf{G}(z)\mathbf{D}(z)$$

d. Find the characteristic equation and the eigenvalues of $\boldsymbol{\phi}(T)$.

5-10

a. Draw a state diagram for the discrete-data system modeled by the following dynamic equations:

$$\mathbf{x}(k+1) = \mathbf{A}\mathbf{x}(k) + \mathbf{B}u(k)$$
$$c(k) = \mathbf{D}\mathbf{x}(k)$$

b. Find the transfer function $C(z)/U(z)$.
c. Find the characteristic equation of the system.

$$\text{(i)}\ \mathbf{A} = \begin{bmatrix} -1 & 1 \\ -0.5 & 0.2 \end{bmatrix} \quad \mathbf{B} = \begin{bmatrix} 0 \\ 1 \end{bmatrix} \quad \mathbf{D} = [1 \quad 0]$$

$$\text{(ii)}\ \mathbf{A} = \begin{bmatrix} 0 & 2 & -1 \\ 0 & 1 & 1 \\ 3 & 3 & -1 \end{bmatrix} \quad \mathbf{B} = \begin{bmatrix} 0 \\ 0 \\ 1 \end{bmatrix} \quad \mathbf{D} = [1 \quad 0 \quad 0]$$

$$\text{(iii)}\ \mathbf{A} = \begin{bmatrix} 0 & 1 & 0 \\ 0 & 0 & 1 \\ 0.2 & -1 & 0.5 \end{bmatrix} \quad \mathbf{B} = \begin{bmatrix} 1 \\ 0 \\ 1 \end{bmatrix} \quad \mathbf{D} = [1 \quad 1 \quad 0]$$

5-11 Find the state transition equations of the following systems by means of the state diagram method.

$$\mathbf{x}(k+1) = \mathbf{A}\mathbf{x}(k) + \mathbf{B}u(k)$$

The initial states are given as $\mathbf{x}(0)$.

$$\mathbf{A} = \begin{bmatrix} 0 & 1 \\ 0.5 & 0.3 \end{bmatrix} \qquad \mathbf{B} = \begin{bmatrix} 0 \\ 1 \end{bmatrix}$$

5-12 The input-output transfer functions of linear discrete-data systems are given below.
a. Draw state diagrams for the systems.
b. Write the dynamic equations of the systems.

(i) $$\frac{C(z)}{R(z)} = \frac{z+0.5}{z^2+0.2z+0.1}$$

(ii) $$\frac{C(z)}{R(z)} = \frac{z^2}{z^3-z^2+0.5z-0.5}$$

5-13 For the digital controller block diagrams shown in Fig. P4-5, assign the outputs of the time-delay units as state variables and write the dynamic equations for the systems.

5-14 Consider the flight control system described in Problem 4-6, Fig. P4-6.
a. Draw a state diagram of the system for the time interval $kT \le t \le (k+1)T$. Use the system parameters given in Problem 4-6, and set $T = 0.1$ s. Define $\theta_m(t) = x_1(t)$ and $\omega_m(t) = x_2(t)$ as the state variables.
b. Write the discrete state equations of the system in the form of

$$\mathbf{x}[(k+1)T] = \boldsymbol{\phi}(T)\mathbf{x}(kT) + \boldsymbol{\theta}(T)\theta_i(kT)$$

c. Find the state transition matrix $\boldsymbol{\phi}(kT)$.

5-15 Repeat Problem 5-14 with $L = 0.05$ H.

5-16 Repeat Problem 5-14 for the flight control system described in Problem 4-8, Fig. P4-8. Use $T = 0.1$ s.

5-17 Repeat Problem 5-14 for the flight control system described in Problem 4-8, Fig. P4-8. Use $T = 0.1$ s and $L = 0.05$ H.

5-18 Consider the Large Space Telescope system described in Problem 4-10, Fig. P4-10.
a. Draw a state diagram of the system for the time interval $kT \le t \le (k+1)T$. Define the state variables as outputs of integrators and number these as x_1, x_2, and x_3, from right to left.
b. Write the discrete state equations in the form of

$$\mathbf{x}[(k+1)T] = \boldsymbol{\phi}(T)\mathbf{x}(kT) + \boldsymbol{\theta}(T)r(kT)$$

5-19 The state diagram of a discrete-data control system is shown in Fig. P5-19. Write the dynamic equations of the system.

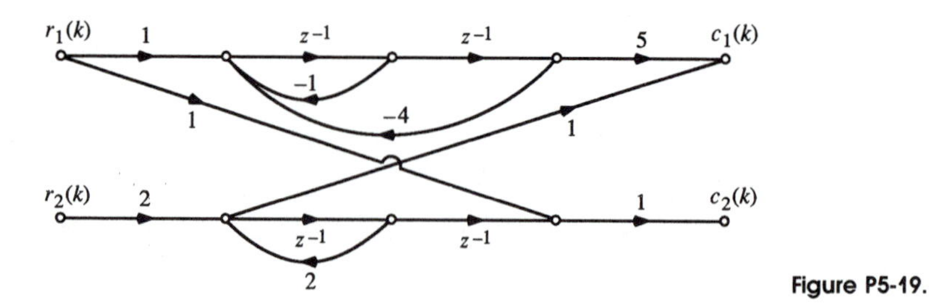

Figure P5-19.

5-20 The block diagram of a discrete-data control system is shown in Fig. P5-20.

a. Draw a state diagram of the system for $kT \leq t \leq (k+1)T$.
b. Write the state equations in the following form:

$$\mathbf{x}[(k+1)T] = \boldsymbol{\phi}(T)\mathbf{x}(kT) + \boldsymbol{\theta}(T)r(kT)$$

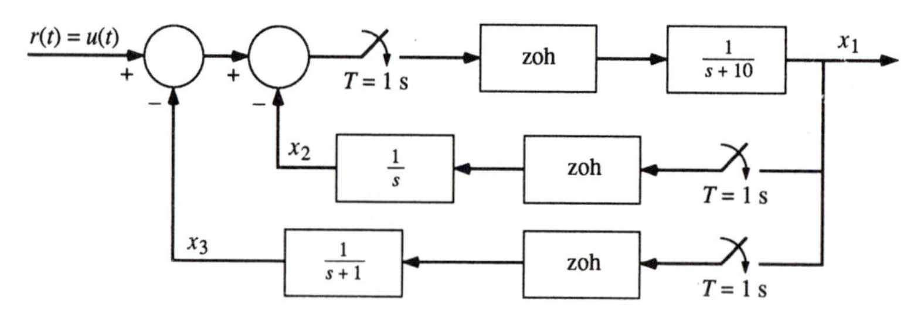

Figure P5-20.

5-21 The block diagram of a feedback control system with discrete data is shown in Fig. P5-21.

$$D(z) = \frac{1 + 0.5z^{-1}}{1 + 0.2z^{-1}} \qquad G(s) = \frac{10(s+5)}{s^2}$$

$T = 0.1$ s.

a. Draw a state diagram for the system.
b. Write the state equations in vector-matrix form. Define the state variables as x_1 and x_2 for $G(s)$ and x_3 for $D(z)$.

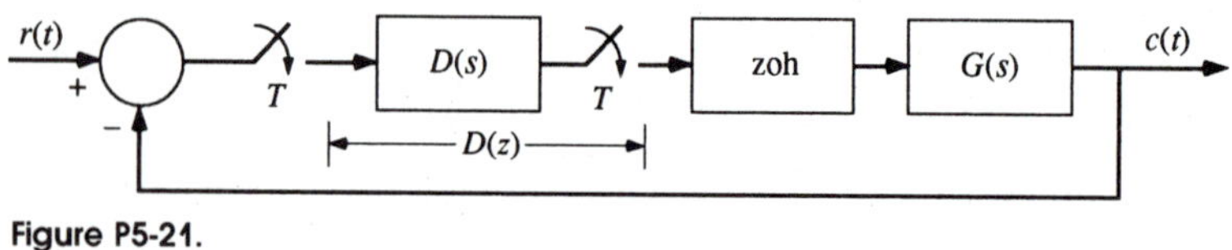

Figure P5-21.

5-22 Decompose the following transfer functions by direct decomposition, and draw the state diagrams. Write the discrete state equations in vector-matrix form.

a. $\dfrac{C(z)}{R(z)} = \dfrac{z + 0.5}{z^3 + 2z^2 + z + 0.5}$

b. $\dfrac{C(z)}{R(z)} = \dfrac{z(z + 0.5)}{z^3 + z^2 + 2z + 0.5}$

5-23 Decompose the following transfer functions by parallel decomposition, and draw the state diagrams. Write the discrete state equations in vector-matrix form.

a. $\dfrac{C(z)}{R(z)} = \dfrac{z - 0.1}{(z - 0.5)(z - 0.8)}$

b. $\dfrac{C(z)}{R(z)} = \dfrac{z}{(z - 0.2)(z - 1)}$

5-24 Decompose the transfer functions in Problem 5-23 by cascade decomposition, and draw the state diagrams. Write the discrete state equations in vector-matrix form.

5-25 Show that the eigenvalues of the digital control system described in Problem 4-10, Fig. P4-10, can be arbitrarily chosen for any sampling period $T \geq 0$ and $J_v > 0$, by choosing K_P, K_I, and K_R appropriately. Find the values of K_P, K_I, and K_R so that the eigenvalues of the system are at $z = 0.5, 0.5$, and 0.8. The sampling period T is 1 s, and $J_v = 0.05$.

5-26 Given the state equations of a digital control system as

$$x_1(k + 1) = x_2(k)$$
$$x_2(k + 1) = x_3(k)$$
$$x_3(k + 1) = 0.4x_2(k) + 0.3x_3(k) + u(k)$$

a. Find the eigenvalues and the corresponding eigenvectors.
b. Find the transformation so that the state equations are transformed into a set of uncoupled state equations.
c. Consider that $u(k) = -[g_1 \quad g_2 \quad g_3]\mathbf{x}(k)$. Find the values of g_1, g_2, and g_3 so that the eigenvalues of the closed-loop system are all equal to zero.

5-27 Consider the automobile idle-speed control system described in Problem 4-11, Fig. P4-11.
a. Draw a state diagram of the system for $kT \leq t \leq (k + 1)T$.
b. Write the state equations in the form of

$$\mathbf{x}[(k + 1)T] = \boldsymbol{\phi}(T)\mathbf{x}(kT) + \boldsymbol{\theta}(T)r(kT)$$

Decompose the continuous-data transfer function by discrete decomposition. Set T_d to zero.

5-28 The block diagram of an open-loop digital control system is shown in Fig. P5-28. The input signal is a unit-step function. Write the state transition equations for the system at $t = kT_2, k = 0, 1, 2, \ldots$.

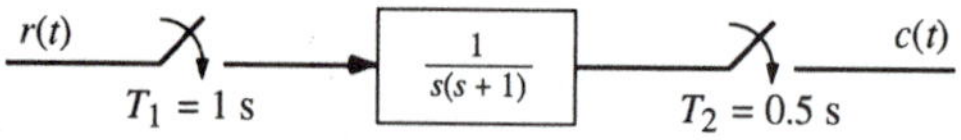

Figure P5-28.

5-29 The block diagram of a digital control system with multirate sampling is shown in Fig. P5-29. Assume that the samplers are synchronized.

a. Write the state equations of the system.

b. Find the state transition equations for the system.

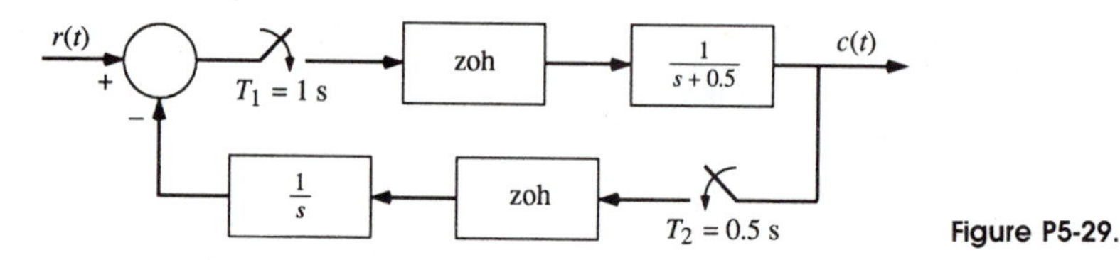

Figure P5-29.

5-30 Figure P5-30 illustrates a linear process whose control signal is the output of a pulsewidth modulator. The control signal $u(t)$ is also shown. The pulsewidth at the ith sampling instant is denoted by α_i, where $\alpha_i \leq T$ for all i. The magnitude of $u(t)$ is designated by $u(kT)$ which is $+1$ or -1 depending on the sign of $r(t)$. The linear process is described by the state equation

$$\frac{d\mathbf{x}(t)}{dt} = \mathbf{A}\mathbf{x}(t) + \mathbf{B}u(t)$$

Show that the system can be described at the sampling instants by the following discrete state equation:

$$\mathbf{x}[(k+1)T] = \boldsymbol{\phi}(T)\mathbf{x}(kT) + \boldsymbol{\phi}(T)\boldsymbol{\phi}(-\alpha_k)\boldsymbol{\theta}(\alpha_k)u(kT)$$

where

$$\boldsymbol{\phi}(T) = e^{\mathbf{A}T}$$

$$\boldsymbol{\theta}(T) = \int_0^T \boldsymbol{\phi}(T-\tau)\mathbf{B}\, d\tau$$

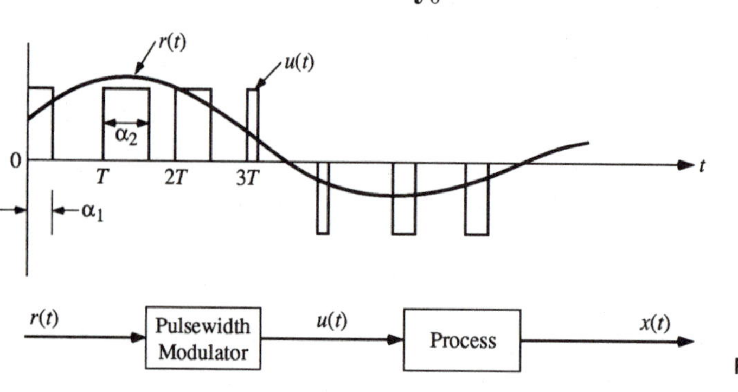

Figure P5-30.

References

1. Kuo, B. C., *Automatic Control Systems*, 6th ed., Prentice-Hall, Englewood Cliffs, N.J., 1991.
2. Zadeh, L. A., "An Introduction to State Space Techniques," Workshop on State Space Techniques for Control Systems, *Proc. Joint Automatic Control Conf.*, Boulder, Colo., 1962.
3. Kamen, E., *Introduction to Signals and Systems*, Macmillan, New York, 1987.
4. Gabel, R. A., and R. A. Roberts, *Signals and Linear Systems*, John Wiley & Sons, New York, 1987.
5. Kalman, R. E., and J. E. Bertram, "A Unified Approach to the Theory of Sampling Systems," *Journal of Franklin Inst.*, vol. 267, pp. 405–436, May 1959.
6. Vogt, W. G., and M. H. Mickle, "A Stochastic Model for a Dynamic Leontief System," *Proc. Tenth Allerton Conf. Circuit and System Theory*, pp. 351–356, October 1972.

Controllability, Observability, and Stability

KEYWORDS AND TOPICS

controllability • state controllability • output controllability • observability • stability • BIBS • BIBO stability • stability tests

6-1 INTRODUCTION

The concepts of *controllability*, *observability*, and *stability* play important roles in the analysis and design of control systems. The significance of the controllability of a system can be stated with reference to the block diagram of Fig. 6-1. *The controlled process* **G** *is said to be controllable if every state of* **G** *can be affected or controlled in finite time by some unconstrained control signal* $\mathbf{u}(k)$. Intuitively, we understand that if any one of the state variables of the system is not accessible from the control $\mathbf{u}(k)$, then there is no way of driving this particular state to a desired state in finite time by means of some control effort. This particular state variable is said to be *uncontrollable*, and the system is said to be uncontrollable. Thus, controllability is an important prerequisite before attempting to design the controls of any process.

The concept of observability is the complement of controllability. Essentially, *the process* **G** *is said to be observable if every state variable of the process eventually affects some of the outputs of the process.* It is often desirable to obtain information on the state variables from measurements of the outputs and inputs. For example, we may want to estimate those state variables that are not accessible for feedback control from measurements of the outputs and inputs. If any one of the states cannot be observed from the measurements of

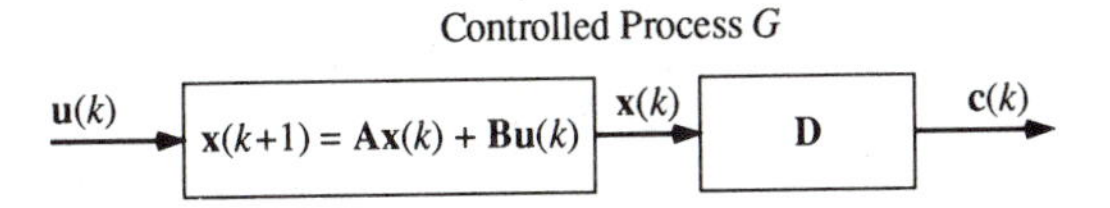

Figure 6-1. Block diagram of a linear discrete-data system.

the outputs, then the state is said to be *unobservable*, and the system is unobservable. Therefore, with reference to Fig. 6-1, observability usually deals with the relationship between the state variables $\mathbf{x}(k)$ and the outputs $\mathbf{c}(k)$, whereas controllability deals with the relationships between the control $\mathbf{u}(k)$ and the states $\mathbf{x}(k)$.

Among the many forms of performance specification used in the design of control systems, the most important requirement is that the system be stable at all times. Relative to the conditions of controllability and observability, stability is an end condition that the system must have, since an unstable system is virtually useless. In many situations, we may start out with an unstable system with one of the design objectives being to achieve stability. Even linear systems have many different types of stability, but to introduce the concept, we can state that stability is the condition wherein *the system output must be bounded for any bounded input.*

Before embarking on the formal discussions of controllability, observability, and stability, let us first provide simple illustrations of the ideas. Figure 6-2 shows the state diagram of a linear discrete-data system with two state variables. Since the input $u(k)$ affects only the state variable $x_1(k)$, we say that $x_2(k)$ is uncontrollable, and the system is *not completely controllable* or simply *uncontrollable.*

Figure 6-3 shows a linear discrete-data system in which the state variable $x_2(k)$ is not connected to the output $c(k)$ in any way. Therefore, we can observe the state variable $x_1(k)$ from measurements of the output $c(k)$, since $c(k) = x_1(k)$. However, the state variable $x_2(k)$ cannot be observed from the information on $c(k)$. Thus, the system is *not completely observable*, or simply *unobservable.* Note that the two examples just given are overly simplified cases devised to illustrate the concepts of controllability and observability. In general, the conditions of controllability and observability usually cannot be detected by mere inspection of the system's state diagram or block diagram.

We shall show that the stability of linear discrete-data systems depends purely on the roots of the characteristic equation. That is, for stability, the roots of the characteristic equation must all lie inside the unit circle $|z| = 1$ in the

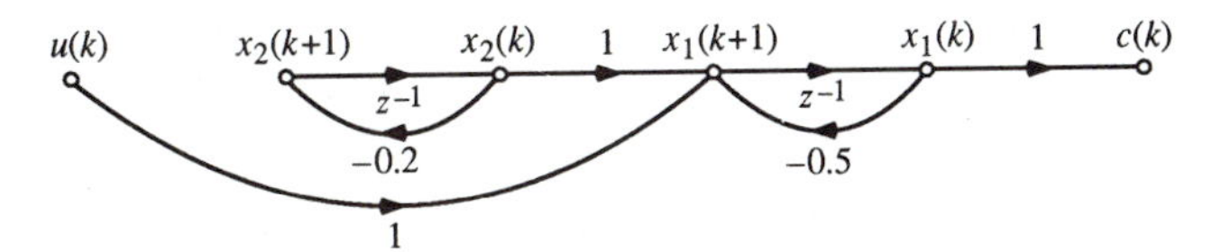

Figure 6-2. State diagram of a process that is not completely controllable.

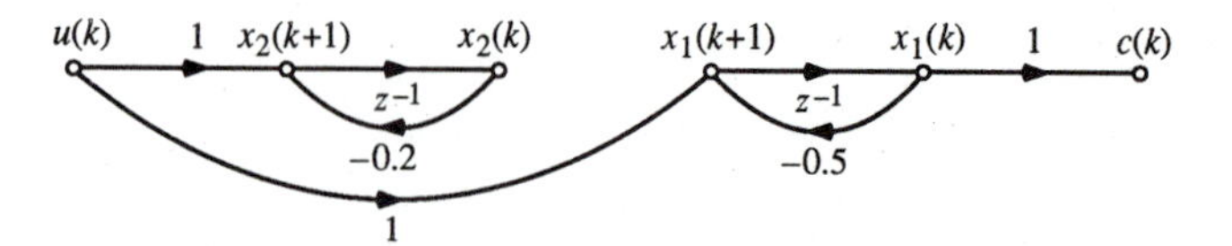

Figure 6-3. State diagram of a process that is not completely observable.

z-plane. From earlier discussions on signal flow graphs and transfer functions, it should be clear that the characteristic equation of a linear system depends only on the loops of the system.

Note also that controllability, observability, and stability are *mutually exclusive* conditions. That is, any system can be controllable and observable, but unstable; it can be stable and controllable, but unobservable; etc.

6-2 CONTROLLABILITY OF LINEAR TIME-INVARIANT DISCRETE-DATA SYSTEMS

6-2-1 Definitions of Controllability

Consider that a linear discrete-data or digital system is described by the following dynamic equations:

$$\mathbf{x}(k+1) = \mathbf{A}\mathbf{x}(k) + \mathbf{B}\mathbf{u}(k) \tag{6-1}$$

$$\mathbf{c}(k) = \mathbf{D}\mathbf{x}(k) + \mathbf{E}\mathbf{u}(k) \tag{6-2}$$

where $\mathbf{x}(k)$ is an n-vector, $\mathbf{u}(k)$ is an r-vector, $\mathbf{c}(k)$ is a p-vector; **A**, **B**, **D**, and **E** are coefficient matrices with appropriate dimensions.

□ **Definition 6-1 Complete State Controllability**

The system described by Eq. (6-1) is said to be completely state controllable if for any initial time (stage) $k = 0$, there exists a set of unconstrained controls $\mathbf{u}(k)$, $k = 0, 1, 2, \ldots, N-1$, which transfers each initial state $\mathbf{x}(0)$ to any final state $\mathbf{x}(N)$ for some finite N. ∎

□ **Definition 6-2 Complete Output Controllability**

The system described by Eqs. (6-1) and (6-2) is said to be completely output controllable if for any initial time (stage) $k = 0$, there exists a set of unconstrained controls $\mathbf{u}(k)$, $k = 0, 1, 2, \ldots, N-1$, such that any final output $\mathbf{c}(N)$ can be reached from arbitrary initial states in finite time (stage) N. ∎

6-2-2 Theorems on Controllability

The following theorems are given on state and output controllability of linear time-invariant discrete-data systems. Some of the theorems are stated without proofs.

□ Theorem 6-1 State Controllability

The system described by Eq. (6-1) is completely state controllable if and only if the $n \times Nr$ matrix

$$\mathbf{S} = [\mathbf{B} \quad \mathbf{AB} \quad \mathbf{A}^2\mathbf{B} \quad \cdots \quad \mathbf{A}^{N-1}\mathbf{B}] \tag{6-3}$$

is of rank n. The matrix $\mathbf{S}$ is called the *state controllability matrix.*

If the system is completely state controllable, then we also say that the pair $[\mathbf{A}, \mathbf{B}]$ is controllable. ■

Proof: The state transition equation of Eq. (6-1) is [from Eq. (5-67)]

$$\mathbf{x}(N) = \mathbf{A}^N\mathbf{x}(0) + \sum_{i=0}^{N-1} \mathbf{A}^{N-i-1}\mathbf{Bu}(i) \tag{6-4}$$

The last equation is rewritten as

$$\mathbf{x}(N) - \mathbf{A}^N\mathbf{x}(0) = \sum_{i=0}^{N-1} \mathbf{A}^{N-i-1}\mathbf{Bu}(i) \tag{6-5}$$

The left side of Eq. (6-5) can be represented by an $n \times 1$ vector $\mathbf{X}(N)$; then the equation is written as

$$\mathbf{X}(N) = \mathbf{SU} \tag{6-6}$$

where $\mathbf{S}$ is given by Eq. (6-3), and $\mathbf{U}$ is the $Nr \times 1$ vector

$$\mathbf{U} = \begin{bmatrix} u(0) \\ u(1) \\ \vdots \\ u(N-1) \end{bmatrix} \tag{6-7}$$

For complete state controllability, every initial state $\mathbf{x}(0)$ must be transferred by unconstrained controls $\mathbf{u}(i)$, $i = 0, 1, 2, \ldots, N-1$, to any final state $\mathbf{x}(N)$ for finite N. Thus, the problem is that of given $\mathbf{S}$ and every vector $\mathbf{X}(N)$ in the n-dimensional state space, solve for the controls from Eq. (6-6). Since Eq. (6-6) represents n simultaneous linear equations, from the theory of linear equations, these equations must be linearly independent for solutions to exist, and the necessary and sufficient condition for this to be true is that the matrix $\mathbf{S}$ has a rank of n. In other words, $\mathbf{S}$ must have at least n independent columns.

If the system in Eq. (6-1) has only one input, $r = 1$, then $\mathbf{S}$ is an $n \times n$ square matrix; the condition of complete state controllability is that $\mathbf{S}$ must be nonsingular.

If the state equations of the linear system are represented as

$$\mathbf{x}[(k+1)T] = \boldsymbol{\phi}(T)\mathbf{x}(kT) + \boldsymbol{\theta}(T)\mathbf{u}(T) \tag{6-8}$$

then the state controllability matrix is given by

$$\mathbf{S} = [\boldsymbol{\theta}(T) \quad \boldsymbol{\phi}(T)\boldsymbol{\theta}(T) \quad \boldsymbol{\phi}^2(T)\boldsymbol{\theta}(T) \quad \cdots \quad \boldsymbol{\phi}^{N-1}(T)\boldsymbol{\theta}(T)] \tag{6-9}$$

□ Theorem 6-2 State Controllability

The system described by Eq. (6-1) is completely state controllable if the $n \times n$ matrix $\mathbf{SS}'$ is nonsingular. ■

Proof: Since if the state controllability matrix $\mathbf{S}$ is of rank n, its transpose $\mathbf{S}'$ has the same rank. Thus, $\mathbf{SS}'$ must also have a rank of n or be nonsingular.

The matrix $\mathbf{SS}'$ is known as the Gramian matrix and is expressed as

$$\mathbf{SS}' = \sum_{i=0}^{N-1} \mathbf{A}^{N-i-1}\mathbf{BB}'(\mathbf{A}^{N-i-1})' \tag{6-10}$$

●

□ Theorem 6-3 State Controllability versus Phase-Variable Canonical Form

If the system represented by Eq. (6-1) has a single input and is in phase-variable canonical form (PVCF), then it is completely state controllable. ■

Proof: The proof of this theorem follows directly from the condition of phase-variable canonical form transformation discussed in Sec. 5-13. Notice that the state controllability matrix $\mathbf{S}$ is identical to the matrix $\mathbf{S}$ given in Eq. (5-252) for testing the condition for PVCF transformation. ●

□ Theorem 6-4 State Controllability of Systems with Distinct Eigenvalues

For the system described in Eq. (6-1), if $\mathbf{A}$ is a diagonal matrix and has distinct eigenvalues of z_i, $i = 1, 2, \ldots, n$, then the system is completely state controllable if and only if $\mathbf{B}$ does not have any rows containing all zeros. ■

Proof: Since $\mathbf{A}$ is a diagonal matrix, the state variables of the system are decoupled from each other. Therefore, if any one row of $\mathbf{B}$ contains all zeros, the corresponding state will not be affected by any one of the inputs, and that state will be uncontrollable. ●

□ Theorem 6-5 State Controllability of Systems with Multiple-Order Eigenvalues

For the system described in Eq. (6-1), if $\mathbf{A}$ has multiple-order eigenvalues, and $\mathbf{A}$ is in Jordan canonical form, then the system is completely state controllable if and only if

1. each Jordan block corresponds to one distinct eigenvalue, and
2. the elements of $\mathbf{B}$ that correspond to the last row of each Jordan block are not all zeros. ■

Proof: Since the last row of each Jordan block corresponds to the state equation that is completely uncoupled from the other state equations, if the elements of the rows of **B** that correspond to these uncoupled states are all zero, these states would not be controllable by any of the inputs. The elements in the rows of **B** that correspond to the other rows of the Jordan blocks could indeed contain all zeros, as the ones in the off-diagonal positions of **A** form couplings between the states.

Notice that the conditions in this theorem also cover the distinct eigenvalues case in Theorem 6-4, since in this case the last row of each Jordan block is the only row.

We can use the following **A** and **B** matrices as an illustration.

$$\mathbf{A} = \begin{bmatrix} z_1 & 0 & 0 & 0 & 0 \\ 0 & z_2 & 0 & 0 & 0 \\ 0 & 0 & z_3 & 0 & 0 \\ 0 & 0 & 0 & z_4 & 1 \\ 0 & 0 & 0 & 0 & z_5 \end{bmatrix} \qquad \mathbf{B} = \begin{bmatrix} b_{11} & b_{12} \\ b_{21} & b_{22} \\ b_{31} & b_{32} \\ b_{41} & b_{42} \\ b_{51} & b_{52} \end{bmatrix} \tag{6-11}$$

where $z_4 = z_5$. Each of the Jordan blocks associated with the eigenvalues z_1, z_2, and z_3 contains only one row. Thus, for controllability, none of the rows of **B** that correspond to these Jordan blocks can contain all zeros. The Jordan block associated with z_4 and z_5 contains two rows. The last row indicates that the state x_5 is not coupled to any of the other four states, and thus, for it to be controllable, the elements b_{51} and b_{52} cannot all be zero. On the other hand, the state x_4 is coupled through x_5, as indicated by the one in the off-diagonal position of the **A** matrix. Thus, the corresponding row of **B** which contains the elements b_{41} and b_{42} can indeed be all zeros, and the system could still be controllable. ●

□ Theorem 6-6 Invariant Theorem on State Controllability

An extension to the state controllability conditions in Theorems 6-4 and 6-5 is that if the matrix **A** is not diagonal for distinct eigenvalues or not in Jordan canonical form for multiple-order eigenvalues, then, as in Sec. 5-9, we can use a similarity transformation **P** to transform the system into canonical forms. That is,

$$\mathbf{\Lambda} = \mathbf{P}^{-1}\mathbf{A}\mathbf{P} \tag{6-12}$$

and

$$\mathbf{\Gamma} = \mathbf{P}^{-1}\mathbf{B} \tag{6-13}$$

Then, the pair $[\mathbf{A}, \mathbf{B}]$ is completely controllable if the pair $[\mathbf{\Lambda}, \mathbf{\Gamma}]$ is completely controllable; i.e., the rows of $\mathbf{\Gamma}$ that correspond to the last rows of the Jordan blocks in $\mathbf{\Lambda}$ do not contain all zeros. Since the controllability matrix of the pair $[\mathbf{\Lambda}, \mathbf{\Gamma}]$ is

$$[\mathbf{\Gamma} \quad \mathbf{\Lambda\Gamma} \quad \mathbf{\Lambda}^2\mathbf{\Gamma} \quad \cdots \quad \mathbf{\Lambda}^{N-1}\mathbf{\Gamma}] = \mathbf{P}^{-1}[\mathbf{B} \quad \mathbf{AB} \quad \mathbf{A}^2\mathbf{B} \quad \cdots \quad \mathbf{A}^{N-1}\mathbf{B}] \tag{6-14}$$

the controllability of $[\mathbf{\Lambda}, \mathbf{\Gamma}]$ implies the same for $[\mathbf{A}, \mathbf{B}]$.

In general, the transformation is not limited to just the similarity transformation described in Eqs. (6-12) and (6-13). The PVCF transformation would also apply. ■

□ Theorem 6-7 Output Controllability

The system described by Eqs. (6-1) and (6-2) is completely output controllable if and only if the following $p \times (N+1)r$ matrix is of rank p, where p is the dimension of the output vector $\mathbf{c}(t)$.

$$\mathbf{T} = [\mathbf{DA}^{N-1}\mathbf{B} \quad \mathbf{DA}^{N-2}\mathbf{B} \quad \cdots \quad \mathbf{DAB} \quad \mathbf{DB} \quad \mathbf{E}] \tag{6-15}$$

The matrix $\mathbf{T}$ is referred to as the *output controllability matrix.* ■

Proof: Substituting the state transition equation of Eq. (6-4) into the output equation of Eq. (6-2) with $k = N$ and rearranging, we have

$$\mathbf{c}(N) - \mathbf{DA}^{N-1}\mathbf{x}(0) = \sum_{i=0}^{N-1} \mathbf{DA}^{N-i-1}\mathbf{Bu}(i) + \mathbf{Eu}(N) \tag{6-16}$$

Since the last equation is similar to Eq. (6-5), the proof of this theorem is parallel to that of Theorem 6-1. ●

The following examples are provided to illustrate the concepts and applications of controllability.

Example 6-1

Consider a linear discrete-data control system whose input-output relation is described by the difference equation

$$c(k+2) + 2c(k+1) + c(k) = u(k+1) + u(k) \tag{6-17}$$

This difference equation is decomposed by direct decomposition into the following dynamic equations:

$$\mathbf{x}(k+1) = \mathbf{Ax}(k) + \mathbf{B}u(k) \tag{6-18}$$

$$c(k) = \mathbf{Dx}(k) \tag{6-19}$$

where

$$\mathbf{A} = \begin{bmatrix} 0 & 1 \\ -1 & -2 \end{bmatrix} \qquad \mathbf{B} = \begin{bmatrix} 0 \\ 1 \end{bmatrix}$$

$$\mathbf{D} = [1 \quad 1] \tag{6-20}$$

The state controllability matrix is

$$\mathbf{S} = [\mathbf{B} \quad \mathbf{AB}] = \begin{bmatrix} 0 & 1 \\ 1 & -2 \end{bmatrix} \tag{6-21}$$

which is nonsingular. Thus, the system is completely state controllable for the state variables $x_1(k)$ and $x_2(k)$. We should have expected this result, since **A** and **B** in Eq. (6-20) are in the phase-variable canonical form.

The output controllability is tested by forming the matrix

$$\mathbf{T} = [\mathbf{DAB} \quad \mathbf{DB}] = [-1 \quad 1] \tag{6-22}$$

which has a rank of one. Since the dimension of $c(t)$ is one, the system is completely output controllable.

As an alternative, we can define the state variables of the system represented by Eq. (6-17) so that the dynamic equations in Eqs. (6-18) and (6-19) have the following coefficient matrices:

$$\mathbf{A} = \begin{bmatrix} 0 & 1 \\ -1 & -2 \end{bmatrix} \qquad \mathbf{B} = \begin{bmatrix} 1 \\ -1 \end{bmatrix}$$

$$\mathbf{D} = [1 \quad 0] \tag{6-23}$$

The state and output controllability matrices are

$$\mathbf{S} = [\mathbf{B} \quad \mathbf{AB}] = \begin{bmatrix} 1 & -1 \\ -1 & 1 \end{bmatrix} \tag{6-24}$$

$$\mathbf{T} = [\mathbf{DAB} \quad \mathbf{DB}] = [1 \quad 1] \tag{6-25}$$

Since **S** in Eq. (6-24) is singular, the state variables represented by the coefficient matrices in Eq. (6-23) are not all controllable, and the system is not state controllable. Thus, we have demonstrated by this example that state controllability depends on how the state variables are defined. The output controllability is not affected by the choice of state variables.

One can easily observe that the system of Eq. (6-17) would result in a transfer function with pole-zero cancellation. In practice, state representation of this type of system would result in system models with controllability and/or observability problems.

Now let us transform **A** and **B** into Jordan canonical form and test the state controllability using matrices $\mathbf{\Lambda}$ and $\mathbf{\Gamma}$. Using the generalized eigenvectors discussed in Sec. 5-10, the transformation matrix **P** is determined to be

$$\mathbf{P} = \begin{bmatrix} 1 & 0 \\ -1 & 1 \end{bmatrix} \tag{6-26}$$

Thus, for the coefficient matrices in Eq. (6-20),

$$\mathbf{\Gamma} = \mathbf{P}^{-1}\mathbf{B} = \begin{bmatrix} 1 & 0 \\ 1 & 1 \end{bmatrix} \begin{bmatrix} 0 \\ 1 \end{bmatrix} = \begin{bmatrix} 0 \\ 1 \end{bmatrix} \tag{6-27}$$

Since $\mathbf{A}$ has a second-order eigenvalue, $\mathbf{\Lambda}$ has only one Jordan block. In this case the last row of $\mathbf{\Gamma}$ that corresponds to the last row of the Jordan block in $\mathbf{\Lambda}$ is nonzero. Thus, the state model is state controllable.

Using the state model described in Eq. (6-23),

$$\mathbf{\Gamma} = \mathbf{P}^{-1}\mathbf{B} = \begin{bmatrix} 1 & 0 \\ 1 & 1 \end{bmatrix}\begin{bmatrix} 1 \\ -1 \end{bmatrix} = \begin{bmatrix} 1 \\ 0 \end{bmatrix} \tag{6-28}$$

Since the last row of $\mathbf{\Gamma}$ that corresponds to the last row of the Jordan block of $\mathbf{\Lambda}$ is zero, the state model of Eq. (6-23) is not completely state controllable.

Example 6-2

In this illustrative example we demonstrate certain practical considerations in the application of controllability. Most practical control systems found in the process and aerospace industries are of high order, so that the application of the controllability criterion may not be as straightforward as that demonstrated in the previous example. The formal definition of controllability (and observability) is a "yes" or "no" proposition; that is, the system is either controllable or uncontrollable. However, in practice, we often encounter systems that are clearly controllable under the strict definition, though the numerical parameters of the system may render it very difficult to control with realistic controller parameters.

The digital model of the dynamics of a space vehicle is represented by the vector matrix equation in Eq. (6-1), where $\mathbf{A}$ is an 11×11 matrix, and $\mathbf{B}$ is 11×1. For convenience, we express $\mathbf{A}$ as a partitioned matrix as follows:

$$\mathbf{A} = \begin{bmatrix} \mathbf{A}_{11} & \mathbf{A}_{12} \\ \mathbf{A}_{21} & \mathbf{A}_{22} \end{bmatrix} \tag{6-29}$$

where $\mathbf{A}_{11} = \mathbf{0}$ (5×5),

$$\mathbf{A}_{12} = \begin{bmatrix} 1 & 0 & 0 & 0 & 0 & 0 \\ 0 & 1 & 0 & 0 & 0 & 0 \\ 0 & 0 & 0 & 1 & 0 & 0 \\ 0 & 0 & 0 & 0 & 1 & 0 \\ 0 & 0 & 0 & 0 & 0 & 1 \end{bmatrix} \tag{6-30}$$

$$\mathbf{A}_{21} = \begin{bmatrix} 0.176 & 0 & 0 & -85.26 & -2.56 \\ 0 & -134.6 & -1.06 & 0 & 0 \\ 0 & -0.69 & 10.17 & 0 & 0 \\ 0 & 19.91 & -293.0 & 0 & 0 \\ -0.366 & 0 & 0 & -52.86 & -0.168 \\ 6.02 & 0 & 0 & -85.4 & -18.84 \end{bmatrix} \tag{6-31}$$

$$\mathbf{A}_{22} = \begin{bmatrix} 0 & -6.1 & -0.188 & 0 & -0.113 & -0.077 \\ -24.0 & 0 & 0 & 0.02 & 0 & 0 \\ 0.69 & 0 & 0 & 0.304 & 0 & 0 \\ -19.9 & 0 & 0 & -8.76 & 2.0 & 0 \\ 0 & -3.66 & 2.36 & -11.67 & -7.0 & -0.005 \\ 0 & -6.1 & -0.188 & 0 & -0.113 & -0.388 \end{bmatrix} \tag{6-32}$$

$$\mathbf{B} = [0 \quad 0 \quad 0 \quad 0 \quad 0 \quad -7.28 \quad 0 \quad 0 \quad 0 \quad -0.478 \quad -7.28] \tag{6-33}$$

Since the system has only one input, we can investigate the controllability of the system by evaluating the determinant of the 11×11 matrix

$$\mathbf{S} = [\mathbf{B} \quad \mathbf{AB} \quad \mathbf{A}^2\mathbf{B} \quad \cdots \quad \mathbf{A}^{10}\mathbf{B}] \tag{6-34}$$

and the result is $|\mathbf{S}| = 4.46 \times 10^{18}$. Therefore, the system is completely state controllable. However, during the process of designing a state-feedback control to place the eigenvalues of the closed-loop system at appropriate locations in the z-plane, numerical difficulties were encountered. It was suspected that certain peculiarities of the system may not have been uncovered by the controllability condition of the system tested by using the **S** matrix in Eq. (6-34). Using an alternative method of testing controllability, we transform **A** into a diagonal matrix by a similarity transformation, according to Theorem 6-6. The eigenvalues of **A** are distinct and are tabulated as follows:

$$\begin{aligned} -0.603 &\pm j30.15 \\ -0.0328 &\pm j1.76 \\ -0.0276 &\pm j1.36 \\ &\pm j1.0 \\ -0.00056 &\pm j0.29 \\ 0 & \end{aligned}$$

The $\boldsymbol{\Gamma}$ matrix is obtained as

$$\boldsymbol{\Gamma} = \mathbf{P}^{-1}B = \begin{bmatrix} -8.16 \times 10^{-2} \\ 1.67 \times 10^{-3} \\ -9.32 \times 10^{-1} \\ 2.99 \times 10^{-2} \\ -1.32 \times 10^{-3} \\ 9.00 \times 10^{-2} \\ -1.76 \times 10^{-1} \\ 1.54 \times 10^{-10} \\ 1.97 \times 10^{-3} \\ -3.19 \\ -2.79 \times 10^{-16} \end{bmatrix} \tag{6-35}$$

Notice that, theoretically, since the matrix $\mathbf{S}$ in Eq. (6-34) has a rank of 11, or $|\mathbf{S}| \neq 0$, or, none of the elements of $\mathbf{\Gamma}$ is zero, the system should be completely state controllable. However, Eq. (6-35) shows that the last element is nearly zero, and we may consider that the system is "nearly uncontrollable." From the practical standpoint, since the elements of $\mathbf{\Gamma}$ do correspond to the coupling of the individual states to the input, the magnitudes of these elements give indication of the degree of controllability of the states. In the present case, the last element of $\mathbf{\Gamma}$ corresponds to the eigenvalue of $\mathbf{A}$ at $z = 0$, and as it turns out it is practically impossible to move this eigenvalue by state feedback.

The purpose of this example is to illustrate that, while the alternate methods of testing controllability may seem to be equivalent, in solving practical problems the numerical properties of a system may render one method more informative than the others. In the present case, while the check on the rank of $\mathbf{S}$ indicates that the system is completely state controllable, the similarity transformation method reveals the difficulty of controlling certain states.

□ Theorem 6-8 Controllability of Closed-Loop Systems with State Feedback

If the digital control system

$$\mathbf{x}(k+1) = \mathbf{A}\mathbf{x}(k) + \mathbf{B}\mathbf{u}(k) \tag{6-36}$$

is completely state controllable, then the closed-loop system derived through state feedback,

$$\mathbf{u}(k) = \mathbf{r}(k) - \mathbf{G}\mathbf{x}(k) \tag{6-37}$$

so that the state equations of the closed-loop system are described by

$$\mathbf{x}(k+1) = (\mathbf{A} - \mathbf{B}\mathbf{G})\mathbf{x}(k) + \mathbf{B}\mathbf{r}(k) \tag{6-38}$$

is also completely state controllable. On the other hand, if the pair $[\mathbf{A}, \mathbf{B}]$ is uncontrollable, then there is no $\mathbf{G}$ that will make the pair $[\mathbf{A} - \mathbf{B}\mathbf{G}, \mathbf{B}]$ controllable. In other words, if the open-loop system is uncontrollable, then it cannot be made controllable through state feedback. ■

Proof: By controllability of the pair $[\mathbf{A}, \mathbf{B}]$ we mean that there exists a control $\mathbf{u}(k)$ over the interval $[0, N]$ such that the initial state $\mathbf{x}(0)$ is driven to $\mathbf{x}(N)$ for the finite time interval N. We can write Eq. (6-37) as

$$\mathbf{r}(k) = \mathbf{u}(k) + \mathbf{G}\mathbf{x}(k) \tag{6-39}$$

which is the control of the closed-loop system. Thus, if $\mathbf{u}(k)$ exists which can drive $\mathbf{x}(0)$ to $\mathbf{x}(N)$ in finite time, Eq. (6-39) implies that $\mathbf{r}(k)$ also exists, and the closed-loop system is also controllable.

Conversely, if the pair $[\mathbf{A}, \mathbf{B}]$ is uncontrollable, which means that no $\mathbf{u}(k)$ exists that will drive any $\mathbf{x}(0)$ to any $\mathbf{x}(N)$ in finite N, then we cannot find an input $\mathbf{r}(k)$ that will do the same to $\mathbf{x}(k)$, since otherwise we can set $\mathbf{u}(k)$ as in Eq. (6-37) to control the open-loop system. ●

Example 6-3

For the system in Eq. (6-36) let

$$\mathbf{A} = \begin{bmatrix} 0 & 1 \\ -1 & -2 \end{bmatrix} \qquad \mathbf{B} = \begin{bmatrix} 0 \\ 1 \end{bmatrix} \tag{6-40}$$

It was shown in Example 6-1 that the system is completely controllable.

Let the state feedback be defined as in Eq. (6-37); the feedback matrix is given by

$$\mathbf{G} = [g_1 \quad g_2] \tag{6-41}$$

where g_1 and g_2 are real constants. The closed-loop system with state feedback is described by Eq. (6-38), with

$$\mathbf{A} - \mathbf{BG} = \begin{bmatrix} 0 & 1 \\ -1 - g_1 & -2 - g_2 \end{bmatrix} \tag{6-42}$$

The characteristic equation of the closed-loop system is

$$|z\mathbf{I} - \mathbf{A} + \mathbf{BG}| = z^2 + (2 + g_2)z + (1 + g_1) = 0 \tag{6-43}$$

Since the feedback gains g_1 and g_2 are isolated in each coefficient of the characteristic equation, they can be arbitrarily set.

Now let the system matrices be

$$\mathbf{A} = \begin{bmatrix} 0 & 1 \\ -1 & -2 \end{bmatrix} \qquad \mathbf{B} = \begin{bmatrix} 1 \\ -1 \end{bmatrix} \tag{6-44}$$

Going through with the state-feedback design, the characteristic equation of the closed-loop system is

$$|z\mathbf{I} - \mathbf{A} + \mathbf{BG}| = z^2 + (2 + g_1 - g_2)z + (1 + g_1 - g_2) = 0 \tag{6-45}$$

Since the two coefficients of the last equation are dependent on each other, the roots of the equation cannot be arbitrarily assigned. It was shown in Example 6-1 that the [**A**, **B**] pair in Eq. (6-44) is not controllable.

6-3 OBSERVABILITY OF LINEAR TIME-INVARIANT DISCRETE-DATA SYSTEMS

6-3-1 Definition of Observability

□ **Definition 6-3 Complete Observability**

The digital system described in Eqs. (6-1) and (6-2) is said to be completely observable if for any initial time (stage) $k = 0$, any state $\mathbf{x}(0)$ can be determined from knowledge of the output $\mathbf{c}(k)$ and input $\mathbf{u}(k)$ for $0 \leq k < N$, where N is some finite time (stage). ■

6-3-2 Theorems on Observability

□ Theorem 6-9 Complete Observability

The linear digital system described by Eqs. (6-1) and (6-2) is completely observable if and only if the following $n \times pN$ matrix is of rank n:

$$\mathbf{L} = [\mathbf{D}' \quad \mathbf{A}'\mathbf{D}' \quad (\mathbf{A}')^2\mathbf{D}' \quad \cdots \quad (\mathbf{A}')^{N-1}\mathbf{D}'] \tag{6-46}$$

where n is the dimension of $\mathbf{x}(k)$, and p is the dimension of $\mathbf{c}(k)$. The matrix $\mathbf{L}$ is known as the observability matrix. ■

Proof: Following Eq. (5-64), the state transition equation of Eq. (6-1) is written as

$$\mathbf{x}(k+j) = \mathbf{A}^j\mathbf{x}(k) + \sum_{i=0}^{j-1} A^{j-i-1}\mathbf{B}\mathbf{u}(k+i) \tag{6-47}$$

for $j = 1, 2, \ldots, N-1$.

Substituting Eq. (6-47) into the output equation, Eq. (6-2), we get

$$\mathbf{c}(k+j) = \mathbf{D}\mathbf{A}^j\mathbf{x}(k) + \sum_{i=0}^{j-1} \mathbf{D}\mathbf{A}^{j-i-1}\mathbf{B}\mathbf{u}(k+i) + \mathbf{E}\mathbf{u}(k+j) \tag{6-48}$$

for $j = 1, 2, \ldots, N-1$.

When j assumes values from 1 through $N-1$, Eq. (6-48) together with Eq. (6-2) represents pN linear algebraic equations that can be put in a matrix form as follows:

$$\begin{bmatrix} \mathbf{c}(k) \\ \mathbf{c}(k+1) \\ \mathbf{c}(k+2) \\ \vdots \\ \mathbf{c}(k+N-1) \end{bmatrix} = \begin{bmatrix} \mathbf{D} \\ \mathbf{DA} \\ \mathbf{DA}^2 \\ \vdots \\ \mathbf{DA}^{N-1} \end{bmatrix} \mathbf{x}(k)$$

$$+ \begin{bmatrix} \mathbf{E} & \mathbf{0} & \mathbf{0} & \cdots & \mathbf{0} & \mathbf{0} \\ \mathbf{DB} & \mathbf{E} & \mathbf{0} & \cdots & \mathbf{0} & \mathbf{0} \\ \mathbf{DAB} & \mathbf{DB} & \mathbf{E} & \cdots & \mathbf{0} & \mathbf{0} \\ \vdots & \vdots & \vdots & & \vdots & \vdots \\ \mathbf{DA}^{N-2}\mathbf{B} & \mathbf{DA}^{N-3}\mathbf{B} & \mathbf{D}^{N-4}\mathbf{AB} & \cdots & \mathbf{DB} & \mathbf{E} \end{bmatrix} \begin{bmatrix} \mathbf{u}(k) \\ \mathbf{u}(k+1) \\ \mathbf{u}(k+2) \\ \vdots \\ \mathbf{u}(k+N-1) \end{bmatrix} \tag{6-49}$$

To determine $\mathbf{x}(k)$ from the last equation, given $\mathbf{c}(k+j)$ and $\mathbf{u}(k+j)$, for $j = 0, 1, 2, \ldots, N-1$, the $pN \times n$ matrix

$$\mathbf{L}' = \begin{bmatrix} \mathbf{D} \\ \mathbf{DA} \\ \mathbf{DA}^2 \\ \vdots \\ \mathbf{DA}^{N-1} \end{bmatrix} \tag{6-50}$$

must have n independent rows (assuming that $pN \geq n$). The condition in Eq. (6-50) is equivalent to the matrix $\mathbf{L}$ in Eq. (6-46) must have a rank n.

For a discrete-data system that is described by the state equations in Eq. (6-8) and the output equations of Eq. (6-2), the condition of complete observability is that the matrix

$$\mathbf{L} = [\mathbf{D}' \quad \boldsymbol{\phi}(T)'\mathbf{D}' \quad \boldsymbol{\phi}(2T)'\mathbf{D}' \quad \cdots \quad \boldsymbol{\phi}[(N-1)T]'\mathbf{D}'] \tag{6-51}$$

must be of rank n. ●

□ Theorem 6-10 Complete Observability

The linear time-invariant system described by Eqs. (6-1) and (6-2) is completely observable if and only if the Gramian matrix

$$\mathbf{V} = \sum_{i=0}^{N-1} (\mathbf{A}^{N-i-1})'\mathbf{D}'\mathbf{D}\mathbf{A}^{N-i-1} \tag{6-52}$$

is nonsingular. ■

□ Theorem 6-11 Complete Observability of Systems with Distinct Eigenvalues

For the digital system described by Eqs. (6-1) and (6-2), if $\mathbf{A}$ has distinct eigenvalues, the dynamic equations are transformed by the similarity transformation

$$\mathbf{x}(k) = \mathbf{P}\mathbf{y}(k) \tag{6-53}$$

to

$$\mathbf{y}(k+1) = \boldsymbol{\Lambda}\mathbf{y}(k) + \boldsymbol{\Gamma}\mathbf{u}(k) \tag{6-54}$$

$$\mathbf{c}(k) = \mathbf{F}\mathbf{y}(k) + \mathbf{E}\mathbf{u}(k) \tag{6-55}$$

where

$$\boldsymbol{\Lambda} = \mathbf{P}^{-1}\mathbf{A}\mathbf{P} = \text{diag}\,[z_i] \tag{6-56}$$

z_i, $i = 1, 2, \ldots, n$, represents the ith eigenvalue of $\mathbf{A}$;

$$\mathbf{\Gamma} = \mathbf{P}^{-1}\mathbf{B} \tag{6-57}$$

$$\mathbf{F} = \mathbf{DP} \tag{6-58}$$

The system is completely observable if and only if the matrix **F** does not have any columns containing all zeros. ■

Proof: The reason behind this theorem is that, since $\mathbf{\Gamma}$ is a diagonal matrix, the system of state equations in Eq. (6-54) is decoupled. Therefore, if any one column of **F** contains all zeros, the corresponding state variable will not be observed by any one of the outputs. ●

□ Theorem 6-12 Complete Observability of Systems with Multiple Eigenvalues

For the system described by Eqs. (6-1) and (6-2), if **A** has multiple eigenvalues, then the similarity transformation of Eq. (6-53) transforms **A** into a Jordan canonical form $\mathbf{\Lambda}$. The conditions of complete observability are

1. each Jordan block corresponds to one distinct eigenvalue, and
2. the elements of the columns of **F** that correspond to the first column of each Jordan block are not all zeros. ■

Proof: The proof of this theorem is similar to that of Theorem 6-5 on the controllability of systems with multiple eigenvalues. ●

□ Theorem 6-13 Invariant Theorem on Nonsingular Transformation, Observability

Given the nth-order digital system as described by Eqs. (6-1) and (6-2), where the pair $[\mathbf{A}, \mathbf{D}]$ is completely observable. The transformation $\mathbf{x}(k) = \mathbf{P}\mathbf{y}(k)$, where **P** is nonsingular, transforms the system equations to

$$\mathbf{y}(k+1) = \mathbf{\Lambda}\mathbf{y}(k) + \mathbf{\Gamma}\mathbf{u}(k) \tag{6-59}$$

and

$$\mathbf{c}(k) = \mathbf{DP}\mathbf{y}(k) + \mathbf{E}\mathbf{u}(k) \tag{6-60}$$

Then, the pair $[\mathbf{\Lambda}, \mathbf{DP}]$ is also observable. ■

Proof: The proof of this theorem is similar to that of Theorem 6-6. We form the observability matrix

$$\mathbf{V}_1 = [(\mathbf{DP})' \quad \mathbf{\Lambda}'(\mathbf{DP})' \quad (\mathbf{\Lambda}^2)'(\mathbf{DP})' \quad \cdots \quad (\mathbf{\Lambda}^{N-1})'(\mathbf{DP})'] \tag{6-61}$$

Substituting $\mathbf{\Lambda} = \mathbf{P}^{-1}\mathbf{AP}$ into the last equation, and simplifying, we can show that $\mathbf{V}_1$ is of rank n if $[\mathbf{A}, \mathbf{D}]$ is observable. ●

Example 6-4

Consider the system described by Eqs. (6-18) through (6-20) of Example 6-1. Applying the observability matrix in Eq. (6-46), we have

$$\mathbf{L} = [\mathbf{D}' \quad \mathbf{A}'\mathbf{D}'] = \begin{bmatrix} 1 & -1 \\ 1 & -1 \end{bmatrix} \tag{6-62}$$

Since $\mathbf{L}$ is singular, the system is unobservable; i.e., not all the states $x_1(k)$ and $x_2(k)$ can be determined from the knowledge of the output $\mathbf{c}(k)$ over the finite interval $[0, N]$, $N \geq 2$.

Now let us consider the system described by the coefficient matrices in Eq. (6-23). Then, the observability matrix is

$$\mathbf{L} = [\mathbf{D}' \quad \mathbf{A}'\mathbf{D}'] = \begin{bmatrix} 1 & 0 \\ 0 & 1 \end{bmatrix} \tag{6-63}$$

which is nonsingular. Thus, the system is completely observable.

Based on the findings here and in Example 6-1, we can conclude that the system described by the coefficient matrices in Eq. (6-20) is *controllable but unobservable*, whereas the system described by the coefficient matrices in Eq. (6-23) is *observable but uncontrollable.*

Continuing with the test of observability of the system in Example 6-1, use the Jordan canonical form. The similarity transformation matrix for the system with the coefficient matrices in Eq. (6-20) was obtained in Eq. (6-26),

$$\mathbf{P} = \begin{bmatrix} 1 & 0 \\ -1 & 1 \end{bmatrix} \tag{6-64}$$

From Eq. (6-58),

$$\mathbf{F} = \mathbf{DP} = [1 \quad 1]\begin{bmatrix} 1 & 0 \\ -1 & 1 \end{bmatrix} = [0 \quad 1] \tag{6-65}$$

Since the first column of $\mathbf{F}$ that corresponds to the first column of the Jordan block of $\mathbf{\Lambda}$ is a zero, the system is unobservable.

Now for the system described by the coefficient matrices of Eq. (6-23),

$$\mathbf{F} = \mathbf{DP} = [1 \quad 0]\begin{bmatrix} 1 & 0 \\ -1 & 1 \end{bmatrix} = [1 \quad 0] \tag{6-66}$$

In this case, the first element of $\mathbf{F}$ is nonzero, so the system is completely observable.

□ Theorem 6-14 Theorem on Observability of Closed-Loop Systems with State Feedback

If the system described by Eqs. (6-1) and (6-2) is controllable and observable, then state feedback of the form of $\mathbf{u}(k) = \mathbf{r}(k) - \mathbf{Gx}(k)$ could destroy observability. In other words, the observability of open-loop and closed-loop systems with state feedback is unrelated. ■

The following example illustrates the relationship between observability and state feedback.

Example 6-5

Consider that the coefficient matrices of a digital control system are

$$\mathbf{A} = \begin{bmatrix} 0 & 1 \\ -2 & -3 \end{bmatrix} \qquad \mathbf{B} = \begin{bmatrix} 1 \\ 1 \end{bmatrix} \qquad \mathbf{D} = [1 \quad 2] \tag{6-67}$$

We can show that the pair $[\mathbf{A}, \mathbf{B}]$ is controllable, and the pair $[\mathbf{A}, \mathbf{D}]$ is observable. Let the state feedback be defined as

$$u(k) = r(k) - \mathbf{Gx}(k) \tag{6-68}$$

where

$$\mathbf{G} = [g_1 \quad g_2] \tag{6-69}$$

Then, the closed-loop system is described by the state equation,

$$\mathbf{x}(k+1) = (\mathbf{A} - \mathbf{BG})\mathbf{x}(k) + \mathbf{B}r(k) \tag{6-70}$$

where

$$\mathbf{A} - \mathbf{BG} = \begin{bmatrix} -g_1 & 1 - g_2 \\ -2 - g_1 & -3 - g_2 \end{bmatrix} \tag{6-71}$$

The observability matrix of the closed-loop system is

$$\mathbf{V} = [\mathbf{D}' \quad (\mathbf{A} - \mathbf{BG})'\mathbf{D}'] = \begin{bmatrix} 1 & -3g_1 - 4 \\ 2 & -3g_2 - 5 \end{bmatrix} \tag{6-72}$$

The determinant of $\mathbf{V}$ is

$$|\mathbf{V}| = 6g_1 - 3g_2 + 3 \tag{6-73}$$

Thus, if g_1 and g_2 are chosen so that $|\mathbf{V}| = 0$, the closed-loop system would be unobservable.

6-4 RELATIONSHIPS BETWEEN CONTROLLABILITY, OBSERVABILITY, AND TRANSFER FUNCTIONS

Classical analysis and design of control systems rely heavily on the use of transfer functions for system modeling. One advantage of using transfer functions is that state controllability and observability are directly related to the minimum order of the transfer function. The following theorem gives the relationship between controllability and observability and the pole-zero cancellation of a transfer function.

□ Theorem 6-15 Controllability, Observability, and Transfer Functions

If the input-output transfer function of a linear time-invariant digital system has pole-zero cancellation, the system will be either not state controllable, unobservable, or both, depending on how the state variables are defined. If the input-output transfer function does not have pole-zero cancellation, then the system can always be represented by dynamic equations as a completely controllable and observable system. ■

Proof: Consider an nth-order digital system with single input, single output, and distinct eigenvalues represented by the dynamic equations

$$\mathbf{x}(k+1) = \mathbf{A}\mathbf{x}(k) + \mathbf{B}u(k) \tag{6-74}$$

$$c(k) = \mathbf{D}\mathbf{x}(k) \tag{6-75}$$

Let the matrix $\mathbf{A}$ be diagonalized by an $n \times n$ nonsingular matrix $\mathbf{P}$, such that $\mathbf{x}(k) = \mathbf{P}\mathbf{y}(k)$. The transformed state equations are represented by

$$\mathbf{y}(k+1) = \mathbf{\Lambda}\mathbf{y}(k) + \mathbf{\Gamma}u(k) \tag{6-76}$$

$$c(k) = \mathbf{F}\mathbf{y}(k) \tag{6-77}$$

where $\mathbf{\Lambda} = \mathbf{P}^{-1}\mathbf{A}\mathbf{P}$, $\mathbf{\Gamma} = \mathbf{P}^{-1}\mathbf{B}$, and $\mathbf{F} = \mathbf{D}\mathbf{P}$.

Since $\mathbf{\Lambda}$ is a diagonal matrix, the ith ($i = 1, 2, \ldots, n$) equation of Eq. (6-76) is

$$y_i(k+1) = z_i y_i(k) + \gamma_i u(k) \tag{6-78}$$

where z_i is the ith eigenvalue of $\mathbf{A}$, and γ_i is the ith element of $\mathbf{\Gamma}$ ($n \times 1$). Taking the z-transform on both sides of Eq. (6-78) and assuming zero initial conditions, the transfer function between $Y_i(z)$ and $U(z)$ is obtained as

$$Y_i(z) = \frac{\gamma_i}{z - z_i} U(z) \tag{6-79}$$

The z-transform of Eq. (6-77) is

$$C(z) = \mathbf{F}\mathbf{Y}(z) = \mathbf{D}\mathbf{P}\mathbf{Y}(z) \tag{6-80}$$

Let the $1 \times n$ matrix $\mathbf{D}$ be represented by

$$\mathbf{D} = [d_1 \quad d_2 \quad \cdots \quad d_n] \tag{6-81}$$

Then,

$$\mathbf{F} = \mathbf{DP} = [f_1 \quad f_2 \quad \cdots \quad f_n] \tag{6-82}$$

where

$$f_i = d_1 + d_2 z_i + \cdots + d_n z_i^{n-1} \tag{6-83}$$

for $i = 1, 2, \ldots, n$. Using Eqs. (6-79) and (6-82), Eq. (6-80) is written as

$$C(z) = [f_1 \quad f_2 \quad \cdots \quad f_n] \begin{bmatrix} \dfrac{\gamma_1}{z - z_1} \\ \dfrac{\gamma_2}{z - z_2} \\ \vdots \\ \dfrac{\gamma_n}{z - z_n} \end{bmatrix} U(z) \tag{6-84}$$

or in the transfer function form,

$$\frac{C(z)}{U(z)} = \sum_{i=1}^{n} \frac{f_i \gamma_i}{z - z_i} \tag{6-85}$$

If the transfer function in Eq. (6-85) has a cancellation of pole and zero, the corresponding coefficient on the right side of the equation would be zero. Assuming that the pole at $z = z_j$ is cancelled by a zero, then

$$f_j \gamma_j = 0 \tag{6-86}$$

which implies that $f_j = 0$, or $\gamma_j = 0$, or both. Since γ_j is the jth element of the matrix $\mathbf{\Gamma}$, $\gamma_j = 0$ would mean that the system is uncontrollable. On the other hand, if $f_j = 0$, where f_j is the jth element of $\mathbf{F}$, the system would be unobservable. ●

Example 6-6

We can show that the input-output transfer functions of the systems in Figs. 6-2 and 6-3 are both equal to

$$\frac{C(z)}{U(z)} = \frac{z + 0.2}{(z + 0.5)(z + 0.2)} = \frac{1}{z + 0.5} \tag{6-87}$$

6-5 CONTROLLABILITY AND OBSERVABILITY VERSUS SAMPLING PERIOD IN A DISCRETE-DATA SYSTEM

When a discrete-data system has sampling operations, the conditions of controllability and observability can be dependent on the sampling period T. Consider an nth-order discrete-data system described by the state equations

$$\mathbf{x}(k+1)T = \boldsymbol{\phi}(T)\mathbf{x}(kT) + \boldsymbol{\theta}(T)\mathbf{u}(kT) \tag{6-88}$$

The condition of state controllability of the system is that the $\mathbf{S}$ matrix of Eq. (6-9) is of rank n. This means that if

$$\mathbf{S}_i = \boldsymbol{\phi}(iT)\boldsymbol{\theta}(T) = \mathbf{S}_j = \boldsymbol{\phi}(jT)\boldsymbol{\theta}(T) \tag{6-89}$$

for $i, j = 1, 2, \ldots, N-1$, $i \neq j$, and $T \neq 0$, the system will be uncontrollable.

Similarly, the condition of observability in Eq. (6-46) indicates that if

$$\mathbf{L}_i = \boldsymbol{\phi}(iT)'\mathbf{D}' = \mathbf{L}_j = \boldsymbol{\phi}(jT)'\mathbf{D}' \tag{6-90}$$

for $i, j = 1, 2, \ldots, N-1$, $i \neq j$, and $T \neq 0$, the system will be unobservable. Therefore, the last two conditions lead to the conclusion that if, for some nonzero sampling period T,

$$\boldsymbol{\phi}(iT) = \boldsymbol{\phi}(jT) \tag{6-91}$$

$i, j = 1, 2, \ldots, N-1$, $i \neq j$, the system will be neither controllable nor observable.

Example 6-7

Consider the transfer function of a linear process described by

$$G(s) = \frac{C(s)}{U(s)} = \frac{\omega}{s^2 + \omega^2} \tag{6-92}$$

It is apparent that the continuous-data process is controllable and observable since $G(s)$ does not have any pole-zero cancellation.

The transfer function $G(s)$ is decomposed into the following state equations:

$$\frac{dx_1(t)}{dt} = x_2(t) \tag{6-93}$$

$$\frac{dx_2(t)}{dt} = -\omega^2 x_1(t) + u(t) \tag{6-94}$$

The state transition matrix $\boldsymbol{\phi}(T)$ is written

$$\boldsymbol{\phi}(T) = \begin{bmatrix} \cos \omega T & \sin \omega T/\omega \\ -\omega \sin \omega T & \cos \omega T \end{bmatrix} \tag{6-95}$$

Since $\cos \omega T = \pm 1$ and $\sin \omega T = 0$ for $\omega T = n\pi, n = 0, \pm 1, \pm 2, \ldots,$ the system is uncontrollable and unobservable for $T = n\pi/\omega$.

6-6 STABILITY OF LINEAR DIGITAL CONTROL SYSTEMS—DEFINITIONS AND THEOREM

One of the most important requirements in the performance of control systems is *stability*. This is true for continuous-data systems as well as discrete-data and digital control systems. In this chapter we shall give definitions of stability and investigate the methods of testing the stability of linear discrete-data and digital control systems.

Before embarking on the methods of testing stability, let us first give definitions of the stability of linear discrete-data systems.

6-6-1 Definitions on Stability

□ Definition 6-4 Zero-State Response

The output response of a discrete-data system that is due to the input only is called the zero-state response; all the initial conditions of the system are set to zero. ■

□ Definition 6-5 Zero-Input Response

The output response of a discrete-data system that is due to the initial conditions only is called the zero-input response; all the inputs of the system are set to zero. ■

From the principle of superposition, when a system is subject to both inputs and initial conditions, the total output response is

$$\text{total response} = \text{zero-state response} + \text{zero-input response} \tag{6-96}$$

□ Definition 6-6 Bounded-Input-Bounded-State Stability

Consider a linear time-invariant discrete-data system that is described by the following dynamic equations:

$$\mathbf{x}(k+1) = \mathbf{A}\mathbf{x}(k) + \mathbf{B}\mathbf{u}(k) \tag{6-97}$$

$$\mathbf{c}(k) = \mathbf{D}\mathbf{x}(k) + \mathbf{E}\mathbf{u}(k) \tag{6-98}$$

The system is said to be bounded-input-bounded-state (BIBS) stable if for any bounded input $\mathbf{u}(k)$, the state $\mathbf{x}(k)$ is also bounded. ■

The norm of an n-vector $\mathbf{x}(k)$ is denoted $\|\mathbf{x}(k)\|$ and is defined as

$$\|\mathbf{x}(k)\| = \left(\sum_{i=1}^{n} x_i^2(k) \right)^{1/2} \tag{6-99}$$

The bounded state implies that the state vector $\mathbf{x}(k)$ satisfies

$$\|\mathbf{x}(k)\| \leq M \qquad \text{for all } k \tag{6-100}$$

where M is a finite number.

□ Definition 6-7 Bounded-Input-Bounded-Output Stability

The system described by Eqs. (6-97) and (6-98) is bounded-input-bounded-output (BIBO) stable if for any bounded input, the output $\mathbf{c}(k)$ is also bounded. ■

Since the output of a system is a linear combination of the state variables, a system that is BIBS stable must also be BIBO stable. However, if the system is BIBO stable, it may or may not be BIBS stable. The latter condition usually happens when the transfer function of the system has an unstable pole that is cancelled by a zero, so that the system is BIBO stable but not BIBS stable.

□ Definition 6-8 Zero-Input Stability

The system described by the dynamic equations of Eqs. (6-97) and (6-98) is said to be zero-input stable or simply stable if the zero-input response $\mathbf{c}(k)$, subject to the finite initial conditions, reaches zero as k approaches infinity; otherwise, the system is unstable. Mathematically, zero-input stability requires that, first,

$$\|\mathbf{c}(k)\| \leq M < \infty \tag{6-101}$$

and, second,

$$\lim_{k \to \infty} |c_i(k)| = 0 \tag{6-102}$$

where $i = 1, 2, \ldots, p$. ■

□ Definition 6-9 Asymptotic Stability

The conditions given in Eqs. (6-101) and (6-102) are also the requirements for asymptotic stability. Thus, zero-input stability implies asymptotic stability. ■

□ **Theorem 6-16 Zero-Input Stability, Asymptotic Stability, and Characteristic Equation Roots**

For the linear discrete-data system described by Eqs. (6-97) and (6-98), BIBO, zero-input, and asymptotic stability all require that the roots of the characteristic equation be inside the unit circle $|z| = 1$ in the z-plane. ■

Proof: From Eq. (4-15), the ith output $c_i(k)$ due to the jth input $u_j(k)$ of the system can be written as

$$c_i(k) = \sum_{m=0}^{k} g_{ji}(k - m)u_j(k) \tag{6-103}$$

where $g_{ji}(k)$ denotes the impulse sequence between $u_j(k)$ and $c_i(k)$. Taking the absolute value on both sides of Eq. (6-103), we get

$$|c_i(k)| = \left| \sum_{m=0}^{k} g_{ji}(k - m)u_j(k) \right| \le \sum_{m=0}^{k} |g_{ji}(k - m)||u_j(k)| \tag{6-104}$$

Since the input is assumed to be bounded, the stability condition in Eq. (6-101) requires that

$$\sum_{k=0}^{\infty} |g_{ji}(k)| < \infty \tag{6-105}$$

for all i and j.

Let the n characteristic equation roots of the system be represented by z_i, $i = 1, 2, \ldots, n$. If m of the n roots are simple, and the rest are of multiple order, $c_i(k)$ will be of the following form:

$$c_i(k) = \sum_{j=1}^{m} K_j(z_j)^k + \sum_{j=m+1}^{n} K_j k^j (z_j)^k \tag{6-106}$$

where K_j are constants. Since the terms $(z_j)^k$ control the response $c_i(k)$ as $k \to \infty$, to satisfy the two conditions in Eqs. (6-101) and (6-102), the magnitude of z_i must be less than one. In other words, *the roots of the characteristic equation must all be inside the unit circle* $|z| = 1$ *in the z-plane.*

From this theorem we see that, for linear time-invariant discrete-data systems, BIBO, zero-input, and asymptotic stability all have the same requirement: the roots of the characteristic equation must all lie inside the unit circle in the z-plane. For this reason, we shall simply refer to the stability condition of a linear discrete-data system as *stable* or *unstable*. The latter refers to the condition that at least one of the characteristic equation roots is not inside the unit circle. For practical reasons we often refer to the situation when the characteristic equation has at least one simple root on the unit circle as

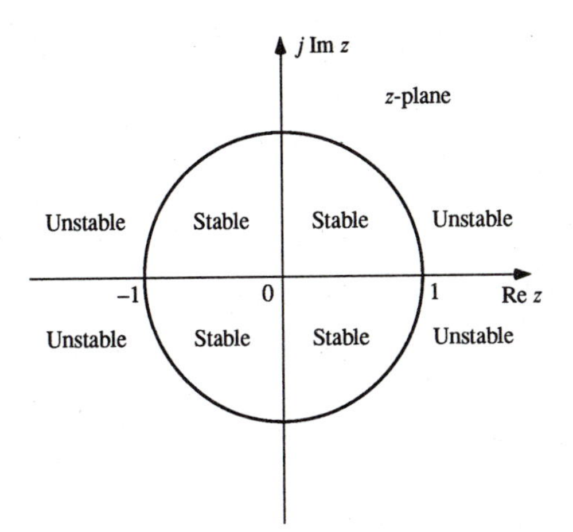

Figure 6-4. The unit circle and stable and unstable regions in the z-plane.

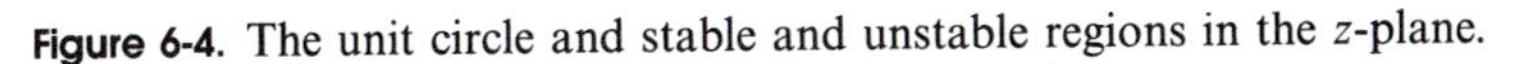

marginally stable or *marginally unstable.* The regions of stability and instability for discrete-data systems in the z-plane are shown with reference to the unit circle in the z-plane in Fig. 6-4. The following example illustrates the relationship between the system transfer function poles which are the roots of the characteristic equation and the stability condition of the system. ●

Example 6-8

The transfer functions of several discrete-data systems and the conditions of stability are stated as follows.

$$M(z) = \frac{10(z - 0.5)}{z(z - 0.15)(z - 0.82)} \tag{6-107}$$

The poles are all inside the unit circle in the z-plane. The system is asymptotically stable, or simply stable.

$$M(z) = \frac{z}{(z + 1.5)(z - 0.5)} \tag{6-108}$$

One pole is outside the unit circle in the z-plane. The system is unstable.

$$M(z) = \frac{K(z + 1)}{(z - 1)(z^2 - 0.25z + 0.1)} \tag{6-109}$$

One pole is on the unit circle in the z-plane. The system is unstable, or marginally stable.

Example 6-9

Consider a second-order discrete-data system described by the dynamic equations in Eqs. (6-97) and (6-98). The inputs to the system are zero, so we need to know only the matrices **A** and **D**,

$$\mathbf{A} = \begin{bmatrix} 0 & 1 \\ -1 & 0 \end{bmatrix} \qquad \mathbf{D} = [1 \quad 0] \tag{6-110}$$

Taking the z-transform on both sides of the state equation in Eq. (6-97) and solving for $\mathbf{X}(z)$ in terms of the initial states $\mathbf{x}(0)$, we get

$$\mathbf{X}(z) = (z\mathbf{I} - \mathbf{A})^{-1} z\mathbf{x}(0) \tag{6-111}$$

or

$$\begin{bmatrix} X_1(z) \\ X_2(z) \end{bmatrix} = \frac{1}{z^2 + 1} \begin{bmatrix} z^2 & z \\ -z & z^2 \end{bmatrix} \begin{bmatrix} x_1(0) \\ x_2(0) \end{bmatrix} \tag{6-112}$$

The responses of $x_1(k)$ and $x_2(k)$ are obtained by taking the inverse z-transform on both sides of Eq. (6-112). We have

$$\begin{bmatrix} x_1(k) \\ x_2(k) \end{bmatrix} = \begin{bmatrix} \cos\frac{k\pi}{2} & \sin\frac{k\pi}{2} \\ -\sin\frac{k\pi}{2} & \cos\frac{k\pi}{2} \end{bmatrix} \begin{bmatrix} x_1(0) \\ x_2(0) \end{bmatrix} \tag{6-113}$$

Since the zero-input states of the system are functions of sine and cosine, they do not go to zero as k approaches infinity. Thus, the system is unstable or not asymptotically stable.

The characteristic equation of the system is

$$|z\mathbf{I} - \mathbf{A}| = z^2 + 1 = 0 \tag{6-114}$$

The roots of the characteristic equation are at $z = j$ and $z = -j$, which are on the unit circle in the z-plane. Thus, the system is also referred to as marginally stable or marginally unstable.

Example 6-10

Now consider that the matrix **A** of the state equations in Eq. (6-97) is

$$\mathbf{A} = \begin{bmatrix} -0.5 & 0 \\ 0 & 0.5 \end{bmatrix} \tag{6-115}$$

The state transition equations expressed in the z-domain are

$$\mathbf{X}(z) = (z\mathbf{I} - \mathbf{A})^{-1}z\mathbf{x}(0) = \begin{bmatrix} \dfrac{z}{z+0.5} & 0 \\ 0 & \dfrac{z}{z-0.5} \end{bmatrix}\mathbf{x}(0) \tag{6-116}$$

The inverse z-transform of $\mathbf{X}(z)$ is

$$\mathbf{x}(k) = \begin{bmatrix} (-0.5)^k & 0 \\ 0 & (0.5)^k \end{bmatrix}\mathbf{x}(0) \tag{6-117}$$

For arbitrary finite initial states, the states $x_1(k)$ and $x_2(k)$ approach zero as k approaches infinity. Thus, the system is asymptotically stable.

The characteristic equation of the system is

$$|z\mathbf{I} - \mathbf{A}| = (z - 0.5)(z + 0.5) = 0 \tag{6-118}$$

which has two real roots at $z = -0.5$ and 0.5. These roots are all inside the unit circle $|z| = 1$.

Example 6-11

Now consider that the matrix $\mathbf{A}$ of the state equations in Eq. (6-97) is

$$\mathbf{A} = \begin{bmatrix} -1.5 & 0 \\ 0 & 0.5 \end{bmatrix} \tag{6-119}$$

The state transition equations in the z-domain are expressed as

$$\mathbf{X}(z) = \begin{bmatrix} \dfrac{z}{z+1.5} & 0 \\ 0 & \dfrac{z}{z-0.5} \end{bmatrix}\mathbf{x}(0) \tag{6-120}$$

The inverse z-transform of $\mathbf{X}(z)$ is

$$\mathbf{x}(k) = \begin{bmatrix} (-1.5)^k & 0 \\ 0 & (0.5)^k \end{bmatrix}\mathbf{x}(0) \tag{6-121}$$

Since the response of $x_1(k)$ will increase without bound for any finite initial state, the system in unstable. This is supported by the characteristic equation,

$$|z\mathbf{I} - \mathbf{A}| = (z + 1.5)(z - 0.5) = 0 \tag{6-122}$$

which has a root at $z = -1.5$ which is outside the unit circle.

6-6-2 Stabilizability

We have pointed out in the preceding sections that controllability and stability are entirely independent of each other. However, controllability does affect whether an unstable system can be stabilized or not. From the definition of controllability, the following theorem is apparent.

□ Theorem 6-17 Stabilizability

A linear discrete-data system that is represented by the dynamic equations of Eqs. (6-97) and (6-98) is stabilizable if and only if all the unstable states are controllable. Thus, a completely controllable system is also stabilizable. ■

Example 6-12

Consider the following discrete-data system:

$$\begin{bmatrix} x_1(k+1) \\ x_2(k+1) \\ x_3(k+1) \end{bmatrix} = \begin{bmatrix} 0.55 & 0 & 0 \\ 0 & -0.12 & 0 \\ 0 & 0 & 2 \end{bmatrix} \begin{bmatrix} x_1(k) \\ x_2(k) \\ x_3(k) \end{bmatrix} + \begin{bmatrix} 0 \\ 1 \\ 2 \end{bmatrix} u(k) \tag{6-123}$$

For simplicity, we have chosen a diagonal matrix for **A** so that the states are decoupled from each other. Apparently, the system is uncontrollable, since the first row of **B** is zero and is unstable due to the eigenvalue at $z = 2$. However, the system is stabilizable, since the unstable state $x_3(k)$ is controllable, and the uncontrollable state $x_1(k)$ is stable. On the other hand, if the eigenvalue of the state $x_1(k)$, which is uncontrollable, were on or outside the unit circle $|z| = 1$, then the system would be unstabilizable.

6-7 STABILITY TESTS OF DISCRETE-DATA SYSTEMS

It has been established in the preceding section that a linear time-invariant discrete-data system is asymptotically stable if the roots of the characteristic equation lie inside the unit circle in the z-plane. The Routh-Hurwitz criterion, which is useful for the stability test of linear continuous-data systems with respect to the characteristic equation roots in the s-plane, *cannot* be applied directly to the z-plane, due to the difference in the stability boundary. In this section we shall introduce a bilinear transformation that transforms the unit circle in the z-plane onto the imaginary axis in another complex plane so that the Routh-Hurwitz criterion can be applied to discrete-data systems. Direct methods of testing the location of the characteristic equation roots with respect to the unit circle in the z-plane also will be discussed.

It should be pointed out that *when the coefficients of the characteristic equation are all known, solving the equation with a root-finding program such as the RL of the DCSP software package on a digital computer would be the easiest and the most preferable method.* When the characteristic equation has unknown parameters, the extended Routh-Hurwitz criterion and the direct methods are more valuable, though they could be quite tedious for high-order systems. We shall present these in the following, and let the reader judge which is more preferable.

6-7-1 The Bilinear Transformation Method—Extension of the Routh-Hurwitz Criterion

The Routh-Hurwitz criterion [7], or the Routh tabulation of the coefficients of the characteristic equation, represents a convenient way of finding the location of the roots with respect to the left and right halves of the s-plane for stability studies of linear continuous-data systems. The z-transformation $z = \exp(Ts)$ transforms the imaginary axis of the s-plane onto the unit circle $|z| = 1$ in the z-plane, and thus renders the Routh-Hurwitz criterion ineffective for discrete-data systems. We can attempt to transform the unit circle in the z-plane back to the imaginary axis of another complex plane, but we cannot use the inverse z-transform, since it would transform an algebraic characteristic equation in z into a transcendental equation of s, which still cannot be tested with the Routh-Hurwitz criterion. Transformations that are algebraic and transform circles in the z-plane onto vertical lines in a complex-variable plane, say the r-plane, are of the following form:

$$z = \frac{ar + b}{cr + d} \tag{6-124}$$

where a, b, c, and d are real constants. One such transformation that transforms the interior of the unit circle $|z| = 1$ onto the left half of the r-plane is

$$z = \frac{r + 1}{r - 1} \tag{6-125}$$

which is referred to as the *r-transformation*. Another possibility is to let $a = b = d = 1$, and $c = -1$, so that

$$z = \frac{r + 1}{-r + 1} \tag{6-126}$$

As it turns out, the amount of algebraic work involved in applying Eq. (6-124) in stability studies when the constants a, b, c, and d vary is relatively the same. The computer program RWT of the DCSP software package can be used to carry out the r-transformation.

Figure 6-5 illustrates the mapping of the unit circle in the z-plane onto the imaginary axis of the r-plane. Notice that the interior of the unit circle corresponds to the left half of the r-plane. Thus, once the characteristic equation in z is transformed into the r-domain using Eq. (6-125), the Routh-Hurwitz criterion can again be applied to the equation in r in the normal fashion.

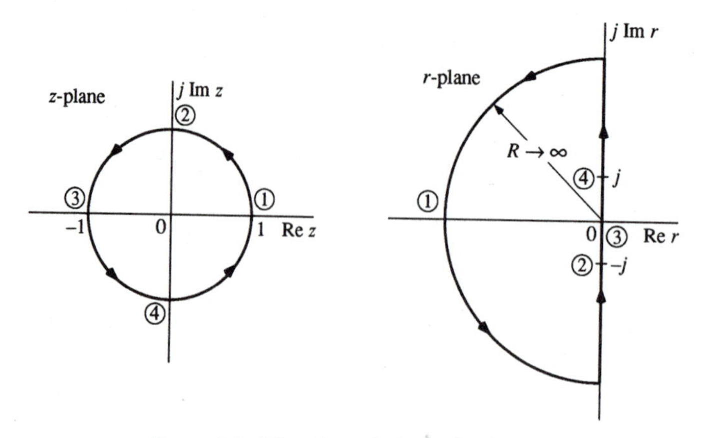

Figure 6-5. Mapping of the unit circle $|z| = 1$ onto the imaginary axis of the r-plane through the r-transformation, $z = (r + 1)/(r - 1)$.

As a quick reference, the Routh tabulation of a polynomial in s is reviewed in Appendix E. For more details on the Routh-Hurwitz criterion, the reader may refer to [7].

Example 6-13

Consider the characteristic equation of the system in Example 6-10, given by Eq. (6-118),

$$F(z) = z^2 - 0.25 = (z - 0.5)(z + 0.5) = 0 \tag{6-127}$$

Transforming $F(z)$ into a polynomial in r using Eq. (6-125), we get

$$0.75r^2 + 2.5r + 0.75 = 0 \tag{6-128}$$

We observe that all the coefficients are of the same sign, and none of the coefficients is zero, which are the necessary and sufficient conditions for all the roots of the second-order equation in Eq. (6-128) to be in the left half of the r-plane. Thus, all the roots of Eq. (6-127) are inside the unit circle.

Example 6-14

Consider that the characteristic equation of a third-order discrete-data control system is given as

$$F(z) = z^3 - 1.25z^2 - 1.375z - 0.25 = 0 \tag{6-129}$$

which is known to have roots at $z = 0.25, 0.5$, and -2.0.

Transforming $F(z)$ into the r-domain using Eq. (6-125), we get

$$-1.875r^3 + 3.875r^2 + 4.875r + 1.125 = 0 \tag{6-130}$$

Since the coefficients of the last equation do not all have the same sign, we know that the system has at least one root in the right half of the r-plane, or $F(z)$ has at least one zero outside the unit circle in the z-plane. Nevertheless, applying the Routh-Hurwitz criterion to Eq. (6-130), we have the following Routh tabulation:

	r^3	-1.875	4.875
Sign change			
	r^2	3.875	1.125
	r^1	5.419	0
	r^0	1.125	

Since there is one sign change in the first column of the Routh tabulation, Eq. (6-130) has one root in the right-half r-plane, or Eq. (6-129) has one root outside the unit circle in the z-plane, which in this case is known to be at $z = -2$.

Singular Cases

In the Routh tabulation of a polynomial in the s-domain, the following two singular cases are often encountered:

1. The first element in any one row of the tabulation is zero, but the others are not, before the tabulation is completed.
2. The elements in one row of the tabulation are all zero, before the tabulation is completed.

The first case is corrected by replacing the zero element in the tabulation by an arbitrarily small positive (or negative) number ε, and then proceeding with the tabulation.

In the second singular case, all the elements in one row of the Routh tabulation being zeros indicates that one or more of the following conditions exist:

1. pairs of real roots with opposite signs (symmetry of roots about the $j\omega$ axis of the s-plane)
2. pairs of imaginary roots
3. complex-conjugate roots forming symmetry about the origin of the s-plane.

The auxiliary equation $A(s) = 0$ is formed by using the coefficients of the row just above the row of zeros in the Routh tabulation. The auxiliary equation *also satisfies the characteristic equation, and the tabulation is continued by replacing the row of zeros by the coefficients of $dA(s)/ds$.*

For the first singular case in the r-plane, the same remedy is applied by replacing the zero element in the first row by an arbitrarily small number ε.

When a row of zeros occurs in the Routh tabulation of the equation in w, looking from the transformation relationships between s and z and r, we conclude that the following conditions may exist:

1. pairs of real roots in the z-plane that are the inverse of each other; e.g., $z_1 = 2$ and $z_2 = 1/z_1 = 0.5$
2. pairs of roots on the unit circle $|z| = 1$, including at $z = 1$ and $z = -1$ simultaneously.

The second condition corresponds to roots on the imaginary axis in the r-plane. The remedy to the singular case is similar to that for the s-domain.

1. Form the auxiliary equation $A(r) = 0$ using the coefficients of the row in the Routh tabulation just above the row of zeros.
2. Perform $dA(r)/dr = 0$.
3. Replace the row of zeros in the tabulation with the coefficients of $dA(r)/dr$.
4. Continue with the Routh tabulation in the usual manner.

Example 6-15

Consider the equation

$$F(z) = z^3 + 3.3z^2 + 3z + 0.8 = 0 \tag{6-131}$$

which has roots at $z = -0.5, -0.8$, and -2.0. The roots at -0.5 and -2.0 form an inverse symmetry about $z = 1$, and should create a singular case for the Routh test.

Transforming $F(z)$ into the r-plane using Eq. (6-125), we get

$$8.1r^3 + 0.9r^2 - 0.9r - 0.1 = 0 \tag{6-132}$$

The Routh tabulation of the last equation is

$$\begin{array}{lll} r^3 & 8.1 & -0.9 \\ r^2 & 0.9 & -0.1 \\ r^1 & 0 & 0 \end{array}$$

Thus, the r^1 row contains all zeros, and the Routh test could not be continued in the usual manner. The auxiliary equation is formed by using the coefficients in the r^2 row:

$$A(r) = 0.9r^2 - 0.1 = 0 \tag{6-133}$$

Taking the derivative of $A(r)$ with respect to r, we get

$$\frac{dA(r)}{dr} = 1.8r \tag{6-134}$$

The coefficients in the r^1 row are now filled with the coefficient of $dA(r)/dr$, and the remaining Routh tabulation is

$$\begin{array}{lll} r^1 & 1.8 & 0 \\ r^0 & -0.1 & \end{array}$$

Since one sign change occurs in the first column of the Routh tabulation, the characteristic equation has one root in the right half of the r-plane, or $F(z)$ has one root outside the unit circle in the z-plane. Solving for the roots of the auxiliary equation $A(r) = 0$, we have

$$r = 0.333 \quad \text{and} \quad -0.333 \tag{6-135}$$

which are two of the roots of the characteristic equation. These two roots correspond to the roots at $z = -2.0$ and -0.5, respectively.

Example 6-16

Consider a discrete-data system that has the following characteristic equation:

$$F(z) = z^3 + z^3 + z + 1 = 0 \tag{6-136}$$

The roots of $F(z) = 0$ are at $z = 1, z = j$, and $z = -j$, which are all on the unit circle. Thus, the system is unstable.

Applying the r-transformation of Eq. (6-125), we get

$$r^3 + r = 0 \tag{6-137}$$

Since the left-hand side of the last equation is an odd polynomial, the r^2 row of its Routh tabulation would contain all zeros. In this case, Eq. (6-137) is also its own auxiliary equation, $A(r) = 0$. The complete Routh tabulation is now

$$\begin{array}{lll} r^3 & 1 & 1 \\ r^2 & 3 & 1 \\ r^1 & 2/3 & 0 \\ r^0 & 1 & \end{array}$$

Since the coefficients in the first column of the tabulation are all of the same sign, Eq. (6-137) does not have any roots in the right half of the r-plane, or Eq. (6-136) does not have any roots outside the unit circle in the z-plane. In this case, we know that all three zeros of $F(z)$ are on the unit circle.

Example 6-17

We pointed out earlier that the true value of the Routh-Hurwitz criterion and the direct methods is seen in the situation where the characteristic equation has variable parameters. In this example we demonstrate the application of the extended Routh-Hurwitz criterion to the design of a discrete-data control system from the stability standpoint.

The characteristic equation of a discrete-data control system is given as

$$F(z) = z^3 + a_2 z^2 + a_1 z + a_0 = 0 \tag{6-138}$$

with

$$a_2 = 111.6T^2 + 16.74T - 3 \tag{6-139}$$

$$a_1 = 3 - 33.48T + 1.395 \times 10^{-4} KT^3 \tag{6-140}$$

$$a_0 = 1.395 \times 10^{-4} KT^3 + 16.74T - 111.6T^3 - 1 \tag{6-141}$$

where T denotes the sampling period in seconds, and K is the gain constant. We want to determine the range of K and T so that the system is asymptotically stable.

Applying the r-transform of Eq. (6-125) to $F(z)$, we have the characteristic equation in the r-domain as

$$A_3 r^3 + A_2 r^2 + A_1 r + A_0 = 0 \tag{6-142}$$

where

$$A_3 = a_2 + a_1 + a_0 + 1 = 2.79 \times 10^{-4} KT^3 \tag{6-143}$$

$$\begin{aligned} A_2 &= a_2 - a_1 + 3a_0 + 3 \\ &= 446.4T^2 - 5.58 \times 10^{-4} KT^3 \end{aligned} \tag{6-144}$$

$$A_1 = -446.4T^2 + 66.96T + 2.79 \times 10^{-4} KT^3 \tag{6-145}$$

$$\begin{aligned} A_0 &= -a_2 + a_1 - a_0 + 1 \\ &= 8 - 66.96T \end{aligned} \tag{6-146}$$

The Routh tabulation in terms of the coefficients of Eq. (6-142) is

$$\begin{array}{lcc} r^3 & A_3 & A_1 \\ r^2 & A_2 & A_0 \\ r^1 & \dfrac{A_1 A_2 - A_0 A_3}{A_2} & 0 \\ r^0 & A_0 & \end{array}$$

For stability, the coefficients in the first column of the Routh tabulation must all be of the same sign. We can show that noncontradictory results can be obtained only if these coefficients are all positive. Thus,

$$A_0 = 8 - 66.96T > 0 \tag{6-147}$$

$$A_2 = 446.4T^2 - 5.58 \times 10^{-4} KT^3 > 0 \tag{6-148}$$

$$A_3 = 2.79 \times 10^{-4} K T^3 > 0 \tag{6-149}$$

$$A_1 A_2 - A_0 A_3 = -15.568 \times 10^{-8} T^3 K^2 + (0.3736T^2 - 0.01868T - 0.002232)K + (29890.94 - 199272.96T) > 0 \tag{6-150}$$

Since the sampling period T must be positive, the condition in Eq. (6-149) gives

$$K > 0 \tag{6-151}$$

The condition in Eq. (6-147) gives

$$T < 0.1195 \text{ s} \tag{6-152}$$

The condition in Eq. (6-148) leads to

$$K < 800000/T \tag{6-153}$$

Thus, the stability of the system is governed by the inequality conditions in Eqs. (6-150) through (6-153). As it turns out, the condition in Eq. (6-150) is more stringent than that in Eq. (6-153). Figure 6-6 shows the stable and unstable regions of the system in the K-versus-T plane, with the vertical axis in logarithmic scale.

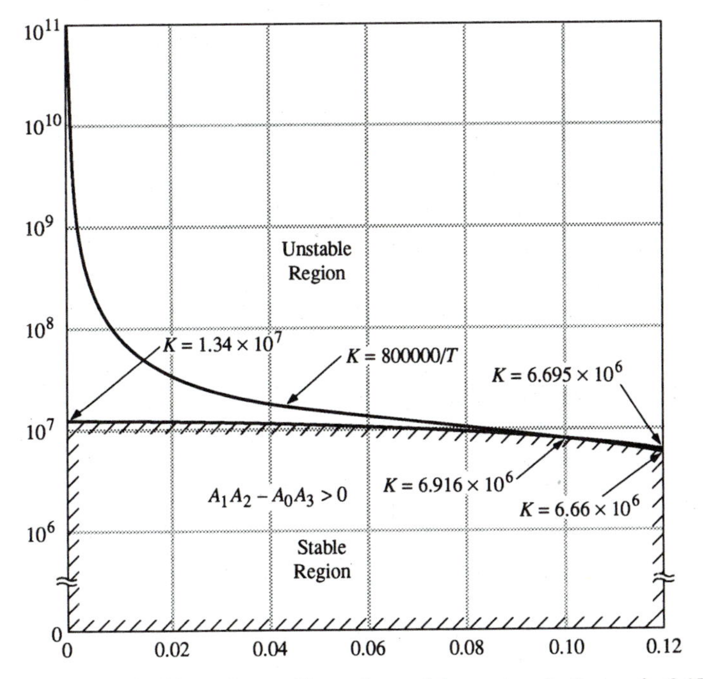

Figure 6-6. Stable and unstable regions of the system in Example 6-17 in the K-versus-T plane.

6-7-2 Jury's Stability Test

One of the first direct methods devised for testing the location of the roots of a polynomial in z with respect to the unit circle in the z-plane is the *Schur-Cohn criterion* [13]. The criterion gives the necessary and sufficient conditions for the roots to lie inside the unit circle in terms of the signs of the *Schur-Cohn determinants.* The Schur-Cohn criterion will not be covered in this text simply because it is very cumbersome for equations higher than the second order. Instead, *Jury's stability test* [9]–[12] is described.

Given an nth-order equation in z,

$$F(z) = a_n z^n + a_{n-1} z^{n-1} + \cdots + a_2 z^2 + a_1 z + a_0 = 0 \tag{6-154}$$

where $a_0, a_1, \ldots, a_n$ are real coefficients. Assume that a_n is positive, or that it can be made positive by changing the signs of all the coefficients of $F(z)$. The following table is made using the coefficients of $F(z)$:

Row	z^0	z^1	z^2	$\cdots$	z^{n-k}	$\cdots$	z^{n-1}	z^n
1	a_0	a_1	a_2	$\cdots$	a_{n-k}	$\cdots$	a_{n-1}	a_n
2	a_n	a_{n-1}	a_{n-2}	$\cdots$	a_k	$\cdots$	a_1	a_0
3	b_0	b_1	b_2	$\cdots$	b_{n-k}	$\cdots$	b_{n-1}	
4	b_{n-1}	b_{n-2}	b_{n-3}	$\cdots$	b_k	$\cdots$	b_0	
5	c_0	c_1	c_2	$\cdots$		c_{n-2}		
6	c_{n-2}	c_{n-3}	c_{n-4}	$\cdots$		c_0		
$\vdots$	$\vdots$	$\vdots$	$\vdots$	$\cdots$				
$2n-5$	p_0	p_1	p_2	p_3				
$2n-4$	p_3	p_2	p_1	p_0				
$2n-3$	q_0	q_1	q_2					

Note that the elements of the $(2k+2)$th row $(k = 0, 1, 2, \ldots)$ consist of the coefficients of the $(2k+1)$th row written in the reverse order. The elements in the table are defined as

$$b_k = \begin{vmatrix} a_0 & a_{n-k} \\ a_n & a_k \end{vmatrix} \qquad c_k = \begin{vmatrix} b_0 & b_{n-1-k} \\ b_{n-1} & b_k \end{vmatrix}$$

$$d_k = \begin{vmatrix} c_0 & c_{n-2-k} \\ c_{n-2} & c_k \end{vmatrix} \qquad \cdots \tag{6-155}$$

$$q_0 = \begin{vmatrix} p_0 & p_3 \\ p_3 & p_0 \end{vmatrix} \qquad q_2 = \begin{vmatrix} p_0 & p_1 \\ p_3 & p_2 \end{vmatrix}$$

The necessary and sufficient conditions for the polynomial $F(z)$ to have no roots on or outside the unit circle in the z-plane are

$$F(1) > 0$$

$$F(-1)\begin{cases} >0 & n \text{ even} \\ <0 & n \text{ odd} \end{cases}$$

$$\left.\begin{array}{l} |a_0| < a_n \\ |b_0| > |b_{n-1}| \\ |c_0| > |c_{n-2}| \\ |d_0| > |d_{n-3}| \\ \vdots \qquad \vdots \\ |q_0| > |q_2| \end{array}\right\} \quad (n-1) \text{ constraints} \tag{6-156}$$

For a second-order system, $n = 2$, the Jury tabulation contains only one row. Therefore, the requirements listed in Eq. (6-156) are reduced to

$$\begin{gathered} F(1) > 0 \\ F(-1) > 0 \\ |a_0| < a_2 \end{gathered} \tag{6-157}$$

One advantage of the Jury stability test is that, for systems of any order, the conditions on $F(1)$, $F(-1)$, *and between* a_0 *and* a_n *in Eq.* (6-156) *form necessary conditions of stability that are very simple to check without carrying out the Jury tabulation. For second-order systems, the conditions in Eq.* (6-157) *are necessary and sufficient.*

For equations that pass the necessary conditions and would require carrying out Jury's tabulation, the inequality tests of the calculated coefficients in Eq. (6-156) *could be quite tedious if the equation contains variable parameters.*

There are other variations of the Jury tabulation, such as *Raible's tabulation* [14]. However, these do not improve the difficulties of equations with variable parameters.

Note that the stability test given in Eq. (6-156) is valid only if the inequality conditions provide conclusive results. As in the Routh-Hurwitz criterion, occasionally the Jury test may encounter singular cases that would require remedial measures to complete the tabulation.

Singular Cases

When some or all of the elements of a row in the Jury tabulation are zero, the tabulation ends prematurely. We refer to this situation as a *singular case.* A singular case can be remedied by expanding or contracting the unit circle $|z| = 1$ infinitesimally [16], which is equivalent to moving the zeros of $F(z)$ off the unit circle. The transformation for this purpose is

$$z = (1 + \varepsilon)z \tag{6-158}$$

where ε is a very small real number. When ε is a positive number in Eq. (6-158), the radius of the unit circle is expanded to $1 + \varepsilon$, and when ε is negative, the radius of the unit circle is reduced to $1 + \varepsilon$. The difference between the number of zeros found inside (or outside) the unit circle when the circle is expanded or contracted by ε is the number of zeros on the circle.

The transformation in Eq. (6-158) is actually quite simple to apply, since

$$(1 + \varepsilon)^n z^n \cong (1 + n\varepsilon)z^n \tag{6-159}$$

for positive or negative ε. This means that the coefficient of the z^n term is multiplied by $(1 + n\varepsilon)$.

Example 6-18

Consider the characteristic equation given in Eq. (6-127),

$$F(z) = z^2 - 0.25 = (z - 0.5)(z + 0.5) = 0 \tag{6-160}$$

With reference to Eq. (6-154), the coefficients of the equation are

$$a_2 = 1 \qquad a_1 = 0 \qquad a_0 = -0.25 \tag{6-161}$$

Since the equation is of the second order, under Jury's test the conditions in Eq. (6-157) are necessary and sufficient for the system to be stable. Thus,

$\boldsymbol{F(1) > 0}$

$$F(1) = 1 - 0.25 = 0.75 > 0 \tag{6-162}$$

$\boldsymbol{F(-1) > 0}$

$$F(-1) = 1 - 0.25 = 0.75 > 0 \tag{6-163}$$

$\boldsymbol{|a_0| > a_2}$

$$|a_0| = 0.25 \qquad a_2 = 1 \tag{6-164}$$

Thus, $|a_0| < a_2$. The system is stable as expected.

Example 6-19

Consider the characteristic equation in Eq. (6-129):

$$F(z) = z^3 - 1.25z^2 - 1.375z - 0.25 = 0 \tag{6-165}$$

We have

$$a_3 = 1 \qquad a_2 = -1.25 \qquad a_1 = -1.375 \qquad a_0 = -0.25 \tag{6-166}$$

Checking the first three conditions in Eq. (6-156),

$\boldsymbol{F(1) > 0}$

$$F(1) = 1 - 1.25 - 1.375 - 0.25 = -1.875 \tag{6-167}$$

$\boldsymbol{F(-1) < 0}$

$$F(-1) = -1 - 1.25 + 1.375 - 0.25 = -1.125 \tag{6-168}$$

$\boldsymbol{|a_0| < a_3}$

$$|a_0| = 0.25 < 1 \tag{6-169}$$

Since $F(1)$ is negative, not all the roots of Eq. (6-165) are inside the unit circle, and the system is unstable. There is no need to carry out the Jury tabulation.

Example 6-20

Consider the following characteristic equation of a discrete-data system:

$$z^3 + 3.3z^2 + 4z + 0.8 = 0 \tag{6-170}$$

The equation has roots at $z = -0.2463$, $z = -1.5268 + j0.9574$, and $z = -1.5268 - j0.9574$. Thus, the system is unstable.

Applying the necessary conditions of the Jury test in Eq. (6-156), we have

$\boldsymbol{F(1) > 0}$

$$F(1) = 9.1 \tag{6-171}$$

$\boldsymbol{F(-1) < 0}$

$$F(-1) = -0.9 \tag{6-172}$$

$\boldsymbol{|a_0| < a_3}$

$$|a_0| = 0.8 < 1 \tag{6-173}$$

Thus, we must carry out the Jury tabulation as follows to determine the stability of the system.

Row	z^0	z^1	z^2	z^3
1	0.8	4.0	3.3	1.0
2	1.0	3.3	4.0	0.8
3	b_0	b_1	b_2	

where

$$b_0 = \begin{vmatrix} a_0 & a_3 \\ a_3 & a_0 \end{vmatrix} = a_0^2 - a_3^2 = -0.36 \tag{6-174}$$

$$b_1 = \begin{vmatrix} a_0 & a_2 \\ a_3 & a_1 \end{vmatrix} = a_0 a_1 - a_2 a_3 = -0.1 \tag{6-175}$$

$$b_2 = \begin{vmatrix} a_0 & a_1 \\ a_3 & a_2 \end{vmatrix} = a_0 a_2 - a_1 a_3 = -1.36 \tag{6-176}$$

Since $|b_0|$ is not greater than $|b_2|$, Eq. (6-170) has at least one root outside the unit circle.

Example 6-21

This example illustrates the application of Jury's stability test to an equation with variable parameters. Consider the system described in Example 6-17, Eq. (6-138). The coefficients of the equation are given in Eqs. (6-139) through (6-141). Since the variable parameters K and T are contained in the coefficients, the stability conditions in Eq. (6-156) would result in the comparison of the absolute values of equations containing K and T. Thus, the procedures would be quite complex. In this case, the r-transform/Routh-Hurwitz criterion method carried out in Example 6-17 would be simpler to use.

Let us solve a simpler version of the problem by assigning a value to the sampling period T and then find the values of K for system stability.

Let $T = 0.1$ s; the coefficients of the equation in Eq. (6-138) become

$$\begin{aligned} a_3 &= 1 \\ a_2 &= -0.21 \\ a_1 &= -0.348 + 1.395 \times 10^{-7}K \\ a_0 &= -0.442 + 1.395 \times 10^{-7}K \end{aligned} \tag{6-177}$$

Applying the conditions in Eq. (6-156), we have

$\boldsymbol{F(1) > 0}$

$$\begin{aligned} F(1) &= 1 + a_2 + a_1 + a_0 \\ &= 2.79 \times 10^{-7}K > 0 \end{aligned} \tag{6-178}$$

$\boldsymbol{F(-1) < 0}$

$$\begin{aligned} F(-1) &= -1 + a_2 - a_1 + a_0 \\ &= -1.304 < 0 \end{aligned} \tag{6-179}$$

$\boldsymbol{|a_0| < a_3}$

$$|1.395 \times 10^{-7}K - 0.442| < 1 \tag{6-180}$$

The coefficients in the last row of the Jury tabulation are

$$b_0 = a_0^2 - a_3^2 = 1.946 \times 10^{-14}K^2 - 1.2332 \times 10^{-7}K - 0.8064 \quad (6\text{-}181)$$

$$b_1 = a_0 a_1 - a_2 a_3 = 1.946 \times 10^{-14}K^2 - 1.1021 \times 10^{-7}K + 0.3638 \quad (6\text{-}182)$$

$$b_2 = a_0 a_2 - a_1 a_3 = -1.688 \times 10^{-7}K + 0.4408 \quad (6\text{-}183)$$

From Eq. (6-178) we have

$$K > 0 \quad (6\text{-}184)$$

From Eq. (6-180),

$$K < 1.0337 \times 10^7 \quad (6\text{-}185)$$

The condition of $|b_0| > |b_2|$ leads to

$$|1.946 \times 10^{-14}K^2 - 1.2332 \times 10^{-7}K - 0.8064| > |-1.688 \times 10^{-7}K + 0.4408| \quad (6\text{-}186)$$

which is best solved graphically, as shown in Fig. 6-7.

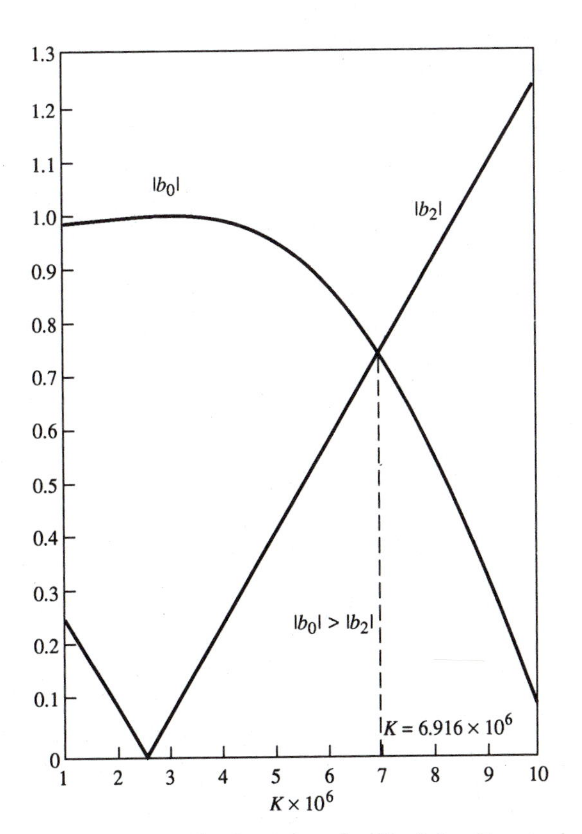

Figure 6-7. Stability boundary for K of the discrete-data system described in Example 6-21.

Thus, the maximum value of K for stability is 6.916×10^6, which is a more stringent requirement than that of Eq. (6-185). The range of K for system stability when $T = 0.1$ s is

$$0 < K < 6.916 \times 10^6 \tag{6-187}$$

which agrees with the result shown in Fig. 6-6.

6-7-3 The Second Method of Liapunov

A powerful method of determining the stability of linear and nonlinear systems with continuous data is the second method of Laipunov [17]. Much has been published about this subject, and it is readily extended into the study of the stability of discrete-data systems. The second method of Liapunov is based on the determination of a V function called the *Liapunov function.* From the properties of the V function, we can show the stability or instability of the system. However, the main disadvantage of Liapunov's stability criterion is that it gives only the sufficient conditions for stability, not the necessary conditions. Furthermore, there are no unique methods of determining the V function for a wide class of systems.

The Liapunov Function

Any function $V = V[\mathbf{x}(k)]$ of definite sign (positive definite or negative definite) is called a Liapunov function if $V(\mathbf{0}) = 0$ and $\mathbf{x}(k)$ is the solution of the state equation

$$\mathbf{x}(k+1) = \mathbf{f}[\mathbf{x}(k)] \tag{6-188}$$

We define a difference operation Δ so that $\Delta V[\mathbf{x}(k)]$ is defined as

$$\Delta V[\mathbf{x}(k)] = V[\mathbf{x}(k+1)] - V[\mathbf{x}(k)] \tag{6-189}$$

□ Stability Theorem of Liapunov

Consider a discrete-data system described by

$$\mathbf{x}(k+1) = \mathbf{f}[\mathbf{x}(k)] \tag{6-190}$$

where $\mathbf{x}(k)$ is the $n \times 1$ state vector and $\mathbf{f}[\mathbf{x}(k)]$ is an $n \times 1$ function vector with the property that

$$\mathbf{f}[\mathbf{x}(k) = \mathbf{0}] = \mathbf{0} \qquad \text{for all } k \tag{6-191}$$

Suppose that there exists a scalar function $V[\mathbf{x}(k)]$ continuous in $\mathbf{x}(k)$ such that

1. $V[\mathbf{x}(k) = \mathbf{0}] = V(\mathbf{0}) = 0$
2. $V[\mathbf{x}(k)] = V(\mathbf{x}) > 0$ *for* $\mathbf{x} \neq 0$ (6-192)
3. $V(\mathbf{x})$ *approaches infinity as* $\|\mathbf{x}\| \to \infty$
4. $\Delta V(\mathbf{x}) < 0$ *for* $\mathbf{x} \neq 0$

Then the equilibrium state $\mathbf{x}(k) = \mathbf{0}$ *(for all k) is asymptotically stable in the large and* $V(\mathbf{x})$ *is a Liapunov function* [17]. ■

□ Instability Theorem of Liapunov

For the system described by Eq. (6-188), if there exists a scalar function $V(\mathbf{x})$ continuous in $\mathbf{x}(k)$ such that

$$\Delta V(\mathbf{x}) < 0 \tag{6-193}$$

then

1. the system is unstable in the finite region for which V is not positive semidefinite ($\geqslant 0$).
2. the response is unbounded as k approaches infinity if V is not positive semidefinite for all $\mathbf{x}(k)$. ■

Example 6-22

Consider the discrete-data system described by

$$x_1(k+1) = -0.5x_1(k) \tag{6-194}$$

$$x_2(k+1) = -0.5x_2(k) \tag{6-195}$$

Let us assign the Liapunov function to be

$$V(\mathbf{x}) = x_1^2(k) + x_2^2(k) \tag{6-196}$$

which is positive for all values of $x_1(k)$ and $x_2(k)$ not equal to zero. Then, the function $\Delta V(\mathbf{x})$ is evaluated using Eq. (6-189),

$$\begin{aligned} \Delta V(\mathbf{x}) &= V[\mathbf{x}(k+1)] - V[\mathbf{x}(k)] \\ &= x_1^2(k+1) + x_2^2(k+1) - x_1^2(k) - x_2^2(k) \\ &= -0.75x_1^2(k) - 0.75x_2^2(k) \end{aligned} \tag{6-197}$$

Since $\Delta V(\mathbf{x})$ is negative for all $\mathbf{x}(k) \neq 0$, the system is asymptotically stable.

Example 6-23

Consider the digital system described by the state equations

$$x_1(k+1) = -1.5x_1(k) \tag{6-198}$$

$$x_2(k+1) = -0.5x_2(k) \tag{6-199}$$

It is simple to show that the eigenvalues of the $\mathbf{A}$ matrix are at -1.5 and -0.5 and that the system is unstable. However, without prior knowledge of the stability conditions of the system, the stability theorem of Liapunov is applied.

Let the Liapunov function be

$$V(\mathbf{x}) = x_1^2(k) + x_2^2(k) \tag{6-200}$$

Then,

$$\begin{aligned}\Delta V(\mathbf{x}) &= V[\mathbf{x}(k+1)] - V[\mathbf{x}(k)] \\ &= 1.25x_1^2(k) - 0.75x_2^2(k)\end{aligned} \tag{6-201}$$

Since $\Delta V(\mathbf{x})$ is indefinite in sign, the test using the Liapunov function of Eq. (6-200) fails, and no conclusion on the stability condition of the system can be reached.

Now let us turn to the instability theorem of Liapunov. Let the Liapunov function be defined as

$$V(\mathbf{x}) = a_1x_1^2(k) + 2a_2x_1(k)x_2(k) + a_3x_2^2(k) \tag{6-202}$$

and let the function $\Delta V(\mathbf{x})$ be of the form

$$\Delta V(\mathbf{x}) = -x_1^2(k) - x_2^2(k) \tag{6-203}$$

so that it is negative definite for all $x_1(k)$ and $x_2(k) \neq 0$.

Forming $\Delta V(\mathbf{x})$ according to Eq. (6-189), we have

$$\begin{aligned}\Delta V(\mathbf{x}) &= V[\mathbf{x}(k+1)] - V[\mathbf{x}(k)] \\ &= 1.25a_1x_1^2(k) - 0.5a_2x_1(k)x_2(k) - 0.75a_3x_2^2(k)\end{aligned} \tag{6-204}$$

Comparing Eqs. (6-203) and (6-204), we have

$$a_1 = -0.8 \qquad a_2 = 0 \qquad a_3 = 1.333 \tag{6-205}$$

Thus, from Eq. (6-202),

$$V(\mathbf{x}) = -0.8x_1^2(k) + 1.333x_2^2(k) \tag{6-206}$$

Since $V(\mathbf{x})$ is indefinite there is again no conclusion of the stability condition.

□ Liapunov Stability Theorem for Linear Digital Systems

The stability and instability theorems of Liapunov are valid for both linear and nonlinear systems. The last two examples illustrate that a successful execution of the Liapunov tests depends upon the correct guess or assignment of $V(\mathbf{x})$ or $\Delta V(\mathbf{x})$, which is always a difficult task. However, for linear digital systems a simple test procedure is available.

Consider that a linear time-invariant digital system is described by the difference equation

$$\mathbf{x}(k+1) = \mathbf{A}\mathbf{x}(k) \tag{6-207}$$

where $\mathbf{x}(k)$ *is* $n \times 1$ *and* $\mathbf{A}$ *is an* $n \times n$ *matrix. The equilibrium state* $\mathbf{x}_e = \mathbf{0}$ *is asymptotically stable if and only if, given any positive-definite real symmetric matrix* $\mathbf{Q}$, *there exists a positive-definite real symmetric matrix* $\mathbf{P}$ *such that*

$$\mathbf{A}'\mathbf{P}\mathbf{A} - \mathbf{P} = -\mathbf{Q} \tag{6-208}$$

Then

$$V(\mathbf{x}) = \mathbf{x}'(k)\mathbf{P}\mathbf{x}(k) \tag{6-209}$$

is a Liapunov function for the system, and further,

$$\Delta V(\mathbf{x}) = -\mathbf{x}'(k)\mathbf{Q}\mathbf{x}(k) \tag{6-210}$$

where $\Delta V(\mathbf{x})$ *is as defined in Eq.* (6-189). ■

Proof: The proof of this theorem is based on Sylvester's theorem, which states that if $\mathbf{P}$ is a positive-definite matrix, then $V(\mathbf{x}) = \mathbf{x}'\mathbf{P}\mathbf{x}$ is positive definite. Using Eq. (6-209) as the Liapunov function,

$$\begin{aligned} \Delta V(\mathbf{x}) &= V[\mathbf{x}(k+1)] - V[\mathbf{x}(k)] \\ &= \mathbf{x}'(k+1)\mathbf{P}\mathbf{x}(k+1) - \mathbf{x}'(k)\mathbf{P}\mathbf{x}(k) \end{aligned} \tag{6-211}$$

Now substituting the state equation of Eq. (6-207) into Eq. (6-211), we have

$$\begin{aligned} \Delta V(\mathbf{x}) &= \mathbf{x}'(k)[\mathbf{A}'\mathbf{P}\mathbf{A} - \mathbf{P}]\mathbf{x}(k) \\ &= -\mathbf{x}'(k)\mathbf{Q}\mathbf{x}(k) \end{aligned} \tag{6-212}$$

Thus, from Sylvester's theorem, if $\Delta V(\mathbf{x})$ is to be negative definite, $\mathbf{Q}$ has to be positive definite. ●

Example 6-24

Consider the same system as in Example 6-22. The state equations are given in Eqs. (6-194) and (6-195). The coefficient matrix is

$$\mathbf{A} = \begin{bmatrix} -0.5 & 0 \\ 0 & -0.5 \end{bmatrix} \tag{6-213}$$

The equilibrium state is $\mathbf{x}_e = 0$. Let $\mathbf{Q}$ be the identity matrix

$$\mathbf{Q} = \begin{bmatrix} 1 & 0 \\ 0 & 1 \end{bmatrix} \tag{6-214}$$

and let $\mathbf{P}$ be of the form

$$\mathbf{P} = \begin{bmatrix} p_{11} & p_{12} \\ p_{21} & p_{22} \end{bmatrix} \tag{6-215}$$

Then Eq. (6-208) becomes

$$\begin{bmatrix} -0.5 & 0 \\ 0 & -0.5 \end{bmatrix}\begin{bmatrix} p_{11} & p_{12} \\ p_{21} & p_{22} \end{bmatrix}\begin{bmatrix} -0.5 & 0 \\ 0 & -0.5 \end{bmatrix} - \begin{bmatrix} p_{11} & p_{12} \\ p_{21} & p_{22} \end{bmatrix} = -\begin{bmatrix} 1 & 0 \\ 0 & 1 \end{bmatrix} \tag{6-216}$$

Solving for the elements of the $\mathbf{P}$ matrix from the last equation yields

$$\mathbf{P} = \begin{bmatrix} 1.33 & 0 \\ 0 & 1.33 \end{bmatrix} \tag{6-217}$$

which is positive definite. Therefore,

$$V(\mathbf{x}) = \mathbf{x}'(k)\mathbf{P}\mathbf{x}(k) \tag{6-218}$$

is the Liapunov function and is positive definite. The function $\Delta V(\mathbf{x})$ is given by Eq. (6-210), which is negative definite, and the equilibrium state is asymptotically stable.

Optimal State-Feedback Design by the Liapunov Method

The stability criterion of Liapunov can be used for the design of a limited class of systems with state feedback. Figure 6-8 shows the block diagram of a linear digital system with state feedback. The state equations of the system are expressed as

$$\mathbf{x}(k+1) = \mathbf{A}\mathbf{x}(k) + \mathbf{B}\mathbf{u}(k) \tag{6-219}$$

The control $\mathbf{u}(k)$ is defined as

$$\mathbf{u}(k) = -\mathbf{G}\mathbf{x}(k) \tag{6-220}$$

where $\mathbf{x}(k)$ is an n-vector, $\mathbf{u}(k)$ an r-vector, and $\mathbf{G}$ is an $r \times n$ constant matrix.

The design objective is to find the feedback matrix $\mathbf{G}$ that will bring the state $\mathbf{x}(k)$ from any initial state $\mathbf{x}(0)$ to the equilibrium state $\mathbf{x} = \mathbf{0}$ in some optimal sense.

The basic assumption is that the system of Eq. (6-219) is asymptotically stable with $\mathbf{u}(k) = \mathbf{0}$. This guarantees that given a positive-definite real symmetric matrix $\mathbf{Q}$, there exists a positive-definite real symmetric matrix $\mathbf{P}$ such that

$$\mathbf{A}'\mathbf{P}\mathbf{A} - \mathbf{P} = -\mathbf{Q} \tag{6-221}$$

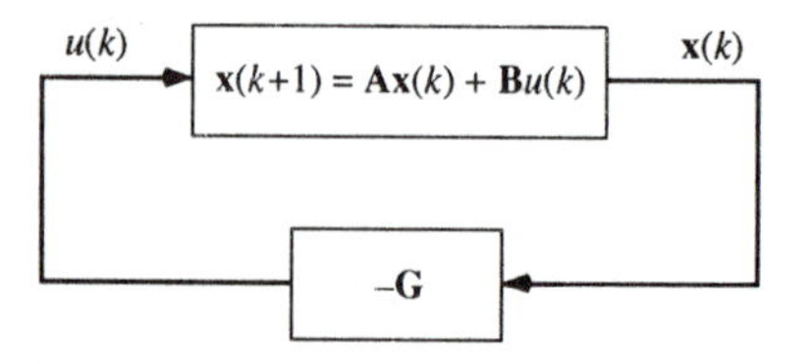

Figure 6-8. A digital control system with state feedback.

The Liapunov function is defined as

$$V[\mathbf{x}(k)] = \mathbf{x}'(k)\mathbf{P}\mathbf{x}(k) \tag{6-222}$$

and

$$\Delta V[\mathbf{x}(k)] = V[\mathbf{x}(k+1)] - V[\mathbf{x}(k)] = -\mathbf{x}(k)\mathbf{Q}\mathbf{x}(k) \tag{6-223}$$

as in Eqs. (6-209) and (6-210), respectively.

As an optimal control problem, we seek an optimal control $\mathbf{u}^o(k)$ which minimizes the performance index

$$J = \Delta V[\mathbf{x}(k)] \tag{6-224}$$

at the instant k.

Since $\Delta V[\mathbf{x}(k)]$ represents the discrete rate of change of $V[\mathbf{x}(k)]$, we can define $V[\mathbf{x}(k)]$ so that minimization of the performance index of Eq. (6-224) will carry a physical meaning in optimal control. For instance, $\Delta V[\mathbf{x}(k)]$ may represent the rate of change of distance or energy along the trajectory $\mathbf{x}(k)$. For a second-order system, if we select $V[\mathbf{x}(k)]$ to be

$$V[\mathbf{x}(k)] = x_1^2(k) + x_2^2(k) \tag{6-225}$$

then

$$-\Delta V[\mathbf{x}(k)] = -x_1^2(k+1) - x_2^2(k+1) + x_1^2(k) + x_2^2(k) \tag{6-226}$$

which is in the form of a quadratic performance index over the period k to $k+1$.

For the general form of $V[\mathbf{x}(k)]$ given in Eq. (6-222), we write

$$\begin{aligned}\Delta V[\mathbf{x}(k)] &= \mathbf{x}'(k+1)\mathbf{P}\mathbf{x}(k+1) - \mathbf{x}'(k)\mathbf{P}\mathbf{x}(k)\\ &= \mathbf{x}'(k)(\mathbf{A}-\mathbf{B}\mathbf{G})'\mathbf{P}(\mathbf{A}-\mathbf{B}\mathbf{G})\mathbf{x}(k) - \mathbf{x}'(k)\mathbf{P}\mathbf{x}(k)\end{aligned} \tag{6-227}$$

Using the state equation relationship in Eq. (6-221), the last equation is written

$$\begin{aligned}\Delta V[\mathbf{x}(k)] &= \mathbf{x}'(k)\mathbf{A}'\mathbf{P}\mathbf{A}\mathbf{x}(k) + \mathbf{u}'\mathbf{B}'\mathbf{P}\mathbf{A}\mathbf{x}(k) + \mathbf{x}'(k)\mathbf{A}'\mathbf{P}\mathbf{B}\mathbf{u}(k)\\ &\quad + \mathbf{u}'(k)\mathbf{B}'\mathbf{P}\mathbf{B}\mathbf{u}(k) - \mathbf{x}'(k)\mathbf{P}\mathbf{x}(k)\end{aligned} \tag{6-228}$$

For minimum $\Delta V[\mathbf{x}(k)]$ with respect to $\mathbf{u}(k)$,

$$\frac{\partial \Delta V[\mathbf{x}(k)]}{\partial \mathbf{u}(k)} = \mathbf{0} \tag{6-229}$$

Substitution of Eq. (6-228) in Eq. (6-229) leads to

$$2\mathbf{B}'\mathbf{P}\mathbf{A}\mathbf{x}(k) + 2\mathbf{B}'\mathbf{P}\mathbf{B}\mathbf{u}(k) = \mathbf{0} \tag{6-230}$$

Therefore, the optimal control is

$$\mathbf{u}^o(k) = -(\mathbf{B}'\mathbf{P}\mathbf{B})^{-1}\mathbf{B}'\mathbf{P}\mathbf{A}\mathbf{x}(k) \tag{6-231}$$

which is in the form of *state feedback*.

It should be pointed out that this design technique is restricted by the requirement that the process of Eq. (6-231) must be asymptotically stable, which would exclude continuous-data systems with one or more poles at the origin of the s-plane.

Example 6-25

Consider that the linear digital process shown in Fig. 6-8 is described by the state equations

$$\mathbf{x}(k+1) = \mathbf{A}\mathbf{x}(k) + \mathbf{B}\mathbf{u}(k) \tag{6-232}$$

where

$$\mathbf{A} = \begin{bmatrix} 0.5 & 0 \\ 0 & 0.2 \end{bmatrix} \tag{6-233}$$

$$\mathbf{B} = \begin{bmatrix} 1 \\ 1 \end{bmatrix} \tag{6-234}$$

Find the optimal control $\mathbf{u}^o(k)$ so that, given $\mathbf{Q} = \mathbf{I}$ (identity matrix), with $V(\mathbf{x}) = \mathbf{x}'(k)\mathbf{P}\mathbf{x}(k)$ and $\mathbf{P}$ the solution of

$$-\mathbf{Q} = \mathbf{A}'\mathbf{P}\mathbf{A} - \mathbf{P} \tag{6-235}$$

the performance index

$$\Delta V(\mathbf{x}) = V[\mathbf{x}(k+1)] - V[\mathbf{x}(k)] \tag{6-236}$$

is minimized.

It is simple to show that the process represented by Eq. (6-232) is asymptotically stable, since the eigenvalues of $\mathbf{A}$ are at 0.5 and 0.2.

Let $\mathbf{P}$ be written as

$$\mathbf{P} = \begin{bmatrix} p_{11} & p_{12} \\ p_{12} & p_{22} \end{bmatrix} \tag{6-237}$$

Substituting Eq. (6-237) into Eq. (6-235) and solving for the elements of $\mathbf{P}$, we have

$$p_{11} = \frac{1}{1-(0.5)^2} = 1.333 \tag{6-238}$$

$$p_{12} = 0$$

$$p_{22} = \frac{1}{1-(0.2)^2} \tag{6-239}$$

Thus,

$$\mathbf{P} = \begin{bmatrix} 1.333 & 0 \\ 0 & 1.042 \end{bmatrix} \tag{6-240}$$

which is positive definite.

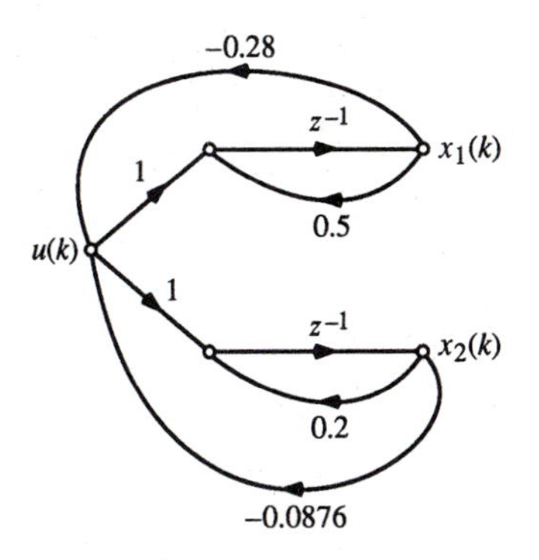

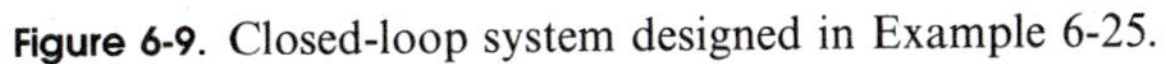
Figure 6-9. Closed-loop system designed in Example 6-25.

The optimal control which minimizes the performance index of Eq. (6-236) is given by Eq. (6-231). The optimal feedback gain is

$$\mathbf{G} = (\mathbf{B'PB})^{-1}\mathbf{B'PA}$$
$$= [0.28 \quad 0.0876] \tag{6-241}$$

The optimal control is

$$\mathbf{u}^o(k) = -0.28x_1(k) - 0.0876x_2(k) \tag{6-242}$$

The state diagram of the closed-loop system is shown in Fig. 6-9. It can be shown that the eigenvalues of the optimal closed-loop system are approximately 0 and 0.3324.

PROBLEMS

6-1 A discrete-data control system is described by the state equation

$$\mathbf{x}(k+1) = \mathbf{Ax}(k) + \mathbf{Bu}(k)$$

a. Determine the controllability of the system.

b. Find a **P** matrix that will diagonalize **A**, or transform **A** into a Jordan canonical form of $\mathbf{\Lambda} = \mathbf{P}^{-1}\mathbf{AP}$. Verify the result obtained in part **a** by checking the elements of $\mathbf{\Gamma} = \mathbf{P}^{-1}\mathbf{B}$.

(i) $\mathbf{A} = \begin{bmatrix} 1 & -2 \\ 1 & -1 \end{bmatrix}$ $\quad \mathbf{B} = \begin{bmatrix} 1 & 0 \\ 0 & -1 \end{bmatrix}$

(ii) $\mathbf{A} = \begin{bmatrix} 0 & 1 \\ -0.25 & 1 \end{bmatrix}$ $\quad \mathbf{B} = \begin{bmatrix} 1 \\ 0.5 \end{bmatrix}$

(iii) $\mathbf{A} = \begin{bmatrix} 0 & 1 & 0 \\ 0 & 0 & 1 \\ 0.04 & -0.53 & 1.4 \end{bmatrix}$ $\quad \mathbf{B} = \begin{bmatrix} 1 & 0 \\ -1 & 1 \\ 0 & 1 \end{bmatrix}$

Show that, in part **b**,

$$\mathbf{P} = \begin{bmatrix} 1 & 1 & 1 \\ \lambda_1 & \lambda_2 & \lambda_3 \\ \lambda_1^2 & \lambda_2^2 & \lambda_3^2 \end{bmatrix}$$

where λ_1, λ_2, and λ_3 are the eigenvalues of $\mathbf{A}$.

6-2 The state diagram of a digital control system is shown in Fig. P6-2. Determine the controllability and observability of the system.

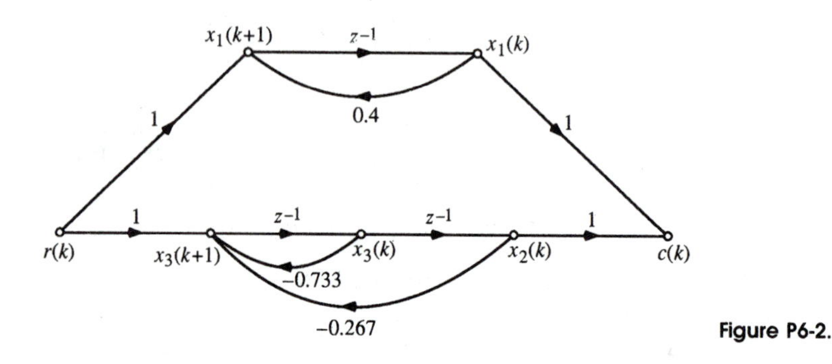

Figure P6-2.

6-3 The block diagram of a digital control system is shown in Fig. P6-3. Determine what value of T must be avoided so that the system will be assured of complete controllability and observability.

$u(t)$ → [sampler T] → $u^*(t)$ → $\dfrac{10(s+5)}{(s^2+5)^2+4}$ → $c(t)$ → [sampler T] → $c^*(t)$

Figure P6-3.

6-4 The linear controlled process of a discrete-data control system is described by the following state equations:

$$\frac{d\mathbf{x}(t)}{dt} = \mathbf{A}\mathbf{x}(t) + \mathbf{B}u(t)$$

where

$$\mathbf{A} = \begin{bmatrix} 0 & 1 \\ -1 & 0 \end{bmatrix} \qquad \mathbf{B} = \begin{bmatrix} 0 \\ 1 \end{bmatrix}$$

$u(t) = u(kT)$ for $kT \le t < (k+1)T$. Determine the values of the sampling period T that make the system uncontrollable; that is, the initial state $\mathbf{x}(0)$ cannot be driven to any state $\mathbf{x}(N)$ for finite N.

6-5 Figure P6-5 shows a stick-balancing system in which the objective is to control the attitude of the stick with the force $u(t)$ applied to the car. The force

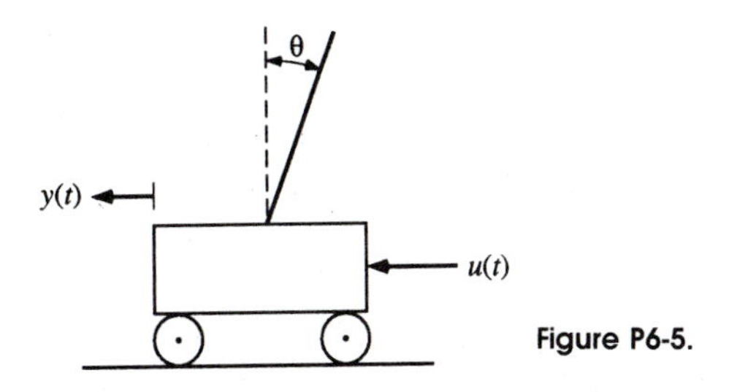

Figure P6-5.

$u(t)$ is sampled and is described by $u(t) = u(kT)$, $kT \leq t < (k+1)T$, where T is the sampling period. The linearized equations that approximate the motion of the stick are

$$\frac{d^2\theta(t)}{dt^2} = \theta(t) + u(t)$$

$$\frac{d^2y(t)}{dt^2} = \theta(t) - u(t)$$

a. Let the state variables be defined as $x_1(t) = \theta(t)$, $x_2(t) = d\theta(t)/dt$, $x_3(t) = y(t)$, and $x_4(t) = dy(t)/dt$. Discretize the system equations and express the state equations in the following form:

$$\mathbf{x}[(k+1)T] = \boldsymbol{\phi}(T)\mathbf{x}(kT) + \boldsymbol{\theta}(T)u(k)$$

b. For $T = 1$ s determine if the discrete-data system in part **a** is completely state controllable.
c. If only one of the state variables can be regarded as the output, that is, $c(k) = x_i(k)$, $i = 1, 2, 3$, or 4, determine which one corresponds to a completely observable system.
d. Find the state feedback $u(k) = -\mathbf{G}\mathbf{x}(k)$ where $\mathbf{G} = [g_1 \quad g_2 \quad g_3 \quad g_4]$ such that the eigenvalues of the closed-loop system are all zero.

6-6 A discrete-data control system is described by the state equation

$$\mathbf{x}(k+1) = \mathbf{A}\mathbf{x}(k) + \mathbf{B}u(k)$$

where

$$\mathbf{A} = \begin{bmatrix} 0 & 0 & 0 \\ 0 & 0.5 & 0 \\ 0 & 0 & 2 \end{bmatrix} \qquad \mathbf{B} = \begin{bmatrix} 1 \\ 0 \\ 1 \end{bmatrix}$$

a. Determine the state controllability of the system.
b. Can the system be stabilized by state feedback of the form

$$u(k) = -[g_1 \quad g_2 \quad g_3]\mathbf{x}(k)$$

where g_1, g_2, and g_3 are real constants?

6-7 The input-output transfer function of a digital control system is

$$\frac{C(z)}{U(z)} = \frac{1.65(z + 0.1)}{z^3 + 0.7z^2 + 0.11z + 0.005}$$

a. Assign state variables to the system so that it is state controllable but not observable.

b. Assign state variables to the system so that it is observable but not state controllable.

6-8 Given the discrete-data control system

$$\mathbf{x}(k + 1) = \mathbf{A}\mathbf{x}(k) + \mathbf{B}u(k)$$
$$c(k) = \mathbf{D}\mathbf{x}(k)$$

where

$$\mathbf{A} = \begin{bmatrix} 0 & 1 \\ -2 & -3 \end{bmatrix} \qquad \mathbf{B} = \begin{bmatrix} 0 \\ 1 \end{bmatrix} \qquad \mathbf{D} = [1 \quad -1]$$

The control is realized through state feedback,

$$u(k) = -\mathbf{G}\mathbf{x}(k) = -[g_1 \quad g_2]\mathbf{x}(k)$$

where g_1 and g_2 are real constants.

Determine the values of g_1 and g_2 that must be avoided for the system to be completely observable.

6-9 Determine the stability conditions of the discrete-data control system that are represented by the following characteristic equations.

a. $z^2 + 0.5z + 0.2 = 0$

b. $z^3 + z^2 + 3z + 0.2 = 0$

c. $z^3 - 1.5z^2 + 1.2z - 0.5 = 0$

d. $z^4 - 1.2z^3 + 0.22z^2 + 0.066z - 0.008 = 0$

e. $z^3 - 1.4z^2 + 0.53z - 0.04 = 0$

f. $z^4 - 2z^3 + z^2 - 2z + 1 = 0$

g. $z^4 - z^3 + z^2 - z + 1 = 0$

6-10 The characteristic equations of linear discrete-data systems are given below. Determine the values of K for the systems to be asymptotically stable.

a. $z^2 + 1.5z + K = 0$

b. $z^2 - Kz + 0.5 = 0$

c. $z^3 + 5z^2 - z + 5K = 0$

d. $z^3 + z^2 - z + K = 0$

e. $z^3 + 0.5z^2 + Kz - K = 0$

f. $z^3 + 0.5z^2 + K = 0$

g. $z^4 + 0.2z^3 - 0.25z^2 - 0.05z + K = 0$

6-11 The block diagram of a discrete-data control system is shown in Fig. P6-11. Determine the range of K for the system to be asymptotically stable.

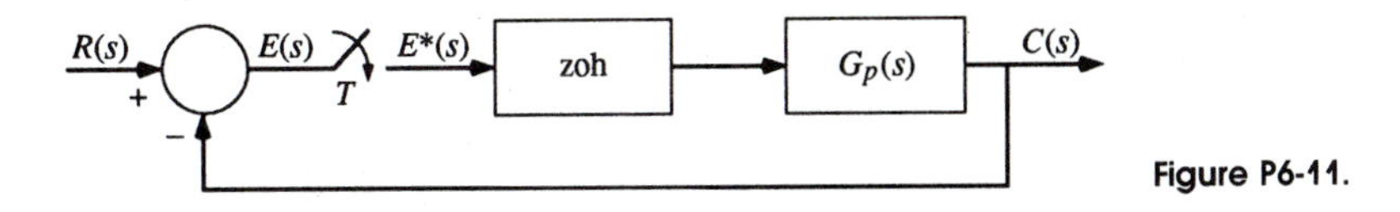

Figure P6-11.

a. $G_p(s) = \dfrac{K}{s(s+5)} \qquad T = 0.5 \text{ s}$

b. $G_p(s) = \dfrac{K(s+1)}{s(s+2)} \qquad T = 0.5 \text{ s}$

c. $G_p(s) = \dfrac{K(s+5)}{s^2} \qquad T = 0.5 \text{ s}$

d. $G_p(s) = \dfrac{K}{s(s+4)(s+8)} \qquad T = 0.5 \text{ s}$

e. $G_p(s) = \dfrac{K}{s^2+s+2} \qquad T = 1 \text{ s}$

6-12 Consider that the amplifier gain K in the flight control system described in Problem 4-8, Fig. P4-8, is variable. Determine the range of K for the system to be stable.

a. $T = 0.01$ s

b. $T = 0.1$ s

6-13 Consider that the amplifier gain K in the flight control system described in Problem 4-9 is variable. Determine the range of K for the system to be stable.

a. $T = 0.01$ s

b. $T = 0.1$ s

6-14 The controlled process of the control system in Fig. P6-11 is described by

$$G_p(s) = \frac{1000e^{-NTs}}{s(s+10)(s+50)}$$

where N is a positive integer. The sampling period is $T = 0.1$ s. Find the maximum value of N so that the system is stable.

6-15 For the Large Space Telescope control system described in Problem 4-10, Fig. P4-10, let $K_I = 0$, $K_R = 10^5$, $K_P = 10^6$, and $J_v = 10^4$. Find the maximum value of the sampling period T so that the system is stable. Find the roots of the characteristic equation when T is set at the marginal value.

6-16 For the Large Space Telescope control system described in Problem 4-10, Fig. P4-10, let $K_I = 0$, $J_v = 10^4$, and $T = 0.1$ s. Find the marginal values of K_R and K_P for the system to be stable. Show the region of stability in the K_R-versus-K_P plane (K_R as the vertical axis).

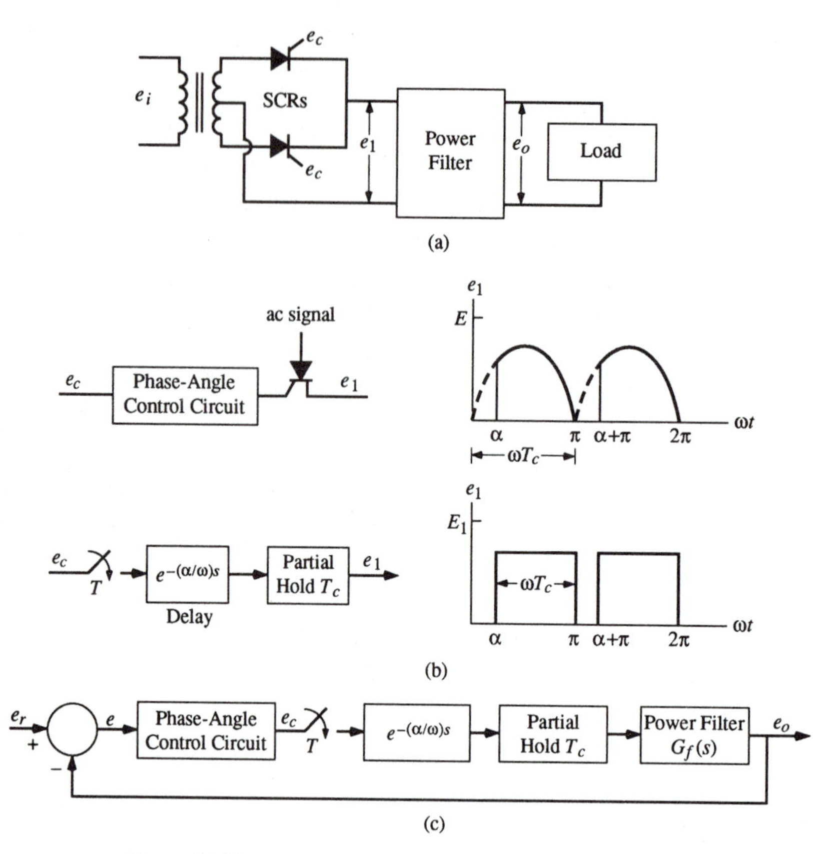

Figure P6-17.

6-17 The circuit diagram of an ac-to-dc silicon-controlled rectifier (SCR) voltage-regulator system is shown in Fig. P6-17(a). Systems of this type are generally difficult to analyze mathematically as the operations are nonlinear. For a sinusoidal input at e_i, the signal at the output of the SCR bridge e_1 is shown in Fig. P6-17(b). The conduction period T_c of the SCRs determines the average value of the output voltage e_o. The triggering angle α is directly related to the conduction period. As an approximation, the SCRs and their phase-angle control circuits can be represented by a sample-and-hold device, as shown in Fig. P6-17(b). In this case the sample-and-hold device has a delay in the sampling of α/ω s and a partial holding period T_c that is less than the sampling period T. The frequency of the input sinusoid is ω rad/s, and therefore, the sampling period is π/ω s.

Figure P6-17(c) shows the block diagram of a closed-loop system that may be used as a steady-state model of an SCR system for the purpose of regulating the voltage e_o. The output voltage e_o is fed back and compared with a reference voltage e_r, and the error $e = e_r - e_o$ is used to regulate the triggering angle α. Since the system shown in Fig. P6-17(c) is still nonlinear, even with the use of the sample-and-hold approximation of Fig. P6-17(b), for linear studies we can only consider the steady-state performance of the system about an equilibrium point for α. Consider that the operating conditions are such that a nominal line voltage of 115 V rms at 60 Hz is applied as e_i. The transfer function for the power filter is

$$G_f(s) = \frac{87}{(s+3)(s+29)}$$

a. Determine the triggering angle α so that the average output voltage e_o is 60 V.

b. For the closed-loop digital control system model shown in Fig. P6-17(c), with the value of α as determined in part **a**, find the range of the gain of the phase-angle control circuit so that the system is stable.

6-18 Given a linear digital control system that is modeled by the state equation,

$$\mathbf{x}(k+1) = \mathbf{A}\mathbf{x}(k) + \mathbf{B}u(k)$$

where

$$\mathbf{A} = \begin{bmatrix} 0 & 1 \\ -1 & 1 \end{bmatrix} \qquad \mathbf{B} = \begin{bmatrix} 0 \\ 1 \end{bmatrix}$$

The control $u(k)$ is implemented through state feedback, $u(k) = -\mathbf{G}\mathbf{x}(k)$, where

$$\mathbf{G} = [g_1 \quad g_2]$$

Find the ranges of g_1 and g_2 for the closed-loop system to be stable. Express the stable region in the g_2-versus-g_1 plane (g_2 as the vertical axis).

6-19 Figure P6-19 shows the schematic diagram of a discrete-data system for proportioning concentrate into a varying flow of forage for cattle feeding. The forage weigher converts the forage flow rate into an electrical signal via the

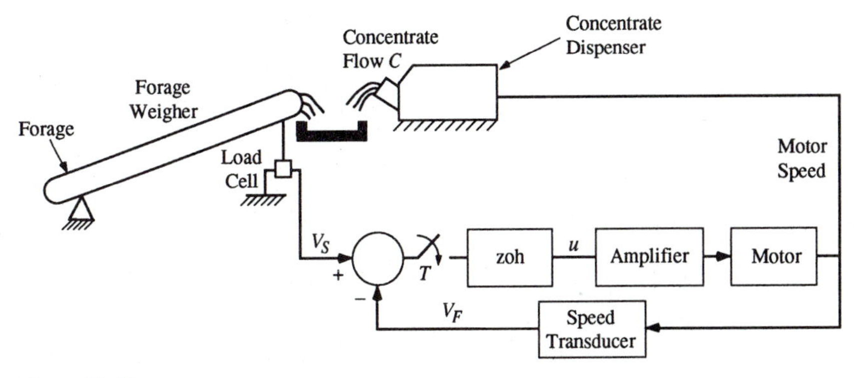

Figure P6-19.

load cell. The transfer function between the input of the amplifier $u(t)$ and the concentrate flow rate $c(t)$ is

$$\frac{C(s)}{U(s)} = \frac{36870}{s^2 + 83.3s + 7780}$$

Let the gain between the motor speed and the concentrate flow rate and that of the speed transducer be unity. For $T = 0.01$ s, find the closed-loop transfer function $C(z)/V_s(z)$ and determine if the system is stable.

6-20 Discrete pattern recognition and adaptive techniques require the use of prediction in digital control systems. Figure P6-20 shows the block diagram of a digital control system with a digital predictor. The transfer function of the ideal predictor is given as $P(z) = z^p$, where p is a positive integer. For $G(s) = K/s^2$, determine the stability conditions of the system for

a. $p = 1$

b. $p = 2$

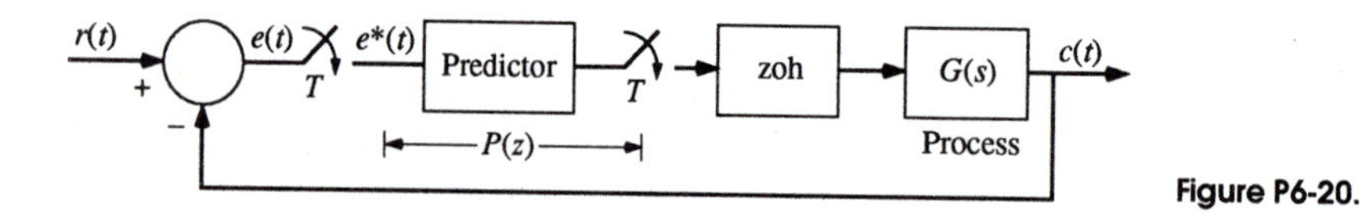

Figure P6-20.

6-21 The block diagram of a digital tracking loop of a missile system is shown in Fig. P6-21. The sampling period is 1/12 s. Determine the values of K for stability as a function of Δ $(0 \leq \Delta \leq 1)$. Plot the maximum value of K for stability as a function of Δ.

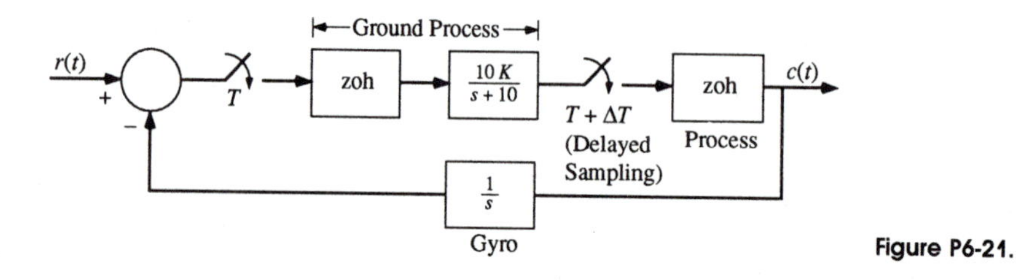

Figure P6-21.

6-22 For the digital control system with nonuniform sampling described in Problem 4-15, Fig. P4-15, determine the region of stability for the system in the K-versus-$1/N$ plane for $0 \leq K < \infty$ and $0 \leq 1/N \leq 1$.

6-23 For the digital control system with multirate sampling described in Problem 4-24, Fig. P4-24, determine the maximum value of K for stability as a function of N for $N \geq 1$.

6-24 A digital process is described by the state equation

$$\mathbf{x}(k+1) = \mathbf{A}\mathbf{x}(k) + \mathbf{B}u(k)$$

where

$$\mathbf{A} = \begin{bmatrix} 0 & 0.81 \\ 1 & 0 \end{bmatrix} \qquad \mathbf{B} = \begin{bmatrix} 0 \\ 1 \end{bmatrix}$$

Find the optimal state-feedback control $u^o(k) = -Gx(k)$ such that the performance index

$$J = [\mathbf{x}(k+1) - \mathbf{x}(k)]'\mathbf{P}[\mathbf{x}(k+1) - \mathbf{x}(k)]$$

is minimized, where $\mathbf{P}$ is the positive-definite solution of the Liapunov equation $\mathbf{A}'\mathbf{P}\mathbf{A} - \mathbf{P} = \mathbf{I}$.

6-25 The state equations of a digital process are expressed as

$$\mathbf{x}(k+1) = \mathbf{A}\mathbf{x}(k) + \mathbf{B}u(k)$$

where

$$\mathbf{A} = \begin{bmatrix} 0 & 1 \\ -1 & -2\zeta \end{bmatrix} \qquad \mathbf{B} = \begin{bmatrix} 0 \\ 1 \end{bmatrix}$$

Find the optimal value of ζ so that the performance index

$$J = \mathbf{x}'(k)\mathbf{P}\mathbf{x}(k) = \text{minimum for all } k \geq 0$$

where $\mathbf{P}$ is the positive-definite solution of the Liapunov equation $\mathbf{A}'\mathbf{P} + \mathbf{P}\mathbf{A} = -\mathbf{I}$, for $x_1(0) = x_0$ and $x_2(0) = 0$.

References

Controllability, Observability, and Stabilizability

1. Kreindler, E., and P. E. Sarachik, "On the Concepts of Controllability and Observability of Linear Systems," *IEEE Trans. Automatic Control*, vol. AC-9, pp. 129–136, April 1964.
2. Kalman, R. E., "Contributions to the Theory of Optimal Control," *Bol. Soc. Mat. Mexicana*, vol. 5, pp. 102–119, 1960.
3. Kalman, R. E., "On the General Theory of Control Systems," *Proc. IFAC*, vol. 1, Butterworths, London, pp. 481–492, 1961.
4. Mullis, C. T., "On the Controllability of Discrete Linear Systems with Output Feedback," *IEEE Trans. Automatic Control*, vol. AC-18, pp. 608–615, December 1973.
5. Hautus, M. L. J., "Controllability and Stabilizability of Sampled Systems," *IEEE Trans. Automatic Control*, vol. AC-17, pp. 528–531, August 1972.
6. Brockett, R. W., "Poles, Zeros, and Feedback: State Space Interpretation," *IEEE Trans. Automatic Control*, vol. AC-10, pp. 129–135, April 1965.

Stability

7. Kuo, B. C., *Automatic Control Systems*, 6th ed., Prentice-Hall, Englewood Cliffs, N.J., 1991.
8. Marden, M., *The Geometry of the Zeros of a Polynomial in a Complex Variable*, chap. 10, American Mathematics Society, New York, 1949.
9. Jury, E. I., and B. H. Bharucha, "Notes on the Stability Criterion for Linear Discrete Systems," *IRE Trans. Automatic Control*, vol. AC-6, pp. 88–90, February 1961.
10. Jury, E. I., and J. Blanchard, "A Stability Test for Linear Discrete Systems in Table Form," *IRE Proc.*, vol. 49, no. 12, pp. 1947–1948, December 1961.
11. Jury, E. I., "The Number of Roots of a Real Polynomial Inside (or Outside) the Unit Circle Using the Determinant Method," *IEEE Trans. Automatic Control*, vol. AC-10, pp. 371–372, July 1965.
12. Jury, E. I., *Theory and Application of the z-Transform Method*, John Wiley & Sons, New York, 1964.
13. Cohen, M. L., "A Set of Stability Constraints on the Denominator of a Sampled-Data Filter," *IEEE Trans. Automatic Control*, vol. AC-11, pp. 327–328, April 1966.
14. Raible, R. H., "A Simplification of Jury's Tabular Form," *IEEE Trans. Automatic Control*, vol. AC-19, pp. 248–250, June 1974.
15. Jury, E. I., and B. D. O. Anderson, "A Simplified Schur-Cohen Test," *IEEE Trans. Automatic Control*, vol. AC-18, pp. 157–163, April 1973.
16. Rao, M. V. C., and A. K. Subramanian, "Elimination of Singular Cases in Jury's Test," *IEEE Trans. Automatic Control*, vol. AC-21, pp. 114–115, February 1976.
17. Kalman, R. E., and J. E. Bertram, "Control Systems Analysis and Design Via the Second Method of Liapunov: II, Discrete Time Systems," *Trans. ASME, J. Basic Eng.*, Series D, no. 3, pp. 371–400, June 1960.

Time-Domain and z-Domain Analysis

KEYWORDS AND TOPICS

transient response • steady-state response • steady-state error • maximum overshoot • rise time • settling time • damping ratio • natural undamped frequency • root-locus techniques • poles and zeros

7-1 INTRODUCTION

Since the outputs of discrete-data control systems are usually functions of the continuous variable t, it is necessary to evaluate the performance of the system in the time domain. However, when the z-transform or the discrete-time state equation is applied as the analysis tool, the outputs of the system are measured only at the sampling instants. Depending on the sampling period chosen and its relation to the time constants of the system, the discrete-data representation may or may not be accurate. In other words, there may be a large discrepancy between the true output $c(t)$ and the fictitiously sampled signal $c^*(t)$, so that the latter is not a realistic representation of the system behavior.

As in the studies of continuous-data control systems, the time response of a discrete-data control system may be characterized by such terms as the *maximum overshoot*, *rise time*, *delay time*, *settling time*, *damping ratio*, *damping factor*, and *natural undamped frequency*. Analysis and design methods using these time-domain specifications are best carried out using the complex s- or z-domain. Analytical tools such as the *root-locus diagram*, which is so popular for linear continuous-data systems, are also applicable for discrete-data control systems.

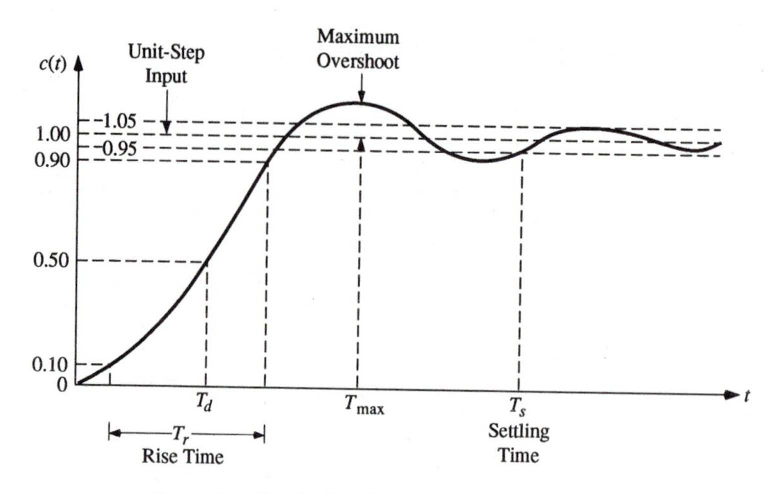

Figure 7-1. Typical unit-step response of a digital control system and illustration of time-domain performance criteria.

The performance of a discrete-data control system in the time domain is often measured by applying a test signal such as a *unit-step function* to the system input. For linear systems the unit-step input can provide valuable information on the transient and steady-state behavior of the system. In fact, maximum overshoot, delay time, rise time, and settling time are all defined with respect to a unit-step function input. The response of a system to a unit-step input is called the *unit-step response.*

Figure 7-1 shows a typical unit-step response of a discrete-data control system. Notice that the unit-step response of a continuous-data system could also be of the same form. Therefore, the time-domain specifications, maximum overshoot, and rise, delay, and settling times all apply equally well to both types of systems when the outputs are in the form of $c(t)$. As a quick review, the definitions of these specifications are summarized as follows.

1. Maximum Overshoot

The maximum overshoot is defined as the difference between the maximum value and the final value of the output response to a unit-step input. Thus, let c_{max} be the maximum value of the output $c(t)$, and c_{ss} be the steady-state value of $c(t)$. Then,

$$\text{maximum overshoot} = c_{max} - c_{ss} \tag{7-1}$$

2. Peak Time

T_{max} is the time at which c_{max} occurs.

3. Percent Maximum Overshoot

The maximum overshoot is often represented as a percentage of the final value of the unit-step response; that is,

$$\text{percent maximum overshoot} = \frac{\text{maximum overshoot}}{\text{final value of } c(t)} \times 100 \text{ percent} \qquad (7\text{-}2)$$

4. Delay Time

The delay time T_d is defined as the time required for the unit-step response to reach 50 percent of its final value.

5. Rise Time

The rise time T_r is defined as the time required for the unit-step response to rise from 10 percent to 90 percent of its final value.

6. Settling Time

The settling time T_s is defined as the time required for the unit-step response to decrease and stay within five percent of its final value.

Figure 7-2 illustrates a typical output of a discrete-data control system that has a maximum value c_{max}. The sampled signal is represented by $c^*(t)$ at the sampling instants. It is apparent that the maximum value of $c^*(t)$, c^*_{max}, is always less than or equal to c_{max}. In the case illustrated in Fig. 7-2, the sampling period is sufficiently small that the discrete-time response $c^*(t)$ gives a sufficiently accurate representation of the true response, and the discrepancy between c_{max} and c^*_{max} is not significant. In general, if the sampling period is too large relative to the oscillations of the response, the discrete-time representation c^* may be entirely erroneous. However, it should be pointed out that the selection of the

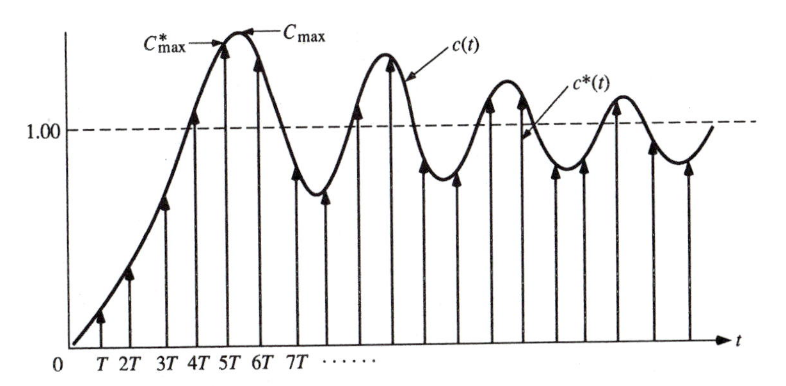

Figure 7-2. Typical unit-step response of a digital control system and its sampled-data representation.

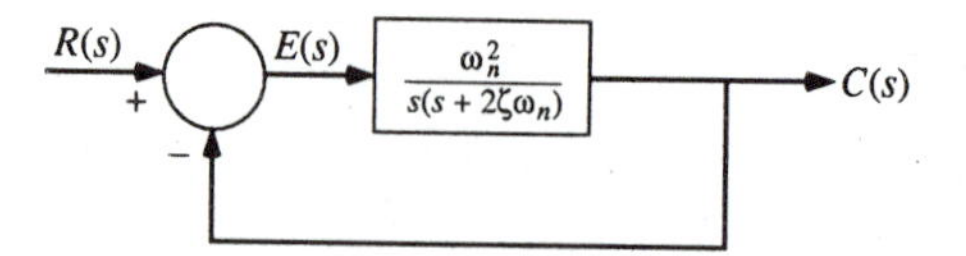

Figure 7-3. Second-order prototype control system.

sampling frequency of a discrete-data control system is usually not based solely on the accuracy of representation of the system responses at the sampling instants; equally important are considerations of system stability, disturbance rejection, system accuracy, and hardware complexity.

7-2 THE PROTOTYPE SECOND-ORDER SYSTEM

Although true second-order systems rarely exist in the real world, most of the s-domain performance parameters are defined with respect to a second-order system model. The study of a second-order system is important since many high-order systems can be approximated by a second-order model if the high-order poles are located so that they provide negligible contribution to the transient response (refer to Sec. 7-8).

Figure 7-3 shows the block diagram of a second-order closed-loop control system which is referred to as the *prototype second-order system* [1]. The open-loop transfer function of the system is

$$G(s) = \frac{C(s)}{E(s)} = \frac{\omega_n^2}{s(s + 2\zeta\omega_n)} \tag{7-3}$$

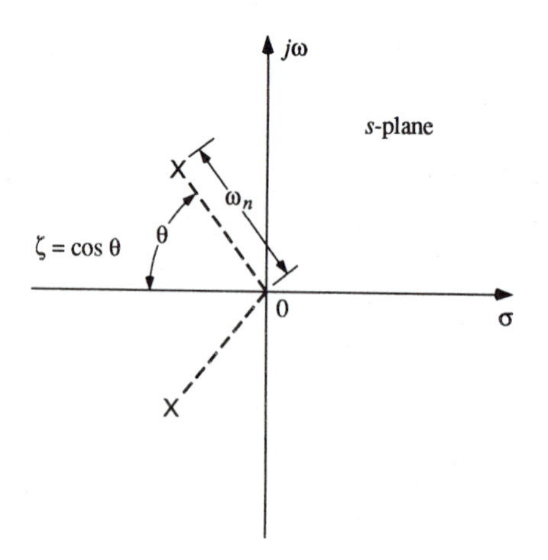

Figure 7-4. Complex roots of characteristic equation of second-order system showing the significance of ζ and ω_n.

The closed-loop transfer function is

$$M(s) = \frac{C(s)}{R(s)} = \frac{\omega_n^2}{s^2 + 2\zeta\omega_n s + \omega_n^2} \tag{7-4}$$

where

$$\zeta = \text{damping ratio} \tag{7-5}$$

$$\omega_n = \text{natural undamped frequency} \tag{7-6}$$

The characteristic equation of the prototype second-order system is

$$s^2 + 2\zeta\omega_n s + \omega_n^2 = 0 \tag{7-7}$$

The roots of Eq. (7-7) are

$$s_1, s_2 = -\zeta\omega_n \pm j\omega_n\sqrt{1 - \zeta^2} \tag{7-8}$$

and are shown in the s-plane in Fig. 7-4. The significance of ζ and ω_n is indicated in terms of the characteristic equation root locations in Fig. 7-4.

7-3 COMPARISON OF TIME RESPONSES OF CONTINUOUS-DATA AND DISCRETE-DATA SYSTEMS

In this section we shall compare the salient features of continuous-data and discrete-data or digital control systems. The objectives are to illustrate the effects of sampling, and how sampling frequency is chosen with respect to the dynamics of the system.

7-3-1 Continuous-Data System

The block diagram of a simplified space vehicle control system is shown in Fig. 7-5. The objective of the system is to control the attitude of the space vehicle in one dimension; for example, pitch. Two similar systems may control the other two space variables if it is a three-coordinate control system. It has been assumed that the dynamics of these coordinates can be decoupled from each other, so that independent controls can be implemented.

Most space vehicles should be modeled as flexible bodies. For simplicity, in Fig. 7-5 the vehicle is modeled as a rigid structure with an inertia J_v. The

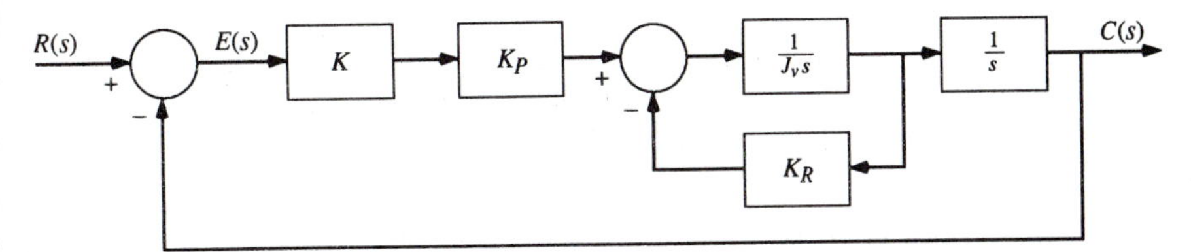

Figure 7-5. Continuous-data space vehicle control system.

transfer function between the applied motor torque and the output position of the vehicle is

$$\frac{C(s)}{T_m(s)} = \frac{1}{J_v s^2} \tag{7-9}$$

The position $c(t)$ and its derivative, the velocity $v(t)$, are fed back by the position and rate sensors, respectively, to form the closed-loop control. The motor that controls the attitude of the space vehicle is actuated by an amplifier with gain K. The parameters of the simple continuous-data control system are

Position sensor gain K_P	$= 1.65 \times 10^6$
Rate sensor gain K_R	$= 3.71 \times 10^5$
Amplifier gain K	= variable
Moment of inertia J_v	= 41822

We assume that all the units are consistent so that, for analytical purposes, we intentionally leave them unspecified.

From Fig. 7-3, the open-loop transfer function of the system is written

$$G(s) = \frac{C(s)}{E(s)} = \frac{KK_P}{s(J_v s + K_R)} \tag{7-10}$$

The closed-loop transfer function is

$$\frac{C(s)}{R(s)} = \frac{G(s)}{1 + G(s)} = \frac{KK_P}{J_v s^2 + K_R s + KK_P} \tag{7-11}$$

Substituting the known values of the system parameters in Eq. (7-10), we have the open-loop transfer function,

$$G(s) = \frac{39.453K}{s(s + 8.871)} \tag{7-12}$$

The closed-loop transfer function becomes

$$\frac{C(s)}{R(s)} = \frac{39.453K}{s^2 + 8.871s + 39.453K} \tag{7-13}$$

The characteristic equation of the system is obtained by setting the denominator of the closed-loop transfer function to zero; we get

$$s^2 + 8.871s + 39.453K = 0 \tag{7-14}$$

Comparing the last equation with the characteristic equation of the prototype second-order system in Eq. (7-7), we have

$$\omega_n = \sqrt{39.453K} \quad \text{rad/s} \tag{7-15}$$

$$\zeta = 8.871/2\omega_n \tag{7-16}$$

Since the system is of the second order, the roots of the characteristic equation will always be in the left half of the s-plane so long as all the system parameters K, K_P, K_R, and J_v are positive and finite. Thus, the continuous-data system will always be asymptotically stable under these conditions.

7-3-2 Digital Control System

Now let us consider that the continuous-data system in Fig. 7-5 is subject to sampled-data control with the position feedback and the input processed by a sample-and-hold device. The rate feedback is not subject to sampling. This would be the case if the position sensor is a digital transducer, and as its output and the input signal are both processed by a microprocessor or digital signal processor. The speed sensor in this case is analog, such as a tachometer or rate gyro. Figure 7-6 is the block diagram of the discrete-data space vehicle control system. The discrete-data operation is represented by the sample-and-hold device, with the sampling period T (seconds).

For the purpose of comparison, we assume that the system parameters K, K_P, K_R, and J_v are the same as those of the continuous-data system. From Fig. 7-6 the transfer function of the process is written

$$G_p(s) = \frac{C(s)}{U(s)} = \frac{KK_P/J_v}{s(s + K_R/J_v)} \tag{7-17}$$

The open-loop transfer function is

$$G(s) = \frac{C(s)}{E(s)} = G_{h0}(s)G_p(s) = \frac{(1 - e^{-Ts})}{s}\frac{KK_P/J_v}{s(s + K_R/J_v)} \tag{7-18}$$

Performing the z-transform of the last equation, we have

$$G(z) = (1 - z^{-1})\frac{KK_P}{K_R}\mathscr{Z}\left(\frac{1}{s^2} - \frac{J_v}{K_R s} + \frac{J_v}{K_R(s + K_R J_v)}\right)$$

$$= (1 - z^{-1})\frac{KK_P}{K_R}\left(\frac{Tz}{(z-1)^2} - \frac{J_v z}{K_R(z-1)} + \frac{J_v z}{K_R(z - e^{-K_R T/J_v})}\right) \tag{7-19}$$

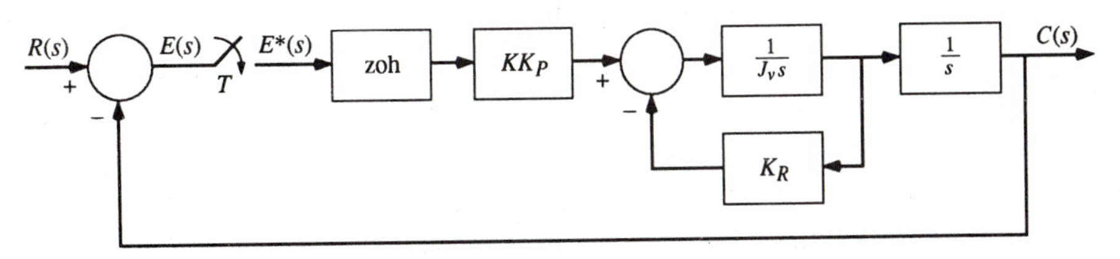

Figure 7-6. Digital space vehicle control system.

After simplification, $G(z)$ becomes

$$G(z) = \frac{KK_P}{K_R^2}\left(\frac{(TK_R - J_v + J_v e^{-K_R T/J_v})z - (TK_R + J_v)e^{-K_R T/J_v} + J_v}{(z-1)(z - e^{-K_R T/J_v})}\right) \tag{7-20}$$

The characteristic equation of the closed-loop system is

$$z^2 + a_1 z + a_0 = 0 \tag{7-21}$$

where

$$a_1 = \frac{KK_P}{K_R^2}(TK_R - J_v + J_v e^{-K_R T/J_v}) - (1 + e^{-K_R T/J_v}) \tag{7-22}$$

$$a_0 = e^{-K_R T/J_v} + \frac{KK_P}{K_R^2}[J_v - (TK_R + J_v)e^{-K_R T/J_v}] \tag{7-23}$$

Substituting the known system parameters into Eqs. (7-22) and (7-23), we have

$$a_1 = 0.000012K(3.71 \times 10^5 T - 41822 + 41822e^{-8.871T}) - 1 - e^{-8.871T} \tag{7-24}$$

$$a_0 = e^{-8.871T} + 0.000012K[41822 - (3.71 \times 10^5 T + 41822)e^{-8.871T}] \tag{7-25}$$

For stability, the roots of Eq. (7-21) must all be inside the unit circle in the z-plane. Applying Jury's stability test to Eq. (7-21), we have the following requirements for stability:

$|a_0| < 1$

$$|e^{-8.871T} + 0.000012K[41822 - (3.71 \times 10^5 T + 41822)e^{-8.871T}]| < 1 \tag{7-26}$$

$F(1) = 1 + a_1 + a_0 > 0$

$$1 - e^{-8.871T} > 0 \tag{7-27}$$

$F(-1) = 1 - a_1 + a_0 > 0$

$$2(1 + e^{-8.871T}) + 0.000012K[83644(1 - e^{-8.871T}) - 3.71 \times 10^5 T(1 + e^{-8.871T})] > 0 \tag{7-28}$$

Since the sampling period T is positive, the condition in Eq. (7-27) is always satisfied. The stability boundaries formed by the conditions in Eqs. (7-26) and (7-28) are shown in the parameter plane of K versus T in Fig. 7-7. Clearly, for small values of T, the stability region is governed by the condition in Eq. (7-26), and for large values of T, Eq. (7-28) gives the dominant condition.

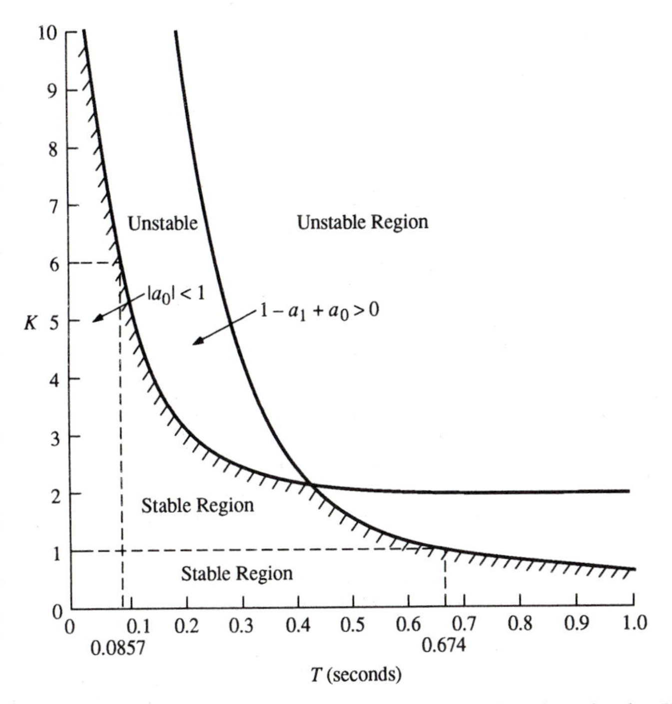

Figure 7-7. Stability boundaries and regions in the K-versus-T parameter plane of the digital space vehicle control system.

Comments on the Selection of *K* and *T*

Figure 7-7 is valuable for the selection of the values of K and T for stable operations of the digital control system. However, for satisfactory system performance, more information is needed for the selection of the proper values of K and T within the stable limits. From studies of continuous-data control systems, it is well known that increasing the value of K generally would reduce the damping ratio, increase the maximum overshoot, increase the natural undamped frequency and thus the bandwidth, and reduce the steady-state error if it is finite and nonzero. These properties also would carry over to discrete-data or digital control systems. As for the sampling period T, Fig. 7-7 shows that for a given K, increasing T would cause instability. On the other hand, a smaller sampling period would require a faster clock rate for the digital computer, which translates into more complex hardware and more cost.

For purposes of illustration, let us first select $K = 1$. Figure 7-7 shows that the maximum sampling period for stable operations is 0.674 s. Figure 7-8 shows the unit-step responses of the digital control system with $T = 0.1, 0.6$, and 0.674 s. These responses are shown as continuous functions by linking the sampled responses by smooth curves, since the true output of the digital system is a function of t. The properties of the unit-step responses of these and several

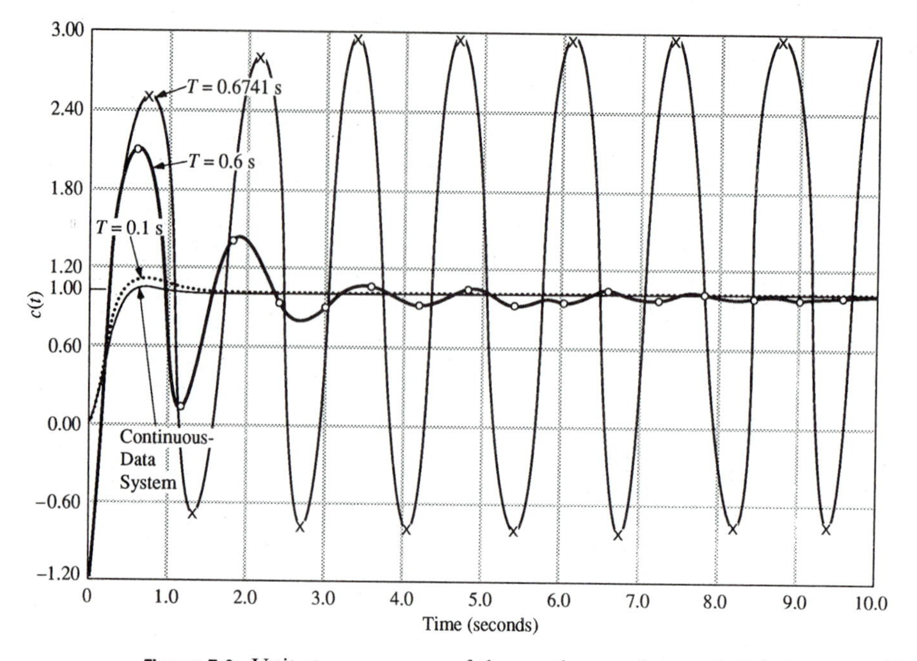

Figure 7-8. Unit-step responses of the continuous-data and digital space vehicle control system with $K = 1$.

other sampling periods with the ratios to the natural undamped frequency and the bandwidth (BW) of the continuous-data system are tabulated in Table 7-1.

The results in Table 7-1 show that the bandwidth or the natural undamped frequency ω_n of the continuous-data system can be used as a guideline for the selection of the sampling period of the digital control system. In the present case, the performance of the digital control system will be close to that of the continuous-data system if the sampling frequency ω_s is at least 15 times the BW or ω_n. For safer measures, sometimes the minimum sampling frequency may be chosen to be 20 times the bandwidth. It should be pointed out that, in the present case, when $K = 1$, the values of BW and ω_n happen to be equal. In general, the value of BW is often greater than ω_n, so that the former is often chosen as the reference guideline.

A better perspective can be gained by plotting the roots of the characteristic equation of the digital control system for the various values of T in the z-plane, as shown in Fig. 7-9. The positions of the roots for the incremental values of T chosen are connected by continuous curves to show that, as the value of T varies continuously, the roots will trace out continuous trajectories. Note that although the root trajectories shown are similar to the conventional root loci when a given system parameter varies, these root loci do not follow the properties of the conventional root loci, since T appears in a nonlinear fashion in the characteristic equation.

Table 7-1 Performance of Digital Space Vehicle Control System for Various Sampling Periods ($K = 1$)

T (s)	Maximum Overshoot	Characteristic Equation Roots	Sampling Frequency ω_s (rad/s)	ω_s/ω_n	ω_s/BW
0.700	unstable	$-0.4148 \quad -1.1961$	8.97		
0.674	1.497	$-0.4947 \quad -1.0000$	9.31	1.48	1.48
0.600	1.173	$-0.5823 \pm j0.3894$	10.47	1.67	1.67
0.200	0.277	$0.3482 \pm j0.5602$	31.40	5.00	5.00
0.100	0.137	$0.6310 \pm j0.3540$	62.80	10.00	10.00
0.067	0.097	$0.7407 \pm j0.2547$	94.30	15.00	15.00
0.050	0.082	$0.7995 \pm j0.1986$	124.00	20.00	20.00
0.033	0.066	$0.8622 \pm j0.1374$	188.60	30.00	30.00
0.010	0.051	$0.9566 \pm j0.0435$	628.00	100.00	100.00

Continuous-data system: $\omega_n = 6.28$ rad/s, BW $= 6.28$ rad/s, maximum overshoot $= 0.0432$.

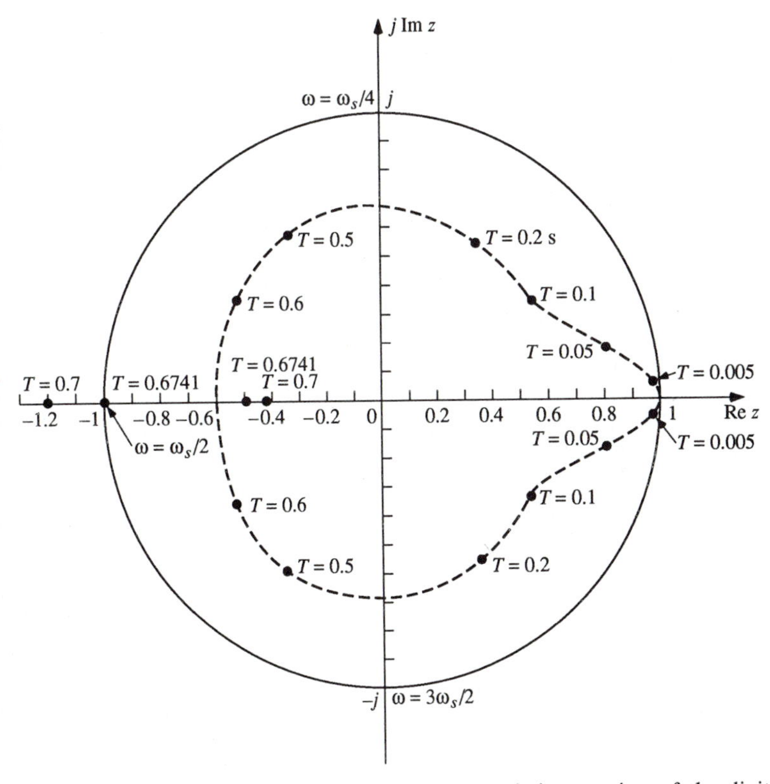

Figure 7-9. Root loci of the characteristic equation of the digital space vehicle control system when T varies. $K = 1$.

The root loci in Fig. 7-9 show that, as the sampling period increases, the roots move from the first and third quadrants into the second and fourth quadrants of the z-plane. From the mapping characteristics between the s-plane and the z-plane discussed in Sec. 3-3, this movement of roots corresponds to an increase in the frequency of the digital control system. When the characteristic equation roots get closer to the negative real axis in the z-plane, the system response becomes very oscillatory. When the roots are all real and negative, the response will be oscillatory with alternate positive and negative values, as shown in Fig. 7-8 for $T = 0.674$ s.

On the other hand, as the sampling period becomes smaller, the roots move closer to the $z = 1$ point in the first and fourth quadrants.

The conclusion from this study on the selection of the sampling period is that the roots of the digital control system should lie in the first and fourth quadrants and very close to the $z = 1$ point for the digital control system to emulate the continuous-data system. Sampling frequencies greater than 30 times the bandwidth of the continuous-data system no longer improve the performance of the digital system in any significant way.

Now let us select a larger value for K so that the continuous-data system is more oscillatory. Let $K = 6$; the continuous-data system has a maximum overshoot of 0.389. The BW is 22.49 rad/s and $\omega_n = 15.38$ rad/s. Figure 7-7 shows that the digital control system with $K = 6$ is stable for T less than 0.0857 s. Figure 7-10 shows the unit-step responses of the continuous-data

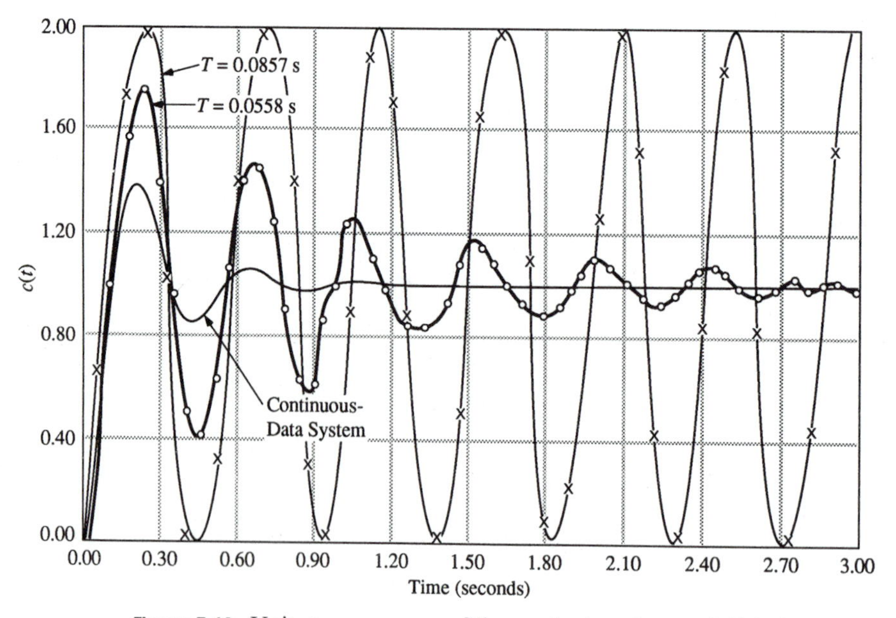

Figure 7-10. Unit-step responses of the continuous-data and digital space vehicle control system with $K = 6$.

Table 7-2 Performance of Digital Space Vehicle Control System with Various Sampling Periods ($K = 6$)

T (s)	Maximum Overshoot	Characteristic Equation Roots	Sampling Frequency ω_s (rad/s)	ω_s/ω_n	ω_s/BW
0.2000	unstable	$-0.8340 \pm j1.0322$	31.40	2.04	1.39
0.1000	unstable	$0.2563 \pm j1.0081$	62.80	4.08	2.78
0.0857	1.000	$0.3930 \pm j0.9203$	73.28	4.76	3.25
0.0558	0.769	$0.6475 \pm j0.6761$	112.45	7.31	5.00
0.0280	0.553	$0.8472 \pm j0.3755$	224.90	14.61	10.00
0.014	0.470	$0.9305 \pm j0.1970$	449.80	29.22	20.00
0.0093	0.440	$0.9554 \pm j0.1329$	674.70	43.84	30.00
0.0056	0.419	$0.9739 \pm j0.0810$	1124.50	73.07	50.00
0.0028	0.404	$0.9873 \pm j0.0409$	2249.00	146.13	100.00

Continuous-data system: $\omega_n = 15.38$ rad/s, BW $= 22.49$ rad/s, maximum overshoot $= 0.389$.

system and the digital control system with $T = 0.0558$ s and 0.0857 s. Table 7-2 gives the performance comparison of the digital control system with various values of T. Figure 7-11 illustrates the root loci of the characteristic equation in Eq. (7-21) for several values of T.

The results in Table 7-2 again show that a sampling frequency of at least 20 times the bandwidth of the continuous-data system is a good choice for the

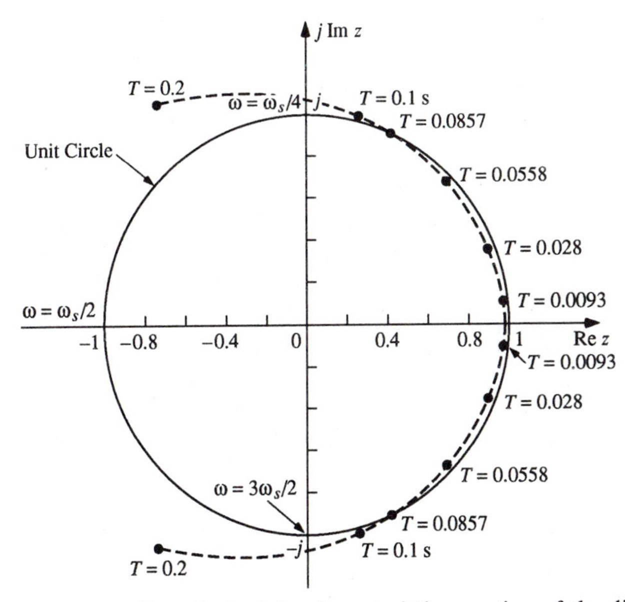

Figure 7-11. Root loci of the characteristic equation of the digital space vehicle control system when T varies. $K = 6$.

digital control system, from the standpoint of emulating the continuous-data system. With the higher value of K, the natural undamped frequency of the continuous-data system is relatively high, which may require the choice of the sampling frequency on the higher end of the 20-to-30-times BW guideline. Any sampling frequency greater than 30 times BW does not justify the improvements made on system performance. The root loci as T varies as shown in Fig. 7-11 indicate that the stable range of T results in characteristic equation roots that stay in the first and fourth quadrants of the z-plane. For $T \geq 0.857$ s the digital system is unstable. The step responses of the digital system shown in Fig. 7-10 support the results of Fig. 7-11. When $T = 0.857$ s, since the roots are complex conjugate on the unit circle, the step response is a sustained oscillation with positive values.

Steady-State Performance

The steady-state performance of the continuous-data control system can be studied by use of the final-value theorem of Laplace transformation. Similarly, the steady-state performance of the digital control system can be evaluated by use of the final-value theorem of z-transformation.

The steady-state error of the continuous-data space vehicle control system in Fig. 7-5 is defined as

$$e_{ss} = \lim_{t \to \infty} e(t) \tag{7-29}$$

where

$$e(t) = r(t) - c(t) \tag{7-30}$$

The Laplace transform of $e(t)$ is written from Fig. 7-5,

$$E(s) = \frac{s(J_v s + K_R)}{J_v s^2 + K_R s + KK_P} R(s) \tag{7-31}$$

Apparently, the steady-state error depends on the input $r(t)$. Applying the final-value theorem of Laplace transformation, Eq. (7-29) is written

$$e_{ss} = \lim_{s \to 0} sE(s) = \lim_{s \to 0} \frac{s^2(J_v s + K_R)}{J_v s^2 + K_R s + KK_P} R(s) \tag{7-32}$$

We must emphasize that the application of the final-value theorem depends on the assumption that the closed-loop system is stable. The values of e_{ss} are tabulated in Table 7-3 for unit-step function, unit-ramp function, and unit-parabolic function inputs.

Since z-transformation is used to analyze the digital control system, the steady-state error is conveniently defined at the sampling instants only. Thus,

$$\text{steady-state error } e_{ss}^* = \lim_{k \to \infty} e(kT) \tag{7-33}$$

Table 7-3 Steady-State Errors of the Continuous-Data Space Vehicle Control System

Types of Input	$R(s)$	e_{ss}
Unit-step function $u_s(t)$	$\frac{1}{s}$	0
Unit-ramp function $tu_s(t)$	$\frac{1}{s^2}$	$\frac{K_R}{KK_P}$
Unit-parabolic function $t^2u_s(t)/2$	$\frac{1}{s^3}$	∞

where $e(kT)$ is defined as the *error sequence*. If the closed-loop system is stable, applying the final-value theorem of Eq. (3-114) to Eq. (7-33), we get

$$e_{ss}^* = \lim_{z \to 1} (1 - z^{-1})E(z) = \lim_{z \to 1} (1 - z^{-1}) \frac{R(z)}{1 + G(z)} \tag{7-34}$$

where $G(z)$ is given in Eq. (7-20). Substituting Eq. (7-20) into the last equation, the steady-state errors of the digital control system at the sampling instants are obtained for the unit-step, unit-ramp, and unit-parabolic inputs and are tabulated in Table 7-4.

Comparing the results in Tables 7-3 and 7-4 we see that the steady-state errors of the continuous-data system and the digital system are identical for the same inputs. In general, the steady-state error of a digital control system may also depend on the sampling period.

Table 7-4 Steady-State Errors of the Digital Space Vehicle Control System

Types of Input	$R(z)$	e_{ss}^*
Unit-step function $u_s(t)$	$\frac{z}{z-1}$	0
Unit-ramp function $tu_s(t)$	$\frac{Tz}{(z-1)^2}$	$\frac{K_R}{KK_P}$
Unit-parabolic function $t^2u_s(t)$	$\frac{T^2z(z+1)}{(z-1)^3}$	∞

SUMMARY

We can summarize the findings from this simple illustrative example on the characteristics of time-domain responses of digital control systems.

1. For the same system configuration and parameters, the digital system with a sample-and-hold device is usually less stable than the continuous-data system. The reader should not have the misconception that continuous-data systems are always more stable than digital systems. Usually, with digital controllers, the digital control systems are more versatile and can have better responses than the continuous-data systems.
2. The performance of the digital control system depends on the value of the sampling period T. Larger sampling periods usually give rise to a higher overshoot in the step response and may eventually cause instability if T becomes too large. A proper way to select the sampling frequency is to use 20 to 30 times the bandwidth of the continuous-data system. A sampling frequency that is too high is wasteful in that a more complex and more costly digital controller may result.
3. The root-locus diagrams in Figs. 7-9 and 7-11 show that, for small values of T, the roots (and in general the dominant roots) of the characteristic equation of the digital control system are located very close to the $z = 1$ point in the z-plane. The concentration of the dominant roots near the $z = 1$ point often represents two practical difficulties in designing digital control systems. One is that controlling the system performance by working in such a small space near the $z = 1$ point becomes difficult. Such parameters as the constant-damping-ratio loci, the constant-frequency loci, etc., all become inaccurate when the roots are clustered together. Second, one may encounter numerical and accuracy problems when working with a relatively large number of nearly identical roots in the z-plane.

7-4 REVIEW OF STEADY-STATE ERROR ANALYSIS OF CONTINUOUS-DATA CONTROL SYSTEMS

Let us first review the steady-state error analysis of linear continuous-data control systems [1]. The system is represented by the block diagram in Fig. 7-12. The transfer function $G_p(s)H(s)$ is assumed to be of the following form:

$$G_p(s)H(s) = \frac{K(1 + T_a s)(1 + T_b s)\cdots(1 + T_m s)}{s^j(1 + T_1 s)(1 + T_2 s)\cdots(1 + T_n s)} e^{-T_d s} \tag{7-35}$$

where K and the T's are nonzero real or complex coefficients; T_d is the time delay in seconds. The *type* of the system is defined to equal j, the power of the s term.

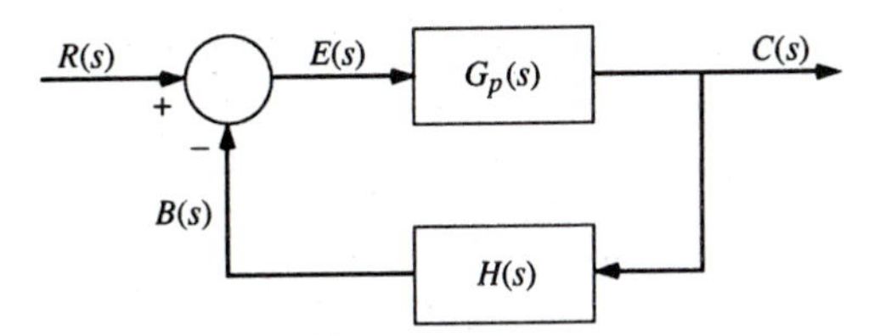

Figure 7-12. Continuous-data control system.

The steady-state error of the system in Fig. 7-12 is defined as

$$e_{ss} = \lim_{t \to \infty} e(t) \tag{7-36}$$

The error constants for the system in Fig. 7-12 are defined as follows.

Step-Error Constant (Step Input)

$$K_p = \lim_{s \to 0} G_p(s)H(s) \tag{7-37}$$

Ramp-Error Constant (Ramp Input)

$$K_v = \lim_{s \to 0} sG_p(s)H(s) \tag{7-38}$$

Parabolic-Error Constant (Parabolic Input)

$$K_a = \lim_{s \to 0} s^2 G_p(s)H(s) \tag{7-39}$$

Recall that these error constants are meaningful only for the individual type of input for which they are intended, and if the system is stable.

The relationships between the error constants and the type of system for step, ramp, and parabolic inputs with magnitude R are summarized in Table 7-5.

Table 7-5 Relationships Between Error Constants and Type of System for Continuous-Data Systems

	Step Input		Ramp Input		Parabolic Input	
Type of System	K_p	e_{ss}	K_v	e_{ss}	K_a	e_{ss}
0	K	$\frac{R}{1+K_p}$	0	$\frac{R}{K_v} = \infty$	0	$\frac{R}{K_a} = \infty$
1	∞	0	K	$\frac{R}{K}$	0	∞
2	∞	0	∞	0	K	$\frac{R}{K}$

7-5 STEADY-STATE ERROR ANALYSIS OF DIGITAL CONTROL SYSTEMS

We have defined and investigated the steady-state error of a digital control system in the last section using an illustrative example. In this section we shall conduct a formal discussion on the subject.

The error signal of a control system may be defined in a number of ways. In control systems, more often than not the difference between the reference input and the output is defined as the error, but there are exceptions. For the digital control system, we may use the block diagram of Fig. 7-13 which is obtained by inserting a sample-and-hold device in the forward path of the continuous-data system of Fig. 7-12. In this case, the signal $e(t)$ is still defined as the error. Thus,

$$e(t) = r(t) - b(t) \tag{7-40}$$

Since $e(t)$ is lost in the z-transform analysis, the sampled error $e^*(t)$ is used, and the steady-state error of the digital system at the sampling instants is defined as

$$e^*_{ss} = \lim_{t \to \infty} e^*(t) = \lim_{k \to \infty} e(kT) \tag{7-41}$$

Applying the final-value theorem of the z-transform, Eq. (3-114), the last equation becomes

$$e^*_{ss} = \lim_{z \to 1} (1 - z^{-1})E(z) \tag{7-42}$$

provided that the function $(1 - z^{-1})E(z)$ does not have any pole on or outside the unit circle $|z| = 1$ in the z-plane.

The steady-state error between the sampling instants can be determined by use of the modified z-transform; that is,

$$\lim_{\substack{k \to \infty \\ 0 \le m \le 1}} e(kT, m) = \lim_{\substack{z \to 1 \\ 0 \le m \le 1}} (1 - z^{-1})E(z, m) \tag{7-43}$$

where $E(z, m)$ is the modified z-transform of $e(t)$.

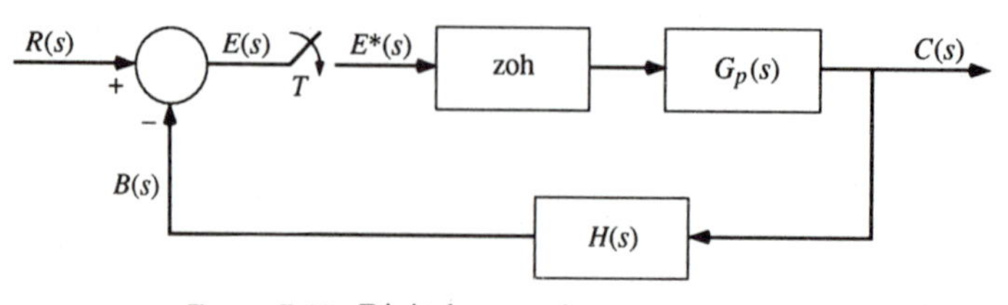

Figure 7-13. Digital control system.

For the system shown in Fig. 7-13, the z-transform of the error signal $e(t)$ is written as

$$E(z) = \frac{R(z)}{1 + GH(z)} \tag{7-44}$$

where

$$GH(z) = (1 - z^{-1})\mathscr{Z}\left[\frac{G_p(s)H(s)}{s}\right] \tag{7-45}$$

Substituting Eq. (7-44) into Eq. (7-42), we have

$$e_{ss}^* = \lim_{t\to\infty} e^*(t) = \lim_{z\to 1} (1 - z^{-1}) \frac{R(z)}{1 + GH(z)} \tag{7-46}$$

This expression shows that the steady-state error depends on the reference input $R(z)$, as well as the loop transfer function $GH(z)$. We shall consider the three basic types of input signal—step function, ramp function, and parabolic function—and define the error constants related to these inputs.

7-5-1 Steady-State Error Due to a Step Function Input—The Discrete Step-Error Constant

Let the reference input to the system of Fig. 7-13 be a step function of magnitude R; that is, $r(t) = Ru_s(t)$. The z-transform of $r(t)$ is

$$R(z) = \frac{Rz}{z - 1} \tag{7-47}$$

Substituting the last equation into Eq. (7-46), we get

$$e_{ss}^* = \lim_{z\to 1} \frac{R}{1 + GH(z)} = \frac{R}{1 + \lim_{z\to 1} GH(z)} \tag{7-48}$$

Let the *discrete step-error constant* be defined as

$$K_p^* = \lim_{z\to 1} GH(z) \tag{7-49}$$

Equation (7-48) becomes

$$e_{ss}^* = \frac{R}{1 + K_p^*} \tag{7-50}$$

Thus, we see that for the steady-state error due to a step function input to be zero, the step-error constant K_p^* must be infinite. This implies that the transfer function $GH(z)$ must have at least one pole at $z = 1$. It should be pointed out that *the discrete step-error constant K_p^* is meaningful only when the reference input is a step function.*

Let $G_p(s)H(s)$ be of the form in Eq. (7-35), then Eq. (7-45) can be written as

$$GH(z) = (1 - z^{-1})\mathscr{Z}\left[\frac{K(1 + T_a s)(1 + T_b s)\cdots(1 + T_m s)}{s^{j+1}(1 + T_1 s)(1 + T_2 s)\cdots(1 + T_n s)} e^{-T_d s}\right] \tag{7-51}$$

Applying partial-fraction expansion to the function inside the bracket, $GH(z)$ is written for a type-0 system as

$$\begin{aligned} GH(z) &= (1 - z^{-1})\mathscr{Z}\left[\frac{K}{s} + \text{terms due to the nonzero poles of } G_p(s)H(s)\right] \\ &= (1 - z^{-1})\left[\frac{Kz}{z-1} + \text{terms due to the nonzero poles of } G_p(s)H(s)\right] \end{aligned} \tag{7-52}$$

Note that the terms due to the nonzero poles of $G_p(s)H(s)$ do not contain the term $(z - 1)$ in the denominator. Thus, substituting Eq. (7-52) into Eq. (7-49), we get

$$K_p^* = \lim_{z \to 1} GH(z) = \lim_{z \to 1} (1 - z^{-1}) \frac{Kz}{z - 1} = K \tag{7-53}$$

Thus, for a type-0 process, the discrete step-error constant is the same as the step-error constant of the continuous-data counterpart.

For a type-1 process, $GH(z)$ is written as

$$GH(z) = (1 - z^{-1})\left[\frac{KTz}{(z-1)^2} + \frac{K_1 z}{z - 1} + \text{terms due to the nonzero poles}\right] \tag{7-54}$$

Thus, clearly, $K_p^* = \infty$. Similarly, we can show that, for all system types greater than 0, $K_p^* = \infty$.

7-5-2 Steady-State Error Due to a Ramp Function Input—The Discrete Ramp-Error Constant

For a ramp function input, $r(t) = Rtu_s(t)$, the z-transform of $r(t)$ is

$$R(z) = \frac{RTz}{(z - 1)^2} \tag{7-55}$$

Substituting $R(z)$ into Eq. (7-46), we have

$$\begin{aligned} e_{ss}^* &= \lim_{z \to 1} \frac{RT}{(z - 1)[1 + GH(z)]} \\ &= \frac{R}{\displaystyle\lim_{z \to 1} \frac{(z - 1)}{T} GH(z)} \end{aligned} \tag{7-56}$$

Let the *discrete ramp-error constant* be defined as

$$K_v^* = \frac{1}{T} \lim_{z \to 1} [(z-1)GH(z)] \tag{7-57}$$

then, Eq. (7-56) becomes

$$e_{ss}^* = \frac{R}{K_v^*} \tag{7-58}$$

The discrete ramp-error constant is meaningful only when the input to the system is a ramp function. Again, Eq. (7-58) is meaningful only if the application of the final-value theorem in Eq. (7-46) is valid. This implies that the closed-loop digital control system must be asymptotically stable.

Equation (7-58) shows that, for e_{ss}^* due to a ramp function input to be zero, K_v^* must equal infinity. From Eq. (7-57) we see that this is equivalent to the requirement that $(z-1)GH(z)$ has at least one pole at $z=1$, or $GH(z)$ has two poles at $z=1$. Again, *the discrete ramp-error constant K_v^* is meaningful only if the input is a ramp function.*

In a similar manner as in the last section, we can show that the discrete ramp-error constant is related to the system type as follows:

Type-0 system:	$K_v^* = 0$
Type-1 system:	$K_v^* = K$
Type-1 and higher type system:	$K_v^* = \infty$

7-5-3 Steady-State Error Due to a Parabolic Function Input—The Discrete Parabolic-Error Constant

For a parabolic function input, $r(t) = Rtu_s(t)/2$. The z-transform of $r(t)$ is

$$R(z) = \frac{RT^2 z(z+1)}{2(z-1)^3} \tag{7-59}$$

From Eq. (7-46), the steady-state error at the sampling instants is written

$$e_{ss}^* = \frac{T^2}{2} \lim_{z \to 1} \frac{R(z+1)}{(z-1)^2[1+GH(z)]} = \frac{R}{\lim_{z \to 1} \frac{(z-1)^2}{T^2} GH(z)} \tag{7-60}$$

provided the final-value theorem is valid.

Let the *discrete parabolic-error constant* be defined as

$$K_a^* = \frac{1}{T^2} \lim_{z \to 1} [(z-1)^2 GH(z)] \tag{7-61}$$

Then, Eq. (7-56) becomes

$$e_{ss}^* = \frac{R}{K_a^*} \tag{7-62}$$

We should point out that *the discrete parabolic-error constant is meaningful only for the parabolic function input* and does not apply to any other types of input. For the steady-state error in Eq. (7-58) to be zero, K_a^* must be infinite. This means that $GH(z)$ must have at least three poles at $z = 1$, while the closed-loop system is still stable.

The relationship between the parabolic-error constant and the system type is

Type-0 system:	$K_a^* = 0$
Type-1 system:	$K_a^* = 0$
Type-2 system:	$K_a^* = K$
Type-3 and higher type system:	$K_a^* = \infty$

7-5-4 Summary of Discrete Error Constants

From the foregoing discussions, we see that when the reference input of the digital control system shown in Fig. 7-13 is of the step, ramp, or parabolic type, the steady-state error of the system at the sampling instants due to each type of input depends on the discrete error constants K_p^*, K_v^*, and K_a^*, respectively. These discrete error constants are summarized as follows.

Step Input: Discrete Step-Error Constant

$$K_p^* = \lim_{z \to 1} GH(z) \tag{7-63}$$

Ramp Input: Discrete Ramp-Error Constant

$$K_v^* = \frac{1}{T} \lim_{z \to 1} (z - 1)GH(z) \tag{7-64}$$

Parabolic Input: Discrete Parabolic-Error Constant

$$K_a^* = \frac{1}{T^2} \lim_{z \to 1} (z - 1)^2 GH(z) \tag{7-65}$$

In a similar manner, the definitions of the discrete error constants can be extended to higher order input types if necessary.

Although Eqs. (7-64) and (7-65) show that K_v^* and K_a^* contain the sampling period T, when $GH(z)$ is substituted into the expressions, T is cancelled by the same parameter in the z-transform of the input. Thus, for the digital control system shown in Fig. 7-13, the steady-state error properties are identical to those of the continuous-data system in Fig. 7-12, and Table 7-5 can be used simply by replacing the error constants by the corresponding discrete error constants.

Let us emphasize that the development given above is for the system configuration of Fig. 7-13. The steady-state error analysis of any other digital control system configuration can be conducted in a straightforward manner using the final-value theorem.

Example 7-1

We can apply the discrete error constants to find the steady-state errors of the digital space vehicle control studied in Sec. 7-2. The open-loop transfer function of the system $G(z)$ is given in Eq. (7-20). From Eq. (7-20) it is apparent that $G(z)$ is of type 1. Thus, the discrete error constants are

$$K_p^* = \lim_{z \to 1} G(z) = \infty \tag{7-66}$$

$$K_v^* = \frac{1}{T} \lim_{z \to 1} (z - 1)G(z) = \frac{KK_P}{K_R} \tag{7-67}$$

$$K_a^* = \frac{1}{T^2} \lim_{z \to 1} (z - 1)^2 G(z) = 0 \tag{7-68}$$

Thus, the steady-state error e_{ss}^* is zero for a step input, K_R/KK_P for a unit ramp input, and ∞ for a parabolic input.

7-6 CORRELATION BETWEEN TIME RESPONSE AND ROOT LOCATIONS IN THE s-PLANE AND THE z-PLANE

The mapping between the s-plane and the z-plane was discussed in Sec. 3-3. In particular, the constant-damping factor, constant-damping ratio, and constant-frequency loci were established in the s- and z-planes. These loci in the z-plane are useful in predicting the performance of a digital control system.

The correlation between the locations of the characteristic equation roots in the s-plane and the transient response is well known for continuous-data control systems. For instance, complex conjugate roots in the left-half s-plane give rise to exponentially decaying sinusoidal responses; roots on the negative real axis correspond to monotonically decaying responses; and simple conjugate roots on the imaginary axis will give rise to undamped constant-amplitude sinusoidal oscillations. Multiple-order roots on the imaginary axis and roots in the right-half s-plane correspond to unstable responses.

Although we have established the relationship between the s-plane and the z-plane, the sampling operation in digital control systems does bring about conditions that require special attention. For instance, when the sampling theorem is not satisfied, the folding effect of sampling (Sec. 2-11) could distort

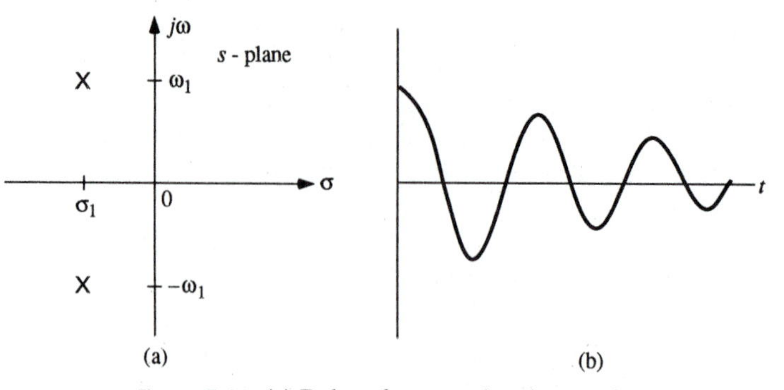

Figure 7-14. (a) Poles of a second-order continuous-data system. (b) Natural response of the system.

the true response of the system. Figure 7-14(a) shows the poles of the transfer function of a second-order continuous-data system. The natural response of the system is shown in Fig. 7-14(b). Now if the system is subjected to sampling with the sampling frequency $\omega_s < 2\omega_1$, the sampling operation generates an infinite number of poles in the s-plane at $s = \sigma_1 \pm j\omega_1 + jn\omega_s$, $n = \pm 1, \pm 2, \ldots$. As shown in Fig. 7-15(a), the sampling operation folds the poles back into the

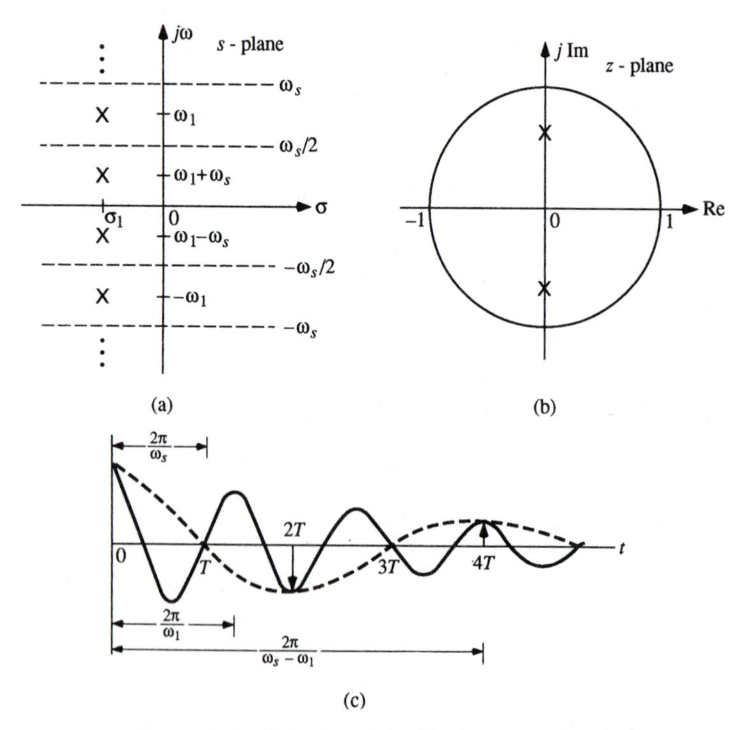

Figure 7-15. Pole locations in the s- and z-planes showing the effect of folding of frequency.

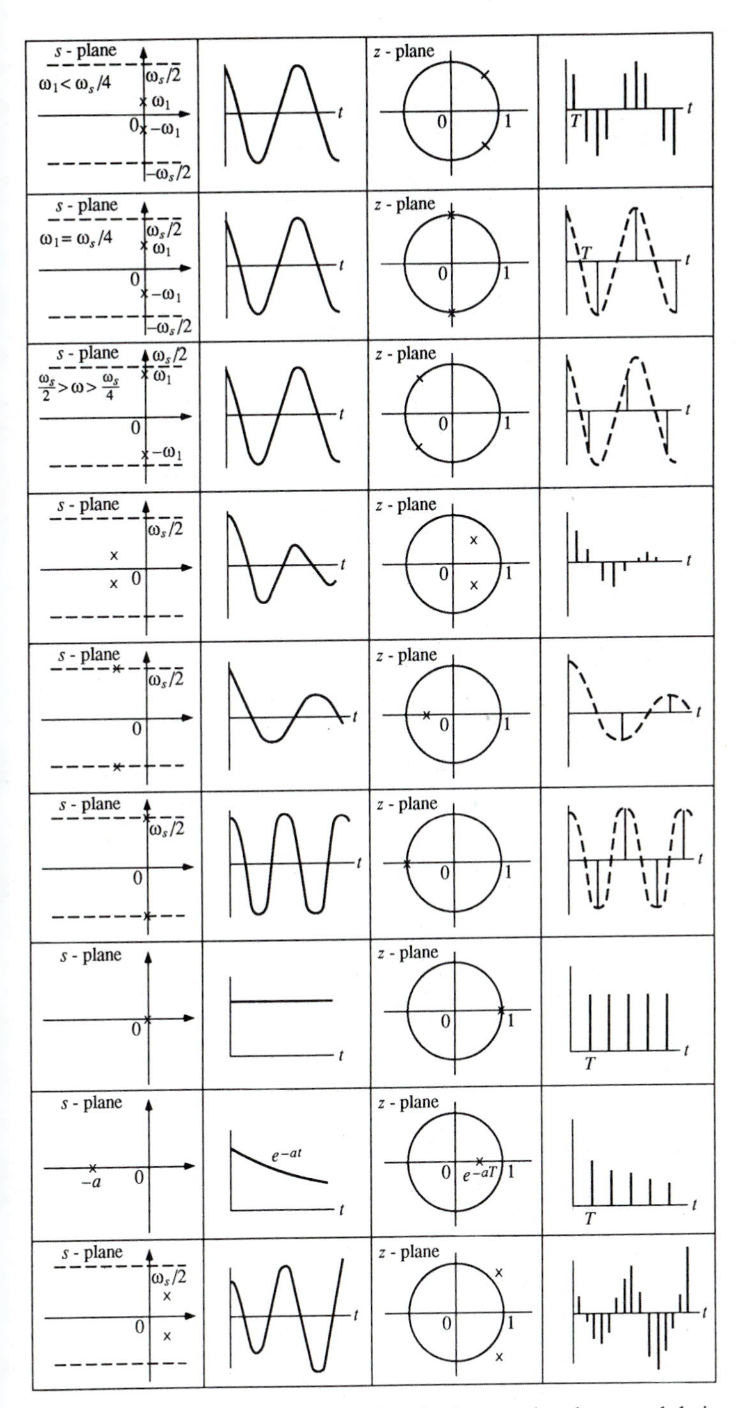

Figure 7-16. Root locations in the s- and z-planes and their corresponding time responses.

primary strip, $-\omega_s/2 < \omega < \omega_s/2$, so that the net effect is equivalent to having a system originally with poles at $s = \sigma_1 \pm j(\omega_s - \omega_1)$. The poles are mapped onto the z-plane as shown in Fig. 7-15(b). Figure 7-15(c) illustrates the sampled signal that demonstrates the effect of frequency folding. The folding effect makes the sampled system appear as if the frequency were equal to $\omega_s - \omega_1$ rather than ω_1.

Figure 7-16 illustrates several cases of root locations of a second-order system in the s- and z-planes and their corresponding time responses.

Although only second-order characteristic equation roots are illustrated in the preceding discussion of the effects of root locations in the z-plane, the correlation between the time response and the second-order roots can generally be applied to high-order systems that have second-order dominant roots.

7-7 CONSTANT-DAMPING-FACTOR AND CONSTANT-DAMPING-RATIO LOCI

The relative stability problem in the z-plane is essentially the study of the location of the characteristic equation roots with respect to the constant-damping-factor and constant-damping-ratio loci. These loci were introduced in Sec. 3-3. The constant-damping-factor loci in the z-plane are a family of concentric circles centered at the origin. The radius of the circle corresponding to a given damping factor σ_1 is $\exp(-\sigma_1 T)$, as shown in Fig. 7-17. If the design requires the system to have a maximum damping factor of σ_1 or a minimum time constant of $1/\sigma_1$, then all the characteristic equation roots of the system must lie inside the circle $|z| = \exp(-\sigma_1 T)$ in the z-plane, as shown in Fig. 7-17.

The constant-damping-ratio loci in the z-plane are a family of logarithmic spirals, except for $\zeta = 0$ and $\zeta = 1$. Typical constant-ζ loci for $\zeta = 0.5$ are shown in Fig. 7-18 for the s- and z-planes. If a certain maximum damping ratio is specified in design, all the characteristic equation roots must lie to the left of

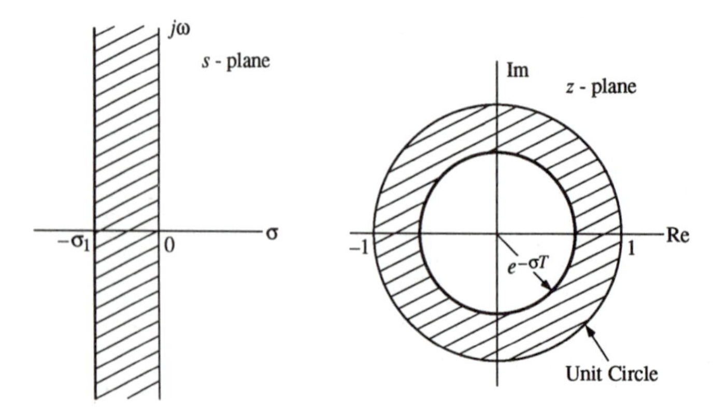

Figure 7-17. Constant-damping loci in the s- and z-planes.

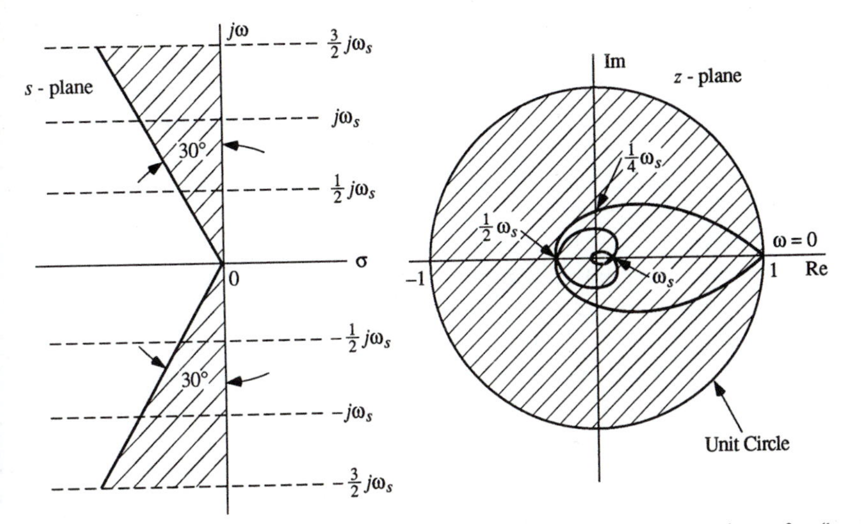

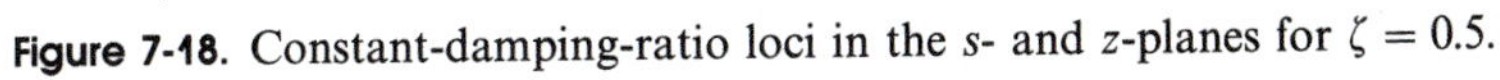

Figure 7-18. Constant-damping-ratio loci in the s- and z-planes for $\zeta = 0.5$.

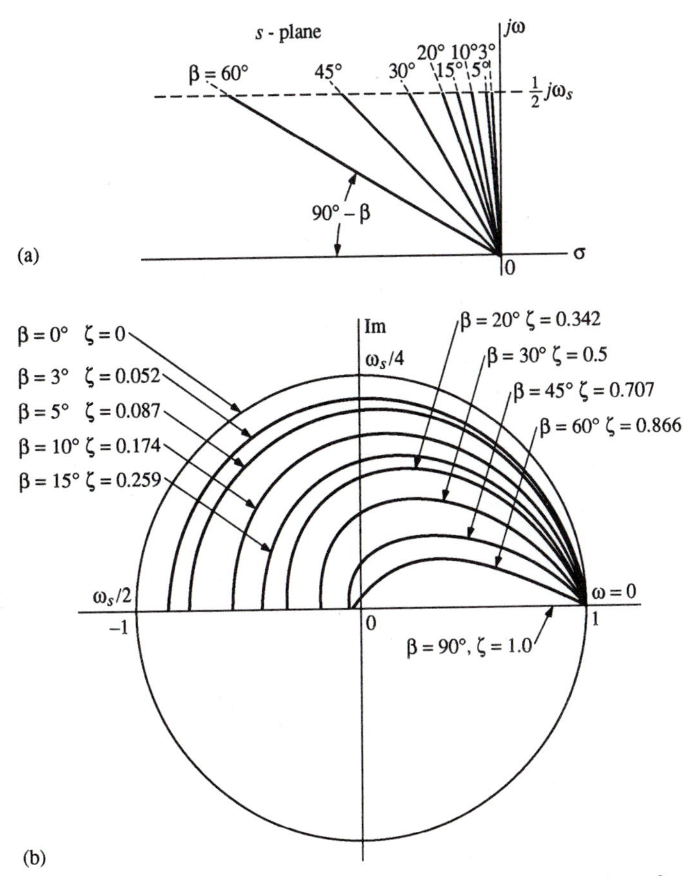

Figure 7-19. (a) The constant-ζ loci in the s-plane for the periodic strip of 0 to $j\omega_s/2$. (b) The constant-ζ loci in the z-plane corresponding to the constant-ζ loci in the s-plane in (a).

the constant-damping-ratio line in the s-plane, or inside the constant-damping-ratio logarithmic spiral in the z-plane, as shown by the shaded region in Fig. 7-18. The entire interior of the unit circle of the z-plane may seem to be excluded, so that the design specification on the damping ratio could never be met. The fact is that most control systems have low-pass filter characteristics, so that in most practical situations working with only the primary strip in the s-plane is sufficient. This assumes that the sampling theorem is satisfied so that the highest frequency component in the system is less than $2\omega_s$. Then, Fig. 7-19 illustrates the constant-damping-ratio loci in the s- and z-planes for several values of ζ for the positive half of the primary strip. The constant-ζ loci for $-\omega_s/2 < \omega < 0$ are just the mirror image of the curves shown in Fig. 7-19, with the mirror placed on the real axis.

7-8 DOMINANT CHARACTERISTIC EQUATION ROOTS

Discussions in the preceding sections show that the location of the characteristic equation roots in the z-plane has pronounced effects on the transient response of digital control systems. In general, some of the roots, due to their location in the z-plane, have more effect on the system response than the others. For analysis and design purposes, it is important to sort out the roots that have a dominant effect on the transient response and call these the *dominant roots.*

In the s-plane, the roots that are closest to the $j\omega$ axis in the left half plane are the dominant ones, since these correspond to time response with the slowest decay. Roots that are far away from the $j\omega$ axis, relative to the dominant ones, correspond to fast decaying time responses. Figure 7-20 shows the preferred locations of the dominant roots and the insignificant roots in the s-plane for continuous-data control systems. Qualitatively, the preferred region for the

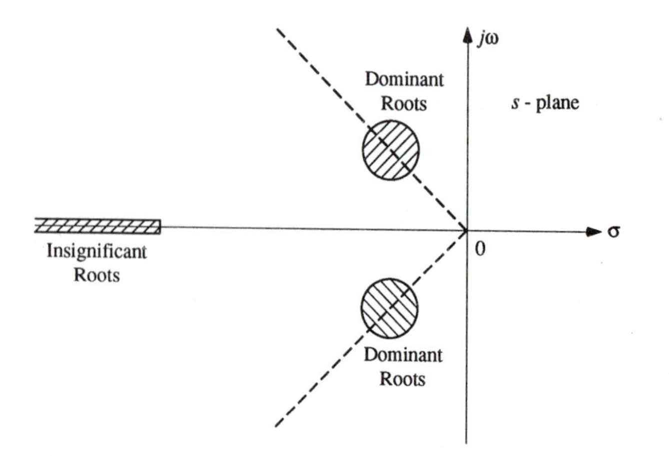

Figure 7-20. Regions for dominant and insignificant roots in the s-plane.

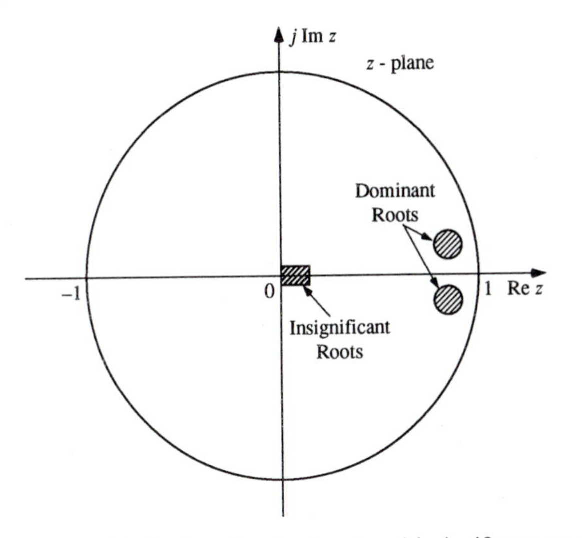

Figure 7-21. Regions for dominant and insignificant roots in the z-plane.

dominant roots is shown arbitrarily as a circle centered at the $\zeta = 0.707$ line. The insignificant roots are shown to be real, although they also could be complex conjugate.

In the z-plane, the dominant roots are the ones that are closest to and inside the unit circle. The roots that are near the origin of the z-plane are the least significant, since as $s \to -\infty$, $z \to 0$. Figure 7-21 shows the preferred locations of the dominant roots and the insignificant roots from the design standpoint. The insignificant roots can also be complex, so long as they are close to the origin. The negative real axis should be avoided, since the corresponding time response is oscillatory with alternate signs.

The difference between neglecting the insignificant roots in the s-plane and the z-plane should be pointed out. In the s-plane, the roots with large negative real parts or negative real roots can be neglected after the zero-frequency gain of the transfer function is adjusted to preserve the steady-state behavior. In the z-domain, the roots near the origin are insignificant only from the standpoint of the maximum overshoot and damping. However, these roots cannot be discarded indiscriminantly, since the excess number of poles over zeros of a closed-loop transfer function has a time-delay effect initially in the time response. For example, adding two poles at $z = 0$ to the closed-loop transfer function of any stable system would not affect the maximum overshoot, but the time response will have an additional delay of two sampling periods.

Therefore, the proper way of simplifying a high-order system with insignificant closed-loop transfer function poles is *not to discard these poles but replace them by poles at* $z = 0$. This will still simplify the analysis considerably, since the poles at $z = 0$ correspond to pure time delays.

7-9 EFFECT OF POLE-ZERO CONFIGURATIONS IN THE z-PLANE ON THE MAXIMUM OVERSHOOT AND PEAK TIME

In the preceding sections we have described the relationship between the characteristic equation roots in the z-plane and the transient response of a second-order digital control system. When the characteristic equation has complex roots inside the unit circle in the z-plane, the unit-step response of the system will be oscillatory with positive damping; that is, the response will decay as time increases. In general, the complex roots that are inside and closest to the unit circle give rise to more oscillatory or less damped responses. It would be useful to determine the relationship between the pole-zero configuration of the closed-loop transfer function and the maximum overshoot and peak time T_{max} of digital control systems.

For the second-order prototype system in Eq. (7-4), the maximum overshoot and peak time are given as follows [1]:

$$\text{maximum overshoot} = 1 + \exp\left[\frac{-\zeta\pi}{\sqrt{1-\zeta^2}}\right] \tag{7-69}$$

and

$$T_{max} = \frac{\pi}{\omega_n\sqrt{1-\zeta^2}} \tag{7-70}$$

For systems of order higher than two, no simple relations can be obtained between the maximum overshoot T_{max} and the pole-zero configuration of the closed-loop transfer function. However, if the system can be characterized by only a pair of dominant poles, and the other poles and zeros are far to the left in the s-plane, the effects of these insignificant poles and zeros on the transient response are negligible. Under this condition, the maximum overshoot and T_{max} can be approximated by Eqs. (7-69) and (7-70) in conjunction with the second-order transfer function in Eq. (7-4). For example, consider the following transfer function of a fourth-order system,

$$\frac{C(s)}{R(s)} = \frac{K}{(s+p_1)(s+p_2)(s^2+2\zeta\omega_n s+\omega_n^2)} \tag{7-71}$$

where p_1 and p_2 are real constants. If p_1 and p_2 are at least six times greater than ω_n, the two poles at $-p_1$ and $-p_2$ will contribute insignificantly to the transient response, and the poles of the second-order term are the dominant poles. However, we cannot simply throw away the two terms, $(s + p_1)$ and $(s + p_2)$ in Eq. (7-71), as *they still affect the steady-state performance of the system.* The proper way to approximate the transfer function in Eq. (7-71) by a second-order transfer function is to write

$$\frac{C(s)}{R(s)} = \frac{K}{p_1p_2(1+s/p_1)(1+s/p_2)(s^2+2\zeta\omega_n s+\omega_n^2)}$$

$$\cong \frac{K}{p_1p_2(s^2+2\zeta\omega_n s+\omega_n^2)} \tag{7-72}$$

For a digital control system, the problem of establishing the relationships between maximum overshoot and peak time and the pole-zero configuration of the closed-loop transfer function is more complex. One reason is that when z-transformation is used, depending on the sampling period T, the output sequence $c(kT)$ may not be an accurate representation of $c(t)$, so that the true value of c_{max} cannot be obtained from $c(kT)$. The second problem is that a simple second-order digital control system usually cannot be modeled by a closed-loop transfer function with just two poles, as in the case of Eq. (7-4) for continuous-data systems.

Let us select the following model as the closed-loop transfer function of a *second-order prototype digital control system*:

$$M(z) = \frac{C(z)}{R(z)} = \frac{K(z - z_1)}{(z - p_1)(z - p_1^*)} \tag{7-73}$$

where z_1 is a real zero, and p_1 and p_1^* are complex conjugate poles. The value of K is chosen to be

$$K = \frac{(1 - p_1)(1 - p_1^*)}{(1 - z_1)} \tag{7-74}$$

so that the steady-state error e_{ss}^* is zero for a unit-step input.

The transfer function in Eq. (7-73) contains a zero at z_1, since most closed-loop continuous-data system transfer functions without zeros when sampled will produce at least one zero in the z-transform transfer functions. A typical example is given in Eq. (7-20), or the reader may simply refer to the z-transform pairs in Appendix C. We also specify that the poles in Eq. (7-73) are complex, since positive real poles in the z-plane would not produce any overshoot.

When the closed-loop digital control system represented by Eq. (7-73) is subject to a unit-step input, the output transform $C(z)$ is written

$$C(z) = \frac{Kz(z - z_1)}{(z - 1)(z - p_1)(z - p_1^*)} \tag{7-75}$$

The output sequence is obtained by applying the inversion formula of the z-transform, Eq. (3-74). We have

$$c(kT) = \frac{1}{2\pi j} \oint_{\Gamma} \frac{Kz(z - z_1)}{(z - 1)(z - p_1)(z - p_1^*)} z^{k-1}\, dz \tag{7-76}$$

where Γ is a closed contour that encloses all the singularities of the integrand. Applying the residues theorem of complex variables to Eq. (7-76), $c(kT)$ is written

$$c(kT) = 1 + 2\left|\frac{K(p_1 - z_1)}{(p_1 - 1)(p_1 - p_1^*)}\right| |p_1|^k \cos(k\phi_1 + \theta_1) \tag{7-77}$$

where

$$\phi_1 = \angle p_1 \tag{7-78}$$

$$\theta_1 = \angle(p_1 - z_1) - \angle(p_1 - 1) - \pi/2 \tag{7-79}$$

The pole-zero configuration of Eq. (7-73) is shown in Fig. 7-22. From the location of the poles and zeros, we define the angle α, as shown in Fig. 7-22, as

$$\pm\alpha = \angle(p_1 - z_1) - \angle(p_1 - 1) + \pi/2 \tag{7-80}$$

where the $\pm$ sign in front of α is selected according to the situations illustrated in Fig. 7-22. That is, *α has a $+$ sign if it is measured in the counterclockwise direction from the perpendicular drop line from z_1 as shown in Fig. 7-22(a) and a negative sign if it is measured in the clockwise direction as shown in Fig. 7-22(b).* From Eq. (7-79), we see that θ_1 is related to α through

$$\theta_1 = \pm\alpha - \pi \tag{7-81}$$

Also, we can readily show that the following relationship holds between α and the closed-loop pole-zero locations:

$$|\sec \alpha| = 2\left|\frac{K(p_1 - z_1)}{(p_1 - 1)(p_1 - p_1^*)}\right| \tag{7-82}$$

where K is given in Eq. (7-74). Substituting Eqs. (7-81) and (7-82) into Eq. (7-77), the output sequence is written

$$c(kT) = 1 + |\sec \alpha| \times |p_1|^k \cos(k\phi_1 \pm \alpha - \pi) \tag{7-83}$$

Equations (7-77) and (7-83) give the response of $c(t)$ only at the sampling instants, and in principle, once $c(t)$ is sampled, information between the sampling instants is lost. In other words, strictly, we cannot reconstruct $c(t)$ from $c(kT)$. However, we can approximate $c(t)$ by a function that passes through all the points of $c(kT)$, and if the sampling period is sufficiently small, the approximation will be a good one.

Let $t = kT$, then

$$|p_1|^k = |p_1|^{t/T} = e^{-\zeta\omega_n t} \tag{7-84}$$

and

$$\phi_1 = \arg(p_1) = \omega T = \omega_n\sqrt{1-\zeta^2}\,T = \tan^{-1}\left(\frac{\text{Im}(p_1)}{\text{Re}(p_1)}\right) \tag{7-85}$$

where $\text{Re}(p_1)$ and $\text{Im}(p_1)$ are the real and imaginary parts of p_1, respectively. Therefore, a continuous-time function that passes through the points of $c(kT)$ is

$$c(t) = 1 + |\sec \alpha| e^{-\zeta\omega_n t} \cos(\omega_n\sqrt{1-\zeta^2}\,t \pm \alpha - \pi) \tag{7-86}$$

The maximum overshoot of the time response $c(t)$ in Eq. (7-86) is now used to approximate the maximum overshoot of the digital control system whose closed-loop transfer function is given by Eq. (7-73). The maximum overshoot

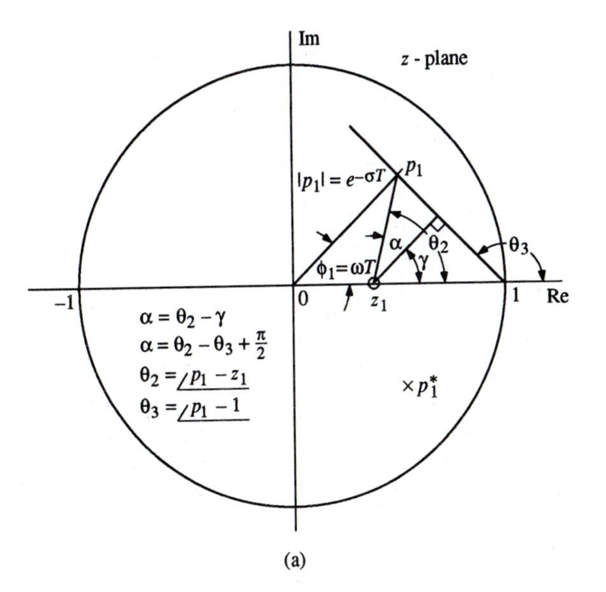

(a)

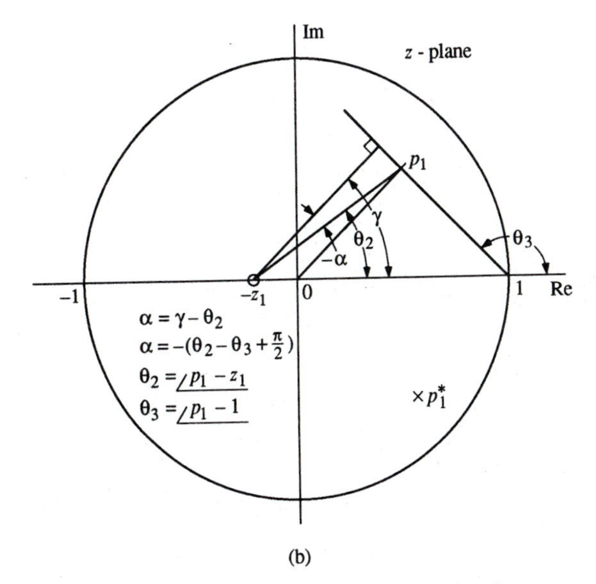

(b)

Figure 7-22. Geometrical representation of α for a second-order sampled-data system with closed-loop transfer function given by Eq. (7-73). (a) α has positive sign. (b) α has negative sign.

of $c(t)$ in Eq. (7-86) is determined by taking the derivative on both sides of the equation with respect to t and setting $dc(t)/dt$ to zero. This leads to the following relationships:

$$\tan\left(\omega_n\sqrt{1-\zeta^2}\,t \pm \alpha - \pi\right) = \frac{-\zeta}{\sqrt{1-\zeta^2}} \tag{7-87}$$

The time t solved from the last equation is the peak time T_{max} of $c(t)$,

$$T_{max} = \frac{1}{\omega_n\sqrt{1-\zeta^2}}\left(\tan^{-1}\frac{-\zeta}{\sqrt{1-\zeta^2}} \mp \alpha + \pi\right) \tag{7-88}$$

Note that the sign of α is now reversed, which means that if α has a positive sign, as determined from the situation in Fig. 7-22(a), then the negative sign should be used in front of α in Eq. (7-88). On the other hand, if α has a negative sign, then $+\alpha$ should be used.

Substituting Eq. (7-88) into Eq. (7-86) and simplifying, the maximum value of $c(t)$ is

$$c_{max} = 1 + \sqrt{1-\zeta^2}\,|\sec\alpha|\exp\left\{\frac{-\zeta}{\sqrt{1-\zeta^2}}\left(\tan^{-1}\frac{-\zeta}{\sqrt{1-\zeta^2}} \mp \alpha + \pi\right)\right\} \tag{7-89}$$

Hence, the maximum overshoot of $c(t)$, which is used to approximate the maximum overshoot of $c^*(t)$, is

$$\text{maximum overshoot} = \sqrt{1-\zeta^2}\,|\sec\alpha|\exp\left\{\frac{-\zeta}{\sqrt{1-\zeta^2}}\left(\tan^{-1}\frac{-\zeta}{\sqrt{1-\zeta^2}} \mp \alpha + \pi\right)\right\} \tag{7-90}$$

We can readily see from Eq. (7-90) that the maximum overshoot is uniquely specified by the damping ratio ζ and the angle α defined in Fig. 7-22.

The percent of maximum overshoot given by Eq. (7-90) is plotted as a function of ζ and α in Fig. 7-23. The angle α and its sign are determined from the pole-zero location of the closed-loop transfer function as described in Fig. 7-22. The value of ζ can be calculated by using Eq. (7-84) once the complex pole p_1 is known. From Eq. (7-84),

$$|p_1| = e^{-\zeta\omega_n T} = \exp\left(\frac{-\zeta\phi_1}{\sqrt{1-\zeta^2}}\right) = \sqrt{[\text{Re}\,(p_1)]^2 + [\text{Im}\,(p_1)]^2} \tag{7-91}$$

where ϕ_1 is measured in radians.

Thus, given p_1, ϕ_1 is determined from Eq. (7-85), and then ζ is calculated from Eq. (7-91), and the result is

$$\zeta = \left(\frac{(\ln|p_1|)^2}{(\ln|p_1|)^2 + \phi_1^2}\right)^{1/2} \tag{7-92}$$

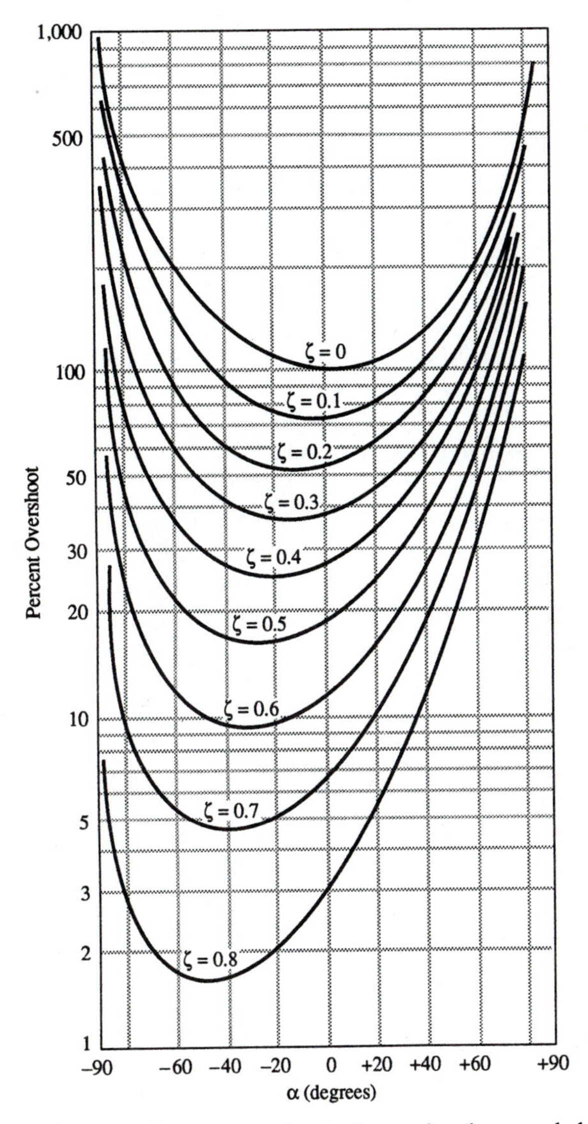

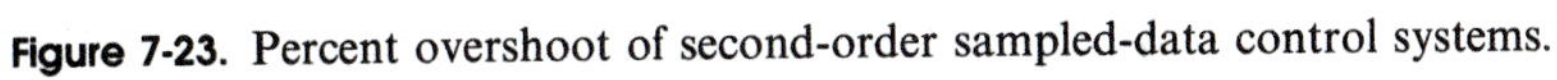
Figure 7-23. Percent overshoot of second-order sampled-data control systems.

The natural undamped frequency ω_n is determined from Eq. (7-91),

$$\omega_n = -\frac{\ln|p_1|}{\zeta T} \tag{7-93}$$

The peak time $T_{\max}$ is found by using Eq. (7-88), or

$$T_{\max} = \frac{T}{\phi_1}\left(\tan^{-1}\frac{-\zeta}{\sqrt{1-\zeta^2}} \mp \alpha + \pi\right) \tag{7-94}$$

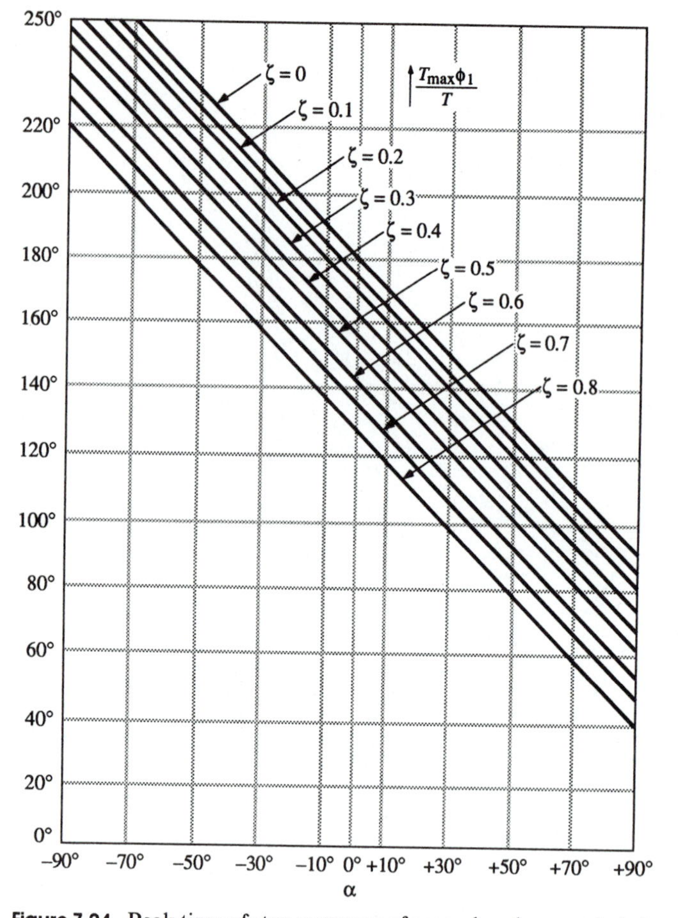

Figure 7-24. Peak time of step response of second-order sampled-data systems versus α.

where the sign of α should be chosen according to Fig. 7-22 and the convention described previously.

Figure 7-24 shows the curves of T_{max}, normalized in $T_{max}\phi_1/T$, plotted as functions of ζ and α.

For better accuracy, Eqs. (7-90) and (7-94) should be used for the calculations of c_{max} and T_{max}. Figures 7-22 and 7-23 are valuable in indicating the effects of moving the poles p_1, p_1^*, and z_1. From Fig. 7-22 we see that the maximum overshoot can be extremely high if α approaches $+90°$, even for large values of ζ. A typical situation would be when the complex conjugate poles are in the second and third quadrants with low damping, and the zero is in the right half plane near the $z = 1$ point. For example, the following closed-loop transfer function has poles at $z = -0.9 + j0.1$ and $-0.9 - j0.1$, and a zero at $z = 0.9$. We can show that the maximum overshoot is 1800 percent and yet the system is stable:

$$\frac{C(z)}{R(z)} = \frac{36.2(z - 0.9)}{z^2 + 1.8z + 0.82} \tag{7-95}$$

It should be emphasized again that the time response of Eq. (7-86) is a closed approximation of $c^*(t)$ only if the sampling frequency is sufficiently high. In terms of pole location, the complex poles p_1 and p_1^* must be located in the first and the fourth quadrants in the z-plane, respectively.

The method described above can be applied to high-order systems having two complex dominant poles and a zero located anywhere on the real axis; all other poles and zeros should be located near the origin in the z-plane, and thus are insignificant as far as the transient response is concerned.

Example 7-2

Consider the digital space vehicle control system shown in Fig. 7-6. The closed-loop transfer function of the system when $K = 1$ and $T = 0.1$ s is

$$\frac{C(z)}{R(z)} = \frac{0.4986(z + 0.7453)}{z^2 - 1.262z + 0.5235} \tag{7-96}$$

where the poles are at $z = 0.631 + j0.354$ and $z = 0.631 - j0.354$. The pole-zero configuration of $C(z)/R(z)$ is shown in Fig. 7-25. In this case, according to the

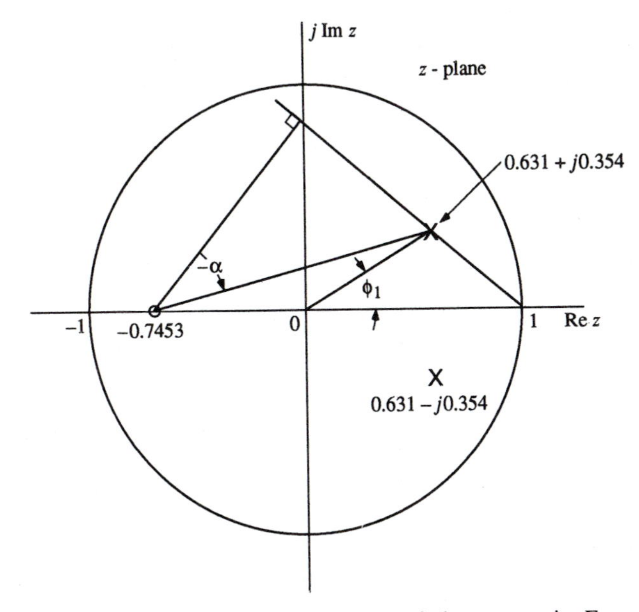

Figure 7-25. Pole-zero configuration of the system in Example 7-2. The sign of α is negative.

convention established in Fig. 7-22, α should have a negative sign in Eq. (7-86), and thus a positive sign in Eq. (7-90). From Eq. (7-85),

$$\phi_1 = \tan^{-1}\left(\frac{\mathrm{Im}\,(p_1)}{\mathrm{Re}\,(p_1)}\right) = \tan^{-1}\left(\frac{0.354}{0.631}\right) = 29.3^\circ = 0.511 \text{ rad} \tag{7-97}$$

From Eq. (7-91),

$$|p_1| = \sqrt{(0.354)^2 + (0.631)^2} = 0.7236 \tag{7-98}$$

Thus, we obtain from Eq. (7-91),

$$\zeta = 0.5347$$

The angle α is calculated from Fig. 7-22:

$$\alpha = -31.75^\circ$$

The maximum overshoot is determined from Eq. (7-90),

$$\text{maximum overshoot} = 0.1378 \tag{7-99}$$

This is extremely close to the value of 0.137 obtained earlier in Table 7-1. From Eq. (7-94) the peak time $T_{\max}$ is determined to be

$$T_{\max} = \frac{T}{\phi_1}\left(\tan^{-1}\frac{-\zeta}{\sqrt{1-\zeta^2}} + \alpha + \pi\right) = 0.6123 \text{ s} \tag{7-100}$$

Earlier we found that $c_{\max}$ occurs at the seventh sampling instant, which is 0.7 s.

Example 7-3

Consider the same digital space vehicle control system as in Example 7-2. Let the sampling period be 0.6 s and $K = 1$. The closed-loop transfer function is

$$\frac{C(z)}{R(z)} = \frac{2.1696(z + 0.2239)}{z^2 + 1.1647z + 0.4908} \tag{7-101}$$

where the poles are at $z = -0.5823 + j0.3894$ and $z = -0.5823 - j0.3894$. The pole-zero configuration of $C(z)/R(z)$ is shown in Fig. 7-26. From Eq. (7-85),

$$\phi_1 = 146.23^\circ = 2.552 \text{ rad} \tag{7-102}$$

Equation (7-91) gives

$$|p_1| = 0.7005 \tag{7-103}$$

The damping ratio ζ is determined using Eq. (7-92),

$$\zeta = 0.138 \tag{7-104}$$

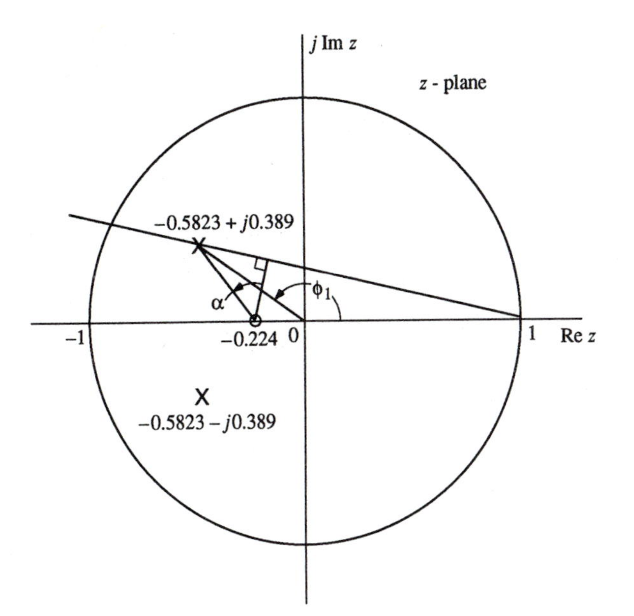

Figure 7-26. Pole-zero configuration of the system in Example 7-3. The sign of α is positive.

From Fig. 7-26, α is calculated to be 56.44°. In the present case, α has a positive sign in Eq. (7-86), and thus a negative sign in Eq. (7-90). The maximum overshoot is obtained from Eq. (7-90):

$$\text{maximum overshoot} = 1.35 \tag{7-105}$$

This compares well with the result obtained in Table 7-1, which is 1.173. The peak time is found from Eq. (7-94),

$$T_{\max} = 0.4745 \text{ s} \tag{7-106}$$

The result obtained in Fig. 7-8 shows that $c_{\max}$ occurs at the first sampling instant after $t = 0$, which is 0.6 s. Thus, with a resolution of $T = 0.6$ s, the result in Eq. (7-106) is the best that can be hoped for.

7-10 DEADBEAT RESPONSE AT THE SAMPLING INSTANTS

Control systems are often designed with the objective that the output response should reach the desired reference value as quickly as possible and without any overshoot. This type of response is generally referred to as a *deadbeat response.* Figure 7-27(a) illustrates a deadbeat response to a unit-step input as a function of t. A typical discrete-data deadbeat response is shown in Fig. 7-27(b). The

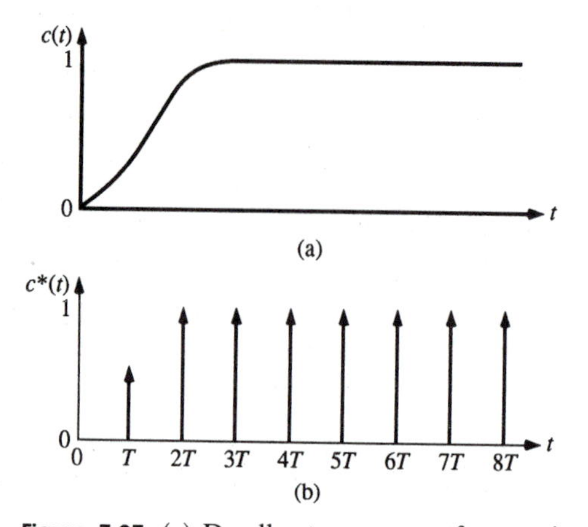

Figure 7-27. (a) Deadbeat response of a continuous-data system to a unit-step input. (b) Deadbeat response of a discrete-data system to a unit-step input.

output sequence is shown to follow the reference unit-step input after two sampling instants. Note that if the response in Fig. 7-27(b) is a sampled representation of a continuous-data response, then the deadbeat characteristics at the sampling instants *do not* guarantee that the continuous-data response is also deadbeat.

For a discrete-data system to present a deadbeat response at the sampling instants, the closed-loop transfer function should be of the form:

$$\frac{C(z)}{R(z)} = \frac{1}{z^N} \tag{7-107}$$

where N is a positive integer. Thus, for a unit-step function input, for example, the output $C(z)$ is expanded as

$$C(z) = z^{-N} + z^{-N-1} + z^{-N-2} + \cdots \tag{7-108}$$

which corresponds to the response shown in Fig. 7-27(b) for $N = 2$.

The design of deadbeat discrete-data control systems will be covered in Chapter 10.

7-11 ROOT LOCI FOR DIGITAL CONTROL SYSTEMS

The root-locus method for the s-plane has been established as a useful tool for the analysis and design of linear time-invariant continuous-data control systems. The root-locus diagram of a continuous-data control system is essentially a plot of the loci of the roots of the characteristic equation in the s-plane as a function of a real parameter K which varies from $-\infty$ to $+\infty$. The root-locus diagram gives an indication of the absolute stability and, to some extent, the relative stability of a control system with respect to the variation of the system parameter K.

Since the characteristic equation of a linear time-invariant digital control system is a rational polynomial in z, the same set of conditions for the construction of the root loci in the s-plane can be applied directly to the z-plane.

Although the root-locus diagram of a digital control system can be constructed in the s-plane using the characteristic equation obtained from the denominator polynomial of the closed-loop transfer function $C^*(s)/R^*(s)$, the root loci will contain an infinite number of branches due to the infinite number of transfer function poles generated by the sampling operation. To illustrate the difficulties of working in the s-plane for digital control systems, let us consider that the closed-loop transfer function of a digital control system is given by

$$\frac{C^*(s)}{R^*(s)} = \frac{G^*(s)}{1 + GH^*(s)} \tag{7-109}$$

where $GH^*(s)$ is the pulse transfer function of

$$G(s)H(s) = \frac{K}{s(s+1)} \tag{7-110}$$

Thus,

$$\begin{aligned} GH^*(s) &= \frac{1}{T} \sum_{n=-\infty}^{\infty} G(s + jn\omega_s)H(s + jn\omega_s) \\ &= \frac{K}{T} \sum_{n=-\infty}^{\infty} \frac{1}{(s + jn\omega_s)(s + jn\omega_s + 1)} \end{aligned} \tag{7-111}$$

The characteristic equation of the digital control system is obtained by equating the numerator polynomial of $1 + GH^*(s)$ to zero. It is well known that the root loci are constructed based on the poles and zeros of the open-loop transfer function $GH^*(s)$. In the present case, the function $GH^*(s)$ has an infinite number of poles, as shown in Fig. 7-28(a). Therefore, the root loci of the digital control system contain an infinite number of branches as shown in Fig. 7-28(b). It is apparent that, for digital control systems with more complex transfer functions, the construction of the root loci in the s-plane will be more tedious.

The use of the z-transform folds the infinite number of poles and zeros into a finite number. Thus, the root loci in the z-plane will contain a finite number of branches. For the system described by Eq. (7-109), the z-transform expression is

$$\frac{C(z)}{R(z)} = \frac{G(z)}{1 + GH(z)} \tag{7-112}$$

The z-transform of $G(s)H(s)$ is written

$$GH(z) = \frac{K(1 - e^{-T})z}{(z-1)(z - e^{-T})} \tag{7-113}$$

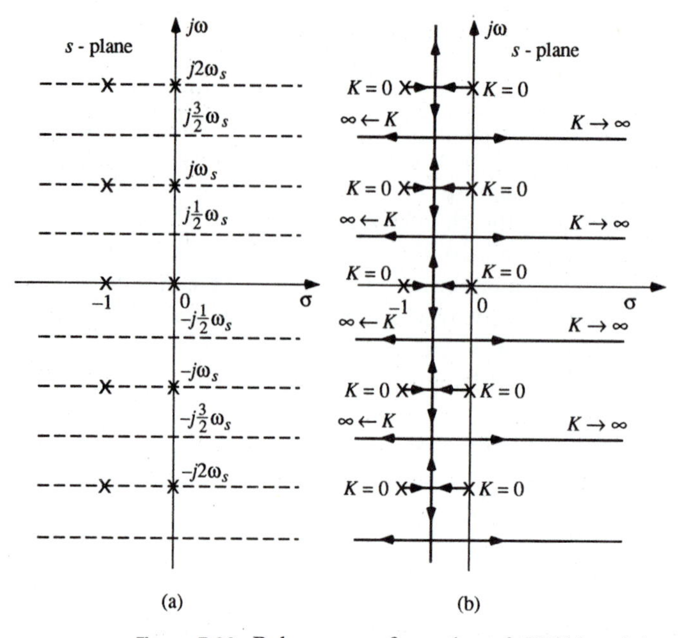

Figure 7-28. Pole-zero configuration of $GH^*(s)$ and the root-locus diagram in the s-plane of the digital control system described by Eqs. (7-107) through (7-109). (a) Poles of $GH^*(s)$. (b) Root loci.

The characteristic equation is

$$(z - 1)(z - e^{-T}) + K(1 - e^{-T})z = 0 \tag{7-114}$$

The root loci of the last equation as K varies from zero to infinity are drawn based on the pole-zero configuration of $GH(z)$, as shown in Fig. 7-29. Since Eq. (7-114) is of the second order, Fig. 7-29 shows only two root loci.

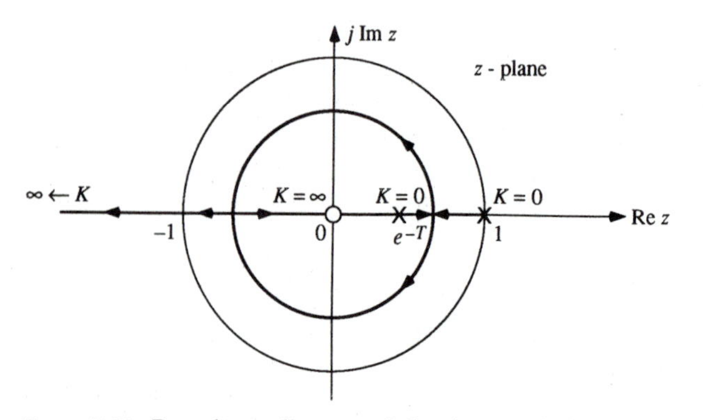

Figure 7-29. Root-locus diagram of the characteristic equation in Eq. (7-112) as K varies from zero to infinity.

7-11-1 Properties of Root Loci in the z-Plane

The properties of the root loci of the characteristic equation of a digital control system with the loop transfer function $GH(z)$ follow directly from those of the continuous-data system with the loop transfer function $G(s)H(s)$ [1].

Consider that the closed-loop transfer function of a digital control system is given by Eq. (7-112). The loop transfer function $GH(z)$ is a rational function with constant coefficients. The variable parameter K is a multiplying factor in $GH(z)$ so that

$$GH(z) = KGH_1(z) \tag{7-115}$$

where $GH_1(z)$ no longer contains K. The roots of the characteristic equation must satisfy the equation

$$1 + KGH_1(z) = 0 \tag{7-116}$$

or

$$GH_1(z) = -\frac{1}{K} \tag{7-117}$$

Equation (7-117) leads to the following two conditions.

Condition on Magnitude

$$|GH_1(z)| = \frac{1}{|K|} \tag{7-118}$$

Conditions on Angle

$$K \geq 0 \qquad \angle GH_1(z) = (2k+1)\pi = \text{odd multiples of } \pi \text{ rad} \tag{7-119}$$

$$K \leq 0 \qquad \angle GH_1(z) = 2k\pi = \text{even multiples of } \pi \text{ rad} \tag{7-120}$$

where $k = 0, \pm 1, \pm 2, \ldots$ (any integer).

Thus, the root loci of a closed-loop system are determined from knowledge of the loop transfer function $GH_1(z)$. When constructing the root loci in the z-plane, the conditions on the angle of $GH_1(z)$ in Eqs. (7-119) and (7-120) are used to find the points on the root loci, whereas the condition on magnitude in Eq. (7-118) is used to find the values of K on the loci.

Table 7-6 summarizes the properties of the root loci in the z-plane for $0 \leq K < \infty$. These properties are useful in sketching the root loci without actually solving for the roots of the characteristic equation. Root-loci programs on computers such as ZRLOCI of the DCSP software package can be used to compute the points and make plots of the root loci. However, familiarity with the properties in Table 7-6 is still essential for the interpretation of the results.

Table 7-6 Properties of Root Loci in the z-Plane

1. $K = 0$ points	The $K = 0$ points on the root loci are at the poles of $GH(z)$. The poles include those at infinity.
2. $K = \infty$ points	The $K = \infty$ points on the root loci are at the zeros of $GH(z)$. The zeros include those at infinity.
3. Number of separate root loci	The total number of root loci is equal to the number of poles or zeros of $GH(z)$, whichever is greater.
4. Symmetry of root loci	The root loci are symmetrical with respect to the axes of symmetry of the poles and zeros of $GH(z)$.
5. Asymptotes of root loci	For large values of z, the root loci are asymptotic to straight lines with angles given by $$\theta_k = \frac{(2k+1)\pi}{\lvert n-m \rvert}$$ where $k = 0, 1, 2, \ldots, \lvert n-m \rvert - 1$; n is the number of finite poles of $GH(z)$, and m is the number of finite zeros of $GH(z)$.
6. Intersection of the asymptotes	a. The intersection of the asymptotes lies only on the real axis of the z-plane. b. The intersect of the asymptotes on the real axis is given by $$\sigma_1 = \frac{\Sigma \text{ real parts of poles of } GH(z) - \Sigma \text{ real parts of zeros of } GH(z)}{n-m}$$
7. Root loci on the real axis	Root loci are found on a given section of the real axis of the z-plane only if the total number of real poles and real zeros of $GH(z)$ to the right of the section is odd.
8. Angles of departure and arrival	The angle of departure of the root loci from a pole or the angle of arrival at a zero of $GH(z)$ can be determined by assuming a point z_1 that is on the root locus associated with the pole, or zero, and is very close to the pole, or zero, and applying Eq. (7-119).
9. Intersection of the root loci with the unit circle $\lvert z \rvert = 1$	The value of z and K at an intersect of the root loci with the unit circle $\lvert z \rvert = 1$ may be determined using one of the stability tests such as Jury's test or the extended Routh-Hurwitz criterion.
10. Breakaway points	The breakaway points on the root loci are points at which multiple-order roots lie. The points are determined from the roots of the following equation: $$\frac{dGH(z)}{dz} = 0$$

Table 7-6 Continued

11. Values of K on the root loci	The value of K at any point z_1 on the root loci is determined from Eq. (7-118), rewritten as $$\lvert K \rvert = \frac{1}{\lvert GH_1(z_1) \rvert} = \frac{\text{product of lengths of vectors drawn from the poles of } GH_1(z) \text{ to } z_1}{\text{product of lengths of vectors drawn from the zeros of } GH_1(z) \text{ to } z_1}$$

7-11-2 Illustrative Examples of Root Loci

In this section we shall use several examples to illustrate the construction of the root loci of digital control systems in the z-plane. The absolute and relative stability of systems will be investigated with respect to the root loci.

Example 7-4

Consider the digital control system described by the loop transfer function of Eq. (7-113). The loop transfer functions $GH(z)$ for $T = 0.1$ s and $T = 1.0$ s are given below, and the root loci are shown in Fig. 7-30.

$\boldsymbol{T = 0.1}$ **s**

$$GH(z) = \frac{0.0952Kz}{(z-1)(z-0.905)} \tag{7-121}$$

$\boldsymbol{T = 1}$ **s**

$$GH(z) = \frac{0.632Kz}{(z-1)(z-0.368)} \tag{7-122}$$

We can show analytically that the complex portions of the root loci are circles with centers at $z = 0$. When $T = 0.1$ s, the circular portion of the root loci is closer to the unit circle than that for $T = 1$ s. This gives the misleading impression that the system with $T = 0.1$ s is less stable than that when $T = 1$ s. Notice that when $T = 0.1$ s, the critical value of K for stability is 40, whereas for $T = 1$ s the critical value of K is 4.33. Thus, as expected, the system with the smaller sampling period has a larger margin of stability in K.

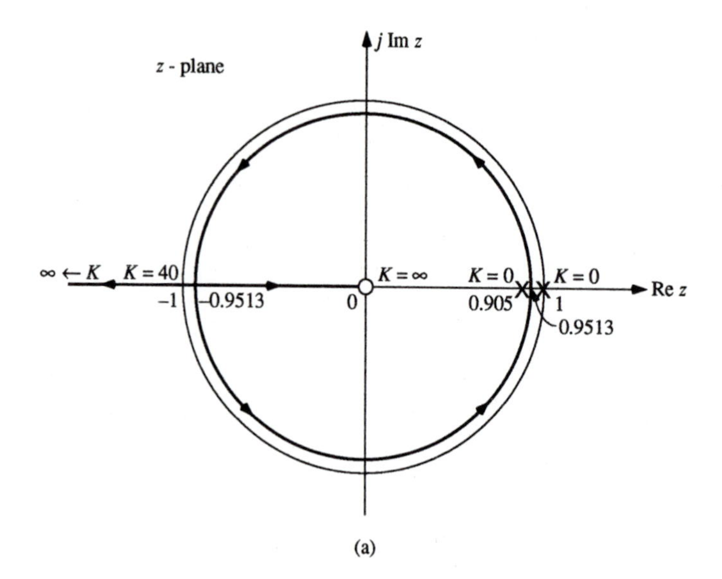

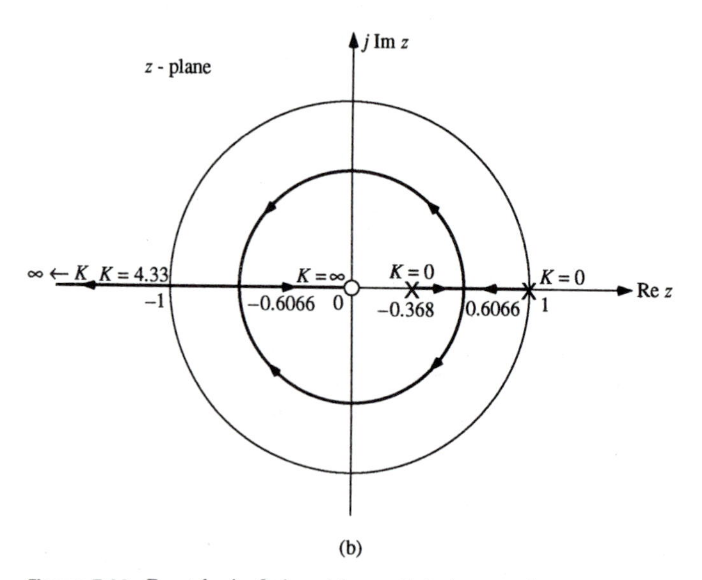

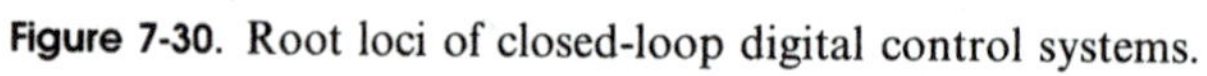

Figure 7-30. Root loci of closed-loop digital control systems.

(a) $GH(z) = \dfrac{0.0952Kz}{(z-1)(z-0.905)}$. (b) $GH(z) = \dfrac{0.632Kz}{(z-1)(z-0.368)}$.

Example 7-5

Consider the space vehicle control system studied in Sec. 7-3-2. The open-loop transfer function of the system is given in Eq. (7-20).

Let $T = 0.1$ s. The open-loop transfer function becomes

$$G(z) = \frac{0.15K(z + 0.7453)}{(z - 1)(z - 0.4119)} \tag{7-123}$$

where K is considered to be a variable parameter.

The root loci of the system are shown in Fig. 7-31 for $0 \le K < \infty$. The closed-loop system is stable for $0 \le K < 5.26$. When $K = 5.26$, the root loci cross the unit circle at $z = 0.3162 + j0.9448$ and $0.3162 - j0.9448$.

If we wish to realize a relative damping ratio of 0.707 for the system, we sketch the constant-damping-ratio locus of $\zeta = 0.707$ in the z-plane, as shown in Fig. 7-31. The intersect between the $\zeta = 0.707$ locus and the root loci gives the desired root location, and the corresponding value of K is approximately 0.7.

Now let $T = 0.5$ s. The open-loop transfer function becomes

$$G(z) = \frac{1.7788K(z + 0.1153)}{(z - 1)(z - 0.1089)} \tag{7-124}$$

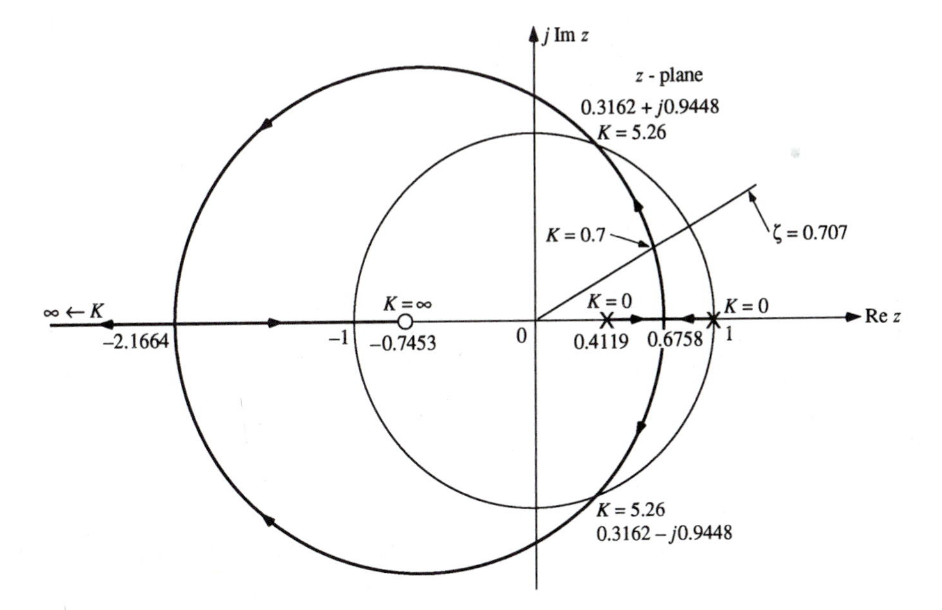

Figure 7-31. Root loci of closed-loop system with

$$G(z) = \frac{0.15K(z + 0.7453)}{(z - 1)(z - 0.4119)}.$$

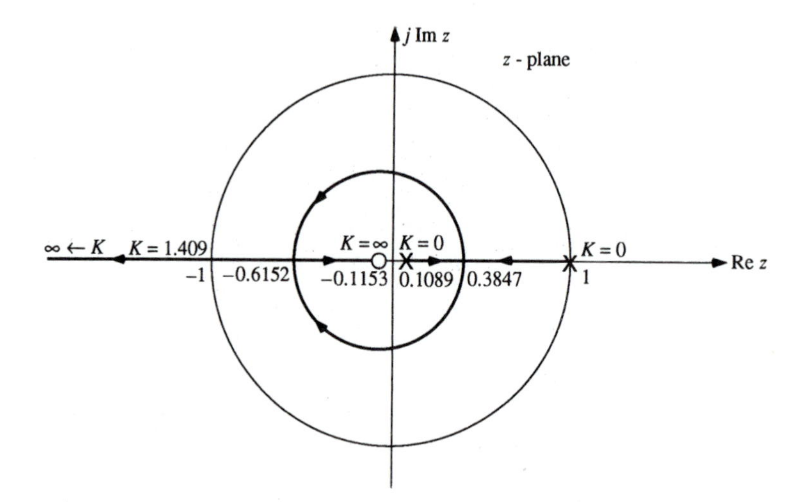

Figure 7-32. Root loci of closed-loop system with

$$G(z) = \frac{1.7788K(z + 0.1153)}{(z - 1)(z - 0.1089)}.$$

The root loci of the system are shown in Fig. 7-32 for $0 \leq K < \infty$. The critical value of K for stability is now reduced to 1.409. The root loci cross the unit circle at $z = -1$. Although the complex portion of the root loci is represented by a smaller circle, the closed-loop system is less stable than when $T = 0.1$ s for the same value of K.

7-11-3 Root Loci of Systems with Pure Time Delays

In continuous-data control systems, if a pure time delay exists in the loop, the construction of the root loci is more complex, since the transfer function $G(s)H(s)$ is no longer a rational function of s. For a digital control system with time delays, if the delay is an integral multiple of the sampling period T, the root loci can be constructed without complications. The following example illustrates the effects of pure time delays in digital control systems.

Example 7-6

Consider that the space vehicle control system analyzed in Example 7-5 now has a pure time delay T_d equal to or capable of being approximated by one sampling period $T = 0.1$ s. The loop transfer function of the system is written

$$G(z) = \frac{0.15K(z + 0.7453)}{z(z - 1)(z - 0.4119)} \qquad \text{(7-125)}$$

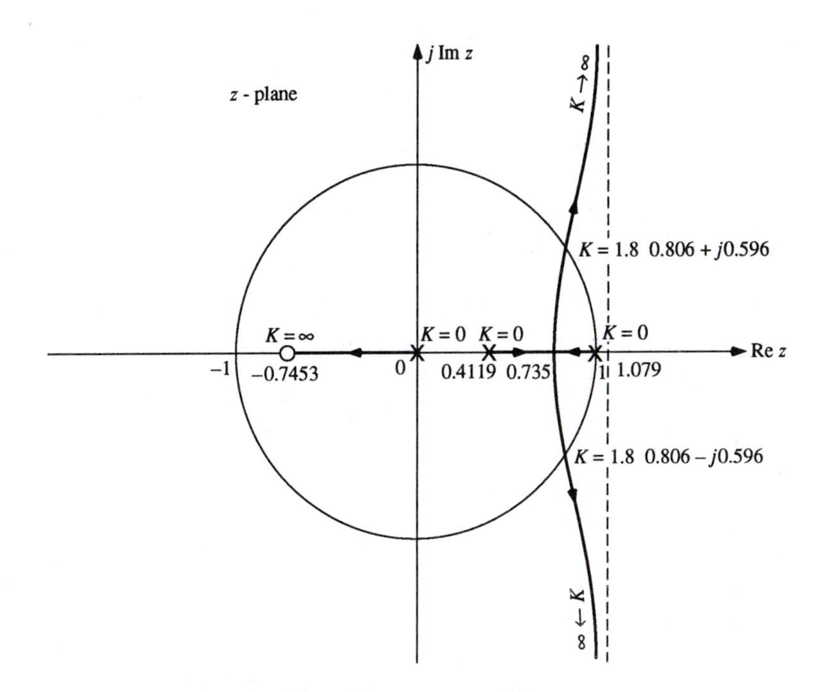

Figure 7-33. Root loci of closed-loop system with

$$G(z) = \frac{0.15K(z + 0.7453)}{(z - 1)(z - 0.4119)}.$$

where the z in the denominator is the result of the pure time delay of one sampling period. Thus, the root loci of the closed-loop system are affected by the addition of the open-loop pole at $z = 0$. Figure 7-33 shows the root loci of the system. Notice that the complex portion of the root loci crosses the unit circle at $0.806 \pm j0.597$, and asymptotically approaches the 90° and 270° asymptotes that intersect at $z = 1.079$. Thus, the time delay has the effect of reducing the stability margin of the system. Now not only is the system of the third order, but one of the poles of the closed-loop transfer function also lies on the negative real axis, which causes oscillations in the system response. Figure 7-34 shows the unit-step response with $K = 1$; due to the time delay, the response has zero values for the first two sampling periods. The maximum overshoot is increased to 47 percent due to the time delay.

Now consider that the pure time delay is increased to two sampling periods, and the loop transfer function becomes

$$G(z) = \frac{0.15K(z + 0.7453)}{z^2(z - 1)(z - 0.4119)} \tag{7-126}$$

The system is now of the fourth order. Figure 7-35 shows the root loci of the system. The critical value of K for stability is now reduced to 1.12, and the

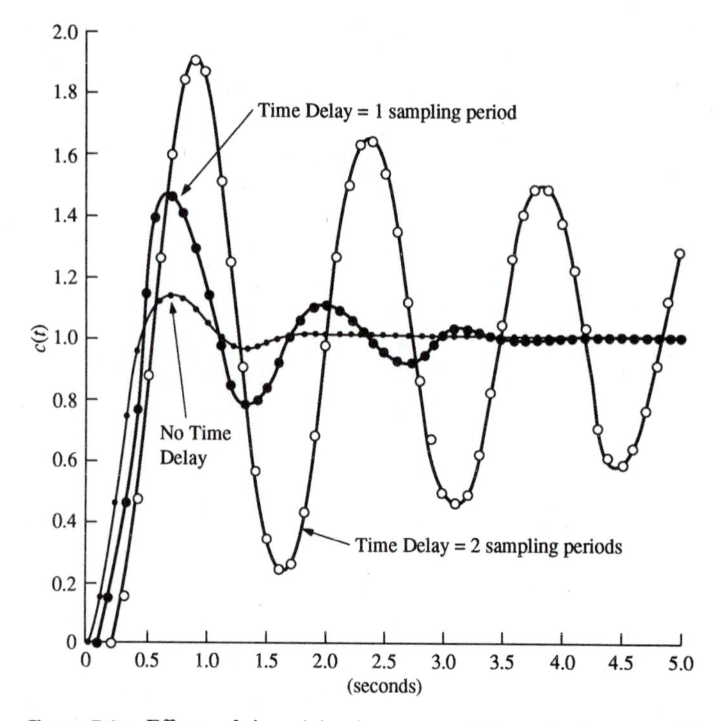

Figure 7-34. Effects of time delay in space vehicle control system: $T = 0.1$ s.

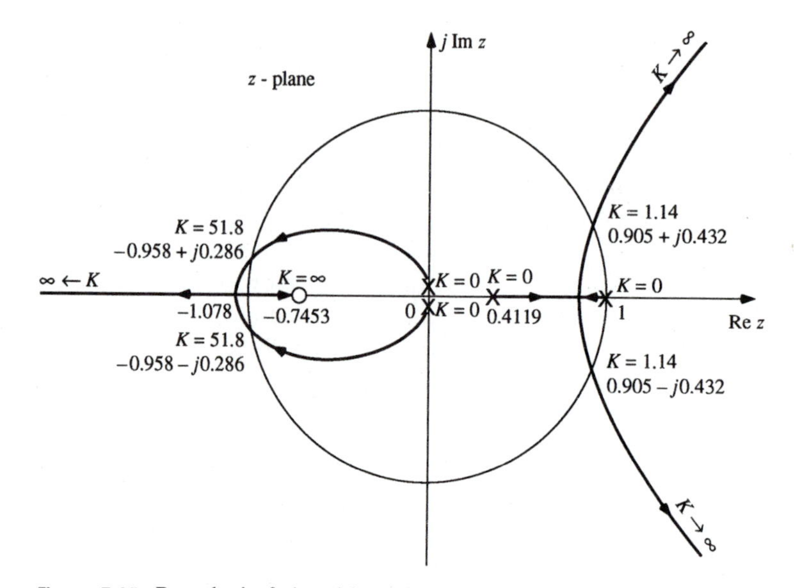

Figure 7-35. Root loci of closed-loop system with

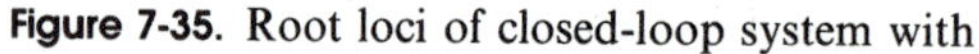

$$G(z) = \frac{0.15K(z + 0.7453)}{z^2(z - 1)(z - 0.4119)}.$$

loci intersect the unit circle at $z = 0.907 + j0.436$ and $0.907 - j0.436$. The asymptotes of the root loci when K approaches infinity are 90°, 180°, and 270°. When the system is stable, the closed-loop transfer function has two poles in the first and fourth quadrants and two poles in the second and third quadrants. The complex poles in the second and third quadrants will cause high-frequency oscillations in the system response. Figure 7-34 shows the unit-step response of the system with $K = 1$. The maximum overshoot is now increased to 89 percent, and the response does not begin until the fourth sampling periods.

7-12 EFFECTS OF ADDING POLES AND ZEROS TO THE OPEN-LOOP TRANSFER FUNCTION

The discussions conducted in the preceding sections are centered on the effects on the system performance of changing the loop gain and the sampling period. In practice, control system design relies to a great extent on the addition of open-loop poles and zeros at desirable locations, and the deletion of undesirable poles and zeros in the z-plane. The root-locus method for the digital control systems described in the previous section is valuable in the qualitative study of system performance when the pole-zero configuration of $GH(z)$ is altered.

7-12-1 Effects of Adding a Zero to the Open-Loop Transfer Function

For continuous-data closed-loop systems, the addition of a zero in the left half s-plane to the open-loop transfer function generally improves the damping, but the maximum overshoot may increase if the zero is too close to the origin in the s-plane, due to the differentiation effect [1].

We shall investigate the effect of adding the term $(z + z_1)$ to the open-loop transfer function of the space vehicle control system given in Eq. (7-123). The study is academic, since the function $(z + z_1)$ is not physically realizable, and most practical digital controllers are represented by transfer functions with at least one pole and one zero.

When $z_1 = 0$, the open-loop transfer function is

$$G(z) = \frac{0.15Kz(z + 0.7453)}{(z - 1)(z - 0.4119)} \tag{7-127}$$

The root loci of the system are shown in Fig. 7-36. The system is now stable for all finite positive values of K. Thus, adding a zero at $z = 0$ to $G(z)$ has the effect of improving the stability of the closed-loop system. Since the zeros of $G(z)$ will appear as zeros of the closed-loop transfer function, adding the zero at $z = 0$ will have a left shift effect on the output response. Table 7-7 gives the values of $c(0)$, c_{max}, and T_{max} when $K = 1$. The maximum overshoot is 0.77 percent, as compared with 13.6 percent when the zero is not added. The step

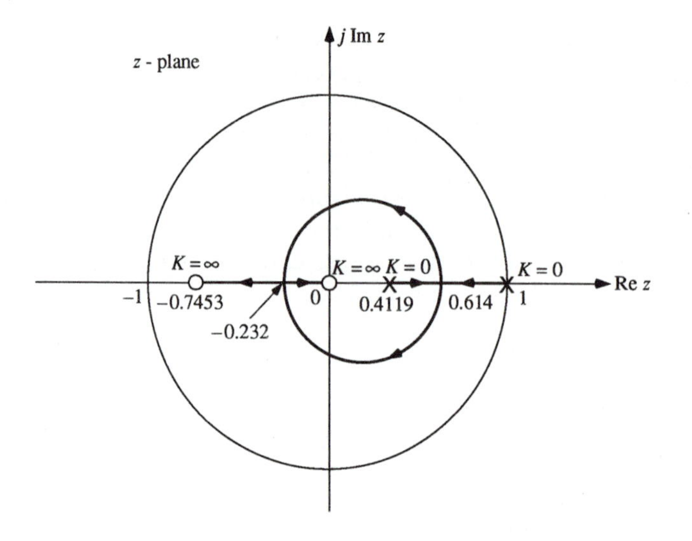

Figure 7-36. Root loci of closed-loop system with

$$G(z) = \frac{0.15K(z + 0.7453)}{(z - 1)(z - 0.4119)}.$$

response has a jump at $t = 0$, since the numbers of poles and zeros of the closed-loop transfer function are the same.

The effect of the added zero on damping is more pronounced as the zero is moved further into the right half z-plane. Table 7-7 gives the results on the unit-step response with $K = 1$ when $z_1 = -0.5$. In this case the response no longer has any overshoot. As the zero at z_1 is moved further to the right, the rise time becomes longer. The zero at $-z_1$ should not be placed outside the unit circle, as it will cause the system to be unstable.

Table 7-7 Effects of Adding a Zero to $G(z)$

Zero $-z_1$	$c(0)$	c_{max}	T_{max} (s)
No zero added	0	1.1366	0.7
0	0.1303	1.0077	0.9
0.1	0.1303	1.0014	1.2
0.5	0.1303	1.0000	∞
0.8	0.1303	1.0000	∞
-0.1	0.1303	1.0191	0.8
-0.5	0.1303	1.0821	0.6
-0.8	0.1303	1.1344	0.5

When the zero at $-z_1$ is placed on the negative real axis, the improvement on damping is reduced, as shown by the results in Table 7-7, but the rise time is still improved.

The conclusion is that adding a zero to the open-loop transfer function will generally improve the damping and reduce the maximum overshoot of the closed-loop system. Adding the zero in the right half plane will generally increase the rise time, and adding the zero in the left half plane tends to reduce the rise time. The steady-state performance of the system due to a step input is not affected, since the system is of type 1.

7-12-2 Effects of Adding a Pole to the Open-Loop Transfer Function

When a pole at $z = -p_1$ is added to the open-loop transfer function of the space vehicle control system, $G(z)$ becomes

$$G(z) = \frac{0.15K(z + 0.7453)}{(z - 1)(z + p_1)(z - 0.4119)} \tag{7-128}$$

In continuous-data control systems, adding a pole in the left half s-plane to the open-loop transfer function generally has the effect of increasing the maximum overshoot and the rise time [1]. From the root-locus standpoint, adding a pole to $G(s)$ tends to push the root loci toward the right half s-plane, thus reducing the relative stability of the closed-loop system.

Table 7-8 summarizes the results on the unit-step responses of the system when the value of p_1 varies and when $K = 1$ in Eq. (7-128).

We may conclude that adding a pole in the right half z-plane to $G(z)$ has the effect of increasing the maximum overshoot, while moderately increasing the rise time. If the pole is too far to the right, the system may become unstable. Adding a pole on the negative real axis may actually reduce the maximum overshoot, especially if the pole is closer to the unit circle, but the rise time is

Table 7-8 Effects of Adding a Pole to $G(z)$

Pole $-p_1$	c_{max}	T_{max} (s)
No pole added	1.1366	0.7
0	1.4703	0.8
0.1	1.5857	0.8
0.2	1.8956	0.8
0.5	unstable	
−0.5	1.1412	0.9
−0.7453	1.0720	1.1
−1.0	1.0350	1.3

increased. The reduction in the maximum overshoot when the added negative pole at $-p_1$ is large is due to the fact that the step response of the digital system is directly affected by the coefficients of the closed-loop transfer function, so that a large coefficient in the denominator has the effect of reducing the value of the output response. Not shown by the results in Table 7-8 is that the step response does not begin until $t = 0.2$ s, since with the added pole, the closed-loop system transfer function has two more poles than zeros. Also not shown in Table 7-8 is that, as the pole $-p_1$ gets larger in the left half plane, the negative real pole of the closed-loop transfer function causes the response to oscillate. For the case when $p_1 = 1.0$, the unit-step response oscillates between 0.996 and 1.003 for a long period of time before diminishing.

PROBLEMS

7-1 The block diagram of a discrete-data control system is shown in Fig. P7-1. Compute and plot the unit-step response $c^*(t)$ of the system. Find c^*_{max} and the sampling instant at which it occurs. Find the step-, ramp-, and parabolic-error constants. Find the final value of $c(kT)$.

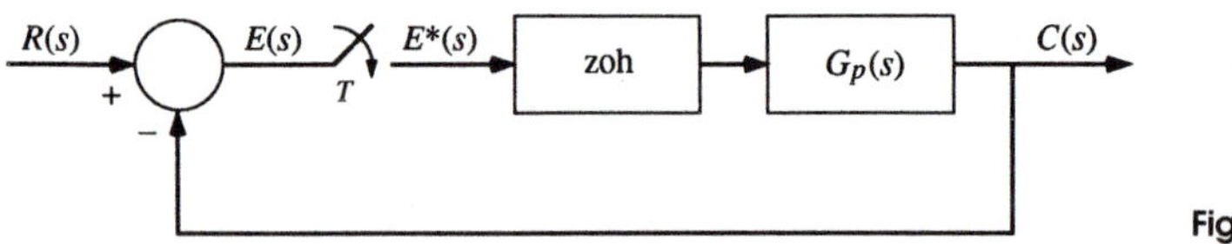

Figure P7-1.

a. $G_p(s) = \dfrac{20}{s(s+5)}$ $\quad T = 0.5$ s

b. $G_p(s) = \dfrac{2(s+1)}{s(s+2)}$ $\quad T = 0.5$ s

c. $G_p(s) = \dfrac{2}{s^2+s+2}$ $\quad T = 1$ s

7-2 Compute and plot the unit-step response of the flight control system described in Problem 4-6, Fig. P4-6. Use the system parameters given in Problem 4-6. Find $(\theta_0^*)_{max}$ and the sampling instant at which it occurs. Find the final value of $\theta_o(kT)$.

a. $T = 0.01$ s

b. $T = 0.1$ s

7-3 Compute and plot the unit-step response of the flight control system described in Problem 4-7. Find $(\theta_0^*)_{max}$ and the sampling instant at which it occurs. Find the final value of $\theta_o(kT)$.

a. $T = 0.01$ s

b. $T = 0.1$ s

7-4 Consider the flight control system described in Problem 4-8, Fig. P4-8. Use the system parameters given in Problem 4-8, except for the values of K which are given below. Compute and plot the unit-step responses of the system. Find $(\theta_0^*)_{\max}$ and the sampling instant at which it occurs.

a. $T = 0.01$ s $\quad K = 1000$

b. $T = 0.1$ s $\quad K = 100$

7-5 The block diagram of the digital control system for the idle-speed control of an automobile engine is shown in Fig. P7-5.

a. Let $T = 0.05$ s. Compute and plot the unit-step response of $\omega^*(t)$. Find $\omega^*_{\max}$ and the sampling instant at which it occurs.

b. Repeat part **a** for $T = 0.1$ s.

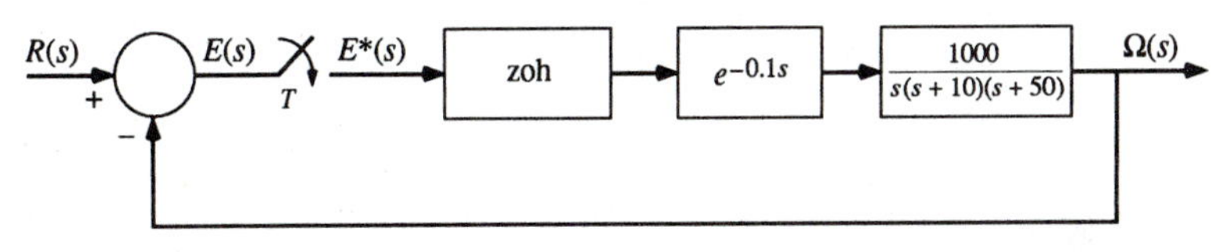

Figure P7-5.

7-6 The schematic diagram of a steel-rolling process is shown in Fig. P7-6. The dc motor is characterized by the following parameters:

Armature resistance	$R = 5\ \Omega$
Armature inductance	negligible
Torque constant	$K_a = 0.721$ Nm/A
Back EMF constant	$K_b = 0.721$ V/rad/s
Lead of lead screw	$L = 0.01$ m/rev $= 0.01/2\pi$ m/rad
Total inertia reflected to motor	$J = 0.001$ kg-m^2
Total viscous friction coefficient reflected to motor	$B = 0.01$ Nm/rad/s
Gain of transducer for measurement of thickness	$K_s = 100$ V/m
Sampling period	$T = 0.01$ s

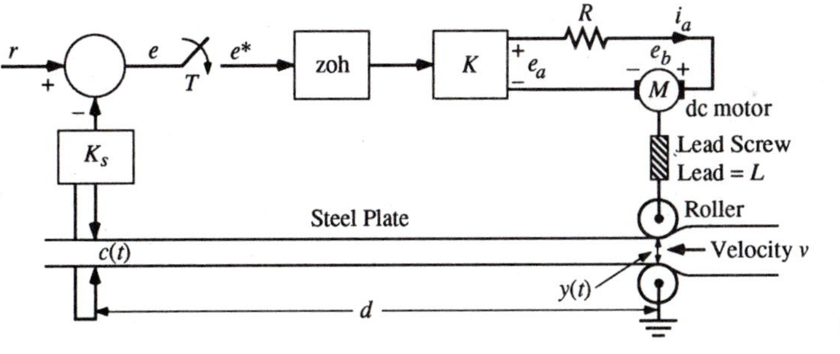

Figure P7-6.

The motor shaft is coupled to a lead screw or linear actuator whose output is connected to the roller to roll the steel plate to the desired thickness. The steel plate is fed through the rollers with a constant speed of v m/s. The distance between the points where the plate is rolled and where the thickness is measured is d meters. Let $v = 10$ m/s and $d = 1$ m. The following equations are written for the system:

$$e_a(t) = Ri_a(t) + e_b(t) \qquad T_m(t) = K_a i_a(t)$$

$$e_b(t) = K_b\omega(t) \qquad T_m(t) = J\frac{d\omega(t)}{dt} + B\omega(t)$$

$$\frac{d\theta(t)}{dt} = \omega(t) \qquad y(t) = L\theta(t)$$

$$c(t) = y(t - T_d) \qquad T_d = d/v$$

$$e(t) = r(t) - K_s c(t)$$

a. Draw a block diagram to represent the transfer function relations of the digital steel-rolling control system.
b. Find the closed-loop transfer function $Y(z)/R(z)$.
c. Determine the range of K for the system to be stable.

7-7 The block diagram of a microprocessor dc motor speed control system is shown in Fig. P7-7(a). The speed of the motor load is controlled to match the desired speed ω_r. The actual speed of the motor load is measured by the microprocessor which processes the signal from the shaft encoder. The speed error is defined as the difference between the desired speed and the actual speed. A controller is programmed into the microprocessor. The output of the microprocessor which is proportional to the speed error is transmitted to the pulse generator. This error signal controls the phase between two pulse trains

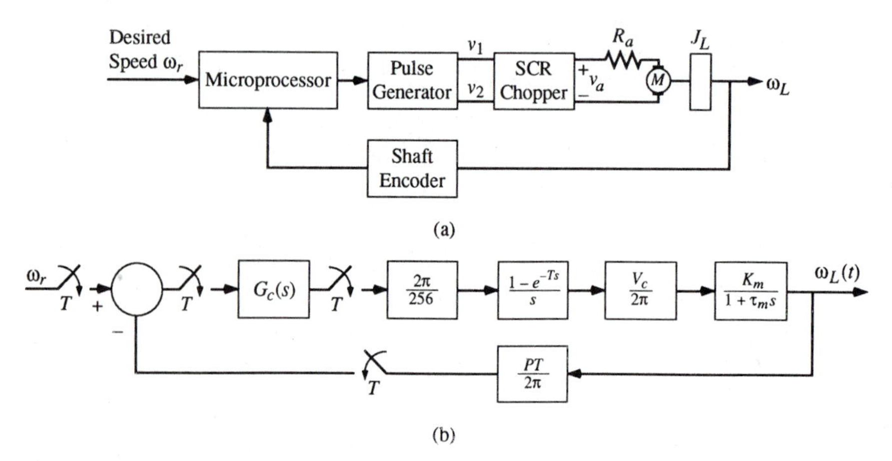

Figure P7-7.

generated by the pulse generator. This error signal controls the phase between two pulse trains generated by the pulse generator. The voltage applied to the dc motor is determined by the duty cycle which is a function of the phase between the two pulse trains.

The speed error is represented by an 8-bit binary number. Let the phase be from zero to 360 degrees. Then, the finest resolution of the digital representation of the phase shift is $2\pi/2^8 = 2\pi/256$ rad. The silicon-controlled rectifier (SCR) chopper outputs a voltage to the dc motor that is determined by the phase between v_1 and v_2. Since the phase information between sampling instants is constant, we can use a sample-and-hold device to model the input-output relation. Thus, the transfer function of the pulse-generator-phase relation is

$$G_p(s) = \frac{2\pi}{256}\,\frac{1 - e^{-Ts}}{s}$$

The transfer function of the SCR chopper is

$$G_d(s) = V_c/2\pi$$

where V_c is the chopper output voltage (full scale). The block diagram of the overall system is shown in Fig. P7-7(b).

a. Let $G_c(s) = K_c =$ constant. Find the characteristic equation of the digital control system shown in Fig. P7-7(b).
b. Find the range of K_c so that the closed-loop digital control system is asymptotically stable.
c. Find the closed-loop transfer function $\Omega_L(z)/\Omega_r(z)$.
d. For $K_c = 1$, find the output response $\omega_L(k)$ when the input ω_r is a unit-step function. What is the steady-state value of $\omega_L(k)$ for this input?

Use $V_c = 24$ V, $K_m = 5$, $\tau_m = 0.05$ s, $T = 0.1$ s, and $P = 100$ pulses/rev.

7-8 For the simplified digital space vehicle control system shown in Fig. P7-8,

a. find the step-error constant K_p;
b. find the ramp-error constant K_v;
c. find the parabolic-error constant K_a.

Express the results in terms of the system parameters.

7-9 For the digital space vehicle control system shown in Fig. P7-8, let $J_v = 41822$ and $T = 0.1$ s.

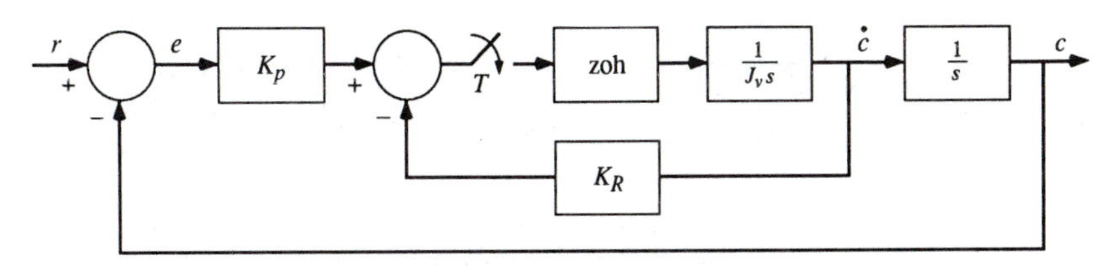

Figure P7-8.

a. Find the relation between K_P and K_R so that the ramp-error constant K_v is equal to 10.

b. Find the loci of the characteristic equation roots when K_P or K_R is varied from zero to infinity under the constraint that the ramp-error constant K_v is 10. Find the ranges of K_P and K_R so that the closed-loop system is stable.

7-10 Given the following closed-loop transfer functions for digital control systems, find the maximum overshoot and the normalized peak time $T_{\max}/T$ of the step response. First use the method of Sec. 7-9, then check the results by evaluating $c(kT)$ for $k = 0, 1, 2, \ldots$.

a. $$\frac{C(z)}{R(z)} = \frac{z + 0.5}{3(z^2 - z + 0.5)}$$

b. $$\frac{C(z)}{R(z)} = \frac{0.5z}{z^2 - z + 0.5}$$

c. $$\frac{C(z)}{R(z)} = \frac{z^{-2}(z + 0.5)}{3(z^2 - z + 0.5)}$$

d. $$\frac{C(z)}{R(z)} = \frac{0.5z^{-2}}{z^2 - z + 0.5}$$

e. $$\frac{C(z)}{R(z)} = \frac{0.316(z + 0.002)(z + 0.5)}{(z^2 - z + 0.5)(z - 0.05)}$$

7-11 The block diagram of a speed control system with a dc motor is shown in Fig. P7-11. The system parameters are given as follows:

K_a = motor torque constant = 0.345
K_b = motor back EMF constant = 0.345
R = motor armature resistance = 1 Ω
L = motor armature inductance = 1 mH
B = viscous friction coefficient of motor and load = 0.25
J = inertia of motor and load = 1.41×10^{-3}
T = sampling period = 0.001 s

It is assumed that the units of these parameters are consistent so that no conversions are necessary. The digital controller is implemented by a micro-

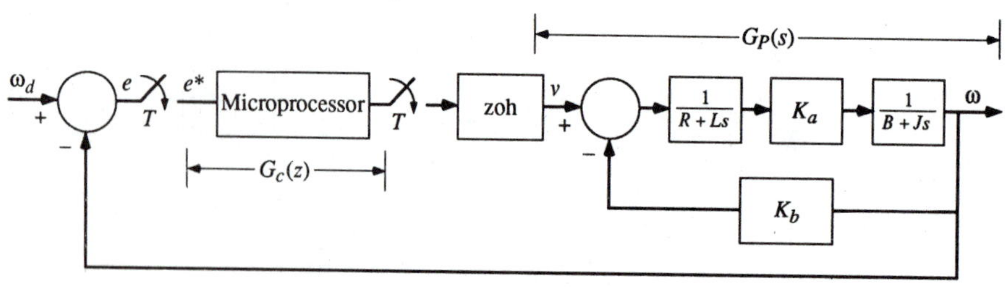

Figure P7-11.

processor and is modeled as shown in the block diagram. The transfer function of the digital controller is

$$G_c(z) = K_P + \frac{K_R T}{2}\left(\frac{z+1}{z-1}\right)$$

where $K_P = 1$ and $K_R = 295.276$.

a. Find the open-loop transfer function $\Omega(z)/E(z)$.

b. Find the closed-loop transfer function $\Omega(z)/\Omega_d(z)$.

c. Find the characteristic equation of the overall system and the roots of the characteristic equation.

d. For $\omega_d =$ unit-step function, find the output response ω at the sampling instants.

7-12 The payload of a space-shuttle pointing control system is modeled as a pure mass M. It is suspended by magnetic bearings so that no friction is encountered in the control. The attitude in the y direction is controlled by magnetic actuators located at the base of the payload. The dynamic system model for the control of the y-axis motion is shown in Fig. P7-12. The controls of the other degrees of motion are independent and are not considered here. Since experiments are located on the payload, electric power must be brought to the payload through wire cables. The linear spring with spring constant K_s is used to model the wire-cable attachment. The total force produced by the magnetic actuators is denoted by $f(t)$. The force equation of motion in the y direction is

$$f(t) - K_s y(t) = M\frac{dy^2(t)}{dt^2}$$

where $K_s = 0.35$ N/m, $M = 600$ kg, and $f(t)$ is in newtons. Let the magnetic actuators be controlled by sampled data, so that

$$f(t) = f(kT) \qquad kT \le t < (k+1)T$$

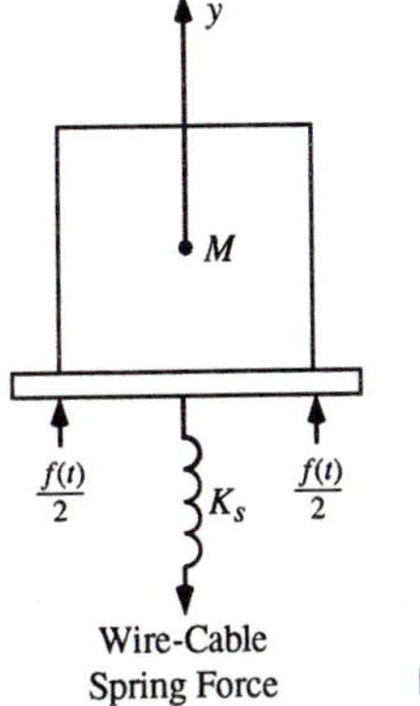

Figure P7-12.

where

$$f(kT) = -K_P y(kT) - K_R \left.\frac{dy(t)}{dt}\right|_{t=kT}$$

$K_P = 37.86$, $K_R = 211$, and T is the sampling period in seconds.

a. Draw a block diagram for the overall system.

b. Find the characteristic equation of the overall system with the feedback control described. Plot the root loci as a function of T. Find the range of T for the system to be stable.

c. Show the region in the parameter plane of K_R versus K_P in which the overall system is stable. Use $T = 1$ s. Find the point in the parameter plane at which the sampled system output is able to follow a step input in a minimum number of sampling periods without overshoot (deadbeat response).

7-13 Given the closed-loop transfer function of a digital control system,

$$\frac{C(z)}{R(z)} = \frac{T^2(K_P z^2 + K_I Tz + K_I T - K_P)}{Az^3 + Bz^2 + Cz + D}$$

where

$$A = 2J_v$$
$$B = T^2 K_P + 2K_R T - 6J_v$$
$$C = 6J_v - 4K_R T + T^3 K_I$$
$$D = 2K_R T + K_I T^3 - 2J_v - K_P T^2$$

$J_v = 41822$. Find the values of K_P, K_I, and K_R as functions of T so that the step response $c(kT)$ follows the step input in a minimum number of sampling periods. What is the maximum overshoot $c(kT)_{\max}$?

7-14 The block diagram of a digital control system with a sample-and-hold device is shown in Fig. P7-14.

a. Sketch the root loci of the characteristic equation and determine the range of K for stability. Find the value of K so that the damping ratio of the system is approximately 70.7 percent. Use $T = 0.1$ s.

$$G(s) = \frac{K}{s(s + 10)}$$

b. Repeat part **a** for $T = 1$ s.

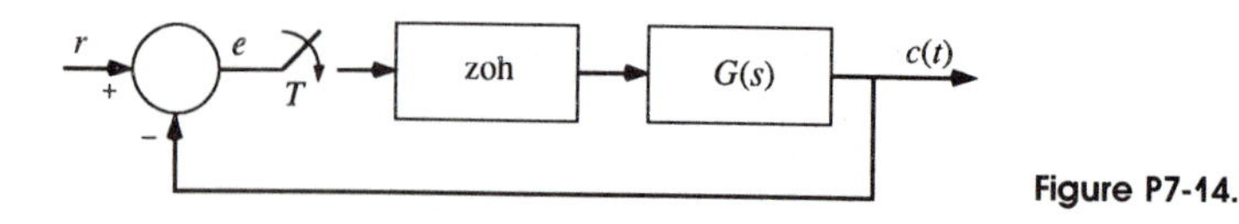

Figure P7-14.

7-15 The characteristic equation of a digital space vehicle control system is

$$2J_v z^2 + (2K_R T - 4J_v + T^2 K_P)z + (2J_v - 2K_R T + T^2 K_P) = 0$$

$T = 0.264$ s, $J_v = 41822$, and $K_P = 1.65 \times 10^6$.

a. Sketch the root loci of the characteristic equation for $0 \le K_R < \infty$.

b. Find the range of K_R so that the system is stable.

c. Find the value of K_R so that the two roots of the characteristic equation are real and equal. Find the unit-step response of the system output for this case.

7-16 For the speed control system described in Problem 7-11, let $K_P = 1$. Sketch the root loci of the overall system characteristic equation for $0 \le K_R < \infty$. Find the values of K_R so that the overall system is stable.

7-17 The schematic diagram shown in Fig. P7-17 represents a control system whose purpose is to hold the liquid in the tank at a fixed level. The level is controlled by a float whose position is connected to the wiper arm of a potentiometer error detector. The error between the reference level and the actual level of the liquid is fed into an analog-to-digital (A/D) converter and then to a digital controller. The transfer functions of the A/D converter, the digital controller, and the D/A converter are lumped together and are represented by $G_c(z)$. The parameters of the system components are given as follows:

DC Motor and Load

Armature resistance	$R = 10\ \Omega$
Armature inductance	$L =$ negligible
Torque constant	$K_a = 10$ lb-ft/A
Back EMF constant	$K_b = 0.075$ V/rad/s
Rotor inertia	$J_m = 0.005$ lb-ft-s^2
Load and rotor friction	negligible
Load inertia	$J_L = 10$ lb-ft-s^2
Gear ratio	$n_1/n_2 = 1/100$

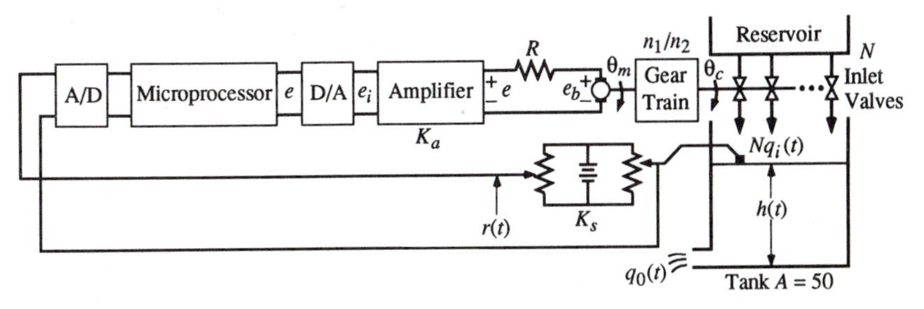

Figure P7-17.

Motor and Load Equations

$$e_a(t) = Ri(t) + e_b(t)$$

$$e_b(t) = K_b \omega_m(t)$$

$$\omega_m(t) = \frac{d\theta_m(t)}{dt}$$

$$T_m(t) = K_a i(t) = J \frac{d\omega_m(t)}{dt} \qquad \text{where } J = J_m + (n_1/n_2)^2 J_L$$

$$\theta_c = n_1 \theta_m / n_2$$

Dynamics of Tank

There are N inlets to the tank from the reservoir. All the inlet valves have the same characteristics and are controlled simultaneously by θ_c. The equations that govern the volume of flow are

Inlet

$$q_i(t) = K_i N \theta_c(t) \qquad K_i = 10 \text{ ft}^3/\text{s-rad}$$

$$q_o(t) = K_o h(t) \qquad K_o = 50 \text{ ft}^2/\text{s}$$

$$h(t) = \frac{\text{volume of tank}}{\text{area of tank}} = \frac{1}{A} \int [q_i(t) - q_o(t)]\, dt$$

Amplifier Gain

$$K = 50$$

Error Detector

$$K_s = 1 \text{ V/ft}$$

a. Draw a block diagram of the control system with the A/D and D/A converters represented by a sample-and-hold device. Model the microprocessor by a digital controller.
b. Find the minimum number of inlets N so that the closed-loop digital control system is stable. The sampling period is 0.05 s. Consider that the microprocessor is represented by a real gain of one.
c. Sketch the root loci of the closed-loop characteristic equation for $0 \le N < \infty$ in the z-plane.
d. For $N = 1$, find the time response $h(t)$ at the sampling instants for a unit-step input $r(t) = u_s(t)$.
e. Repeat part **d** for $N = 3$.
f. Repeat part **d** for $N = 5$.

7-18 The characteristic equation of a digital control system is

$$\Delta(z) = z^3 + (111.6T^2 + 16.74T - 3)z^2 + (3 - 33.48T + 1.395 \times 10^{-4}K_I T^3)z$$
$$+ 1.395 \times 10^{-4}K_I T^3 + 16.74T - 111.6T^2 - 1 = 0$$

Sketch the root loci of the characteristic equation for $T = 0.08$ s and $0 \le K_I < \infty$.

7-19 The schematic diagram of a dc motor controlling a load with inertia, friction, and compliance is shown in Fig. P7-19(a). The system parameters are defined as follows:

Amplifier gain	$K = 150$
Motor armature resistance	$R_a = 4\ \Omega$
Feedback resistance	$R_s = 0.18\ \Omega$
Torque constant	$K_a = 3.5$ oz-in/A
Back EMF constant	$K_b = 0.0247$ V/rad/s
Motor armature inductance	$L_a = 0.002$ H
Motor inertia	$J_m = 2.7 \times 10^{-4}$ oz-in-s^2
Load inertia	$J_L = 12 \times 10^{-4}$ oz-in-s^2
Motor friction	$B_m = 0.1045$ oz-in/rad/s
Spring constant	$K_s = 3453$ oz-in/rad

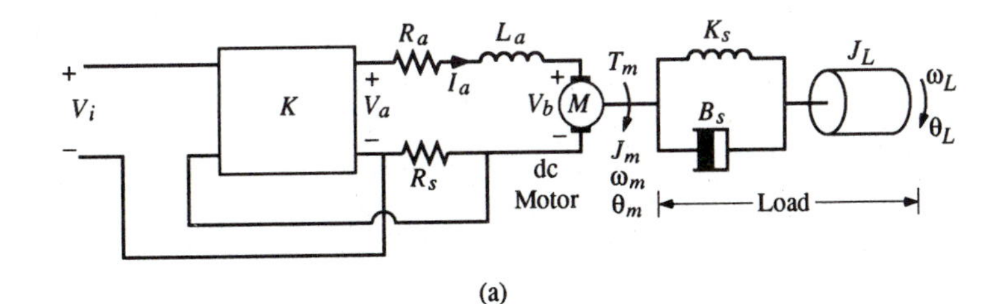

(a)

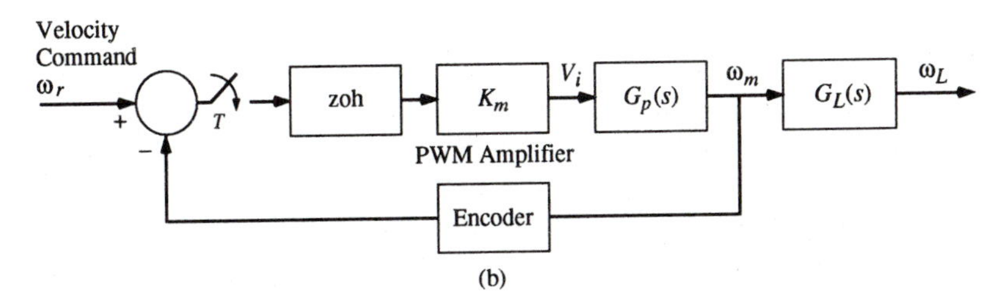

(b)

Figure P7-19.

a. Show that the transfer function between the motor input voltage v_i and the load velocity ω_L is

$$\frac{\Omega_L(s)}{V_i(s)} = \frac{KK_a(B_m s + K_s)/L_a J_m J_L}{\Delta(s)}$$

where

$$\Delta(s) = s^4 + \left(\frac{R_a + KR_s}{L_a} + \frac{B_m}{J_m} + \frac{B_m}{J_L}\right)s^3$$
$$+ \left\{\frac{K_a K_b}{J_m L_a} + \frac{(R_a + KR_s)B_m}{L_a}\left(\frac{1}{J_m} + \frac{1}{J_L}\right) + \left(\frac{1}{J_m} + \frac{1}{J_L}\right)\right\}s^2$$
$$+ \left\{\frac{(R_a + KR_s)K_s}{L_a}\left(\frac{1}{J_m} + \frac{1}{J_L}\right) + \frac{K_a K_b B_m}{L_a J_m J_L}\right\}s + \frac{K_a K_b K_s}{L_a J_m J_L}$$

b. The block diagram of a discrete-data control system using the dc motor for speed control is shown in Fig. P7-19(b). The encoder senses the motor speed, and the output of the encoder is compared with the speed command. For analytical purposes, the pulsewidth-modulation amplifier and the encoder are modeled by constant gains. Find the transfer function $G_p(s)$ and $G_L(s)$. Let the gain of the encoder be unity. Based on the dynamics of the system, determine what is the largest sampling period that can be used without violating the sampling theorem.

c. Let $T = 0.001$ s. Sketch the root loci of the equation

$$1 + K_m \mathscr{Z}[G_{h0}(s)G_p(s)] = 0$$

for $0 \le K_m < \infty$. Comment on the stability condition of the overall discrete-data system.

d. Repeat part **c** for $T = 0.0005$ s.

7-20 The purpose of this problem is to apply digital control theory to the analysis of electronic control of the stoichiometric air-fuel-ratio control problem for automobile engines. A considerable amount of effort is being spent by the automobile manufacturers to meet the exhaust emission performance standards of the various government agencies. Modern automotive powerplant systems consist of an internal combustion engine which has an internal cleanup device called the catalytic converter. Such a system requires control of such variables as engine air-fuel ratio A/F, ignition spark timing, exhaust gas recirculation, and injection air.

The control system problem considered in this exercise deals with the control of the air-to-fuel ratio A/F. In general, depending on fuel composition and other factors, a typical stoichiometric A/F is 14.7 to 1; that is, 14.7 g of air to 1 g of fuel. An A/F greater or less than stoichiometry will cause high hydrocarbons, carbon monoxide, and oxides of nitrogen in the tailpipe emission.

A control system is devised to control the air-to-fuel ratio so that a desired output variable is maintained for a given command signal. Figure P7-20(a) shows a block diagram of such a system. The sensor senses the composition of the exhaust gas mixture entering the catalytic converter. The digital controller detects the difference or the error between the command and the sensor signals and computes the control signal necessary to achieve the desired exhaust gas composition. The disturbance signal is used to represent changing or unknown operating conditions such as variations in temperature, humidity, barometric

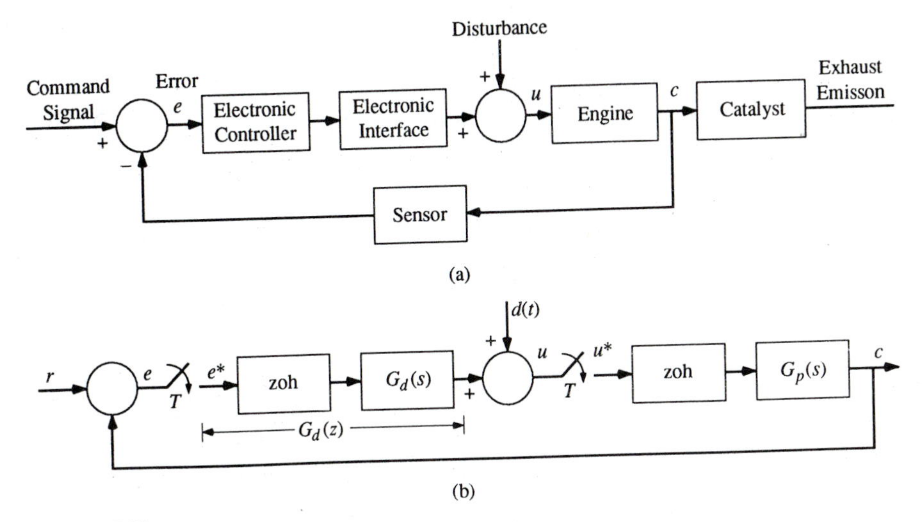

Figure P7-20.

pressure, and fuel composition. The output variable $c(t)$ denotes the effective air-fuel ratio A/F.

A discrete-data model of the system is shown in Fig. P7-20(b). The transfer function of the engine is given by

$$G_p(s) = \frac{e^{-T_d s}}{1 + \tau s}$$

where T_d is a time delay. Let $\tau = 0.25$ s, $T_d = 1$ s, and $T = 0.1$ s.

a. Let $d(t) = 0$ and $G_d(s) = K =$ constant. Find the characteristic equation of the closed-loop system. Sketch the root loci as K varies from zero to infinity. Determine the marginal value of K for stability.

b. Let $d(t) = 0$ and $G_d(s) = K =$ constant. Find the steady-state error between $r(t)$ and $c(t)$ when $r(t)$ is a unit-step function. Evaluate the steady-state error at the sampling instants (in terms of K).

c. Under the conditions given above, obtain the time responses of $c(t)$ at the sampling instants for $K = 1$ when the input $r(t)$ is a unit-step function.

d. Repeat part **c** when $K = 0.5$.

7-21 For the automatic forage and concentrate-proportioning system described in Problem 6-19, let

$$\frac{C(s)}{U(s)} = \frac{36870K}{s^2 + 83.3s + 7780}$$

Sketch the root loci of the characteristic equation for $0 \le K < \infty$ in the z-plane.

a. Find the value of K when the damping ratio ζ is zero.

b. Find the damping ratio ζ when $K = 0.1$.

c. For $K = 0.1$, find the percent overshoot of the output $c(kT)$ when the input v_s is a unit-step function.

7-22 The open-loop transfer function of a unity-feedback digital control system is given as $G(z)$. The characteristic equation of the closed-loop system is obtained by setting the numerator polynomial of $1 + G(z)$ to zero. Sketch the root loci of the characteristic equation for $0 \le K < \infty$. Indicate all important information on the root loci.

a. $G(z) = \dfrac{Kz}{(z-1)(z-0.5)}$

b. $G(z) = \dfrac{K(z+0.5)(z+0.2)}{(z-1)(z-0.5)}$

c. $G(z) = \dfrac{Kz}{(z-1)(z^2-z+0.5)}$

d. $G(z) = \dfrac{K(z+0.5)(z+0.2)}{(z-1)(z^2-z+0.5)}$

e. $G(z) = \dfrac{K(z^2+0.5z+0.2)}{(z-1)(z^2-z+0.5)}$

f. $G(z) = \dfrac{Kz}{(z-1)^2}$

7-23 The closed-loop transfer function of a digital control system is

$$\frac{C(z)}{R(z)} = \frac{D(z)G_p(z)}{1 + D(z)G_p(z)}$$

where

$$G_p(z) = \frac{Kz}{(z-1)^2}$$

$$D(z) = \frac{z+a}{z+b}$$

Sketch the root loci of the characteristic equation for $0 \le K < \infty$ for the following values of a and b.

a. $a = -0.5$, $b = -0.1$

b. $a = -0.7$, $b = -0.1$

c. $a = -0.9$, $b = -0.1$

Comment on the effects of the various values of a and b on the root loci and the system stability.

References

1. Kuo, B. C., *Automatic Control Systems*, 6th ed., Prentice-Hall, Englewood Cliffs, N.J., 1991.
2. Kuo, B. C., *Analysis and Synthesis of Sampled-Data Control Systems*, Prentice-Hall, Englewood Cliffs, N.J., 1963.

Frequency-Domain Analysis

KEYWORDS AND TOPICS

frequency response • Nyquist plot • Bode diagram • gain-phase plot • Nyquist stability criterion • gain margin • phase margin • Nichols chart • bandwidth

8-1 INTRODUCTION

In the preceding chapter time response of digital control systems was studied through the use of the pole-zero configuration of transfer functions in the z-plane. A direct method of time-domain analysis involves the computation of the system response analytically or by a computer. In general, a clear-cut or direct method of designing digital control systems in the time- or the z-domain is lacking. An example of this difficulty is clearly illustrated by the lack of correlation between the maximum overshoot and peak time to pole-zero configuration of high-order systems.

Frequency-domain analysis and design, on the other hand, possess a wealth of graphical and semigraphical techniques that can be applied to linear time-invariant control systems of virtually any complexity. Historically, the analysis and design of continuous-data control systems in the frequency domain have been well developed, and practically all these methods can be extended to digital control systems. Such well-known methods as the Nyquist criterion for stability analysis, the Bode plot, and the Nichols chart can all be extended to the analysis and design of digital control systems without complications.

The basic feature of the frequency-response method is that the description of the performance of a linear time-invariant system is given in terms of its steady-state response to sinusoidally varying input signals. The crux of the problem is that characteristics of the time-domain performance of a linear system can be predicted based on the sinusoidal steady-state analysis information. For instance, the bandwidth parameter in the frequency-domain analysis is linked directly to how "fast" and oscillatory the time response will be. Therefore, in many design problems, instead of specifying the maximum overshoot, the rise time, and the settling time of the time response, the designer often chooses to specify the bandwidth of the system.

Another advantage of frequency-domain analysis is that the controller in a control system can be treated as a filter, and in the case of a digital controller, it is a digital filter. In many situations, the characteristics of the digital filter are better described in terms of its frequency-domain properties.

It is well known that in the Laplace-transform domain the sinusoidal steady-state analysis is carried out by setting $s = j\omega$. Thus, in the z-domain, we replace z by $e^{j\omega T}$. Some of the unique characteristics found in digital control systems make the frequency-domain study more interesting. For instance, the responses of a linear continuous-data control system contain only the same frequency component as the sinusoidal input signal, and only in systems with nonlinear elements could harmonics or subharmonics of the input signal be found. However, in linear digital control systems, the sampler, whether it is a real or fictitious component, acts as a harmonic generator, so that the system response in general may contain higher harmonics of the input signal. Therefore, the main difficulty with studying control systems in the frequency domain stems from the fact that these high-frequency components often make the construction and interpretation of the frequency-loci plots more difficult.

The study of digital control systems in the frequency domain essentially relies on the extension of all the existing techniques devised for the analysis of continuous-data systems. Some of the well-known methods are described as follows.

1. The Nyquist Plot

The Nyquist plot of a transfer function, usually the loop transfer function $GH(z)$, is a mapping of the Nyquist path in the z-plane onto the $GH(z)$ plane which is in polar coordinates. Thus, the Nyquist plot is also known as a polar plot. Absolute and relative stabilities of closed-loop digital control systems can be determined from the Nyquist plot of $GH(z)$.

2. The Bode Diagram

The Bode diagram is a plot of the amplitude in decibels (dB) and the phase angle of a transfer function, usually the open-loop transfer function $G(z)$, as a function of frequency ω. The Bode diagram may be used to investigate the absolute and relative stabilities of a closed-loop digital control system.

3. The Gain-Phase Plot

The gain-phase plot of the open-loop transfer function of a control system is a plot of amplitude in dB versus phase in degrees. The plot can be used to determine the absolute and relative stabilities of the closed-loop system. When the gain-phase plot of $G(z)$ is superimposed on the Nichols chart [1], relative stability and information on the closed-loop frequency response can be obtained.

We shall show in the following sections how these frequency-domain tools are applied to digital control systems.

8-2 POLAR PLOT OF $GH(z)$

Given the loop transfer function $GH(z)$ of a closed-loop digital control system, the polar plot of $GH(z)$ is obtained by setting

$$z = e^{j\omega T} \tag{8-1}$$

and varying ω from zero to infinity. This corresponds to a mapping of the points on the unit circle $|z| = 1$ in the z-plane onto the $GH(z)$ plane. The best method of conducting the calculations of the mapping is to use a digital computer program, such as FREQRPZ of the DCSP software package. The following example illustrates the analytical relations of the polar plot, and some of the important characteristics of the mapping.

Example 8-1

Consider that the loop transfer function of a digital control system is given by

$$G(s)H(s) = \frac{1.57}{s(s+1)} \tag{8-2}$$

The sampling frequency ω_s is arbitrarily chosen to be 4 rad/s, or the sampling period T is $\pi/2$ s. The z-transform of $G(s)H(s)$ is

$$GH(z) = \frac{1.243z}{(z-1)(z-0.208)} \tag{8-3}$$

Substituting Eq. (8-1) into the last equation, we have

$$GH(e^{j\omega T}) = \frac{1.243e^{j\omega T}}{(e^{j\omega T}-1)(e^{j\omega T}-0.208)} \tag{8-4}$$

Since $GH(e^{j\omega T})$ is a complex quantity, it can be written as

$$\begin{aligned} GH(e^{j\omega T}) &= |GH(e^{j\omega T})| \angle GH(e^{j\omega T}) \\ &= \text{Re}\,[GH(e^{j\omega T})] + j\,\text{Im}\,[GH(e^{j\omega T})] \end{aligned} \tag{8-5}$$

Figure 8-1 shows the graphical interpretation of the magnitude and phase of $GH(e^{j\omega T})$ in the z-plane when z takes on any value $z_1 = j\omega_1 T$ on the unit circle. Thus,

$$|GH(e^{j\omega_1 T})| = \frac{1.243A}{B \cdot C} \tag{8-6}$$

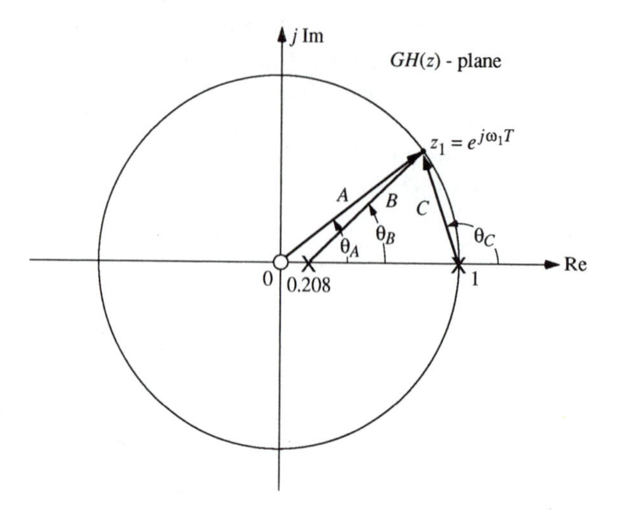

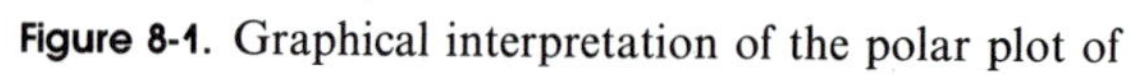
Figure 8-1. Graphical interpretation of the polar plot of

$$GH(z) = \frac{1.243z}{(z-1)(z-0.208)}.$$

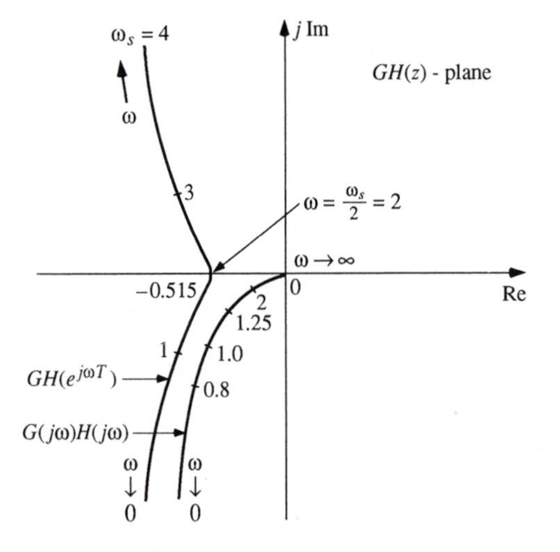

Figure 8-2. Polar plot of

$$G(s)H(s) = \frac{1.57}{s(s+1)}$$

and of

$$GH(z) = \frac{1.243z}{(z-1)(z-0.208)} \qquad \text{with} \quad T = \frac{\pi}{2}\ \text{s}.$$

where A is the length of the phasor drawn from the zero at $z = 0$ to z_1, and B and C are the lengths of the phasors drawn from the poles at $z = 0.208$ and $z = 1$, respectively, to z_1. The angle relationship is given by

$$\angle GH(e^{j\omega_1 T}) = \theta_A - \theta_B - \theta_C \tag{8-7}$$

where θ_A, θ_B, and θ_C are the angles of the three phasors mentioned above.

Since the unit circle is traversed once for every $\omega = n\omega_s$ for $n = 1, 2, \ldots,$ the polar plot of $GH(z)$ repeats over the same frequency range. Figure 8-2 shows the polar plot of $GH(z)$ when z takes on values along the unit circle. The polar plot of $G(j\omega)H(j\omega)$ for positive values of ω is also shown in the figure. In the present case, the polar plot of $GH(z)$ repeats every $\omega = \omega_s = 4$ rad/s. Thus, only the portion for $\omega = 0$ to $\omega = 4$ rad/s is shown; the portions corresponding to the rest of the frequency range are all identical. Since the portion of the unit circle from $\omega = 0$ to $\omega = \omega_s/2$ is symmetrical to that from $\omega = \omega_s/2$ to ω_s, the corresponding portions of the polar plot of $GH(z)$ will also be symmetrical in general.

Example 8-2

As another illustrative example of the polar plots of digital control systems, consider the open-loop transfer function of a digital control system,

$$G(z) = \frac{0.758(z + 1)}{(z - 1)(z - 0.242)} \tag{8-8}$$

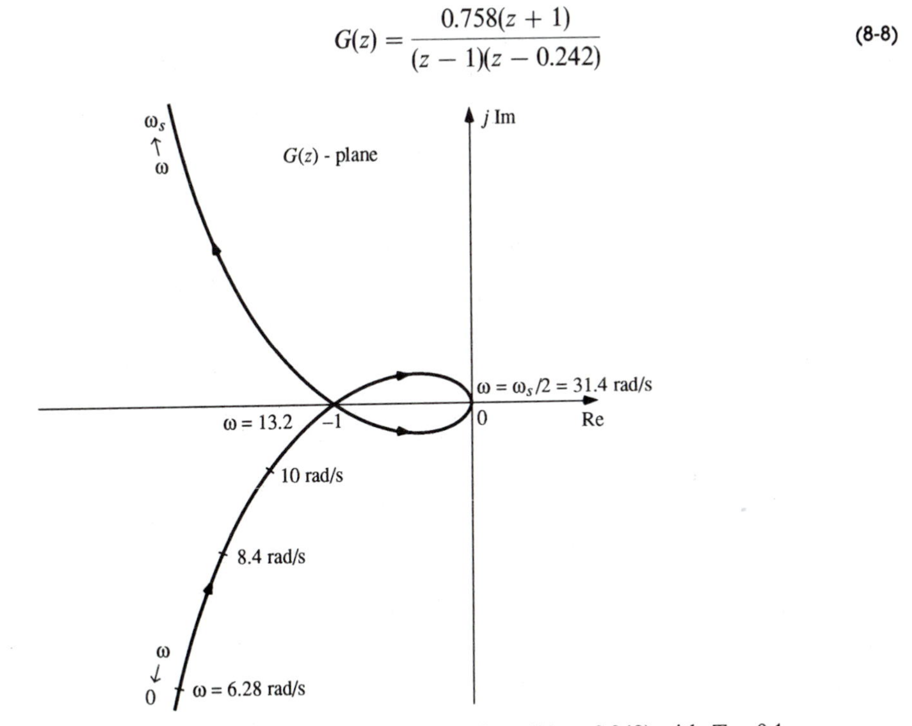

Figure 8-3. Polar plot of $G(z) = 0.758(z + 1)/(z - 1)(z - 0.242)$ with $T = 0.1$ s.

The sampling period of the system is $T = 0.1$ s; thus, $\omega_s = 62.8$ rad/s. Substituting Eq. (8-1) into Eq. (8-8), the polar plot of $G(z)$ is computed by the computer program FREQRPZ and is plotted as shown in Fig. 8-3 for $\omega = 0$ to ω_s.

8-3 THE NYQUIST STABILITY CRITERION

In Chapter 6 we defined the stability condition of digital control systems and introduced several analytical methods of stability testing. The Nyquist criterion is a graphical method of testing the stability of a closed-loop system by sketching the polar plot of the loop transfer function $GH(z)$.

The closed-loop transfer function of a single-loop, single-input, single-output digital control system is written as

$$M(z) = \frac{C(z)}{R(z)} = \frac{G(z)}{1 + GH(z)} \tag{8-9}$$

In Chapter 6 we established that the stability of the system is studied by investigating the zeros of the function $1 + GH(z)$ or the characteristic equation roots satisfying the following equation:

$$1 + GH(z) = 0 \tag{8-10}$$

For the system described by Eq. (8-9) to be stable, all the roots of the characteristic equation must lie inside the unit circle $|z| = 1$ in the z-plane.

As with the Nyquist criterion for linear continuous-data systems [1], the Nyquist criterion for digital control systems involves the following steps.

1. Define the Nyquist path in the z-plane that encloses the exterior of the unit circle.
2. Map the Nyquist path in the z-plane onto the $GH(z)$ plane with the function $GH(z)$. This results in the Nyquist plot for $GH(z)$.
3. The stability condition of the closed-loop system is determined by investigating the behavior of the Nyquist plot of $GH(z)$ with respect to the *critical point* $(-1, j0)$ in the $GH(z)$ plane.

8-3-1 The Nyquist Paths

We shall devise the Nyquist criterion for digital control systems according to the general version introduced by Yeung and Lai [2]. Two Nyquist paths are defined as shown in Fig. 8-4(a) and (b). Assuming that the function $GH(z)$ has poles on the unit circle marked by $\times$'s as shown, *the Nyquist path* Γ_{z1} *in Fig. 8-4(a) encloses the exterior of the unit circle, excluding the poles of GH(z) on the unit circle. The Nyquist path* Γ_{z2} *in Fig. 8-4(b) encloses the exterior of the unit circle, including the poles of the GH(z) on the unit circle. The usual definition of enclosure is that the region to the left of a traversed path is enclosed.*

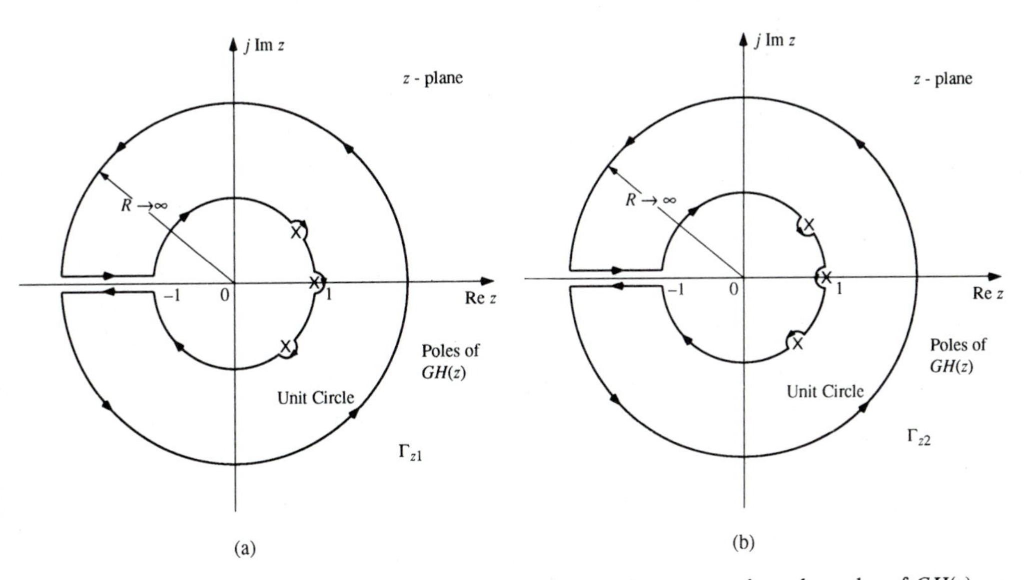

Figure 8-4. Nyquist paths in the z-plane. (a) Γ_{z1}, does not enclose the poles of $GH(z)$ on the unit circle. (b) Γ_{z2}, encloses the poles of $GH(z)$ on the unit circle.

8-3-2 The Nyquist Criterion

Let us define the following quantities with respect to the function $GH(z)$.

- Z_{-1} number of zeros of $1 + GH(z)$ that are outside the unit circle in the z-plane.
- P_{-1} number of poles of $1 + GH(z)$ that are outside the unit circle in the z-plane.
- P_{ω} number of poles of $GH(z)$ [same as the number of poles of $1 + GH(z)$] that are on the unit circle in the z-plane.
- N_1 number of times the $(-1, j0)$ point of $GH(z)$ is encircled by the Nyquist plot of $GH(z)$ corresponding to Γ_{z1}.
- N_2 number of times the $(-1, j0)$ point of $GH(z)$ is encircled by the Nyquist plot of $GH(z)$ corresponding to Γ_{z2}.

Then, with reference to the Nyquist paths in Fig. 8-4 and according to the principle of the argument [1] of complex-variable theory,

$$N_1 = Z_{-1} - P_{-1} \tag{8-11}$$

and

$$N_2 = Z_{-1} - P_{-1} - P_{\omega} \tag{8-12}$$

Let

Φ_1 net angle traversed by the phasor drawn from the $(-1, j0)$ point to the $GH(z)$ plot corresponding to the Nyquist path Γ_{z1}.

Φ_2 net angle traversed by the phasor drawn from the $(-1, j0)$ point to the $GH(z)$ plot corresponding to the Nyquist path Γ_{z2}.

Alternatively, in terms of N_1, N_2, Z_{-1}, P_{-1}, and P_ω,

$$\Phi_1 = N_1 \times 360° = (Z_{-1} - P_{-1})360° \tag{8-13}$$

$$\Phi_2 = N_2 \times 360° = (Z_{-1} - P_{-1} - P_\omega)360° \tag{8-14}$$

The angles Φ_1 and Φ_2 are positive if the net angle traversed by the phasor is counterclockwise and negative if clockwise.

Let us consider that each of the Nyquist paths in the z-plane is composed of three sections:

1. the section along the circle with infinite radius
2. the section on the unit circle, excluding all the small indentations around the poles of $GH(z)$
3. all the small indentations on the unit circle.

Since the Nyquist paths are symmetrical about the real axis in the z-plane, the angles traversed by the Nyquist plots corresponding to these Nyquist paths are identical for positives and negatives of z. Thus, the angles Φ_1 and Φ_2 are expressed as

$$\Phi_1 = 2\Phi_{11} + \Phi_{12} + \Phi_{13} \tag{8-15}$$

$$\Phi_2 = 2\Phi_{11} - \Phi_{12} + \Phi_{13} \tag{8-16}$$

where

Φ_{11} angle traversed by the phasor drawn to the $GH(z)$ plot from the $(-1, j0)$ point corresponding to the positive half from $\omega = \omega_s/2$ to $\omega = 0$, or to the negative half from $\omega = \omega_s/2$ to ω_s, on the unit circle of Γ_{z1} or Γ_{z2}, excluding the small indentations.

Φ_{12} angle traversed by the phasor drawn to the $GH(z)$ plot from the $(-1, j0)$ point corresponding to the small indentations on the unit circle on Γ_{z1}. Since the directions of the small indentations on Γ_{z2} with respect to the poles on the unit circle are opposite to the direction of those on Γ_{z1}, the sign in front of Φ_{12} in Eq. (8-16) is negative.

Φ_{13} angle traversed by the phasor drawn to the $GH(z)$ plot from the $(-1, j0)$ point corresponding to the circle with infinite radius on Γ_{z1} and Γ_{z2}.

For physically realizable transfer functions, $GH(z)$ cannot have more zeros than poles. This means that the Nyquist plot of $GH(z)$ that corresponds to the circle with infinite radius must either be a point on the real axis or a trajectory around the origin of the $GH(z)$ plane. Thus, the angle Φ_{13} is always zero.

Now adding Eq. (8-15) to Eq. (8-16), we have

$$\begin{aligned}\Phi_1 + \Phi_2 &= 4\Phi_{11} \\ &= (2Z_{-1} - 2P_{-1} - P_{\omega})360^\circ\end{aligned} \tag{8-17}$$

Solving for Φ_{11} from the last equation, we get

$$\Phi_{11} = (Z_{-1} - P_{-1} - 0.5P_{\omega})180^\circ \tag{8-18}$$

The last equation represents the Nyquist criterion, which is stated as follows.

The total angle traversed by the phasor drawn from the $(-1, j0)$ point to the Nyquist plot of $GH(z)$ that corresponds to the upper half of the unit circle of the z-plane from $\omega = \omega_s/2$ to $\omega = 0$, excluding the small indentations, if any, equals

> [number of zeros of $1 + GH(z)$ that lie outside the unit circle $- 0.5 \times$ (number of poles of $GH(z)$ that lie on the unit circle) $-$ number of poles of $GH(z)$ that lie outside the unit circle] $\times$ 180 degrees

Thus, *the Nyquist stability criterion for linear digital control systems can be carried out by constructing only the Nyquist plot of GH(z) that corresponds to the* $\omega = \omega_s/2$ *to* $\omega = 0$ *portion on the Nyquist paths, excluding the small indentations, if any. Furthermore, if the closed-loop system is unstable* ($Z_{-1} \neq 0$), *knowing the values of* Φ_{11}, P_{ω}, *and* P_{-1}, *Eq.* (8-18) *gives the value of* Z_{-1} *which is the number of roots of the characteristic equation that are outside the unit circle in the z-plane.*

For the closed-loop digital control system to be stable, Z_{-1} must equal zero. Thus, *the Nyquist criterion for stability of the closed-loop system is stated as*

$$\Phi_{11} = -(0.5P_{\omega} + P_{-1})180^\circ \tag{8-19}$$

Since P_{ω} and P_{-1} cannot be negative, the last equation leads to the following observation.

For the closed-loop digital control system to be stable, the angle traversed by the phasor from the $(-1, j0)$ *point to the Nyquist plot of GH(z) as* ω *varies from* $\omega_s/2$ *to* 0 *cannot be negative.*

The crux of the Nyquist criterion is that the application of the criterion does not require a detailed plot of $GH(z)$, as long as the value of Φ_{11} can be determined from the plot.

8-3-3 Illustrative Examples of the Nyquist Criterion

The following examples illustrate the application of the Nyquist criterion to the stability of digital control systems.

Example 8-3

Consider the digital control system described in Example 7-4. For $T = 0.1$ s, the loop transfer function is

$$GH(z) = \frac{0.0952Kz}{(z-1)(z-0.905)} \tag{8-20}$$

With reference to the Nyquist criterion, the following quantities are obtained for $GH(z)$:

$P_{-1} = 0$ $GH(z)$ does not have any poles outside the unit circle.

$P_{\omega} = 1$ $GH(z)$ has one pole on the unit circle (at $z = 1$).

Thus, the Nyquist path must have a small indentation at $z = 1$ on the unit circle. The Nyquist plot of $GH(z)$ is sketched as shown in Fig. 8-5 for points on the top half of the unit circle, excluding the small indentation at $z = 1$ on the Nyquist path. The Nyquist plot of $GH(z)$ intersects the negative real axis of the $GH(z)$-plane at $-0.025K$ when $\omega = \omega_s/2 = 31.4$ rad/s. Depending on the value of K, the critical $(-1, j0)$ point can be to the left or the right of the critical point. Applying the Nyquist criterion in Eq. (8-19), the closed-loop system is stable if

$$\Phi_{11} = -(0.5P_{\omega} + P_{-1})180° = -90° \tag{8-21}$$

Thus, we see from Fig. 8-5 that for Φ_{11} to be -90 degrees, the $(-1, j0)$ point must be located to the left of the $-0.025K$ point. Thus, $K < 1/0.025 =$

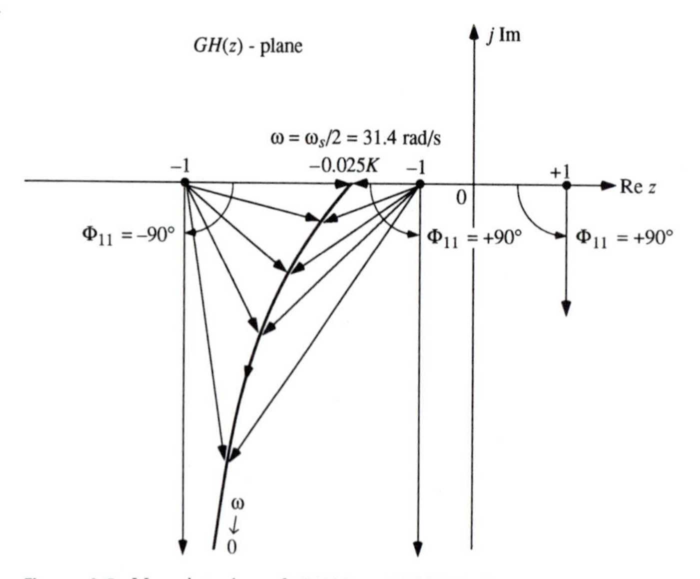

Figure 8-5. Nyquist plot of $GH(z) = 0.0952Kz/(z-1)(z-0.905)$ and the interpretation of stability of the closed-loop system with respect to the value of K.

40 for stability. If K is greater than 40, the critical point will be to the right of $-0.025K$, and Fig. 8-5 shows that $\Phi_{11} = +90$ degrees. From Eq. (8-18), we have

$$Z_{-1} = \frac{\Phi_{11}}{180°} + P_{-1} + 0.5P_{\omega} = 1 \tag{8-22}$$

Thus, for $K > 40$, the characteristic equation of the closed-loop system would have one root outside the unit circle.

If K is negative, we can still use the same Nyquist plot of $GH(z)$ in Fig. 8-5, but refer to the $(+1, j0)$ point as the critical point. From Fig. 8-5, it is apparent that Φ_{11} still equals $+90$ degrees and the system is unstable. Thus, the stability requirement is

$$0 \leq K < 40 \tag{8-23}$$

which agrees with the result in Example 7-4.

Example 8-4

The following transfer function from Example 7-4 is obtained when the sampling period is increased to 1 s:

$$GH(z) = \frac{0.632Kz}{(z-1)(z-0.368)} \tag{8-24}$$

Thus, $P_{-1} = 0$ and $P_{\omega} = 1$. The Nyquist plot of $GH(z)$ is shown in Fig. 8-6.

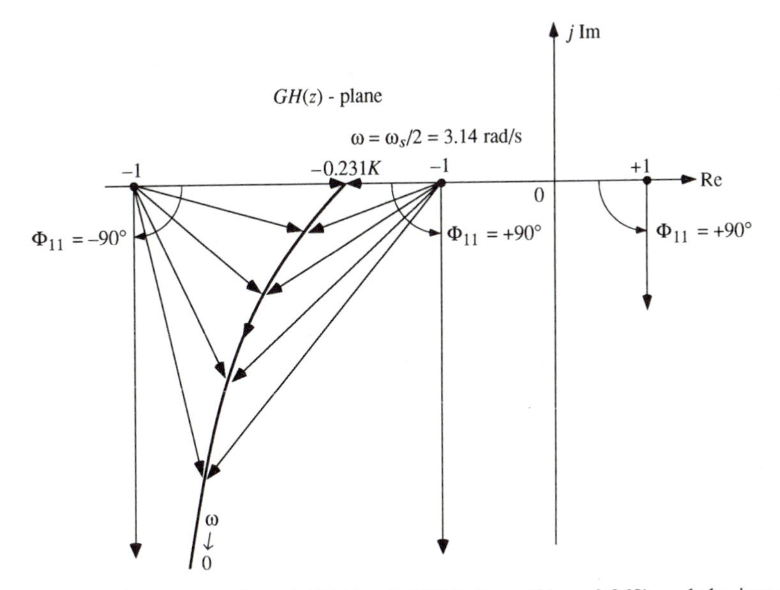

Figure 8-6. Nyquist plot of $GH(z) = 0.632Kz/(z-1)(z-0.368)$ and the interpretation of stability of the closed-loop system with response to the value of K.

The intersect of $GH(z)$ on the negative real axis of the $GH(z)$-plane is $-0.231K$, where $\omega = \omega_s/2 = 3.14$ rad/s. Following the same analysis as in Example 8-3, the stability requirement is

$$0 \le K < 4.33 \tag{8-25}$$

which agrees with the result obtained in Example 7-4.

Example 8-5

Consider the digital space vehicle control system described in Example 7-5. The open-loop transfer function of the system when $T = 0.1$ s is

$$G(z) = \frac{0.15K(z + 0.7453)}{(z - 1)(z - 0.4119)} \tag{8-26}$$

Since $GH(z)$ has one pole on the unit circle at $z = 1$ and no poles outside the unit circle, $P_\omega = 1$ and $P_{-1} = 0$. From Eq. (8-19), the closed-loop system is stable if

$$\begin{aligned}\Phi_{11} &= -(0.5P_\omega + P_{-1})180^\circ \\ &= -90^\circ\end{aligned} \tag{8-27}$$

Figure 8-7 shows the Nyquist plot of $GH(z)$ with K as a variable parameter for the top half of the unit circle on the Nyquist path. The $GH(z)$ plot intersects

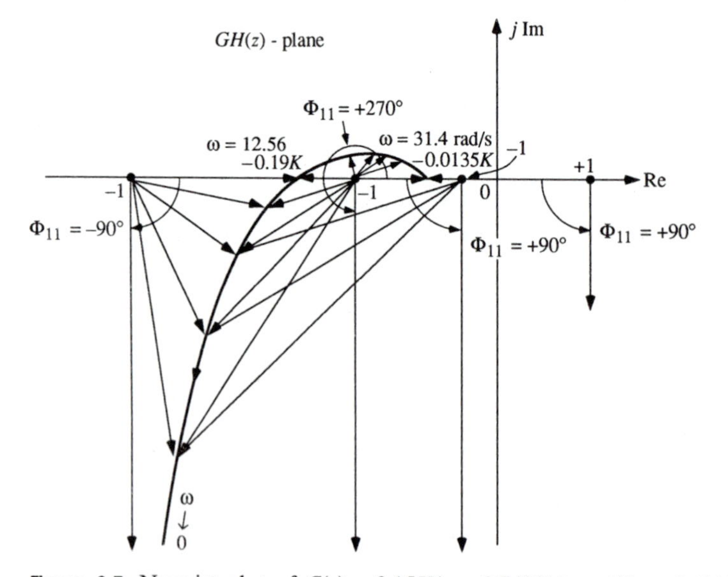

Figure 8-7. Nyquist plot of $G(z) = 0.15K(z + 0.7433)/(z - 1)(z - 0.4119)$ and the interpretation of stability of the closed-loop system with respect to the value of K.

the negative real axis at $-0.19K$ and $-0.0135K$ where $\omega = 12.56$ rad/s and $\omega_s/2 = 31.4$ rad/s, respectively. As shown in Fig. 8-7, for $K > 0$, the critical point $(-1, j0)$ can lie in one of three possible regions, depending on the value of K. The stability conditions are tabulated as follows:

Critical Point $(-1, j0)$	Φ_{11} (degrees)	Z_{-1}	Stability Condition
To the left of $-0.19K$	-90	0	stable
Between $-0.19K$ and $-0.0135K$	$+270$	2	unstable
Between $-0.0135K$ and 0	$+90$	1	unstable

When K is negative, the critical point is at $(1, j0)$, and $\Phi_{11} = +90$ degrees, the closed-loop system is unstable. Thus, for stability, the requirement on K is

$$0 \leq K < 5.26 \tag{8-28}$$

which agrees with the results obtained in Example 7-5.

Example 8-6

For the same space vehicle control system described in Example 7-5, when $T = 0.5$ s, the open-loop transfer function becomes

$$G(z) = \frac{1.7788K(z + 0.1153)}{(z - 1)(z - 0.1089)} \tag{8-29}$$

The open-loop transfer function $G(z)$ has one pole on the unit circle at $z = 1$, and no poles outside the unit circle. Thus, $P_\omega = 1$ and $P_{-1} = 0$. Using Eq. (8-19), the stability requirement is that Φ_{11} must equal -90 degrees.

The Nyquist plot of $G(z)$ for the top half of the unit circle of the Nyquist path, minus the small indentation at $z = 1$, is sketched as shown in Fig. 8-8. The $G(z)$ plot intersects the negative real axis at $-0.7096K$. The stability conditions with respect to positive values of K are tabulated below:

Critical Point $(-1, j0)$	Φ_{11} (degrees)	Z_{-1}	Stability Condition
To the left of $-0.7096K$	-90	0	stable
Between $-0.7096K$ and 0	$+90$	1	unstable

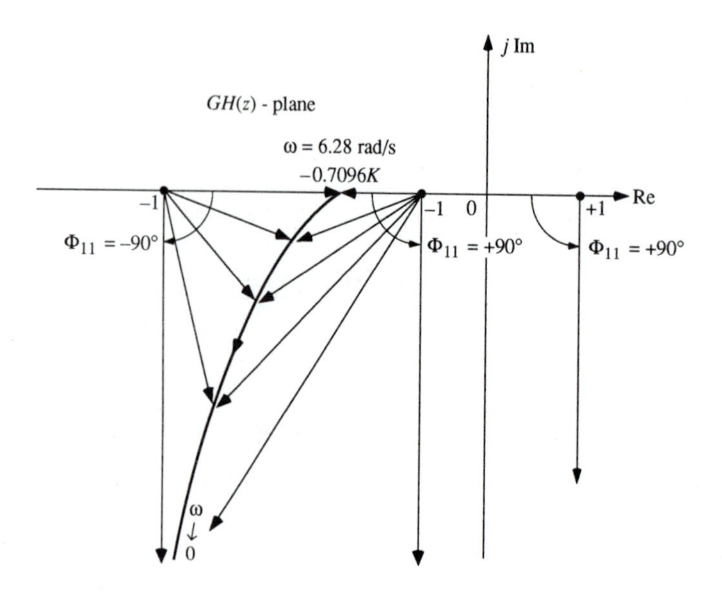

Figure 8-8. Nyquist plot of $G(z) = 1.7788K(z + 0.1153)/(z - 1)(z - 0.1089)$ and the interpretation of stability of the closed-loop system with respect to the value of K.

When K is negative, $\Phi_{11} = +90$ degrees, and the system is unstable. Thus, the stability requirement is

$$0 \leq K < 1.409 \tag{8-30}$$

which agrees with the results obtained in Example 7-5.

Example 8-7

Consider that the open-loop transfer function of a unity-feedback digital control system is given as

$$G(z) = \frac{0.01Kz}{(z-1)^2(z-0.905)} \tag{8-31}$$

The sampling period is 0.1 s. The function $G(z)$ has two poles on the unit circle at $z = 1$ and no poles outside the unit circle. Thus, $P_\omega = 2$ and $P_{-1} = 0$. From Eq. (8-19), we have

$$\Phi_{11} = -(0.5P_\omega + P_{-1})180° = -180° \tag{8-32}$$

Thus, Φ_{11} must equal -180 degrees for the closed-loop system to be stable.

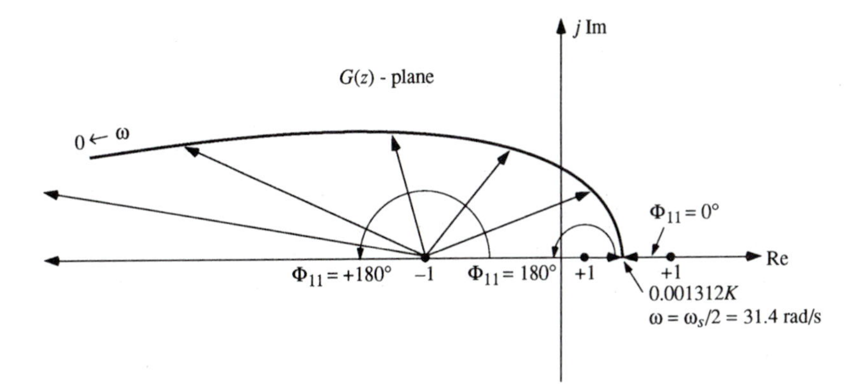

Figure 8-9. Nyquist plot of $G(z) = 0.01Kz/(z-1)(z-0.905)$ and the interpretation of stability of the closed-loop system with respect to the value of K.

Figure 8-9 shows the Nyquist plot of $G(z)$ that corresponds to the top half of the unit circle of the Nyquist path. The Nyquist plot intersects the real axis at $0.001312K$ when $\omega = \omega_s/2 = 31.4$ rad/s. For all positive values of K, the angle Φ_{11} associated with the $(-1, j0)$ point is $+180$ degrees. Thus, the system is unstable for all positive values of K. From Eq. (8-18), we have

$$Z_{-1} = \frac{\Phi_{11}}{180^\circ} + P_{-1} + 0.5P_\omega = 1 + 1 = 2 \tag{8-33}$$

Thus, when K is positive, the characteristic equation of the closed-loop system has two roots outside the unit circle.

When K is negative, the critical point is at $(+1, j0)$. From Fig. 8-9 we see that the following two conditions exist as K varies.

Critical Point $(+1, j0)$	Φ_{11} (degrees)	Z_{-1}	Stability Condition
Between 0 and $0.001312K$	$+180$	2	unstable
To the right of $0.001312K$	0	1	unstable

Thus, the system is unstable for all values of K, and the characteristic equation can have one or two roots outside the unit circle as K varies. That is,

$0 \le K < \infty$	two roots outside the unit circle
$-\infty < K < -762.2$	two roots outside the unit circle
$-762.2 < K < 0$	one root outside the unit circle

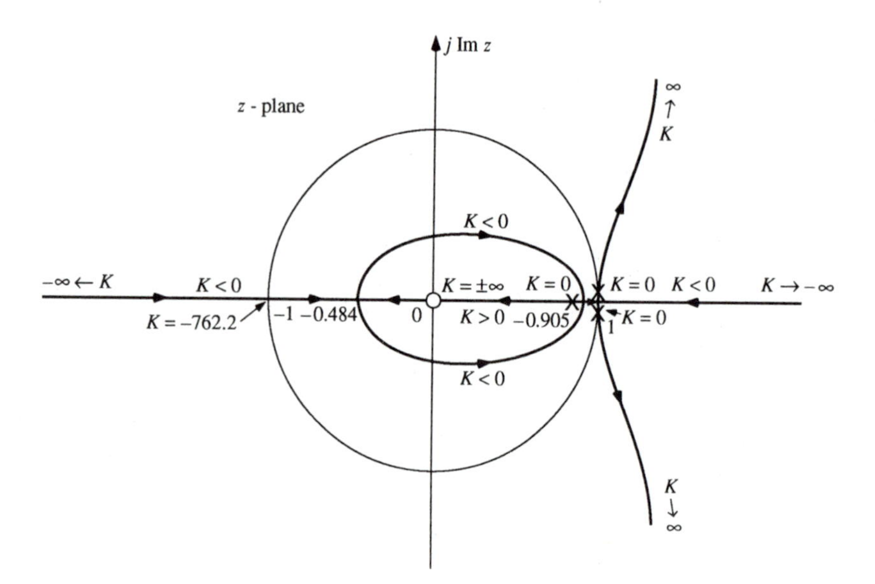

Figure 8-10. Root-locus diagram of $G(z) = 0.01Kz/(z-1)^2(z-0.905)$ for $-\infty < K < \infty$.

The root-locus diagram of the system is shown in Fig. 8-10. The correlation between the results of the Nyquist criterion and the root loci should be quite enlightening.

Example 8-8

Consider that the open-loop transfer function of a unity-feedback digital control system is given as

$$G(z) = \frac{K(z-0.5)(z-0.7)}{(z-1)(z^2-1.414z+1)} \tag{8-34}$$

The sampling period is 0.1 s. The poles of $G(z)$ are at $z = 1$, $z = 0.707 + j0.707$, and $z = 0.707 - j0.707$, which are all on the unit circle. Thus, $P_\omega = 3$ and $P_{-1} = 0$. From Eq. (8-19), for the closed-loop system to be stable, Φ_{11} must satisfy

$$\Phi_{11} = -(0.5P_\omega + P_{-1})180^\circ = -270^\circ \tag{8-35}$$

The Nyquist plot of $G(z)$ that corresponds to the top half of the unit circle of the Nyquist path, excluding the small indentations, is sketched as shown in Fig. 8-11. The sketch is not drawn to scale since the exact plot of $G(z)$ is not necessary as long as the value of Φ_{11} can be determined accurately. As shown in Fig. 8-11, the $G(z)$ plot has a discontinuity when z takes on the value of

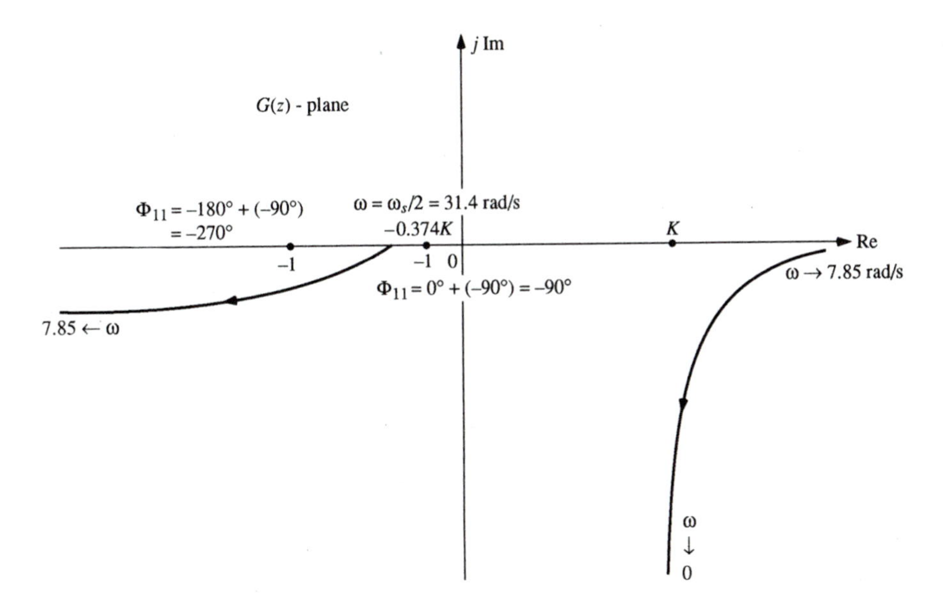

Figure 8-11. Nyquist plot of $G(z) = K(z - 0.5)(z - 0.7)/(z - 1)(z^2 - 1.414z + 1)$ and the interpretation of stability of the closed-loop system with respect to the value of K.

$0.707 + j0.707$. The corresponding value of ω at this point is found from

$$z = 1\angle\pi/4 = 1\angle\omega T = 1\angle 0.1\omega \tag{8-36}$$

Solving for ω from the last equation, we get $\omega = 7.85$ rad/s.

Figure 8-11 shows that when $\omega = \omega_s/2 = 31.4$ rad/s, the Nyquist plot of $G(z)$ intersects the negative real axis at $-0.374K$. When ω approaches 7.85 rad/s, the real part of $G(z)$ approaches $-\infty$, and the angle of $G(z)$ approaches 180 degrees. The $G(z)$ plot then jumps to 0 degrees and then approaches -90 degrees when ω goes to zero. If we had used the entire Nyquist path, including the three small indentations and the large circle around infinity, the Nyquist plot of $G(z)$ would have traced out a continuous trajectory. However, the entire Nyquist plot of $G(z)$ is not necessary, as we shall see that the stability of the closed-loop system can be determined from the plot shown in Fig. 8-11.

From Fig. 8-11, we see that for $K > 0$, the critical point $(-1, j0)$ can be located either to the right or to the left of the point $-0.375K$. For $K < 2.67$, the critical point is to the left of $-0.375K$. The angle Φ_{11} of the phasor drawn from the $(-1, j0)$ point to the Nyquist plot as ω varies from 31.4 rad/s to 0 consists of the portion from $\omega = 31.4$ rad/s to 7.85 rad/s which is -180 degrees, and the portion from 7.85 rad/s to zero which is -90 degrees. Thus,

$0 < K < 2.67$

$$\Phi_{11} = -180^\circ + (-90^\circ) = -270^\circ \tag{8-37}$$

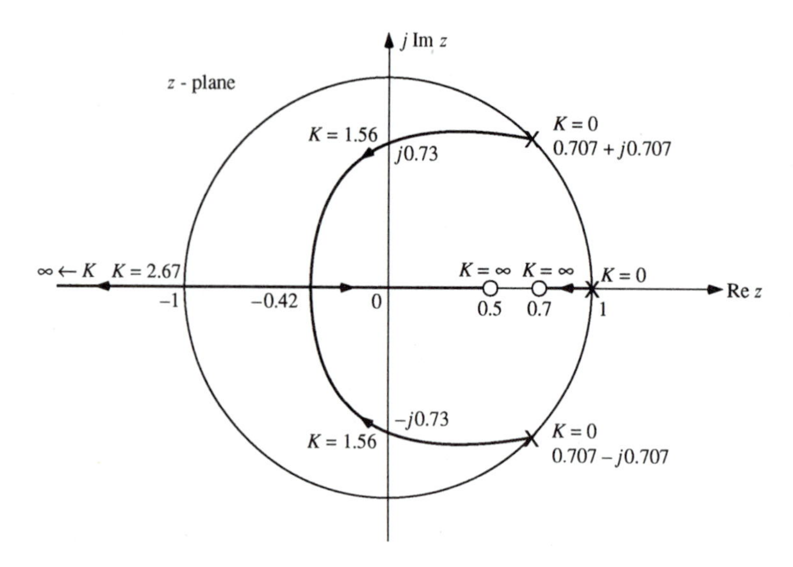

Figure 8-12. Root-locus diagram of $G(z) = K(z - 0.5)(z - 0.7)/(z - 1)(z^2 - 1.414z + 1)$ for $K > 0$.

Since the value of Φ_{11} agrees with that required for stability, the system is stable for $K < 2.67$. Similarly,

$\infty > \boldsymbol{K} > \mathbf{2.67}$

$$\Phi_{11} = 0° + (-90°) = -90° \tag{8-38}$$

and thus the system is unstable. We can show by using the $(+1, j0)$ point as the critical point that the system is unstable for all negative values of K.

The root-locus diagram of the system for $0 \leq K < \infty$ is shown in Fig. 8-12. The root-loci results corroborate those obtained from the Nyquist plot. The reader may verify that when K is greater than 2.67, the characteristic equation of the closed-loop system has one root outside the unit circle.

8-4 BODE PLOT

Given a z-transform transfer function $G(z)$, the Bode plot consists of a magnitude plot and a phase plot. The magnitude plot is the plot of $20|G(e^{j\omega T})|$ in decibels versus ω. The phase plot is $\angle G(e^{j\omega T})$ in degrees versus ω.

The simplest way of making the Bode plot is by means of a digital computer program, such as FREQRPZ of the DCSP software package. As pointed out earlier in Sec. 8-2, the frequency response of $G(e^{j\omega T})$ repeats every ω_s, and the polar plot is symmetrical about the real axis. Thus, the magnitude plot of the

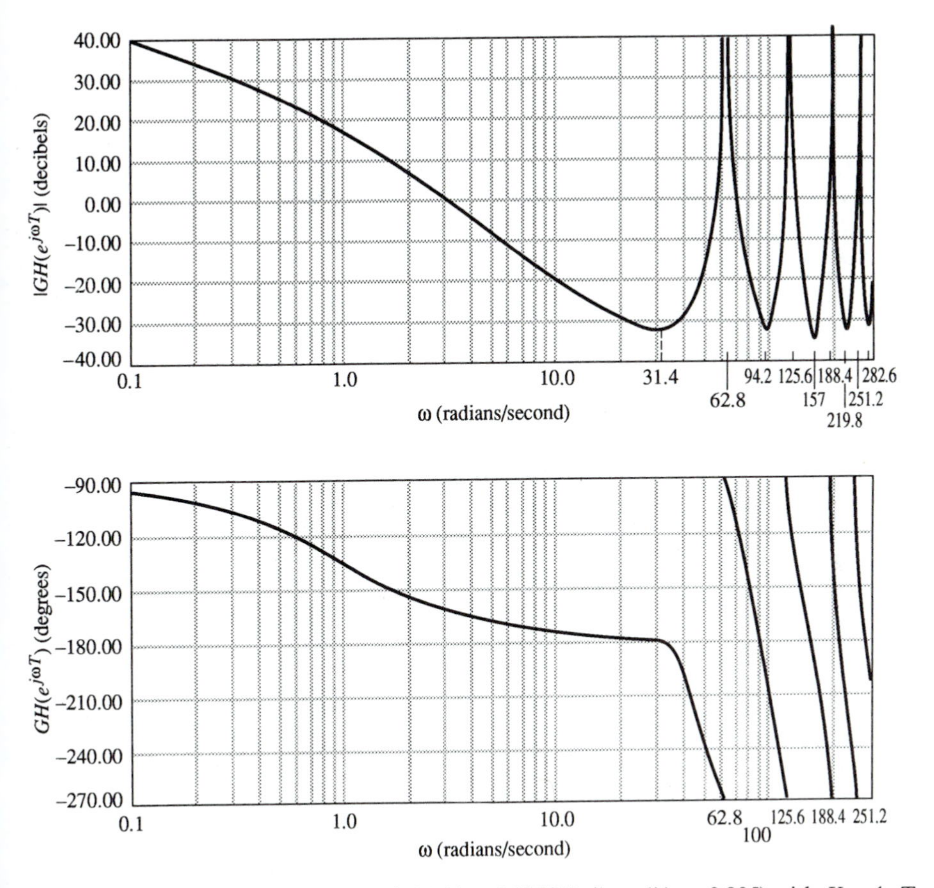

Figure 8-13. Bode plot of $GH(z) = 0.0952Kz/(z - 1)(z - 0.905)$ with $K = 1$, $T = 0.1$ s.

Bode diagram is identical for every ω_s. As an illustrative example, the Bode plot of $GH(z)$ given in Eq. (8-20) is shown in Fig. 8-13. Due to the logarithmic ω axis, the magnitude plot of $GH(e^{j\omega T})$ is distorted, but the magnitude repeats for every ω_s. The shape of the phase plot also repeats for the same frequency range. The Nyquist plot and the root-locus plot show that the absolute and the relative stability of the closed-loop system are governed only by the Bode plot from $\omega = 0$ to $\omega_s/2$. Thus, for all practical purposes, it is necessary only to plot the Bode plot over this frequency range. As an example, the Bode plot of $G(z)$ of Eq. (8-8) is shown in Fig. 8-14 for the frequency range of $\omega = 0$ to $\omega_s/2 = 31.4$ rad/s.

One of the advantages of the Bode plot is that, since the magnitude curve is in decibels, changing the gain factor of $G(z)$ simply causes the curve to move up or down without distortion, and multiplying $G(z)$ by another transfer function $H(z)$ simply involves the addition of the magnitude curve of $H(z)$ to

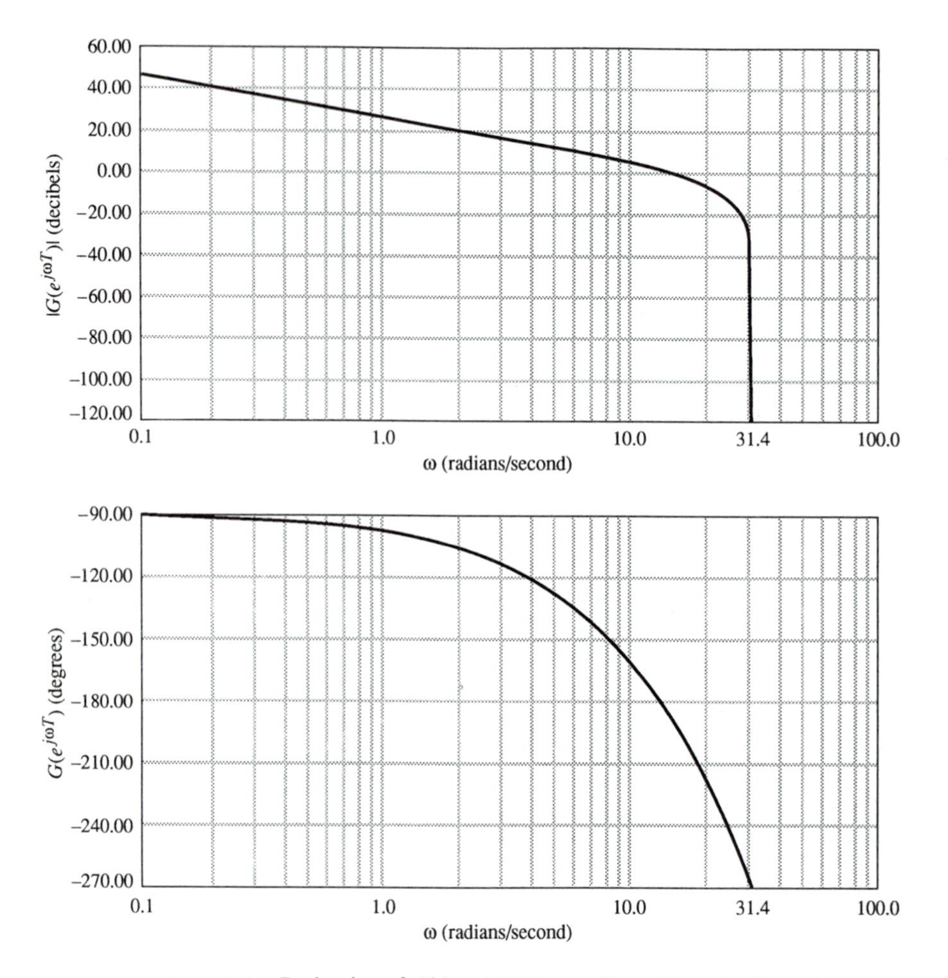

Figure 8-14. Bode plot of $G(z) = 0.758(z + 1)/(z - 1)(z - 0.242)$ with $K = 1$, $T = 0.1$ s.

that of $G(z)$ or

$$20 \log_{10} |G(e^{j\omega T})H(e^{j\omega T})| = 20 \log_{10} |G(e^{j\omega T})| + 20 \log_{10} |H(e^{j\omega T})| \tag{8-39}$$

Similarly, for the phase curve,

$$\angle [G(e^{j\omega T})H(e^{j\omega T})] = \angle G(e^{j\omega T}) + \angle H(e^{j\omega T}) \tag{8-40}$$

This means that when $G(z)$ has a pure time delay nT, which is equal to an integral multiple of the sampling period, the magnitude curve of the Bode plot of $G(z)$ is not affected, but the phase is reduced by $n\omega T$ at the frequency ω. Thus, representing a transfer function by a Bode plot facilitates design in the frequency domain.

Another advantage of the Bode plot is that, in the s-domain, the magnitude curve can be approximated by straight lines [1]. This allows the sketching of the magnitude curve without exact computation. However, this convenient feature is lost in the z-domain, since z is related to $j\omega$ through $e^{j\omega T}$. We can again use a bilinear transformation to transform the unit circle of the z-plane to the imaginary axis of another complex plane. The transformation introduced in Chapter 6 for stability studies, Eq. (6-125), may be used for this purpose. While Eq. (6-125) is suitable for extending the Routh-Hurwitz criterion, as an alternative we can find a bilinear transformation which transforms z into, say, w, such that the imaginary axis of the w-plane closely resembles the $j\omega$ axis of the s-plane. Let w be related to z through the z-transform relation; i.e.,

$$w = \frac{1}{T} \ln z \tag{8-41}$$

Expanding ln z into a power series and taking the first term of the expansion, we get

$$w = \frac{2}{T} \frac{z-1}{z+1} \tag{8-42}$$

Solving for z from the last equation, we get

$$z = \frac{(2/T) + w}{(2/T) - w} \tag{8-43}$$

We shall define Eqs. (8-42) and (8-43) as the *w-transformation.*

Substituting $z = e^{j\omega T} = \cos \omega T + j \sin \omega T$ into Eq. (8-42), we get

$$w = \frac{2}{T} \frac{\cos \omega T + j \sin \omega T - 1}{\cos \omega T + j \sin \omega T + 1} = j \frac{2}{T} \frac{\sin \omega T}{1 + \cos \omega T} = j \frac{2}{T} \tan \frac{\omega T}{2} \tag{8-44}$$

Thus, when z takes on values on the unit circle, w is imaginary and can be written as

$$w = j\omega_w = j \frac{\omega_s}{\pi} \tan \frac{\pi\omega}{\omega_s} \tag{8-45}$$

The last equation establishes the relationship between the transformed "frequency" ω_w and the real frequency ω, and the sampling frequency ω_s.

For frequency-domain analysis, we can substitute Eq. (8-43) into the transfer function $G(z)$. If $G(z)$ is a rational function of z, consisting of first-order and/or second-order factors in the numerator and the denominator, then the transformed function $G(w)$ will also be a rational function with first-order and/or second-order factors. The advantage, of course, is that the function $G(w)$, or rather $G(j\omega_w)$, can now be sketched using the straight-line approximation method, similar to the s-domain techniques.

The disadvantage with the w-transformation is that the sampling period T appears in the transform equation in Eq. (8-43), whereas the r-transformation in Eq. (6-125) is simpler to apply algebraically. Both the r-transformation and the w-transformation can be performed for $G(z)$ using the program RWT of the DCSP software package.

As an illustrative example of using the w-transformation for Bode plots, the transfer function $GH(z)$ in Eq. (8-20) is transformed through Eq. (8-43) into the following form:

$$GH(w) = \frac{10.02K(1 - 0.0025w^2)}{w(1 + 1.0026w)} \tag{8-46}$$

Figure 8-15 shows the Bode diagram of $GH(w)$ in magnitude (dB) and phase versus ω_w. The straight-line components and approximation of $G(w)$ are also

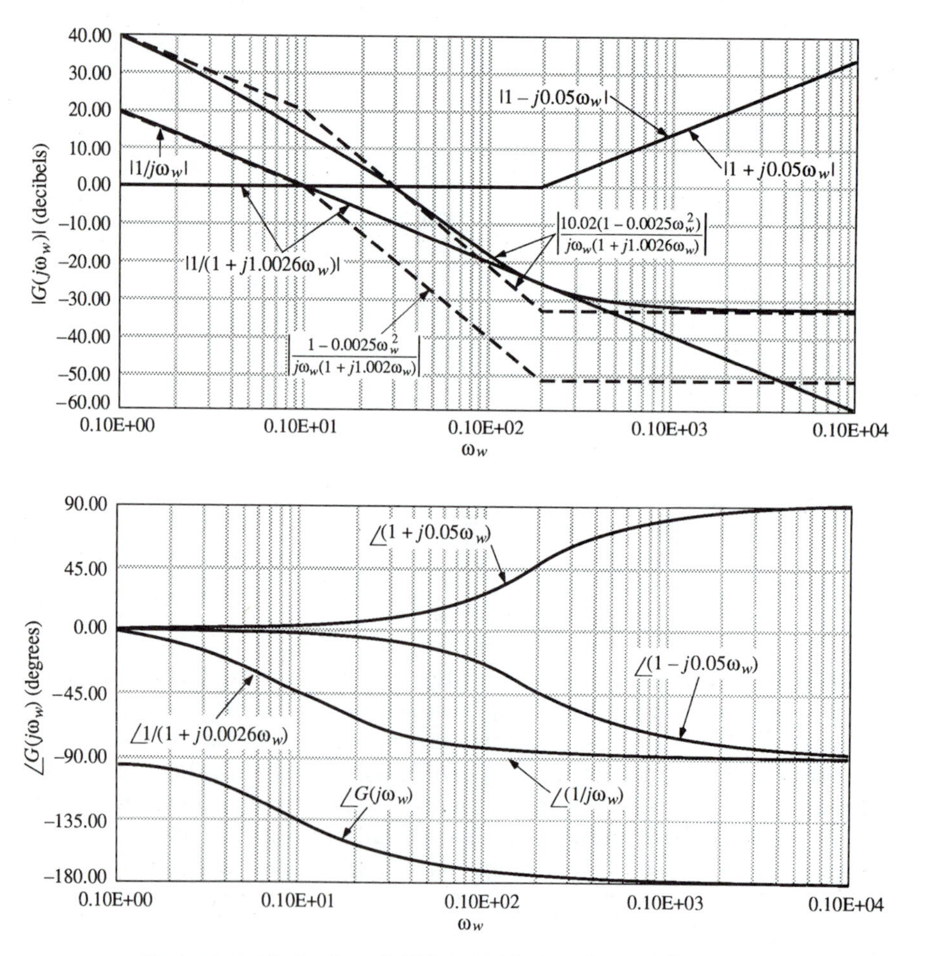

Figure 8-15. Bode plot of $G(z) = 10.02K(1 - 0.0025w^2)/(1 + 1.0026w)$ with $K = 1$, $T = 0.1$ s.

shown for the magnitude curve. Note that the abscissa of Fig. 8-15 is ω_w, whereas that of Fig. 8-13 is ω. Thus, we should not compare the two plots, although they compare very well at low frequencies. The fact is that the plot in Fig. 8-13 terminates at $\omega = \omega_s/2$ which corresponds to $\omega_w = \infty$ in Fig. 8-15.

8-5 GAIN MARGIN AND PHASE MARGIN

Two of the performance specifications in the frequency domain on the relative stability of closed-loop control systems are the *gain margin* and the *phase margin*. These specifications, which have been quite useful for the design of linear continuous-data control systems [1], can be easily extended to digital control systems.

8-5-1 Gain Margin

Let us refer to the Nyquist plot of the loop transfer function $GH(z)$ shown in Fig. 8-5. The intersect of the plot with the negative real axis of the $GH(z)$-plane is at $-0.025K$. The conclusion reached from applying the Nyquist criterion in Example 8-3 is that the closed-loop system would become unstable if K is $\geq 1/0.025$ or 40. Thus, $K = 40$ is the "safety factor" to which the gain K can be increased before the system reaches instability. When $K = 1$, the *gain margin* (GM) of the system is 40. In terms of the Bode diagram of the same system, shown in Fig. 8-13, when the value of K increases, the magnitude curve is raised by the amount $20 \log_{10} K$, while the phase curve is unchanged. When $K = 1$, the magnitude of $G(e^{j\omega T})$ at $\omega = \omega_s/2 = 31.4$ rad/s where the phase of $G(e^{j\omega T})$ is 180 degrees is -32 dB. Thus, we say that the gain margin is 32 dB (0.025), which means that the gain K can be increased by 32 dB before reaching instability. Thus, it is customary to express gain margin in decibels which can be easily read off the Bode diagram. In general, the definition of GM can be written as

$$\text{GM} = 20 \log_{10} \frac{1}{|GH(e^{j\omega_c T})|} \text{ dB} \tag{8-47}$$

$|GH(e^{j\omega_c T})|$ is the magnitude of $GH(e^{j\omega T})$ measured at the point where the Nyquist plot intersects the negative real axis, and the frequency at the point is denoted by ω_c. The frequency ω_c is defined at the point where the phase of $G(e^{j\omega T})$ is 180 degrees and is called the *phase-crossover frequency*.

Figure 8-16 shows the condensed versions of the Nyquist and Bode plots of Eq. (8-20) when $K = 1$. The phase-crossover point and the gain margin are indicated on the plots. Figure 8-17 shows the same for the transfer function in Eq. (8-8) for $K = 1$. In this case, when $K = 1$, the Nyquist plot of $G(z)$ intersects the negative real axis at the critical point $(-1, j0)$. The phase-crossover frequency is 13.2 rad/s. Since the critical value of K is 1, the gain margin for $K = 1$ is 0 dB.

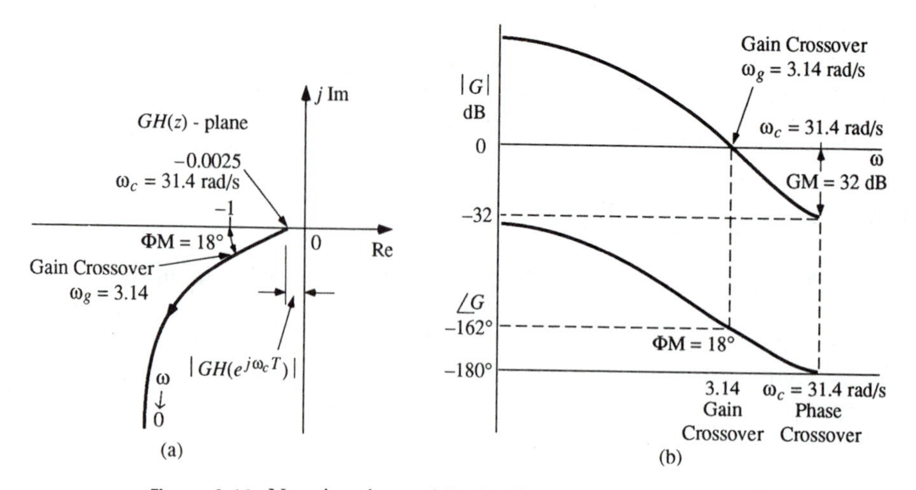

Figure 8-16. Nyquist plot and Bode diagram of $GH(z) = 0.0952z/(z-1)(z-0.905)$ with $T = 0.1$ s.

8-5-2 Phase Margin

Gain margin alone is inadequate to represent the relative stability of a closed-loop control system. Since the gain factor is not the only parameter that is subject to variation, a given system may have a large gain margin and yet still have a low stability margin due to parameter variations that affect the phase of the transfer function. Referring to the Nyquist plot in Fig. 8-16, if the

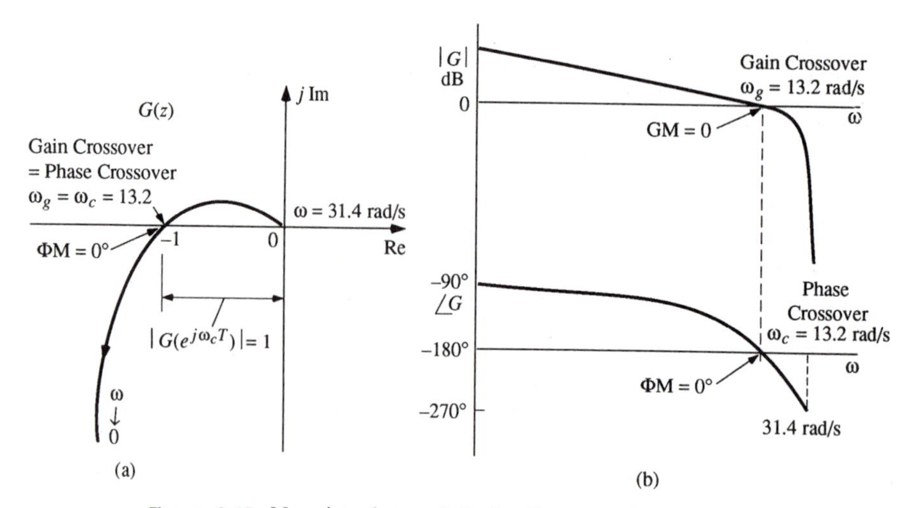

Figure 8-17. Nyquist plot and Bode diagram of $G(z) = 0.758(z+1)/(z-1)(z-0.242)$ with $T = 0.1$ s.

locus is held rigid and is rotated about the intersect on the real axis in the clockwise direction, it can intersect the $(-1, j0)$ point after a rotation of 18 degrees. Thus, this angle is defined as the *phase margin* (ΦM) of the system.

In general, to find the phase margin, we first locate the point at which the magnitude of the transfer function $GH(e^{j\omega T})$ is unity. The frequency at this point is called the *gain-crossover frequency* ω_g. The phase margin is defined as

$$\Phi M = \angle GH(e^{j\omega_g T}) - 180^\circ \tag{8-48}$$

Figure 8-16 illustrates the gain crossover and the phase margin of the system described by Eq. (8-20). In Fig. 8-17(a), the locus passes through the $(-1, j0)$ point. Thus, the gain- and phase-crossover points are the same. The phase margin is 0 degrees.

We conclude from the above discussions that *the gain margin is measured at the phase-crossover frequency, whereas the phase margin is measured at the gain-crossover frequency.*

It should be pointed out that care should be taken in the determination of gain and phase margins when $GH(z)$ has poles outside the unit circle. In general, one should first determine where the $(-1, j0)$ point should be with respect to the Nyquist plot of $GH(z)$ and then determine the gain and phase margins according to the definitions. Thus, there is a danger of using the Bode diagram indiscriminantly without first checking with the Nyquist criterion.

Example 8-9

Consider the open-loop transfer function of a unity-feedback digital control system,

$$G(z) = \frac{Kz(z + 0.5)}{(z - 1)(z - 1.5)} \tag{8-49}$$

The root-locus diagram of the closed-loop system for positive K is shown in Fig. 8-18. Notice that since one of the poles of $G(z)$ is outside the unit circle, the closed-loop system is stable only for $K > 0.5$.

To apply the Nyquist criterion, $G(z)$ has one pole on the unit circle, and one pole outside the unit circle. Thus, $P_\omega = 1$ and $P_{-1} = 1$. For stability, the angle Φ_{11} must satisfy Eq. (8-19),

$$\Phi_{11} = -(0.5P_\omega + P_{-1})180^\circ = -270^\circ \tag{8-50}$$

The Nyquist plot of $G(z)$ is sketched as shown in Fig. 8-19 for $K = 1$. The plot intersects the negative real axis at -2 where $\omega = 7.25$ rad/s. From Fig. 8-19, we found that, when $K > 0.5$, the intersect of the Nyquist plot of $G(z)$ with the real axis is to the left of the $(-1, j0)$ point; $\Phi_{11} = -270$ degrees, and the system is stable. If $0 < K < 0.5$, the intersect is to the right of the critical point; $\Phi_{11} = +90$ degrees, and the system is unstable.

The phase-crossover frequency is at $\omega_c = 7.25$ rad/s, and the gain-crossover frequency is $\omega_g = 10.47$ rad/s. Since the magnitude of $G(e^{j\omega T})$ at the phase

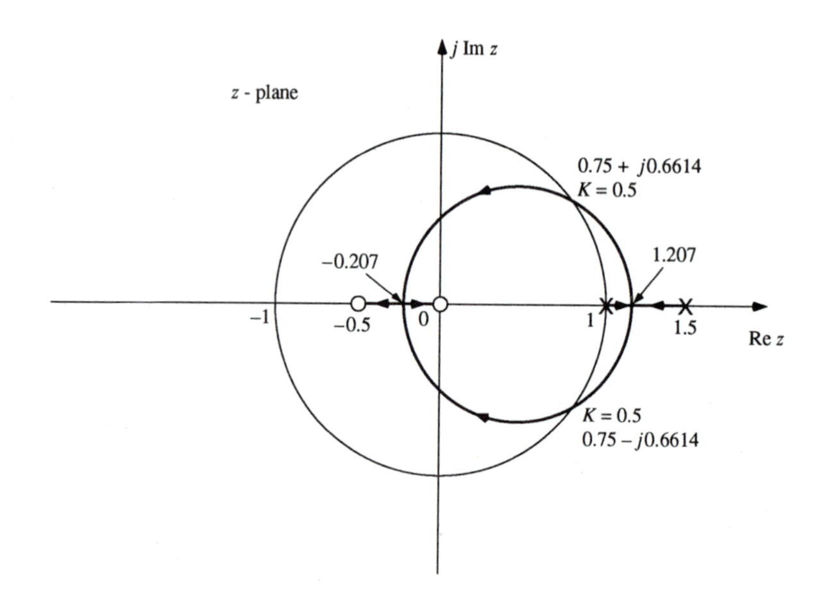

Figure 8-18. Root loci of $G(z) = Kz(z + 0.5)/(z - 1)(z - 1.5)$.

crossover is greater than one, Eq. (8-47) gives a gain margin that is negative; i.e.,

$$\text{GM} = 20 \log_{10} (1/2) = -6.02 \text{ dB} \tag{8-51}$$

Ordinarily, this would correspond to an unstable situation; but in the present case, the system is stable, and the negative gain margin simply means that the gain has to be reduced by 6.02 dB before instability.

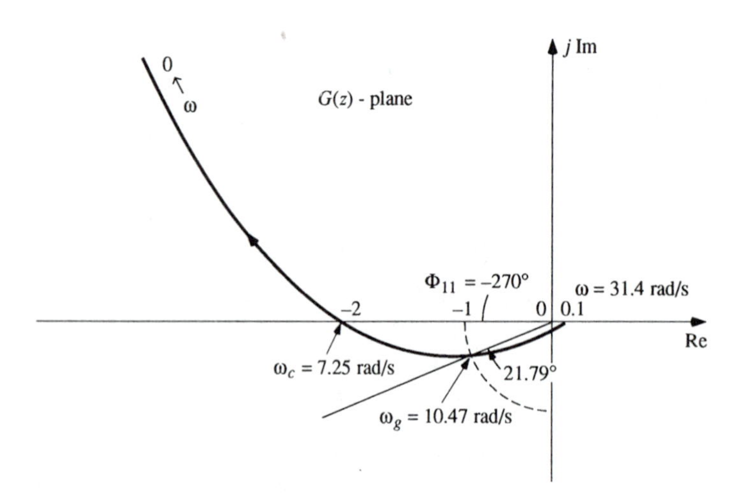

Figure 8-19. Nyquist plot of $G(z) = Kz(z + 0.5)/(z - 1)(z - 1.5)$.

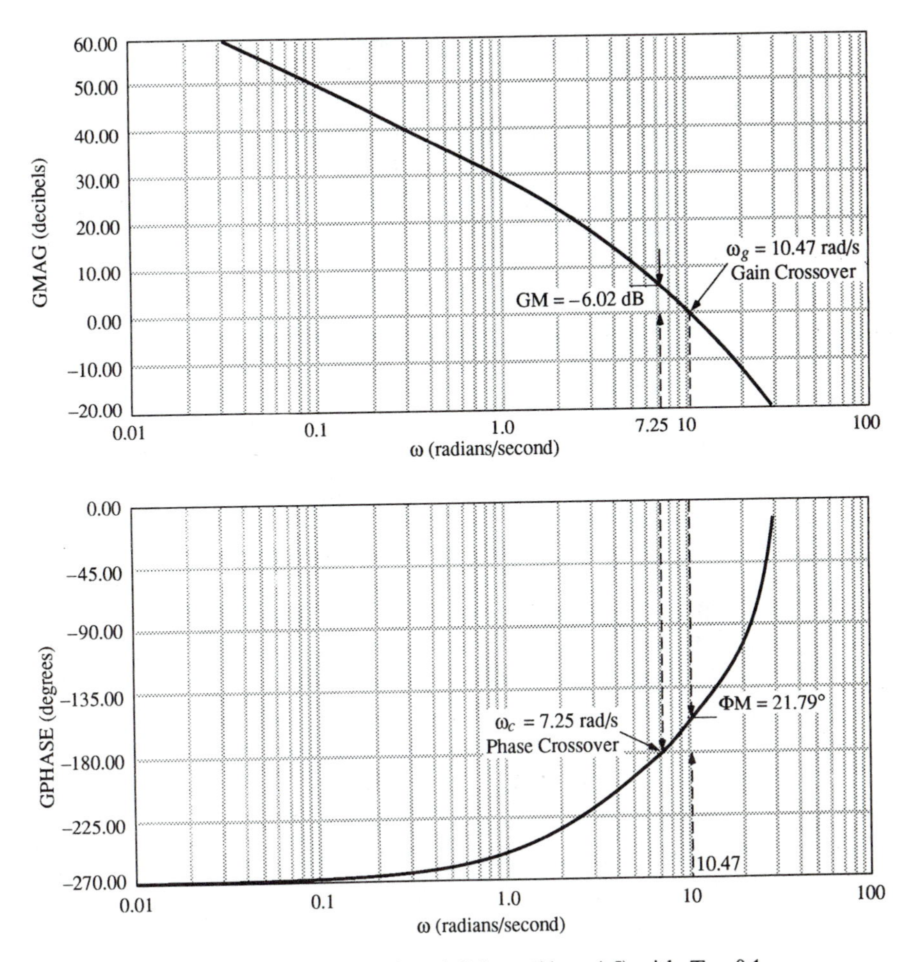

Figure 8-20. Bode plot of $G(z) = Kz(z + 0.5)/(z - 1)(z - 1.5)$ with $T = 0.1$ s.

The phase of the Nyquist plot of $G(z)$ at the gain-crossover frequency $\omega_g = 10.47$ rad/s is -158.21 or 201.79 degrees. Thus, from Eq. (8-48), the phase margin is

$$\Phi M = 201.79^\circ - 180^\circ = 21.79^\circ \tag{8-52}$$

From the Nyquist plot of Fig. 8-19, this phase margin value is interpreted such that if the plot is rotated clockwise by 21.79 degrees, the system would be unstable.

The Bode diagram of $G(z)$ is shown in Fig. 8-20. The gain and phase crossovers and phase and gain margins are all shown on the diagram in the usual manner. However, without the knowledge already established by the root loci or the Nyquist plot, the Bode diagram in this case does not give conclusive results on either the absolute or the relative stability of the system.

8-6 GAIN-PHASE PLOT AND THE NICHOLS CHART

Still another means of representing frequency-domain characteristics is the gain-versus-phase plot. In this case, the magnitude of $G(e^{j\omega T})$ [or $G(j\omega_r)$] is plotted as a function of the phase of the same function with frequency ω (or ω_r) as a varying parameter on the locus. One advantage of using the gain-phase plot is that when superposed with the Nichols chart [1], the plot gives information on the frequency response of the closed-loop system. Information on the gain and phase margins and the bandwidth of the closed-loop system is also easily determined from the gain-phase plot.

To illustrate the gain-phase plot, the frequency-domain data of the transfer function in Eq. (8-20) for $K = 1$ are transferred from the Nyquist plot or the Bode plot, as shown in Fig. 8-21. The Nichols chart provides information on the magnitude and phase of the closed-loop transfer function $M(z) = C(z)/R(z)$ in Eq. (8-9) for a unity-feedback system as a function of frequency. The values of $M = |C(e^{j\omega T})/R(e^{j\omega T})|$ at various frequencies are found at the intersects between the $G(e^{j\omega T})$ locus and the Nichols chart loci. For $G(z)$ that does not have poles outside the unit circle, the Nichols locus that is tangent to the $G(e^{j\omega T})$ curve from above gives the maximum value of the closed-loop frequency response M_p, the *resonance peak*; the frequency at which it occurs is called the *resonant frequency* ω_p. From Fig. 8-21, we see that with $K = 1$, $M_p = 3.24$, and $\omega_p = 3.14$ rad/s.

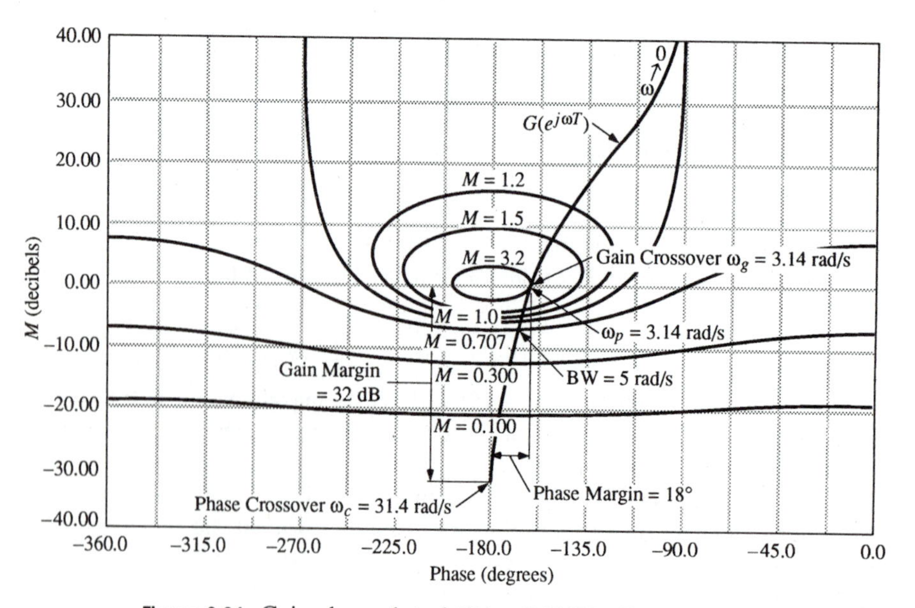

Figure 8-21. Gain-phase plot of $G(z) = 0.0952Kz/(z - 1)(z - 0.905)$ with $K = 1$.

The intersect of the 0-dB and the -180-degree axes in the gain-phase coordinates corresponds to the critical point $(-1, j0)$ of the Nyquist plot. As shown in Fig. 8-21, the gain-crossover point is the intersect of the $G(e^{j\omega T})$ curve and the 0-dB axis, and the phase-crossover point is the intersect of the $G(e^{j\omega T})$ curve and the -180-degree axis. Thus, the gain and phase margins are easily determined as shown in Fig. 8-21. Positive phase margin is measured to the right of the critical point, and positive gain margin is measured below the critical point. However, the absolute and relative stability of the system relative to the signs and magnitudes of the gain and phase margins depend on whether the transfer function $G(z)$ has any poles outside the unit circle.

8-7 BANDWIDTH CONSIDERATIONS

It is common to use bandwidth as a qualitative measure of the system performance when conducting control system design in the frequency domain. For instance, the practicing engineer often specifies that the bandwidth of a control system should be, say, 2 Hz. The use of bandwidth as one of the system specifications applies to continuous-data as well as digital control systems, since from the input-output standpoint, it does not matter what type of system it is.

It is important to clarify what bandwidth means in general terms. As it turns out, for the continuous-data second-order prototype system described by Eq. (7-4) a direct correlation exists between bandwidth (BW), damping ratio ζ, and natural-undamped frequency ω_n. Specifically, for a second-order prototype system with the transfer function

$$\frac{C(s)}{R(s)} = \frac{\omega_n^2}{s^2 + 2\zeta\omega_n s + \omega_n^2} \tag{8-53}$$

the bandwidth is given by [1]

$$\mathrm{BW} = \omega_n\left[(1 - 2\zeta^2) + \sqrt{4\zeta^4 - 4\zeta^2 + 2}\right]^{1/2} \tag{8-54}$$

which is determined by finding the frequency ω at which the magnitude of $C(j\omega)/R(j\omega)$ is equal to 0.707.

For digital control systems, the simplest second-order transfer function usually has a first-order term $(z - z_1)$ in the numerator, and z is related to ω through the relation $z = e^{j\omega T}$ in the frequency domain. It becomes difficult to arrive at a clean analytical expression such as Eq. (8-54) which relates the bandwidth to the pole-zero location of $C(z)/R(z)$. Therefore, the most convenient way of finding the bandwidth of a digital control system is to find it graphically using the gain-phase plot and the Nichols chart. Using Fig. 8-21, we see that bandwidth is found at the intersect of the $G(e^{j\omega T})$ curve and the $M = 0.707$ locus. In the case of Fig. 8-21, BW is found to be 5 rad/s. This is also marked as shown in Fig. 8-22.

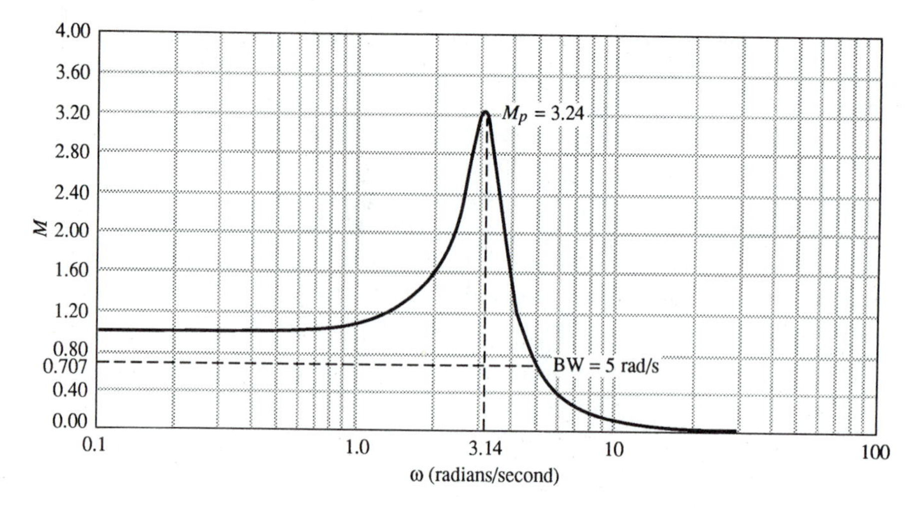

Figure 8-22. Magnitude plot of closed-loop frequency response of $G(z) = 0.0952Kz/(z-1)(z-0.905)$ with $K = 1$.

If the w-transformation of Eq. (8-43) is used, the bandwidth found in terms of $G(w)$ is designated as $(\mathrm{BW})_w$. The true bandwidth in the real ω-domain is determined by use of Eq. (8-45). Then,

$$\mathrm{BW} = \frac{\omega_s}{\pi} \tan^{-1} \frac{\pi(\mathrm{BW})_w}{\omega_s} \tag{8-55}$$

8-8 SENSITIVITY ANALYSIS

In many control system designs, the system not only must satisfy the relative stability and accuracy requirements, but the design must also yield performance that is robust (insensitive) to external disturbances and parameter variations. Thus, sensitivity analysis and design with regard to system sensitivity to parameter variations and disturbances are important for continuous-data as well as discrete-data control systems.

System sensitivity is easily interpreted using the frequency-domain plots such as the Nyquist plot and the Nichols chart. In this section we define the sensitivity function $S_G^M(z)$ which is defined as the sensitivity of the closed-loop transfer function $M(z)$ with respect to the open-loop transfer function $G(z)$ or, rather, a gain parameter in $G(z)$. The properties of $S_G^M(z)$ will be investigated in the frequency domain.

Consider that a linear discrete-data control system with unity feedback has the closed-loop transfer function

$$M(z) = \frac{C(z)}{R(z)} = \frac{G(z)}{1 + G(z)} \tag{8-56}$$

where $G(z)$ represents the open-loop transfer function. The sensitivity of $M(z)$ with respect to $G(z)$ is defined as

$$S_G^M(z) = \frac{dM(z)/M(z)}{dG(z)/G(z)} = \frac{dM(z)}{dG(z)} \frac{G(z)}{M(z)} \tag{8-57}$$

Substituting Eq. (8-56) into the last equation and simplifying, we have

$$S_G^M(z) = \frac{1}{1 + G(z)} = \frac{1/G(z)}{1 + [1/G(z)]} \tag{8-58}$$

Clearly, when $z = e^{j\omega T}$ is substituted in the last equation, $S_G^M(e^{j\omega T})$ is shown to be a function of frequency. From the design standpoint, a performance criterion can be that of limiting the magnitude of the sensitivity function to a certain maximum value over a certain frequency range. Equation (8-58) shows that $S_G^M(z)$ is analogous to the closed-loop transfer function $M(z)$ simply by replacing $G(z)$ by $1/G(z)$ in the latter. Thus, $S_G^M(z)$ in Eq. (8-58) can be determined by plotting $1/G(z)$, with $z = e^{j\omega T}$, on the Nichols chart. As an illustrative example, the open-loop transfer function $G(z)$ of the digital space vehicle control system in Eq. (8-26) and $1/G(z)$ are plotted on the Nichols chart as shown in Fig. 8-23. The frequency response plots of $|M(e^{j\omega T})|$ and $|S_G^M(e^{j\omega T})|$ are shown in Fig. 8-24. Without feedback, the sensitivity S_G^M of the open-loop system is unity for all frequencies. With feedback, the magnitude of S_G^M is less than unity only for frequencies up to 5.3 rad/s, beyond which the sensitivity of the closed-loop

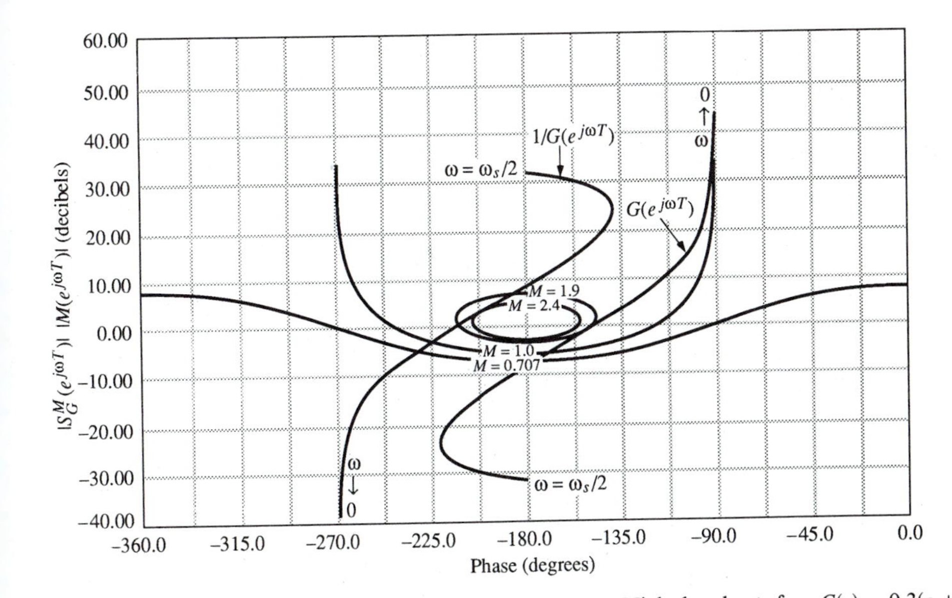

Figure 8-23. $G(z)$ and $1/G(z)$ plots in the Nichols chart for $G(z) = 0.3(z + 0.7433)/(z - 1)(z - 0.4119)$.

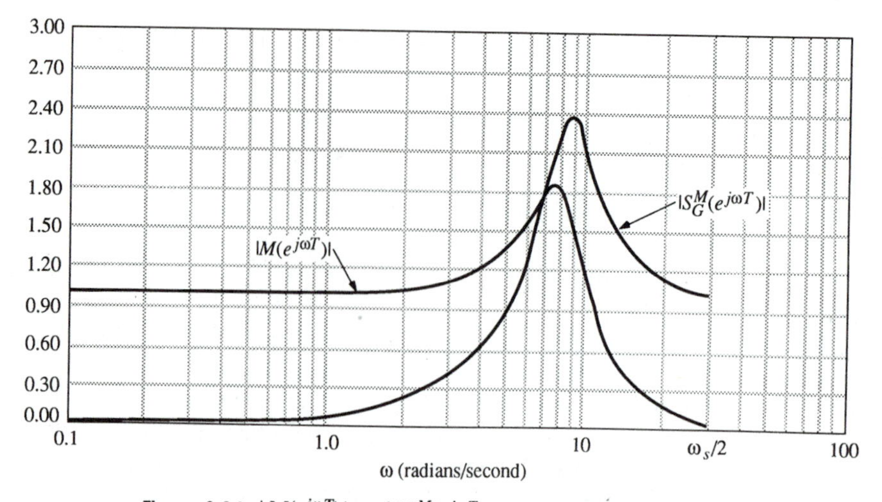

Figure 8-24. $|M(e^{j\omega T})|$ and $|S_G^M(e^{j\omega T})|$ versus frequency of the discrete-data control system with $G(z) = 0.3(z + 0.7453)/(z - 1)(z - 0.4119)$.

system is actually inferior. It is not difficult to see from this example that, in general, the sensitivity function would be low over a wider frequency range if the system bandwidth is wider. Equation (8-58) also shows that for low sensitivity, the gain of $G(z)$ must be high, which may cause instability to the closed-loop system. Thus, the conditions of high stability and low sensitivity are not naturally compatible. The design of robust or low-sensitivity digital control systems is discussed in Chapter 10.

PROBLEMS

8-1 For the discrete-data control systems described in Problem 6-11, sketch the Nyquist plot of $G(z)$, where

$$G(z) = (1 - z^{-1})\mathscr{Z}[G_p(s)]$$

Determine the range of K for the system to be stable using the Nyquist criterion.

8-2 Determine the values of K for stability for the systems in Problem 7-22 by means of the Nyquist criterion.

8-3 Determine the values of K for stability for the systems in Problem 6-10 by means of the Nyquist criterion. [*Hint*: Create an equivalent $G(z)$ by dividing both sides of the characteristic equation by terms that do not contain K. This creates an equivalent $1 + G(z) = 0$.]

8-4 For the digital space vehicle control system shown in Fig. P7-8, let $J_v = 41822$ and $T = 0.1$ s. Set the relation between K_R and K_P so that the ramp-error constant is 10. Sketch the Nyquist plot of an equivalent $G(z)$ that has K_R as a multiplying factor. Determine the ranges of $K_P K_R$ so that the system is stable using the Nyquist criterion.

8-5 The open-loop z-transfer function of the liquid-level control system described in Problem 7-17, Fig. P7-17, is represented as $G(z)$, where

$$G(s) = \frac{1 - e^{-Ts}}{s}\left(\frac{16.67N}{s(s+1)(s+12.5)}\right) \qquad T = 0.05 \text{ s}$$

Construct a Nyquist plot for $G(z)/N$ for $z = e^{j\omega T}$, $0 \le \omega < \omega_s/2$. Determine the values of N (positive integer) for the system to be stable.

8-6 Consider that the amplifier gain K in the flight control system described in Problem 4-9 is variable. Determine the range of K for the system to be stable by means of the Nyquist criterion.

a. $T = 0.01$ s
b. $T = 0.1$ s

8-7 The controlled process of a unity-feedback digital control system with a zero-order hold is

$$G_p(s) = \frac{1000e^{-NTs}}{s(s+10)(s+50)}$$

where N is a positive integer. The sampling period T is 0.1 s. Apply the Nyquist criterion to find the maximum value of N so that the system is stable.

8-8 The stoichiometric air-fuel ratio control system described in Problem 7-20 has the open-loop transfer function

$$G(z) = \frac{0.33Kz^{-10}}{z - 0.67}$$

The sampling period is 0.1 s. Sketch the Nyquist plot of $G(z)$ with $z = e^{j\omega T}$, $0 \le \omega < \omega_s/2$, and determine the marginal value of K for stability.

8-9 Consider the multirate discrete-data control system described in Problem 4-15, Fig. P4-15. Sketch the Nyquist plots of the normalized open-loop transfer function $G(z)/K = C(z)/KE(z)$, for $z = e^{j\omega T}$, $0 \le \omega < \omega_s/2$, $N = 1, 2, 3, 5, 10$, and ∞. Find the critical value of K for the closed-loop system to be stable for each of the given values of N.

8-10 Plot the Bode diagrams for $G(z)/K$, with $z = e^{j\omega T}$, $0 \le \omega \le \omega_s/2$, for the open-loop transfer functions in Problem 6-11. Determine the positive values of K for system stability from the Bode diagrams.

8-11 Transform the open-loop transfer functions in Problem 6-11 into $G(w)$ by the w-transformation given in Eq. (8-43). Plot the Bode diagrams of $G(w)$ with $w = j\omega_w$, and determine the values of K for system stability from the Bode diagrams.

8-12 Construct the Bode diagrams of the open-loop transfer function $G(z) = C(z)/E(z)$ with $z = e^{j\omega T}$, $0 \le \omega \le \omega_s/2$, for the systems described in Problem 7-1. Determine the gain margin and phase margin of the system.

8-13 Consider the flight control system described in Problem 4-8, Fig. P4-8. Use the system parameters given in Problem 4-8, except for the values of K given below. Plot the Bode diagram of $G(z)$ with $z = e^{j\omega T}$ for $0 \le \omega \le \omega_s/2$. Find the gain and phase margins of the system.

a. $T = 0.01$ s, $K = 1000$
b. $T = 0.1$ s, $K = 100$

8-14 For the digital idle-speed control system modeled in Fig. P7-5, set the sampling period T at 0.1 s.

a. Construct the Bode diagram of $G(z)$ with $z = e^{j\omega T}$ for $0 \le \omega \le \omega_s/2$. Determine the gain and phase margins of the system.
b. Construct the Bode diagram of $G(z)$ with the w-transformation of Eq. (8-43), $w = j\omega_w$. Determine the gain and phase margins of the system.
c. Plot $G(z)$ on the Nichols chart and determine the bandwidth and M_p. Plot $|M|$ versus ω for ω up to $\omega_s/2$.

8-15 A unity-feedback digital control system has the open-loop transfer function

$$G(z) = \frac{K(z + 0.5)}{3(z - 1)(z - 0.333)} \qquad T = 1 \text{ s}$$

a. Plot the Bode diagram of $G(z)/K$ with $z = e^{j\omega T}$, for $0 \le \omega \le \omega_s/2$. Plot $G(z)/K$ on the Nichols chart.
b. Find the gain margin, phase margin, M_p, ω_p, and BW from the Nichols chart when $K = 1$.
c. Find the value of K so that the gain margin is 0 dB.
d. Find the value of K so that the gain margin is 20 dB.
e. Find the value of K so that the phase margin is 60 degrees.
f. Find the value of K so that $M_p = 1.25$.
g. Find the value of K so that $M_p = 1.0$ with $\omega_p \neq 0$.

8-16 For the Large Space Telescope control system described in Problem 4-10, Fig. P4-10, let $K_R = 731{,}885$, $K_P = 10{,}455{,}500$, $K_I = 41{,}822{,}000$, $J_v = 41{,}822$, and $T = 0.1$ s. Construct the Bode diagram of $G(z) = C(z)/E(z)$ for $z = e^{j\omega T}$ for ω up to $\omega_s/2$. Determine the gain and phase margins of the system.

8-17 The open-loop transfer function of the space vehicle control system described in Problem 7-8, Fig. P7-8, has the open-loop transfer function

$$G(z) = \frac{1.2 \times 10^{-7} K_P(z + 1)}{(z - 1)(z - 0.242)} \qquad T = 0.1 \text{ s}$$

a. Draw the Nyquist plot of $G(z)$ with $z = e^{j\omega T}$ for $0 \le \omega \le \omega_s/2$. Determine the marginal value of K_P for stability.

b. Transform $G(z)$ to $G(r)$ by the r-transformation of Eq. (6-126) and draw the Bode diagram of $G(r)$ normalized by the gain factor; that is, $G(r)/K$. Find the value of K_P so that the gain margin is 20 dB.

c. From the Bode diagram in part **b**, find the value of K_P so that the phase margin is 45 degrees.

d. Plot the normalized $G(w)$ on the Nichols chart. Find the value of K_P so that $M_p = 1.01$. Find the bandwidth of the system when K_p is set at the value for $M_p = 1.01$.

References

1. Kuo, B. C., *Automatic Control Systems*, 6th ed., Prentice-Hall, Englewood Cliffs, N.J., 1991.
2. Yeung, K. S., and H. M. Lai, "A Reformulation of Nyquist Criterion for Discrete Systems," *IEEE Trans. Educ.*, vol. 31, no. 1, February 1988.
3. Whitbeck, R. F., *Analysis of Digital Flight Control Systems with Flying Qualities Applications*, vol. I, Executive Summary, Tech. Report AFFDL-TR-78-115, Air Force Flight Dynamics Lab., Wright-Patterson Air Force Base, Ohio, September 1978.
4. Whitbeck, R. F., and L. G. Hoffmann, *Analysis of Digital Flight Control Systems with Flying Qualities Applications*, vol. II, Executive Summary, Tech. Report AFFDL-TR-78-115, Air Force Flight Dynamics Lab., Wright-Patterson Air Force Base, Ohio, September 1978.
5. Whitbeck, R. F., and D. G. Didaleusky, *Multi-Rate Digital Control Systems with Simulation Applications*, vol. I, *Technical Report*, Tech. Report AFWAL-TR-80-3101, Flight Dynamics Lab., Wright-Patterson Air Force Base, Ohio, May 1980.

Digital Simulation and Digital Redesign

KEYWORDS AND TOPICS

digital simulation • numerical integration • digital redesign

9-1 INTRODUCTION

It is well known that digital computers play an important role in the analysis and design of feedback control systems. Digital computers are used not only in the computation and simulation of control system performance, but also in direct on-line control of processes. Since continuous-data control systems are still in abundance and most control systems contain analog components, it is a common practice to simulate these systems with a digital computer. The procedure of finding the digital equivalent of an analog controller is called *digital redesign.* Although a digital controller may be designed independently, if a continuous-data system with satisfactory performance already exists, it may be feasible simply by finding the digital equivalent of the analog controller. In this chapter various methods of digital simulation and a digital redesign method are discussed.

A continuous-data control system can be simulated on a digital computer once its system dynamics are approximated by a transfer function in the z-domain, or by difference state equations. The analysis usually consists of the following two steps.

1. Represent the continuous-data system by a digital model.
2. Simulate the digital model on a digital computer.

Of course, the digital model can also be studied analytically. In general, digital modeling can be conducted by the following methods.

1. Insert sample-and-hold devices in the continuous-data system.
2. Approximate analog integration numerically.
3. Discrete state equation.

These topics will be discussed in the following sections.

9-2 DIGITAL SIMULATION—DIGITAL MODELING WITH SAMPLE-AND-HOLD DEVICES

The simplest way of approximating a continuous-data system by a digital model is to insert fictitious sample-and-hold devices at strategic locations in the system. Then the system can be described by z-transform transfer functions or difference state equations. For instance, the continuous-data system shown in Fig. 9-1(a) is approximated by the digital model shown in Fig. 9-1(b). The ideal samplers have a sampling period of T s, and the hold device can be of any type at the discretion of the analyst. It can even be a polygonal hold (Sec. 2-15) which is not physically realizable but nevertheless quite acceptable for analytical purposes. In Fig. 9-1(b), let $G_h(s)$ denote the transfer function of the hold device. Then the z-transform transfer function of the digital model is

$$\frac{C(z)}{R(z)} = \frac{G_hG(z)}{1 + G_hG(z)G_hH_1(z) + G_hG(z)G_hH_2(z)} \tag{9-1}$$

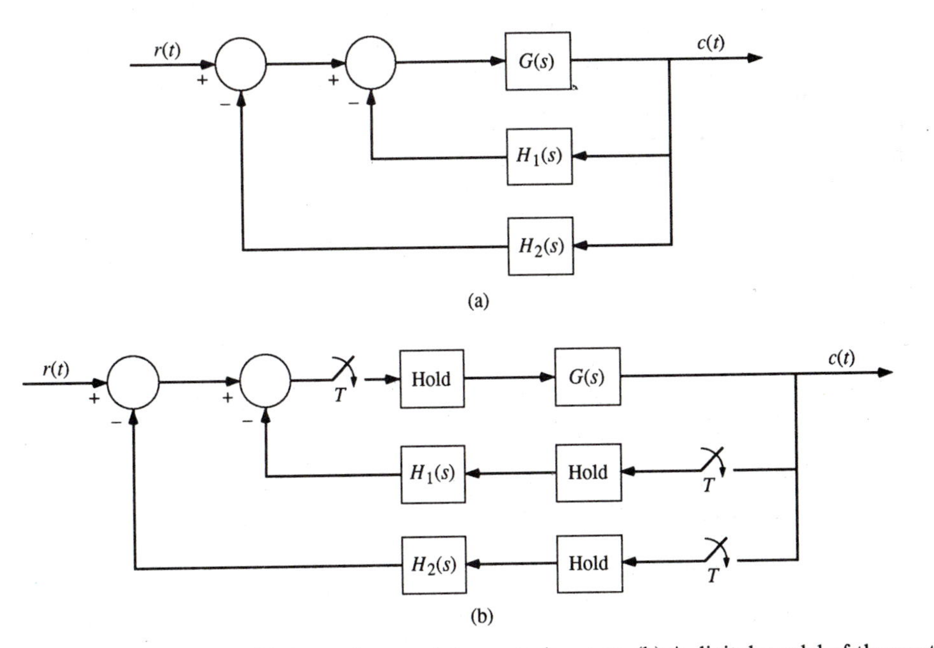

Figure 9-1. (a) A continuous-data control system. (b) A digital model of the continuous-data system with fictitious sample-and-hold devices.

where

$$G_hG(z) = \mathscr{Z}[G_h(s)G(s)] \tag{9-2}$$

$$G_hH_1(z) = \mathscr{Z}[G_h(s)H_1(s)] \tag{9-3}$$

$$G_hH_2(z) = \mathscr{Z}[G_h(s)H_2(s)] \tag{9-4}$$

Although in principle the digital modeling method just described is straightforward, there are two important considerations when selecting the value of the sampling period T. The first concerns the accuracy and the time required of the digital simulation. The second consideration is that the sampling period should be chosen so that the digital system is stable. In fact, sample-and-hold devices generally have an adverse effect on system stability.

As an illustrative example, let us refer to the continuous-data system shown in Fig. 9-2(a). A digital approximation of the system may be obtained by inserting a sample-and-hold device in the feedback path, as shown in Fig. 9-2(b). The z-transform of the output of the digital model is

$$C(z) = \frac{RG(z)}{1 + G_{h0}G(z)} \tag{9-5}$$

where

$$RG(z) = \mathscr{Z}[R(s)G(s)] \tag{9-6}$$

and

$$G_{h0}G(z) = \mathscr{Z}[G_{h0}(s)G(s)] \tag{9-7}$$

where $G_{h0}(s)$ is the transfer function of the zero-order hold (zoh).

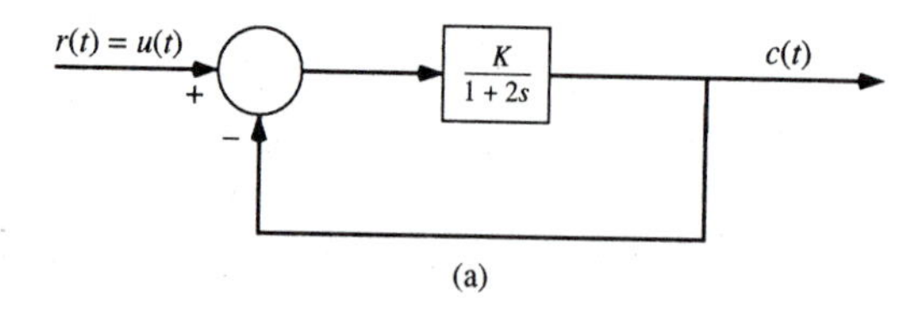

(a)

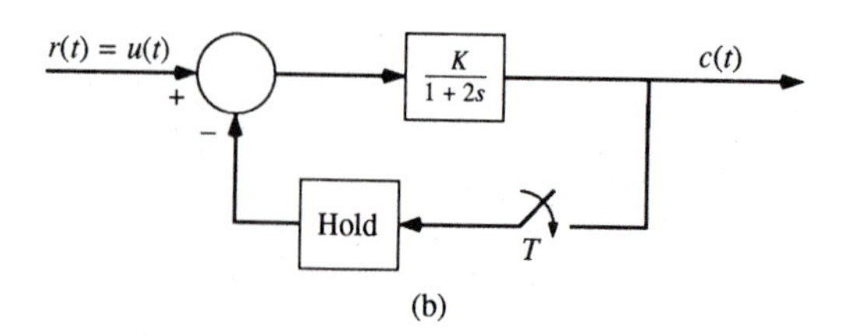

(b)

Figure 9-2. (a) A continuous-data control system. (b) A digital model of the continuous-data system with sample-and-hold devices inserted in the feedback path.

For a unit-step input, Eq. (9-6) becomes

$$RG(z) = \mathscr{Z}\left(\frac{K}{s(1+2s)}\right) = \frac{K(1-e^{-0.5T})z}{(z-1)(z-e^{-0.5T})} \tag{9-8}$$

Equation (9-7) gives

$$G_{h0}G(z) = \mathscr{Z}\left(\frac{1-e^{-Ts}}{s}\,\frac{K}{1+2s}\right) = \frac{K(1-e^{-0.5T})}{(z-e^{-0.5T})} \tag{9-9}$$

Substituting Eqs. (9-8) and (9-9) into Eq. (9-5), we have

$$C(z) = \frac{K(1-e^{-0.5T})z}{(z-1)[z-e^{-0.5T}+K(1-e^{-0.5T})]} \tag{9-10}$$

The sampling period T must be chosen so that the digital system is stable. In addition, T should be sufficiently small that the digital model is a good approximation of the original continuous-data system. The characteristic equation of the digital system is

$$z - e^{-0.5T} + K(1-e^{-0.5T}) = 0 \tag{9-11}$$

Figure 9-3 shows the stable region in the K-versus-T plane. Notice that when $K = 1$, for stability, T can lie anywhere between zero and infinity. When $K = 2$, for instance, T must be less than 2.2 s. To select the value of T so that the digital control system is a good approximation of the continuous-data system, we use the guideline established in Sec. 7-2. That is, the sampling

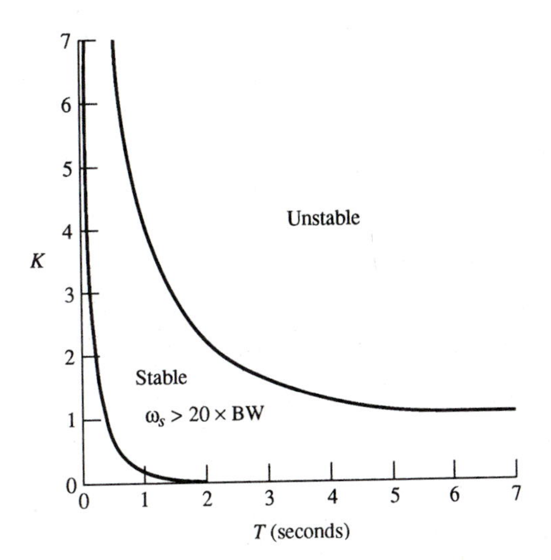

Figure 9-3. Stable and unstable regions and region in which $\omega_s > 20 \times \text{BW}$ of the digital control system in Fig. 9-2.

frequency should be at least 20 times the bandwidth of the system. When $K = 1$, the bandwidth of the continuous-data system is 1 rad/s. Thus, the sampling frequency ω_s should be at least 20 rad/s, or the maximum sampling period T should be 0.314 s. When $K = 2$, BW = 1.49 rad/s; thus the minimum value for ω_s is 29.8 rad/s, and the maximum T is 0.21 s. Figure 9-3 shows the region for good digital approximation in the K-versus-T plane. In general, the requirement of good digital approximation is far more stringent than the requirement of stability.

With $K = 1$, let the sampling period be 0.314 s. Equation (9-10) becomes

$$C(z) = \frac{0.145z}{(z-1)(z-0.709)} \tag{9-12}$$

The output sequence $c(kT)$ is obtained by expanding $C(z)$ into a power series by long division. The result is

$$\begin{aligned} C(z) = {} & 0.145z^{-1} + 0.247z^{-2} + 0.32z^{-3} + 0.372z^{-4} + 0.409z^{-5} + 0.435z^{-6} \\ & + 0.453z^{-7} + 0.466z^{-8} + 0.475z^{-9} + 0.482z^{-10} + 0.487z^{-11} \\ & + 0.490z^{-12} + 0.492z^{-13} + 0.494z^{-14} + 0.495z^{-15} + \cdots \end{aligned} \tag{9-13}$$

The final value of $c(kT)$ is 0.5. Figure 9-4 shows the unit-step responses of the continuous-data system and the digital control system with $K = 1$ and $T = 0.314$ s. Figure 9-4 also shows the unit-step response of the digital control system when $K = 1$ and $T = 1$ s. Naturally, the approximation is inferior to that when $T = 0.314$ s. For sampling periods less than 0.314 s, the improvement becomes insignificant, and the expense of a higher sampling frequency is not justified.

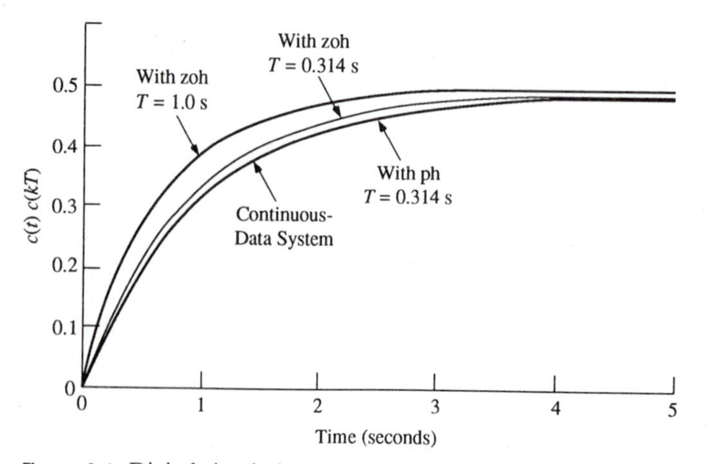

Figure 9-4. Digital simulation with zero-order hold and polygonal hold.

Let us consider that the hold device is a polygonal hold with the transfer function given in Eq. (2-107),

$$G_{hp}(s) = \frac{e^{Ts} - 2 + e^{-Ts}}{Ts^2} \tag{9-14}$$

Then,

$$\begin{aligned}\mathscr{Z}[G_{hp}(s)G(s)] = G_{hp}G(z) &= \frac{(z-1)^2}{Tz}\mathscr{Z}\left(\frac{K}{s^2(1+2s)}\right)\\ &= \frac{(0.5T - 1 + e^{-0.5T})z + 1 - (1 + 0.5T)e^{-0.5T}}{0.5T(z - e^{-0.5T})}\end{aligned} \tag{9-15}$$

The z-transform of the output of the digital model is

$$\begin{aligned}C(z) &= \frac{RG(z)}{1 + G_{hp}G(z)}\\ &= \frac{0.5T(1 - e^{-0.5T})z}{(z-1)[T - 1 + e^{-0.5T})z + 1 - (1 + T)e^{-0.5T}]}\end{aligned} \tag{9-16}$$

For $K = 1$ and $T = 0.314$, $C(z)$ becomes

$$C(z) = \frac{0.1352z}{(z-1)(z-0.7297)} \tag{9-17}$$

which is expanded to

$$\begin{aligned}C(z) = {} & 0.1352z^{-1} + 0.2339z^{-2} + 0.3058z^{-3} + 0.3584z^{-4} + 0.3967z^{-5}\\ & + 0.4247z^{-6} + 0.4451z^{-7} + 0.4600z^{-8} + 0.4708z^{-9} + 0.4788z^{-10}\\ & + 0.4846z^{-11} + 0.4888z^{-12} + 0.4919z^{-13} + 0.4941z^{-14} + 0.4958z^{-15}\\ & + 0.4970z^{-16} + 0.4978z^{-17} + 0.4985z^{-18} + 0.4989z^{-19} + \cdots\end{aligned} \tag{9-18}$$

The unit-step response of the digital model with the polygonal hold is shown in Fig. 9-4. The results show that, with the same values of K and T, the system with the polygonal hold is a better approximation than that with the zero-order hold.

9-3 DIGITAL SIMULATION—STATE-VARIABLE FORMULATION

Once the digital model is obtained by inserting sample-and-hold devices in a continuous-data system, instead of using transfer functions, discrete dynamic equations can be written in the following form:

$$\mathbf{x}[(k+1)T] = \boldsymbol{\phi}(T)\mathbf{x}(kT) + \boldsymbol{\theta}(T)\mathbf{r}(kT) \tag{9-19}$$

$$c(kT) = \mathbf{D}\mathbf{x}(kT) \tag{9-20}$$

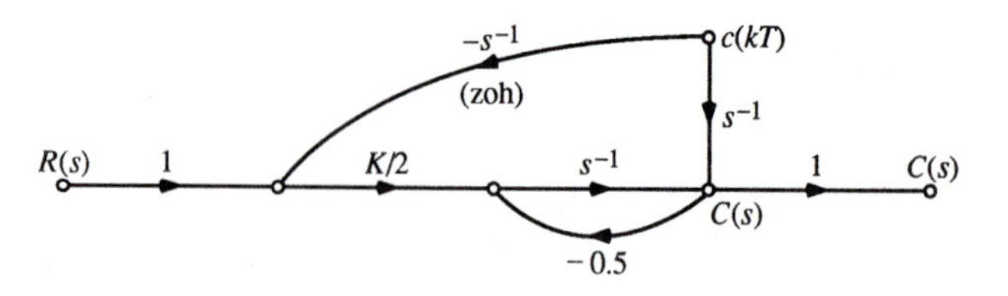

Figure 9-5. State diagram of the digital control system in Fig. 9-2(b).

As an illustrative example, the state diagram of the digital control system shown in Fig. 9-2(b) is shown in Fig. 9-5 when the hold device is a zero-order hold. Writing the output transform in terms of the input $R(s)$ and the initial state $c(kT)$, we have

$$C(s) = \left(\frac{-K}{s} + \frac{K+1}{s+0.5}\right)c(kT) + \frac{KR(s)}{2(s+0.5)} \tag{9-21}$$

For a unit-step input, $R(s) = 1/s$, the last equation becomes

$$C(s) = \left(\frac{-K}{s} + \frac{K+1}{s+0.5}\right)c(kT) + K\left(\frac{1}{s} - \frac{1}{s+0.5}\right) \tag{9-22}$$

Taking the inverse Laplace transform on both sides of the last equation, and letting $t = (k+1)T$, we have

$$c[(k+1)T] = [-K + (K+1)e^{-0.5T}]c(kT) + K(1 - e^{-0.5T}) \tag{9-23}$$

Setting $K = 1$, $T = 0.314$ s, and $c(0) = 0$, we can show that the same results as given by Eq. (9-13) are obtained for the output sequence.

9-4 DIGITAL SIMULATION—NUMERICAL INTEGRATION

Another popular method for digital simulation of continuous-data systems is the use of numerical integration. Since integration is the most time-consuming and difficult basic mathematical operation on a digital computer, its digital simulation plays an important role here. Instead of inserting sample-and-hold devices at strategic locations in a continuous-data system, the approach now is to approximate the continuous integration operation by numerical methods. The problem can also be stated as the simulation of the integrator s^{-1} in a continuous-data state diagram by a digital model. Shown in Fig. 9-6 is the integrator element of a state diagram. The input-output relationship is written as

$$x(t) = \int_0^t r(\tau)\,d\tau \tag{9-24}$$

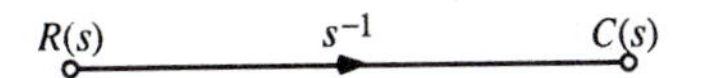

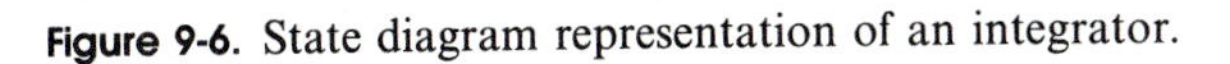
Figure 9-6. State diagram representation of an integrator.

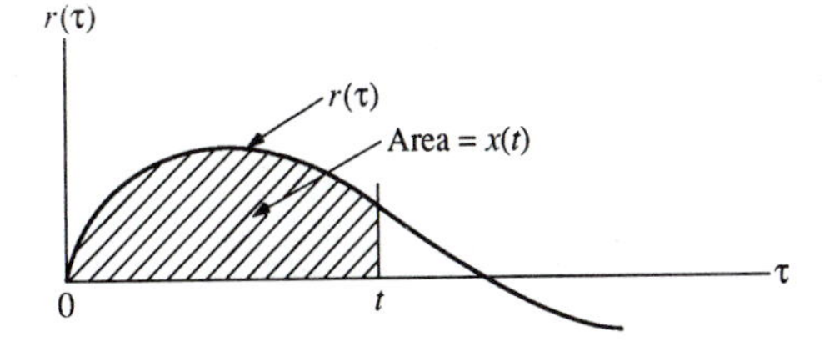

Figure 9-7. Input-output relation of an integrator.

or, in the transform domain,

$$\frac{X(s)}{R(s)} = \frac{1}{s} \tag{9-25}$$

Suppose that the input $r(\tau)$ to the integrator is as shown in Fig. 9-7, the output $x(t)$ is equal to the area under the curve $r(\tau)$ between $\tau = 0$ and $\tau = t$. We shall investigate several types of numerical integration in the following sections.

9-4-1 Rectangular Integration

One of the standard methods of numerical integration is rectangular integration, which is further divided into forward-rectangular integration and backward-rectangular integration, as illustrated in Fig. 9-8. As shown in Fig. 9-8, the

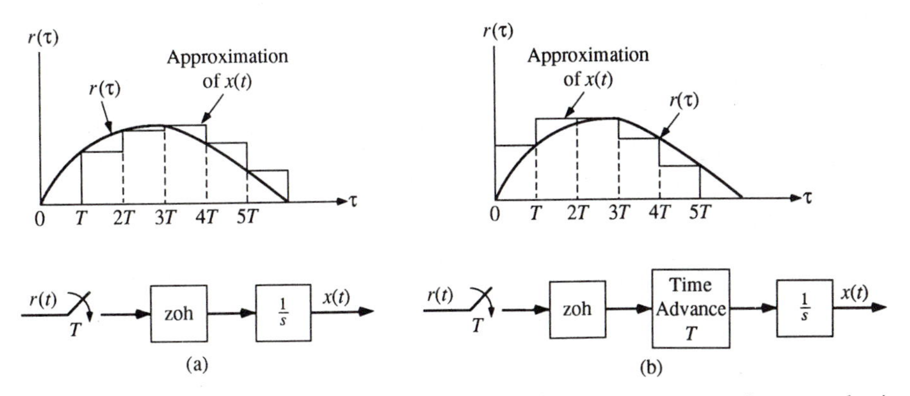

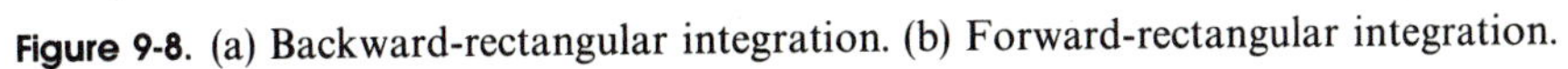
Figure 9-8. (a) Backward-rectangular integration. (b) Forward-rectangular integration.

rectangular integration schemes are all equivalent to inserting a zero-order hold in front of the integrator. The forward-rectangular integration also has a time-advance unit which is noncausal.

Referring to Fig. 9-8(a), the z-transfer function of the backward-rectangular integrator is

$$\frac{X(z)}{R(z)} = (1 - z^{-1})\mathscr{Z}\left(\frac{1}{s^2}\right) = \frac{T}{z-1} \tag{9-26}$$

The state equation is

$$x[(k+1)T] = x(kT) + Tr(kT) \tag{9-27}$$

Similarly, the z-transfer function of the forward-rectangular integrator is

$$\frac{X(s)}{R(z)} = z(1 - z^{-1})\mathscr{Z}\left(\frac{1}{s^2}\right) = \frac{Tz}{z-1} \tag{9-28}$$

and the state equation is

$$x[(k+1)T] = x(kT) + Tr[(k+1)T] \tag{9-29}$$

As an illustrative example, the continuous-data system of Fig. 9-2(a) is first approximated by backward-rectangular integration. The state diagram of the system of Fig. 9-2(a) is shown in Fig. 9-9(a). Figure 9-9(b) shows the system

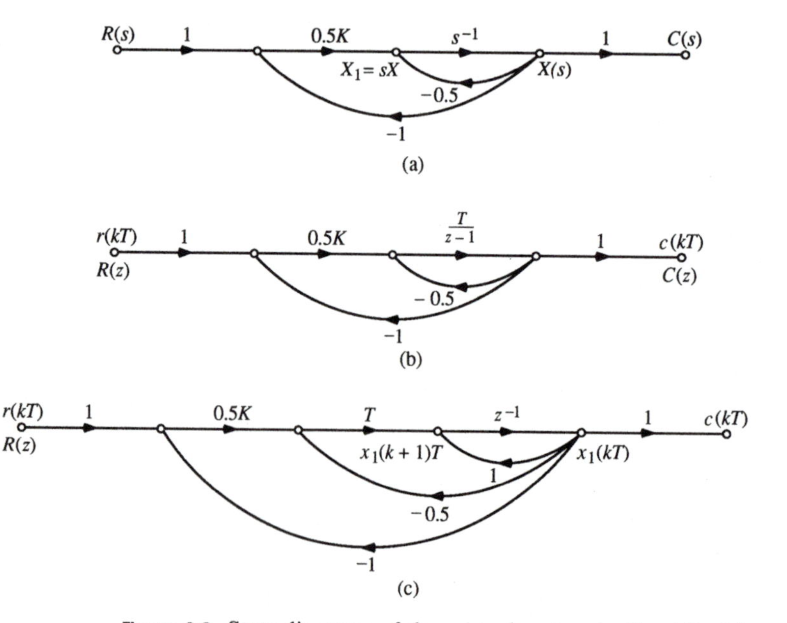

Figure 9-9. State diagrams of the control system in Fig. 9-2 with rectangular-integration approximation. (a) Continuous-data system. (b) Integrator replaced by backward-rectangular integration. (c) State diagram of digital model in (b).

diagram with the integrator $1/s$ replaced by the transfer function $T/(z-1)$. Figure 9-9(c) shows the state diagram of the digital system in Fig. 9-9(b). The input-output transfer function of the digital system is obtained from Fig. 9-9(b),

$$\frac{C(z)}{R(z)} = \frac{0.5KT}{z + (0.5T + 0.5KT - 1)} \tag{9-30}$$

For $K = 1$, $T = 0.314$ s, and a unit-step function input, the output transform is written

$$C(z) = \frac{0.157z}{(z-1)(z-0.686)} \tag{9-31}$$

The dynamic equations of the system in Fig. 9-9(c) are written

$$x_1[(k+1)T] = (1 - 0.5T - 0.5KT)x_1(kT) + 0.5KTr(kT) \tag{9-32}$$

$$c(kT) = x_1(kT) \tag{9-33}$$

Expanding $C(z)$ in Eq. (9-31) into a power series in inverse powers of z, we have the output sequence as

$$\begin{aligned} C(z) = {} & 0.1570z^{-1} + 0.2647z^{-2} + 0.3386z^{-3} + 0.3893z^{-4} + 0.4240z^{-5} \\ & + 0.4479z^{-6} + 0.4642z^{-7} + 0.4755z^{-8} + 0.4832z^{-9} + 0.4885z^{-10} \\ & + 0.4921z^{-11} + 0.4946z^{-12} + 0.4963z^{-13} + 0.4974z^{-14} + \cdots \end{aligned} \tag{9-34}$$

The unit-step response is shown in Fig. 9-10.

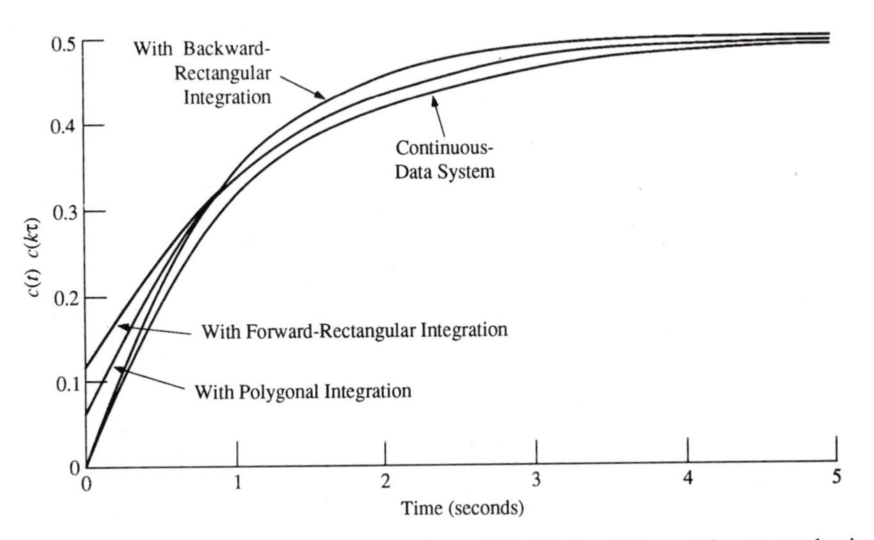

Figure 9-10. Unit-step responses of digital simulation with rectangular integrations and polygonal integration.

With the forward-rectangular integration, the integrator in Fig. 9-9(a) is replaced by $Tz/(z-1)$. The output transform of the system for $K = 1$, $T = 0.314$ s, and a unit-step input is

$$C(z) = \frac{0.1195z^2}{(z-1)(z-0.761)} \tag{9-35}$$

Expanding $C(z)$ into a power series of inverse powers of z, we have

$$\begin{aligned} C(z) = {} & 0.1195 + 0.2104z^{-1} + 0.2796z^{-2} + 0.3323z^{-3} + 0.3724z^{-4} \\ & + 0.4029z^{-5} + 0.4261z^{-6} + 0.4438z^{-7} + 0.4572z^{-8} + 0.4674z^{-9} \\ & + 0.4811z^{-10} + 0.4856z^{-11} + 0.4856z^{-12} + 0.4891z^{-13} + 0.4917z^{-14} \\ & + 0.4937z^{-15} + 0.4952z^{-16} + \cdots \end{aligned} \tag{9-36}$$

The unit-step response is shown in Fig. 9-10. Notice that the forward-rectangular integration approximation gives a unit-step response that jumps at $t = 0$. As time increases, the forward-rectangular integration gives a better approximation than that of the backward-rectangular integration.

9-4-2 Polygonal (Trapezoidal) Integration–Bilinear Transformation

A generally more accurate numerical integration scheme is achieved by using the polygonal hold concept discussed in Chapter 2. As shown in Fig. 9-11(a), the area under the curve $r(\tau)$ can be approximated by summing up the areas under the polygons of widths T. It is apparent that the approximation gets better with the reduction of the sampling period T. This approximation is often referred to as the *polygonal integration* or *trapezoidal integration.* As shown in Fig. 9-11(b), this type of approximation is equivalent to inserting a sample-and-polygonal-hold device before each integrator. Since the transfer function of the polygonal hold is [Eq. (2-107)]

$$G_{hp}(s) = \frac{e^{Ts} + e^{-Ts} - 2}{Ts^2} \tag{9-37}$$

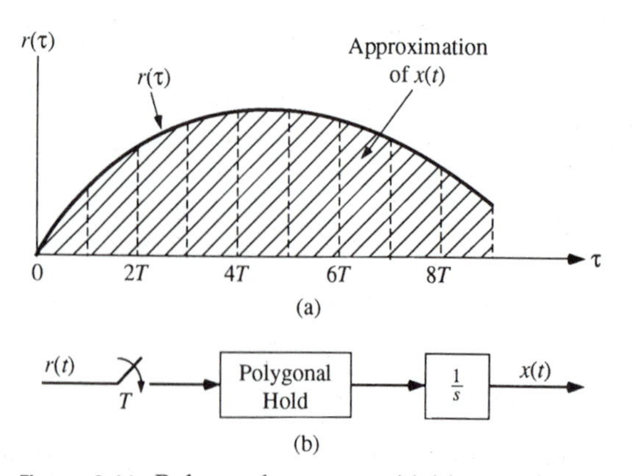

Figure 9-11. Polygonal or trapezoidal-integration approximation.

the transfer function of the polygonal integrator is written as

$$\frac{X(z)}{R(z)} = \frac{z + z^{-1} - 2}{T} \mathscr{Z}\left(\frac{1}{s^3}\right) = \frac{T}{2}\left(\frac{z+1}{z-1}\right) \tag{9-38}$$

It is interesting to note that the polygonal-integration approximation to integration is identical to the r-transformation discussed in Sec. 8-4; that is, r corresponds to s. Thus, the polygonal integrator is also referred to as the *bilinear-transformation* approximation.

With the polygonal integration, the integrator in Fig. 9-9(a) is replaced by $T(z+1)/2(z-1)$. The output transform of the system with $K = 1$, $T = 0.314$ s, and a unit-step input is

$$C(z) = \frac{0.0678z(z+1)}{(z-1)(z-0.7286)} \tag{9-39}$$

Expanding $C(z)$ into a power series in inverse powers of z, we have

$$\begin{aligned} C(z) = 0.0678 &+ 0.1850z^{-1} + 0.2704z^{-2} + 0.3326z^{-3} + 0.3779z^{-4} \\ &+ 0.4110z^{-5} + 0.4350z^{-6} + 0.4526z^{-7} + 0.4653z^{-8} + 0.4746z^{-9} \\ &+ 0.4814z^{-10} + 0.4864z^{-11} + 0.4900z^{-12} + 0.4926z^{-13} + 0.4945z^{-14} \\ &+ 0.4959z^{-15} + 0.4969z^{-16} + \cdots \end{aligned} \tag{9-40}$$

The unit-step response is shown in Fig. 9-10. Comparing the results of the polygonal integration with the forward-rectangular integration, we see that the initial jump is smaller, and as time progresses, the two responses are very similar.

In practice, the bilinear-transformation approximation is preferable for digital simulation in control systems, due to the fact that it maps the $j\omega$ axis of the s-plane onto the unit circle of the z-plane. This means that, given any rational transfer function $G(s)$, poles and zeros in the left and right halves of the s-plane will map onto corresponding poles and zeros inside and outside of the z-plane, respectively, for $G(z)$, when the following bilinear transformation is applied.

$$s = \frac{2}{T}\frac{z-1}{z+1} \tag{9-41}$$

None of the other numerical integration schemes has this property.

9-5 FREQUENCY-DOMAIN CHARACTERISTICS—FREQUENCY WARPING

It would be enlightening to investigate the frequency-domain characteristics of the digital simulation techniques described in the preceding sections. The Bode plot of the closed-loop transfer function of the continuous-data system shown in Fig. 9-2 is shown in Fig. 9-12 for $K = 1$. The Bode plots of closed-loop transfer functions $C(z)/R(z)$ of the digital equivalent systems derived from

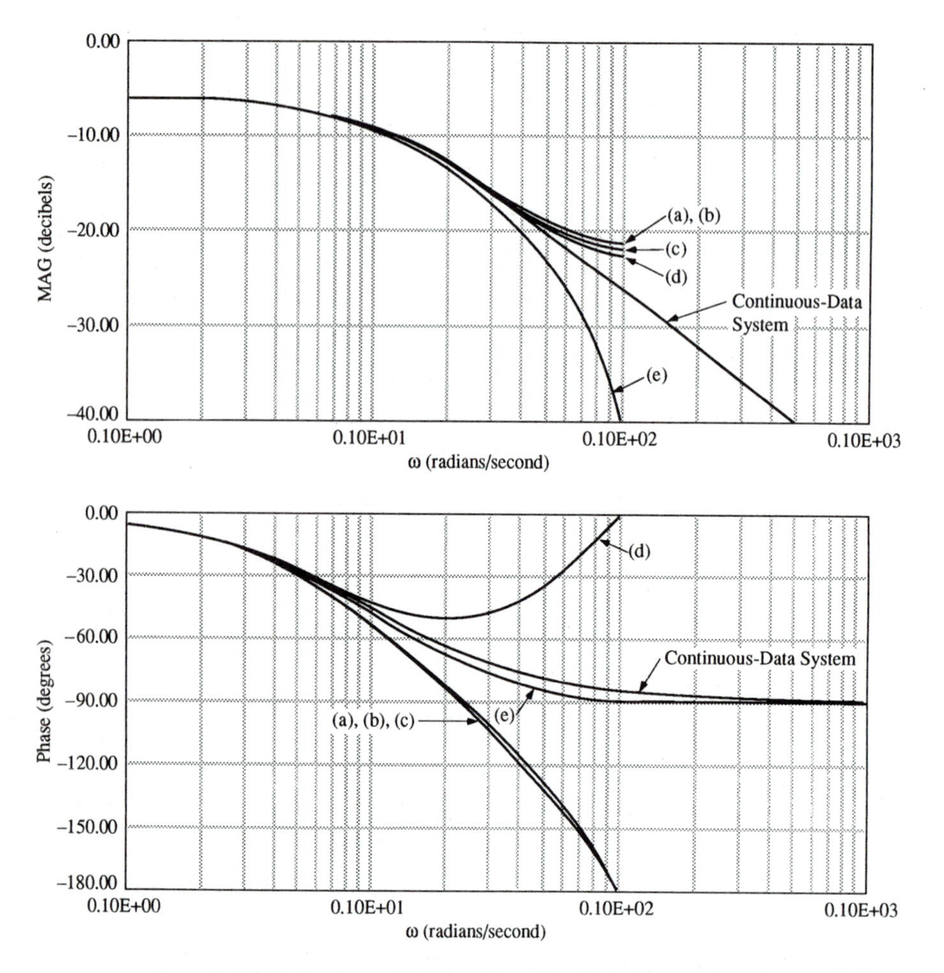

Figure 9-12. Bode plots of $0.5/(s + 1)$ and its digital approximations. (a) With sample-and-zoh. $C(z)/R(z) = 0.145/(z - 0.709)$. (b) With sample-and-polygonal-hold. $C(z)/R(z) = 0.157/(z - 0.7297)$. (c) With backward-rectangular integration. $C(z)/R(z) = 0.157/(z - 0.686)$. (d) With forward-rectangular integration. $C(z)/R(z) = 0.1195z/(z - 0.761)$. (e) With polygonal integration. $C(z)/R(z) = 0.0678(z + 1)/(z - 0.7286)$.

sample and zoh, sample-and-polygonal-hold, backward-rectangular integration, forward-rectangular integration, and polygonal integration are all shown in the same figure. Notice that the Bode plots of the digital systems are all shown up to $\omega = \omega_s/2 = 10$ rad/s.

Comparing the Bode plot of the continuous-data system with those of the digital models, we see that the digital approximations are close at frequencies only slightly beyond the bandwidth of the continuous-data system which is 1 rad/s. At high frequencies, especially beyond $\omega_s/2$, the differences in the magnitude plots and phase plots between the continuous-data and digital systems are quite drastic. The time-domain simulation results shown in Figs. 9-4 and 9-10 indicate that matching the frequency responses at very high frequencies is not important.

One important property of the digital simulation schemes revealed by the Bode plots is the "warping" of the gain and phase characteristics of the digital systems near the frequency $\omega = \omega_s/2$. The magnitude curve of the bilinear-transformation approximation in Fig. 9-12 warps downward, whereas the others warp upward. Apparently, the extent of warping depends on the sampling period T. If the sampling period is large, the sampling frequency ω_s is low. The magnitude curve in Fig. 9-12(e) will warp down more severely relative to the magnitude curve of the continuous-data system.

9-5-1 Frequency Prewarping

The discussions in the preceding section show that when the sampling period of a digital simulation using the bilinear transformation of Eq. (9-41) is not sufficiently small, the warping of the frequency response, relative to that of the continuous-data system, can cause inaccuracy in digital simulations. In many practical situations, using a relatively large sampling period may have economic advantages, or there are situations under which the sampling period cannot be randomly selected. Several textbooks [1]–[3] have suggested the "prewarping" of the transfer function of the continuous-data system so that the effect of warping due to sampling can be reduced. We shall show in the following that the matching of the frequency response characteristics at one point does not guarantee a better simulation model for the time response. The prewarping principle is explained by using the relationship between the frequencies of the s-domain and that of the bilinear transformation in the z-domain. Using Eq. (8-45), the relationship between the frequency ω of the continuous-data system and frequency ω_d of the digital approximating system is

$$\omega = \frac{2}{T} \tan\left(\frac{\omega_d T}{2}\right) \tag{9-42}$$

The relationship between the frequencies of the two systems in Eq. (9-42) explains the warping of the frequency characteristics of the digital equivalent system, since when $\omega_d = \pi/T$, ω approaches infinity. The prewarping principle suggests that if we prewarp the transfer function of the continuous-data system by setting the critical frequency or the bandwidth of the system equal to that of the digital system, then the frequency responses of the two systems can be matched at least at this one point. Let us again use the continuous-data system in Fig. 9-2(a) as an example. When the sampling period is 0.314 s, the sampling frequency is 20 times the bandwidth of the system, which is quite adequate. Thus, with $T = 0.314$ s, the digital system obtained with the bilinear transformation of Eq. (9-41) gives a fairly good approximation of the continuous-data system. However, when $T = 1.57$ s, the transfer function of the digital approximating system using the bilinear transformation is

$$\frac{C(z)}{R(z)} = \frac{0.1946(z+1)}{z-0.2218} \tag{9-43}$$

The Bode diagram in Fig. 9-13 shows that the magnitude of $C(e^{j\omega T})/R(e^{j\omega T})$ at $\omega = 1$ rad/s is -11.42 dB, as compared with -9.03 dB of the continuous-data system.

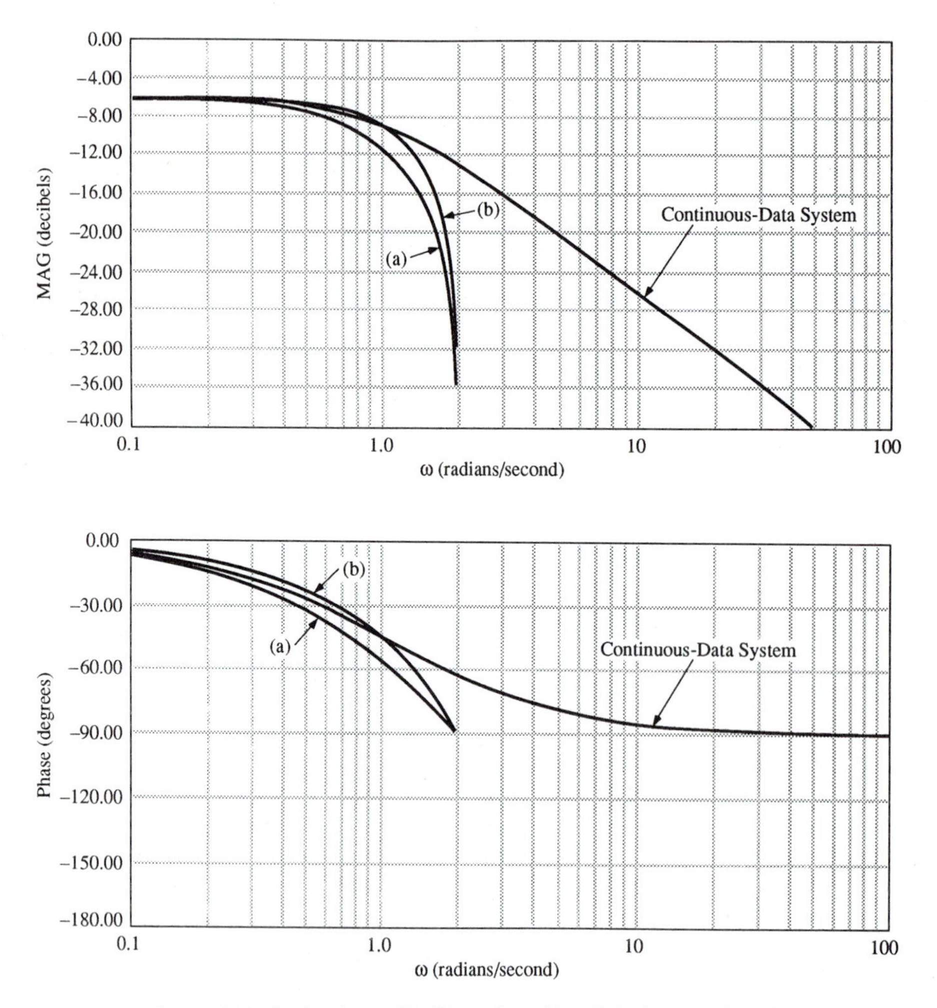

Figure 9-13. Bode plots of $0.5/(s+1)$ and its digital approximations, $T = 1.57$ s. (a) With bilinear approximation. $C(z)/R(z) = 0.1946(z+1)/(z-0.2218)$. (b) With bilinear transformation and prewarping. $C(z)/R(z) = 0.25(z+1)/z$.

Let us set the bandwidth of the digital system at 1 rad/s also, which will bring the magnitude up from -11.42 to -9.03 dB. Setting $\omega_d = 1$ rad/s in Eq. (9-42), we get $\omega = 1.274$ rad/s. Thus, the prewarped transfer function of the continuous-data system becomes

$$\frac{C(s)}{R(s)} = \frac{0.637}{s + 1.274} \tag{9-44}$$

where the gain is adjusted to give the same zero-frequency gain of 0.5. Now substituting the bilinear transformation of Eq. (9-41) with $T = 1.57$ s into the last equation, we get

$$\frac{C(z)}{R(z)} = \frac{0.25(z + 1)}{z} \tag{9-45}$$

The Bode plot of $C(z)/R(z)$ in Eq. (9-45) is shown in Fig. 9-13(b), where the digital system based on the prewarped continuous-data model of Eq. (9-44) is shown to match the original system at $\omega = 1$ rad/s. At $\omega = 1$ rad/s, the magnitude and phase of the digital and the continuous-data systems are equal. It is interesting to compare the unit-step responses of the systems with the prewarped provision. Figure 9-14 shows that the unit-step response of the digital system described by Eq. (9-45) actually deviates more from that of the continuous-data system than the digital system described by Eq. (9-43) which is determined without prewarping. Careful examination of the Bode diagram of Fig. 9-13 reveals that, although the digital system obtained from the prewarped model matches the continuous-data system at $\omega = 1$ rad/s, the characteristics at other frequencies are still mismatched. This may lead to an inferior simulation in the time domain. Thus, it is important to realize that, for a control system problem,

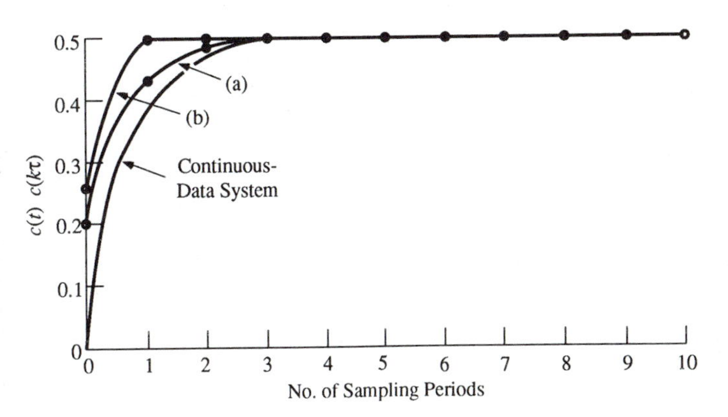

Figure 9-14. Unit-step responses of a continuous-data system and its digital approximations. $C(s)/R(s) = 0.5/(s + 1)$. (a) With bilinear transformation. $C(z)/R(z) = 0.1946(z + 1)/(z - 0.2218)$. (b) With bilinear transformation and prewarping. $C(z)/R(z) = 0.25(z + 1)/z$.

the time-domain characteristics of digital simulation are perhaps more important than the frequency-domain characteristics which often govern the performance of digital filters used in communication systems.

9-6 DIGITAL REDESIGN

A large number of control systems in operation in industry are continuous-data systems. From the performance standpoint, these systems are quite satisfactory. However, as the technology of digital computers and digital signal processors becomes more advanced, refitting these systems with digital transducers and digital controllers is often desirable. Rather than carry out a completely new design using digital control theory, it is possible to apply the digital redesign technique to arrive at an equivalent digital system. The digital system is said to be equivalent to the continuous-data system if the responses of the two systems are closely matched for the same input and initial conditions. The digital redesign technique can also be regarded as digital simulation, since the digital equivalent of a continuous-data system is found. The method is more elegant than the simple scheme of adding sample-and-hold devices in a continuous-data system, since the system parameters are altered to allow an arbitrarily selected sampling period.

The block diagram of the continuous-data system under consideration is shown in Fig. 9-15. As a matter of convenience, the state-feedback model is used. It is conceivable that any linear feedback control system could be modeled in this form.

The controlled process of the continuous-data system is represented by the state equation

$$\frac{d\mathbf{x}_c(t)}{dt} = \mathbf{A}\mathbf{x}_c(t) + \mathbf{B}\mathbf{u}(t) \tag{9-46}$$

The control vector $\mathbf{u}(t)$ is dependent on the state vector $\mathbf{x}_c(t)$ and the input vector $\mathbf{r}(t)$ through the relation

$$\mathbf{u}(t) = \mathbf{E}(0)\mathbf{r}(t) - \mathbf{G}(0)\mathbf{x}_c(t) \tag{9-47}$$

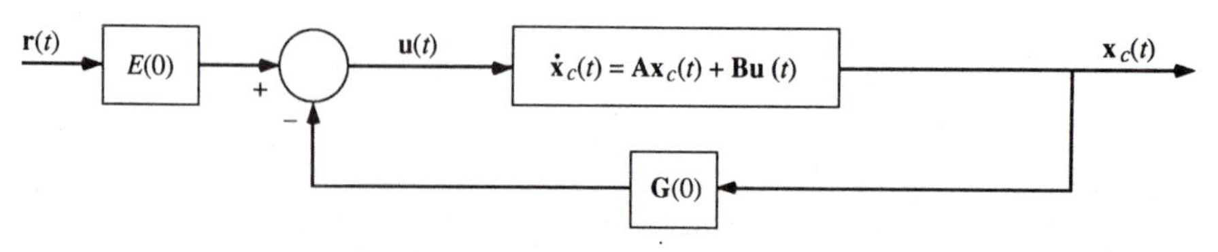

Figure 9-15. Continuous-data system.

The following vectors and matrices are defined:

$\mathbf{x}_c(t)$ state vector ($n \times 1$)
$\mathbf{u}(t)$ control vector ($m \times 1$)
$\mathbf{r}(t)$ input vector ($m \times 1$)
$\mathbf{A}$ coefficient matrix ($n \times n$)
$\mathbf{B}$ coefficient matrix ($n \times m$)
$\mathbf{E}(0)$ input matrix ($m \times m$)
$\mathbf{G}(0)$ feedback matrix ($m \times n$)

Substituting Eq. (9-47) into Eq. (9-46) yields the state equations of the closed-loop system:

$$\frac{d\mathbf{x}_c(t)}{dt} = [\mathbf{A} - \mathbf{B}\mathbf{G}(0)]\mathbf{x}_c(t) + \mathbf{B}\mathbf{E}(0)\mathbf{r}(t) \tag{9-48}$$

The solution of the last equation for $t \geq t_0$ is

$$\mathbf{x}_c(t) = \boldsymbol{\phi}_c(t - t_0)\mathbf{x}_c(t_0) + \int_{t_0}^{t} \boldsymbol{\phi}_c(t - \tau)\mathbf{B}\mathbf{E}(0)r(\tau)\, d\tau \tag{9-49}$$

where $\mathbf{x}_c(t_0)$ is the initial state of $\mathbf{x}_c(t)$ at $t = t_0$, and

$$\boldsymbol{\phi}_c(t - t_0) = e^{[\mathbf{A} - \mathbf{B}\mathbf{G}(0)](t - t_0)} \tag{9-50}$$

As defined in Chapter 4, $\boldsymbol{\phi}_c(t - t_0)$ is the state transition matrix of $\mathbf{A} - \mathbf{B}\mathbf{G}(0)$ and is represented by the power series

$$\boldsymbol{\phi}_c(t - t_0) = \sum_{j=0}^{\infty} \frac{1}{j!} [\mathbf{A} - \mathbf{B}\mathbf{G}(0)]^j (t - t_0)^j \tag{9-51}$$

The block diagram of the digital control system which is to approximate the system of Fig. 9-15 is shown in Fig. 9-16. Here the digital system may be considered as a digital simulation model of the continuous-data system, or it may be regarded as the digitally redesigned model of the continuous-data system in Fig. 9-15.

The outputs of the sample-and-hold devices in Fig. 9-16 are a series of step functions with the amplitudes denoted by the elements of the vector $\mathbf{u}_s(kT)$ for $kT \leq t < (k + 1)T$. The matrices $\mathbf{G}(T)$ and $\mathbf{E}(T)$ denote the feedback gain and the forward gain matrices of the digital system, respectively.

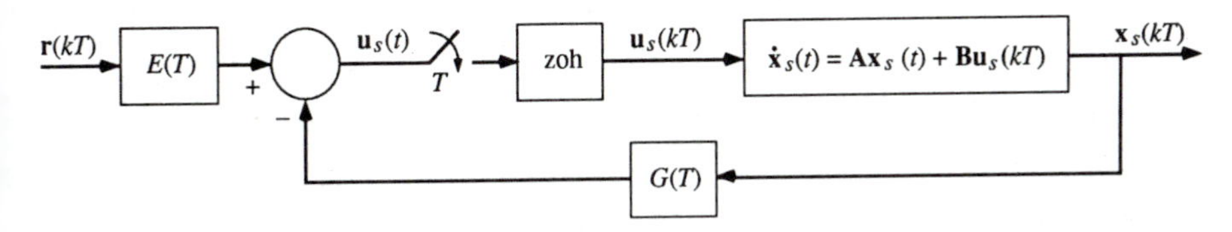

Figure 9-16. Digital control system that approximates the continuous-data system in Fig. 9-15.

The state equations of the digital system are denoted by

$$\frac{d\mathbf{x}_s(t)}{dt} = \mathbf{A}\mathbf{x}_s(t) + \mathbf{B}\mathbf{u}_s(kT) \tag{9-52}$$

for $kT \le t \le (k+1)T$, where

$$\mathbf{u}_s(kT) = \mathbf{E}(T)\mathbf{r}(kT) - \mathbf{G}(T)\mathbf{x}_s(kT) \tag{9-53}$$

It is important to note that the matrices **A** and **B** are identical to those of Eq. (9-46). Substituting Eq. (9-53) into Eq. (9-52), we get

$$\frac{d\mathbf{x}_s(t)}{dt} = \mathbf{A}\mathbf{x}_s(t) + \mathbf{B}[\mathbf{E}(T)\mathbf{r}(kT) - \mathbf{G}(T)\mathbf{x}_s(kT)] \tag{9-54}$$

for $kT \le t \le (k+1)T$.

The solution of Eq. (9-54) with $t = (k+1)T$ and $t_0 = kT$ is

$$\mathbf{x}_s[(k+1)T] = \left(\boldsymbol{\phi}(T) - \int_{kT}^{(k+1)T} \boldsymbol{\phi}(kT + T - \tau)\, d\tau \mathbf{B}\mathbf{G}(T)\right)\mathbf{x}_s(kT) + \int_{kT}^{(k+1)T} \boldsymbol{\phi}(kT + T - \tau)\, d\tau \mathbf{B}\mathbf{E}(T)\mathbf{r}(kT) \tag{9-55}$$

where

$$\boldsymbol{\phi}(T) = e^{\mathbf{A}T} = \sum_{j=0}^{\infty} \frac{(\mathbf{A}T)^j}{j!} \tag{9-56}$$

The problem is to find the matrices $\mathbf{E}(T)$ and $\mathbf{G}(T)$ so that the state variables of the digital system in Fig. 9-16 are as close as possible to those of the continuous-data system at the sampling instants for a given input $\mathbf{r}(t)$.

It is necessary to assume that $\mathbf{r}(t) = \mathbf{r}(kT)$ for $kT \le t < (k+1)T$ so that the solution of $\mathbf{E}(T)$ is independent of $\mathbf{r}(t)$. Therefore, effectively, the input of the continuous-data system of Fig. 9-15 is assumed to pass through sample-and-hold devices. This assumption would not affect the solution if $\mathbf{r}(t)$ had step functions as its elements. If the inputs were other than step functions, the approximation would be a good one for small sampling periods.

Now letting $t_0 = kT$ and $t = (k+1)T$ in Eq. (9-49), and assuming $\mathbf{r}(\tau)$ is $\mathbf{r}(kT)$ over one sampling period, we have

$$\mathbf{x}_c[(k+1)T] = \boldsymbol{\phi}_c(T)\mathbf{x}_c(kT) + \int_{kT}^{(k+1)T} \boldsymbol{\phi}_c(kT + T - \tau)\mathbf{B}\mathbf{E}(0)\, d\tau \mathbf{r}(kT) \tag{9-57}$$

for $kT \le t \le (k+1)T$.

The responses of Eqs. (9-55) and (9-57) will match at $t = (k+1)T$ for an arbitrary initial state $\mathbf{x}_c(kT) = \mathbf{x}_s(kT)$ and an arbitrary input $\mathbf{r}(\tau)$ if and only if the following two equations are satisfied:

$$\boldsymbol{\phi}_c(T) = \boldsymbol{\phi}(T) - \int_{kT}^{(k+1)T} \boldsymbol{\phi}(kT + T - \tau)\, d\tau \mathbf{B}\mathbf{G}(T) \tag{9-58}$$

and

$$\int_{kT}^{(k+1)T} \boldsymbol{\phi}(kT + T - \tau)\, d\tau \mathbf{B}\mathbf{E}(T) = \int_{kT}^{(k+1)T} \boldsymbol{\phi}_c(kT + T - \tau)\mathbf{B}\mathbf{E}(0)\, d\tau \tag{9-59}$$

First, referring to Eq. (9-58) and letting $\lambda = (k + 1)T - \tau$, we have

$$\boldsymbol{\phi}_c(T) = \boldsymbol{\phi}(T) - \boldsymbol{\theta}(T)\mathbf{G}(T) \tag{9-60}$$

where

$$\boldsymbol{\theta}(T) = \int_0^T e^{\mathbf{A}\lambda}\mathbf{B}\, d\lambda \tag{9-61}$$

In principle, the feedback matrix $\mathbf{G}(T)$ of the digital system can be determined from Eq. (9-60). However, it is subject to the limitations discussed in the following section.

9-6-1 A Closed-Form Solution for $G(T)$

For all n states of the digital system $\mathbf{x}_s(kT)$ to match those of the continuous-data system $\mathbf{x}_c(kT)$ at each sampling instant, it is sufficient that Eq. (9-60) be satisfied. However, Eq. (9-60) consists of n^2 scalar equations with mn unknown elements in $\mathbf{G}(T)$. If the number of state variables n is equal to the number of inputs m and the matrix $\boldsymbol{\theta}(T)$ is nonsingular, the feedback matrix $\mathbf{G}(T)$ is solved from Eq. (9-60) as

$$\mathbf{G}(T) = [\boldsymbol{\theta}(T)]^{-1}[\boldsymbol{\phi}(T) - \boldsymbol{\phi}_c(T)] \tag{9-62}$$

For most control systems, $n > m$; that is, there are more states than inputs, and Eq. (9-60) will generally not have a solution. However, if the following rank condition is satisfied, the system of equations in Eq. (9-60) is consistent, and a solution still exists.

Let

$$\mathbf{G}(T) = [\mathbf{g}_1 \quad \mathbf{g}_2 \quad \cdots \quad \mathbf{g}_n] \tag{9-63}$$

$$\boldsymbol{\phi}(T) - \boldsymbol{\phi}_c(T) = [\mathbf{d}_1 \quad \mathbf{d}_2 \quad \cdots \quad \mathbf{d}_n] \tag{9-64}$$

where $\mathbf{g}_i$, $i = 1, 2, \ldots, n$, are m-dimensional vectors, and $\mathbf{d}_i$, $i = 1, 2, \ldots, n$, are n-dimensional vectors. Then if

$$\text{rank}[\boldsymbol{\theta}] = \text{rank}[\boldsymbol{\theta}, \mathbf{d}_i] \tag{9-65}$$

for all $i = 1, 2, \ldots, n$, the system of equations in Eq. (9-60) has at least one solution. However, if the above rank condition is not satisfied, the equations are inconsistent, and no solution exists.

9-6-2 Partial Matching of States

In general, the rank condition of Eq. (9-65) is rarely satisfied by the matrices $\boldsymbol{\theta}(T)$ and $\boldsymbol{\phi}(T) - \boldsymbol{\phi}_c(T)$. Thus, when $n > m$, not all of the states of the continuous-data and the digital systems can be made to match at the sampling instants.

Although it is not possible to match all of the states exactly, it can be shown that it is possible to match some of the states or the algebraic sums of the states at each sampling instant. We introduce a weighting matrix $\mathbf{H}$ which allows a partial matching of the states. Equation (9-60) is rewritten as

$$\mathbf{D}(T) = \boldsymbol{\phi}(T) - \boldsymbol{\phi}_c(T) = \boldsymbol{\theta}(T)\mathbf{G}(T) \tag{9-66}$$

Both sides of the last equation are premultiplied by an $m \times n$ *weighting matrix* $\mathbf{H}$ to give

$$\mathbf{HD}(T) = \mathbf{H}\boldsymbol{\theta}(T)\mathbf{G}(T) \tag{9-67}$$

If $\mathbf{H}$ is chosen such that the $m \times n$ matrix $\mathbf{H}\boldsymbol{\theta}(T)$ is nonsingular, Eq. (9-67) may be solved for $\mathbf{G}(T)$ to give a partial matching of the states. Let the feedback matrix for partial matching be designated by $\mathbf{G}_w(T)$. Then, from Eq. (9-67),

$$\mathbf{G}_w(T) = [\mathbf{H}\boldsymbol{\theta}(T)]^{-1}\mathbf{HD}(T) \tag{9-68}$$

Note that the solution of $\mathbf{G}_w(T)$ from Eq. (9-68) does not satisfy Eq. (9-66) except when $\mathbf{H}$ is an identity matrix. The reason is that when Eq. (9-66) is premultiplied by the matrix $\mathbf{H}$, it reduces the system of n^2 equations to a system of mn equations for $n > m$. Therefore, the solution will not satisfy the original equations.

To explore the physical meaning of the transformation of Eq. (9-67) and of the solution in Eq. (9-68), let us again equate Eq. (9-55) to Eq. (9-57), and then premultiply both sides of the equation by $\mathbf{H}$. We get

$$\begin{aligned}\mathbf{Hx}_c[(k+1)T] = \mathbf{Hx}_s[(k+1)T] &= \mathbf{H}\boldsymbol{\phi}_c(T)\mathbf{x}_c(kT) + \mathbf{H}\boldsymbol{\theta}_c(T)\mathbf{E}(0)\mathbf{r}(kT)\\ &= \mathbf{H}[\boldsymbol{\phi}(T) - \boldsymbol{\theta}(T)\mathbf{G}(T)]\mathbf{x}_s(kT) + \mathbf{H}\boldsymbol{\theta}(T)\mathbf{E}(T)\mathbf{r}(kT)\end{aligned} \tag{9-69}$$

where

$$\boldsymbol{\theta}_c(T) = \int_0^T \boldsymbol{\phi}_c(\lambda)\mathbf{B}\, d\lambda \tag{9-70}$$

For arbitrary $\mathbf{x}_c(kT)$, $\mathbf{x}_s(kT)$, and $\mathbf{r}(kT)$, Eq. (9-69) leads to

$$\mathbf{H}\boldsymbol{\phi}_c(T)\mathbf{x}_c(kT) = \mathbf{H}[\boldsymbol{\phi}(T) - \boldsymbol{\theta}(T)\mathbf{G}(T)]\mathbf{x}_s(kT) \tag{9-71}$$

and

$$\mathbf{H}\boldsymbol{\theta}_c(T)\mathbf{E}(0)\mathbf{r}(kT) = \mathbf{H}\boldsymbol{\theta}(T)\mathbf{E}(T)\mathbf{r}(kT) \tag{9-72}$$

The significant point is that the solution $\mathbf{G}_w(T)$ of Eq. (9-68) satisfies Eq. (9-71) for any arbitrary initial state $\mathbf{x}_c(kT)$. Premultiplying both sides of the state transition equation of Eqs. (9-55) and (9-57) by the weighting matrix $\mathbf{H}$ simply transforms the n-dimensional state vectors $\mathbf{x}_c[(k+1)T]$ and $\mathbf{x}_s[(k+1)T]$ into a new m-dimensional vector $\mathbf{y}[(k+1)T]$, such that

$$\mathbf{y}[(k+1)T] = \mathbf{H}\mathbf{x}_c[(k+1)T] = \mathbf{H}\mathbf{x}_s[(k+1)T] \tag{9-73}$$

Equation (9-73) indicates that the m new states, $y_i[(k+1)T]$, $i = 1, 2, \ldots, m$, are algebraic sums of the original n state variables; that is,

$$y_i[(k+1)T] = \sum_{j=1}^{n} h_{ij} x_j[(k+1)T] \qquad i = 1, 2, \ldots, m \tag{9-74}$$

Therefore, the solution $\mathbf{G}_w(T)$ from Eq. (9-68) represents a matching by the n states of the digital system, the weighted algebraic sum of the n states of the continuous-data system, at the sampling instants.

9-6-3 Solution of the Feedback Matrix by Series Expansion

Although Eq. (9-68) gives the exact solution of $\mathbf{G}_w(T)$, it is possible to simplify the procedure by expanding $\mathbf{G}(T)$ into a Taylor series about $T = 0$. In general, if the series converges, $\mathbf{G}(T)$ can be approximated by taking a finite number of terms of the series expansion.

Let $\mathbf{G}(T)$ be expanded into a Taylor series about $T = 0$,

$$\mathbf{G}(T) = \lim_{K \to \infty} \mathbf{G}_k(T) = \lim_{K \to \infty} \sum_{j=0}^{K-1} \frac{1}{j!} \mathbf{G}^{(j)}(T) T^j \tag{9-75}$$

where

$$\mathbf{G}^{(j)}(T) = \left. \frac{\partial^j \mathbf{G}(T)}{\partial T^j} \right|_{T=0} \tag{9-76}$$

Substituting $\mathbf{G}(T)$ from Eq. (9-75) into Eq. (9-60) gives

$$\boldsymbol{\phi}_c(T) = \boldsymbol{\phi}(T) - \boldsymbol{\theta}(T) \sum_{j=0}^{\infty} \frac{1}{j!} \mathbf{G}^{(j)}(T) T^j \tag{9-77}$$

The last equation is written as

$$\sum_{j=0}^{\infty} \frac{[\mathbf{A} - \mathbf{B}\mathbf{G}(0)]^j T^j}{j!} = \sum_{j=0}^{\infty} \frac{\mathbf{A}^j T^j}{j!} - \sum_{j=0}^{\infty} \frac{\mathbf{A}^j T^{j+1}}{(j+1)!} \mathbf{B} \sum_{k=0}^{\infty} \frac{\mathbf{G}^{(k)}(T) T^k}{k!} \tag{9-78}$$

or

$$\sum_{j=0}^{\infty} \left(\frac{[\mathbf{A} - \mathbf{B}\mathbf{G}(0)]^j T^j}{j!} - \frac{\mathbf{A}^j T^j}{j!} + \frac{\mathbf{A}^j \mathbf{B}}{(j+1)!} \sum_{k=0}^{\infty} \frac{\mathbf{G}^{(k)}(T) T^{k+j+1}}{k!} \right) = \mathbf{0} \tag{9-79}$$

Now equating the coefficients of T^i, $i = 1, 2, \ldots$, to zero, we have

$$\frac{[\mathbf{A} - \mathbf{B}\mathbf{G}(0)]^i}{i!} - \frac{\mathbf{A}^i}{i!} + \sum_{j=0}^{i-1} \frac{\mathbf{A}^{i-j-1} \mathbf{B} \mathbf{G}^{(j)}(T)}{(i-j)!\, j!} = \mathbf{0} \tag{9-80}$$

In general, it is possible to express $\mathbf{G}^{(i-1)}(T)$ in terms of $\mathbf{G}^{(i-2)}(T)$, $\mathbf{G}^{(i-3)}(T)$, ..., $\mathbf{G}^{(1)}(T)$, and $\mathbf{G}^{(0)}(T)$, where $\mathbf{G}^{(0)}(T) = \mathbf{G}(0)$.

Equation (9-80) is written

$$\frac{[\mathbf{A} - \mathbf{BG}(0)]^i}{i!} - \frac{\mathbf{A}^i}{i!} + \sum_{j=0}^{i-2} \frac{\mathbf{A}^{i-j-1}\mathbf{BG}^{(j)}(T)}{(i-j)!j!} + \frac{\mathbf{BG}^{(i-1)}(T)}{(i-1)!} = \mathbf{0} \tag{9-81}$$

$i = 1, 2, \ldots$. Solving for the last term in the last equation, we get

$$\frac{\mathbf{BG}^{(i-1)}(T)}{(i-1)!} = \frac{\mathbf{A}^i}{i!} - \frac{[\mathbf{A} - \mathbf{BG}(0)]^i}{i!} - \sum_{j=0}^{i-2} \frac{\mathbf{A}^{i-j-1}\mathbf{BG}^{(j)}(T)}{(i-j)!j!} \tag{9-82}$$

To solve for $\mathbf{G}^{(i-1)}(T)$ from the last equation, since $\mathbf{B}$ is generally not a square matrix, we again introduce the weighting matrix $\mathbf{H}$ ($m \times n$) such that $\mathbf{HB}$ ($m \times m$) is nonsingular. Premultiplying both sides of Eq. (9-82) by $\mathbf{H}$ and solving for $\mathbf{G}^{(i-1)}(T)$, we have

$$\mathbf{G}_w^{(i-1)}(T) = (\mathbf{HB})^{-1}\mathbf{H}\left(\frac{\mathbf{A}^i}{i} - \frac{[\mathbf{A} - \mathbf{BG}(0)]^i}{i} - (i-1)! \sum_{j=0}^{i-2} \frac{\mathbf{A}^{i-j-1}\mathbf{BG}^{(j)}(T)}{(i-j)!j!}\right) \tag{9-83}$$

$i = 1, 2, \ldots$. Table 9-1 gives the expressions for $\mathbf{G}_w^{(i-1)}(T)$ for $i = 1, 2$, and 3.

The results in Table 9-1 allow the approximation of $\mathbf{G}(T)$ by using up to three terms in the series expansion of Eq. (9-75). In Eq. (9-75),

$K = 1$

$$\mathbf{G}_1(T) = \mathbf{G}^{(0)}(T) = \mathbf{G}(0) \tag{9-84}$$

$K = 2$

$$\mathbf{G}_2(T) = \mathbf{G}(0) + T\mathbf{G}^{(1)}(T) \tag{9-85}$$

$K = 3$

$$\mathbf{G}_3(T) = \mathbf{G}(0) + T\mathbf{G}^{(1)}(T) + \frac{T^2}{2}\mathbf{G}^{(2)}(T) \tag{9-86}$$

Table 9-1 Expressions for $\mathbf{G}_w^{(i-1)}(T)$

i	$\mathbf{G}_w^{(i-1)}(T)$
1	$\mathbf{G}(0)$
2	$\frac{1}{2}\mathbf{G}(0)[\mathbf{A} - \mathbf{BG}(0)]$
3	$(\mathbf{HB})^{-1}\mathbf{H}\{-\frac{1}{6}\mathbf{ABG}(0)[\mathbf{A} - \mathbf{BG}(0)] + \frac{1}{3}\mathbf{BG}(0)[\mathbf{A} - \mathbf{BG}(0)]^2\}$

In reality, we can use only $\mathbf{G}_w^{(0)}(T)$, $\mathbf{G}_w^{(1)}(T)$, and $\mathbf{G}_w^{(2)}(T)$ as approximations to their exact counterparts. However, it is interesting to note from Table 9-1 that $\mathbf{G}_w^{(0)}(T)$ and $\mathbf{G}_w^{(1)}(T)$ are not dependent upon $\mathbf{H}$. Therefore, the one- and two-term approximations of $\mathbf{G}(T)$, $\mathbf{G}_1(T)$, and $\mathbf{G}_2(T)$, respectively, attempt to match all the states of the continuous-data and the digital systems. Beyond two terms, the weighting matrix $\mathbf{H}$ should be used, since $\mathbf{G}_w^{(2)}(T) \neq \mathbf{G}^{(2)}(T)$, and only certain states and combinations of states are matched, depending on the $\mathbf{H}$ selected, when $\mathbf{G}_w^{(2)}(T)$ is used in place of $\mathbf{G}^{(2)}(T)$ in Eq. (9-86).

In general, as more terms are used in the series approximation of $\mathbf{G}(T)$ in Eq. (9-75), the solution will approach that of $\mathbf{G}_w(T)$ in Eq. (9-68).

9-6-4 An Exact Solution for $E(T)$

We now turn our attention to the determination of the feedforward matrix $\mathbf{E}(T)$. The condition of state matching in Eq. (9-59) is written as

$$\boldsymbol{\theta}(T)\mathbf{E}(T) = \boldsymbol{\theta}_c(T)\mathbf{E}(0) \tag{9-87}$$

$$\mathbf{E}(T) = [\boldsymbol{\theta}(T)]^{-1}\boldsymbol{\theta}_c(T)\mathbf{E}(0) \tag{9-88}$$

In general, when $n > m$, the solution of $\mathbf{E}(T)$ that corresponds to a partial matching of states is

$$\mathbf{E}_w(T) = [\mathbf{H}\boldsymbol{\theta}(T)]^{-1}\mathbf{H}\boldsymbol{\theta}_c(T)\mathbf{E}(0) \tag{9-89}$$

where it is assumed that $\mathbf{H}\boldsymbol{\theta}(T)$ is nonsingular.

9-6-5 Solution of $E(T)$ by Series Expansion

As in the solution of $\mathbf{G}(T)$, the matrix $\mathbf{E}(T)$ may be expanded into a Taylor series about $T = 0$:

$$\mathbf{E}(T) = \lim_{K\to\infty} \mathbf{E}_K(T) = \lim_{K\to\infty} \sum_{j=0}^{K-1} \frac{1}{j!}\mathbf{E}^{(j)}(T)T^j \tag{9-90}$$

where

$$\mathbf{E}^{(j)}(T) = \left.\frac{\partial^j \mathbf{E}(T)}{\partial T^j}\right|_{T=0} \tag{9-91}$$

Substituting the series expansion of $\mathbf{E}(T)$ into Eq. (9-87) and carrying out steps similar to those in Eqs. (9-78) to (9-83), we have

$$\mathbf{E}_w^{(i-1)}(T) = (\mathbf{HB})^{-1}\mathbf{H}\left(\frac{[\mathbf{A} - \mathbf{BG}(0)]^{i-1}\mathbf{BE}(0)}{i} - (i-1)!\sum_{j=0}^{i-2}\frac{\mathbf{A}^{i-j-1}\mathbf{BE}^{(j)}(T)}{(i-j)!\,j!}\right) \tag{9-92}$$

$i = 1, 2, \ldots$. Table 9-2 gives the results of $\mathbf{E}_w^{(i-1)}(T)$ for $i = 1, 2$, and 3.

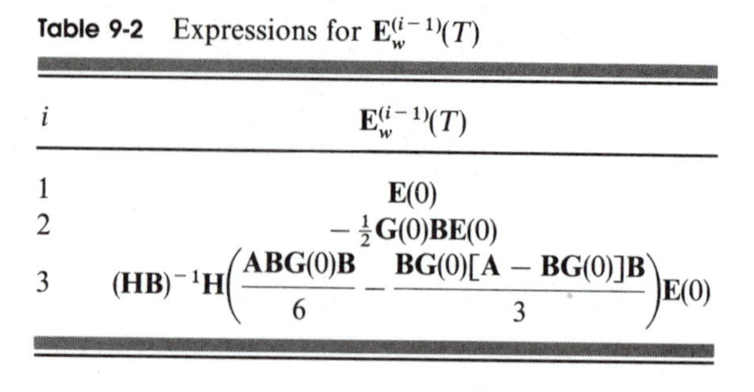

Table 9-2 Expressions for $\mathbf{E}_w^{(i-1)}(T)$

i	$\mathbf{E}_w^{(i-1)}(T)$
1	$\mathbf{E}(0)$
2	$-\frac{1}{2}\mathbf{G}(0)\mathbf{B}\mathbf{E}(0)$
3	$(\mathbf{HB})^{-1}\mathbf{H}\left(\frac{\mathbf{ABG}(0)\mathbf{B}}{6} - \frac{\mathbf{BG}(0)[\mathbf{A} - \mathbf{BG}(0)]\mathbf{B}}{3}\right)\mathbf{E}(0)$

Like the situation for $\mathbf{G}(T)$, if only up to two terms are used for the series approximation for $\mathbf{E}(T)$, the weighting matrix $\mathbf{H}$ is not needed, and an attempt is made to match all the states of the digital and the continuous-data systems.

9-6-6 Stability Considerations and Constraints on the Selection of the Weighting Matrix H

In the preceding sections the weighting matrix $\mathbf{H}$ was introduced to achieve a partial matching of states in the digital redesign problem. While the selection of the elements of $\mathbf{H}$ is not an exact science, one important requirement is that the closed-loop digital control system be asymptotically stable. The problem is to find the constraints on $\mathbf{H}$ such that the stability requirement is met.

Solving the state equations in Eq. (9-52) and evaluating the states at the sampling instants, we get

$$\mathbf{x}_s[(k+1)T] = \boldsymbol{\phi}(T)\mathbf{x}_s(kT) + \boldsymbol{\theta}(T)\mathbf{u}_s(kT) \tag{9-93}$$

where $\boldsymbol{\phi}(T)$ and $\boldsymbol{\theta}(T)$ are defined in Eqs. (9-56) and (9-61), respectively. For $\mathbf{r}(kT) = 0$, the state-feedback control is

$$\mathbf{u}_s(kT) = -\mathbf{G}(T)\mathbf{x}_s(kT) \tag{9-94}$$

Substitution of Eq. (9-94) into Eq. (9-93) yields

$$\mathbf{x}_s[(k+1)T] = [\boldsymbol{\phi}(T) - \boldsymbol{\theta}(T)\mathbf{G}(T)]\mathbf{x}_s(kT) \tag{9-95}$$

The digital closed-loop system described by Eq. (9-95) is asymptotically stable if all the eigenvalues of $[\boldsymbol{\phi}(T) - \boldsymbol{\theta}(T)\mathbf{G}(T)]$ are located inside the unit circle $|z| = 1$. Since $\boldsymbol{\phi}(T)$ and $\boldsymbol{\theta}(T)$ are known once the sampling period T is specified, the constraints on $\mathbf{G}(T)$ for stability can be established using standard stability test criteria.

Replacing $\mathbf{G}(T)$ by $\mathbf{G}_w(T)$ in Eq. (9-67), we have

$$\mathbf{HD}(T) = \mathbf{H}\boldsymbol{\theta}(T)\mathbf{G}_w(T) \tag{9-96}$$

Taking the matrix transpose on both sides of the last equation, we get

$$\mathbf{D}'(T)\mathbf{H}' = \mathbf{G}_w'(T)\boldsymbol{\theta}'(T)\mathbf{H}' \tag{9-97}$$

or, rearranging,

$$[\mathbf{G}'_w(T)\boldsymbol{\theta}'(T) - \mathbf{D}'(T)]\mathbf{H}' = \mathbf{0} \tag{9-98}$$

This matrix equation represents a set of linear homogeneous equations with nontrivial solutions if and only if the following condition is satisfied:

$$|\mathbf{G}'_w(T)\boldsymbol{\theta}'(T) - \mathbf{D}'(T)| = 0 \tag{9-99}$$

which is equivalent to

$$|\boldsymbol{\theta}(T)\mathbf{G}_w(T) - \mathbf{D}'(T)| = \mathbf{0} \tag{9-100}$$

Thus, if Eq. (9-100) is satisfied, a nonzero $\mathbf{H}$ always exists that will satisfy

$$\mathbf{G}_w(T) = [\mathbf{H}\boldsymbol{\theta}(T)]^{-1}\mathbf{H}\mathbf{D}(T) \tag{9-101}$$

9-6-7 Digital Redesign of the Simplified Skylab Satellite

An illustrative example of the digital redesign technique is presented in this section. The block diagram in Fig. 9-17 represents the dynamics of a simplified one-axis Skylab satellite system. The continuous-data Skylab system can be shown to have a damping ratio of approximately 0.707.

The state equations of the system in Fig. 9-17 can be expressed in the form of Eq. (9-46) with

$$\mathbf{A} = \begin{bmatrix} 0 & 1 \\ 0 & 0 \end{bmatrix} \qquad \mathbf{B} = \begin{bmatrix} 0 \\ \dfrac{1}{970741} \end{bmatrix} \tag{9-102}$$

The feedback matrix is

$$\mathbf{G}(0) = [11800 \quad 151800] \tag{9-103}$$

and the feedforward matrix is

$$E(0) = 11800 \tag{9-104}$$

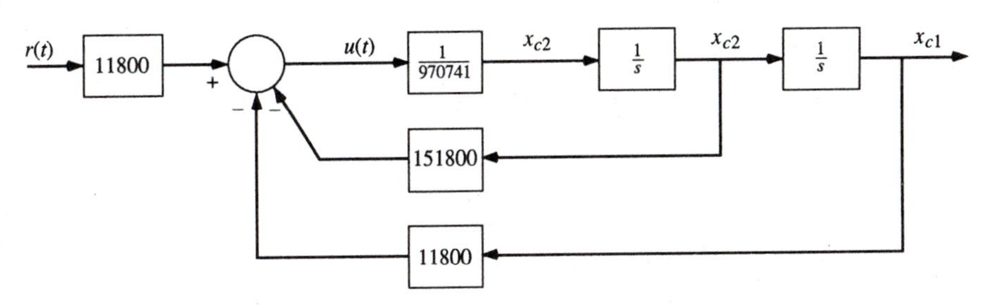

Figure 9-17. Simplified one-axis Skylab satellite system.

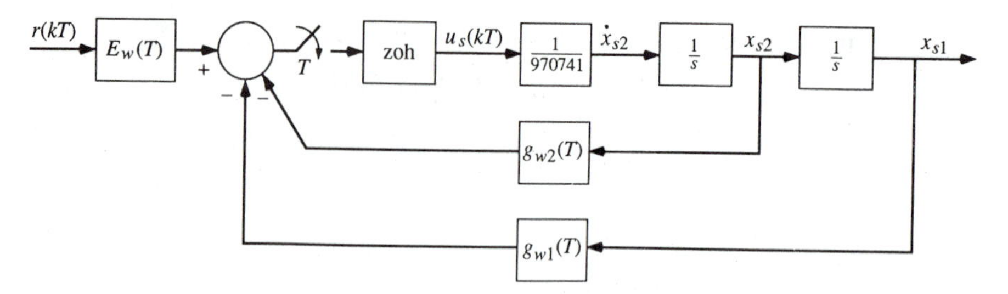

Figure 9-18. Equivalent digital system of the Skylab system.

We now wish to find an equivalent digital system with the block diagram in Fig. 9-18 whose states will match those of the continuous-data Skylab at the sampling instants for the same input and initial states.

The gain matrices $E_w(T)$ and

$$\mathbf{G}_w(T) = [g_{w1}(T) \quad g_{w2}(T)] \tag{9-105}$$

are to be determined using the digital redesign method discussed in the preceding sections.

When only one term is used in the series approximation of $\mathbf{G}(T)$ and $E(T)$, the results are $\mathbf{G}(0)$ and $E(0)$, respectively. In this case, the sampling period of the digital system shown in Fig. 9-18 must be chosen so that the system is asymptotically stable. For a two-term approximation,

$$\begin{aligned}\mathbf{G}_w(T) &\cong \mathbf{G}(0) + \frac{T}{2}\mathbf{G}(0)[\mathbf{A} - \mathbf{B}\mathbf{G}(0)] \\ &= [11800 - 922.6T \quad 151800 - 5968.9T]\end{aligned} \tag{9-106}$$

$$E_w(T) \cong E(0) - \frac{T}{2}\mathbf{G}(0)\mathbf{B}E(0) = 11800 - 922.6T \tag{9-107}$$

These results show that the gains of the feedback and feedforward matrices decrease as the sampling period increases. Table 9-3 gives the results of the elements of $\mathbf{G}_w(T)$ and $E_w(T)$ with the two-term approximation for $T = 1$ through 5 s.

Table 9-3 Elements of $\mathbf{G}_w(T)$ and $E_w(T)$ with Two-Term Approximation

T (s)	$g_{w1}(T)$	$g_{w2}(T)$	$E_w(T)$
1	10877.4	145831	10877.4
2	9954.8	139862	9954.8
3	9032.2	133893	9032.2
4	8109.6	127924	8109.6
5	7187.0	121955	7187.0

Table 9-4 Exact $\mathbf{G}_w(T)$ and $E_w(T)$ for $T = 1$ through 5 s

T (s)	$\mathbf{H}$	$g_{w1}(T)$	$g_{w2}(T)$	$E_w(T)$
1	[0 1]	10901.5	145840	10901.5
2	[0 1]	10051.2	139921	10051.2
3	[0 1]	9248.4	134071	9248.4
4	[0 1]	8492.5	128315	8492.5
5	[0 1]	7782.3	122674	7782.3
1	[1 0]	11197.0	147825	11197.0
2	[1 0]	10618.1	143867	10618.1
3	[1 0]	10063.1	139937	10063.1
4	[1 0]	9531.8	136048	9531.8
5	[1 0]	9023.7	132207	9023.7

The exact matrices $\mathbf{G}_w(T)$ and $E_w(T)$ are computed using Eqs. (9-68) and (9-89), respectively. Table 9-4 gives the results for the exact $\mathbf{G}_w(T)$ and $E_w(T)$ for $T = 1$ through 5 s, and $\mathbf{H} = [0 \quad 1]$ and $\mathbf{H} = [1 \quad 0]$. With $\mathbf{H} = [0 \quad 1]$, the state x_{c2} is to be matched, and with $\mathbf{H} = [1 \quad 0]$, the state x_{c1} is to be matched.

The digital control system with the matrices tabulated in Table 9-4 is simulated on the digital computer with $T = 2$ s. The step responses of the continuous-data Skylab system are also computed, with the values of $\mathbf{G}(0)$ and $E(0)$ given in Eqs. (9-103) and (9-104), respectively. Figures 9-19 through 9-21

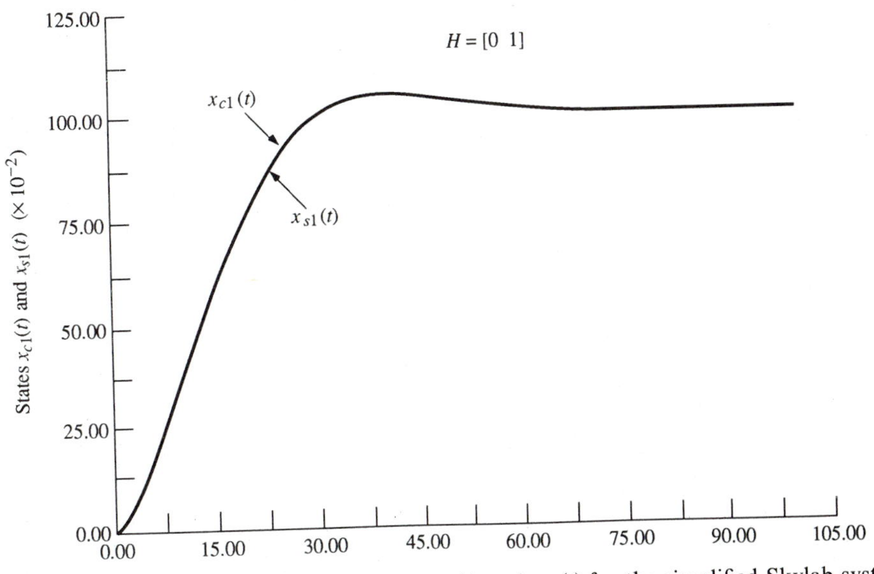

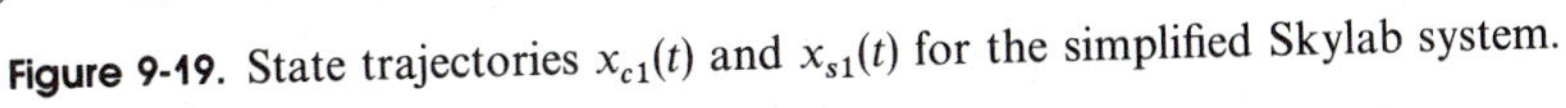

Figure 9-19. State trajectories $x_{c1}(t)$ and $x_{s1}(t)$ for the simplified Skylab system.

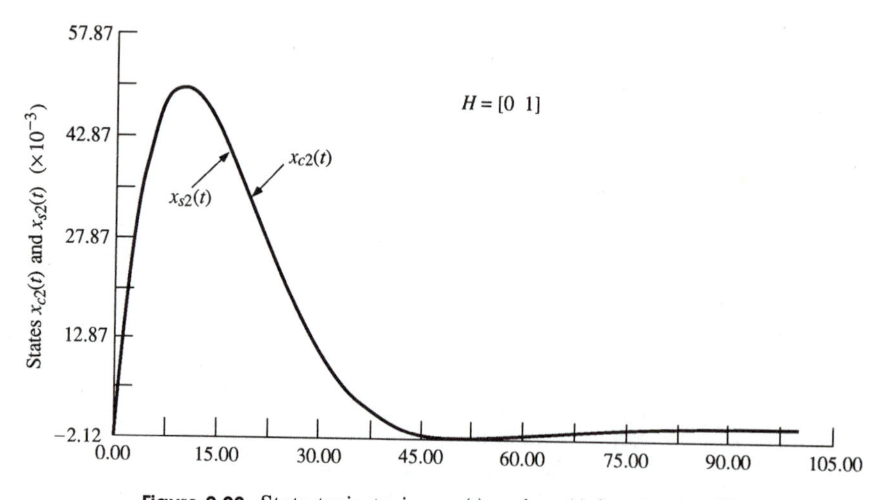

Figure 9-20. State trajectories $x_{c2}(t)$ and $x_{s2}(t)$ for the simplified Skylab system.

show the comparison for $T = 2$ s and $\mathbf{H} = [0 \quad 1]$. Figures 9-22 through 9-24 give the results for $\mathbf{H} = [1 \quad 0]$. These figures show that the digital redesign technique leads to a digital Skylab system with responses that match the continuous-data Skylab system very closely. The sampling period of 2 s appears to be quite adequate. The results are better with $\mathbf{H} = [0 \quad 1]$ than with $\mathbf{H} = [1 \quad 0]$. This means that the responses of the two systems are better matched if the matching of $x_{c2}(t)$ is emphasized rather than that of $x_{c1}(t)$.

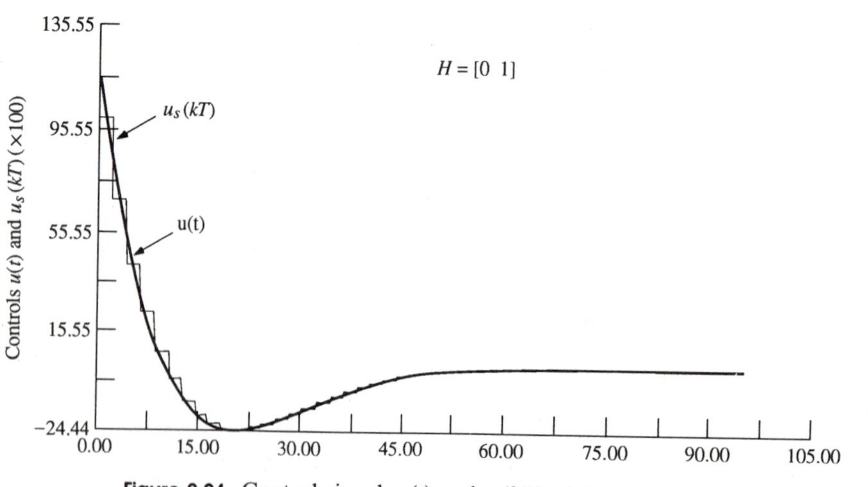

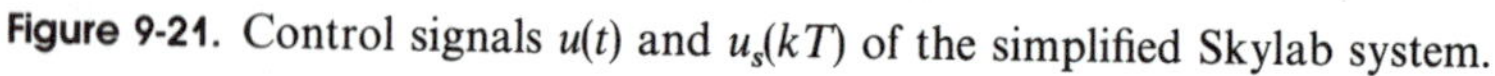
Figure 9-21. Control signals $u(t)$ and $u_s(kT)$ of the simplified Skylab system.

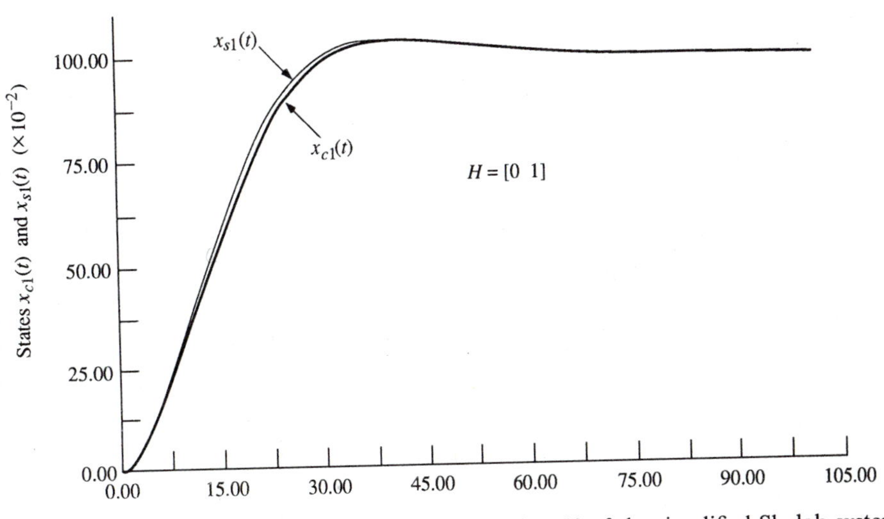

Figure 9-22. State trajectories $x_{c1}(t)$ and $x_{s1}(t)$ of the simplified Skylab system.

We can investigate the general stability condition in the parameter plane of $g_{w2}(T)$ versus $g_{w1}(T)$. The characteristic equation of the closed-loop digital system is

$$F(z) = |z\mathbf{I} - \boldsymbol{\phi}(T) + \boldsymbol{\theta}(T)\mathbf{G}_w(T)|$$

$$= z^2 + 2\left(\frac{g_{w1}}{970741} + \frac{g_{w2}}{970741} - 1\right)z - \frac{2g_{w1}}{970741} - \frac{2g_{w2}}{970741} + 1 = 0 \qquad \textbf{(9-108)}$$

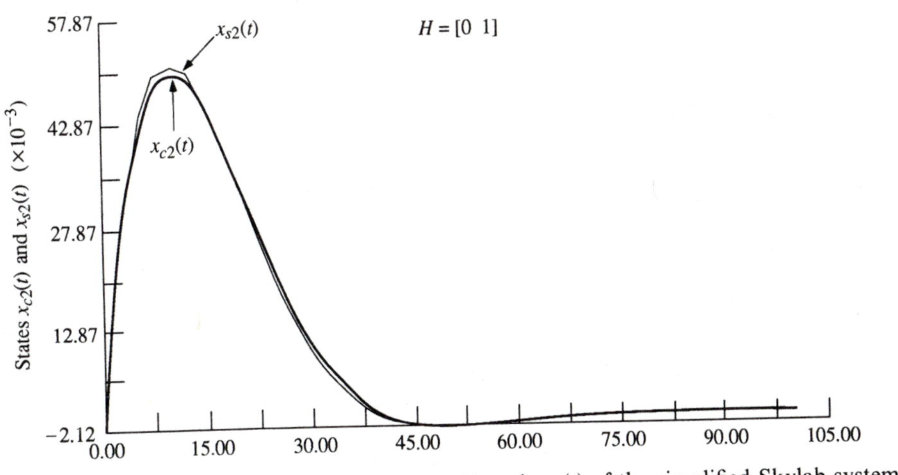

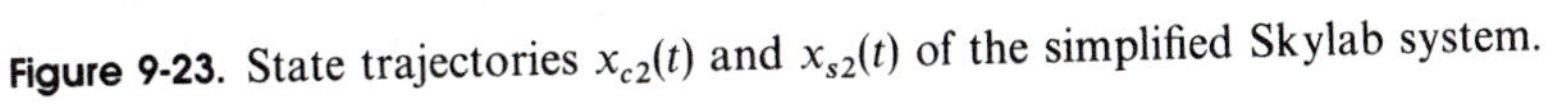
Figure 9-23. State trajectories $x_{c2}(t)$ and $x_{s2}(t)$ of the simplified Skylab system.

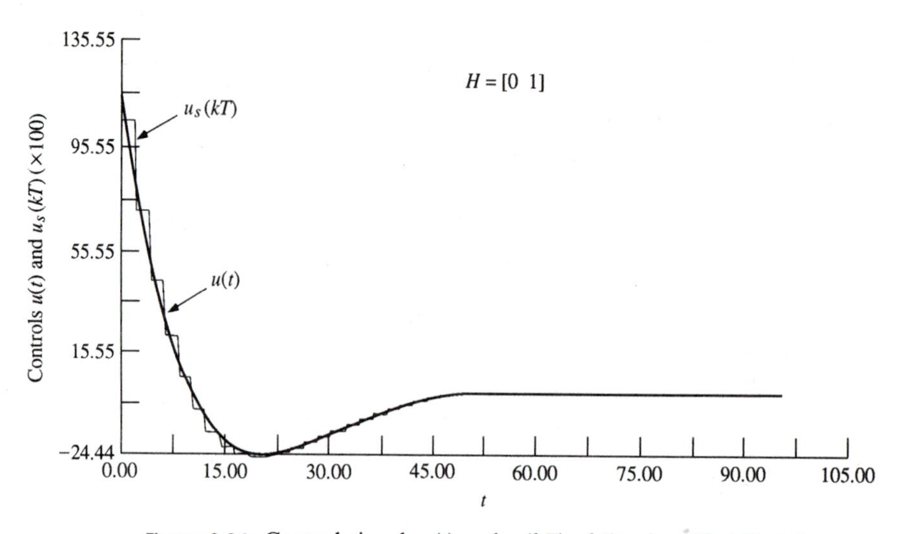

Figure 9-24. Control signals $u(t)$ and $u_s(kT)$ of the simplified Skylab system.

Applying the stability criterion, we have the following three requirements:

$$
\begin{aligned}
F(0) &< 1 \qquad g_{w1} < g_{w2} \\
F(1) &> 0 \qquad g_{w1} > 0 \\
F(-1) &> 0 \qquad g_{w2} < 970741
\end{aligned}
\tag{9-109}
$$

The stable region is bounded by these straight-line boundaries in the g_{w2}-versus-g_{w1} plane, as shown in Fig. 9-25. All the results in Tables 9-3 and 9-4 can be shown to lie inside the stable region.

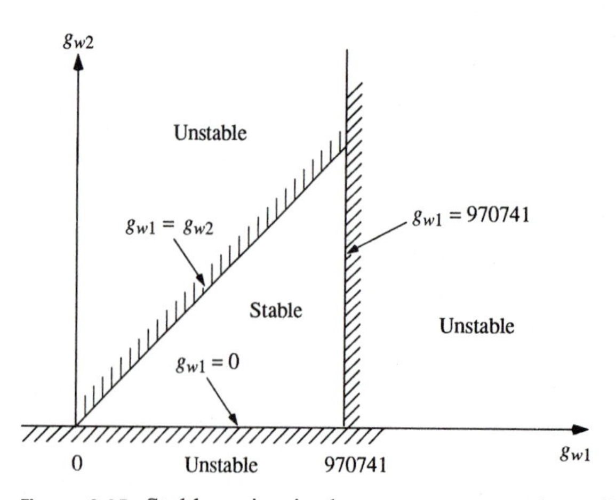

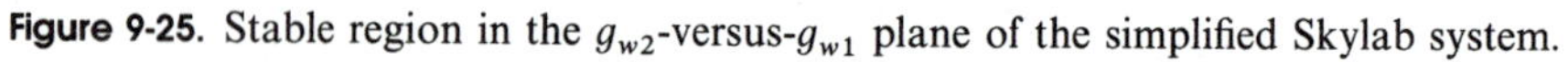
Figure 9-25. Stable region in the g_{w2}-versus-g_{w1} plane of the simplified Skylab system.

In this section we have introduced the digital redesign method using point-by-point comparison. In general, there are other realistic ways of matching the states between a continuous-data system and its equivalent digital model. For instance, the states can be matched at the sampling instants by using a higher order hold with the sampler, instead of the zero-order hold used in this section. Furthermore, it has been shown that states can be matched at multiples of sampling instants or by varying the feedback gain matrix $\mathbf{G}(T)$ and feedforward gain matrix $\mathbf{E}(T)$ between sampling instants, without using the weighting matrix $\mathbf{H}$. These and other methods of digital redesign may be found in the literature listed in the reference section of this chapter.

PROBLEMS

9-1 The block diagram in Fig. P9-1 represents a discrete-data model of a continuous-data control system by inserting a sample-and-hold device in the forward path. Select the sampling frequency ω_s of the discrete-data system so that it is five times the bandwidth of the continuous-data system. Compare the unit-step responses of the discrete-data and the continuous-data systems. Repeat the problem by using a sampling frequency that is 20 times the bandwidth. Do not obtain the unit-step response if the discrete-data system is unstable.

a. $G_p(s) = \dfrac{100}{s(s+0.1)} \qquad H(s) = 0.1s$

b. $G_p(s) = \dfrac{10}{s(s+1)(s+10)} \qquad H(s) = 0$

c. $G_p(s) = \dfrac{10}{s^2} \qquad H(s) = 0$

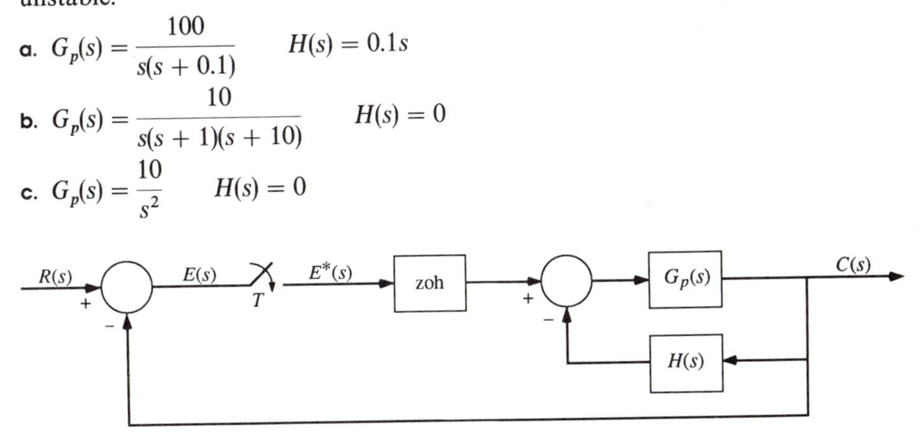

Figure P9-1.

9-2 The block diagram in Fig. P9-2 represents a discrete-data model of a continuous-data system by the insertion of a sample-and-hold device in the forward path.

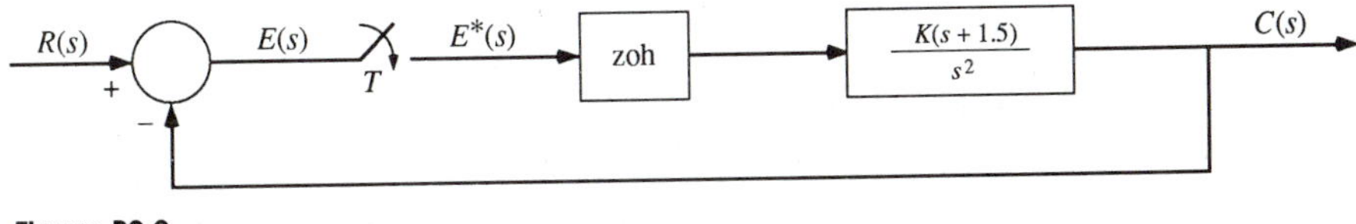

Figure P9-2.

a. Find the region of stability in the K-versus-T plane (K in the vertical axis) for the discrete-data system.

b. Find the region in the K-versus-T plane so that the sampling frequency is greater than 20 times the bandwidth of the continuous-data system.

c. Let $K = 10$. Compare the unit-step responses of the discrete-data system and the continuous-data system when the sampling frequency ω_s is 20 times the bandwidth of the continuous-data system. Repeat when ω_s is five times BW.

d. Let $K = 10$. Replace the zero-order hold by a slewer hold [Eq. (2-108)]. Compare the unit-step responses of the discrete-data system and the continuous-data system when the sampling frequency is 20 times the bandwidth of the continuous-data system.

9-3 Figure P9-3 shows a state diagram of the continuous-data control system described in Problem 9-2.

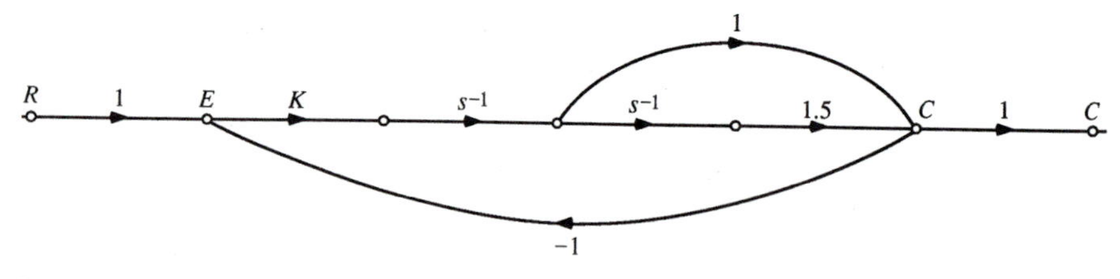

Figure P9-3.

a. Approximate the integrators by backward-rectangular integration of Eq. (9-26). Find the region of stability in the K-versus-T plane. For $K = 10$ and $T = 0.0272$ s, compare the unit-step responses of the discrete-data and the continuous-data systems.

b. Repeat part **a** by approximating the integrators by forward-rectangular integration of Eq. (9-28).

c. Repeat part **a** by approximating the integrators by the trapezoidal integration of Eq. (9-38).

d. Sketch the root loci of the discrete-data systems in parts **a**, **b**, and **c** for $T = 0.0272$ s and $0 \leq K < \infty$.

9-4

a. Construct a Bode diagram for the open-loop transfer function of the continuous-data control system described in Problem 9-2. Find the phase margin, gain margin, M_p, ω_p, and bandwidth of the system.

b. For $K = 10$ and $T = 0.0272$ s, construct the Bode diagrams of the open-loop transfer functions of the discrete-data systems obtained with the three numerical integrations in Problem 9-3. Find the phase margin, gain margin, M_p, ω_p, and bandwidth for each case. Compare the results with those of the continuous-data system and comment on the results.

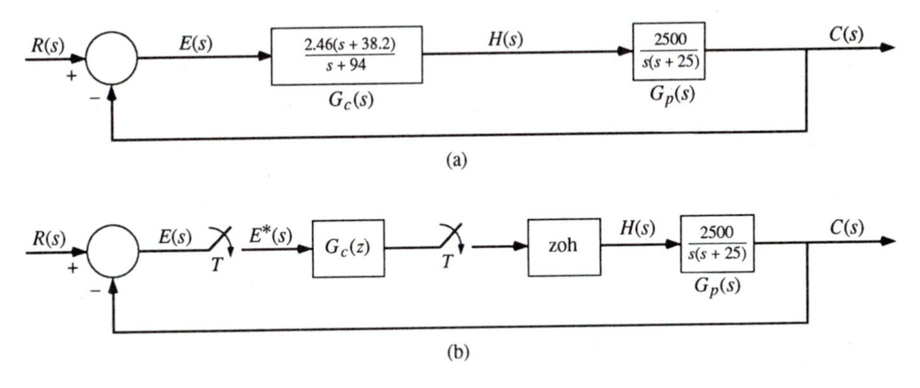

Figure P9-5.

9-5 A continuous-data control system with a phase-lead controller is shown in Fig. P9-5(a).

a. Compute and plot the unit-step response of the system. Determine the maximum overshoot $c_{\max}$. Find the bandwidth of the system.

b. Discretize the controller $G_c(s)$ using the backward-rectangular integration approximation for $1/s$. Find the equivalent transfer function $G_c(z)$.

c. Repeat part **b** with the forward-rectangular integration approximation.

d. Repeat part **b** with the trapezoidal-integration approximation.

e. A discrete-data equivalent of the continuous-data system is shown in Fig. P9-5(b). Set the sampling frequency ω_s to be approximately 20 times the bandwidth of the continuous-data system. Round off ω_s to the form of $X \cdot 10^{-N}$ rad/s, where X and N are positive integers. Compute the unit-step responses of the discrete-data systems obtained in parts **b**, **c**, and **d**, and compare these responses with that of the continuous-data system.

9-6

a. For the continuous-data system in Fig. P9-5(a), plot the Bode diagram of the open-loop transfer function and determine the gain margin, phase margin, M_p, ω_p, and bandwidth.

b. Repeat part **a** with the controller $G_c(s)$ approximated by backward-rectangular integration. Set the sampling frequency ω_s to be approximately 20 times the bandwidth of the continuous-data system. Round off ω_s to the form of $X \cdot 10^{-N}$ rad/s, where X and N are positive integers.

c. Repeat part **b** with $G_c(s)$ approximated by forward-rectangular integration.

d. Repeat part **b** with $G_c(s)$ approximated by the trapezoidal integration.

9-7 For the discrete-data system in Fig. P9-5(b), set the sampling period T to 0.01 s, which corresponds to a sampling frequency approximately six times the bandwidth of the continuous-data system in Fig. P9-5(a). Compute and plot the unit-step responses of the discrete-data systems obtained in parts **b**, **c**, and **d** of Problem 9-5, and compare these responses with that of the continuous-data system.

9-8 For the discrete-data system in Fig. P9-5(b), set the sampling period T to 0.01 s, which corresponds to a sampling frequency approximately six times the bandwidth of the continuous-data system in Fig. P9-5(a).

a. Plot the Bode diagram of the open-loop transfer function of the discrete-data system with the controller $G_c(s)$ approximated by backward-rectangular integration. Find the gain margin, phase margin, M_p, ω_p, and bandwidth. Compare these quantities with those of the continuous-data system in Fig. P9-5(a).

b. Repeat part **a** with $G_c(s)$ approximated by forward-rectangular integration.

c. Repeat part **a** with $G_c(s)$ approximated by trapezoidal integration.

9-9 A continuous-data control system with state feedback is described by the following state equations:

$$\frac{d\mathbf{x}(t)}{dt} = \mathbf{A}\mathbf{x}(t) + \mathbf{B}u(t)$$

where

$$u(t) = E(0)r(t) - \mathbf{G}(0)\mathbf{x}(t)$$

$E(0) = 1$ and $\mathbf{G}(0) = [1 \quad 1]$.

$$\mathbf{A} = \begin{bmatrix} 0 & 0 \\ 0 & -0.6 \end{bmatrix} \qquad \mathbf{B} = \begin{bmatrix} 0 \\ 1 \end{bmatrix}$$

The system is to be approximated by a sampled-data model

$$\frac{d\mathbf{x}_s(t)}{dt} = \mathbf{A}\mathbf{x}_s(t) + \mathbf{B}u_s(kT)$$

where

$$u_s(kT) = E(T)r(kT) + \mathbf{G}(T)\mathbf{x}_s(kT)$$

Determine $E(T)$ and $\mathbf{G}(T)$ using the first two terms of the Taylor series expansion for $T = 0.3$ s and $T = 0.7$ s. Compare $\mathbf{x}_s(kT)$ with $\mathbf{x}(t)$ for $r(t)$, the unit-step function.

9-10 Given a continuous-data system

$$\frac{d\mathbf{x}(t)}{dt} = \mathbf{A}\mathbf{x}(t) + \mathbf{B}u(t)$$

where

$$u(t) = -\mathbf{G}(0)\mathbf{x}(t)$$

$$\mathbf{A} = \begin{bmatrix} 0 & 1 \\ 0 & 0 \end{bmatrix} \qquad \mathbf{B} = \begin{bmatrix} 0 \\ 1 \end{bmatrix} \qquad \mathbf{G}(0) = [2 \quad 3]$$

The following discrete-data system is used to approximate the continuous-data system by matching the response at the sampling instants.

$$\frac{d\mathbf{x}_s(t)}{dt} = \mathbf{A}\mathbf{x}_s(t) + \mathbf{B}u_s(t)$$

$$u_s(kT) = -\mathbf{G}(T)\mathbf{x}_s(kT)$$

$$\mathbf{G}(T) = [g_1(T) \quad g_2(T)]$$

The sampling period is 1 s. Using the weighted feedback gain matrix $\mathbf{G}_w(T)$ for $\mathbf{G}(T)$, find the stable region in the $g_2(T)$-versus-$g_1(T)$ plane. Find the corresponding region in the h_2-versus-h_1 plane, where $\mathbf{H} = [h_1 \quad h_2]$ is the weighting matrix for $\mathbf{G}_w(T)$.

9-11 The dynamics of a closed-loop continuous-data system are described by

$$\frac{d\mathbf{x}(t)}{dt} = \mathbf{A}\mathbf{x}(t) + \mathbf{B}u(t)$$

where

$$\mathbf{A} = \begin{bmatrix} 0 & 1 \\ 0 & 0 \end{bmatrix} \qquad \mathbf{B} = \begin{bmatrix} 0 \\ 1 \end{bmatrix}$$

$$u(t) = -\mathbf{G}_c\mathbf{x}(t) + E_c r(t)$$

$$\mathbf{G}_c = [100 \quad 20] \qquad E_c = -100$$

The system is to be approximated by a discrete-data model

$$\frac{d\mathbf{x}_s(t)}{dt} = \mathbf{A}\mathbf{x}_s(t) + \mathbf{B}u_s(kT) \qquad T = 0.01 \text{ s}$$

where $u_s(kT)$ is the output of a first-order hold; i.e.,

$$u_s(kT) = [E_0 + (t - kT)E_1]r(kT) - [\mathbf{G}_0 + (t - kT)\mathbf{G}_1]\mathbf{x}_s(kT)$$

E_0 and E_1 are scalars; $\mathbf{G}_0$ and $\mathbf{G}_1$ are 1×2 row matrices. For $r(t) \cong r(kT)$, $kT \le t < (k+1)T$, find $E_0, E_1, \mathbf{G}_0$ and $\mathbf{G}_1$ so that the responses $\mathbf{x}_s(kT)$ and $\mathbf{x}(t)$ are matched at the sampling instants. Compute and plot the responses of $\mathbf{x}(t)$ and $\mathbf{x}_s(t)$ for a unit-step input with zero initial states.

9-12 Given the continuous-data control system

$$\frac{d\mathbf{x}_c(t)}{dt} = \mathbf{A}\mathbf{x}_c(t) + \mathbf{B}u_c(t)$$

where $\mathbf{x}_c(t)$ is an n-vector, $u_c(t)$ is a scalar input, and

$$u_c(t) = E_c r(t) - \mathbf{G}_c\mathbf{x}_c(t)$$

a. Show that the response of the continuous-data system can be uniquely matched at every $n = N$ sampling instant by that of the discrete-data system

$$\frac{d\mathbf{x}_s(t)}{dt} = \mathbf{A}\mathbf{x}_s(t) + \mathbf{B}u_s(kT)$$

where

$$u_s(kT) = E(T)r(kT) - \mathbf{G}(T)\mathbf{x}_s(kT) \qquad kT \le t < (k+1)T$$

$\mathbf{x}_s(kT)$ is an n-vector, if the discretized system

$$\mathbf{x}_s[(k+1)T] = \boldsymbol{\phi}(T)\mathbf{x}_s(kT) + \boldsymbol{\theta}(T)u_s(kT)$$

is completely state controllable. Assume that $r(t) \cong r(kT)$ for $kT \le t \le (k+1)T$.

b. For

$$\mathbf{A} = \begin{bmatrix} 0 & 1 \\ 0 & 0 \end{bmatrix} \qquad \mathbf{B} = \begin{bmatrix} 0 \\ 1 \end{bmatrix}$$

$$\mathbf{G}_c = [100 \quad 20] \qquad E_c = -100$$

find $\mathbf{G}(T)$ and $E(T)$ such that the state responses of the continuous-data system are matched by the discrete-data system every two sampling instants; $T = 0.01$ s.

References

1. Astrom, K. J., and B. Wittenmark, *Computer Controlled Systems, Theory and Design*, Prentice-Hall, Englewood Cliffs, N.J., 1984.
2. Ogata, K., *Discrete-Time Control Systems*, Prentice-Hall, Englewood Cliffs, N.J., 1987.
3. Franklin, G. F., J. D. Powell, and M. L. Workman, *Digital Control of Dynamic Systems*, 2nd ed., Addison-Wesley, Reading, Mass., 1990.
4. Tustin, A., "A Method of Analyzing the Behavior of Linear Systems in Terms of Time Series," *Journal IEE, London*, vol. 94, II-A, pp. 130–142, May 1947.
5. Singh, G., B. C. Kuo, and R. A. Yackel, "Digital Approximation by Point-to-Point State Matching with Higher-Order Holds," *International J. Control*, vol. 20, no. 1, pp. 81–90, 1974.
6. Kuo, B. C., G. Singh, and R. A. Yackel, "Digital Approximation of Continuous-Data Control Systems by Point-to-Point State Comparison," *Computers & Elec. Eng.*, vol. 1, Pergamon Press, pp. 155–170, 1973.
7. Kuo, B. C., and D. W. Peterson, "Optimal Discretization of Continuous-Data Control Systems," *IFAC Automatica*, vol. 9, no. 1, pp. 125–129, 1973.

8. Yackel, R. A., B. C. Kuo, and G. Singh, "Digital Redesign of Continuous Systems by Matching of States at Multiple Sampling Periods," *IFAC Automatica*, vol. 10, Pergamon Press, pp. 105–111, 1974.
9. Moore, B. C., and L. M. Silverman, "Model Matching by State Feedback and Dynamic Compensation," *IEEE Trans. Automatic Control*, vol. AC-17, pp. 491–497, 1972.

Design of Discrete-Data Control Systems

KEYWORDS AND TOPICS

continuous-data controllers, • time-delay approximation • design by bilinear transformation • digital controllers • digital programming • digital PID controller • phase-lead • phase-lag controllers • root-locus design • forward-controller design • noise and disturbance rejection • robust control • deadbeat-response design • state-feedback design • pole-placement design • incomplete state feedback • dynamic output feedback • dynamic controllers

10-1 INTRODUCTION

The design problems encountered in discrete-data control systems are essentially similar to those found in the design of continuous-data control systems. Basically, a process needs to be controlled so that its output will behave according to some prescribed performance specifications. In the conventional design approach we decide at the outset that there should be feedback from the outputs to the reference inputs, so that errors can be formed between these signals for control efforts. Then, in general, we find that a controller is needed to operate on the error signals in such a way that the design specifications are satisfied by the outputs. In discrete-data control systems, the design problems have more variations and flexibilities. For instance, the discrete data may be due to the use of certain digital or incremental transducers that naturally put out digital or discrete data, or the designer may simply decide to use a digital controller. On the other hand, it is also feasible to use a continuous-data controller on a digital signal after the signal has been smoothed out by a data hold. Thus, we see that there are a great variety of possible configurations and schemes for the design of a discrete-data control system.

In this chapter we shall introduce several methods of designing discrete-data control systems. Design methods in the s-domain, the frequency domain, as well as with state variables will be discussed.

The conventional design of control systems is characterized by the *fixed-configuration* concept, in that the designer first fixes the composition of the overall system, including the controlled process and the controller. Figures 10-1 through 10-4 illustrate some of the configurations most frequently encountered in practice. Figure 10-1 is the block diagram of a sampled-data system in which the controller is analog. The sampler may represent the fact that digital or sampled data exist at the input and the feedback channels, due to the use of

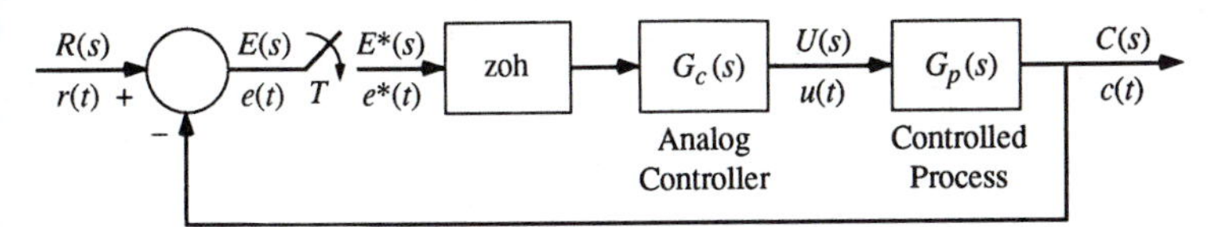

Figure 10-1. Digital control system with cascade analog controller.

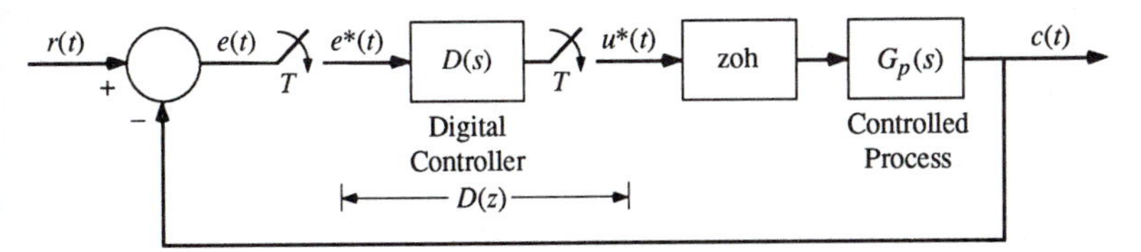

Figure 10-2. Digital control system with cascade digital controller.

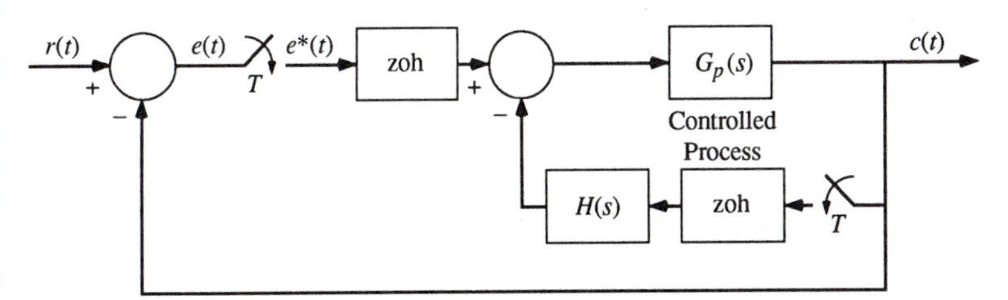

Figure 10-3. Digital control system with analog controller in minor feedback path.

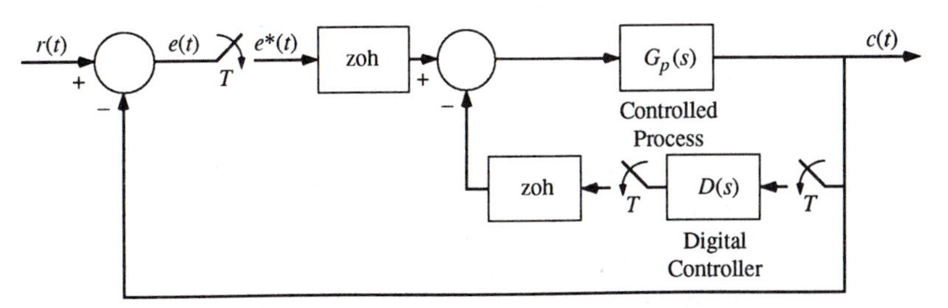

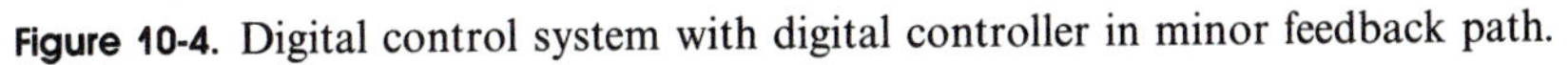
Figure 10-4. Digital control system with digital controller in minor feedback path.

digital transducers. In this case, a continuous-data controller is selected to operate on the sampled signal after it is decoded and smoothed out by the data hold.

Figure 10-2 shows the classical case of a digital control system in which a digital controller is located in the forward path. The digital controller operates on the digital signal $e^*(t)$, which is represented as the output of a sampler, and outputs the digital signal $u^*(t)$, which in turn is filtered by the usual data hold before being applied to the controlled process. Figure 10-3 shows the case with the continuous-data controller located in the minor feedback loop. Figure 10-4 gives the counterpart of the system in Fig. 10-3, with the digital controller replacing the analog controller.

A powerful design method in the state-variable domain is the state-feedback or output-feedback design. Figure 10-5 is the block diagram of a multivariable digital control system with state-variable feedback, assuming that all the state variables are accessible. In practice, not all the state variables are accessible, so that we either have to use an *observer* to estimate some or all of the state variables, or simply use output feedback. Figure 10-6 shows the block diagram of a multivariable digital control system with state feedback and an observer. In this case the observer estimates the state variables from information it receives from the output vector $\mathbf{c}(k)$. The output of the observer is the estimated state vector $\mathbf{x}(k)$. Figure 10-7 shows a system with output feedback in the multivariable configuration. In this case, the output $\mathbf{c}(k)$ is fed directly to the

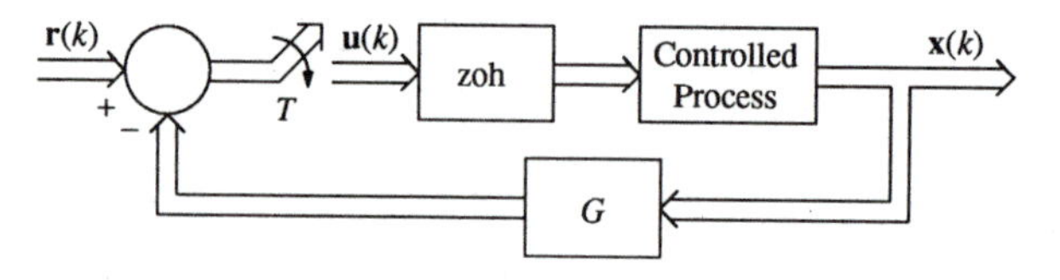

Figure 10-5. Digital control system with state feedback.

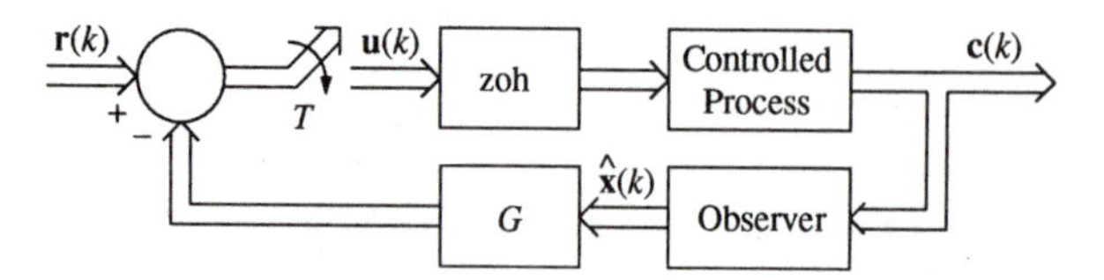

Figure 10-6. Digital control system with state feedback and observer.

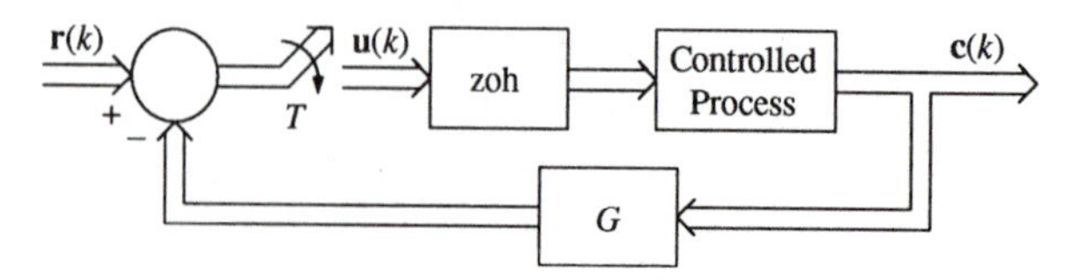

Figure 10-7. Digital control system with output feedback.

feedback gain matrix **G**. Since, in general, there are fewer output variables than state variables, the output-feedback scheme would have more constraints than the state-feedback design.

10-2 CASCADE COMPENSATION BY CONTINUOUS-DATA CONTROLLERS

In this section we shall consider the design of a discrete-data control system with a cascade continuous-data controller, as shown in Fig. 10-1. Since the controller is located between the zero-order hold and the controlled process, the z-transform of the open-loop transfer function is written

$$G(z) = \frac{C(z)}{E(z)} = \mathscr{Z}[G_{h0}(s)G_c(s)G_p(s)] = (1 - z^{-1})\mathscr{Z}\left(\frac{G_c(s)G_p(s)}{s}\right) \tag{10-1}$$

The design objective is to find a physically realizable transfer function $G_c(s)$ of the continuous-data controller which will cause the discrete-data control system to perform according to specifications. Unfortunately, as shown by Eq. (10-1), the transfer function $G_c(s)$ is imbedded in the transfer function of the process, $G_p(s)$, so that the effects of the controller cannot be investigated independently. In the following section we shall introduce a simple approximation method that has been used often in industry, and show why it must be applied with great care. A design method using bilinear transformation is also introduced.

10-2-1 Time-Delay Approximation of Sample-and-Hold Process

A method that has been used frequently by practicing engineers to approximate a sampled-data system by a continuous-data system relies on the approximation of the sample-and-hold operation by means of a pure time delay. Although we are reviewing this method here, the approximation is not generally regarded as accurate, and great care should be exercised when applying the method to design problems.

Let us refer to the sinusoidal steady-state transfer function of the zero-order hold given in Eq. (2-89),

$$G_{h0}(j\omega) = T\,\frac{\sin(\omega T/2)}{\omega T/2}\,e^{-j\omega T/2} \tag{10-2}$$

The sinusoidal steady-state pulsed transfer function of the open-loop system is written

$$G_{h0}G_cG_p^*(j\omega) = \frac{1}{T}\sum_{n=-\infty}^{\infty} G_{h0}(j\omega + jn\omega_s)G_c(j\omega + jn\omega_s)G_p(j\omega + jn\omega_s) \tag{10-3}$$

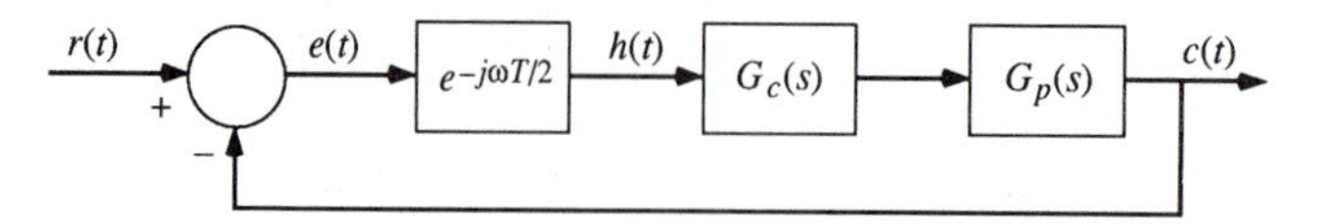

Figure 10-8. Continuous-data system approximation of the sampled-data system in Fig. 10-1.

Since in most control systems $G_p(j\omega)$ has low-pass filter characteristics, and at low frequencies the magnitude of $\sin(\omega T/2)/(\omega T/2)$ is approximately unity, we can approximate the right-hand side of Eq. (10-3) by just the $n = 0$ term; that is,

$$G_{h0} G_c G_p^*(j\omega) \cong G_c(j\omega) G_p(j\omega) e^{-j\omega T/2} \tag{10-4}$$

which means that the sample-and-hold operation can be approximated by a pure time delay of one-half the sampling period T. Figure 10-8 shows the continuous-data system with the pure time delay approximating the sample-and-hold device.

Another perspective of the pure time-delay approximation of the sample-and-hold operation is illustrated by the signal waveform shown in Fig. 10-9. It appears that if we approximate the output of the zero-order hold by a smooth curve passing through the center of the flat-top signal during each sampling period, we have a continuous-data signal which is approximately the same as the input signal of the sampler, except that it is delayed by one-half of the sampling period. However, this type of approximation is not applicable to a broad class of sampled-data control systems. In general, the one-term approximation of the infinite series in Eq. (10-3) may not be adequate, and the ramifications of the approximation as illustrated by Fig. 10-9 are not at all clear. Therefore, the best we can say is that the continuous-data system with pure time delay shown in Fig. 10-8 can be used to approximate the sampled-data system in Fig. 10-1 only if the sampling period is relatively *very small*. Otherwise, if all sampled-data systems could be approximated by continuous-data systems with pure time delays, there would be no need to devise and study the analysis and design of discrete-data systems.

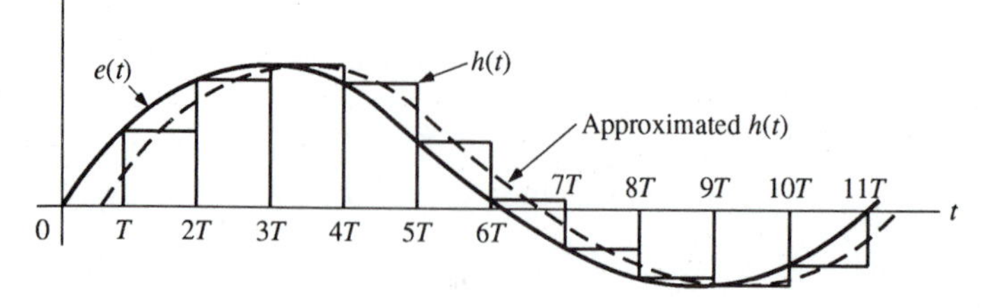

Figure 10-9. Typical waveform showing the approximation of the sample-and-hold operation by a pure time delay of $T/2$.

We shall use the following example to illustrate the time-delay approximation method.

Example 10-1

Consider that the transfer function of the controlled process of the system shown in Fig. 10-1 is that of the space vehicle control system given in Eq. (7-12) and is repeated as follows.

$$G_p(s) = \frac{39.453K}{s(s + 8.871)} \tag{10-5}$$

The block diagram of the discrete-data space vehicle control system is shown in Fig. 7-6. The open-loop pulse transfer function of the approximating continuous-data system with time delay shown in Fig. 10-8 is

$$G^*(s) = G_{h0}G_cG_p^*(s) \cong G_c(s)G_p(s)e^{-Ts/2} \tag{10-6}$$

Substituting Eq. (10-5) into the last equation, and with $G_c(s) = 1$, we have

$$G_{h0}G_cG_p^*(s) \cong \frac{39.453K}{s(s + 8.871)}\, e^{-Ts/2} \tag{10-7}$$

Let the sampling period T be 0.01 s. The open-loop transfer function of the discrete-data system is

$$G(z) = (1 - z^{-1})\mathscr{Z}\left(\frac{39.453K}{s^2(s + 8.871)}\right) \tag{10-8}$$

or

$$G(z) = \frac{0.001916K(z + 0.9708)}{(z - 1)(z - 0.915)} \tag{10-9}$$

Let us set K at 22.5 which corresponds to a ramp-error constant K_v of 100, and the sampling period T is 0.01 s. Figure 10-10 (a) shows the Bode diagram of Eq. (10-7) with $K = 22.5$ and $T = 0.01$ s. The phase margin of the approximating continuous data with time delay is 8.60 degrees, and the gain margin is 6.08 dB. The Bode diagram of the $G(z)$ in Eq. (10-9) which is the open-loop transfer function of the discrete-data system is drawn in Fig. 10-10 (b). We see that except for the warping near the frequency of $\omega_s/2$, the time-delay approximation is quite good. In fact, the phase margin of the closed-loop discrete-data system is 8.63 degrees, and the gain margin is 6.15 dB.

Based on the continuous-data system with time-delay model of Eq. (10-7), a phase-lead controller is designed using the method described in [1] to yield a phase margin of 45 degrees. The transfer function of the cascade controller is

$$G_c(s) = \frac{1 + 0.05473s}{1 + 0.008s} \tag{10-10}$$

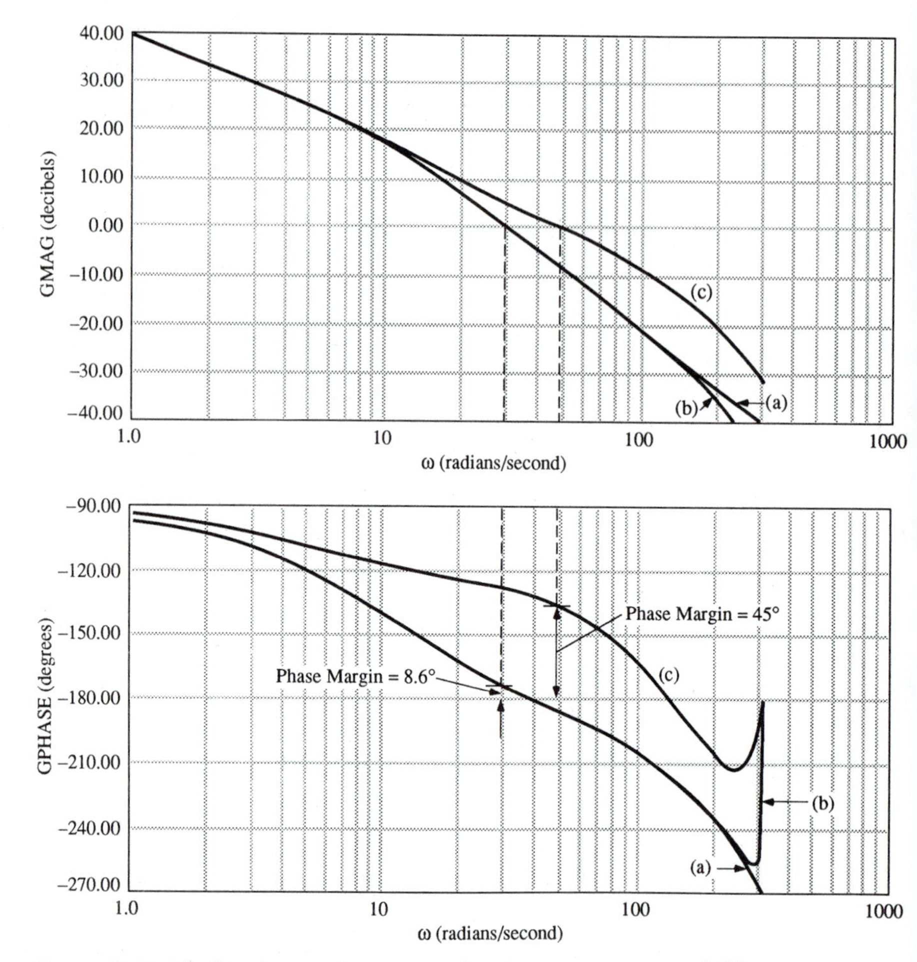

Figure 10-10. (a) Continuous-data approximation. $G_p(s) = 100e^{-0.005s}/s(1 + 0.1127s)$. (b) Discrete-data system. $G_p(s) = 100/s(1 + 0.1127s)$, $T = 0.01$ s. (c) Discrete-data system with phase-lead controller. $G_p(s) = 100(1 + 0.05473s)/s(1 + 0.1127s)(1 + 0.008s)$, $T = 0.01$ s.

The Bode diagram of the discrete-data system with the open-loop transfer function

$$G_c(s)G_p(s) = \frac{100(1 + 0.05473s)}{s(1 + 0.1127s)(1 + 0.008s)} \tag{10-11}$$

is plotted as shown in Fig. 10-10 (c). As shown in Fig. 10-10 (c), the phase margin of the discrete-data system with the phase-lead controller is 45 degrees, which is the desired value. Thus we see that, when the time-delay approximation gives an adequate approximation of the discrete-data system, the approximating continuous-data system can be used for the design of the discrete-data system.

10-2-2 Design by Bilinear Transformation

In the preceding chapters we introduced two forms of bilinear transformation: the r-transformation [Eq. (6-125) or (6-126)] and the w-transformation [Eq. (8-43)]. These bilinear transforms may be used to transform a z-transfer function $G(z)$ of a discrete-data system into a new function $G(r)$ or $G(w)$ which is a rational function in the complex variable r or w, respectively. As shown in Chapter 8, the Bode diagram of $G(w)$ can be drawn in decibels of magnitude and phase in degrees versus ω_w which is the imaginary part of w. Then, information on stability and relative stability can be expressed in terms of gain and phase margins, and all the design tools of the continuous-data systems in the frequency domain can be applied.

Since bilinear transformation is simply a convenient analytical tool, we shall select the r-transformation of Eq. (6-126) for design purposes here. By using the r-transformation, the transfer function $G(z)$ is transformed into $G(r)$, and all the design techniques involving the Bode diagram and the Nichols chart [1] can be utilized.

With reference to the block diagram in Fig. 10-1, the design steps are outlined as follows:

1. Obtain the r-transfer function $G_{h0}G_cG_p(r)$ of the system with $G_c(s) = 1$, i.e., without compensation, where

$$G_{h0}G_cG_p(r) = G_{h0}G_p(r) = G_{h0}G_p(z)|_{z=(1+r)/(1-r)} \tag{10-12}$$

2. Construct the Bode diagram of $G_{h0}G_p(r)$ by setting $r = j\omega_r$, and if necessary, transfer the data of the Bode diagram onto the Nichols chart. Predict the dynamic behavior of the uncompensated system by evaluating the phase margin, gain margin, peak resonance M_p, and bandwidth from the Bode diagram and the Nichols chart. The real frequency ω is related to ω_r through

$$\omega_r = \tan\left(\frac{\omega T}{2}\right) \tag{10-13}$$

3. Should the system need improvement, a cascade controller with the transfer function $G_c'(r)$ is multiplied to $G_{h0}G_p(r)$. The objective of the controller $G_c'(r)$ is to reshape the transfer function loci of $G_{h0}G_p(r)$ in the Bode plot or the Nichols chart. The transfer function of the controller in the r-domain can be classified as phase-lead, phase-lag, lead-lag, or lag-lead, according to the filtering characteristics in the r-domain. It is important to point out that the filtering characteristics in the r-domain do not have the same correlations in the s-domain. The purpose of identifying $G_c'(r)$ in the r-domain according to filtering properties is to use the established knowledge of the design of continuous-data systems in the frequency domain. Once the transfer function $G_r'(r)G_{h0}G_p(r)$ giving the desired system performance is determined, the transfer function of the actual controller in the s-domain, $G_c(s)$, is found by first transforming $G_r'(r)G_{h0}G_p(r)$ to

$G_{h0}(s)G_c(s)G_p(s)$, and then solving $G_c(s)$ from the latter expression, since $G_{h0}(s)G_p(s)$ is already known. Since $G_c(s)$ must be physically realizable, and furthermore, the transfer function is preferably realizable by a network with only resistors and capacitors, it must satisfy the following requirements.

a. The poles of $G_c(s)$ must lie in the left half of the s-plane and must be simple and real.

b. The number of zeros of $G_c(s)$ must not exceed the number of poles of $G_c(s)$.

In general, the zeros of $G_c(s)$ may lie anywhere in the s-plane. From **a**, the poles of $G_c(s)$ are restricted to the negative real axis of the s-plane. Since the poles of $G_c(s)$ are generated by the poles of $G_c'(r)$ and the negative real axis of the s-plane corresponds to the portion $-1 \leq r \leq 0$ in the r-plane, as shown in Fig. 10-11, it follows that $G_c'(r)$ can have only simple poles that lie in the range of -1 and 0 on the real axis of the r-plane.

4. Once $G_c'(r)G_{h0}G_p(r)$ is determined, take the inverse transform from r to s to obtain $G_{h0}(s)G_c(s)G_p(s)$. Note that since

$$G_c'(r)G_{h0}G_p(r) = G_{h0}G_cG_p(z)|_{z=(1+r)/(1-r)} \tag{10-14}$$

$$G_c'(r) \neq \mathscr{Z}[G_c(z)]|_{z=(1+r)/(1-r)} \tag{10-15}$$

The z-transform of the uncompensated open-loop transfer function is written as

$$G_{h0}G_p(z) = (1 - z^{-1})\mathscr{Z}\left(\frac{G_p(s)}{s}\right) \tag{10-16}$$

In terms of the r-transform, we have

$$\begin{aligned} G_{h0}G_p(r) &= (1 - z^{-1})\mathscr{Z}\left(\frac{G_p(s)}{s}\right)\bigg|_{z=(1+r)/(1-r)} \\ &= \frac{2r}{1+r}\mathscr{Z}\left(\frac{G_p(s)}{s}\right)\bigg|_{z=(1+r)/(1-r)} \end{aligned} \tag{10-17}$$

With the cascade controller $G_c'(r)$, the r-transform open-loop transfer function is written

$$\begin{aligned} G_c'(r)G_{h0}G_p(r) &= G_{h0}G_cG_p(r) \\ &= \frac{2r}{r+1}\mathscr{Z}\left(\frac{G_c(s)G_p(s)}{s}\right)\bigg|_{z=(1+r)/(1-r)} \end{aligned} \tag{10-18}$$

or

$$\mathscr{Z}\left(\frac{G_c(s)G_p(s)}{s}\right)\bigg|_{z=(1+r)/(1-r)} = \frac{r+1}{2r}G_c'(r)G_{h0}G_p(r) \tag{10-19}$$

The transfer function $G_c(s)G_p(s)$ is found from Eq. (10-19) by taking the partial-fraction expansion of $G_c'(r)G_{h0}G_p(r)$ and finding the corresponding pairs of transfer functions in s. A table of the Laplace z-r transfer functions is given in Table 10-1 for this purpose. Since the function $G_p(s)/s$ must

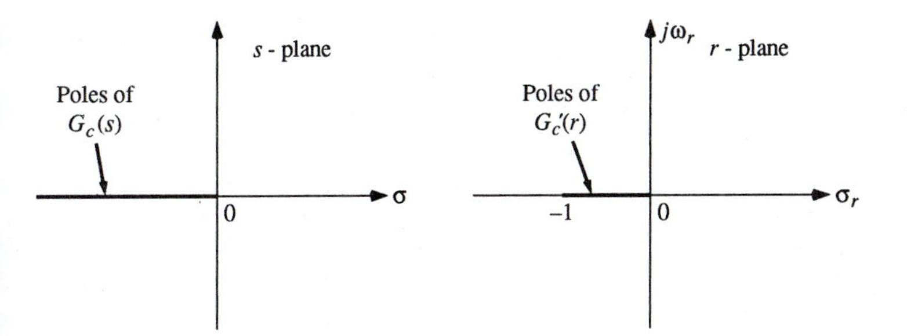

Figure 10-11. Locations of the poles of $G_c(s)$ in the s-plane and $G_c'(r)$ in the r-plane for RC realizability.

Table 10-1 s-z-r Transformations

Laplace Transform $G(s)$	z-Transform $G(z)$	r-Transform $G(r)$
$\dfrac{1}{s}$	$\dfrac{z}{z-1}$	$\dfrac{r+1}{2}$
$\dfrac{1}{s^2}$	$\dfrac{Tz}{(z-1)^2}$	$\dfrac{T(1+r)(1-r)}{4r^2}$
$\dfrac{1}{s^3}$	$\dfrac{T^2z(z+1)}{2(z-1)^3}$	$\dfrac{T^2(1+r)(1-r)}{8r^3}$
$\dfrac{1}{s+a}$	$\dfrac{z}{z-e^{-aT}}$	$\dfrac{1+r}{(1-e^{-aT})\left[1+\dfrac{1+e^{-aT}}{1-e^{-aT}}r\right]}$
$\dfrac{1}{(s+a)^2}$	$\dfrac{Tze^{-aT}}{(z-e^{-aT})^2}$	$\dfrac{(1+r)(1-r)Te^{-aT}}{(1-e^{-aT})^2\left[1+\dfrac{1+e^{-aT}}{1-e^{-aT}}r\right]^2}$
$\dfrac{a}{s(s+a)}$	$\dfrac{(1-e^{-aT})z}{(z-1)(z-e^{-aT})}$	$\dfrac{(1+r)(1-r)}{2r\left[1+\dfrac{1+e^{-aT}}{1-e^{-aT}}r\right]}$
$\dfrac{a}{s^2(s+a)}$	$\dfrac{Tz}{(z-1)^2}-\dfrac{1-e^{-aT}}{a(z-1)(z-e^{-aT})}$	$\dfrac{T(1-r)(1+r)}{4r^2}-\dfrac{(1-r)(1+r)}{2ar\left[1+\dfrac{1+e^{-aT}}{1-e^{-aT}}r\right]}$
$\dfrac{\omega}{s^2+\omega^2}$	$\dfrac{z\sin\omega T}{z^2-2z\cos\omega T+1}$	$\dfrac{(1-r)(1+r)\sin\omega T}{2[(1+r^2)-(1-r^2)\cos\omega T]}$
$\dfrac{\omega}{(s+a)^2+\omega^2}$	$\dfrac{ze^{-aT}\sin\omega T}{z^2-2ze^{-aT}\cos\omega T+e^{-2aT}}$	$\dfrac{r\dfrac{(1+r)(1-r)e^{-aT}\sin\omega T}{(1+r)^2-2(1+r)(1-r)}}{e^{-aT}\cos\omega T+(1-r)e^{-2aT}}$

eventually be factored out of $G_c(s)G_p(s)/s$ to find $G_c(s)$, care must be taken in performing the partial-fraction expansion of $G_c'(r)G_{h0}G_p(r)$.

5. Once $G_c(s)G_p(s)/s$ is determined, $G_c(s)$ is readily obtained. However, because of the lack of correspondence between the number of s-plane zeros and r-plane zeros, the function $G_c(s)$ obtained from the previous steps will in many cases have more zeros than poles. This can be circumvented by adding one or more remote poles to $G_c(s)$ on the negative real axis in the s-plane to make $G_c(s)$ physically realizable. These additional poles should have negligible effect on the overall transient and steady-state performance of the system.

Example 10-2

In this example we repeat the design problem in Example 10-1, using the r-transform method. The transfer function of the controlled process is given in Eq. (10-5), with $K = 22.5$ and $T = 0.01$ s. The design objective is to attain a phase margin of 45 degrees while maintaining the ramp-error constant K_v at 100.

The z-transform of the open-loop transfer function without compensation is

$$\mathscr{Z}[G_{h0}(s)G_p(s)] = G_{h0}G_p(z) = (1 - z^{-1})\mathscr{Z}\left(\frac{887.3}{s^2(s + 8.871)}\right)$$

$$= \frac{0.04308(z + 0.9708)}{(z - 1)(z - 0.915)} \tag{10-20}$$

Applying the r-transformation of $z = (1 + r)/(1 - r)$ to the last equation, we have

$$G_{h0}G_p(r) = G_{h0}G_p(z)|_{z=(1+r)/(1-r)} = \frac{0.4995(1 + 0.0148r)(1 - r)}{r(1 + 22.53r)} \tag{10-21}$$

The Bode diagrams of $G_{h0}G_p(r)$ are constructed in Fig. 10-12. The phase margin of the uncompensated system is shown to be 8.66 degrees, which compares closely with the result obtained in Example 10-1. Since $G_{h0}G_p(r)$ in Eq. (10-21) is of the form of a continuous-data transfer function, the design follows the frequency-domain design established for continuous-data control systems [1]. A phase-lead controller for the system to attain a phase margin of 45 degrees is

$$G_c'(r) = \frac{1 + 10.767r}{1 + 1.5887r} \tag{10-22}$$

The transfer function of the compensated system in the r-domain is

$$G_c'(r)G_{h0}G_p(r) = \frac{0.4995(1 + 0.0148r)(1 - r)(1 + 10.767r)}{r(1 + 22.53r)(1 + 1.5887r)} \tag{10-23}$$

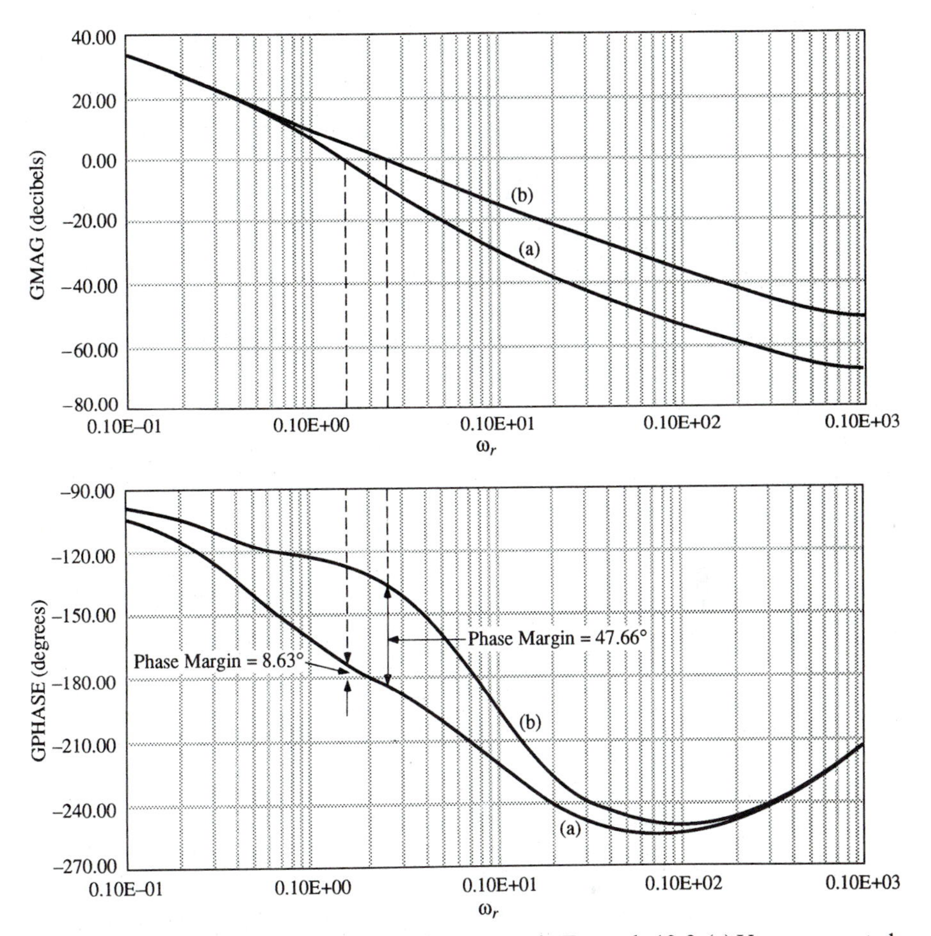

Figure 10-12. Bode diagrams of discrete-data system in Example 10-2. (a) Uncompensated system.

$$G_{h0}G_p(r) = \frac{0.4995(1 + 0.0148r)(1 - r)}{r(1 + 22.53r)}.$$

(b) With phase-lead controller.

$$G'_c(r)G_{h0}(r)G_p(r) = \frac{0.4995(1 + 0.0148r)(1 - r)(1 + 10.767r)}{r(1 + 22.53r)(1 + 1.588r)}.$$

The remaining task in this design problem involves the determination of the transfer function $G_c(s)$ of the continuous-data controller. Since most of the r-transform expressions in Table 10-1 contain the term $1 - r$ in the numerator, we should perform the partial-fraction expansion of $G'_c(r)G_{h0}G_p(r)/(1 - r)$. Thus,

$$\frac{G'_c(r)G_{h0}G_p(r)}{1 - r} = \frac{0.4995}{r} - \frac{0.2804}{r + 0.04439} - \frac{0.2169}{r + 0.6295} \tag{10-24}$$

and

$$G_c'(r)G_{h0}G_p(r) = \frac{0.4995(1-r)}{r} - \frac{0.2804(1-r)}{r+0.04439} - \frac{0.2169(1-r)}{r+0.6295} \tag{10-25}$$

Substituting the last equation into Eq. (10-19), we get

$$\mathscr{Z}\left(\frac{G_c(s)G_p(s)}{s}\right)\Bigg|_{z=(1+r)/(1-r)} = \frac{0.4995(1+r)(1-r)}{2r^2} - \frac{0.2804(1+r)(1-r)}{2r(r+0.04439)} - \frac{0.2169(1+r)(1-r)}{2r(r+0.6295)} \tag{10-26}$$

The transform pairs for the terms in the last equation are identified in Table 10-1, and we have

$$\begin{aligned}\frac{G_c(s)G_p(s)}{s} &= \frac{100}{s^2} - \frac{56.04}{s(s+8.871)} - \frac{51}{s(s+148)} \\ &= \frac{7.04(s+18.566)(-s+1004.5)}{s^2(s+8.871)(s+148)}\end{aligned} \tag{10-27}$$

Now cancelling out the transfer function $G_p(s)$, the transfer function of the phase-lead controller is written

$$G_c(s) = \frac{0.007934(s+18.566)(-s+1004.5)}{s+148} \tag{10-28}$$

For a physically realizable controller, $G_c(s)$ must not have more zeros than poles. Ordinarily, we can add one more pole to $G_c(s)$ in Eq. (10-28) that is far to the left on the negative real axis. In the present case, since the zero at $s = 1004.5$ is quite large, it can be neglected. Thus, $G_c(s)$ can be approximated by

$$G_c(s) = \frac{7.97(s+18.566)}{s+148} \tag{10-29}$$

Note that in arriving at the approximation $G_c(s)$ by neglecting the zero at $s = 1004.5$, the zero-frequency gain is maintained so that the ramp-error constant is not altered. The transfer function $G_c(s)G_p(s)$ of the discrete-data system with phase-lead compensation is

$$G_c(s)G_p(s) = \frac{7071.68(s+18.566)}{s(s+8.871)(s+148)} \tag{10-30}$$

Figure 10-13 shows the Bode diagram of $(1-z^{-1})\mathscr{Z}[G_c(s)G_p(s)/s]$. It is shown that the actual phase margin of the designed system is 47.66 degrees.

The unit-step responses of the compensated and the uncompensated systems are shown in Fig. 10-14. As shown, the unit-step response of the uncompensated system is quite oscillatory, and the maximum overshoot is 78.7 percent. The maximum overshoot of the compensated system is 24 percent.

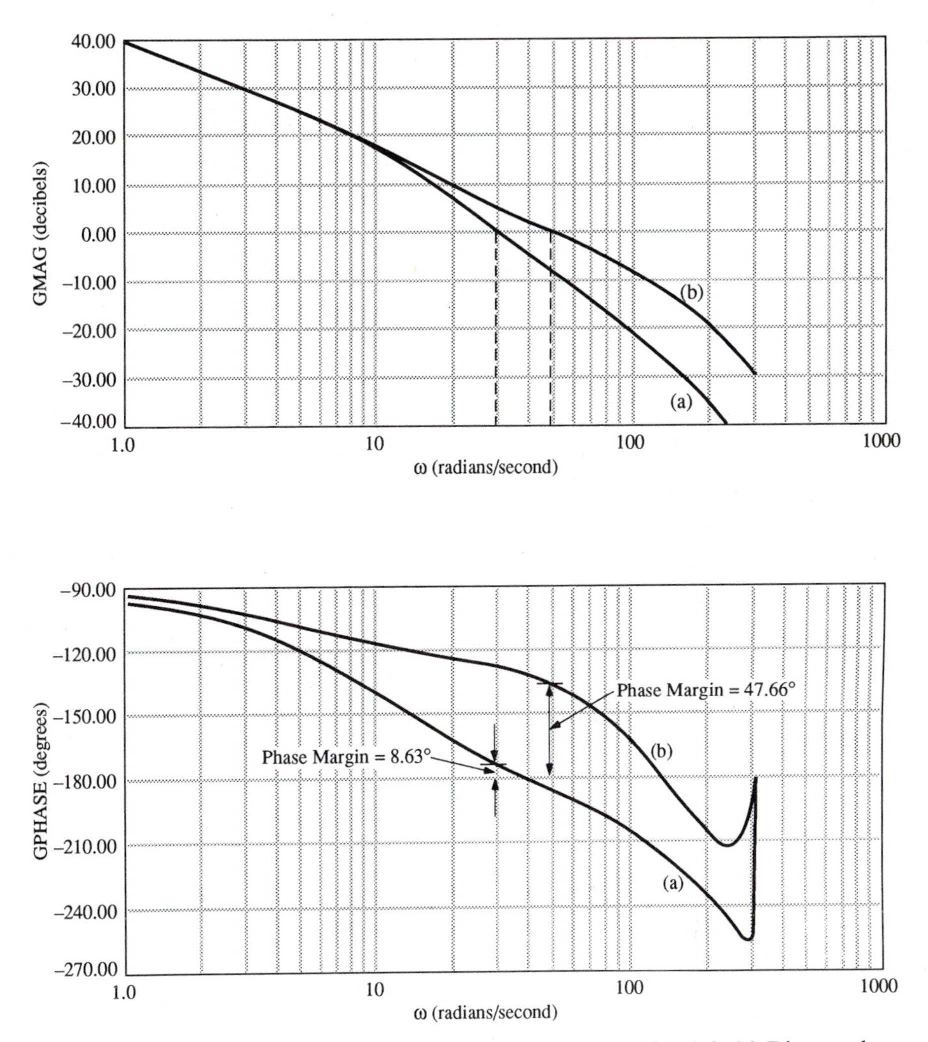

Figure 10-13. Bode diagram of discrete-data system in Example 10-2. (a) Discrete-data system (uncompensated). $G_p(s) = 100/s(1 + 0.1127s)$, $T = 0.01$ s. (b) Discrete-data system with phase-lead controller. $G_c(s)G_p(s) = 7071.68(s + 18.566)/s(s + 8.871)(s + 148)$, $T = 0.01$ s.

Further improvement of the system can be made by selecting a higher phase margin. However, it should be cautioned that, given the controlled process with the designated sampling period, there is a limit to how much the system stability can be improved by using a phase-lead controller. For the present case, the maximum phase margin obtainable with a phase-lead controller of the form $G_c(s) = (1 + T_1 s)/(1 + T_2 s)$, $T_1 > T_2$, can be shown to be approximately 58 degrees, provided that the poles of $G_c'(r)$ stay within the range of $-1 \leq r \leq 0$.

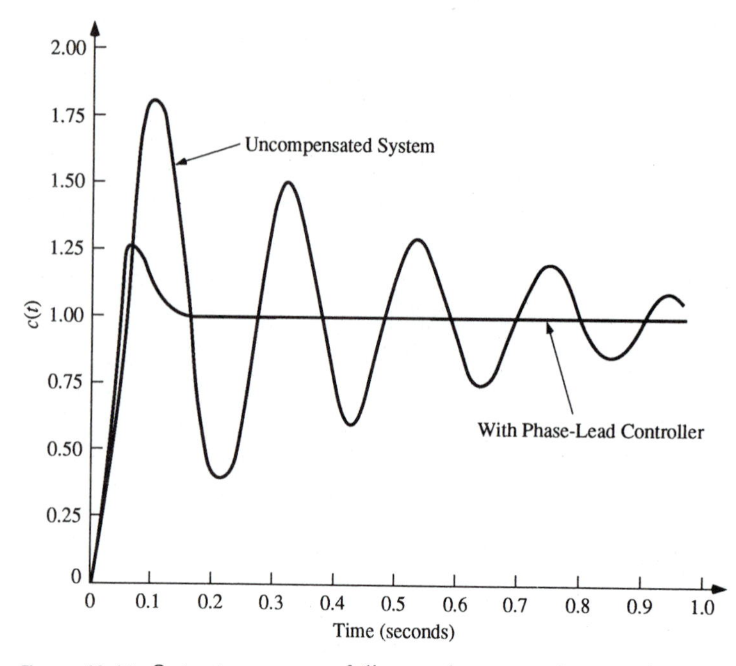

Figure 10-14. Output responses of discrete-data control system in Example 10-2.

From this illustrative example we see that the r-transformation method of designing a discrete-data control system with a continuous-data controller is analytically more tedious than the time-delay approximation method illustrated in Example 10-1. However, the time-delay method is subject to the validity and accuracy of the approximation model.

10-3 DESIGN OF CONTINUOUS-DATA CONTROLLERS WITH EQUIVALENT DIGITAL CONTROLLERS

We shall show in the following sections that the discrete-data system with a cascade digital controller $D(z)$ is the simplest to design analytically, since $D(z)$ is separated from the transfer functions of the zero-order hold and controlled process. Thus, the design of the continuous-data controller of the system in Fig. 10-1 can be carried out by first designing the digital controller $D(z)$ of the system configuration in Fig. 10-2, and then finding the equivalent $G_c(s)$. In fact, we can design the digital controller $D(z)$ using the system configuration of Fig. 10-2 and then find the other equivalent controller configurations using Fig. 10-1 or 10-3 or others.

10-3-1 Cascade Continuous-Data Controller

For the discrete-data control system with digital controller shown in Fig. 10-2, the open-loop transfer function is

$$G(z) = \frac{C(z)}{E(z)} = D(z)\mathscr{Z}[G_{h0}(s)G_p(s)]$$

$$= (1 - z^{-1})D(z)\mathscr{Z}\left(\frac{G_p(s)}{s}\right) \quad (10\text{-}31)$$

The open-loop transfer function of the discrete-data control system with continuous-data controller in Fig. 10-1 is

$$G(z) = \frac{C(z)}{E(z)} = (1 - z^{-1})\mathscr{Z}\left(\frac{G_c(s)G_p(s)}{s}\right) \quad (10\text{-}32)$$

For the two systems to be equivalent, we equate the last two equations. Thus,

$$\mathscr{Z}\left(\frac{G_c(s)G_p(s)}{s}\right) = D(z)\mathscr{Z}\left(\frac{G_p(s)}{s}\right) \quad (10\text{-}33)$$

Given $G_p(s)$ and $D(z)$, $G_c(s)$ can be solved from the last equation.

Example 10-3

Consider the space vehicle control system studied in Examples 10-1 and 10-2. The transfer function of the controlled process $G_p(s)$ is given in Eq. (10-5). Let the transfer function of the digital controller be

$$D(z) = \frac{10.5882(z - 0.915)}{z - 0.1} \quad (10\text{-}34)$$

We do not have to elaborate on how $D(z)$ is determined at this stage. Substituting Eq. (10-34) into Eq. (10-33), we have

$$\mathscr{Z}\left(\frac{G_c(s)G_p(s)}{s}\right) = \frac{0.4561z(z + 0.9708)}{(z - 1)^2(z - 0.1)} \quad (10\text{-}35)$$

Performing partial-fraction expansion to the right-hand side of the last equation, we get

$$\mathscr{Z}\left(\frac{G_c(s)G_p(s)}{s}\right) = \frac{0.6030z}{z - 0.1} - \frac{0.6030z}{z - 1} + \frac{0.9988z}{(z - 1)^2} \quad (10\text{-}36)$$

Taking the inverse z-transform on both sides of the last equation, we get

$$\frac{G_c(s)G_p(s)}{s} = \frac{0.6030}{s+230} - \frac{0.6030}{s} + \frac{99.876}{s^2}$$

$$= \frac{-38.8(s-592)}{s^2(s+230)} \tag{10-37}$$

Substituting the transfer function $G_p(s)$ from Eq. (10-5) into Eq. (10-37) and solving for $G_c(s)$, we have

$$G_c(s) = \frac{-38.8(s+8.871)(s-592)}{s+230} \tag{10-38}$$

For a physically realizable $G_c(s)$, we can add a pole far to the left on the negative real axis of the s-plane. However, a better solution in this case is to delete the zero at $s = 592$. The result for $G_c(s)$ is

$$G_c(s) = \frac{25.89(s+8.871)}{s+230} \tag{10-39}$$

We can show that the controller in Eq. (10-38) gives a unit-step response that has a maximum overshoot of 42 percent, whereas the controller in Eq. (10-39), which is an approximation of Eq. (10-38), gives a maximum overshoot of 25 percent.

10-3-2 Feedback Continuous-Data Controller

In this section the continuous-data controller $H(s)$ in the feedback path of Fig. 10-3 is to be determined based on the digital controller $D(z)$ in Fig. 10-2.

The open-loop transfer function of the discrete-data control system in Fig. 10-3 is

$$G(z) = \frac{C(z)}{E(z)} = \frac{\mathscr{Z}[G_{ho}(s)G_p(s)]}{1 + \mathscr{Z}[G_{ho}(s)H(s)G_p(s)]} \tag{10-40}$$

Equating the last equation to Eq. (10-31) and solving for $\mathscr{Z}[G_{ho}(s)H(s)G_p(s)]$, we get

$$\mathscr{Z}[G_{ho}(s)H(s)G_p(s)] = \frac{1 - D(z)}{D(z)} \tag{10-41}$$

or

$$\mathscr{Z}[H(s)G_p(s)] = \frac{z}{z-1}\,\frac{1 - D(z)}{D(z)} \tag{10-42}$$

Thus, knowing $G_p(s)$ and $D(z)$, we can solve for $H(s)$ from the last equation.

Example 10-4

The continuous-data feedback controller in Fig. 10-3 is to be determined so that the system is equivalent to the one in Fig. 10-2 with $D(z)$ given in Eq. (10-34).

Substituting Eq. (10-34) into Eq. (10-42), we get

$$\mathscr{Z}[H(s)G_p(s)] = \frac{-0.9055z}{z - 0.915} \tag{10-43}$$

Now substituting Eq. (10-5) in the last equation for $G_p(s)$ and taking the inverse z-transform on both sides of the equation, we get

$$H(s) = -0.001021s^2 \tag{10-44}$$

10-4 DIGITAL CONTROLLERS

The most versatile way of compensating a discrete-data control system is to use a digital controller. In general, digital controllers can be implemented by digital networks, digital computers, microprocessors, or digital signal processors (DSPs). Compared with continuous-data controllers, better performance can be realized for a control system with digital controllers. Another advantage of using digital controllers is that the control algorithm can be easily changed by changing the program of the controller, whereas changing the components of a continuous-data controller is rather difficult once the controller has been implemented.

Before entering into the design of digital controllers, we shall first discuss the physical realizability and the composition of the digital controller. The block diagram in Fig. 10-15 can be used to represent a single-input-single-output digital controller. The input of the controller, $e_1^*(t)$, is in the form of a sequence of numbers, $e_1(kT)$, $k = 0, 1, 2, \ldots$. Analytically, $e_1^*(t)$ is represented as the output of an ideal sampler. The digital controller performs certain linear operations on the sequence $e_1(kT)$ and delivers the output sequence $e_2(kT)$ which is portrayed as the output of another ideal sampler, $e_2^*(t)$. The transfer function of the digital controller is described by

$$D(z) = \frac{E_2(z)}{E_1(z)} = \frac{b_m z^m + b_{m-1}z^{m-1} + \cdots + b_0}{a_n z^n + a_{n-1}z^{n-1} + \cdots + a_0} \tag{10-45}$$

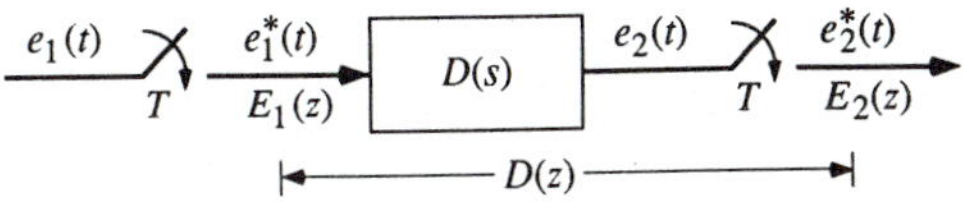

Figure 10-15. The block diagram of a digital controller.

10-4-1 Physical Realizability Considerations

The *a priori* requirement on the design of the digital controller is that the transfer function $D(z)$ be physically realizable. The condition of physical realizability implies that no output signal of the system will appear before an input signal is applied. Expanding $D(z)$ in Eq. (10-45) into a power series in z^{-1}, the coefficients of the series represent the values of the weighting sequence of the digital controller. The coefficient of the z^{-k} term, $k = 0, 1, 2, \ldots$, corresponds to the value of the weighting sequence at $t = kT$. Clearly, for the digital controller to be physically realizable, the power-series expansion of $D(z)$ must not contain any positive power in z. Any positive power in z in the series expansion of $D(z)$ would indicate *prediction* or simply that the output precedes the input. Therefore, for the transfer function in Eq. (10-45) to be physically realizable, the highest power of the denominator must be equal to or greater than that of the numerator, or simply $n \geq m$.

If the digital controller has the same number of poles and zeros, it is expressed as

$$D(z) = \frac{b_0 + b_1 z^{-1} + \cdots + b_m z^{-m}}{a_0 + a_1 z^{-1} + \cdots + a_n z^{-n}} \tag{10-46}$$

where n and m are any positive integers. In this case, in order that the power-series expansion of $D(z)$ will not contain positive powers of z, the denominator must not contain any factor of z^{-1}, if $b_0 \neq 0$. Therefore, *if $b_0 \neq 0$, the condition for $D(z)$ to be physically realizable is that $a_0 \neq 0$.*

10-4-2 Realization of Digital Controllers by Digital Programming

The most versatile way of implementing a digital controller is to use a digital computer, microprocessor, or DSP. Because of the advantages in computing speed, storage capacity, and flexibility, the use of digital computers as controllers has become increasingly important. In recent years the advancements made in the field of microcomputers and DSPs have made these devices more attractive as controllers.

In general, the transfer function of a digital controller can be realized by a digital program. Three basic methods of programming are discussed in the following. These are *direct programming*, *cascade programming*, and *parallel programming*. From the analytical standpoint, these programming methods are very similar to the decomposition techniques presented in Chapter 5.

We shall consider that a digital computer is capable of performing specific arithmetic operations of addition, multiplication, subtraction, storage, and shifting. Some microcomputers cannot multiply two numbers directly, and a subprogram must be written to perform simple multiplication. The discussions in this section are intended only for analytical purposes, and no specific machine characteristics are considered.

1. Direct Digital Programming

A transfer function of a physically realizable digital controller may be written as

$$D(z) = \frac{E_2(z)}{E_1(z)} = \frac{b_0 + b_1 z^{-1} + b_2 z^{-2} + \cdots + b_m z^{-m}}{a_0 + a_1 z^{-1} + a_2 z^{-2} + \cdots + a_n z^{-n}} \tag{10-47}$$

where $a_0 \neq 0$ if $b_0 \neq 0$; m and n are positive integers. $E_1(z)$ and $E_2(z)$ represent the z-transforms of the input and output of the controller, respectively. To conduct a direct digital programming of $D(z)$, we first perform cross multiplication in Eq. (10-47), and then take the inverse z-transform; we get the following equation:

$$a_0 e_2^*(t) + \sum_{k=1}^{n} a_k e_2^*(t - kT) = \sum_{k=0}^{m} b_k e_1^*(t - kT) \tag{10-48}$$

Solving for $e_2^*(t)$ from the last equation, we get

$$e_2^*(t) = \frac{1}{a_0} \sum_{k=0}^{m} b_k e_1^*(t - kT) - \frac{1}{a_0} \sum_{k=1}^{n} a_k e_2^*(t - kT) \tag{10-49}$$

The last equation shows that the present value of the output $e_2^*(t)$ depends on the present and past values of the input $e_1^*(t)$ as well as the past information of the output. To implement the digital program using Eq. (10-49), two basic mathematical operations are required. The first operation is *data storage* which stores the past samples of the output and the input so that they may be used in the computation of the output $e_2^*(t)$. The second operation involves arithmetic manipulations such as multiplication of the stored input and output data by constants, addition, and subtraction.

Let

$$x^*(t) = \frac{1}{a_0} \sum_{k=0}^{m} b_k e_1^*(t - kT) \tag{10-50}$$

and

$$y^*(t) = \frac{1}{a_0} \sum_{k=1}^{n} a_k e_2^*(t - kT) \tag{10-51}$$

Equation (10-49) becomes

$$e_2^*(t) = x^*(t) - y^*(t) \tag{10-52}$$

The block diagram representation of the direct digital programming of Eqs. (10-50), (10-51), and (10-52) is shown in Fig. 10-16. Note that the data storages are simply time-delay units which delay the input data for one sampling period.

The direct digital program implemented in Fig. 10-16 requires a total of $n + m$ data storages. As an alternative, we can use the direct-decomposition scheme discussed in Chapter 5 as a direct digital program implementation of

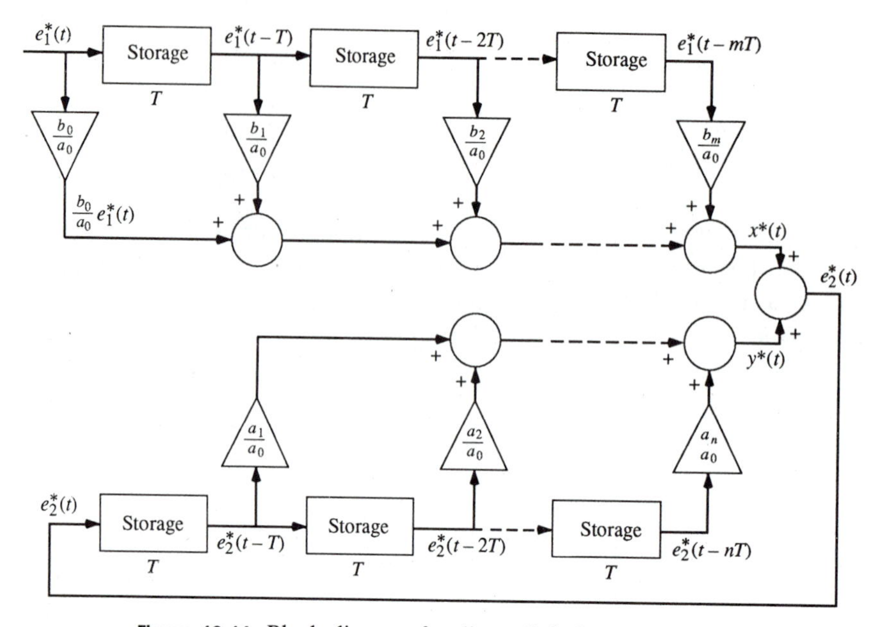

Figure 10-16. Block diagram for direct digital programming of $D(z)$ in Eq. (10-47).

Eq. (10-47). Applying direct decomposition to Eq. (10-47), we have the following equations:

$$E_2(z) = \frac{1}{a_0}(b_0 + b_1 z^{-1} + \cdots + b_m z^{-m})X(z) \tag{10-53}$$

$$X(z) = \frac{1}{a_0} E_1(z) - \frac{1}{a_0}(a_1 z^{-1} + a_2 z^{-2} + \cdots + a_n z^{-n})X(z) \tag{10-54}$$

where $X(z)$ is simply a dummy variable. Figure 10-17 shows the block diagram of direct digital programming by direct decomposition. It is assumed in the illustrated case that $n = m$. The block diagram can be easily modified if $n \neq m$. In general, the number of data-storage units is equal to the greater of n and m.

2. Cascade Digital Programming

The transfer function $D(z)$ may be written as a product of a number of simple transfer functions each realizable by a simple digital program. Then, the digital programming of $D(z)$ may be represented by a series of cascaded digital programs of the simple transfer functions. Writing Eq. (10-47) in factored form, we have

$$D(z) = \prod_{k=1}^{p} D_k(z) \tag{10-55}$$

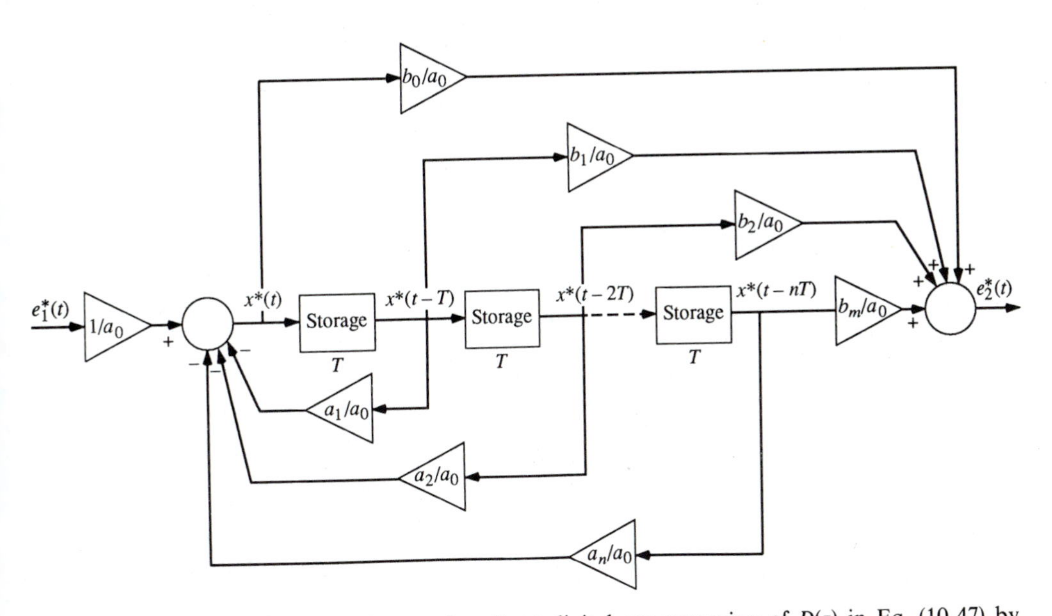

Figure 10-17. Block diagram for direct digital programming of $D(z)$ in Eq. (10-47) by direct decomposition, $n = m$.

where p is the greater of n and m. Figure 10-18 shows the block diagram representation of the cascade digital programming of $D(z)$. In general, the transfer function $D_k(z)$ may assume the following forms depending on the poles and zeros of $D(z)$ and the relative magnitudes of m and n.

Real Pole and Zero

$$D_k(z) = K_k \frac{1 + c_k z^{-1}}{1 + d_k z^{-1}} \tag{10-56}$$

Two Complex Conjugate Poles

$$D_k(z) = K_k \frac{1}{1 + d_k z^{-1} + f_k z^{-2}} \tag{10-57}$$

One Real Zero and Two Complex Conjugate Poles

$$D_k(z) = K_k \frac{1 + c_k z^{-1}}{1 + d_k z^{-1} + f_k z^{-2}} \tag{10-58}$$

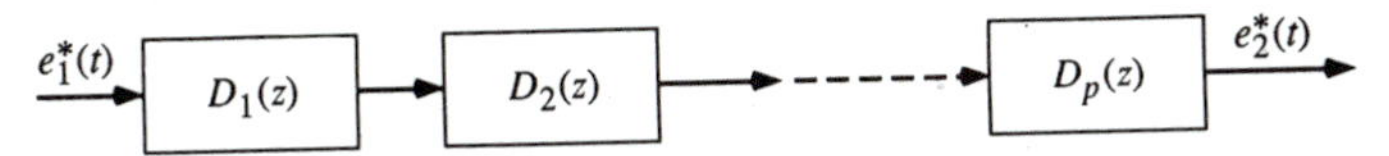

Figure 10-18. Block diagram of cascade digital programming of $D(z)$.

Complex Conjugate Poles and Zeros

$$D_k(z) = K_k \frac{1 + g_k z^{-1} + h_k z^{-2}}{1 + d_k z^{-1} + f_k z^{-2}} \tag{10-59}$$

Real Zero ($m > n$)

$$D_k(z) = K_k(1 + c_k z^{-1}) \tag{10-60}$$

Complex Conjugate Zeros ($m > n$)

$$D_k(z) = K_k(1 + g_k z^{-1} + h_k z^{-2}) \tag{10-61}$$

One Real Pole and Two Complex Conjugate Zeros ($m > n$)

$$D_k(z) = K_k \frac{1 + g_k z^{-1} + h_k z^{-2}}{1 + d_k z^{-1}} \tag{10-62}$$

These transfer functions can all be realized by the direct digital programming method discussed in the last section. The case illustrated by Eq. (10-56) is perhaps the most common, as the real pole and zero of $D(z)$ are represented. The case in Eq. (10-57) is for two complex conjugate poles of $D(z)$, since we want to avoid dealing with complex numbers in a digital program. The cases in Eqs. (10-58) through (10-62) are just some other possible configurations depending on the relative magnitudes of n and m.

3. Parallel Digital Programming

Another method of implementing digital programming is *parallel programming*, in which case the transfer function $D(z)$ is first expanded by partial-fraction expansion into a sum of simple first- or second-order transfer functions.

In general, the transfer function $D(z)$ in Eq. (10-47) can be written as

$$D(z) = \sum_{k=1}^{p} D_k(z) \tag{10-63}$$

where p is the greater of m and n. Depending on the nature of $D(z)$, $D_k(z)$ can be of the following forms:

a. $D_k(z) = K_k/(1 + d_k z^{-1})^j$, real pole of multiplicity j; $j = 1, 2, \ldots, N$

b. $D_k(z) = K_k(1 + c_k z^{-1})/(1 + d_k z^{-1} + f_k z^{-2})^j$, complex poles of multiplicity j; $j = 1, 2, \ldots, N$

c. $D_k(z) = K_k/z^j$, pole at $z = 0$; $j = 1, 2, \ldots, N$; $N = (m - n) > 0$

Each of these transfer functions can again be realized by the direct digital programming method. The block diagram of parallel digital programming is illustrated in Fig. 10-19.

The following example illustrates the three methods of digital programming presented in this section.

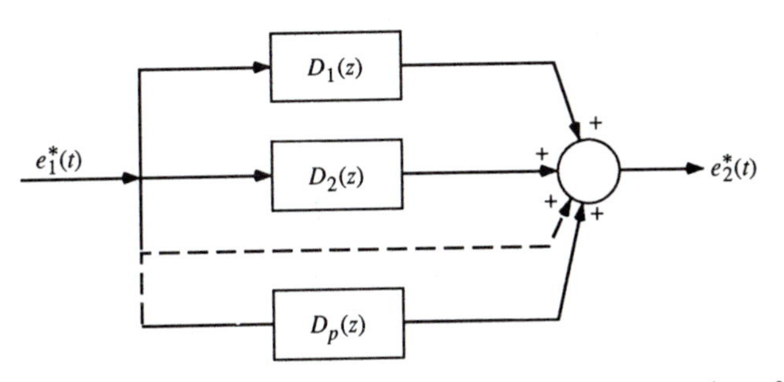

Figure 10-19. Block diagram of parallel digital programming of $D(z)$.

Example 10-5

Consider the following transfer function of a digital controller.

$$\frac{E_2(z)}{E_1(z)} = D(z) = \frac{5(1 + 0.25z^{-1})}{(1 - 0.5z^{-1})(1 - 0.1z^{-1})} \tag{10-64}$$

It is apparent that the transfer function is physically realizable.

1. Direct Digital Programming

Expanding the denominator factors and cross multiplying, Eq. (10-64) is written

$$(1 - 0.6z^{-1} + 0.05z^{-2})E_2(z) = 5(1 + 0.25z^{-1})E_1(z) \tag{10-65}$$

Taking the inverse z-transform on both sides of the last equation and solving for $e_2^*(t)$, we get

$$e_2^*(t) = 5e_1^*(t) + 1.25e_1^*(t - T) + 0.6e_2^*(t - T) - 0.05e_2^*(t - 2T) \tag{10-66}$$

where T is the sampling period. The block diagram of the direct digital programming of Eq. (10-64) according to Eq. (10-66) is shown in Fig. 10-20.

Alternatively, we may apply direct decomposition to Eq. (10-64) and get

$$E_2(z) = (5 + 1.25z^{-1})X(z) \tag{10-67}$$

$$X(z) = E_1(z) + 0.6z^{-1}X(z) - 0.05z^{-2}X(z) \tag{10-68}$$

These last two equations are realized by the digital program shown in Fig. 10-21.

2. Cascade Digital Programming

The right-hand side of Eq. (10-64) is arbitrarily divided and written as the product of two functions:

$$\frac{E_2(z)}{E_1(z)} = \frac{1 + 0.25z^{-1}}{1 - 0.5z^{-1}} \frac{5}{1 - 0.1z^{-1}} \tag{10-69}$$

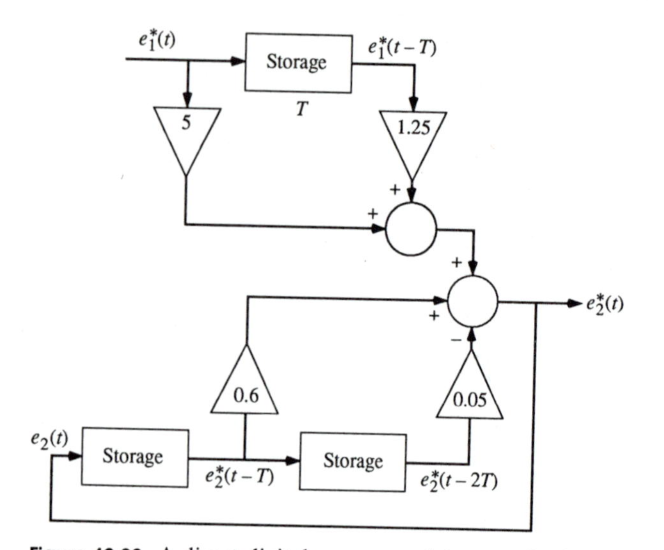

Figure 10-20. A direct digital program of the transfer function $D(z)$ in Eq. (10-64).

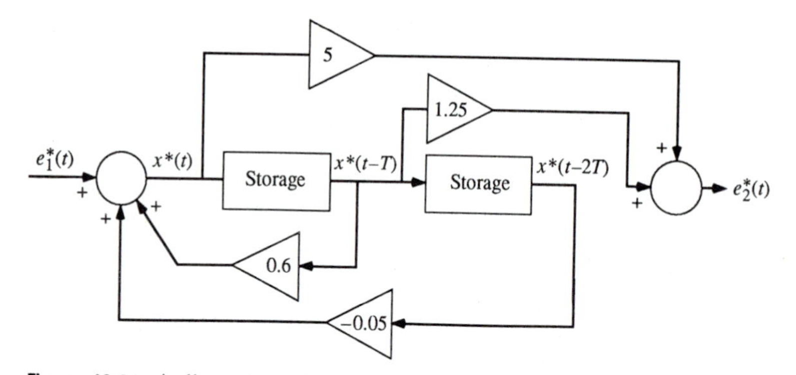

Figure 10-21. A direct digital program of the transfer function $D(z)$ in Eq. (10-64).

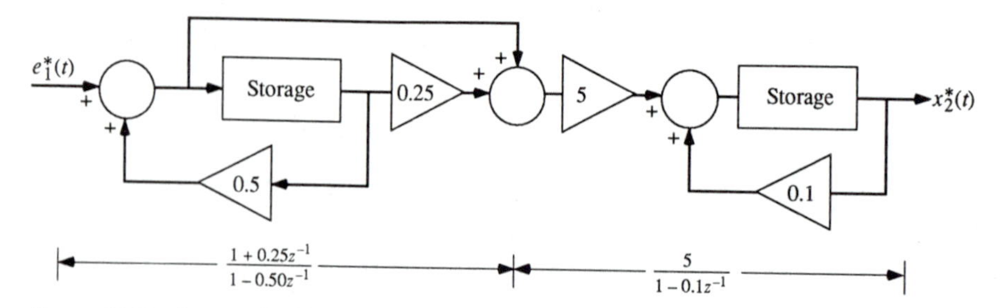

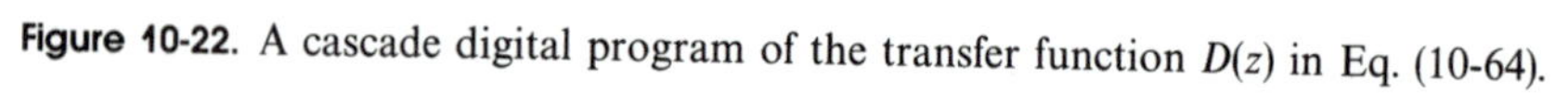

Figure 10-22. A cascade digital program of the transfer function $D(z)$ in Eq. (10-64).

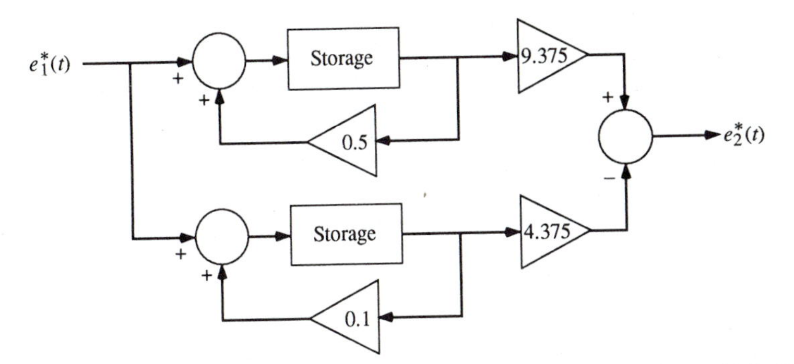

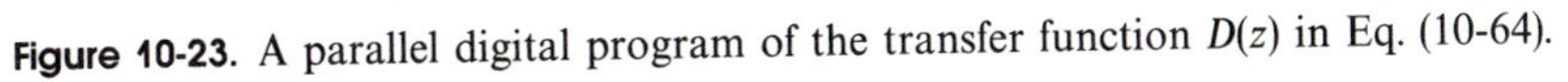
Figure 10-23. A parallel digital program of the transfer function $D(z)$ in Eq. (10-64).

The digital program shown in Fig. 10-22 is drawn according to the division of the transfer function in Eq. (10-69).

3. Parallel Digital Programming

The right-hand side of Eq. (10-64) is expanded by partial-fraction expansion into the following form:

$$\frac{E_2(z)}{E_1(z)} = \frac{9.375}{1 - 0.5z^{-1}} - \frac{4.375}{1 - 0.1z^{-1}} \tag{10-70}$$

Figure 10-23 shows the parallel digital programming of Eq. (10-64) using Eq. (10-70).

10-4-3 The Digital PID Controller

One of the most widely used controllers in the design of continuous-data control systems is the proportional-integral-derivative (PID) controller [1]. Figure 10-24 shows the block diagram of a continuous-data PID controller acting on an error signal $e(t)$. The proportional control simply multiplies $e(t)$

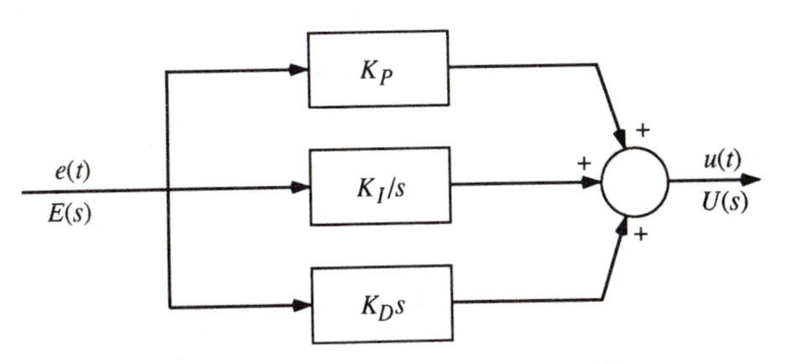

Figure 10-24. A continuous-data PID controller.

by a constant K_P, the integral control multiplies the time integral of $e(t)$ by K_I, and the derivative control generates a signal equal to K_D times the time derivative of $e(t)$. The function of the integral control is to provide action to reduce the area under $e(t)$, which leads to the reduction of the steady-state error. The derivative control provides an anticipatory action to reduce the overshoots and oscillations in the time response. The same principle of PID control can be applied to digital control. In digital control, the proportional control is still implemented by a proportional constant K_P. In general, there are a number of ways of implementing integration and derivatives digitally. The rectangular integration schemes discussed in Chapter 9 can be used for this purpose. The transfer functions of these integration schemes including the proportional constant K_I are summarized as follows:

Backward-Rectangular Integration

$$D_I(z) = K_I \frac{T}{z-1} \tag{10-71}$$

Forward-Rectangular Integration

$$D_I(z) = K_I \frac{Tz}{z-1} \tag{10-72}$$

Bilinear-Transformation Integration

$$D_I(z) = K_I \frac{T}{2} \frac{z+1}{z-1} \tag{10-73}$$

The most common method of approximating the derivative of $e(t)$ at $t = T$ that results in a physically realizable transfer function is

$$\left.\frac{de(t)}{dt}\right|_{t=T} = \frac{e(kT) - e[(k-1)T]}{T} \tag{10-74}$$

Taking the z-transform on both sides of the last equation and including the proportional constant K_D, we have the transfer function of the digital derivative controller,

$$D_D(z) = K_P \frac{z-1}{Tz} \tag{10-75}$$

The block diagram of the digital PID controller is shown in Fig. 10-25. Digital-program implementation of the PID controller can be conducted by any one of the methods discussed in the preceding sections.

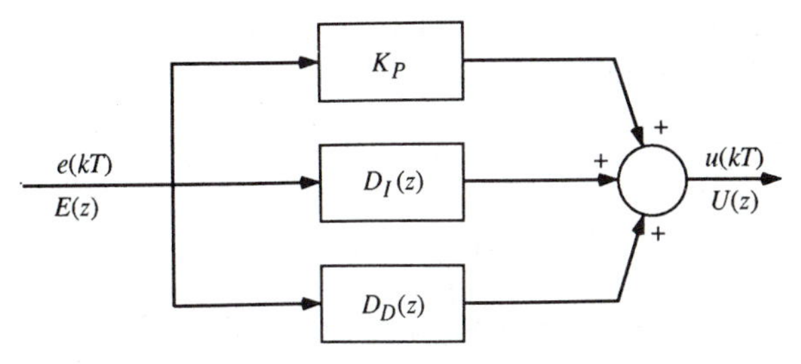

Figure 10-25. A digital PID controller.

10-5 DESIGN OF DIGITAL CONTROL SYSTEMS WITH DIGITAL CONTROLLERS THROUGH BILINEAR TRANSFORMATION

In this section we shall consider the design of a digital control system with a digital controller using the frequency-domain technique. This type of design problem is generally simpler to carry out than the design of discrete-data systems with continuous-data controllers. The reason is because the transfer function of the digital controller is isolated from that of the controlled process, so that the effects of varying the controller parameters may be investigated by means of the Bode diagram using either the r-transformation of Eq. (6-126) with $r = j\omega_r$ or the w-transformation of Eq. (8-43) with $w = j\omega_w$. The principle of design is outlined below with reference to the block diagram in Fig. 10-26 using the r-transformation. A similar design procedure can be formulated using the w-transformation.

1. Evaluate the z-transform of the zero-order hold (zoh) and controlled process combination, $G_{h0}G_p(z)$. Apply the r-transformation, $z = (1 + r)/(1 - r)$, to obtain $G_{h0}G_p(r)$.
2. Construct the Bode diagram of $G_{h0}G_p(r)$ in magnitude (dB) and phase (degrees) versus ω_r. Transfer the data from the Bode plot to the Nichols chart if necessary. Determine the performance characteristics of the uncompensated system by finding the gain margin, phase margin, bandwidth, resonance peak, and resonant frequency from the Bode plot and the Nichols chart.
3. Should the system need compensation, the open-loop transfer function of the system with a digital controller becomes $D(z)G_{h0}G_p(z)$ or, in the

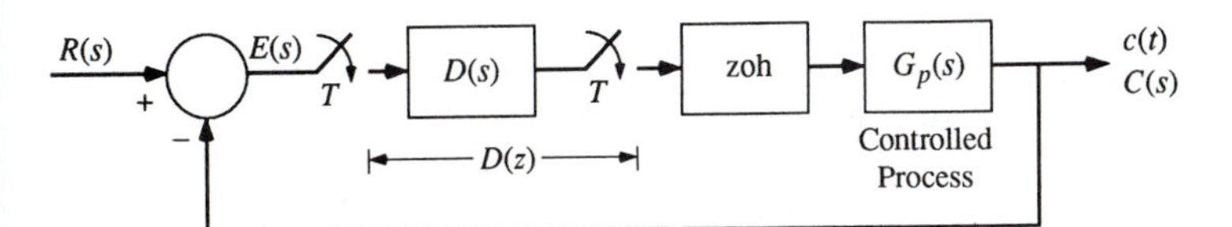

Figure 10-26. A control system with digital controller.

r-domain, $D(r)G_{h0}G_p(r)$. The digital controller transfer function $D(r)$ is to be determined so that the desired system performance specifications are satisfied. The selection of $D(r)$ may follow the design principle of continuous-data systems, which involves a trial-and-error procedure and, to some extent, is based on the experience and imagination of the designer.

Since r is the working domain, the transfer function $D(r)$ must be chosen so that the corresponding z-domain transfer function of the digital controller, $D(z)$, is physically realizable. Let $D(r)$ be of the following form:

$$D(r) = \frac{c_m r^m + c_{m-1} r^{m-1} + \cdots + c_1 r + c_0}{d_n r^n + d_{n-1} r^{n-1} + \cdots + d_1 r + d_0} \tag{10-76}$$

where n and m are positive integers. Let us investigate the constraints on the relative magnitudes of n and m and the coefficients of $D(r)$ so that the corresponding transfer function $D(z)$, obtained through the transformation $r = (z-1)/(z+1)$, is physically realizable. Substituting $r = (z-1)/(z+1)$ into Eq. (10-76), we have

$$D(z) = \frac{c_m(z-1)^m + c_{m-1}(z-1)^{m-1}(z+1) + \cdots + c_1(z-1)(z+1)^{m-1} + c_0(z+1)^m}{d_n(z-1)^n + d_{n-1}(z-1)^{n-1}(z+1) + \cdots + d_1(z-1)(z+1)^{n-1} + d_0(z+1)^n} \times (z+1)^{n-m} \tag{10-77}$$

This expression for $D(z)$ indicates that if $D(r)$ is as given in Eq. (10-76), regardless of the relative magnitudes of m and n, $D(z)$ will always have the same number of poles and zeros. This means that $D(z)$ will always be physically realizable so long as $D(r)$ is of the form of Eq. (10-76). However, if we want the digital controller to be stable, then all the poles of $D(z)$ must lie inside the unit circle $|z| = 1$ in the z-plane. This means that all the poles of $D(r)$ must lie inside the left half of the r-plane. In addition, we note from Eq. (10-77) that if $m > n$, $D(z)$ will have $m - n$ poles at $z = -1$ which correspond to an unstable condition. Therefore, for a stable digital controller, we must also require that $D(r)$ does not have more zeros than poles. In practice, it is permissible to have an unstable digital controller so long as the overall system is stable, although, if possible, it is desirable to keep both the open-loop as well as the closed-loop systems stable.

4. Once $D(r)$ is determined, $D(z)$ is obtained by substituting $r = (z-1)/(z+1)$ in $D(r)$. The final step in the design involves the realization of $D(z)$ by one of the digital programming methods discussed in Sec. 10-4. If $D(z)$ is to be implemented by a microprocessor or DSP, then the designer should be aware of the limitations and constraints of these devices and take them into consideration when carrying out the design.

Let us first review the basics of phase-lead and phase-lag controllers in the r-plane, based on the properties in the s-domain.

10-5-1 Basic Properties of a Phase-Lead Controller for $D(r)$

The transfer function of a single-stage phase-lead controller for $D(r)$ is of the form

$$D(r) = \frac{1 + \alpha\tau r}{1 + \tau r} \qquad \alpha > 1 \tag{10-78}$$

where α and τ are positive real constants. The Bode plot of $D(r)$ for $r = j\omega_r$ is shown in Fig. 10-27. The maximum phase of $D(r)$ is given by

$$\phi_m = \sin^{-1}\left(\frac{\alpha - 1}{\alpha + 1}\right) \tag{10-79}$$

and it occurs at

$$\omega_{rm} = \frac{1}{\sqrt{\alpha\tau}} \tag{10-80}$$

The phase-lead design is carried out by making use of the maximum phase lead ϕ_m. If the design is based on a specified phase margin, the designer should first determine the desired ϕ_m and place the phase curve of $D(r)$ so that ϕ_m occurs at the new gain-crossover frequency. In view of Fig. 10-27, since a gain of $20 \log_{10} \alpha$ is always provided by the phase-lead controller at high frequencies, the new gain-crossover frequency is always higher than that of the uncompensated system. In addition, the phase curve of the uncompensated process usually has a negative slope, so that the phase decreases as the frequency increases. Thus, the design trick is first to set a desired value for ϕ_m and then locate ω_m at the new gain-crossover frequency in ω_r.

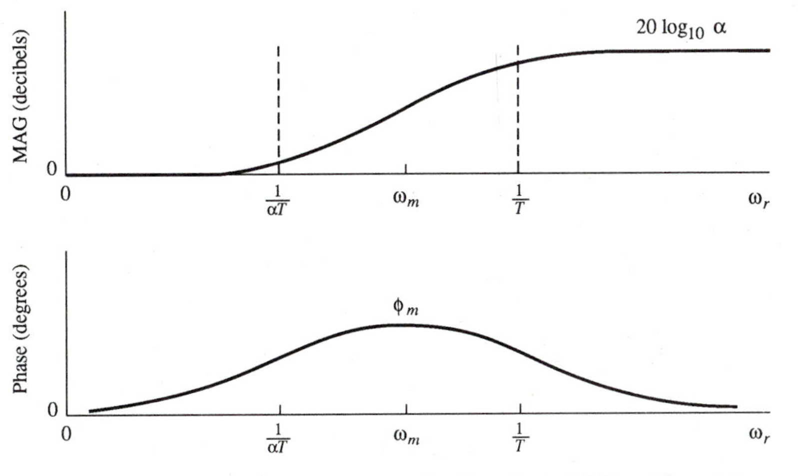

Figure 10-27. Bode diagram of $D(r) = (1 + \alpha T_r)/(1 + T_r)$, $\alpha > 1$.

In general, if one stage of the phase-lead controller in the form of Eq. (10-78) is inadequate in achieving the desired relative stability, two or more stages can be connected in cascade, if necessary, to satisfy the design objectives.

10-5-2 Basic Properties of a Phase-Lag Controller for $D(r)$

The transfer function of a single-stage phase-lag controller for $D(r)$ has the same form as Eq. (10-78), except that $\alpha < 1$. The Bode plot of $D(j\omega_r)$ is shown in Fig. 10-28. In this case, the phase of $D(j\omega_r)$ is negative for all ω_r. The minimum value of the phase is still given by Eq. (10-79), and it occurs at ω_{rm} given by Eq. (10-80).

The phase-lag design is carried out by using the attenuation provided by the phase-lag controller which is $20 \log_{10} \alpha$ dB.

Based on the desired phase margin or gain margin, the designer should first determine how much attenuation is needed to bring down the magnitude curve of the uncompensated process so that the gain-crossover frequency occurs at the desired phase margin, or the desired gain margin is obtained at the phase crossover. This establishes the value of α. However, these are carried out based on the assumption that the phase curve of the process is not appreciably affected by the phase-lag controller. As seen from Fig. 10-28, the phase-lag controller always contributes a negative phase which is detrimental to the design effort. Thus, the strategy is to place the lower corner frequency of the controller, $\omega_r = 1/\alpha\tau$, so that the negative phase does not appreciably affect the phase curve of the compensated system. A rule of thumb is to place $1/\alpha\tau$ at approximately 1/10 of the gain-crossover frequency. Again, for more stringent design requirements, several stages of the phase-lag controller can be placed in cascade to form the overall controller.

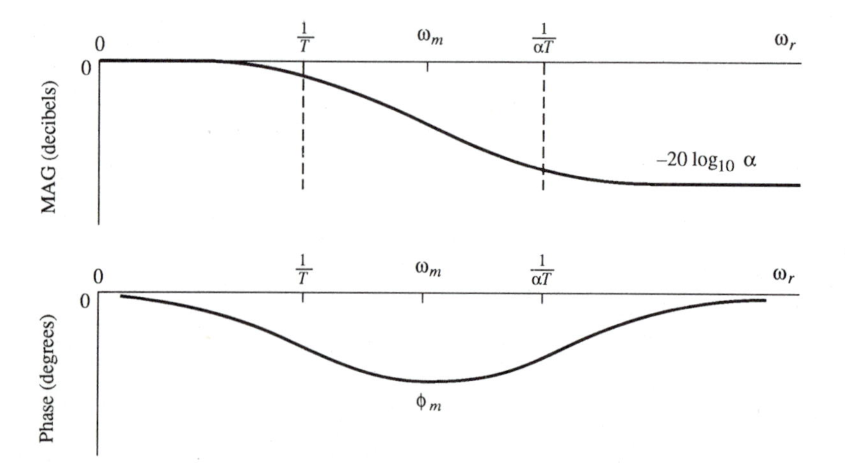

Figure 10-28. Bode diagram of $D(r) = (1 + \alpha T_r)/(1 + T_r)$, $\alpha < 1$.

In general, the combined effects of lag-lead or lead-lag controllers and the pole-zero-cancellation controller [1] can be used if necessary.

To illustrate the frequency-domain design in the r-plane, the following illustrative examples are given.

Example 10-6

Consider that the transfer function of the controlled process of the system shown in Fig. 10-26 is that of the space vehicle control system given in Eq. (10-5). As in Example 10-1, let us select K to be 22.5, which corresponds to a ramp-error constant K_v of 100, and the sampling period T is 0.01 s. The open-loop transfer function of the system without compensation is

$$G_{h0}G_p(z) = (1 - z^{-1})\mathscr{Z}\left(\frac{887.69}{s^2(s + 8.871)}\right)$$

$$= \frac{0.0431(z + 0.9708)}{(z - 1)(z - 0.915)} \tag{10-81}$$

Design Using the r-Transformation

Applying the r-transformation of $z = (1 + r)/(1 - r)$ to the last equation, we have

$$G_{h0}G_p(r) = G_{h0}G_p(z)|_{z=(1+r)/(1-r)} = \frac{0.4995(1 + 0.0148r)(1 - r)}{r(1 + 22.53r)} \tag{10-82}$$

Figure 10-29 shows the Bode plot of $G_{h0}G_p(r)$. The phase margin of the uncompensated system is 8.63 degrees, and the gain margin is 6.15 dB. From the Nichols chart in Fig. 10-30, the resonance peak M_p is found to be 6.7, and the bandwidth is 47.24 rad/s.

Let us set the desired phase margin at 45 degrees. We shall design various types of controller in the following.

1. Phase-Lead Controller for $D(r)$

Using the transfer function of the phase-lead controller in Eq. (10-78), the design is carried out by making use of the maximum phase lead ϕ_m. Since the uncompensated system has a phase margin of 8.63 degrees, to achieve a phase margin of 45 degrees, theoretically, an additional phase lead of 36.37 degrees is needed. However, as shown by the Bode diagram in Fig. 10-27, the magnitude curve of $D(r)$ will inevitably move the gain crossover to a higher frequency, causing the maximum phase ϕ_m to be added to a lower phase of $G_p(j\omega_r)$. Thus, let us set the desired value of ϕ_m at 48 degrees. From Eq. (10-79), the value of α is determined to be 6.777.

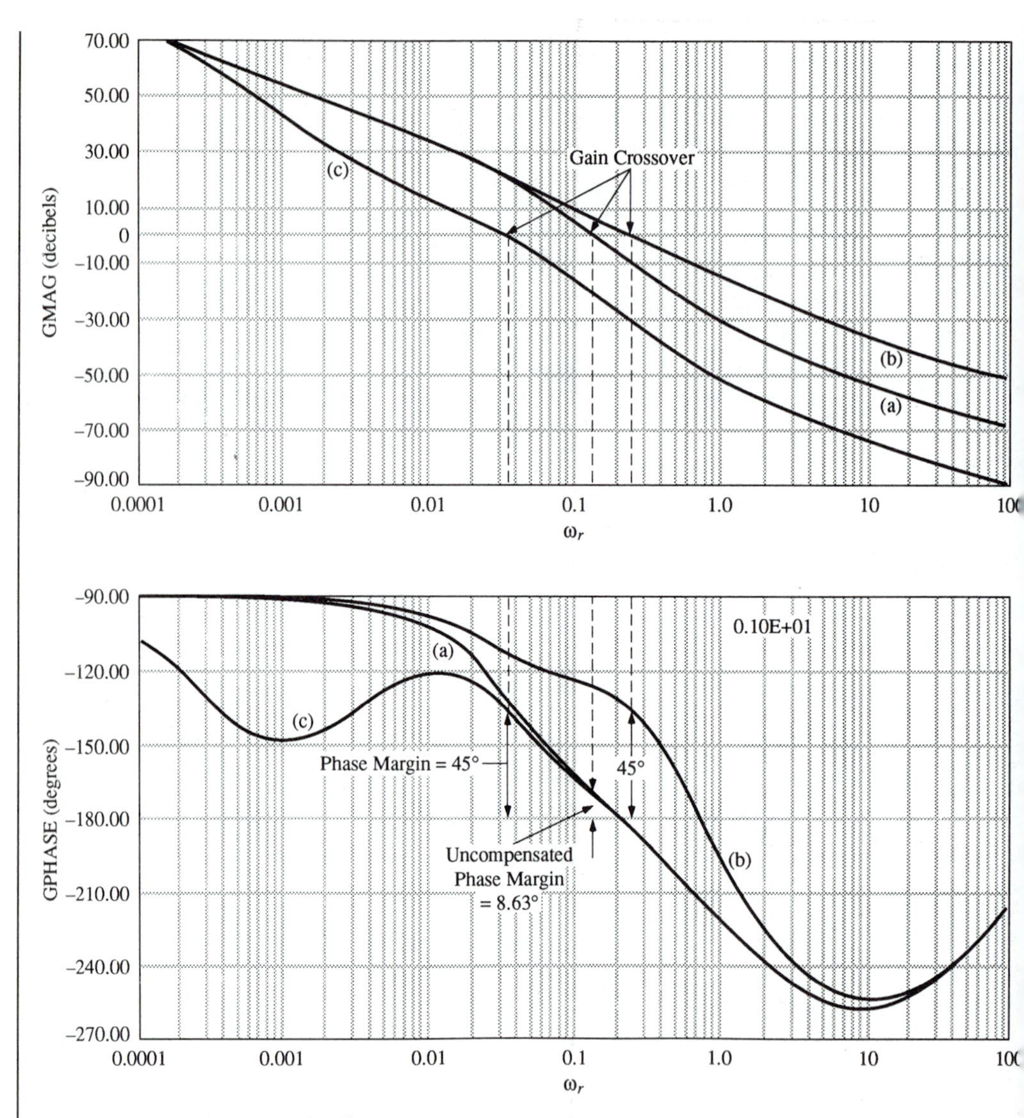

Figure 10-29. Bode diagram of open-loop transfer function in Example 10-6. (a) Uncompensated system.

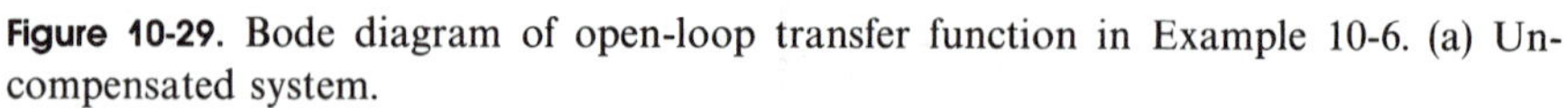

$$G_{h0}G_p(r) = \frac{0.4995(1 + 0.0148r)(1 - r)}{r(1 + 22.53r)}.$$

(b) With phase-lead controller.

$$D(r) = \frac{1 + 10.76r}{1 + 1.5887r}.$$

(c) With phase-lag controller.

$$D(r) = \frac{1 + 294.11r}{1 + 3333.33r}.$$

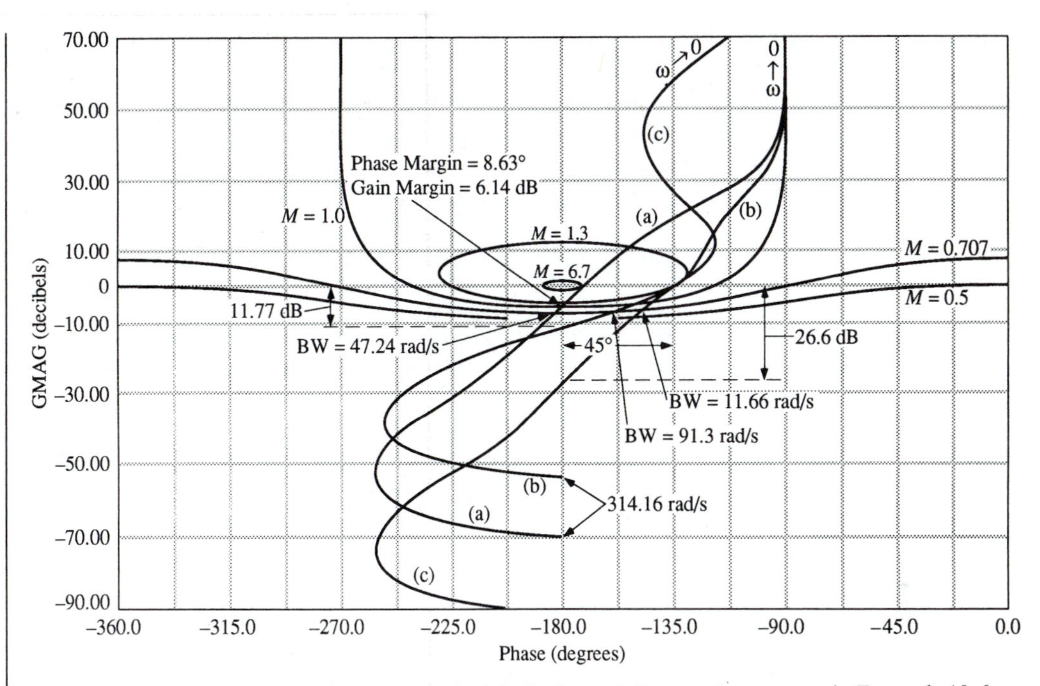

Figure 10-30. Gain-phase plot and Nichols chart of discrete-data system in Example 10-6. (a) Uncompensated system.

$$G_{h0}G_p(r) = \frac{0.4995(1 + 0.0148r)(1 - r)}{r(1 + 22.53r)}.$$

(b) With phase-lead controller.

$$D(r) = \frac{1 + 10.767r}{1 + 1.5887r}.$$

(c) With phase-lag controller.

$$D(r) = \frac{1 + 294.11r}{1 + 3333.33r}.$$

The new gain crossover where ω_{rm} should be located is placed at the point where the gain of the uncompensated Bode plot of $G_p(j\omega_r)$ is $-10\log_{10}\alpha = -8.3$ dB, which is found from Fig. 10-29 to be 0.242. Now using Eq. (10-80), τ is found to be

$$\tau = \frac{1}{\sqrt{\alpha}\omega_{rm}} = \frac{1}{\sqrt{6.777}\,0.242} = 1.5887 \tag{10-83}$$

Thus,

$$D(r) = \frac{1 + \alpha\tau r}{1 + \tau r} = \frac{1 + 10.767r}{1 + 1.5887r} \tag{10-84}$$

Notice that this transfer function is the same as that obtained in Example 10-2, which has the same controlled process with the same phase margin requirement. The compensated Bode diagram is shown in Fig. 10-29 with the controller transfer function as given in Eq. (10-84). The transfer function of the digital controller in the z-domain is obtained by substituting $r = (z-1)/(z+1)$ in Eq. (10-84). We have

$$D(z) = \frac{(\alpha\tau + 1)z + 1 - \alpha\tau}{(1+\tau)z + 1 - \tau} = \frac{4.5455(z - 0.83)}{z - 0.2274} \tag{10-85}$$

Figure 10-30 shows the Nichols chart with the compensated frequency locus, and the value of M_p is reduced from 6.7 to 1.31. The bandwidth is increased to 91.3 rad/s.

The unit-step responses of the uncompensated system and the system with the digital controller in Eq. (10-85) are shown in Fig. 10-31. Notice that the maximum overshoot of the uncompensated system is 78.8 percent, whereas that of the system with the phase-lead controller is 27.8 percent. The phase-lead compensated system also has a faster rise time, since the bandwidth is higher.

2. Phase-Lag Controller for $D(r)$

Instead of using a phase-lead controller, the phase-lag controller described by Eq. (10-78) but with $\alpha < 1$ is now applied to compensate the discrete-data system. From the Bode diagram in Fig. 10-29 we see that a phase margin of

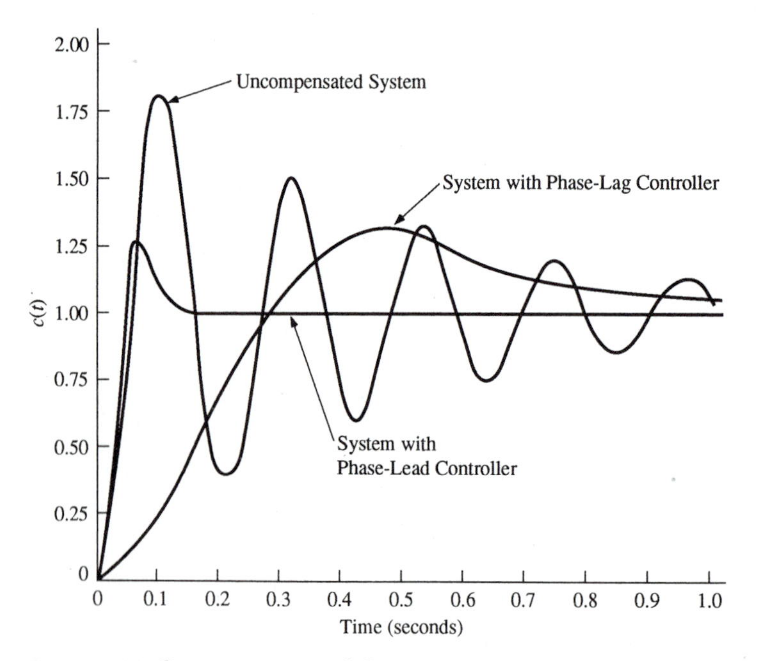

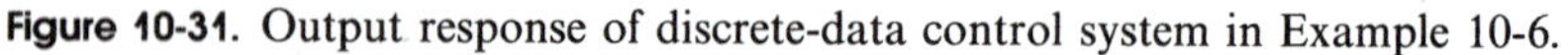
Figure 10-31. Output response of discrete-data control system in Example 10-6.

45 degrees is obtained if the gain crossover is moved from 0.12 to 0.04 on the ω_r axis, if the phase curve is not affected by the phase-lag controller $D(r)$. This requires an attenuation of -20 dB for the open-loop transfer function. However, since a phase lag accompanies the phase-lag controller which will degrade the final phase margin, we arbitrarily select the desired attenuation to be -21 dB. From Fig. 10-28, the corresponding value of α is

$$\alpha = 10^{-21/20} = 0.089 \tag{10-86}$$

This places the new gain crossover at $\omega_r = 0.034$. Now, setting the value of $1/\alpha\tau$ at 1/10 of 0.034, we have $1/\alpha\tau = 0.0034$. The value of $1/\tau$ is determined to be 0.0003. Thus, the transfer function of the digital controller in the r-domain is

$$D(r) = \frac{1 + \alpha\tau s}{1 + \tau s} = \frac{1 + 294.11r}{1 + 3333.33r} \tag{10-87}$$

The transfer function of the digital controller in the z-domain is

$$D(z) = 0.0872 \frac{z - 0.9931}{z - 0.9994} \tag{10-88}$$

Figure 10-29 shows the Bode diagram of the compensated system with the phase-lag controller. Figure 10-30 shows the gain-phase plot of the open-loop transfer function with the phase-lead controller. The peak resonance is reduced to 1.3. The bandwidth is now 11.66 rad/s. The corresponding unit-step response in Fig. 10-31 shows that although the maximum overshoot is reduced to 27 percent, the phase-lag controller gives a much longer rise time, and the step response approaches the final value of unity very slowly.

Design Using the w-Transformation

We shall illustrate the frequency-domain design of the same system using the w-transformation. Applying the w-transformation of Eq. (8-43), $z = (2 + wT)/(2 - wT)$, $T = 0.01$ s, to Eq. (10-81), we have

$$G_{h0}G_p(w) = \frac{99.93(1 + 0.000074w)(1 - 0.005w)}{w(1 + 0.11265w)} \tag{10-89}$$

The Bode plot of $G_{h0}G_p(w)$ is shown in Fig. 10-32. The same gain margin of 6.15 dB and phase margin of 8.63 degrees are shown for the uncompensated system.

Applying the usual frequency-domain design techniques, the single-stage phase-lag and phase-lead controllers are designed in the w-domain for a 45-degree phase margin requirement. The results are

1. Phase-Lead Controller for $D(w)$

$$D(w) = \frac{1 + 0.0538w}{1 + 0.00794w} \tag{10-90}$$

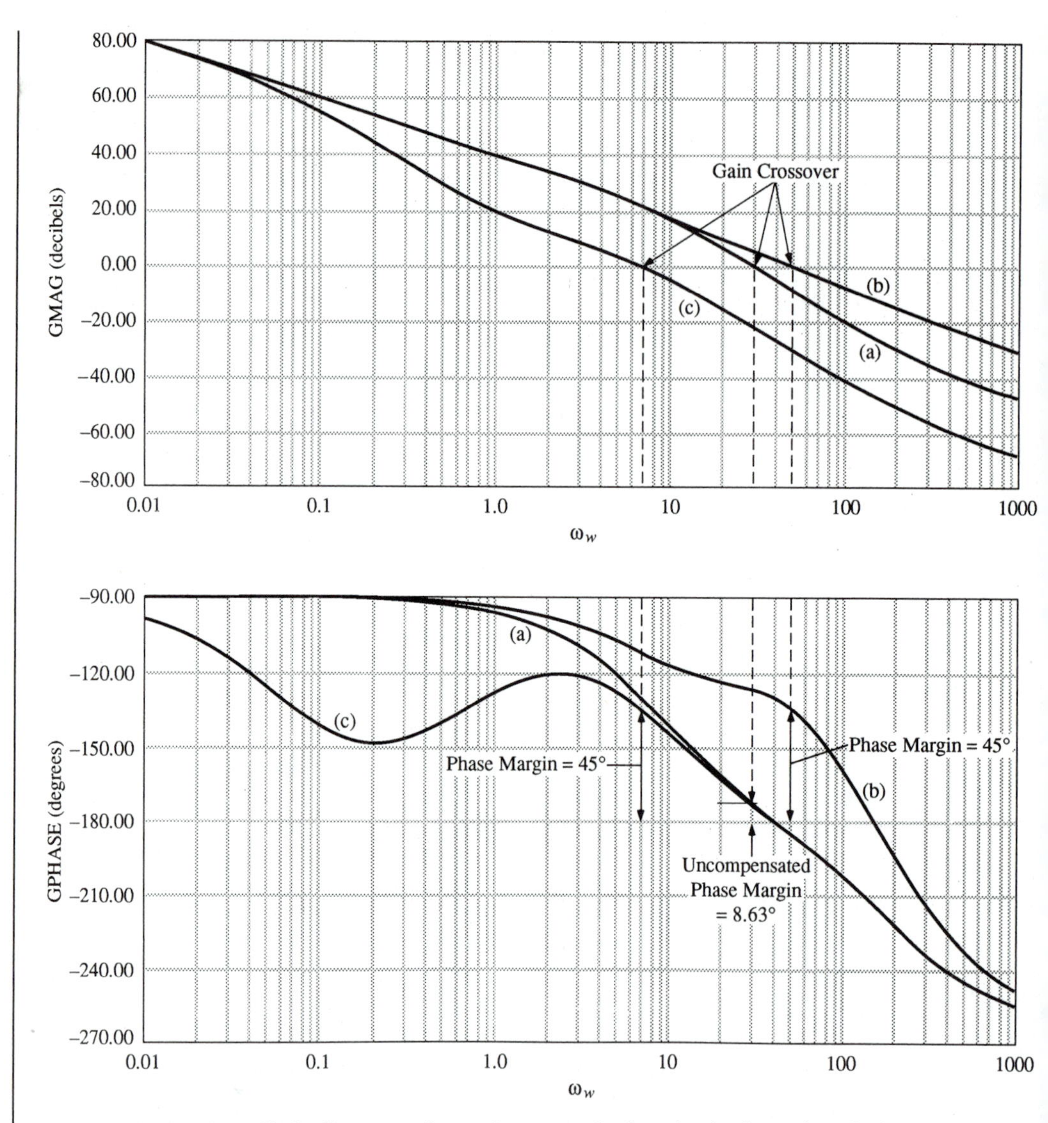

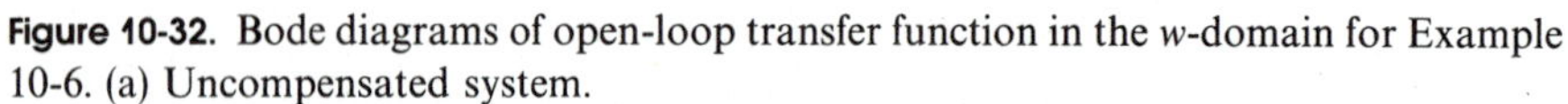

Figure 10-32. Bode diagrams of open-loop transfer function in the w-domain for Example 10-6. (a) Uncompensated system.

$$G_{h0}G_p(w) = \frac{99.93(1 + 0.000074w)(1 - 0.005w)}{w(1 + 0.11265w)}.$$

(b) With phase-lead controller.

$$D(w) = \frac{1 + 0.0538w}{1 + 0.00794w}.$$

(c) With phase-lag controller.

$$D(w) = \frac{1 + 1.45w}{1 + 16.683w}.$$

2. Phase-Lag Controller for $D(w)$

$$D(w) = \frac{1 + 1.45w}{1 + 16.683w} \tag{10-91}$$

Applying the w-transformation relation, we can show that Eqs. (10-90) and (10-91) essentially lead to the same z-domain transfer functions for $D(z)$ in Eqs. (10-85) and (10-88), respectively. The Bode diagrams of the compensated system transfer functions are shown in Fig. 10-32. As shown, the desired phase margin of 45 degrees is realized.

Example 10-7

As another illustrative example on the design of discrete-data control systems with the r-transformation and w-transformation, let us consider the system shown in Fig. 10-33 which models the idle-speed control system of an automobile. The sampling period of the system is 0.1 s.

The z-transform of the open-loop transfer function is

$$G_{h0}G_p(z) = \frac{0.0125(z + 0.195)(z + 2.821)}{z(z - 1)(z - 0.368)(z - 0.8187)} \tag{10-92}$$

Design Using the r-Transformation

Applying the r-transformation to the last expression, we have

$$G_{h0}G_p(r) = \frac{0.25(1 - r)^2(1 + 0.6736r)(1 - 0.4766r)}{r(1 + 2.1646r)(1 + 10.03r)(1 + r)} \tag{10-93}$$

Figure 10-34 shows the Bode diagram of the last equation. The system is unstable with a phase margin of -4.03 degrees and gain margin of -0.93 dB. The gain-phase plot is shown in Fig. 10-35.

1. Phase-Lag Controller

Since the uncompensated system in unstable and the phase curve has a steep negative slope beyond the gain crossover, a phase-lead controller will not be effective. We can show that a single-stage phase-lead controller of the form of Eq. (10-78) can achieve a maximum phase margin of only 6.6 degrees. Let us attempt a phase-lag controller design, with the objective of improving the

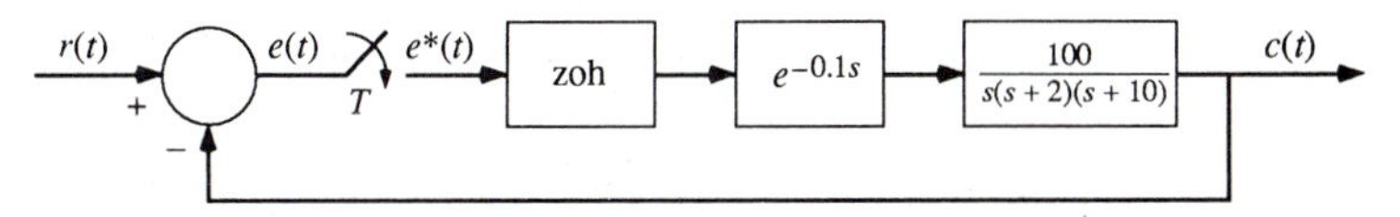

Figure 10-33. Discrete-data control system in Example 10-7.

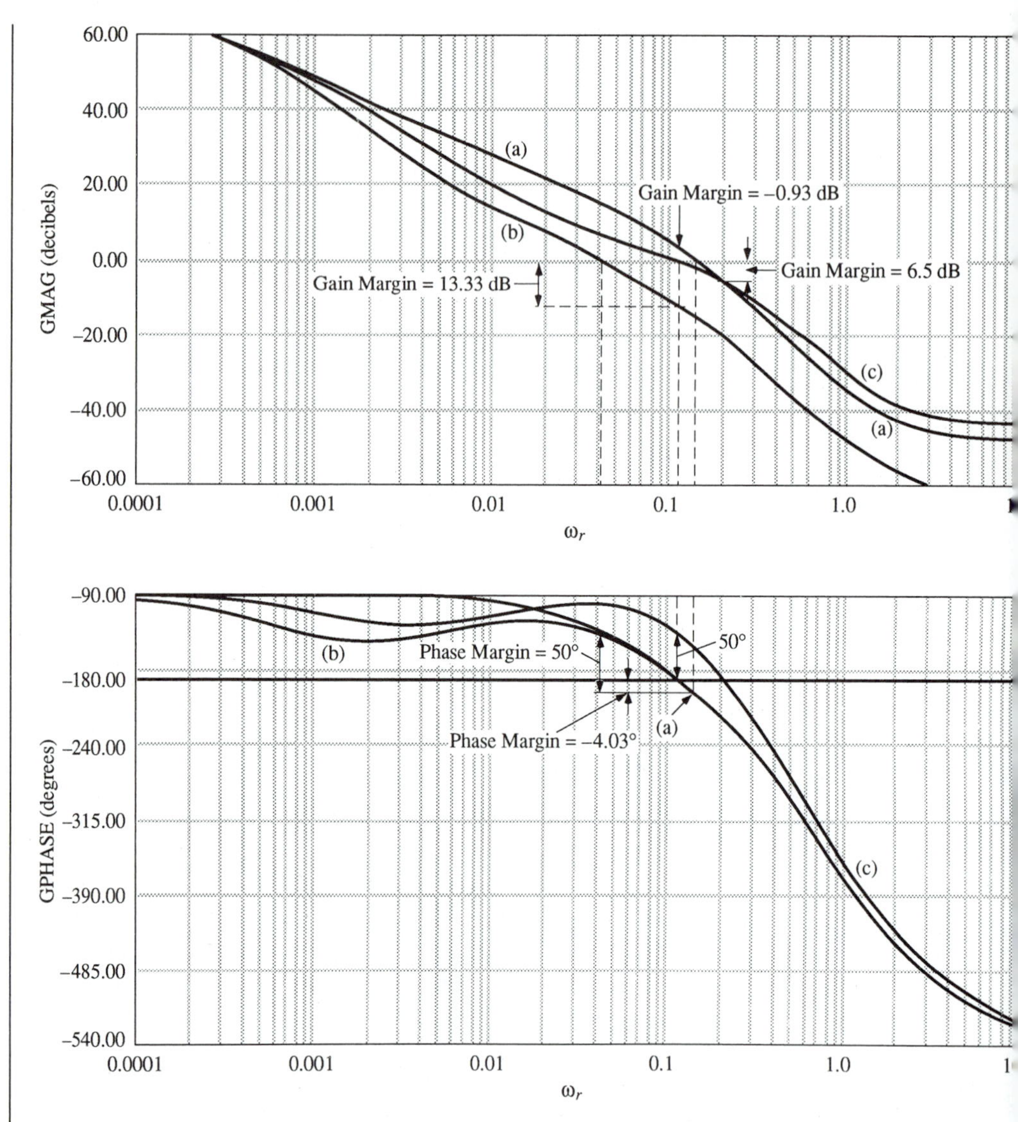

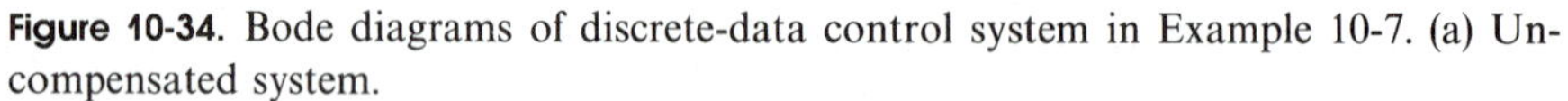

Figure 10-34. Bode diagrams of discrete-data control system in Example 10-7. (a) Uncompensated system.

$$G_p(s) = \frac{100e^{-0.1s}}{s(s+2)(s+10)}.$$

(b) With phase-lag controller.

$$D(r) = \frac{1 + 233.94r}{1 + 1260.37r}.$$

(c) With lag-lead controller.

$$D(r) = \frac{(1 + 155.546r)(1 + 20.39r)}{(1 + 507.62r)(1 + 3.6365r)}.$$

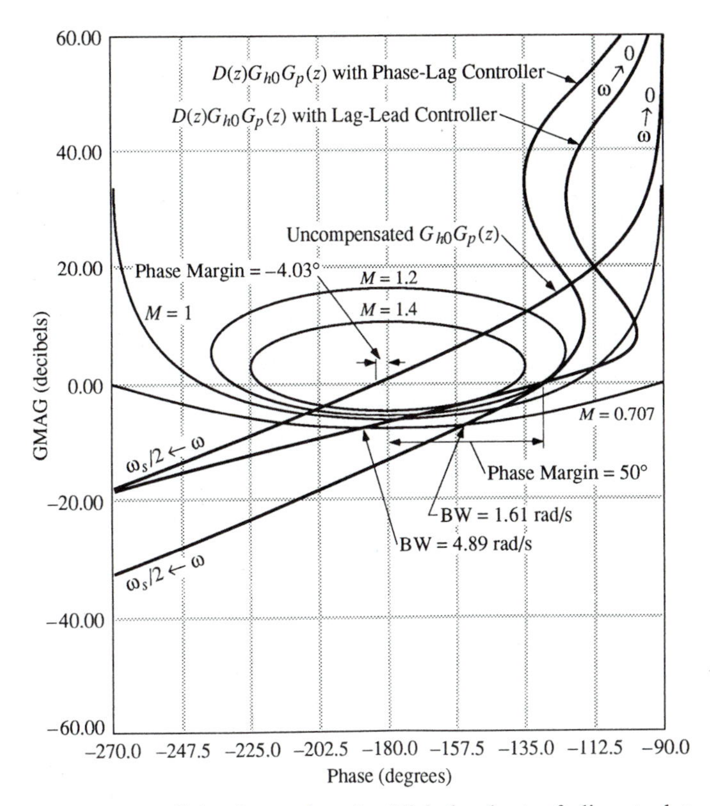

Figure 10-35. Gain-phase plots in Nichols chart of discrete-data control system in Example 10-7.

phase margin to 50 degrees. From the Bode diagram we see that a phase margin of 50 degrees is achieved if the gain crossover is moved to $\omega_r = 0.0427$. The magnitude of $G_{h0}G_p(j\omega_r)$ at this point is 14.628 dB. Thus, the phase-lag controller should have an attenuation of -14.628 dB at high ω_r. Thus,

$$\alpha = 10^{-14.628/20} = 0.1856 \tag{10-94}$$

Setting $1/\alpha\tau = 0.00427$, which is one-tenth of the new gain crossover, we have $\tau = 1260.37$. Thus, the transfer function of the phase-lag controller in the r-domain is

$$D(r) = \frac{1 + \alpha\tau r}{1 + \tau r} = \frac{1 + 233.94r}{1 + 1260.37r} \tag{10-95}$$

In the z-domain, the phase-lag controller is described by

$$D(z) = 0.18626\,\frac{z - 0.9915}{z - 0.9984} \tag{10-96}$$

Figure 10-34 shows the Bode diagram of the phase-lag compensated system in the r-domain, and the gain-phase plot is given in Fig. 10-35 in terms

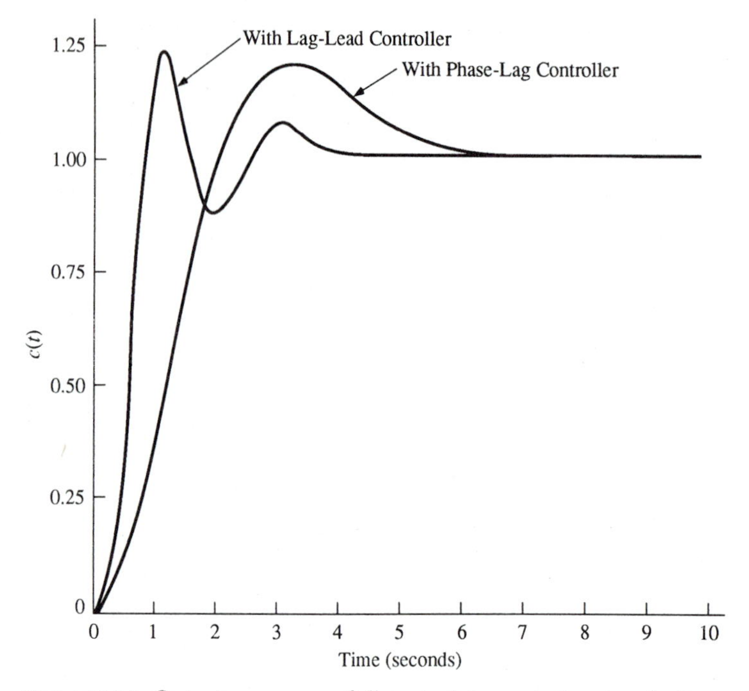

Figure 10-36. Output responses of discrete-data control system in Example 10-7.

of ω. As seen, the phase-lag compensated system has an M_p of 1.18, and bandwidth = 1.61 rad/s. The unit-step response of the system is shown in Fig. 10-36. As in the last example, the phase-lag controller stabilizes the system, but at the expense of a longer rise time and settling time.

2. Lag-Lead Controller

We can use a controller with the combination of the phase-lead and phase-lag effects. This is called the lag-lead or lead-lag controller [1]. By first realizing a phase margin of 35 degrees with a phase-lag controller and then the remaining 15 degrees with a phase-lead stage, the transfer function of a lag-lead controller is

$$D(r) = \frac{(1 + 155.546r)(1 + 20.39r)}{(1 + 507.623r)(1 + 3.6365r)} \tag{10-97}$$

The Bode diagram and the gain-phase plots of the system with the lag-lead controller are shown in Figs. 10-34 and 10-35, respectively. Figure 10-36 shows that the rise and settling times of the system are substantially improved.

The transfer function of the lag-lead controller in the z-domain is

$$D(z) = 1.4199\,\frac{(z - 0.9872)(z - 0.9065)}{(z - 0.9961)(z - 0.5686)} \tag{10-98}$$

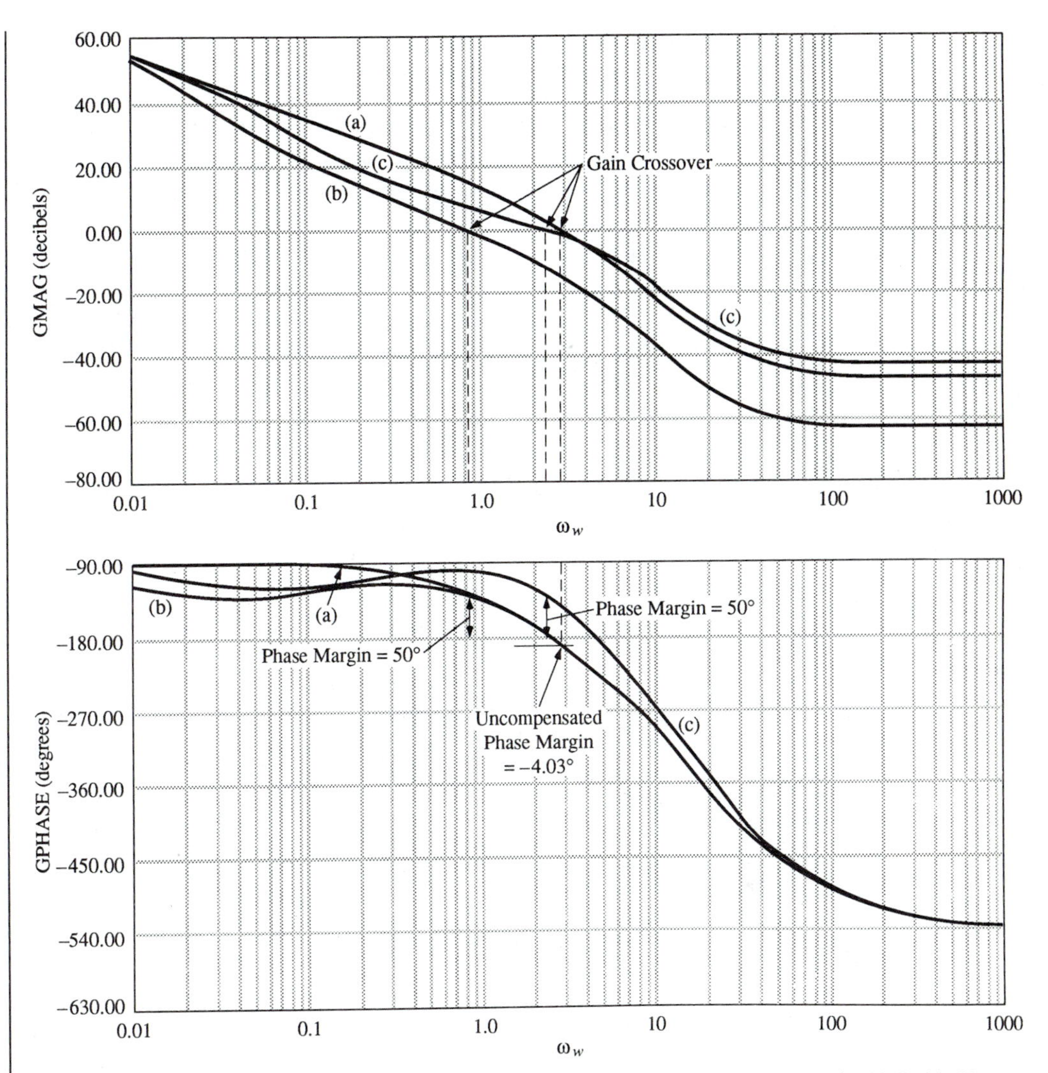

Figure 10-37. Bode diagram of discrete-data control system in Example 10-7. (a) Uncompensated system.

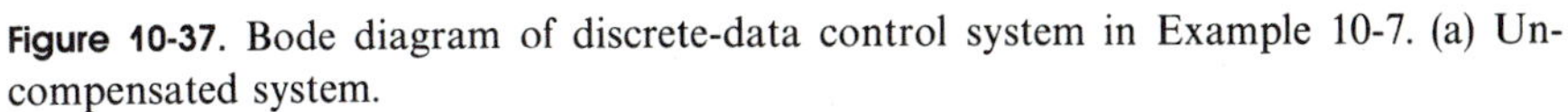

$$G_p(s) = \frac{100}{s(s+1)(s+2)} e^{-0.1s}.$$

(b) With phase-lag controller.

$$D(w) = \frac{1 + 11.698w}{1 + 62.985w}.$$

(c) With lag-lead controller.

$$D(w) = \frac{(1 + 7.778w)(1 + 1.0192w)}{(1 + 25.367w)(1 + 0.1819w)}.$$

Design Using the *w*-Transformation

The design in this example can be carried out using the *w*-transformation of Eq. (8-43). Applying $z = (20 + w)/(20 - w)$ to Eq. (10-92), we have

$$G_{h0}G_p(w) = \frac{4.997(1 + 0.03368w)(1 - 0.02383w)(1 - 0.05w)^2}{w(1 + 0.05w)(1 + 0.1082w)(1 + 0.5016w)} \tag{10-99}$$

The Bode plot of $G_{h0}G_p(w)$ is shown in Fig. 10-37. As shown, the gain and phase margins of the uncompensated system are -0.93 dB and -0.403 degrees, respectively.

Repeating the phase-lag controller and the lag-lead controller designs in the *w*-domain for a phase margin of 50 degrees, the following results are obtained.

1. Phase-Lag Controller

$$D(w) = \frac{1 + 11.698w}{1 + 62.985w} \tag{10-100}$$

2. Lag-Lead Controller

$$D(w) = \frac{(1 + 7.778w)(1 + 1.0192w)}{(1 + 25.367w)(1 + 0.1819w)} \tag{10-101}$$

The lag-lead controller is designed with 35 degrees contributed from the phase-lag section and 15 degrees from the phase-lead section.

We can show that the transfer functions in Eqs. (10-100) and (10-101) lead to essentially the same transfer functions in Eqs. (10-96) and (10-98), respectively, in the *z*-domain.

The Bode plots of the compensated systems are shown in Fig. 10-37.

10-6 DESIGN IN THE *z*-PLANE USING THE ROOT-LOCUS DIAGRAM

In Chapter 7 the root-locus technique was extended to the analysis of discrete-data control systems. It was shown that the properties of the construction of the root loci in the *z*-plane are essentially the same as those for the root loci of continuous-data systems which are drawn in the *s*-plane. However, in interpreting the system performance from the *z*-plane roots, the unit circle $|z| = 1$ is the stability boundary, and all information on relative stability should be evaluated accordingly. One may regard the design of a control system by the root-locus method as a pole-placement problem solved by trial-and-error. In other words, the root-locus design essentially involves the determination of the system and controller parameters so that the roots of the characteristic equation are at the desired locations. For systems with an order higher than third, it is generally very difficult to establish the relation between the controller parameters and the characteristic equation roots. Furthermore, the conventional root-locus diagram allows only one parameter to vary at a given time. Therefore, the design of a discrete-data system in the z-plane using a root-locus diagram is essentially a trial-and-error method. Alternatively, the designer may rely on the digital computer to plot out a large number of root loci by scanning through a wide range of possible values of the controller parameters, and select the best solution. However, the experienced designer can still make proper and intelligent initial "guesses" so that the amount of trial-and-error effort is kept to a minimum. Therefore, it is useful to investigate the effect of the various pole-zero configurations of the digital controller on the overall system performance and the characteristic equation roots.

10-6-1 Phase-Lead and Phase-Lag Controllers

Let a first-order digital controller be described by the transfer function,

$$D(z) = K_c \frac{z - z_1}{z - p_1} \tag{10-102}$$

where z_1 is a real zero and p_1 is a real pole. For practical design reasons, we limit z_1 and p_1 to being inside the unit circle $|z| = 1$. In general, if the digital controller is not to affect the steady-state performance of the system, we set

$$\lim_{z \to 1} D(z) = 1 \tag{10-103}$$

Thus, K_c in Eq. (10-102) must be set to

$$K_c = \frac{1 - p_1}{1 - z_1} \tag{10-104}$$

We can classify $D(z)$ as a low-pass or high-pass controller, depending on the relative magnitudes of z_1 and p_1. Substituting $z = e^{Ts} \cong 1 + Ts$ in Eq. (10-102), the equation can be approximated as

$$D^*(s) \cong K_c \frac{s + (1 - z_1)/T}{s + (1 - p_1)/T} \tag{10-105}$$

If the magnitudes of z_1 and p_1 are less than one, as required earlier, the zero and pole of $D^*(s)$ are necessarily in the left half of the s-plane. Thus, according to the classification of continuous-data controllers, the following classification for $D(z)$ can be made.

Phase-Lag (Low Pass)

$$(1 - z_1)/T > (1 - p_1)/T \qquad \text{or} \qquad p_1 > z_1 \tag{10-106}$$

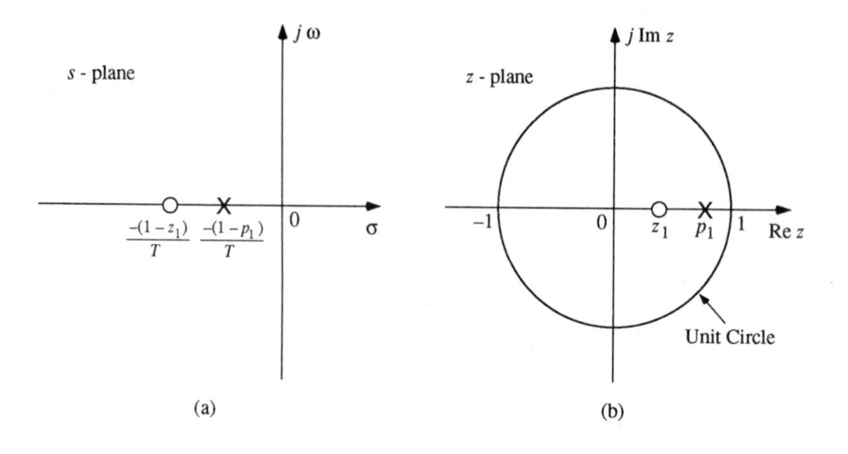

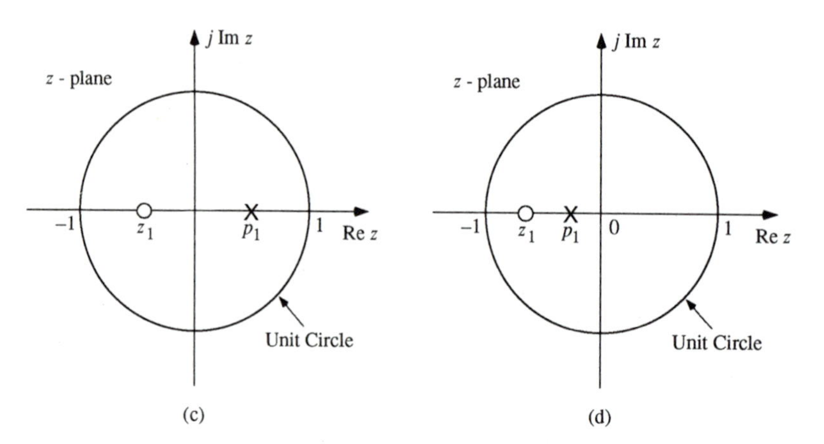

Figure 10-38. (a) Pole-zero configuration in the s-plane of the phase-lag controller in Eq. (10-105). (b), (c), (d) Pole-zero configuration in the z-plane of the phase-lag controller in Eq. (10-102).

Phase-Lead (High Pass)

$$(1 - z_1)/T < (1 - p_1)/T \qquad \text{or} \qquad p_1 < z_1 \tag{10-107}$$

Typical pole-zero configurations in the s-plane and the z-plane for a phase-lag digital controller based on the transfer functions in Eqs. (10-105) and (10-102), respectively, are shown in Fig. 10-38. Similarly, Fig. 10-39 shows the pole-zero configurations for phase-lead controllers.

The problem of designing discrete-data controller systems in the z-plane involves essentially the selection of a controller configuration and the placing of their poles and zeros in such a way that the design objectives are achieved.

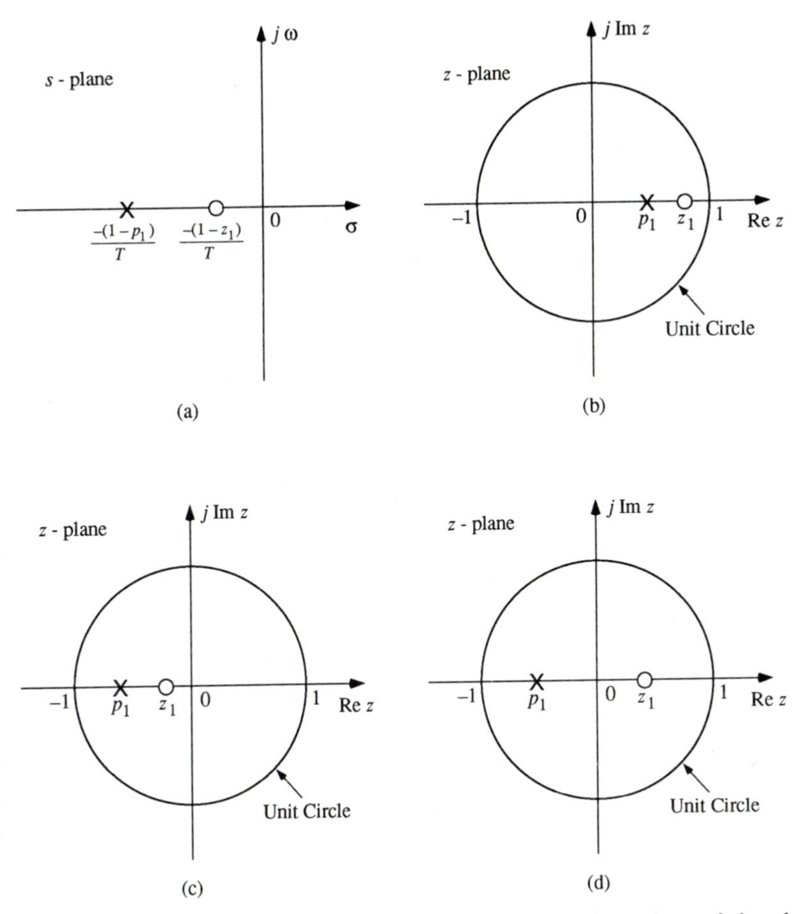

Figure 10-39. (a) Pole-zero configuration in the s-plane of the phase-lead controller in Eq. (10-105). (b), (c), (d) Pole-zero configuration in the z-plane of the phase-lead controller in Eq. (10-102).

10-6-2 The Digital PID Controller

The digital PID controller discussed in Sec. 10-4-3 is represented by the following transfer function,

$$D(z) = K_P + D_I(z) + D_D(z) \tag{10-108}$$

where K_P is the constant of proportional control, $D_I(z)$ is the integral control which can be modeled by one of the transfer functions in Eqs. (10-71) through (10-73), and $D_D(z)$ is the derivative control modeled in Eq. (10-75). Using the three rectangular integration schemes, the transfer functions of the digital PID controller are summarized as follows.

Backward-Rectangular Integration

$$D(z) = \frac{(K_P T + K_D)z^2 + (K_I T^2 - K_P T - 2K_D)z + K_D}{Tz(z-1)} \tag{10-109}$$

Forward-Rectangular Integration

$$D(z) = \frac{(K_P T + K_D + K_I T^2)z^2 - (K_P T + 2K_D)z + K_D}{Tz(z-1)} \tag{10-110}$$

Bilinear-Transformation Integration

$$D(z) = \frac{(2K_P T + K_I T^2 + 2K_D)z^2 + (K_I T^2 - 2K_P T - 4K_D)z + 2K_D}{2Tz(z-1)} \tag{10-111}$$

Thus, the digital PID controller has a pole at $z = 0$ and one at $z = 1$. There are two zeros which can be real or in complex conjugate pairs.

10-6-3 The Digital PD Controller

Depending on the design requirements, frequently only the proportional and derivative components of the PID controller are needed. Setting $K_I = 0$ in any one of the transfer functions in Eqs. (10-109) through (10-111), we get the transfer function of the digital PD controller,

$$D(z) = \frac{(K_P T + K_D)z - K_D}{Tz} \tag{10-112}$$

Thus, the digital PD controller has a pole at $z = 0$ and a zero at $K_D/(K_P T + K_D)$ which lies on the positive real axis inside the unit circle. Comparing Eq. (10-112) with Eq. (10-102), we see that the PD controller is a special case of the phase-lead controller of Eq. (10-102) with $K_c = (K_P T + K_D)/T$, $z_1 = K_D/(K_P T + K_D)$, and $p_1 = 0$.

10-6-4 The Digital PI Controller

Under certain conditions using only the proportional and integral components of the PID controller is adequate for design purposes. Setting $K_D = 0$ in the transfer functions in Eqs. (10-109) through (10-111) for the three cases of integration, we get the following results.

Backward-Rectangular Integration

$$D(z) = \frac{K_P z - (K_P - K_I T)}{z - 1} \tag{10-113}$$

Apparently this PI controller is a special case of the phase-lag controller modeled in Eq. (10-102) with $K_c = K_P$, $z_1 = (K_P - K_I T)/K_P$, and $p_1 = 1$. The zero z_1 can lie anywhere on the real axis inside the unit circle.

Forward-Rectangular Integration

$$D(z) = \frac{(K_P + K_I T)z - K_P}{z - 1} \tag{10-114}$$

This PI controller is a special case of the phase-lag controller modeled in Eq. (10-102) with $K_c = K_P + K_I T$, $p_1 = 1$, and $z_1 = K_P/(K_P + K_I T)$ which lies on the positive real axis inside the unit circle.

Bilinear-Transformation Integration

$$D(z) = \frac{(2K_P + K_I T)z + (K_I T - 2K_P)}{2(z - 1)} \tag{10-115}$$

Comparing the last equation with the phase-lag controller in Eq. (10-102), we have $K_c = (2K_P + K_I T)/2$, $p_1 = 1$, and $z_1 = (K_I T - 2K_P)/(K_I T + 2K_P)$, which lies inside on the real axis inside the unit circle.

10-6-5 Pole-Zero Cancellation Design

In the design of control systems using the s-plane or the z-plane, a common practice has been to attempt to cancel the undesirable poles and zeros of the controlled process transfer function by poles and zeros of the controller, and new open-loop poles and zeros are added at more advantageous locations to satisfy the design specifications. The cancellation compensation is used often in systems that have complex poles in the controlled process and a severe stability problem. Cancelling the undesirable poles and adding new ones at any place one wishes would seem the simplest control strategy. However, keep in mind that the pole-zero cancellation compensation scheme does not always provide a satisfactory solution to a broad class of design problems. The resulting controller could involve excessive complexity; and if the undesirable poles of the controlled process are close to the unit circle in the z-plane, inexact cancellation, which almost always occurs in practice, could result in a conditionally stable system. As a simple example to illustrate the effect of inexact cancellation, let us refer to the root-locus diagram shown in Fig. 10-40. In Fig. 10-40(a), the pole-zero configuration of the open-loop transfer function of a certain discrete-data system is shown. A digital controller with the transfer function

$$D(z) = K_c \frac{(z - p_1)(z - p_2)}{(z - a)(z - b)} \tag{10-116}$$

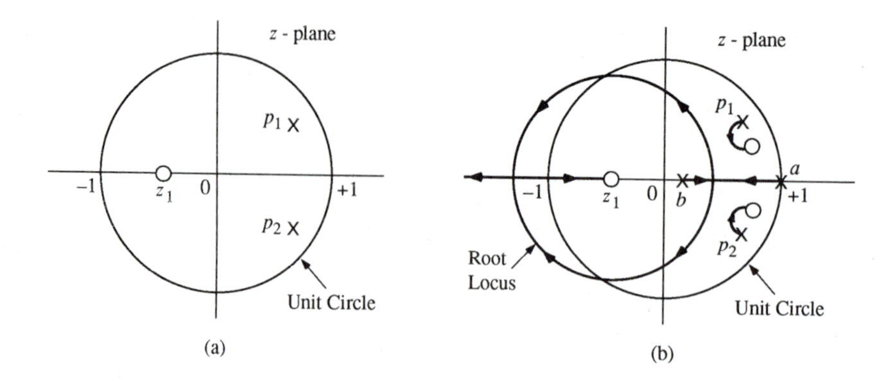

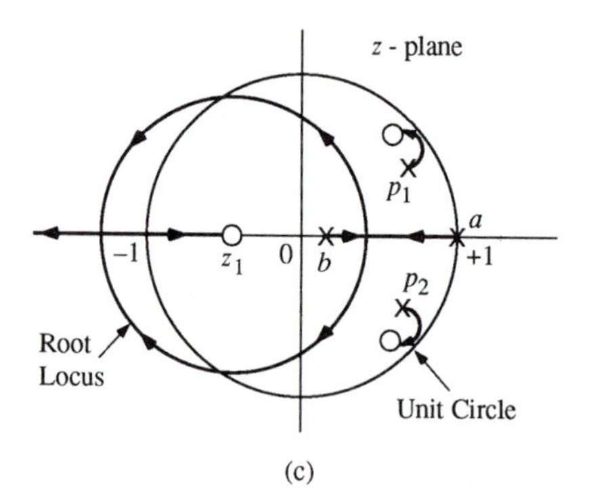

Figure 10-40. Root-locus diagrams illustrating the effects of inexact cancellation of undesirable poles.

is devised to cancel the complex poles p_1 and p_2 and to add new poles at $z = a$ and $z = b$. Figure 10-40(b) shows an assumed case of inexact cancellation, where the complex zeros of the controller are not exactly equal to the poles p_1 and p_2 of the controlled process. In this case, the relative positions of the poles and zeros involved are such that the inexact cancellation does not have any adverse effect on the system performance, since the closed-loop poles near p_1 and p_2 are stable and are very close to the two zeros of the controller, so that the transient responses due to these closed-loop poles would have negligible magnitudes. On the other hand, if the poles and zeros of the inexact cancellation are arranged as shown in Fig. 10-40(c), a portion of the root loci between the uncancelled pole and zero pairs may lie outside the unit circle. The system is called a *conditionally stable system*, since it is stable only for low and high values of the loop gain. Thus, in applying pole-zero cancellation in control system design, inexact cancellation is almost inevitable, and care must be taken so that the inexact cancellation does not result in conditionally stable systems.

10-6-6 Design Examples

Example 10-8

Consider the discrete-data control system described in Example 10-6. The open-loop z-transfer function with K unspecified is

$$G_{h0}G_p(z) = \frac{0.00192K(z + 0.9708)}{(z - 1)(z - 0.915)} \tag{10-117}$$

The root loci of the characteristic equation are plotted as shown in Fig. 10-41. Only the portion relevant to system stability is shown. The critical value of K for stability is 45.6, and when $K = 22.5$, which corresponds to a ramp-error constant of 100, the roots are at $0.936 + j0.285$ and $0.936 - j0.285$. It was indicated in Example 10-6 that, when $K = 22.5$, the system has a very low stability margin, and the maximum overshoot of the unit-step response is 78.8 percent.

Phase-Lead Controller Design

Let us start with the phase-lead controller designed in Example 10-6 which yielded a phase margin of 45 degrees. The transfer function of $D(z)$ is

$$D(z) = 4.5455\frac{z - 0.83}{z - 0.2274} \tag{10-118}$$

Comparing the last equation with Eq. (10-102), we have $K_c = 4.5455$, $z_1 = 0.83$, and $p_1 = 0.2274$. The pole-zero configuration of $D(z)$ is that of Fig. 10-39(b). The value of K_c is chosen so that Eq. (10-103) is satisfied. The root loci of the system with the controller in Eq. (10-118) are shown in Fig. 10-41. The marginal value of K for stability is improved to 87.5, and as indicated in Example 10-6, the maximum overshoot is reduced to 27.8 percent. The root loci of the compensated system in Fig. 10-41 show that the effect of the zero $z_1 = 0.83$ of $D(z)$ is to pull the complex portion of the root loci to the left in the z-plane, thus improving relative stability. The design carried out in Example 10-6 in the frequency domain was to achieve a phase margin of 45 degrees. Certainly, the system performance can be improved further by rearranging the pole and zero of $D(z)$. Without elaboration, we can rule out the pole-zero configuration of $D(z)$ shown in Fig. 10-39(c), since the zero z_1 is too far to the left. The situation shown in Fig. 10-39(d) may render further improvement, since the general effect of the pole at p_1 is to push the root loci to the right; thus, placing it in the left half plane may be preferable. Figure 10-41 shows the root loci when $K_c = 8.8286$, $z_1 = 0.83$, and $p_1 = -0.5$. The marginal value of K is now 94, and when $K = 22.5$, the three roots of the characteristic equation are all real: -0.328, 0.606, and 0.756. The maximum overshoot is reduced to 10.8 percent.

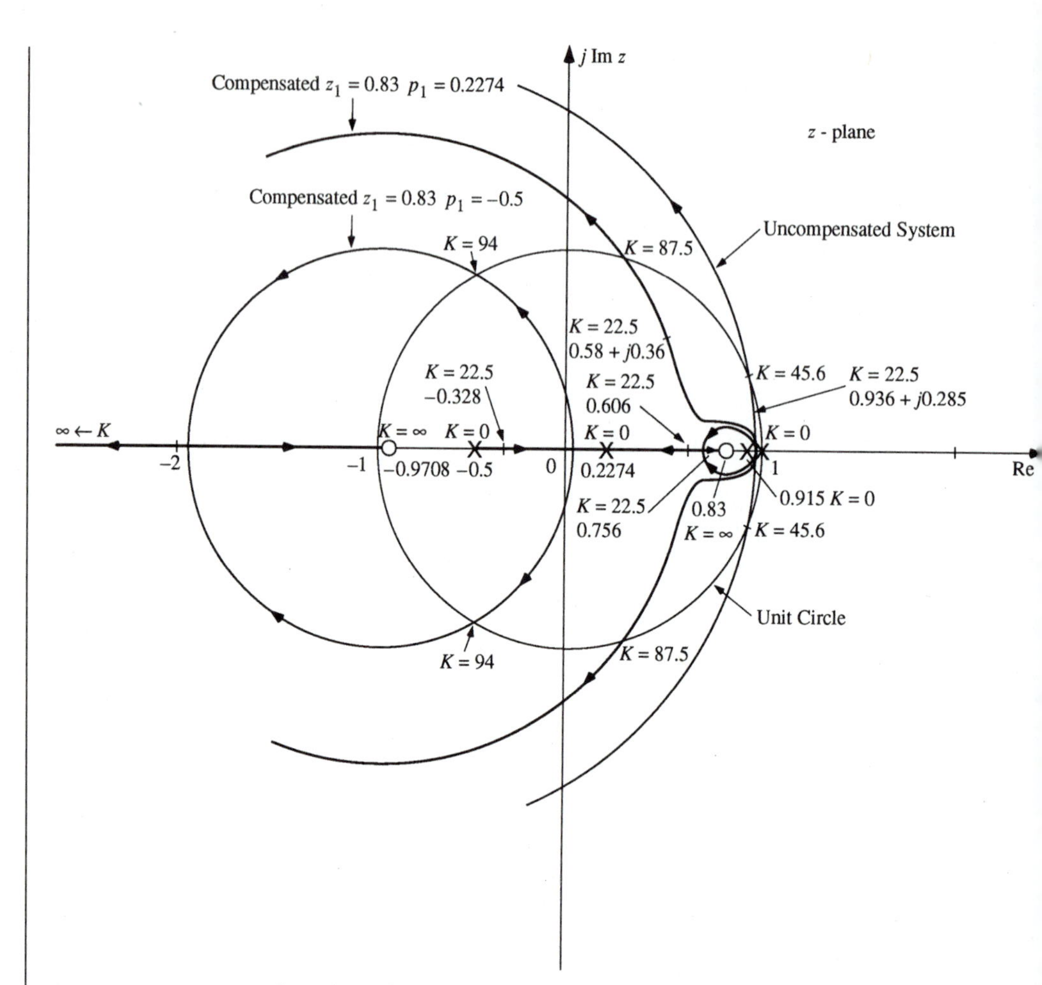

Figure 10-41. Root loci of discrete-data control system in Example 10-8.

$$G_{h0}G_p(z) = \frac{0.00192K(z + 0.9708)}{(z - 1)(z - 0.915)}.$$

Phase-lead controllers:

$$D(z) = \frac{4.5455(z - 0.83)}{z - 0.2274} \quad \text{and} \quad D(z) = \frac{8.8236(z - 0.83)}{z + 0.5}.$$

It should be pointed out that the effects of moving the pole and zero of $D(z)$ on stability are not always clear-cut since the value of K_c must also be altered accordingly to satisfy Eq. (10-103).

Table 10-2 summarizes some of the results of the phase-lead controller with varied controller parameters. The results indicate the general effects of the values of z_1 and p_1, but the study is by no means exhaustive. The choice of

Table 10-2 Phase-Lead Controller with Varied Parameters

K_c	z_1	p_1	Marginal K for Stability	Closed-Loop Poles at $K = 22.5$	c_{max}
1.00	0	0	45.6	$0.936 \pm j0.285$	1.788
4.5455	0.83	0.2274	87.5	$0.785, 0.581 \pm j0.36$	1.278
5.8824	0.83	0	90.4	$0.775, 0.443 \pm j0.261$	1.190
8.8236	0.83	−0.5	94.0	−0.328, 0.606, 0.756	1.108
11.593	0.83	−0.9708	94.06	0.681, 0.733	1.094
5.00	0.80	0	107.0	$0.349, 0.675 \pm j0.159$	1.214
6.6667	0.85	0	80.0	$0.825, 0.401 \pm j0.357$	1.190
10.00	0.90	0	53.6	$0.898, 0.292 \pm j0.579$	1.313

$z_1 = 0.83$ is probably near optimum from the maximum overshoot standpoint. Table 10-2 shows that, when z_1 is moved to either side of 0.83, the maximum overshoot increases, although when $z_1 = 0.8$ and $p_1 = 0$, the marginal value of K for stability is increased to 107.

Phase-Lag Controller Design

The phase-lag controller designed in Example 10-6 based on a phase margin of 45 degrees is

$$D(z) = K_c \frac{z - z_1}{z - p_1} \tag{10-119}$$

where $K_c = 0.0872$, $z_1 = 0.9931$, and $p_1 = 0.9994$. The pole-zero configuration of $D(z)$ in the last equation fits the situation shown in Fig. 10-38(b). In fact, p_1 and z_1 are very close to each other in this case and are close to unity. These characteristics are by no means accidental, and this pole-zero combination is known as a *dipole* phase-lag controller. Thus, effectively the phase-lag controller is equivalent to introducing a pure attenuation of K_c in the control loop. The open-loop transfer function of the compensated system with the controller described in Eq. (10-118) can be approximated as

$$D(z)G_{h0}G_p(z) \cong K_c G_{h0} G_p(z) = 0.0872 G_{h0} G_p(z) \tag{10-120}$$

Figure 10-42 shows the root loci in the neighborhood of $z = 1$ of the uncompensated and the compensated systems with the phase-lag controller. The dipole phase-lag controller creates a small loop near $z = 1$ and the shape of the important portion of the compensated root loci is essentially the same as that of the uncompensated system, except that the gains along the loci are different. Notice that when $K = 22.5$ for the compensated system, the two complex conjugate roots are at $0.956 + j0.0737$ and $0.956 - j0.0737$. For the uncompensated system, when $K = 1.962$ ($= 0.0872 \times 22.5$) the roots are at approximately the same place.

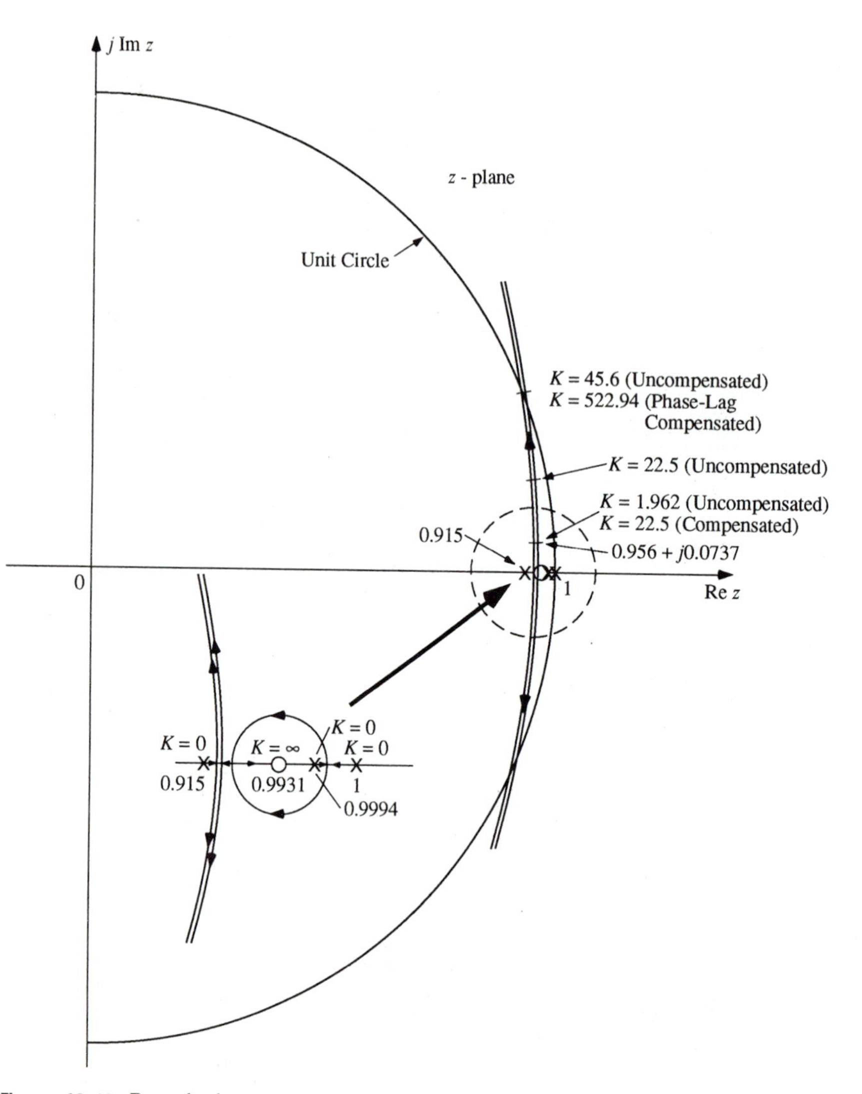

Figure 10-42. Root loci near $z = 1$ for the discrete-data system in Example 10-8 with a phase-lag digital controller.

From the root-locus design standpoint, the dipole phase-lag controller can be designed using the following steps.

1. Set the value of K based on the requirement of the error constant. Refer to this value of K as K_{old}.
2. Plot the root-locus diagram of the uncompensated system.
3. Find the value of K on the root-locus diagram where the relative stability requirement is satisfied. This value of K is referred to as K_{new}.

4. Set the gain of the phase-lag controller as

$$K_c = K_{new}/K_{old} \tag{10-121}$$

5. Set z_1 to be arbitrarily close to unity. The value of p_1 is determined from Eq. (10-104), or

$$p_1 = 1 - K_c(1 - z_1) \tag{10-122}$$

PD Controller Design

Since the process transfer function in Eq. (10-117) is of type 1, for the design requirement of $K_v = 100$, there is no need to bring in another pole at $z = 1$. Thus, a PD controller would be adequate for this system. Referring to Eq. (10-112), $T = 0.1$ s, we must set K_P so that Eq. (10-103) is satisfied; thus $K_P = 1$. Since the PD controller is just a special case of the phase-lead controller, we can select one of the designs listed in Table 10-2. Choosing $z_1 = 0.83$ and $p_1 = 0$, we have $K_D = 0.4882$, and the transfer function of the PD controller becomes

$$D(z) = \frac{5.882(z - 0.83)}{z} \tag{10-123}$$

Table 10-2 shows that, when $K = 22.5$, the closed-loop roots are at 0.775, and $0.443 \pm j0.261$, and the maximum overshoot is 19 percent.

PI Controller Design

If applied properly, the PI controller can be used to improve the steady-state performance by increasing the type of the system and, at the same time, improve the relative stability. This is achieved by using the dipole phase-lag principle discussed earlier. In the present case, the pole of the PI controller is fixed at $z = 1$, and the design strategy is to place the zero of $D(z)$ near the pole. Any one of the three PI controllers modeled in Eqs. (10-113) through (10-115) can be used, since the effective design restricts the zero to be in the right half z-plane. Let us select Eq. (10-113), which has the simplest form. Since the ramp-error constant K_v is now infinite, the value of K_P does not matter. Realistically, however, since the zero of $D(z)$ is to be placed very close to $z = 1$, the value of K_P should be chosen to yield a satisfactory relative stability.

Let us set the zero of $D(z)$ in Eq. (10-113) at 0.995, so that

$$D(z) = 0.002\,\frac{z - 0.995}{z - 1} \tag{10-124}$$

Thus, $K_P = 0.002$ and $K_I = 0.0001$. The discrete-data control system with the PI controller in Eq. (10-124) can be shown to have a maximum overshoot of 16 percent. However, the PI controller is still subject to long rise and settling times, so that its main function is to improve the steady-state error of the system.

PID Controller Design

The PID controller acts as a lag-lead controller, since it contains both phase-lead and phase-lag components. Let us consider the digital PID controller modeled by Eq. (10-109). With $T = 0.1$ s, $D(z)$ is written

$$D(z) = \frac{(0.1K_P + K_D)z^2 + (0.01K_I - 0.1K_P - 2K_D)z + K_D}{0.1z(z-1)} \tag{10-125}$$

We can set a zero of $D(z)$ at $z = 0.995$, together with the pole at $z = 1$, forming the phase-lag portion. The other zero of $D(z)$ can be placed between

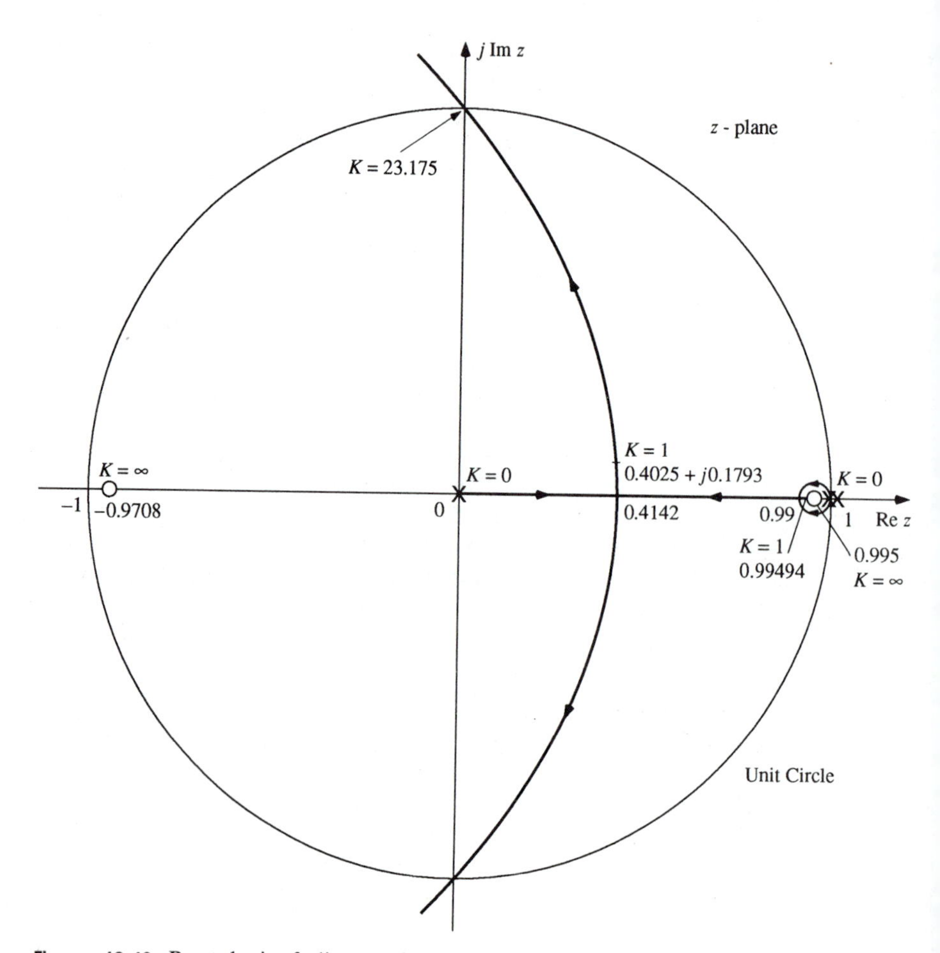

Figure 10-43. Root loci of discrete-data system in Example 10-8 with digital PID controller.

$$G_{h0}G_p(z) = \frac{0.00192K(z + 0.9708)}{(z-1)(z-0.915)} \quad \text{and} \quad D(z) = \frac{4.63(z-0.915)(z-0.995)}{z(z-1)}.$$

$z = 0$ and $z = 0.915$, say at $z = 0.83$ and, together with the pole at $z = 0$, form the phase-lead portion. The controller should improve the system with the PI controller designed earlier due to the added zero at $z = 0.83$ and the pole at $z = 0$. The gain constant of $D(z)$, which is $K_P + 10K_D$, should be selected so that the relative stability or the maximum overshoot is satisfied. Let the open-loop transfer function of the system with the PID controller be

$$D(z)G_{h0}G_p(z) = \frac{0.00192K(K_P + 10K_D)(z + 0.9708)(z - 0.995)}{z(z - 1)^2} \tag{10-126}$$

where we have chosen one zero of $D(z)$ to be at 0.915 so that it cancels the pole of the controlled process at the same point. The value of K can be set arbitrarily, since the open-loop gain can still be set by the value of K_P.

Figure 10-43 shows the root loci of the PID controller compensated system with $0.00192K(K_P + 10K_D)$ as the variable parameter. Notice that the significant portion of the complex root loci is not moved further to the left, on account of the cancellation of the open-loop pole at 0.915. We can show that, for minimum overshoot in the step response, there is an optimal value of the gain. For small values of the gain, the closed-loop roots are very close to the $z = 1$ point, and the overshoot can be excessive. By trial-and-error, the optimal value of $0.00192K(K_P + 10K_D)$ is 0.2. The maximum overshoot is 47 percent, and the peak time is 0.9 s.

Setting $K = 1$ arbitrarily, and equating Eq. (10-125) to the designed digital controller, we have $K_I = 0.426$, $K_P = 9.33$, and $K_D = 9.484$.

We can show that setting a zero of $D(z)$ not too close to $z = 1$, say at $z = 0.9$, while the other zero cancelling the pole of the process is at 0.915 would not be particularly effective. For instance, setting the zero at 0.9, the optimal gain is still around 0.2, but the best maximum overshoot that can be achieved is 23 percent. This is because the complex portion of the root loci will be pushed further to the left, making the design less effective.

Example 10-9

In this example the transfer function of the controlled process of the system shown in Fig. 10-26 is given as

$$G_p(s) = \frac{10}{(s + 1)(s + 2)} \tag{10-127}$$

The sampling period is 0.1 s. The open-loop transfer function of the uncompensated system is

$$G_{h0}G_p(z) = (1 - z^{-1})\mathscr{Z}\left(\frac{10}{s(s + 1)(s + 2)}\right)$$

$$= \frac{0.0453(z + 0.905)}{(z - 0.905)(z - 0.819)} \tag{10-128}$$

Since the uncompensated system is type 0, the steady-state error due to a step input is not zero. Solving for the step-error constant, we have

$$K_P^* = \lim_{z \to 1} G_{h0} G_p(z) = 5.0187 \quad \text{(10-129)}$$

Thus, for a unit-step input, the steady-state error is

$$e_{ss}^* = \frac{1}{1 + K_P^*} = \frac{1}{6.0197} = 0.166 \quad \text{(10-130)}$$

Let us assume that the performance specifications of the system are as follows:

1. Ramp-error constant $K_v^* \geq 5$.
2. Maximum overshoot of $c_{max}^* \leq 5$ percent.
3. T_{max} (time at which c_{max}^* occurs) ≤ 1 s.

PI Controller Design

To eliminate the steady-state error when the input is a step function, we need the integral control to increase the type of the system by one. Let us first apply only the PI control with the controller transfer function given in Eq. (10-113). The open-loop transfer function with the PI controller is

$$D(z)G_{h0}G_p(z) = \frac{0.0453[K_P z - (K_P - K_I T)](z + 0.905)}{(z - 1)(z - 0.905)(z - 0.819)} \quad \text{(10-131)}$$

Applying the requirement on K_v^*, we get

$$K_v^* = \frac{1}{T} \lim_{z \to 1} (z - 1)D(z)G_{h0}G_p(z) = 5.0187K_I \geq 5 \quad \text{(10-132)}$$

Thus, $K_I \geq 0.9963$. We let $K_I = 1$.

The effects of K_P on system performance can be investigated by using the root contour method [1]. With $K_I = 1$, the characteristic equation of the compensated system is

$$(z - 1)(z - 0.905)(z - 0.819) + 0.00453(z + 0.905) + 0.0453K_P(z - 1)(z + 0.905) = 0 \quad \text{(10-133)}$$

Dividing both sides of the last equation by the terms that do not contain K_P, we get

$$1 + G_{eq}(z) = 1 + \frac{0.0453K_P(z - 1)(z + 0.905)}{z^3 - 2.7236z^2 + 2.469z - 0.7367} = 0 \quad \text{(10-134)}$$

Thus, with K_P appearing only as a multiplying factor of $G_{eq}(z)$, we can construct the root contours of Eq. (10-133) using the pole-zero configuration of $G_{eq}(z)$, as shown in Fig. 10-44. The root contours show that the closed-loop system with the PI controller, $K_I = 1$, is unstable for $0.25 < K_P < 6$. However, it is apparent that the relative stability of the compensated system with $K_I = 1$ is

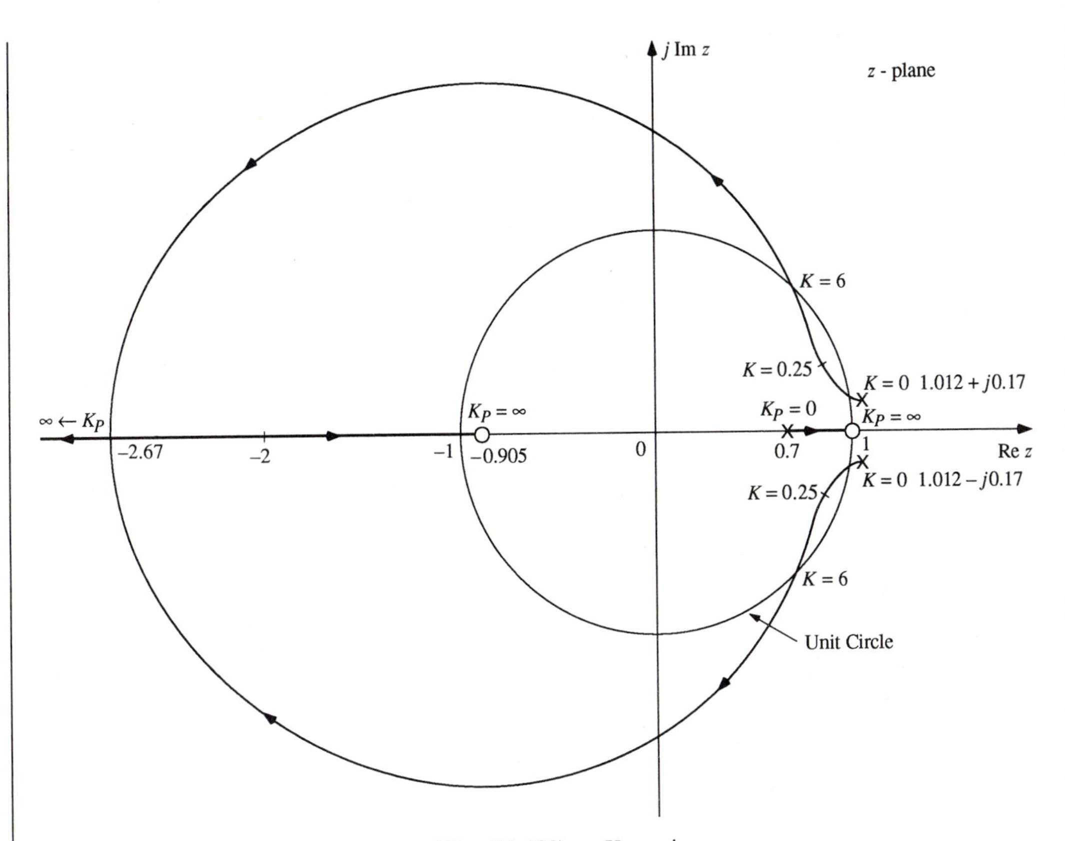

Figure 10-44. Root contours of Eq. (10-133) as K_P varies.

poor, since the portion of the root contours inside the unit circle does not dip down close enough to the real axis. The following tabulation shows that the best overshoot that can be obtained with various values of K_P is 1.464.

K_P	c_{max}	T_{max} (s)
1.5	1.476	9
1.4	1.473	10
1.0	1.464	12
0.9	1.476	12
0.5	1.617	16

Thus, while the PI controller does eliminate the steady-state error due to a step input, it cannot meet the requirements on K_v^* and T_{max}.

PID Controller Design

To eliminate the steady-state error and simultaneously realizing the requirements on K_v^* and $T_{\max}$, a PID controller is applied. Let the PID controller be described by the transfer function in Eq. (10-111). Now we have three unknown parameters in K_P, K_I, and K_D to be determined.

We can show that the ramp-error constant requirement still leads to the condition in Eq. (10-132). Thus, we can set $K_I = 1$. As a simple design trial, we can set the two zeros of $D(z)$ to cancel the poles of $G_{h0}G_p(z)$ at $z = 0.905$ and 0.819. This is,

$$z^2 + \frac{K_I T^2 - 2K_P T - 4K_D}{2K_P T + K_I T^2 + 2K_D} z + \frac{2K_D}{2K_P T + K_I T^2 + 2K_D} = (z - 0.905)(z - 0.819) \tag{10-135}$$

With $T = 0.1$ s, and $K_I = 1$, the values of K_P and K_D are found from the last equation to be $K_P = 1.454$ and $K_D = 0.431$. Substituting the controller parameters in Eq. (10-111), we have

$$D(z) = 5.811 \frac{(z - 0.905)(z - 0.819)}{z(z - 1)} \tag{10-136}$$

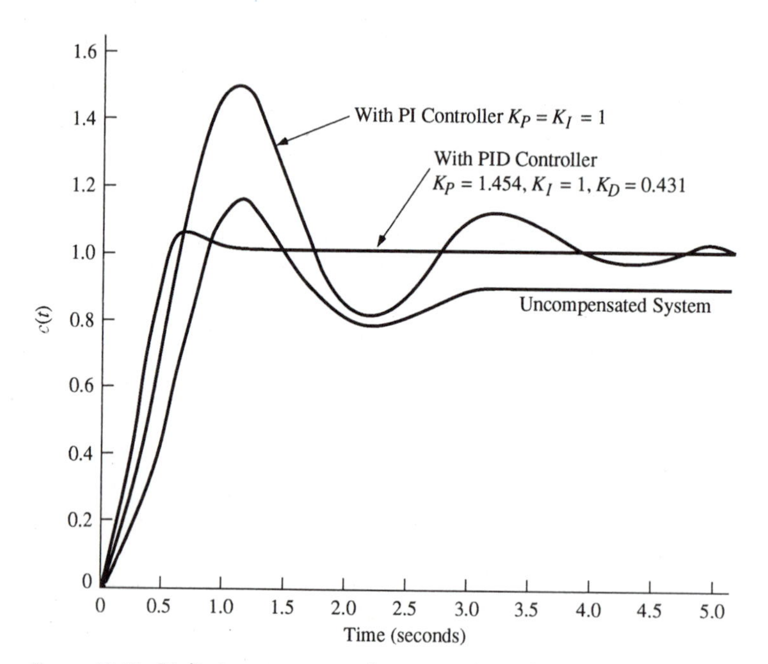

Figure 10-45. Unit-step responses of uncompensated and compensated systems in Example 10-9.

The open-loop transfer function of the system with the PID controller is

$$D(z)G_{h0}G_p(z) = \frac{0.263(z+0.905)}{z(z-1)} \tag{10-137}$$

Figure 10-45 shows the unit-step responses of the uncompensated system and the system with the PI and the PID controllers. With the PID controller designed to cancel the open-loop poles, the maximum overshoot is 3.8 percent and T_{max} is 0.6 s, which satisfy the transient response requirements.

Example 10-10

Consider that the transfer function of the controlled process of the system shown in Fig. 10-26 is given as

$$G_p(s) = \frac{5}{s^2+s+10} \tag{10-138}$$

The sampling period is 0.5 s. The open-loop transfer function of the uncompensated system is

$$G_{h0}G_p(z) = (1-z^{-1})\mathscr{Z}\left(\frac{5}{s(s^2+s+10)}\right)$$

$$= \frac{0.434(z+0.834)}{z^2-0.01487z+0.6065} \tag{10-139}$$

Since the poles of $G_{h0}G_p(z)$ are complex at $0.007435 \pm j0.7788$, the closed-loop system is lightly damped. We can show that the roots of the characteristic equation are at $-0.2095 \pm j0.9615$. Furthermore, the system is of type 0; so the steady-state error due to a step input is nonzero.

The performance requirements are

1. Ramp-error constant $K_v^* = 10$.
2. Maximum overshoot ≤ 15 percent.

The design strategy is to apply a digital PID controller so that the system type is increased to one. In addition, the controller parameters are selected to satisfy the K_v^* requirement and to cancel the complex poles of $G_{h0}G_p(z)$. Let us select the PID controller modeled by Eq. (10-109). First applying the requirement that $K_v^* = 10$, we get $K_I = 20$. Equating the numerator coefficients of Eq. (10-109) to the denominator coefficients of Eq. (10-139), we get $K_P = 2.472$ and $K_D = 1.9054$. The transfer function of the PID controller is

$$D(z) = \frac{6.283(z^2-0.01487z+0.6065)}{z(z-1)} \tag{10-140}$$

The open-loop transfer function of the compensated system is

$$D(z)G_{h0}G_p(z) = \frac{2.7263(z + 0.834)}{z(z - 1)} \tag{10-141}$$

It can be shown that, while the PID-compensated system is now of type 1, the system is unstable. The next step is to add another stage to the PID controller for stabilization. Since the system is unstable, a phase-lead controller would not be effective. Let us use the phase-lag controller modeled in Eq. (10-102). We can show that the relative stability requirement is satisfied if an attenuation of 0.1 is applied to Eq. (10-141). Thus, we let $K_c = 0.1$. Selecting $z_1 = 0.95$, we have from Eq. (10-104), $p_1 = 0.995$. The transfer function of the PID-phase-lead controller is

$$D(z) = \frac{0.6283(z^2 - 0.01487z + 0.6065)(z - 0.95)}{z(z - 1)(z - 0.995)} \tag{10-142}$$

The open-loop transfer function of the final system is

$$D(z)G_{h0}G_p(z) = \frac{0.2726(z + 0.834)(z - 0.95)}{z(z - 1)(z - 0.995)} \tag{10-143}$$

The maximum overshoot of the compensated system is 12 percent and $T_{\max} = 3$ s.

Let us now examine the effects of inexact cancellation. Consider that the values of the parameters of the PID controller are slightly changed to $K_I = 20$, $K_P = 2.35$, and $K_D = 1.8$. The transfer function of the controller becomes

$$D(z) = \frac{0.59(z^2 + 0.0847z + 0.6102)(z - 0.95)}{z(z - 1)(z - 0.995)} \tag{10-144}$$

We can show that the maximum overshoot of the compensated system is now 12.2 percent and $T_{\max} = 2.5$ s. The small change in the system performance is due to the fact that, although the open-loop poles at $0.007435 \pm j0.7788$ are not cancelled, the closed-loop poles generated by these poles are still very close to the zeros of $z^2 + 0.0847z + 0.6102$ so that their effect on the transient response is negligible.

10-7 TWO-DEGREE-OF-FREEDOM COMPENSATION

The compensation schemes presented in the preceding sections all have one degree of freedom in that the system has only one controller although that controller can contain several design parameters. The disadvantage of a one-degree-of-freedom controller is that the performance criteria that can be realized are limited. For example, if the parameters of a single controller are designed to realize a certain degree of relative stability, the noise-rejection property of the system may be poor, or it may have poor sensitivity to parameter variations. If the roots of the characteristic equation are selected to realize good relative damping, the maximum overshoot of the step response may still be excessive, owing to the bad locations of the zeros of the closed-loop transfer function.

Figure 10-46 shows typical compensation schemes that have two degrees of freedom. The block diagram shown in Fig. 10-46(a) contains series-feedback compensation. Figure 10-46(b) and (c) show systems with forward and feedforward controllers, respectively. In Fig. 10-46(b), the controller $D_f(z)$ is placed in

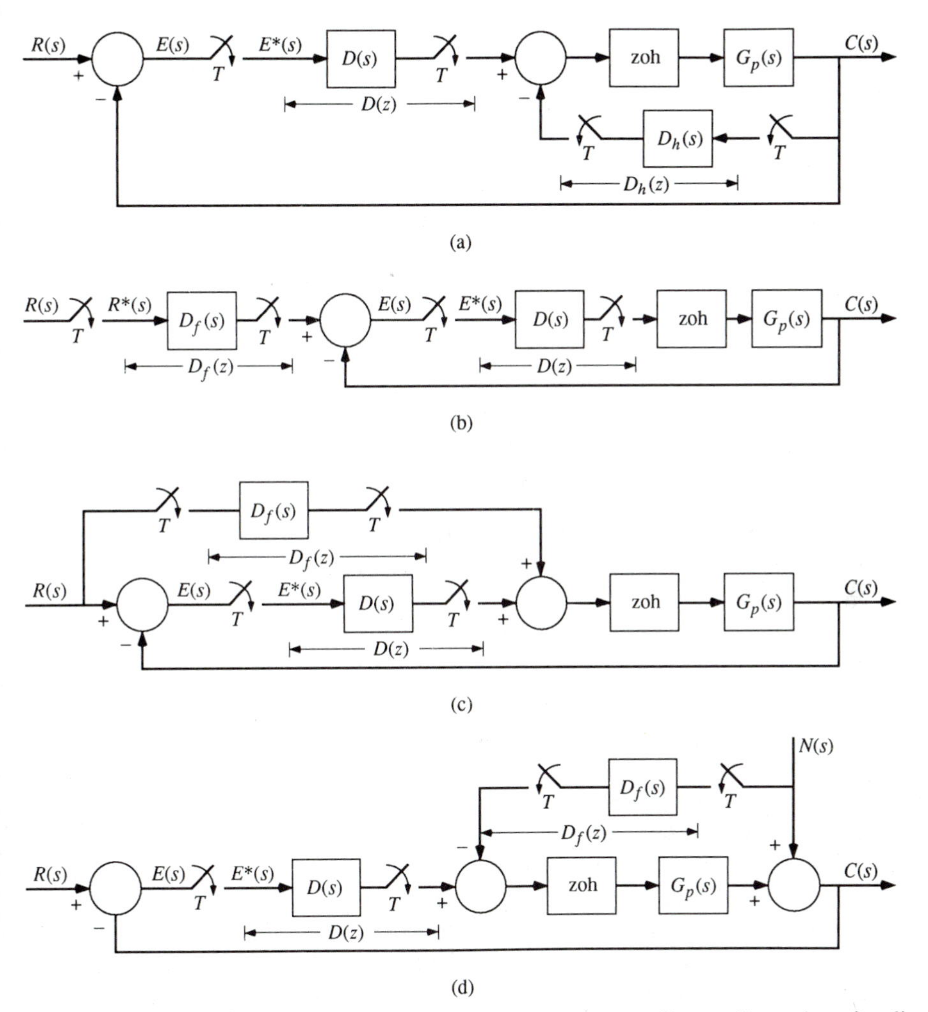

Figure 10-46. Various two-degree-of-freedom controller configurations for discrete-data control systems. (a) Series-feedback compensation. (b) Forward compensation. (c) Feedforward compensation. (d) Forward load disturbance with series compensation.

series with the closed-loop system which has a controller $D(z)$ in the forward path. In Fig. 10-46(c) the feedforward controller $D_f(z)$ is in parallel with the forward path. Analytically, the two configurations in Fig. 10-46(b) and (c) are equivalent, and knowing the $D_f(z)$ in one case one can find the equivalent $D_f(z)$ in the other. The essence of the feedforward compensation is that the controller $D_f(z)$ is not in the system loop, so that it does not affect the roots of the characteristic equation of the original system. The poles and zeros of $D_f(z)$ may be selected to add or cancel the unwanted poles and zeros of the closed-loop transfer function.

Figure 10-46(d) shows a control system with a two-degree-of-freedom controller to minimize the effects of the disturbance $N(s)$. The controller $D(z)$ is designed to achieve a desired closed-loop transfer function, whereas the feedforward controller $D_f(z)$ is designed to reduce or eliminate the effect of $N(s)$ on the output.

10-7-1 Feedforward Controller Design

The folowing example illustrates how a feedforward controller can be used to move a zero of the closed-loop transfer function for better system performance.

Example 10-11

Consider that the open-loop transfer function of a conditionally stable system is given as

$$G_{h0}G_p(z) = \frac{0.8(z-0.9)}{(z-1)(z-1.4)} \tag{10-145}$$

The closed-loop transfer function is

$$\frac{C(z)}{R(z)} = \frac{0.8(z-0.9)}{z^2-1.6z+0.68} \tag{10-146}$$

The roots of the characteristic equation are $0.8+j0.2$ and $0.8-j0.2$ which correspond to good damping, but the zero at $z=0.9$ which is carried over from the open-loop transfer function causes the overshoot to be very high. Figure 10-47 shows the unit-step response of the system without compensation. A logical design strategy is to cancel the zero of the closed-loop transfer function at $z=0.9$, while maintaining the positions of the poles. We can utilize the feedforward compensation configuration of Fig. 10-46(b) and select the transfer function of the controller as

$$D_f(z) = \frac{0.1z}{z-0.9} \tag{10-147}$$

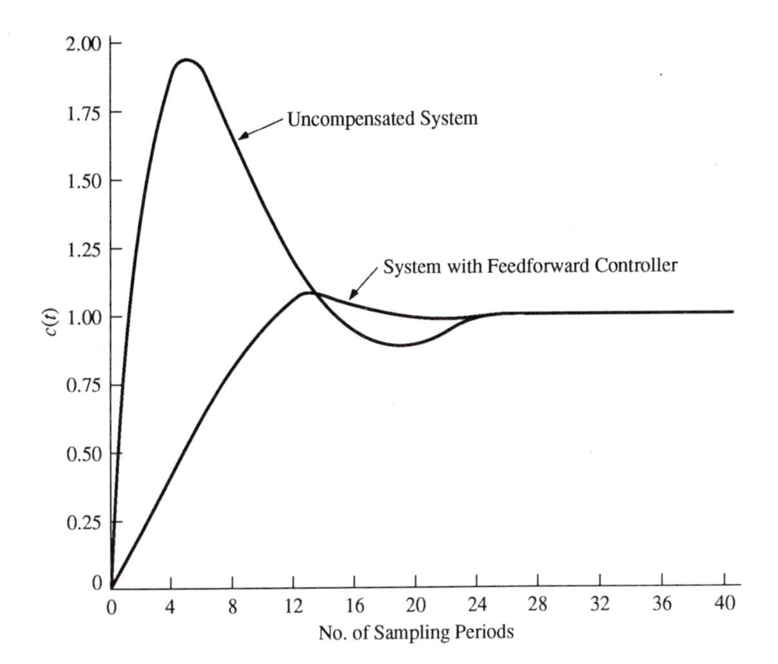

Figure 10-47. Unit-step responses of discrete-data system in Example 10-11.

Thus, the closed-loop transfer function of the compensated system is

$$\frac{C(z)}{R(z)} = \frac{0.08z}{z^2 - 1.6z + 0.68} \tag{10-148}$$

Figure 10-47 shows that the maximum overshoot of the system with the feedforward controller is approximately 8 percent.

10-7-2 Noise and Disturbance Rejection

All control systems are subject to some form of noise or disturbance during operation. Design considerations should include the reduction or elimination of the effects of the disturbance to the outputs. For the system configuration shown in Fig. 10-46(d), the transfer function between the sampled disturbance $N(z)$ and the output $C(z)$ is $[R(z) = 0]$

$$\frac{C(z)}{N(z)} = \frac{1 - D_f(z)G_{h0}G_p(z)}{1 + D(z)G_{h0}G_p(z)} \tag{10-149}$$

From the last equation we see that, if $D_f(z) = 0$, the basic properties of feedback reduce the effect of $N(z)$ on $C(z)$ if the magnitude of $1 + D(z)G_{h0}G_p(z)$

is greater than unity and the system is stable. Theoretically, the output will be completely unaffected by $N(z)$ if

$$D_f(z) = \frac{1}{G_{h0}G_p(z)} \tag{10-150}$$

In practice, the transfer function $G_{h0}G_p(z)$ usually has more poles than zeros, which makes $D_f(z)$ in Eq. (10-150) physically unrealizable. The problem can be circumvented by adding insignificant poles to $D_f(z)$ at or near $z = 0$ so that the number of poles of $D_f(z)$ is at least equal to the number of zeros.

Example 10-12

Consider that the discrete-data control system described in Example 10-9 now has a disturbance input added directly to the output, as shown in Fig. 10-46(d). The transfer function of the controlled process $G_{h0}G_p(z)$ is given in Eq. (10-128), and the open-loop transfer function with the PID controller designed, $D(z)G_{h0}G_p(z)$, is given in Eq. (10-137). Since $G_{h0}G_p(z)$ has two poles and one zero, $D_f(z)$ in Eq. (10-150) would not be physically realizable. Let us add a pole to $D_f(z)$ at $z = 0$ so that

$$D_f(z) = \frac{1}{zG_{h0}G_p(z)} = \frac{22.075(z - 0.905)(z - 0.819)}{z(z + 0.905)} \tag{10-151}$$

The transfer function $C(z)/N(z)$ $[R(z) = 0]$ is given below for $D_f(z) = 0$:

$$\frac{C(z)}{N(z)} = \frac{z(z - 1)}{z^2 - 0.737z + 0.238} \tag{10-152}$$

When $D_f(z)$ is as given in Eq. (10-151),

$$\frac{C(z)}{N(z)} = \frac{(z - 1)^2}{z^2 - 0.737z + 0.238} \tag{10-153}$$

Figure 10-48 shows the Bode plots of Eqs. (10-152) and (10-153) for three decades of ω up to $\omega_s/2$. Notice that with $D_f(z)$ implemented as shown in Fig. 10-46(d), ideally, the magnitude plot of Fig. 10-48 should be $-\infty$ dB. However, with the compromised solution for $D_f(z)$ given in Eq. (10-151), Fig. 10-48 shows that the attenuation to $N(z)$ over a significant portion of the frequency range is much greater than that when $D_f(z)$ is absent.

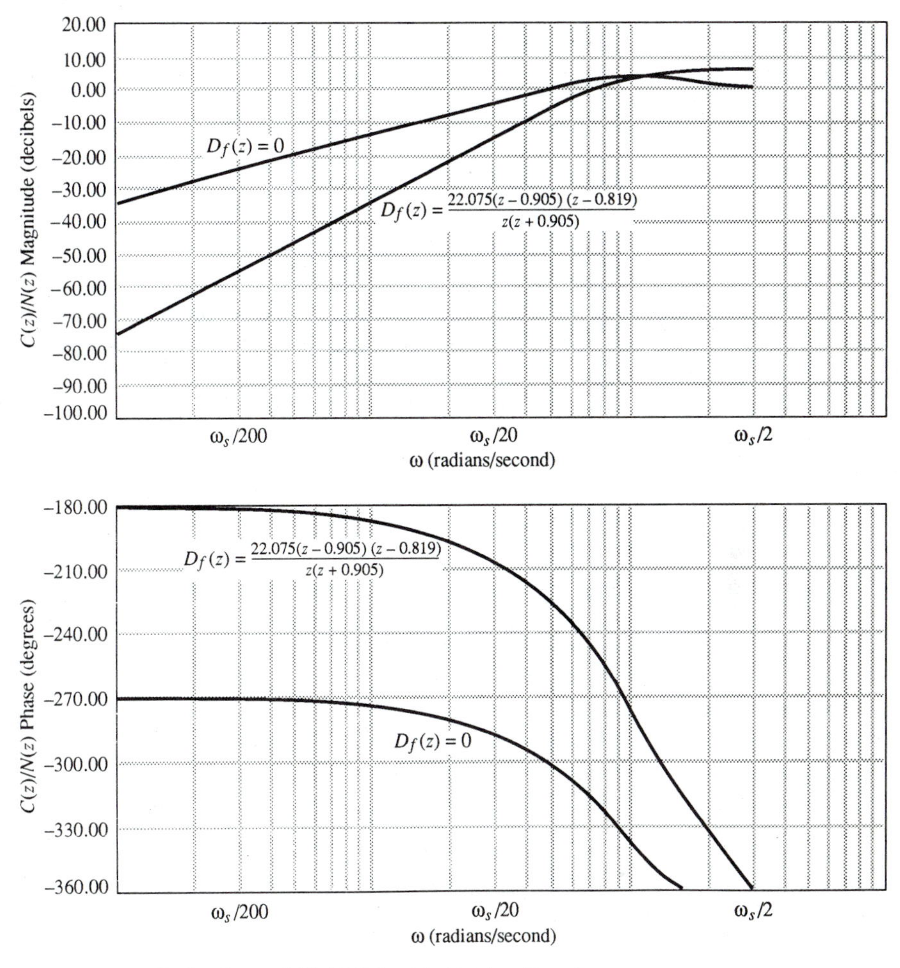

Figure 10-48. Bode plots of $C(z)/N(z)$ with $D_f(z) = 0$ and

$$D_f(z) = \frac{22.075(z - 0.905)(z - 0.819)}{z(z - 0.905)}$$

for the system in Example 10-12.

10-8 DESIGN OF ROBUST CONTROL SYSTEMS

The sensitivity function of a closed-loop discrete-data control system with respect to the open-loop transfer function gain is defined in Sec. 8-7, Eq. (8-57). We have shown that feedback in control systems has the inherent ability to reduce the effects of external disturbance and parameter variations. However,

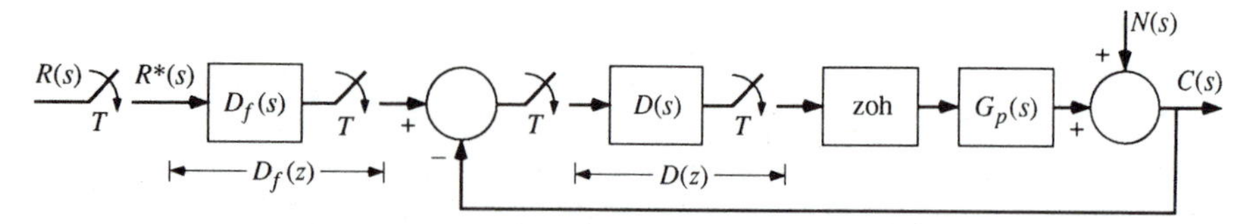

Figure 10-49. Discrete-data system with forward controller and disturbance inputs.

with feedback alone, robustness is achieved only with a high loop gain which may be detrimental to system stability. Let us consider the discrete-data system with external disturbance shown in Fig. 10-49. The digital controller in the loop is $D(z)$, and the forward controller is designated as $D_f(z)$. The input-output transfer function of the system is $[N(s) = 0]$,

$$M(z) = \left.\frac{C(z)}{R(z)}\right|_{N=0} = \frac{D_f(z)D(z)G_{h0}G_p(z)}{1 + D(z)G_{h0}G_p(z)} \tag{10-154}$$

The disturbance-output transfer function is $[R = 0]$,

$$\left.\frac{C(z)}{N(z)}\right|_{R=0} = \frac{1}{1 + D(z)G_{h0}G_p(z)} \tag{10-155}$$

In general, the strategy is to design the controller $D(z)$ so that the output is insensitive to the gain variation in the open loop and to the disturbance over the frequency range in which the latter is significant, and the forward controller $D_f(z)$ is designed to achieve the desired transfer function between the input and the output. From Eq. (8-58), the sensitivity of $M(z)$ due to variations in the magnitude of $G(z) = D(z)G_{h0}G_p(z)$ is written

$$S_G^M = \frac{dM(z)/M(z)}{dG(z)/G(z)} = \frac{1}{1 + D(z)G_{h0}G_p(z)} \tag{10-156}$$

which is identical to Eq. (10-155). Thus, the sensitivity function and the disturbance-output transfer function are identical, which means that disturbance suppression and robustness with respect to parameter variation can be designed with the same control schemes. The following examples illustrate the general approach to the utilization of feedforward control for disturbance suppression and robustness to parameter variations.

Example 10-13

Let us first consider a simple second-order system to illustrate the design ideas presented above. Consider that the controlled process of the system shown in Fig. 10-49 is

$$G_p(s) = \frac{K}{s(s + 10)} \tag{10-157}$$

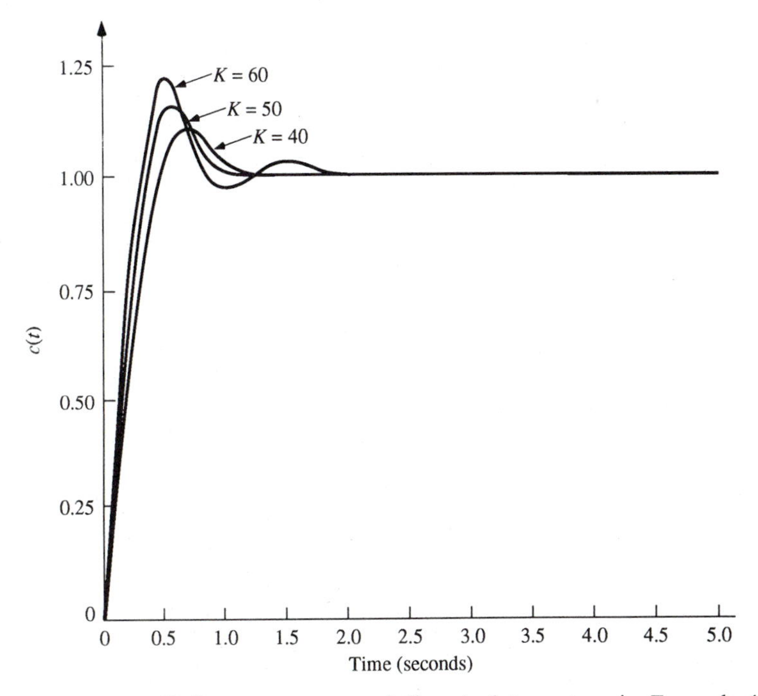

Figure 10-50. Unit-step responses of discrete-data system in Example 10-13 without compensation for $K = 40$, 50, and 60. $N(s) = 0$.

The sampling period is 0.1 s. Without any controller, the open-loop transfer function of the system is

$$G_{h0}G_p(z) = (1 - z^{-1})\mathscr{Z}\left(\frac{K}{s^2(s + 10)}\right) = \frac{0.003678K(z + 0.7183)}{(z - 1)(z - 0.3679)} \qquad \textbf{(10-158)}$$

Figure 10-50 shows the unit-step response of the system without compensation when $K = 40$, 50, and 60. The maximum overshoots for these three gains are 8.35, 14.6, and 21.3 percent, respectively. Figure 10-51 shows the frequency-response plot of $|S_G^M|$, which is the same as that of $|C(e^{j\omega T})/N(e^{j\omega T})|$, for $K = 50$. Notice that $|S_G^M|$ of the uncompensated system is less than unity up to $\omega =$ 4 rad/s.

Figure 10-52 shows the root loci of the uncompensated system and the location of the roots when $K = 40$, 50, and 60. The design strategy of the robust digital controller is to place the two zeros of the controller near the location where the desired dominant characteristic equation roots should be. Since the root loci end at the zeros of the open-loop transfer function, for zeros in the finite z-plane the root sensitivity [1] will be low near these zeros. That is, for large values of K, the movements of the roots will not be large when K varies. Let the two zeros of the controller $D(z)$ be at $0.6 + j0.2$ and $0.6 - j0.2$. The

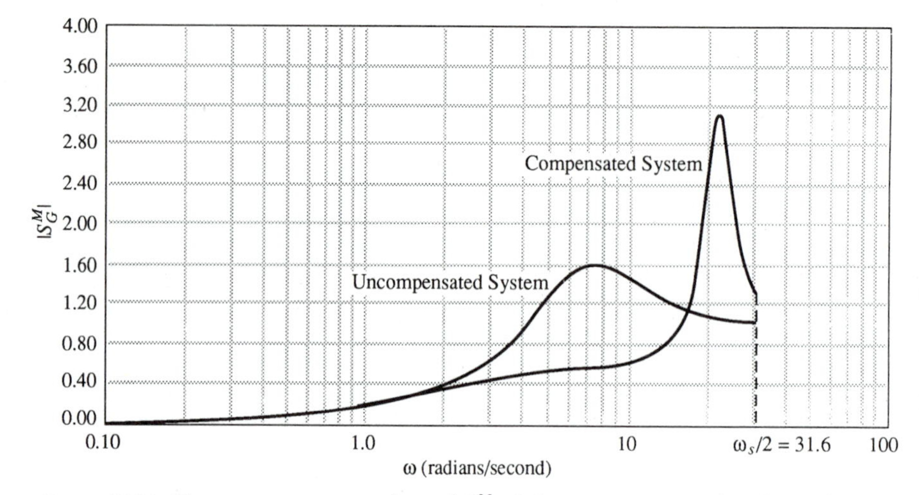

Figure 10-51. Frequency-response plots of S_G^M of the uncompensated and compensated systems in Example 10-13 for $K = 50$.

transfer function of $D(z)$ is

$$D(z) = \frac{5z^2 - 6z + 2}{z^2} \tag{10-159}$$

where the condition of $D(1) = 1$ has been applied, and two poles are added to $D(z)$ for physical realizability requirements. To make the controller $D(z)$

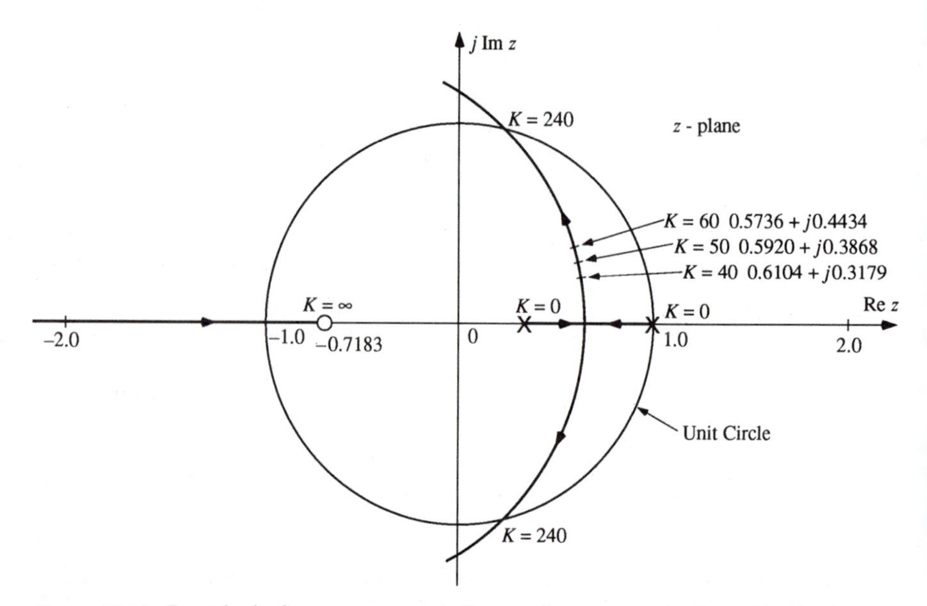

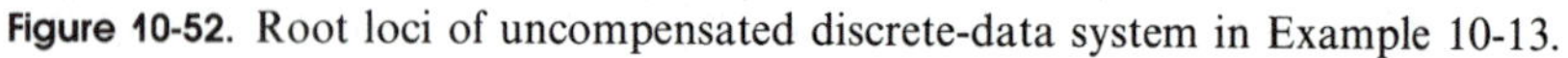

Figure 10-52. Root loci of uncompensated discrete-data system in Example 10-13.

effective, we must eliminate the zeros at $0.6 \pm j0.2$ from the closed-loop transfer function; otherwise, the characteristic equation roots approaching and near these zeros will be nearly cancelled. Thus, we set the transfer function of the forward controller as

$$D_f(z) = \frac{0.2z^2}{z^2 - 1.2z + 0.4} \tag{10-160}$$

where the two zeros at $z = 0$ are added just to speed up the response by two sampling periods. The root loci of the compensated system are shown in Fig. 10-53. Notice that the dominant roots near the two zeros of $D(z)$ are very robust, since they do not move much with the variation of K. The two roots in the second and third quadrants are less dominant to the transient response.

The closed-loop transfer functions and the characteristic equation roots of the compensated system with $D(z)$ and $D_f(z)$ given in Eqs. (10-159) and (10-160), respectively, are given below for $K = 40$, 50, and 60 $[N(s) = 0]$.

$\boldsymbol{K = 40}$

$$\frac{C(z)}{R(z)} = \frac{0.1472(z + 0.7183)}{z^4 - 0.6321z^3 + 0.01345z^2 - 0.3399z + 0.2114} \tag{10-161}$$

Roots: $0.66 \pm j0.02637$, $-0.3438 \pm j0.6055$.

$\boldsymbol{K = 50}$

$$\frac{C(z)}{R(z)} = \frac{0.1839(z + 0.7183)}{z^4 - 0.4482z^3 - 0.07516z^2 - 0.4248z + 0.2642} \tag{10-162}$$

Roots: $0.6525 \pm j0.086$, $-0.4285 \pm j0.653$.

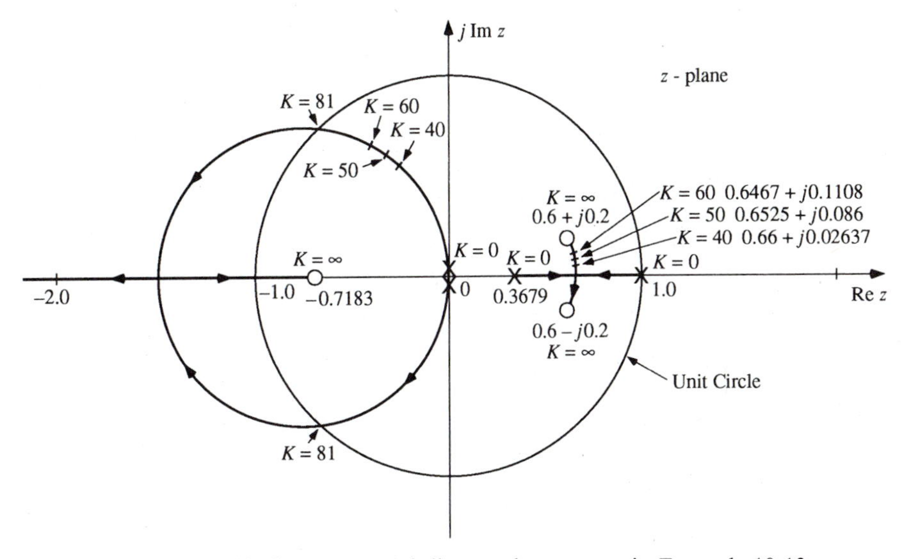

Figure 10-53. Root loci of compensated discrete-data system in Example 10-13.

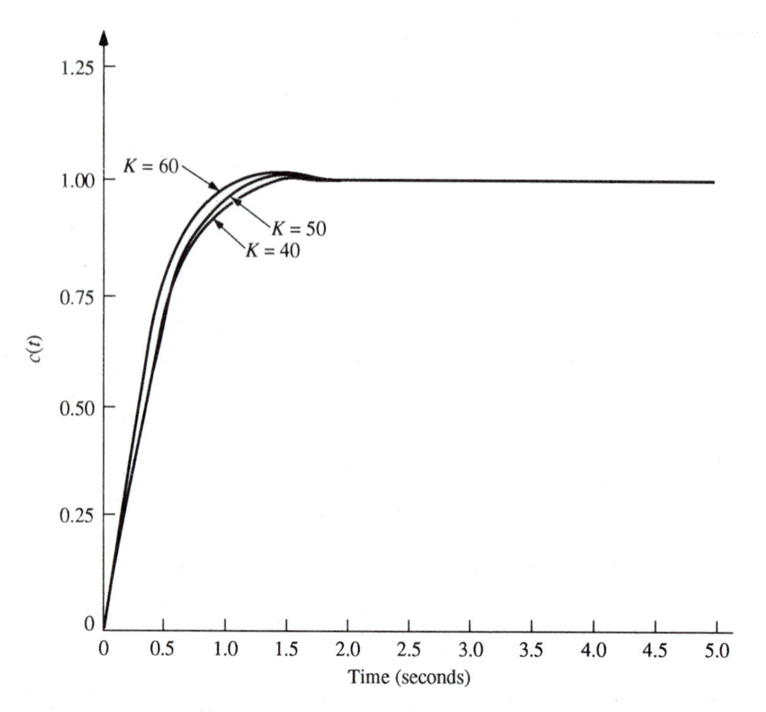

Figure 10-54. Unit-step responses of discrete-data system with robust and forward controllers in Example 10-13 for $K = 40$, 50, and 60. $N(s) = 0$.

$K = 60$

$$\frac{C(z)}{R(z)} = \frac{0.2207(z + 0.7183)}{z^4 - 0.2642z^3 - 0.1638z^2 - 0.5098z + 0.3171} \tag{10-163}$$

Roots: $0.6467 \pm j0.1108$, $-0.5146 \pm j0.687$.

The unit-step responses of the compensated system for $K = 40$, 50, and 60 are shown in Fig. 10-54. The robustness of the system to the variation of K is clearly indicated.

To show the robustness of the compensated system and its ability to reduce the effect of the external disturbance, we obtain the sensitivity function S_G^M by plotting the function $1/[D(z)G_{h0}G_p(z)]$ on the Nichols chart. Figure 10-51 shows the frequency response of S_G^M for $K = 50$ of the compensated system. Notice that the sensitivity is now less than unity for frequencies up to 15 rad/s, although the maximum value is increased substantially at high frequencies. This simply means that external disturbances with frequencies less than 15 rad/s will be suppressed by the robust and forward controllers. In addition, the sensitivity will be less over the same frequency range.

Example 10-14

In this example we apply the robust- and forward-controller configuration of Fig. 10-49 to the automobile idle-speed control system shown in Fig. 10-33. The open-loop transfer function of the uncompensated system is

$$G_{ho}G_p(z) = \frac{0.000125K(z + 0.195)(z + 2.821)}{z(z - 1)(z - 0.368)(z - 0.8187)} \quad (10\text{-}164)$$

where $K = 100$ and the sampling period is 0.1 s. It was shown in Example 10-7 that the uncompensated system is unstable. In Example 10-7, a phase margin of 50 degrees was realized by using a phase-lag controller, Eq. (10-96), and a lag-lead controller, Eq. (10-98). Figure 10-55 shows the unit-step responses of the compensated systems when K takes on values of 80, 100, and 120. The sensitivity functions $|S_G^M|$ of the compensated systems are shown in Fig. 10-56. For the system with the phase-lag controller, $|S_G^M|$ becomes greater than unity at $\omega = 0.85$ rad/s, whereas for the system with the lag-lead controller, this frequency is 2.1 rad/s.

Using the system configuration in Fig. 10-49, we can place two zeros of $D(z)$ at $0.75 + j0.1$ and $0.75 - j0.1$, and two poles at $z = 0$. Thus, the transfer

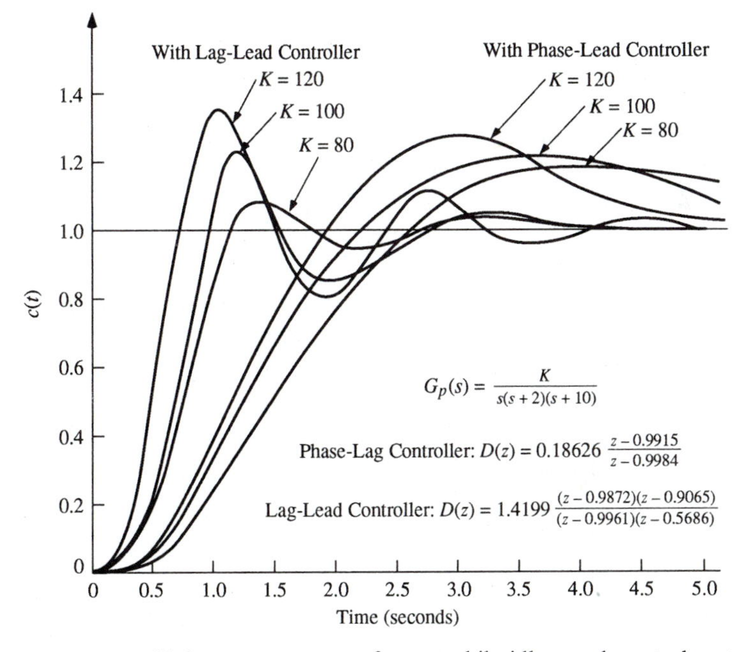

Figure 10-55. Unit-step responses of automobile idle-speed control system designed in Example 10-7 with phase-lag and lag-lead controllers.

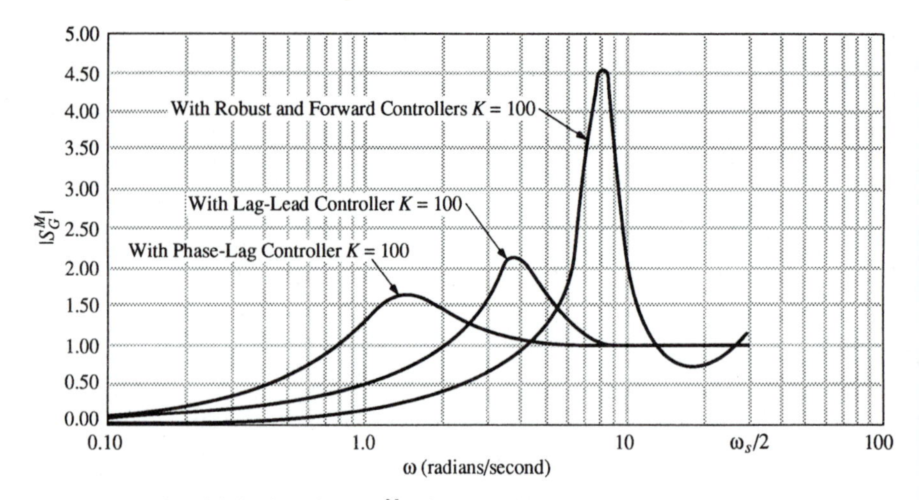

Figure 10-56. Sensitivity functions $|S_G^M|$ of automobile idle-speed control system in Example 10-14, with phase-lag and lag-lead controllers designed in Example 10-7 and the robust-forward controller designed in Example 10-14.

function of $D(z)$ is

$$D(z) = \frac{13.793(z^2 - 1.5z + 0.5725)}{z^2} \tag{10-165}$$

Figure 10-57 shows the root loci of the compensated system. As seen, the roots near the open-loop zeros at $0.75 \pm j0.1$ hardly move as the value of K varies from 80 to 120. However, in the present case, the complex roots that emanate from the poles at $z = 0$ and 0.368 still remain in the first quadrant, so that their effects on the transient response are not negligible. Completing the design with the forward controller $D_f(z)$, we let

$$D_f(z) = \frac{0.0725z^2}{z^2 - 1.5z + 0.5725} \tag{10-166}$$

The closed-loop transfer functions of the compensated system with $D(z)$ and $D_f(z)$ for $K = 80$, 100, and 120 are given below.

$\boldsymbol{K = 80}$

$$\frac{C(z)}{R(z)} = \frac{0.0100(z + 0.195)(z + 2.821)}{z^6 - 2.1866z^5 + 1.626z^4 - 0.0913z^3 - 0.4710z^2 + 0.1249z + 0.0436} \tag{10-167}$$

$\boldsymbol{K = 100}$

$$\frac{C(z)}{R(z)} = \frac{0.1731(z + 0.195)(z + 2.821)}{z^6 - 2.1866z^5 + 1.661z^4 - 0.0388z^3 - 0.5888z^2 + 0.1561z + 0.0545} \tag{10-168}$$

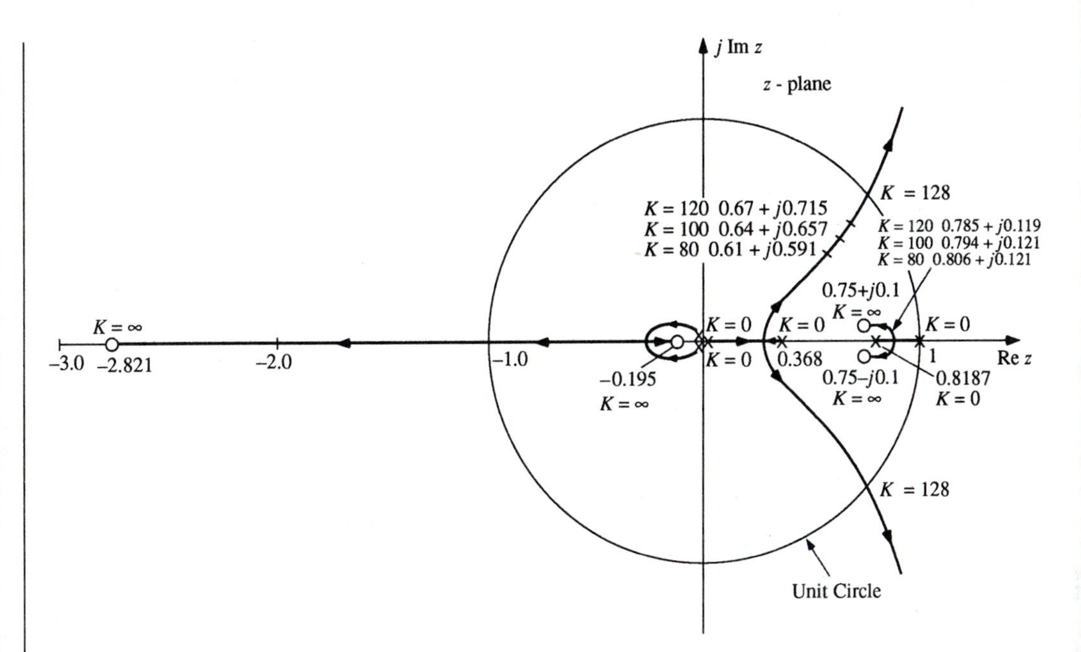

Figure 10-57. Root loci of automobile idle-speed control system with robust controller in Example 10-14.

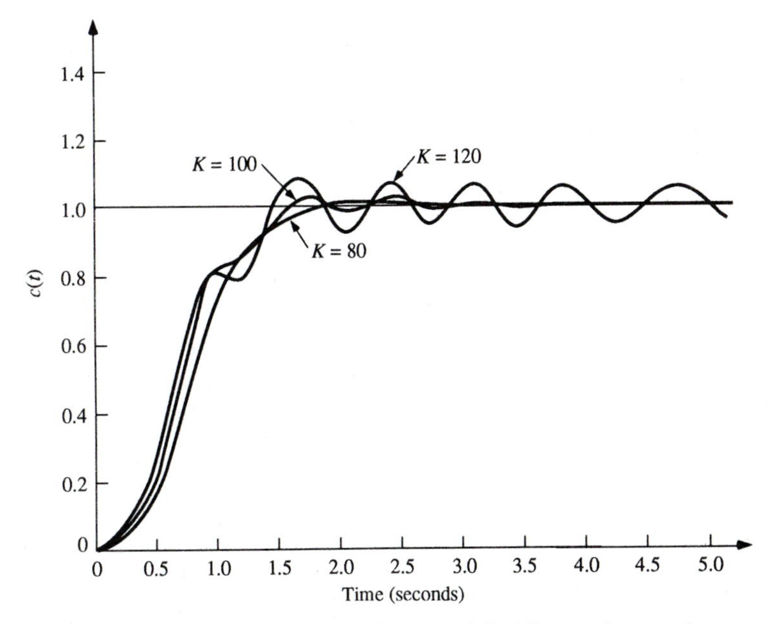

Figure 10-58. Unit-step responses of automobile idle-speed control system with robust and forward controllers in Example 10-14.

$K = 120$

$$\frac{C(z)}{R(z)} = \frac{0.01506(z + 0.195)(z + 2.821)}{z^6 - 2.1866z^5 + 1.696z^4 + 0.0136z^3 - 0.7065z^2 + 0.1874z + 0.0654} \tag{10-169}$$

The sensitivity function $|S_G^M|$ of the compensated system is shown in Fig. 10-56. Now the value of $|S_G^M|$ does not exceed one until $\omega = 4$ rad/s.

The unit-step responses of the compensated system for $K = 80$, 100, and 120 are shown in Fig. 10-58. The oscillations in the response when $K = 120$ are due to the characteristic equation roots at $0.67 \pm j0.715$ which are still in the first and fourth quadrants of the z-plane. The objective of this design example is simply to demonstrate the strategy of robust and disturbance reduction design. For further improvements to the system performance, we can add more elements to $D(z)$ so that the complex root loci in question are pulled toward the second and third quadrants.

10-9 DESIGN OF DISCRETE-DATA SYSTEMS WITH DEADBEAT RESPONSE

As defined in Sec. 7-10, a deadbeat response is one that follows the reference input without error in minimum finite time. Discrete-data systems are capable of having deadbeat responses due to the switching operations on signals in these systems. Without switching or sampling operations, a deadbeat response cannot be achieved in linear continuous-data control systems. As shown in Sec. 7-10, it is necessary (but not sufficient) that the characteristic equation of a discrete-data system with deadbeat response be of the form

$$\Delta(z) = z^N \tag{10-170}$$

where N is a positive integer.

10-9-1 Design of Digital Control Systems with Deadbeat Response

The design methods discussed in the preceding sections all involve the extension of the design experiences acquired in the design of linear continuous-data control systems, e.g., phase-lag, phase-lead, lag-lead, PID controllers, etc. Since the implementation of digital controllers is more versatile, we can come up with independent design methods.

In this section we deal with the deadbeat-response design of a digital control system. Figure 10-59 shows the block diagram of a digital control system in which all the components are subject to only digital data. We must distinguish between the designs of deadbeat response for digital control systems and sampled-data systems because the signals between samples in a digital control system are zero, whereas the outputs of a sampled-data system are continuous-

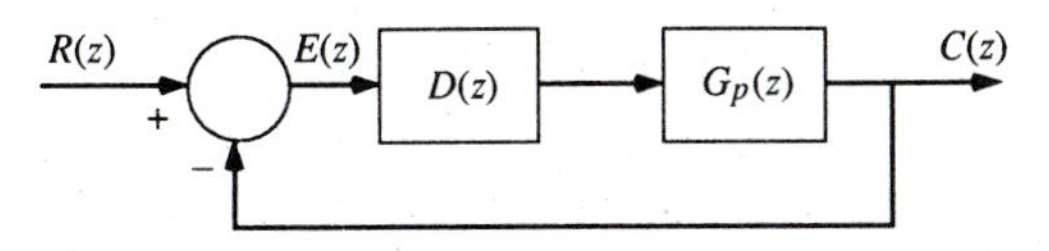

Figure 10-59. An all-digital control system.

time functions. Therefore, a deadbeat response at the sampling instants of a sampled-data system must guårantee that no ripples occur between the sampling instants.

To illustrate the design principle of a deadbeat-response controller for a digital control system, let us consider that the transfer function of a digital process is described by

$$G_p(z) = \frac{z + 0.5}{z^2 - z - 1} \tag{10-171}$$

Let the transfer function of the cascade digital controller be

$$D(z) = \frac{z^2 - z - 1}{(z - 1)(z + 0.5)} \tag{10-172}$$

The open-loop transfer function of the system becomes

$$G(z) = D(z)G_p(z) = \frac{1}{z - 1} \tag{10-173}$$

The closed-loop transfer function is

$$M(z) = \frac{C(z)}{R(z)} = \frac{G(z)}{1 + G(z)} = \frac{1}{z} \tag{10-174}$$

For a unit-step input, the output transform is

$$C(z) = \frac{1}{z}\frac{z}{z - 1} = \frac{1}{z - 1} = z^{-1} + z^{-2} + \cdots \tag{10-175}$$

Thus, the output response $c(k)$ reaches the desired steady-state value in one sampling period without overshoot and stays at that value thereafter. The response $c(k)$ is referred to as a deadbeat response. It should be pointed out that if $G_p(z)$ in Eq. (10-171) were the result of sampling a continuous-data process, then the digital controller $D(z)$ given in Eq. (10-172) does not guarantee that no ripples occur between the sampling instants in the continuous-data output $c(t)$.

Examining $D(z)$ in Eq. (10-172), we see that the deadbeat-response design principle involves the introduction of poles and zeros in $D(z)$ to cancel the zeros and poles in $G_p(z)$. However, this approach requires that all the poles and zeros of $G_p(z)$ are inside the unit circle.

Poles and Zeros of $G_p(z)$ Are All Inside the Unit Circle

We shall first consider the situation where the poles and zeros of $G_p(z)$ are all inside the unit circle. The deadbeat-response design is characterized by the following design criteria.

1. The system must have zero steady-state error at the sampling instants for the specified input signal.
2. The time for the output to reach the steady state should be finite and minimum.
3. The digital controller $D(z)$ must be physically realizable; i.e., it must not have more zeros than poles.

With reference to the system block diagram in Fig. 10-59, the closed-loop transfer function is

$$M(z) = \frac{C(z)}{R(z)} = \frac{D(z)G_p(z)}{1 + D(z)G_p(z)} \tag{10-176}$$

Solving for $D(z)$ from the last equation, we get

$$D(z) = \frac{1}{G_p(z)} \frac{M(z)}{1 - M(z)} \tag{10-177}$$

The z-transform of the error signal is written

$$\begin{aligned} E(z) &= R(z) - C(z) \\ &= R(z)[1 - M(z)] = \frac{R(z)}{1 + D(z)G_p(z)} \end{aligned} \tag{10-178}$$

Let the z-transform of the input be described by the function

$$R(z) = \frac{A(z)}{(1 - z^{-1})^N} \tag{10-179}$$

where N is a positive integer, and $A(z)$ is a polynomial in z^{-1} with no zeros at $z = 1$. For example, for a unit-step input, $A(z) = 1$ and $N = 1$; for a unit-ramp function input, $A(z) = Tz^{-1}$ and $N = 2$; for a parabolic input, $r(kT) = (kT)^2$, $A(z) = T^2z(z + 1)$, and $N = 3$. For zero steady-state error,

$$\begin{aligned} \lim_{k \to \infty} e(kT) &= \lim_{z=1} (1 - z^{-1})E(z) \\ &= \lim_{z=1} (1 - z^{-1}) \frac{A(z)[1 - M(z)]}{(1 - z^{-1})^N} = 0 \end{aligned} \tag{10-180}$$

Since the polynomial $A(z)$ does not contain any zeros at $z = 1$, the necessary condition for the steady-state error to be zero is that $1 - M(z)$ contain the factor $(1 - z^{-1})^N$. Thus, $1 - M(z)$ should have the form

$$1 - M(z) = (1 - z^{-1})^N F(z) \tag{10-181}$$

where $F(z)$ is a polynomial of z^{-1}. Solving for $M(z)$ in the last equation, we have

$$M(z) = 1 - (1 - z^{-1})^N F(z) = \frac{Q(z)}{z^P} \tag{10-182}$$

where $Q(z)$ is a polynomial in z, and $P \geq N$. Thus, the last equation verifies that the characteristic equation of the system with zero steady-state error in finite time is in the form of

$$z^P = 0 \tag{10-183}$$

Now substituting Eq. (10-182) into Eq. (10-178) yields

$$E(z) = A(z)F(z) \tag{10-184}$$

Since $A(z)$ and $F(z)$ are both polynomials of z^{-1}, $E(z)$ will have a finite number of terms in its power-series expansion in inverse powers of z. Thus, *when the characteristic equation of a digital control system is in the form of Eq.* (10-183), *that is, when the characteristic equation roots are all at* $z = 0$, *the error signal will go to zero in a finite number of sampling periods.*

Physical Realizability Considerations for $D(z)$

Equation (10-177) indicates that one way of designing the digital controller $D(z)$ is first to determine the desired closed-loop transfer function, and $F(z)$ and $Q(z)$ are found from Eq. (10-182). Given the process transfer function $G_p(z)$, the physical realizability condition on $D(z)$ places constraints on the form of $M(z)$. Let $G_p(z)$ and $M(z)$ be expressed by the following series expansions in powers of z^{-1}:

$$G_p(z) = g_n z^{-n} + g_{n+1} z^{-n-1} + \cdots \tag{10-185}$$

$$M(z) = m_k z^{-k} + m_{k+1} z^{-k-1} + \cdots \tag{10-186}$$

where n and k are the excess poles over zeros of $G_p(z)$ and $M(z)$, respectively. Substituting the last two equations in Eq. (10-177) and expanding $D(z)$ in a power series of z^{-1}, we get

$$D(z) = d_{k-n} z^{-(k-n)} + d_{k-n+1} z^{-(k-n+1)} + \cdots \tag{10-187}$$

Thus, for $D(z)$ to be physically realizable, $k \geq n$; that is, *the excess of poles over zeros of* $M(z)$ *must be at least equal to the excess of poles over zeros of* $G_p(z)$.

In view of the physical realizability requirement on $D(z)$, if $G_p(z)$ does not have poles or zeros on or outside the unit circle, the closed-loop transfer function $M(z)$ should have the following forms according to the type of input.

Step Input

$$R(z) = \frac{z}{z-1} \qquad M(z) = \frac{1}{z^n} \tag{10-188}$$

Ramp Input

$$R(z) = \frac{Tz}{(z-1)^2} \qquad M(z) = \frac{2z-1}{z^{n+1}} \tag{10-189}$$

Parabolic Input

$$R(z) = \frac{Tz(z+1)}{(z-1)^3} \qquad M(z) = \frac{3z^2-3z+1}{z^{n+2}} \tag{10-190}$$

where n is the excess number of poles over zeros of $G_p(z)$.

Example 10-15

Consider that the digital process of the system shown in Fig. 10-59 is described by the transfer function

$$G_p(z) = \frac{z+0.5}{z^2-z-1} \tag{10-191}$$

The objective is to design digital controllers so that the system will have deadbeat responses when the input is a unit-step, unit-ramp, and unit-parabolic function, respectively.

The poles and zeros of $G_p(z)$ are all inside the unit circle. Since the number of poles exceeds the number of zeros by one for $G_p(z)$, $n = 1$.

Unit-Step Input

When the input is a unit-step function, from Eq. (10-179), $N = 1$. Equation (10-188) gives $M(z) = z^{-1}$. Thus, from Eq. (10-177),

$$D(z) = \frac{1}{G_p(z)}\frac{M(z)}{1-M(z)} = \frac{z^2-z-1}{(z-1)(z+0.5)} \tag{10-192}$$

which is identical to Eq. (10-172). The output transform $C(z)$ is given in Eq. (10-175). Thus, the output follows the unit-step input after one sampling period, as shown in Fig. 10-60(a).

Unit-Ramp Input

When the input is a unit-ramp function,

$$R(z) = \frac{Tz}{(z-1)^2} \tag{10-193}$$

Thus, from Eq. (10-189), the closed-loop transfer function is

$$M(z) = \frac{2z-1}{z^2} \tag{10-194}$$

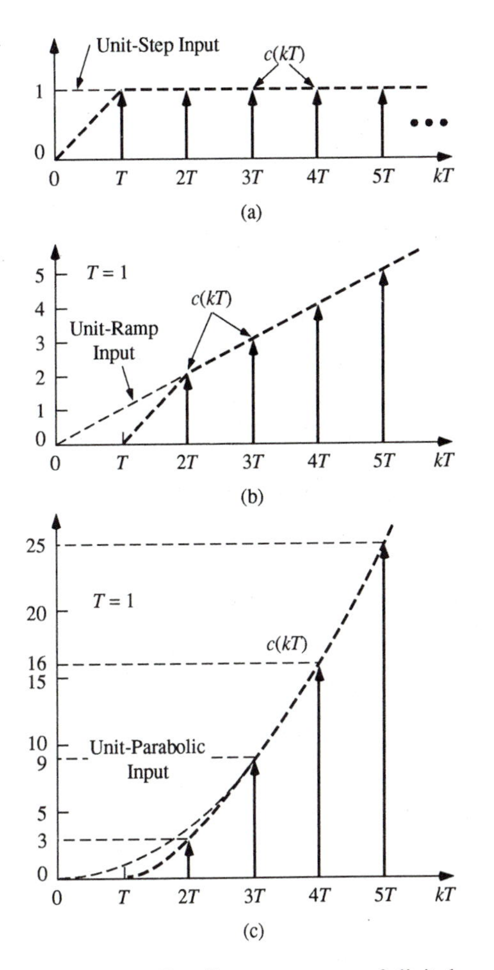

Figure 10-60. Deadbeat responses of digital control system designed in Example 10-15. (a) Unit-step response. (b) Unit-ramp response. (c) Unit-parabolic response.

Substituting the last equation into Eq. (10-177), we get

$$D(z) = \frac{2(z - 0.5)(z^2 - z - 1)}{(z - 1)^2(z + 0.5)} \tag{10-195}$$

For the unit-ramp input defined in Eq. (10-193), the output transform is

$$C(z) = \frac{T(z - 0.5)}{z(z - 1)^2} \tag{10-196}$$

Expanding $C(z)$ in inverse powers of z, we get

$$C(z) = T(2z^{-2} + 3z^{-3} + 4z^{-4} + \cdots) \tag{10-197}$$

Thus, the output sequence follows the input after two sampling periods, as shown in Fig. 10-60(b).

Unit-Parabolic Input

For a unit-parabolic input, the input transform is

$$R(z) = \frac{T^2 z(z+1)}{(z-1)^3} \tag{10-198}$$

Thus, from Eq. (10-190), the closed-loop transfer function is

$$M(z) = \frac{3z^2 - 3z + 1}{z^3} \tag{10-199}$$

Substituting the last equation into Eq. (10-177), we get

$$D(z) = \frac{(z^2 - z - 1)(3z^2 - 3z + 1)}{(z + 0.5)(z - 1)^3} \tag{10-200}$$

The output transform is

$$C(z) = \frac{T(z+1)(3z^2 - 3z + 1)}{z^2(z-1)^3} \tag{10-201}$$

Expanding $C(z)$ in inverse powers of z, we have

$$C(z) = T(3z^{-2} + 9z^{-3} + 16z^{-4} + 25z^{-5} + 36z^{-6} + 49z^{-7} + \cdots) \tag{10-202}$$

Thus, the output sequence follows the input after three sampling periods, as shown in Fig. 10-60(c).

Example 10-16

Consider that the digital process of the system in Fig. 10-49 is described by the transfer function

$$G_p(z) = \frac{0.05(z + 0.5)}{(z - 0.9)(z - 0.8)(z - 0.35)} \tag{10-203}$$

The objective is to design a deadbeat-response controller so that the system output sequence $c(kT)$ will follow a unit-step input in the minimum number of sampling periods.

Since $G_p(z)$ has two more poles than zeros and all the poles and zeros are inside the unit circle, $n = 2$. The closed-loop transfer function is given by Eq. (10-188), $M(z) = 1/z^2$. The transfer function of the digital controller is

$$D(z) = \frac{1}{G_p(z)} \frac{M(z)}{1 - M(z)} = \frac{20(z - 0.9)(z - 0.8)(z - 0.35)}{(z + 0.5)(z + 1)(z - 1)} \tag{10-204}$$

The output transfer is given by

$$C(z) = M(z)R(z) = \frac{1}{z^2 - z} = z^{-2} + z^{-3} + z^{-4} + \cdots \tag{10-205}$$

Thus the output sequence $c(kT)$ follows the unit-step input after two sampling periods.

Process Transfer Functions with Poles or Zeros on or Outside the Unit Circle

The development above shows that the deadbeat-response design of digital control systems depends on the cancellation of the poles and zeros of the process transfer function $G_p(z)$ by the zeros and poles of the digital controller $D(z)$. However, if $G_p(z)$ has poles or zeros on or outside the unit circle, imperfect cancellation, which is very likely to occur in practice, will result in an unstable closed-loop system. Thus, for all practical purposes no attempt should be made to cancel the poles and zeros of $G_p(z)$ that are on or outside the unit circle. This simply imposes additional constraints on the selection of the closed-loop transfer function $M(z)$.

Let the transfer function of the controlled process of the system in Fig. 10-59 be written as

$$G_p(z) = \frac{\prod_{i=1}^{K} (1 - z_i z^{-1})}{\prod_{j=1}^{L} (1 - p_j z^{-1})} A(z) \tag{10-206}$$

where z_i, $i = 1, 2, \ldots, K$, and p_j, $j = 1, 2, \ldots, L$, are the zeros and poles, respectively, of $G_p(z)$ that are *outside* the unit circle, and $A(z)$ is a rational function in z^{-1} with poles and zeros only inside the unit circle. Substituting Eq. (10-206) into Eq. (10-177), we get

$$D(z) = \frac{\prod_{j=1}^{L} (1 - p_j z^{-1})}{\prod_{i=1}^{K} (1 - z_i z^{-1})} \frac{M(z)}{A(z)[1 - M(z)]} \tag{10-207}$$

Since $D(z)$ cannot contain the poles p_j and zeros z_i as its zeros and poles, respectively, they must be cancelled by the zeros and poles of $1 - M(z)$ and $M(z)$. In other words, $M(z)$ must contain the factors

$$\prod_{i=1}^{K} (1 - z_i z^{-1})$$

and $1 - M(z)$ must contain the factors

$$\prod_{j=1}^{L} (1 - p_j z^{-1})$$

Thus, in general, $M(z)$ and $1 - M(z)$ should have the following forms:

$$M(z) = \prod_{i=1}^{K} (1 - z_i z^{-1})(M_k z^{-k} + M_{k+1} z^{-k-1} + \cdots) \tag{10-208}$$

$$1 - M(z) = \prod_{j=1}^{L} (1 - p_j z^{-1})(1 - z^{-1})^P (1 + a_1 z^{-1} + a_2 z^{-2} + \cdots) \tag{10-209}$$

In the last two equations, $k \geq n$, where n is the number of excess poles over zeros of $G_p(z)$. The power of the term $(1 - z^{-1})$, P, should equal either the order of the poles of $R(z)$, N, or the order of the poles of $G_p(z)$ at $z = 1$, whichever is greater. The truncation of the last two equations is determined by the following conditions.

1. The orders of the poles of $M(z)$ and $1 - M(z)$ must be equal.
2. The total number of unknowns in M_k, M_{k+1}, ... and a_1, a_2, ... must equal the order of $M(z)$, so that these coefficients can be solved independently.

To illustrate the deadbeat-response design of a digital control system that has a controlled process with poles or zeros on or outside the unit circle, the following examples are given.

Example 10-17

Consider that the digital control system shown in Fig. 10-59 has the controlled process described by

$$G_p(z) = \frac{0.0125(z + 0.195)(z + 2.821)}{z(z - 1)(z - 0.368)(z - 0.8187)} \tag{10-210}$$

The objective is to design a deadbeat-response controller so that the output sequence $c(kT)$ will follow a unit-step input in minimum time.

Notice that this is the same transfer function of the automobile idle-speed control system studied in Example 10-7. It is assumed here that the process is purely digital, so that the possibility of intersampling ripples as a result of the deadbeat-response design will not be an issue.

Multiplying the numerator and denominator of $G_p(z)$ by z^{-4}, we have

$$G_p(z) = \frac{0.0125 z^{-2}(1 + 0.195 z^{-1})(1 + 2.821 z^{-1})}{(1 - z^{-1})(1 - 0.368 z^{-1})(1 - 0.8187 z^{-1})} \tag{10-211}$$

Since $G_p(z)$ has a zero at -2.821 which is outside the unit circle and a pole at $z = 1$ which is on the unit circle, we cannot cancel the pole and zero by the elements of $D(z)$. The closed-loop transfer functions $M(z)$ and $1 - M(z)$ are selected based on the following considerations.

1. According to Eq. (10-208), $M(z)$ must contain the term $(1 + 2.821 z^{-1})$.

2. Since $G_p(z)$ has two more poles than zeros, k in Eq. (10-208) should equal two, so that $M(z)$ will have the same pole-over-zero excess.
3. The function $1 - M(z)$ should include the term $1 - z^{-1}$, since $G_p(z)$ has one pole at $z = 1$, and the input is a step function.
4. The minimum order of $M(z)$ is three, since this would allow three unknown coefficients in Eqs. (10-208) and (10-209) to be solved uniquely.

Thus, Eqs. (10-208) and (10-209) become

$$M(z) = (1 + 2.821z^{-1})M_2 z^{-2} \tag{10-212}$$

$$1 - M(z) = (1 - z^{-1})(1 + a_1 z^{-1} + a_2 z^{-2}) \tag{10-213}$$

Substituting Eq. (10-212) into Eq. (10-213) and equating the coefficients of like powers of z^{-1}, we get

$$a_1 = 1 \tag{10-214}$$

$$a_1 - a_2 = M_2 \tag{10-215}$$

$$a_2 = 2.821 M_2 \tag{10-216}$$

The solutions of these equations are $a_1 = 1, a_2 = 0.7383$, and $M_2 = 0.2617$. The closed-loop transfer function of the designed system is

$$M(z) = \frac{0.2617(z + 2.821)}{z^3} \tag{10-217}$$

Substituting the last equation in Eq. (10-177) and simplifying, we have

$$D(z) = \frac{20.936z(z - 0.368)(z - 0.8187)}{(z + 0.195)(z^2 - z + 0.7383)} \tag{10-218}$$

The output transform is

$$C(z) = \frac{0.2617(z + 2.821)}{z^2(z - 1)} = 0.2617z^{-2} + z^{-3} + z^{-4} + \cdots \tag{10-219}$$

Thus, the output sequence $c(kT)$ tracks the unit-step input perfectly after three sampling periods. Referring to Eq. (10-188), we see that if $G_p(z)$ did not have any poles or zeros outside the unit circle, the output would be able to track the step input in two sampling periods, since $G_p(z)$ has two more poles than zeros.

Example 10-18

Consider that the transfer function of the digital process in the digital control system shown in Fig. 10-59 is

$$G_p(z) = \frac{0.0003916(z + 2.8276)(z + 0.19)}{(z - 1)^2(z - 0.2865)} \tag{10-220}$$

which has one zero outside the unit circle at $z = -2.8276$ and two poles at $z = 1$. The number of poles exceeds the number of zeros in $G_p(z)$ by one.

The objective is to find the deadbeat-response controllers so that the output sequence $c(kT)$ will track a unit-step input and then a unit-ramp input.

Unit-Step Input

The closed-loop transfer functions $M(z)$ and $1 - M(z)$ are selected based on the following considerations.

1. According to Eq. (10-208), $M(z)$ must contain the term $(1 + 2.8276z^{-1})$.
2. Since $G_p(z)$ has one more pole than zeros, k in Eq. (10-208) should equal one, so that $M(z)$ will have the same pole-over-zero excess.
3. The function $1 - M(z)$ should include the term $(1 - z^{-1})^2$ since $G_p(z)$ has two poles at $z = 1$, and the input is a step function.
4. The minimum order of $M(z)$ is three, since this would allow three unknown coefficients in Eqs. (10-208) and (10-209) to be solved uniquely.

Thus, Eqs. (10-208) and (10-209) become

$$M(z) = (1 + 2.8276z^{-1})(M_1 z^{-1} + M_2 z^{-2}) \tag{10-221}$$

$$1 - M(z) = (1 - z^{-1})^2(1 + a_1 z^{-1}) \tag{10-222}$$

Substituting Eq. (10-221) into Eq. (10-222) and equating the coefficients of like powers of z^{-1}, we get

$$M_1 = 2 - a_1 \tag{10-223}$$

$$2.8276M_1 + M_2 = 2a_1 - 1 \tag{10-224}$$

$$2.8276M_2 = -a_1 \tag{10-225}$$

The solutions of these equations are $M_1 = 0.7155$, $M_2 = -0.4543$, and $a_1 = 1.2845$. Thus, the closed-loop transfer function of the designed system is

$$M(z) = \frac{0.7155z^2 + 1.5688z - 1.2845}{z^3} \tag{10-226}$$

From Eq. (10-177) the transfer function of the digital controller is determined as

$$D(z) = \frac{1827.17(z - 0.6349)(z - 0.2865)}{(z + 0.19)(z + 1.2845)} \tag{10-227}$$

The output transfer is

$$C(z) = \frac{0.7155z^2 + 1.5688z - 1.2845}{z^2(z - 1)}$$

$$= 0.7155z^{-1} + 2.28433z^{-2} + z^{-3} + z^{-4} + z^{-5} \cdots \tag{10-228}$$

The output sequence $c(kT)$ is plotted as shown in Fig. 10-61(a). Notice that although $c(kT)$ reaches the desired steady-state value in three sampling periods,

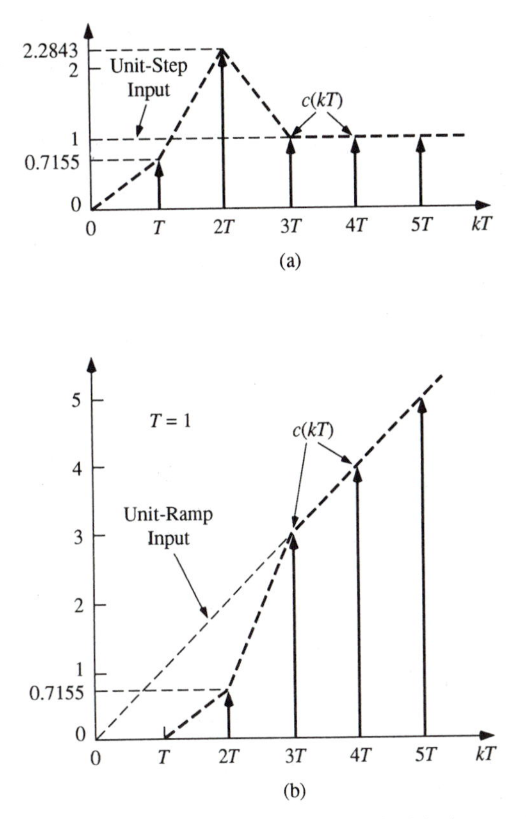

Figure 10-61. Deadbeat responses of digital control system designed in Example 10-18. (a) Unit-step response. (b) Unit-ramp response.

the maximum overshoot is 128.43 percent. The reason for this is that the digital process is a type 2 system, so that a deadbeat response without overshoot cannot be obtained for a step input. Since the two poles at $z = 1$ are not allowed to be cancelled by the zeros of the controller, the output response rises too fast so that the unit-step response will always exhibit an overshoot. Furthermore, the open-loop system is unstable, due to the pole of $D(z)$ at $z = -1.2845$. Thus, the closed-loop system is conditionally stable. This example also shows that, for certain systems, it is not always possible to design a deadbeat response without overshoot for a given input.

Unit-Ramp Input

The design of the deadbeat-response system for a ramp input is identical to that for a step input, since the function $1 - M(z)$ in Eq. (10-222) is still chosen with the term $(1 - z^{-1})^2$. Thus, the digital controller in Eq. (10-227) is still the result for the deadbeat response due to a ramp input. The output transform

$C(z)$ is now

$$C(z) = \frac{T(0.7155z^2 + 1.5585z - 1.2845)}{z^2(z-1)^2}$$

$$= T(0.7155z^{-2} + 3z^{-3} + 4z^{-4} + 5z^{-5} + \cdots) \tag{10-229}$$

As shown in Fig. 10-61(b), the unit-ramp response follows the input without steady-state error after three sampling periods.

SUMMARY ON THE DEADBEAT-RESPONSE DESIGN OF DIGITAL CONTROL SYSTEMS

The following summary is given on the deadbeat-response design of digital control systems.

1. The deadbeat-response design developed for digital control systems does not guarantee zero intersampling ripples if $G_p(z)$ is the z-transform of a sampled continuous-data process.
2. The design principle relies on the cancellation of the stable poles and zeros of $G_p(z)$ by the zeros and poles of $D(z)$, respectively. The unstable zeros of $G_p(z)$ must be carried over as zeros of $M(z)$, and the unstable poles of $G_p(z)$ must be designated as the zeros of $1 - M(z)$. The function $1 - M(z)$ must also contain the term $(1 - z^{-1})^P$, where P is the greater of the order of the pole at $z = 1$ of $G_p(z)$ or of $R(z)$.
3. Depending on the type of $G_p(z)$, a deadbeat response with zero overshoot may not be attainable for a given input.
4. The deadbeat-response design is "tuned" only for the type of input it is designed for and would give inferior or unacceptable responses to other inputs. For example, the system designed in Example 10-17 may be regarded as ideal for a ramp input, but the maximum overshoot to a unit-step input is 128.43 percent.
5. Since the characteristic equation roots of a system with deadbeat response are all at $z = 0$, we show in the following that the sensitivity of these roots due to a change in any system parameter is theoretically infinite.

   Let the characteristic equation of a digital system be written as

$$P(z) + KQ(z) = 0 \tag{10-230}$$

where K is any system parameter; $P(z)$ and $Q(z)$ are polynomials of z with constant coefficients and do not contain K. Consider that K is varied by an increment ΔK; Eq. (10-230) becomes

$$P(z) + (K + \Delta K)Q(z) = 0 \tag{10-231}$$

Dividing both sides of the last equation by $P(z) + KQ(z)$, we get

$$1 + \frac{\Delta K Q(z)}{P(z) + KQ(z)} = 0 \tag{10-232}$$

Notice that the denominator of the last equation contains the characteristic equation of Eq. (10-230). Consider that the characteristic equation in Eq. (10-230) has a root z_i with multiplicity n (>1). The quotient in Eq. (10-232) is expanded by partial-fraction expansion in the neighborhood of z_i. Letting $\Delta z_i = z - z_i$, Eq. (10-232) can be approximated as

$$1 + \frac{\Delta K Q(z)}{P(z) + KQ(z)} \cong 1 + \frac{\Delta K A_i}{(\Delta z)^n} \cong 0 \tag{10-233}$$

where A_i represents the partial-fraction expansion coefficients. The last equation leads to

$$\frac{\Delta K}{\Delta z} = \frac{-(\Delta z)^{n-1}}{A_i} \tag{10-234}$$

Now taking the limit on both sides of Eq. (10-234) as ΔK and Δz approach zero, we have

$$\lim_{\substack{\Delta K \to 0 \\ \Delta z \to 0}} \frac{\Delta K}{\Delta z} = \frac{dK}{dz} = \lim_{\substack{\Delta K \to 0 \\ \Delta z \to 0}} \frac{-(\Delta z)^{n-1}}{A_i} = 0 \tag{10-235}$$

The root sensitivity with respect to change of a system parameter K is defined as

$$S_z^K = \frac{dz/z}{dK/K} = \frac{dz}{dK}\frac{K}{z} \tag{10-236}$$

Since Eq. (10-235) shows that dK/dz is zero at a multiple-order root, this corresponds to S_z^K being infinite at the root. The conclusion is that a system with a deadbeat response is very sensitive to parameter variations.

10-9-2 Design of Sampled-Data Control Systems with Deadbeat Response

The deadbeat-response design presented in the preceding section is based on the condition that the system is purely digital, so that the output response between the sampling instants is zero. When the controller process is continuous, the output $c(t)$ is a function of t; the deadbeat-response design based on the cancellation of the stable poles and zeros of the controlled process *may* result in a system that has intersampling ripples in the output, although the steady-state error at the sampling instants goes to zero in finite time. The reason behind this is that the cancellation of process zeros by controller poles results in controller dynamics that are excited by the input but are not affected by feedback. *Therefore, the deadbeat-response strategy for a sampled-data system with the process transfer function $G_{h0}G_p(z)$ having at least one zero is not to cancel the zeros at all by the poles of the controller, whether these zeros are inside or outside the unit circle.* H. R. Sirisena [3] gives a mathematical solution to the design of ripple-free deadbeat-response discrete-data systems. Let the transfer function of a stable process be expressed as

$$G_{h0}G_p(z^{-1}) = \frac{Q(z^{-1})}{P(z^{-1})} \tag{10-237}$$

where $Q(z^{-1})$ and $P(z^{-1})$ are polynomials of z^{-1}. Sirisena shows that the digital controller for ripple-free deadbeat response to a step input is

$$D(z) = \frac{P(z^{-1})}{Q(1) - Q(z^{-1})} \tag{10-238}$$

The reader may refer to [3] for more details.

The design of discrete-data systems with ripple-free deadbeat responses can still follow the approach discussed in the last section, using Eqs. (10-208) and (10-209), except that in Eq. (10-208) the zeros at $z = z_i$ should include *all* the zeros of $G_{h0}G_p(z)$. Thus, the added restriction will increase the response time of the system.

Example 10-19

Consider that the controlled process of the discrete-data system shown in Fig. 10-62 has the transfer function

$$G_p(s) = \frac{10}{s(s+2)} \tag{10-239}$$

The sampling period is 0.1 s. The objective is to design a ripple-free deadbeat-response when the input is a step function.

The z-transfer function of the open-loop transfer function without compensation is

$$G_{h0}G_p(z) = \frac{0.04683(z+0.9355)}{(z-1)(z-0.8187)} \tag{10-240}$$

Let us first use the design devised for digital control systems, without addressing the intersampling ripples. Since the zero of $G_{h0}G_p(z)$ is inside the unit circle, we can choose $M(z)$ to be z^{-1}. Thus, $1 - M(z) = 1 - z^{-1}$. The digital controller has the transfer function

$$D(z) = \frac{21.355(z-0.8187)}{(z+0.9355)} \tag{10-241}$$

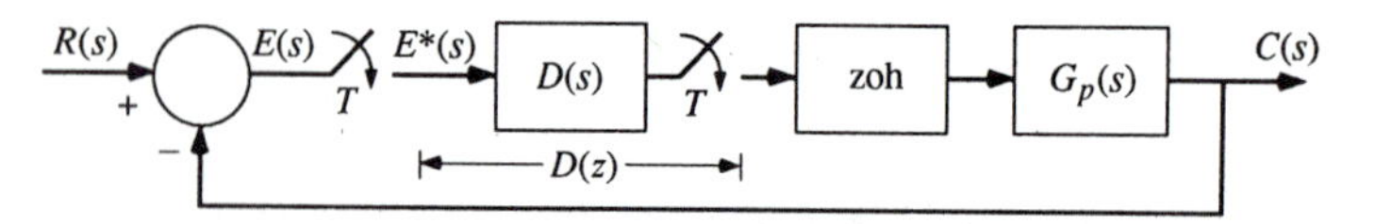

Figure 10-62. A discrete-data control system.

The output transform to a unit-step input is expanded as

$$C(z) = \frac{1}{z-1} = z^{-1} + z^{-2} + z^{-3} + \cdots \qquad (10\text{-}242)$$

However, we can show that the output response is deadbeat only at the sampling instants. The true output $c(t)$ has intersampling ripples so that it takes forever for it to reach the steady-state value. The necessary and sufficient conditions for $c(t)$ to track a unit-step input with finite response time are

$$c(NT) = 1 \qquad \text{and} \qquad \left.\frac{dc(t)}{dt}\right|_{t=NT} = 0 \qquad (10\text{-}243)$$

for finite N, and strictly, all higher derivatives must equal zero.

Let $\omega(t) = dc(t)/dt$. The z-transform of $\omega(t)$ can be written

$$\Omega(z) = \frac{D(z)(1-z^{-1})\mathscr{Z}[G_p(s)]}{1 + D(z)(1-z^{-1})\mathscr{Z}[G_p(s)/s]} R(z) \qquad (10\text{-}244)$$

Substituting Eqs. (10-239) and (10-241) into the last equation, we get

$$\Omega(z) = \frac{3.871(z-1)}{z(z+0.9355)} R(z) \qquad (10\text{-}245)$$

Since the poles of $\Omega(z)/R(z)$ are not all at $z = 0$, the unit-step response of $\Omega(z)$ will not go to zero in finite time. Figure 10-63(a) shows the unit-step response of the system with $D(z)$ as given in Eq. (10-241).

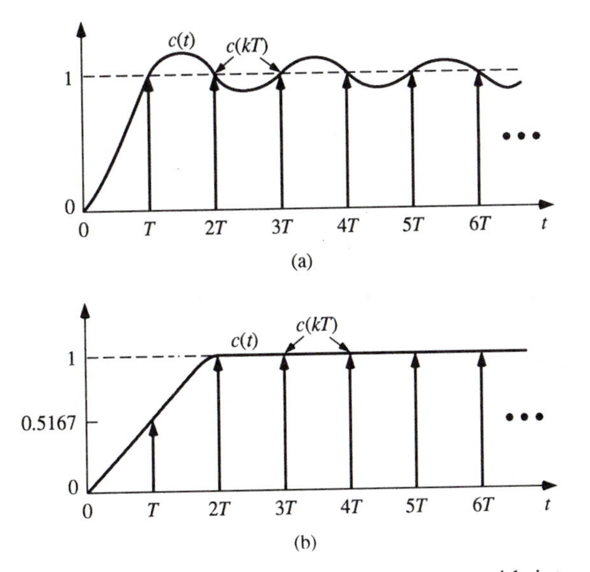

Figure 10-63. (a) Deadbeat unit-step response with intersampling ripples. (b) Deadbeat unit-step response without intersampling ripples.

Now let us apply the condition that the zero of $G_{h0}G_p(z)$ at $z = -0.9355$ should not be cancelled by a pole of $D(z)$. According to Eqs. (10-208) and (10-209), $M(z)$ and $1 - M(z)$ are selected as

$$M(z) = (1 + 0.9355z^{-1})(M_1 z^{-1}) \tag{10-246}$$

$$1 - M(z) = (1 - z^{-1})(1 + a_1 z^{-1}) \tag{10-247}$$

Solving for M_1 and a_1 from the last two equations, we get $M_1 = 0.5166$ and $a_1 = 0.4833$. Thus, the closed-loop transfer function is

$$M(z) = \frac{0.5166(z + 0.9355)}{z^2} \tag{10-248}$$

The transfer function of the digital controller is obtained from Eq. (10-177),

$$D(z) = \frac{11.033(z - 0.8187)}{z + 0.4834} \tag{10-249}$$

We can show that $D(z)$ can also be determined using Eq. (10-238). Comparing the last equation with Eq. (10-241), we see that the gain of $D(z)$ is changed from 21.355 to 11.033, and the pole is moved from -0.9355 to -0.4834. The output transform for a unit-step input is

$$C(z) = \frac{0.5166(z + 0.9355)}{z(z - 1)} = 0.5166z^{-1} + z^{-2} + z^{-3} + \cdots \tag{10-250}$$

Thus, the deadbeat response reaches the steady state in two sampling periods. We can show that the z-transform of the derivative of $c(t)$ when the input is a unit-step function is

$$\Omega(z) = 2z^{-1} \tag{10-251}$$

which verifies that the output $c(t)$ does reach its steady-state value in two sampling periods, as shown in Fig. 10-63(b).

Example 10-20

As another example of the deadbeat-response design of a discrete-data control system, let us consider the automobile idle-speed control system shown in Fig. 10-33. The uncompensated open-loop transfer function is

$$G_{h0}G_p(z) = \frac{0.0125(z + 0.195)(z + 2.821)}{z(z - 1)(z - 0.368)(z - 0.8187)} \tag{10-252}$$

The objective is to design a cascade digital controller so that the output will be a ripple-free deadbeat response when the input is a step function.

Since $G_{h0}G_p(z)$ has two finite zeros and one pole at $z = 1$ and two more poles than zeros, according to Eqs. (10-208) and (10-209), $M(z)$ and $1 - M(z)$ are chosen as follows:

$$M(z) = (1 + 0.195z^{-1})(1 + 2.821z^{-1})M_1z^{-2} \tag{10-253}$$

$$1 - M(z) = (1 - z^{-1})(1 + a_1z^{-1} + a_2z^{-2} + a_3z^{-3}) \tag{10-254}$$

The keys are as follows.

1. $M(z)$ should contain all the zeros of $G_{h0}G_p(z)$.
2. The number of poles over zeros of $M(z)$ should at least equal that of $G_{h0}G_p(z)$, which is two.
3. $1 - M(z)$ must include the term $1 - z^{-1}$, since $G_{h0}G_p(z)$ has one pole at $z = 1$ and $R(z)$ is a step input.
4. The orders of $M(z)$ and $1 - M(z)$ should be the same and should equal the number of unknown coefficients.

Solving for M_1, a_1, a_2, and a_3 from Eqs. (10-253) and (10-254), we get the following simultaneous equations:

$$1 - a_1 = 0 \tag{10-255}$$

$$M_1 = a_1 - a_2 \tag{10-256}$$

$$3.016M_1 = a_2 - a_3 \tag{10-257}$$

$$0.55M_1 = a_3 \tag{10-258}$$

The solutions are $M_1 = 0.219$, $a_1 = 1$, $a_2 = 0.781$, and $a_3 = 0.1205$. The closed-loop transfer function is

$$M(z) = \frac{0.219z^2 + 0.66z + 0.1205}{z^4} \tag{10-259}$$

The transfer function of the digital controller is obtained by substituting $G_{h0}G_p(z)$, $M(z)$, and $1 - M(z)$ into Eq. (10-177),

$$D(z) = \frac{17.52z(z - 0.368)(z - 0.8187)}{(z + 0.1927)(z^2 + 0.8073z + 0.6255)} \tag{10-260}$$

which can also be obtained using Eq. (10-238). The output transform to a unit-step input is written

$$C(z) = \frac{0.219z^2 + 0.66z + 0.1205}{z^2(z - 1)}$$

$$= 0.219z^{-2} + 0.879z^{-3} + z^{-4} + z^{-5} + \cdots \tag{10-261}$$

Thus, the output sequence $c(kT)$ reaches the final steady state in four sampling periods. This is one more sampling period than the deadbeat response designed in Example 10-17 for a pure digital system. If the controller in Eq. (10-218) were applied to the discrete-data system considered in this example,

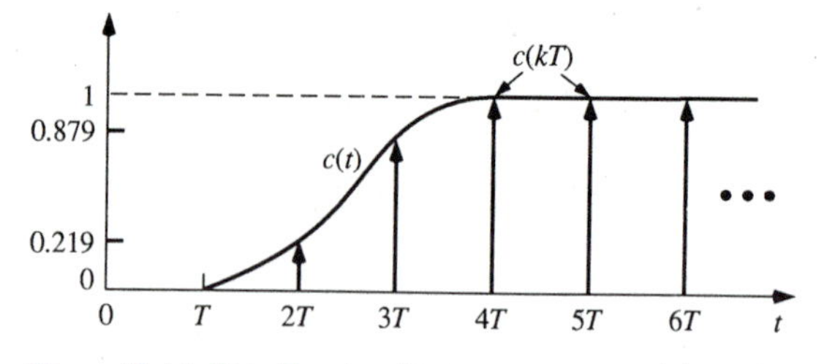

Figure 10-64. Deadbeat unit-step response without intersampling ripples of discrete-data system in Example 10-20.

the output unit-step response would reach steady state in three sampling periods, but only at the sampling instants, and intersampling ripples would occur.

To show that the unit-step response $c(t)$ is indeed deadbeat, we derive the z-transform of the derivative of $c(t)$, $\omega(t)$. We have

$$\Omega(z) = \frac{D(z)(1 - z^{-1})\mathscr{Z}[G_{ho}G_p(s)]}{1 + D(z)(1 - z^{-1})\mathscr{Z}[G_{ho}G_p(s)/s]} R(z) \tag{10-262}$$

$$= \frac{6(z + 0.6714)}{z^3} = 6z^{-2} + 4.028z^{-3} \tag{10-263}$$

Thus, the derivative of $c(t)$ is zero for $kT \geq 4T$, which shows that the step response reaches the final value in four sampling instants, with no intersampling ripples. The unit-step response is shown in Fig. 10-64.

Example 10-21

Consider that the process of the discrete-data control system shown in Fig. 10-62 has the transfer function

$$G_p(s) = \frac{25}{s^2(s + 25)} \tag{10-264}$$

The sampling period is 0.05 s. The open-loop z-transfer function of the uncompensated system is

$$G_{ho}G_p(z) = \frac{0.0003916(z + 2.8276)(z + 0.19)}{(z - 1)^2(z - 0.2865)} \tag{10-265}$$

Notice that this is the same transfer function used in Example 10-18, where it is assumed that the process is purely digital. The result obtained in Example 10-18 for a deadbeat-response design was that the unit-step response has a maximum overshoot of 128.43 percent, and the digital controller is unstable, causing the system to be conditionally stable.

For ripple-free deadbeat response to a step input, we express $G_{h0}G_p(z)$ as

$$G_{h0}G_p(z^{-1}) = \frac{0.0003916z^{-1}(1 + 2.8276z^{-1})(1 + 0.19z^{-1})}{(1 - z^{-1})^2(1 - 0.2865z^{-1})} \tag{10-266}$$

Thus,

$$Q(z^{-1}) = 0.0003916z^{-1}(1 + 2.827z^{-1})(1 + 0.19z^{-1}) \tag{10-267}$$

$$P(z^{-1}) = (1 - z^{-1})^2(1 - 0.2865z^{-1}) \tag{10-268}$$

and $Q(1) = 0.00153$. Substituting the last two equations and the value of $Q(1)$ in Eq. (10-238), we have

$$D(z) = \frac{560.654(z - 1)(z - 0.2865)}{(z + 0.205)(z + 0.5755)} \tag{10-269}$$

The closed-loop transfer function is

$$M(z) = \frac{D(z)G_{h0}G_p(z)}{1 + D(z)G_{h0}G_p(z)} = \frac{0.2196(z + 2.8276)(z + 0.19)}{z^3} \tag{10-270}$$

The same results can be obtained by setting $M(z)$ as

$$M(z) = (1 + 2.8276z^{-1})(1 + 0.19z^{-1})M_1z^{-1} \tag{10-271}$$

and $1 - M(z)$ as

$$1 - M(z) = (1 - z^{-1})^2(1 + a_1z^{-1}) \tag{10-272}$$

and solve for M_1 and a_1.

The output transform due to a unit-step input is

$$C(z) = \frac{0.2196z^2 + 0.6625z + 0.11795}{z^2(z - 1)}$$

$$= 0.2196z^{-1} + 0.8821z^{-2} + z^{-3} + z^{-4} + z^{-5} + \cdots \tag{10-273}$$

Thus, the deadbeat response reaches the steady state in three sampling periods. To show that the response is indeed ripple-free, we can again show that the z-transform of the derivative of $c(t)$ goes to zero starting at the third sampling period. The result is

$$\Omega(z) = 12.03z^{-1} + 7.97z^{-2} \tag{10-274}$$

It is interesting to note that the deadbeat response not only does not have intersampling ripples, but the digital controller is also stable.

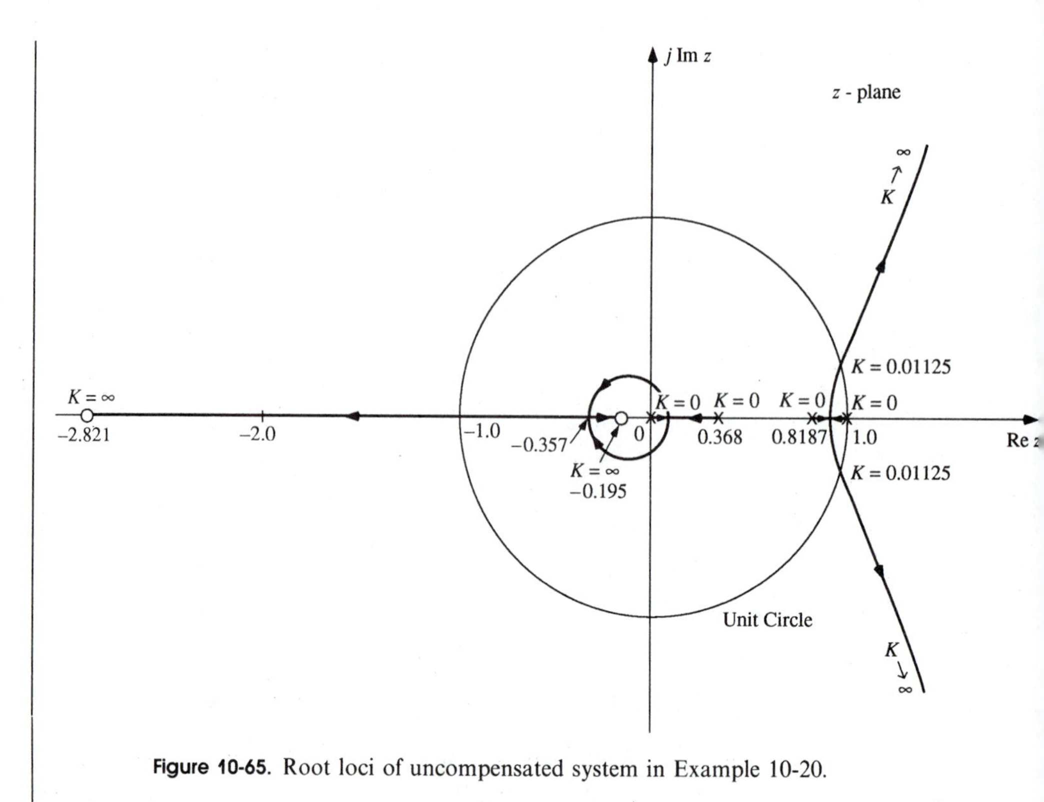

Figure 10-65. Root loci of uncompensated system in Example 10-20.

10-9-3 Root Loci of Deadbeat Discrete-Data Systems

It is enlightening to investigate the root loci of discrete-data control systems with deadbeat responses. Since the characteristic equation of a system with deadbeat response is of the form of Eq. (10-183), the roots of the characteristic equation must all be at the origin of the z-plane.

For the discrete-data system designed in Example 10-20, the open-loop transfer function of the uncompensated system is given in Eq. (10-252). The root loci of the system are drawn as shown in Fig. 10-65 with K as the variable parameter. The open-loop transfer function of the compensated system with deadbeat response to a step input is

$$D(z)G_{ho}G_p(z) = \frac{K(z + 0.195)(z + 2.821)}{(z - 1)(z + 0.1927)(z^2 + 0.8073z + 0.6255)} \tag{10-275}$$

where $K = 0.219$. Figure 10-66 shows that the poles of the uncompensated open-loop system at $z = 0$, 0.368, and 0.8187 are cancelled by the zeros of the digital controller, and new poles are added at -0.1927, $-0.4036 + j0.68$ and $-0.4036 - j0.68$, and the root loci of the characteristic equation. When $K =$ 0.219, the four roots all converge at $z = 0$.

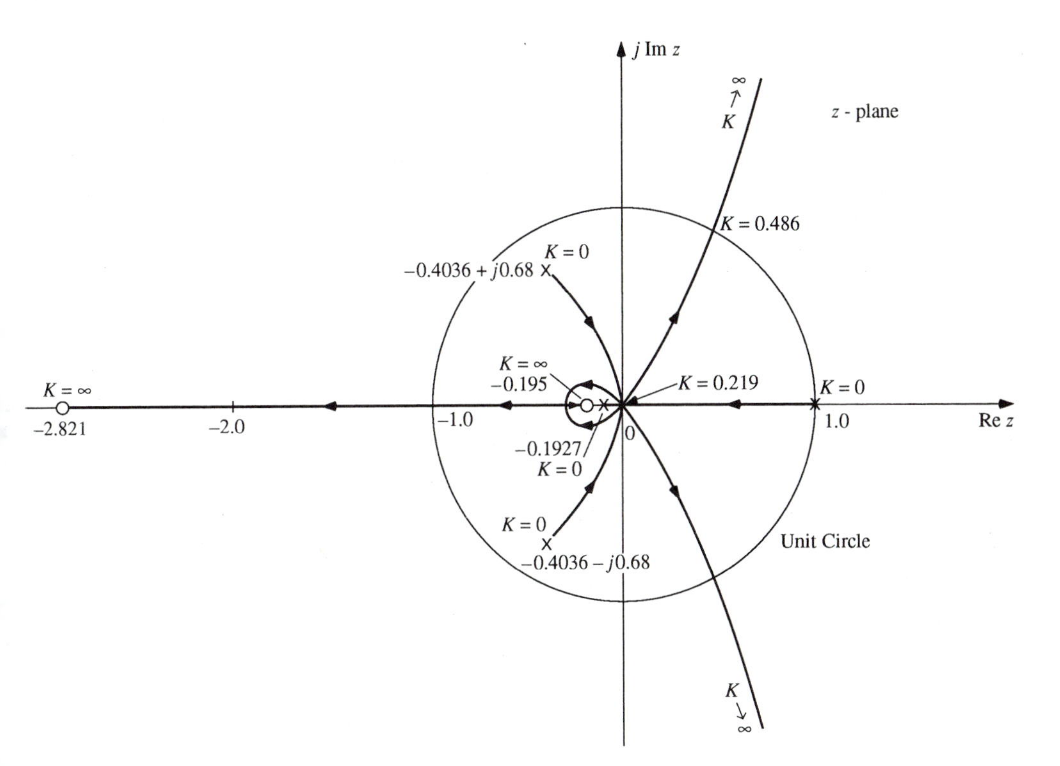

Figure 10-66. Root loci of discrete-data system designed with deadbeat step response in Example 10-20.

10-10 DESIGN WITH NOTCH CONTROLLERS

Notch networks, and notch controllers in general, are used frequently in continuous-data control systems [1] to suppress undesirable resonances and oscillations. The transfer function of a typical continuous-data notch controller is

$$D(s) = \frac{s^2 + 2\zeta_z \omega_n s + \omega_n^2}{s^2 + 2\zeta_p \omega_n s + \omega_n^2} \tag{10-276}$$

where ζ_z and ζ_p are the damping ratios of the numerator and the denominator polynomials, respectively, and ω_n is the resonant frequency of the controller in rad/s. In the s-plane design, the zeros of $D(s)$ are chosen to cancel or placed very close to the complex poles of the controlled process, and the poles of $D(s)$ are placed at appropriate places to yield desirable relative stability. In the frequency domain, the notch controller design involves the positioning of ω_n at the resonant frequency of the process, and the values of ζ_z and ζ_p are determined from the desired attenuation that $D(s)$ must provide to suppress the

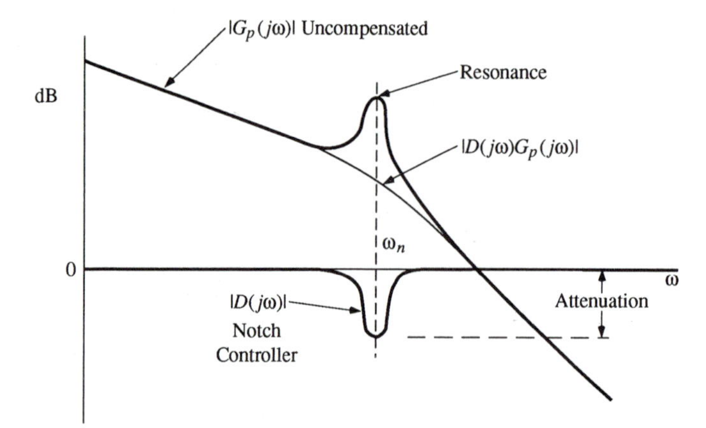

Figure 10-67. A qualitative illustration of the design of a notch controller for a continuous-data system in the frequency domain.

undesirable system resonances. Figure 10-67 gives a qualitative description of the frequency-domain design principle of a continuous-data system with a notch controller.

A digital notch controller can be described by

$$D(z) = K_d \frac{z^2 + az + b}{z^2 + cz + d} \tag{10-277}$$

where K_d is generally chosen so that $D(1) = 1$. The design of the notch controller in the z-plane is straightforward, since it again involves the placement of the poles and zeros of $D(z)$. The deadbeat-response design will naturally lead to the cancellation of the complex poles of the process by the complex zeros of $D(z)$. The design in the frequency domain can be carried out by use of the bilinear transformation, but the task is more complex, since establishing the relationship of the parameters of $D(z)$ to the resonant frequency is difficult. The following example illustrates the design of a discrete-data system that requires a notch controller.

Example 10-22

The block diagram of a dc-motor velocity control system is shown in Fig. 10-68. The compliance between the motor and the load causes the system to have a very oscillatory response. The transfer function of the controlled process is

$$G_p(s) = \frac{\Omega_L(s)}{U(s)} = \frac{1000(s + 10000)}{s^3 + 40s^2 + 300200s + 2000000} \tag{10-278}$$

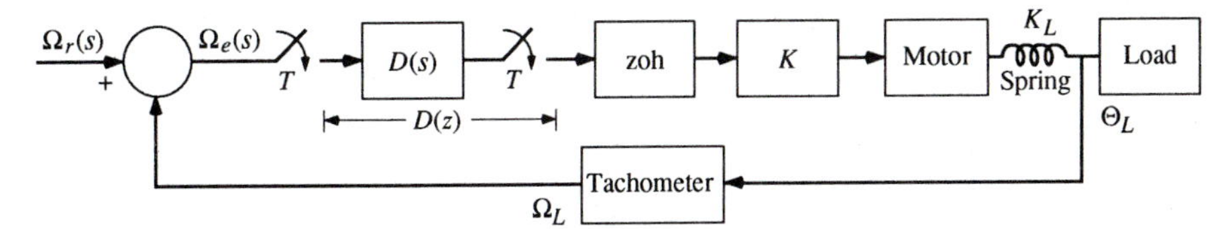

Figure 10-68. A discrete-data dc-motor velocity control system.

where $\Omega_L(s)$ is the output velocity and $U(s)$ is the output of the zero-order hold. The sampling period is chosen to be 0.001 s. To simplify the problem, we assume that the gain of the tachometer is unity.

The z-transform of the uncompensated open-loop system $D(s) = 1$ is

$$G_{ho}G_p(z) = \frac{\Omega_L(z)}{\Omega_e(z)} = \frac{0.002109(z + 0.1887)(z + 2.814)}{(z - 0.9603)(z^2 - 1.7104z + 0.9994)} \tag{10-279}$$

where the two complex poles are at $z = 0.8552 \pm j0.5176$. Figure 10-69 shows the unit-step response of the uncompensated system. Although the system is stable, the compliance between the motor and the load causes the response to oscillate with a frequency of approximately 550 rad/s. Since the system is of type 0, the final value of $c(t)$ is not unity.

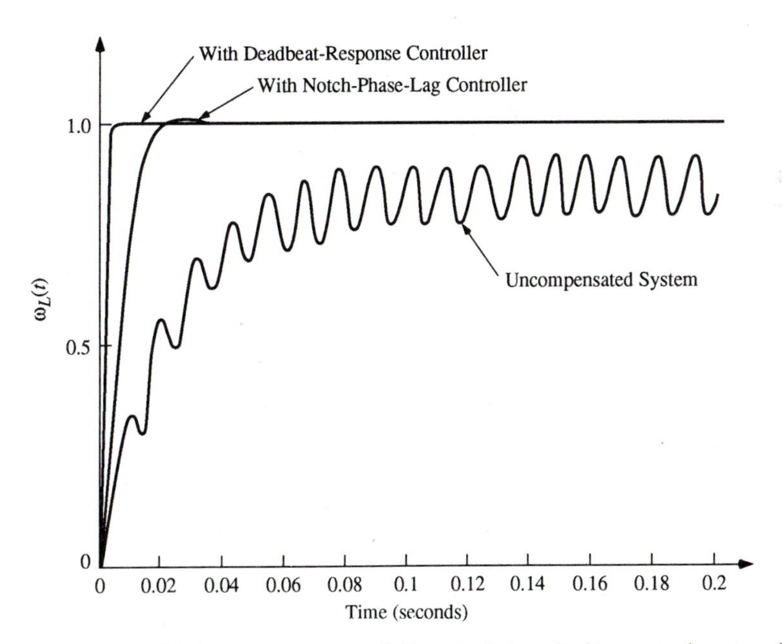

Figure 10-69. Unit-step responses of discrete-data velocity control system in Example 10-22.

We shall conduct two designs in the z-plane, one involving a notch-and-phase-lag controller and the other a deadbeat-response design.

Notch-and-Phase-Lag Controller

Let the digital controller be described by the transfer function

$$D(z) = \frac{4(z^2 - 1.7104z + 0.9994)(z - 0.9603)}{(z - 1)(z - 0.4)(z - 0.5)} \tag{10-280}$$

The two complex zeros of $D(z)$ are introduced to cancel the complex poles of $G_{h0}G_p(z)$. The phase-lag portion of $D(z)$ contains the zero at 0.9603 which is designed to cancel the real pole of $G_{h0}G_p(z)$, and the pole at $z = 1$ makes the compensated system type 1. Finally, the poles at $z = 0.4$ and 0.5 are used to make $D(z)$ physically realizable. The gain of four of $D(z)$ is selected so that the unit-step response of the compensated system has a near zero overshoot. The open-loop transfer function of the compensated system is

$$D(z)G_{h0}G_p(z) = \frac{0.008436(z + 2.814)(z + 0.1887)}{(z - 1)(z - 0.4)(z - 0.5)} \tag{10-281}$$

Figure 10-69 shows the unit-step response of the compensated system with the notch-phase-lag controller of Eq. (10-280). The maximum overshoot is 0.6 percent which occurs at $t = 0.026$ s.

Deadbeat-Response Design (Step Input)

Equation (10-279) is written as

$$G_{h0}G_p(z) = \frac{0.002109z^{-1}(1 + 0.1887z^{-1})(1 + 2.814z^{-1})}{(1 - 0.9603z^{-1})(1 - 1.7104z^{-1} + 0.9994z^{-2})} \tag{10-282}$$

Thus,

$$Q(z^{-1}) = 0.002109z^{-1}(1 + 0.1887z^{-1})(1 + 2.814z^{-1}) \tag{10-283}$$

$$P(z^{-1}) = (1 - 0.9603z^{-1})(1 - 1.7104z^{-1} + 0.9994z^{-2}) \tag{10-284}$$

and $Q(1) = 0.009562$. Applying Eq. (10-238) and simplifying, we have

$$D(z) = \frac{104.585(z - 0.9603)(z^2 - 1.7104z + 0.9994)}{(z - 1)(z^2 + 0.7794z + 0.1171)} \tag{10-285}$$

The open-loop transfer function of the compensated system is

$$D(z)G_{h0}G_p(z) = \frac{0.2206(z + 0.1887)(z + 2.814)}{(z - 1)(z^2 + 0.7794z + 0.1171)} \tag{10-286}$$

The closed-loop transfer function is

$$M(z) = \frac{0.2206(z + 0.1887)(z + 2.814)}{z^3} \tag{10-287}$$

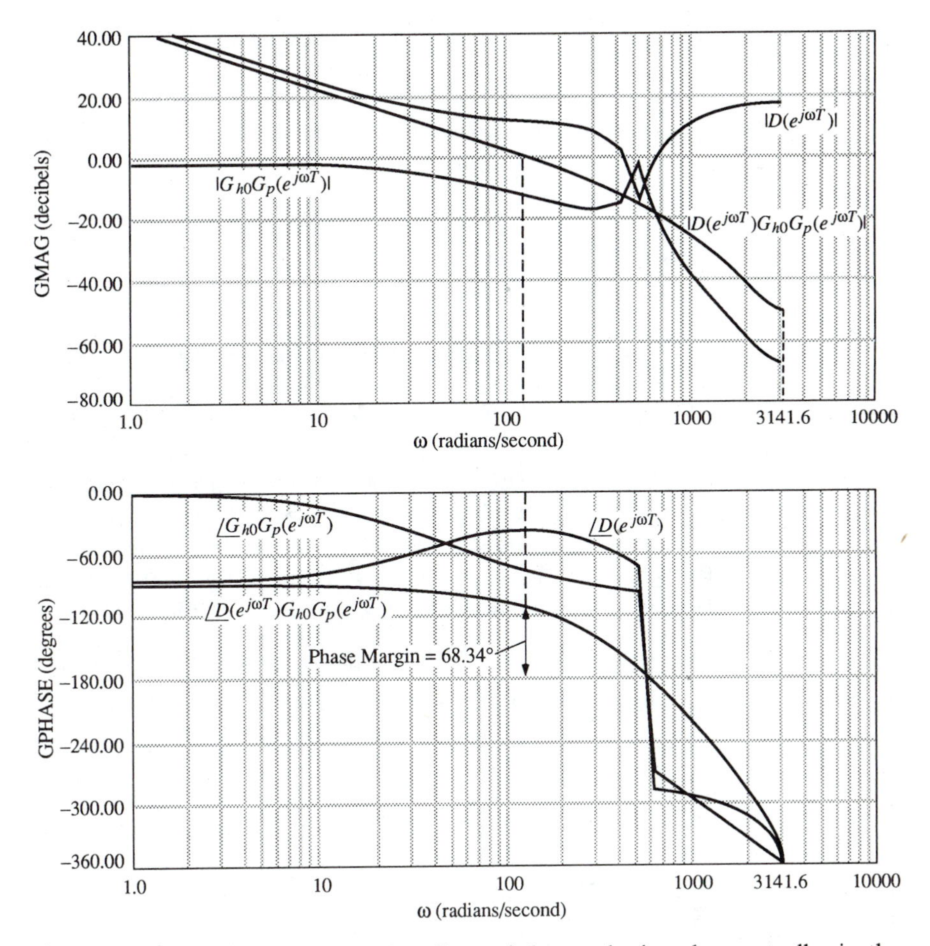

Figure 10-70. Bode plots showing the effects of the notch-phase-lag controller in the discrete-data system in Example 10-22.

The output transform to a unit-step input is

$$C(z) = 0.2206z^{-1} + 0.8829z^{-2} + z^{-3} + z^{-4} + z^{-5} + \cdots \qquad \textbf{(10-288)}$$

Thus, the output $c(t)$ reaches the final steady state without intersampling ripples in three sampling periods, or 0.003 s, as shown in Fig. 10-69.

Figure 10-70 shows the Bode plots of $G_{h0}G_p(z)$ in Eq. (10-279), $D(z)$ in Eq. (10-280), and $D(z)G_{h0}G_p(z)$ in Eq. (10-281). The uncompensated system has a phase margin of infinity and a gain margin of 7.93 dB, but a resonance occurs at 530 rad/s. The controller described by Eq. (10-280) places a "notch" at the resonant frequency and effectively cancels the resonance. The compensated system has a phase margin of 68.34 degrees, and a gain margin of 17.08 dB.

10-11 POLE-PLACEMENT DESIGN BY STATE FEEDBACK (SINGLE INPUT)

In Chapter 5 we considered the design of a single-input digital control system by state feedback. It was shown by means of the phase-variable canonical form (PVCF) transformation that for a single-input system

$$\mathbf{x}[(k+1)T] = \mathbf{A}\mathbf{x}(kT) + \mathbf{B}u(kT) \tag{10-289}$$

with the state-feedback control

$$u(kT) = -\mathbf{G}\mathbf{x}(kT) \tag{10-290}$$

where $\mathbf{G}$ is the $1 \times n$ constant feedback matrix, the eigenvalues of $\mathbf{A} - \mathbf{BG}$ can be arbitrarily assigned if and only if the pair $[\mathbf{A}, \mathbf{B}]$ is completely controllable. As a matter of fact, the property of arbitrary pole placement by state feedback for a controllable system is also true for a system with multiple inputs.

In Chapter 5 the subject of pole-placement design by state feedback was touched upon using the phase-variable canonical form. In this section we shall present more general methods of pole-placement design for single-input systems with state feedback. First we shall establish certain functional relationships as design tools. We define the following functional relations:

$\mathbf{T}_0(z) = -\mathbf{G}(z\mathbf{I} - \mathbf{A})^{-1}\mathbf{B}$ open-loop return signal matrix (10-291)

$\mathbf{T}_c(z) = -\mathbf{G}(z\mathbf{I} - \mathbf{A} + \mathbf{BG})^{-1}\mathbf{B}$ closed-loop return signal matrix (10-292)

$\mathbf{T}(z) = \mathbf{I} - \mathbf{T}_0(z)$ return difference matrix (10-293)

$\Delta_0(z) = |z\mathbf{I} - \mathbf{A}|$ characteristic equation of $\mathbf{A}$ (open-loop system) (10-294)

$\Delta_c(z) = |z\mathbf{I} - \mathbf{A} + \mathbf{BG}|$ characteristic equation of $\mathbf{A} - \mathbf{BG}$ (closed-loop system) (10-295)

$$\Delta(z) = |\mathbf{I} - \mathbf{T}_0(z)| = |\mathbf{T}(z)| \tag{10-296}$$

In the preceding equations $\mathbf{I}$ denotes the identity matrix of the appropriate dimension in a given equation.

First we shall show that

$$\Delta(z) = \frac{\Delta_c(z)}{\Delta_0(z)} \tag{10-297}$$

This is accomplished by writing

$$z\mathbf{I} - \mathbf{A} + \mathbf{BG} = (z\mathbf{I} - \mathbf{A})[\mathbf{I} + (z\mathbf{I} - \mathbf{A})^{-1}\mathbf{BG}] \tag{10-298}$$

Taking the determinant on both sides of the last equation, we get

$$\Delta_c(z) = |z\mathbf{I} - \mathbf{A} + \mathbf{BG}| = \Delta_0(z)|\mathbf{I} + (z\mathbf{I} - \mathbf{A})^{-1}\mathbf{BG}| \tag{10-299}$$

Since

$$|\mathbf{I} + (z\mathbf{I} - \mathbf{A})^{-1}\mathbf{BG}| = |\mathbf{I} + \mathbf{BG}(z\mathbf{I} - \mathbf{A})^{-1}| = |\mathbf{I} + \mathbf{G}(z\mathbf{I} - \mathbf{A})^{-1}\mathbf{B}| \tag{10-300}$$

where the identity matrices are of different dimensions, Eq. (10-299) becomes

$$\Delta_c(z) = \Delta_0(z)\Delta(z) \tag{10-301}$$

Thus, Eq. (10-297) is proven.

The next functional relation of importance is

$$\mathbf{T}_0(z) = \mathbf{T}(z)\mathbf{T}_c(z) \tag{10-302}$$

or

$$\mathbf{G}(z\mathbf{I} - \mathbf{A})^{-1}\mathbf{B} = [\mathbf{I} + \mathbf{G}(z\mathbf{I} - \mathbf{A})^{-1}\mathbf{B}]\mathbf{G}(z\mathbf{I} - \mathbf{A} + \mathbf{BG})^{-1}\mathbf{B} \tag{10-303}$$

Taking the matrix inverse on both sides of Eq. (10-298), we get

$$(z\mathbf{I} - \mathbf{A} + \mathbf{BG})^{-1} = [\mathbf{I} + (z\mathbf{I} - \mathbf{A})^{-1}\mathbf{BG}]^{-1}(z\mathbf{I} - \mathbf{A})^{-1} \tag{10-304}$$

Premultiplication on both sides of the last equation by $\mathbf{I} + (z\mathbf{I} - \mathbf{A})^{-1}\mathbf{BG}$ yields

$$[\mathbf{I} + (z\mathbf{I} - \mathbf{A})^{-1}\mathbf{BG}](z\mathbf{I} - \mathbf{A} + \mathbf{BG})^{-1} = (z\mathbf{I} - \mathbf{A})^{-1} \tag{10-305}$$

Then we premultiply both sides by $\mathbf{G}$ and post-multiply by $\mathbf{B}$; the last equation becomes

$$\mathbf{G}[\mathbf{I} + (z\mathbf{I} - \mathbf{A})^{-1}\mathbf{BG}](z\mathbf{I} - \mathbf{A} + \mathbf{BG})^{-1}\mathbf{B} = \mathbf{G}(z\mathbf{I} - \mathbf{A})^{-1}\mathbf{B} \tag{10-306}$$

The last equation is written

$$[\mathbf{I} + \mathbf{G}(z\mathbf{I} - \mathbf{A})^{-1}\mathbf{B}]\mathbf{G}(z\mathbf{I} - \mathbf{A} + \mathbf{BG})^{-1}\mathbf{B} = \mathbf{G}(z\mathbf{I} - \mathbf{A})^{-1}\mathbf{B} \tag{10-307}$$

Thus, the relation in Eq. (10-302) is proven.

The last relation we need is derived by use of Eqs. (10-297) and (10-304). Equation (10-293) is written as

$$\begin{aligned}\mathbf{T}(z) &= \mathbf{I} + \mathbf{G}(z\mathbf{I} - \mathbf{A})^{-1}\mathbf{B} \\ &= \mathbf{I} + \mathbf{G}\,\frac{\operatorname{adj}(z\mathbf{I} - \mathbf{A})\mathbf{B}}{\Delta_0(z)}\end{aligned} \tag{10-308}$$

where adj $(z\mathbf{I} - \mathbf{A})$ denotes the adjoint of the matrix $z\mathbf{I} - \mathbf{A}$.

Let

$$\mathbf{k}(z) = [\operatorname{adj}(z\mathbf{I} - \mathbf{A})]\mathbf{B} \qquad (n \times 1) \tag{10-309}$$

Then, Eq. (10-308) is written as

$$\mathbf{T}(z) = \frac{\Delta_0(z) + \mathbf{Gk}(z)}{\Delta_0(z)} = \Delta(z) \tag{10-310}$$

where $\mathbf{T}(z)$ is shown to be a scalar.

Using Eq. (10-297), the last equation leads to

$$\mathbf{Gk}(z) = \Delta_c(z) - \Delta_0(z) \tag{10-311}$$

Thus, knowing $\mathbf{k}(z)$, $\Delta_c(z)$, and $\Delta_0(z)$, we should be able to solve for the feedback gain matrix $\mathbf{G}$ from Eq. (10-311) if the pair $(\mathbf{A}, \mathbf{B})$ is completely controllable.

Two expressions for $\mathbf{G}$ can be obtained from Eq. (10-311). Let

$$\Delta_c(z) = z^n + \alpha_n z^{n-1} + \alpha_{n-1} z^{n-2} + \cdots + \alpha_2 z + \alpha_1 \tag{10-312}$$

$$\Delta_0(z) = z^n + a_n z^{n-1} + a_{n-1} z^{n-2} + \cdots + a_2 z + a_1 \tag{10-313}$$

From Eq. (5-209), $\mathbf{k}(z)$ is written

$$\mathbf{k}(z) = [\text{adj }(z\mathbf{I} - \mathbf{A})]\mathbf{B} = \sum_{j=1}^{n} z^{j-1} \sum_{i=j}^{n} a_{i+1}\mathbf{A}^{i-j}\mathbf{B} \tag{10-314}$$

Then, Eq. (10-311) is written

$$\mathbf{G} \sum_{j=1}^{n} z^{j-1} \sum_{i=j}^{n} a_{i+1}\mathbf{A}^{i-j}\mathbf{B} = \sum_{i=0}^{n-1} (\alpha_{i+1} - a_{i+1})z^i \tag{10-315}$$

Equating the coefficients of the like powers of z on both sides of the last equation, we get

$$\begin{aligned}
\mathbf{G}(a_{n+1}\mathbf{B}) &= \alpha_n - a_n \qquad a_{n+1} = 1 \\
\mathbf{G}(a_n + a_{n+1}\mathbf{A})\mathbf{B} &= \alpha_{n-1} - a_{n-1} \\
&\vdots \\
\mathbf{G} \sum_{i=2}^{n} a_{i+1}\mathbf{A}^{i-2}\mathbf{B} &= \alpha_2 - a_2 \\
\mathbf{G} \sum_{i=1}^{n} a_{i+1}\mathbf{A}^{i-1}\mathbf{B} &= \alpha_1 - a_1
\end{aligned} \tag{10-316}$$

or, in matrix form,

$$\begin{bmatrix} a_{n+1} & 0 & 0 & \cdots & 0 \\ a_n & a_{n+1} & 0 & \cdots & 0 \\ a_{n-1} & a_n & a_{n+1} & \cdots & 0 \\ \vdots & \vdots & \vdots & \cdots & \vdots \\ a_2 & a_3 & a_4 & \cdots & a_{n+1} \end{bmatrix} \begin{bmatrix} \mathbf{B}' \\ \mathbf{B}'\mathbf{A}' \\ \mathbf{B}'(\mathbf{A}')^2 \\ \vdots \\ \mathbf{B}'(\mathbf{A}')^{n-1} \end{bmatrix} \mathbf{G}' = \begin{bmatrix} \alpha_n - a_n \\ \alpha_{n-1} - a_{n-1} \\ \alpha_{n-2} - a_{n-2} \\ \vdots \\ \alpha_1 - a_1 \end{bmatrix} \tag{10-317}$$

Let

$$\mathbf{M} = \begin{bmatrix} a_{n+1} & 0 & 0 & \cdots & 0 \\ a_n & a_{n+1} & 0 & \cdots & 0 \\ a_{n-1} & a_n & a_{n+1} & \cdots & 0 \\ \vdots & \vdots & \vdots & & \vdots \\ a_1 & a_2 & a_3 & \cdots & a_{n+1} \end{bmatrix} \tag{10-318}$$

$$\mathbf{S} = [\mathbf{B} \quad \mathbf{AB} \quad \mathbf{A}^2\mathbf{B} \quad \cdots \quad \mathbf{A}^{n-1}\mathbf{B}] \tag{10-319}$$

$$\boldsymbol{\alpha} = [\alpha_n \quad \alpha_{n-1} \quad \cdots \quad \alpha_1]' \tag{10-320}$$

$$\mathbf{a} = [a_n \quad a_{n-1} \quad \cdots \quad a_1]' \tag{10-321}$$

Thus, Eq. (10-317) is written

$$\mathbf{MS'G'} = \boldsymbol{\alpha} - \mathbf{a} \tag{10-322}$$

Solving for $\mathbf{G}$ from the last equation, we get

$$\mathbf{G} = [(\mathbf{MS'})^{-1}(\boldsymbol{\alpha} - \mathbf{a})]' \qquad \text{(10-323)*}$$

Since $\mathbf{M}$ is a triangular matrix with ones on the main diagonal, it is nonsingular. Thus, for $\mathbf{G}$ to have the solution given by Eq. (10-323), the controllability matrix $\mathbf{S}$ must be of rank n, or the pair $(\mathbf{A}, \mathbf{B})$ must be controllable.

The feedback matrix $\mathbf{G}$ is expressed as a function of the coefficients of the closed-loop characteristic equation, α_i, $i = 1, 2, \ldots, n$, in Eq. (10-323). An alternate expression for $\mathbf{G}$ can be obtained in terms of the desired closed-loop eigenvalues. Let these eigenvalues be $z_1, z_2, z_3, \ldots, z_m$, all distinct, and the remaining ones are of multiple order. Then,

$$\Delta_c(z_i) = 0 \qquad i = 1, 2, \ldots, n \tag{10-324}$$

and, thus, Eq. (10-311) leads to

$$\mathbf{Gk}(z_i) = -\Delta_0(z_i) \qquad i = 1, 2, \ldots, m \tag{10-325}$$

Letting

$$\mathbf{k}_i = \mathbf{k}(z_i) \tag{10-326}$$

and

$$\Delta_0(z_i) = \Delta_{0i} \tag{10-327}$$

for $i = 1, 2, \ldots, m$, Eq. (10-325) becomes

$$\mathbf{Gk}_i = -\Delta_{0i} \tag{10-328}$$

For a multiple-order eigenvalue of multiplicity q, we take the derivatives on both sides of Eq. (10-311) with respect to z, and then setting $z = z_{m+j}$, $j = 1, 2, \ldots, q$, we have

$$\mathbf{Gk}_{m+j} = -\Delta_{m+j} \tag{10-329}$$

where

$$\mathbf{k}_{m+j} = \frac{d^j}{dz^j}\mathbf{k}(z)\bigg|_{z=z_{m+j}} \qquad j = 1, 2, \ldots, q \tag{10-330}$$

and

$$\Delta_{m+j} = \frac{d^j}{dz^j}\Delta_0(z)\bigg|_{z=z_{m+j}} \qquad j = 1, 2, \ldots, q \tag{10-331}$$

---

* The expression for $\mathbf{G}$ is simplified and can be expressed in terms of the closed-loop eigenvalues if $\mathbf{A}$ and $\mathbf{B}$ are in phase-variable canonical form (see Problem 10-33).

Now for all n eigenvalues, we have

$$\mathbf{G}[\mathbf{k}_1 \quad \mathbf{k}_2 \quad \cdots \quad \mathbf{k}_n] = -[\Delta_{01} \quad \Delta_{02} \quad \cdots \quad \Delta_{0n}] \tag{10-332}$$

Thus,

$$\mathbf{G} = -[\Delta_{01} \quad \Delta_{02} \quad \cdots \quad \Delta_{0n}]\mathbf{K}^{-1} \tag{10-333}$$

where

$$\mathbf{K} = [\mathbf{k}_1 \quad \mathbf{k}_2 \quad \cdots \quad \mathbf{k}_n] \tag{10-334}$$

Since $(\mathbf{A}, \mathbf{B})$ is controllable, we are assured of a solution for $\mathbf{G}$ from Eq. (10-333); and, under this condition, $\mathbf{K}^{-1}$ exists.

Example 10-23

This example is aimed at illustrating the design of a digital control system with pole placement and state feedback. The controlled process is a dc motor whose block diagram is shown in Fig. 10-71. The dc motor is used to control a pure inertia load in such a way that any nonzero initial values in the armature current i_a, and the motor-load velocity ω, should be reduced to zero as quickly as possible. This type of system belongs to a well-known class of control systems called *regulators*. The system parameters of the dc motor are given as follows:

Armature resistance	$R_a = 1\ \Omega$
Armature inductance	$L_a =$ negligible
Torque constant	$K_a = 0.345$ Nm/A
Back EMF constant	$K_b = 0.345$ Nm/A
Motor and load inertia	$J = 1.41 \times 10^{-3}$ kg-m^2
Viscous friction coefficient	$B = 0.25$ kg-m

Note that the parameters are given in SI units and are consistent. Figure 10-72 shows a state diagram of the dc motor. The input is the armature voltage $e_a(t)$. From the state diagram it is apparent that the state variables are $\theta(t)$ and $\omega(t)$. The state equations of the dc motor system are written as

$$\dot{\mathbf{x}}(t) = \mathbf{A}\mathbf{x}(t) + \mathbf{B}u(t) \tag{10-335}$$

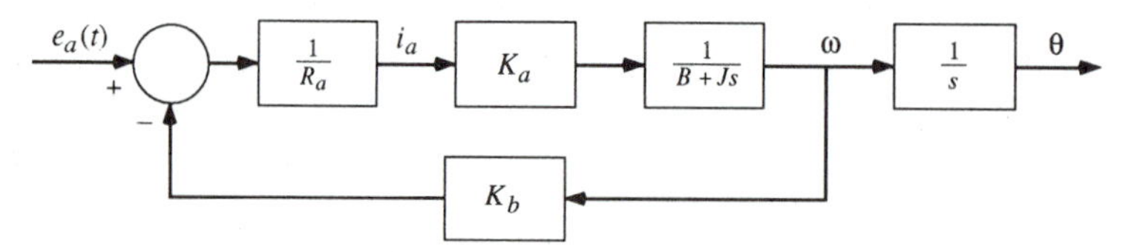

Figure 10-71. Block diagram of a dc motor system.

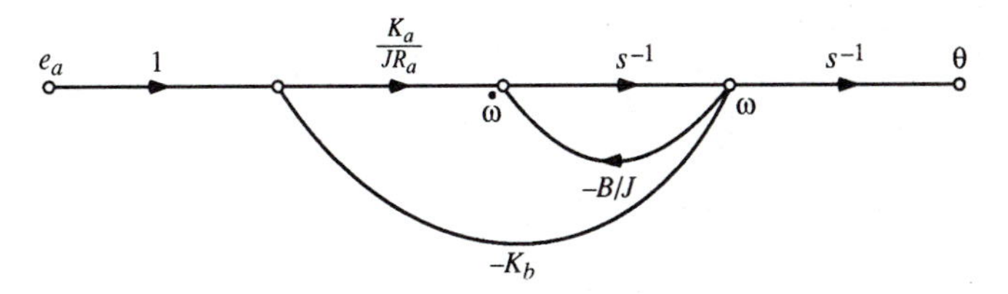

Figure 10-72. A state diagram of the dc motor system shown in Fig. 10-69.

where

$$\mathbf{x}(t) = \begin{bmatrix} \theta(t) \\ \omega(t) \end{bmatrix} \tag{10-336}$$

$$u(t) = e_a(t) \tag{10-337}$$

$$\mathbf{A} = \begin{bmatrix} 0 & 1 \\ 0 & -\dfrac{(BR_a + K_a K_b)}{JR_a} \end{bmatrix} \tag{10-338}$$

$$\mathbf{B} = \begin{bmatrix} 0 \\ \dfrac{K_a}{JR_a} \end{bmatrix} \tag{10-339}$$

Substituting the system parameters in the matrices **A** and **B**, we have

$$\mathbf{A} = \begin{bmatrix} 0 & 1 \\ 0 & -267.19 \end{bmatrix} \tag{10-340}$$

$$\mathbf{B} = \begin{bmatrix} 0 \\ 244.68 \end{bmatrix} \tag{10-341}$$

Let us incorporate state feedback by sensing the displacement θ and the velocity ω, and the control is derived through sample-and-hold; that is,

$$u(kT) = -\mathbf{G}\mathbf{x}(kT) \tag{10-342}$$

where

$$\mathbf{G} = [g_1 \quad g_2] \tag{10-343}$$

and the sampling period T is 0.005 s.

The design problem is to find g_1 and g_2 such that the eigenvalues of the closed-loop digital control system are $\lambda_1 = \lambda_2 = 0$. It is important to point out that the eigenvalues of the closed-loop digital control system are the eigenvalues of $\boldsymbol{\phi}(T) - \boldsymbol{\theta}(T)\mathbf{G}$, where

$$\boldsymbol{\phi}(T) = e^{\mathbf{A}T} \tag{10-344}$$

and

$$\boldsymbol{\theta}(T) = \int_0^T \boldsymbol{\phi}(\lambda)\mathbf{B}\, d\lambda \tag{10-345}$$

The state-transition matrix $\boldsymbol{\phi}(T)$ for $T = 0.005$ s is determined by use of the **A** matrix in Eq. (10-340). Thus,

$$\boldsymbol{\phi}(T) = \mathscr{L}^{-1}[(s\mathbf{I} - \mathbf{A})^{-1}]|_{t=T=0.005} = \begin{bmatrix} 1 & 0.00276 \\ 0 & 0.263 \end{bmatrix} \tag{10-346}$$

$$\boldsymbol{\theta}(T) = \mathscr{L}^{-1}[(s\mathbf{I} - \mathbf{A})^{-1}\mathbf{B}s^{-1}]|_{t=T=0.005} = \begin{bmatrix} 0.00205 \\ 0.675 \end{bmatrix} \tag{10-347}$$

The controlled process is now discretized, and the state equations in the difference equation form are written as

$$\mathbf{x}[(k+1)T] = \boldsymbol{\phi}(T)\mathbf{x}(kT) + \boldsymbol{\theta}(T)u(kT) \tag{10-348}$$

where $\boldsymbol{\phi}(T)$ and $\boldsymbol{\theta}(T)$ are given by Eqs. (10-346) and (10-347), respectively. The state-feedback control is described by

$$u(kT) = -\mathbf{G}\mathbf{x}(kT) \tag{10-349}$$

The characteristic equation of the open-loop system, or of $\boldsymbol{\phi}(T)$, is

$$\Delta_0(z) = |z\mathbf{I} - \boldsymbol{\phi}(T)| = \begin{vmatrix} z-1 & -0.00276 \\ 0 & z-0.263 \end{vmatrix}$$

$$= z^2 - 1.263z + 0.263 = 0 \tag{10-350}$$

Since the pair $[\boldsymbol{\phi}(T), \boldsymbol{\theta}(T)]$ is completely controllable, the eigenvalues of $\boldsymbol{\phi}(T) - \boldsymbol{\theta}(T)\mathbf{G}$ can be arbitrarily placed by state feedback. Let the desired closed-loop characteristic equation be

$$\Delta_c(z) = z^2 \tag{10-351}$$

From Eq. (10-309),

$$\mathbf{k}(z) = \text{adj}\,[z\mathbf{I} - \boldsymbol{\phi}(T)] \cdot \boldsymbol{\theta}(T)$$

$$= \begin{bmatrix} 0.00205z + 0.00132 \\ 0.675(z-1) \end{bmatrix} \tag{10-352}$$

The feedback matrix **G** is determined from Eq. (10-333),

$$\mathbf{G} = -[\Delta_{01} \quad \Delta_{02}]\mathbf{K}^{-1} \tag{10-353}$$

where

$$\Delta_{01} = \Delta_0(z)|_{z=0} = 0.263 \tag{10-354}$$

$$\Delta_{02} = \frac{d}{dz}\Delta_0(z)|_{z=0} = (2z - 1.263)|_{z=0} = -1.263 \tag{10-355}$$

$$\mathbf{K} = [\mathbf{k}_1 \quad \mathbf{k}_2]$$

where

$$\mathbf{k}_1 = \text{adj}\,[z\mathbf{I} - \boldsymbol{\phi}(T)] \cdot \boldsymbol{\theta}(T)|_{z=0}$$

$$= \begin{bmatrix} 0.001323 \\ -0.675 \end{bmatrix} \tag{10-356}$$

$$\mathbf{k}_2 = \frac{d}{dz}\,\text{adj}\,[z\mathbf{I} - \boldsymbol{\phi}(T)] \cdot \boldsymbol{\theta}(T)|_{z=0}$$

$$= \begin{bmatrix} 0.00205 \\ 0.675 \end{bmatrix} \tag{10-357}$$

Substituting Eqs. (10-354) through (10-357) into Eq. (10-353), we get

$$\mathbf{G} = [296.3 \quad 0.970] \tag{10-358}$$

Substituting the feedback gain matrix and the control $u(kT)$ in Eq. (10-348), we get the state equation of the closed-loop system as

$$\begin{bmatrix} x_1[(k+1)T] \\ x_2[(k+1)T] \end{bmatrix} = \begin{bmatrix} 0.395 & 0.00078 \\ -200 & -0.392 \end{bmatrix} \begin{bmatrix} x_1(kT) \\ x_2(kT) \end{bmatrix} \tag{10-359}$$

The solution of the last equation in the z-domain is

$$\mathbf{X}(z) = \begin{bmatrix} z^{-1} + 0.395z^{-2} & 0.00078z^{-2} \\ -200z^{-2} & z^{-1} - 0.392z^{-2} \end{bmatrix} \mathbf{x}(0) \tag{10-360}$$

Thus, we see that the natural responses at the sampling instants of the designed system with state feedback are deadbeat responses, since the two eigenvalues of the closed-loop system are all placed at $z = 0$. For any initial conditions for $x_1(kT)$ and $x_2(kT)$, the responses will become zero after two sampling periods, as indicated by Eq. (10-360).

As an alternate method, the feedback matrix $\mathbf{G}$ can also be found by use of Eq. (10-323).

10-12 POLE-PLACEMENT DESIGN BY STATE FEEDBACK (MULTIPLE INPUTS)

With a slight modification the pole-placement design method for systems with single inputs can be applied to systems with multiple inputs. Consider the system

$$\mathbf{x}[(k+1)T] = \mathbf{A}\mathbf{x}(kT) + \mathbf{B}\mathbf{u}(kT) \tag{10-361}$$

where $\mathbf{x}(kT)$ is an n-vector and $\mathbf{u}(kT)$ is an r-vector. The pair $(\mathbf{A}, \mathbf{B})$ is assumed to be completely controllable. The design problem is stated as follows. Find the feedback matrix $\mathbf{G}$ $(r \times n)$ such that the control

$$\mathbf{u}(kT) = -\mathbf{G}\mathbf{x}(kT) \tag{10-362}$$

places the eigenvalues of $\mathbf{A} - \mathbf{BG}$ at arbitrarily assigned positions in the z-plane.

Let us construct the single-input system

$$\mathbf{x}[(k+1)T] = \mathbf{A}\mathbf{x}(kT) + \mathbf{B}^* u(kT) \tag{10-363}$$

and let the $n \times 1$ matrix $\mathbf{B}^*$ be defined as

$$\mathbf{B}^* = \mathbf{B}\mathbf{w} \tag{10-364}$$

where $\mathbf{w}$ is $r \times 1$. The matrix $\mathbf{w}$ must be chosen so that the pair $(\mathbf{A}, \mathbf{B}^*)$ is controllable. Then, we can apply the feedback

$$u(kT) = -\mathbf{G}^*\mathbf{x}(kT) \tag{10-365}$$

to place the eigenvalues of $\mathbf{A} - \mathbf{B}^*\mathbf{G}^*$ at the same locations as those of $\mathbf{A} - \mathbf{BG}$. Then, the problem becomes that of designing the state feedback for the single-input system of Eq. (10-363). Once the feedback matrix $\mathbf{G}^*$ is determined, $\mathbf{G}$ is given by

$$\mathbf{G} = \mathbf{w}\mathbf{G}^* \tag{10-366}$$

since $\mathbf{BG} = \mathbf{B}^*\mathbf{G}^*$.

It is apparent that in general $\mathbf{w}$ is not unique, and it only has to satisfy the condition that $(\mathbf{A}, \mathbf{Bw})$ is controllable. The feedback gain $\mathbf{G}^*$ of the single-input model can be determined using either Eq. (10-323) or Eq. (10-333).

Example 10-24

Consider the multiple-input digital control system

$$\mathbf{x}[(k+1)T] = \mathbf{A}\mathbf{x}(kT) + \mathbf{B}\mathbf{u}(kT) \tag{10-367}$$

where

$$\mathbf{A} = \begin{bmatrix} 0 & 1 \\ -1 & -2 \end{bmatrix} \qquad \mathbf{B} = \begin{bmatrix} 1 & 0 \\ 0 & 1 \end{bmatrix} \tag{10-368}$$

The pair $(\mathbf{A}, \mathbf{B})$ is controllable. The design problem involves the determination of the feedback matrix $\mathbf{G}$ such that the state feedback

$$\mathbf{u}(kT) = -\mathbf{G}\mathbf{x}(kT) \tag{10-369}$$

places the closed-loop eigenvalues at $z_1 = 0.1$ and $z_2 = 0.2$.

Let us define

$$\mathbf{B}^* = \mathbf{B}\mathbf{w} = \begin{bmatrix} 1 & 0 \\ 0 & 1 \end{bmatrix}\begin{bmatrix} w_1 \\ w_2 \end{bmatrix} = \begin{bmatrix} w_1 \\ w_2 \end{bmatrix} \tag{10-370}$$

For the pair ($\mathbf{A}$, $\mathbf{B}^*$) to be controllable, we require that the matrix $[\mathbf{B}^* \quad \mathbf{AB}^*]$ be nonsingular. Thus,

$$|\mathbf{B}^* \quad \mathbf{AB}^*| = \begin{vmatrix} w_1 & w_2 \\ w_2 & -w_1 - 2w_2 \end{vmatrix} = -(w_1 + w_2)^2 \neq 0 \tag{10-371}$$

or $w_1 \neq -w_2$.

The feedback matrix $\mathbf{G}^*$ of the single-input system is first determined by use of Eq. (10-333). We have

$$\mathbf{G}^* = -[\Delta_{01} \quad \Delta_{02}]\mathbf{K}^{-1} \tag{10-372}$$

where

$$\Delta_{01} = |z\mathbf{I} - \mathbf{A}|_{z=0.1} = 1.21 \tag{10-373}$$

$$\Delta_{02} = |z\mathbf{I} - \mathbf{A}|_{z=0.2} = 1.44 \tag{10-374}$$

$$\begin{aligned} \mathbf{k}(z) &= \text{adj } (z\mathbf{I} - \mathbf{A})\mathbf{B}^* \\ &= \text{adj} \begin{bmatrix} z & -1 \\ 1 & z+2 \end{bmatrix} \begin{bmatrix} w_1 \\ w_2 \end{bmatrix} = \begin{bmatrix} w_1(z+2) + w_2 \\ -w_1 + w_2 z \end{bmatrix} \end{aligned} \tag{10-375}$$

Then,

$$\mathbf{k}_1 = \mathbf{k}(z_1) = \begin{bmatrix} 2.1w_1 + w_2 \\ -w_1 + 0.1w_2 \end{bmatrix} \tag{10-376}$$

$$\mathbf{k}_2 = \mathbf{k}(z_2) = \begin{bmatrix} 2.2w_1 + w_2 \\ -w_1 + 0.2w_2 \end{bmatrix} \tag{10-377}$$

$$\mathbf{K} = [\mathbf{k}_1 \quad \mathbf{k}_2] = \begin{bmatrix} 2.1w_1 + w_2 & 2.2w_1 + w_2 \\ -w_1 + 0.1w_2 & -w_1 + 0.2w_2 \end{bmatrix} \tag{10-378}$$

From the last equation we can again show that for $\mathbf{K}$ to be nonsingular, the condition in Eq. (10-371) must be satisfied. Let us arbitrarily choose $\mathbf{w} = [1 \quad 1]'$, although w_1 and w_2 satisfy the condition $w_1 \neq -w_2$. Then,

$$\mathbf{K} = \begin{bmatrix} 3.1 & 3.2 \\ -0.9 & -0.8 \end{bmatrix} \tag{10-379}$$

Substituting Eqs. (10-373), (10-374), and (10-378) in Eq. (10-372), we get

$$\mathbf{G}^* = -[0.82 \quad 1.48] \tag{10-380}$$

The feedback matrix of the multiple-input system is determined by using Eq. (10-366),

$$\mathbf{G} = \mathbf{w}\mathbf{G}^* = -\begin{bmatrix} 0.82 & 1.48 \\ 0.82 & 1.48 \end{bmatrix} \tag{10-381}$$

Now it is simple to show that

$$|z\mathbf{I} - \mathbf{A} + \mathbf{B}^*\mathbf{G}^*| = |z\mathbf{I} - \mathbf{A} + \mathbf{BG}| = z^2 - 0.3z + 0.02 \tag{10-382}$$

which has the desired roots at $z = 0.1$ and 0.2.

An alternate method of finding $\mathbf{G}^*$ is to use Eq. (10-323). In the present case,

$$\mathbf{M} = \begin{bmatrix} 1 & 0 \\ 2 & 1 \end{bmatrix} \tag{10-383}$$

$$\mathbf{S} = [\mathbf{B}^* \quad \mathbf{AB}^*] = \begin{bmatrix} w_1 & w_2 \\ w_2 & -w_1 - 2w_2 \end{bmatrix} \tag{10-384}$$

For the closed-loop eigenvalues to be at 0.1 and 0.2,

$$\boldsymbol{\alpha} = [-0.3 \quad 0.02]' \tag{10-385}$$

Also,

$$\mathbf{a} = [2 \quad 1]' \tag{10-386}$$

Then, Eq. (10-323) gives

$$\begin{aligned} \mathbf{G}^* &= [(\mathbf{MS}')^{-1}(\boldsymbol{\alpha} - \mathbf{a})]' \\ &= \frac{1}{(w_1 + w_2)^2}[-2.3w_1 - 0.98w_2 \quad -3.62w_1 - 2.3w_2] \end{aligned} \tag{10-387}$$

which shows that w_1 cannot be equal to $-w_2$. If, as in the above, we choose $w_1 = w_2 = 1$, we have

$$\mathbf{G}^* = -[0.82 \quad 1.48] \tag{10-388}$$

which is the same answer as obtained in Eq. (10-380).

Weighted- or Incomplete-Input Feedback

The method introduced in the foregoing discussions showed that the pole-placement design of multiple-input systems can be conducted in lieu of an "equivalent" single-input system. *The two systems are "equivalent" only in the sense that they have the same eigenvalues.*

The matrix $\mathbf{w}$ used to convert the multiple-input system to a single-input system has only the constraint that the latter must be controllable. However, although $\mathbf{w}$ can otherwise be arbitrarily chosen, in practice, we can impose other useful constraints on the selection of its elements. Since the input vector $\mathbf{u}(kT)$ is multiplied by $\mathbf{w}$, the significance of $\mathbf{w}$ is that its elements place various weights on the state-feedback control. For instance, if we choose $w_1 = 2w_2$, the gain in the feedback to u_1 would be twice as much as compared to that for u_2.

10-13 POLE-PLACEMENT DESIGN BY INCOMPLETE STATE FEEDBACK OR OUTPUT FEEDBACK

In practice not all the state variables are accessible. For economical reasons, it may not be feasible to feed back all the state variables, especially for high-order systems. Thus, incomplete state feedback or output feedback must be considered. The design of digital control systems with incomplete state feedback and output feedback for pole placement is discussed in the following.

Incomplete State Feedback

Consider the digital control system

$$\mathbf{x}[(k+1)T] = \mathbf{A}\mathbf{x}(kT) + \mathbf{B}\mathbf{u}(kT) \tag{10-389}$$

where $\mathbf{x}(kT)$ is an n-vector and $\mathbf{u}(kT)$ is an r-vector. The state feedback is described by

$$\mathbf{u}(kT) = -\mathbf{G}\mathbf{x}(kT) \tag{10-390}$$

Let us assume that $x_i(kT)$ is not available for feedback, where i can be one or more values from 1 to n. *This requires that the corresponding columns of* **G** *must contain all zeros*. Using the design procedure introduced in the last section, we let

$$\mathbf{G} = \mathbf{w}\mathbf{G}^* \tag{10-391}$$

where $\mathbf{w}$ is $r \times 1$ and $\mathbf{G}^*$ is $1 \times r$. The matrix $\mathbf{w}$ should be chosen so that $(\mathbf{A}, \mathbf{Bw})$ is completely controllable. *With incomplete feedback, the columns of* $\mathbf{G}^*$ *that correspond to the zero columns of* **G** *must equal zero*. Since the feedback matrix $\mathbf{G}^*$ for the single-input model is related to the desired closed-loop eigenvalue, the system parameters, and $\mathbf{w}$, through the relationship

$$\mathbf{G}^* = -[\Delta_{01} \quad \Delta_{02} \quad \cdots \quad \Delta_{0n}]\mathbf{K}^{-1} \tag{10-392}$$

with one or more columns of $\mathbf{G}^*$ forced to be zero, constraints are placed on the values of the desired closed-loop eigenvalues. The following example illustrates the design of a digital control system with incomplete state feedback.

Example 10-25

Consider the system

$$\mathbf{x}[(k+1)T] = \mathbf{A}\mathbf{x}(kT) + \mathbf{B}u(kT) \tag{10-393}$$

where

$$\mathbf{A} = \begin{bmatrix} 0 & 1 \\ -1 & -2 \end{bmatrix} \qquad \mathbf{B} = \begin{bmatrix} 0 \\ 1 \end{bmatrix}$$

The state feedback is defined as

$$u(kT) = -\mathbf{G}\mathbf{x}(kT) \tag{10-394}$$

where

$$\mathbf{G} = [g_1 \quad g_2] \tag{10-395}$$

First let us assume that only $\mathbf{x}_1(kT)$ is available for feedback; thus, $g_2 = 0$. The characteristic equation of the closed-loop system is

$$|z\mathbf{I} - \mathbf{A} + \mathbf{B}\mathbf{G}| = z^2 + (g_1 + 2)z + (1 - g_1) = 0 \tag{10-396}$$

Since we have only one parameter in g_1, the two eigenvalues of the closed-loop system cannot be arbitrarily assigned. Dividing both sides of Eq. (10-396) by the terms that do not contain g_1, we have

$$1 + \frac{g_1(z-1)}{z^2 + 2z + 1} = 0 \tag{10-397}$$

The root loci of Eq. (10-396) are constructed based on the pole-zero configuration of $(z-1)/(z^2+2z+1)$, as shown in Fig. 10-73(a). Notice that for negative values of g_1 both roots are outside the unit circle, and for positive values of g_1 one root always stays to the left of the -1 point in the z-plane. Thus, the conclusion is that with only $x_1(kT)$ available for feedback, the trajectories of the roots of Eq. (10-396) are restricted to those shown in Fig. 10-73(a), and the system cannot be stabilized for any value of g_1.

Now let us consider that only $x_2(kT)$ is available for feedback, so that

$$\mathbf{G} = [0 \quad g_2] \tag{10-398}$$

The characteristic equation of the closed-loop system is

$$|z\mathbf{I} - \mathbf{A} + \mathbf{B}\mathbf{G}| = z^2 + (g_2 + 2)z + (1 + g_2) = 0 \tag{10-399}$$

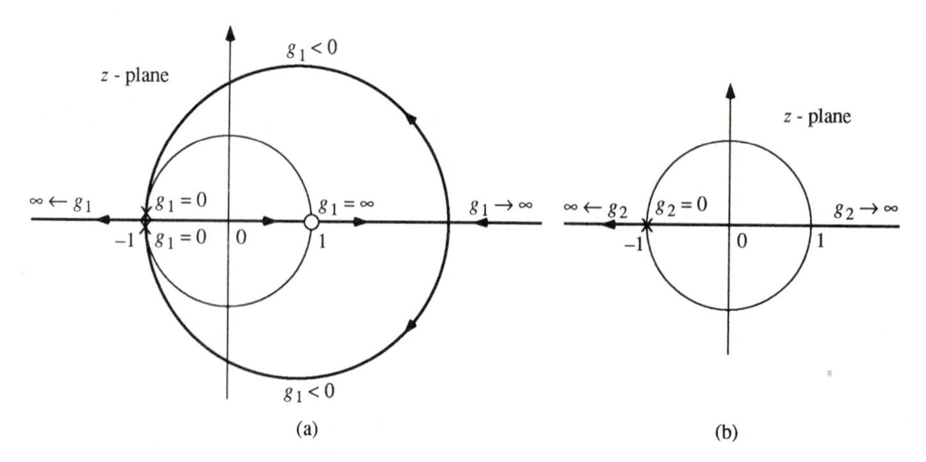

Figure 10-73. Root loci of the system in Example 10-25.

The root loci of the last equation are constructed with the help of the following equation:

$$1 + \frac{g_2(z + 1)}{z^2 + 2z + 1} = 0 \tag{10-400}$$

as shown in Fig. 10-73(b). In this case, the open-loop pole at $z = -1$ cannot be moved. As g_2 varies between $-\infty$ and ∞, the other root moves from $z = \infty$ to $z = -\infty$ on the real axis in the z-plane. Thus, we see that for the single-input system not only can the eigenvalues of the closed-loop system not be arbitrarily placed by incomplete state feedback, but the system is not stabilizable.

Let us modify the system by changing **B** to

$$\mathbf{B} = \begin{bmatrix} 1 & 0 \\ 0 & 1 \end{bmatrix}$$

The feedback matrix is

$$\mathbf{G} = \begin{bmatrix} g_{11} & g_{12} \\ g_{21} & g_{22} \end{bmatrix}$$

Then,

$$|z\mathbf{I} - \mathbf{A} + \mathbf{BG}| = z^2 + (2 + g_{11} + g_{22})z + g_{11}(2 + g_{22}) + (1 - g_{12})(1 + g_{21}) = 0 \tag{10-401}$$

When $x_1(kT)$ is not available for feedback, $g_{11} = g_{21} = 0$. Equation (10-401) becomes

$$z^2 + (2 + g_{22})z + (1 - g_{12}) = 0 \tag{10-402}$$

Since we have two independent parameters in g_{12} and g_{22} in the last equation, the two eigenvalues of $\mathbf{A} - \mathbf{BG}$ can be arbitrarily assigned. Similarly, for $g_{12} = g_{22} = 0$, we have

$$z^2 + (2 + g_{11})z + 2g_{11} + 1 + g_{21} = 0 \tag{10-403}$$

Again, the eigenvalues can be arbitrarily assigned by choosing the values of g_{11} and g_{21}.

Output Feedback

Since the outputs of a system are always measurable, we can always count on feeding back these signals through constant gains for control purposes. Thus, output feedback may be considered as an alternative to incomplete feedback.

Consider the system

$$\mathbf{x}[(k + 1)T] = \mathbf{Ax}(kT) + \mathbf{Bu}(kT) \tag{10-404}$$

$$\mathbf{c}(kT) = \mathbf{Dx}(kT) \tag{10-405}$$

where $\mathbf{x}(kT)$ is an n-vector, $\mathbf{u}(kT)$ is an r-vector, and $\mathbf{c}(kT)$ is a p-vector. The output feedback is defined as

$$\mathbf{u}(kT) = -\mathbf{Gc}(kT) \tag{10-406}$$

where $\mathbf{G}$ is the $r \times p$ output-feedback matrix. The design objective is to find $\mathbf{G}$ so that the eigenvalues of the closed-loop system are at the desired values. However, since in general $p \leq r \leq n$, not all the n eigenvalues can be arbitrarily assigned. We shall show that the number of eigenvalues that can be arbitrarily assigned depends on the ranks of $\mathbf{D}$ and $\mathbf{B}$.

The design of the output-feedback control can be carried out in a manner similar to that of the state-feedback control. Let us consider first the single-input case. Substitution of Eq. (10-405) in Eq. (10-406) and then in Eq. (10-404) gives us

$$\mathbf{x}[(k+1)T] = (\mathbf{A} - \mathbf{BGD})\mathbf{x}(kT) \tag{10-407}$$

Since the last equation is equivalent to the state equation of a closed-loop system with state-feedback gain $\mathbf{GD}$, the latter can be solved directly from Eq. (10-323) or Eq. (10-333) if the pair $(\mathbf{A}, \mathbf{B})$ is completely controllable. Thus, we have

$$\mathbf{GD} = [(\mathbf{MS}')^{-1}(\boldsymbol{\alpha} - \mathbf{a})]' \tag{10-408}$$

or

$$\mathbf{GD} = [\Delta_{01} \quad \Delta_{02} \quad \cdots \quad \Delta_{0n}]\mathbf{K}^{-1} \tag{10-409}$$

However, since $\mathbf{D}$ and $\mathbf{DK}$ are generally not square matrices, we cannot solve for $\mathbf{G}$ directly from the last two equations.

For the single-input case, $\mathbf{G}$ is $1 \times p$, $\mathbf{D}$ is $p \times n$, and $\mathbf{B}$ is $n \times 1$; so $\mathbf{GD}$ is always a $1 \times n$ row matrix. There are p gain elements in $\mathbf{G}$, but only m of these are free or independent parameters that may be used for design purposes, where m is the rank of $\mathbf{D}$, and $m \leq p$. For example,

$$\mathbf{D} = \begin{bmatrix} 1 & 0 & 0 \\ 0 & 1 & 0 \\ 0 & 2 & 0 \end{bmatrix}$$

which has a rank of two, and $\mathbf{G} = [g_1 \quad g_2 \quad g_3]$. Then,

$$\mathbf{GD} = [g_1 \quad g_2 \quad 2g_2]$$

which has only two independent gain parameters. This means that only two of the three eigenvalues of the system can be placed arbitrarily by the output feedback. For the single-input case, if the rank of $\mathbf{D}$ is equal to n, the order of the system, then output feedback is the same as complete state feedback, and if $(\mathbf{A}, \mathbf{B})$ is controllable, all the eigenvalues can be arbitrarily assigned.

For multiple-input systems, $\mathbf{B}$ is $n \times r$; we again form

$$\mathbf{B}^* = \mathbf{Bw} \tag{10-410}$$

where $\mathbf{w}$ is an $r \times 1$ matrix with r parameters; thus, $\mathbf{B}^*$ is $n \times 1$. Similarly,

$$\mathbf{G} = \mathbf{w}\mathbf{G}^* \tag{10-411}$$

where

$$\mathbf{G}^* = [g_1^* \quad g_2^* \quad \cdots \quad g_n^*] \qquad (1 \times n) \tag{10-412}$$

Then,

$$\mathbf{BGD} = \mathbf{BwG}^*\mathbf{D} = \mathbf{B}^*\mathbf{G}^*\mathbf{D} \tag{10-413}$$

and the characteristic equation of the closed-loop system is

$$|z\mathbf{I} - \mathbf{A} + \mathbf{BGD}| = |z\mathbf{I} - \mathbf{A} + \mathbf{B}^*\mathbf{G}^*\mathbf{D}| = 0 \tag{10-414}$$

Thus, $\mathbf{G}^*\mathbf{D}$ can be determined using Eq. (10-323) or Eq. (10-333).

Unlike in the single-input case with output feedback, the solution of the feedback gain now depends on the ranks of $\mathbf{D}$ and $\mathbf{B}$. In general, if the rank of $\mathbf{D}$ is greater than or equal to that of $\mathbf{B}$, the elements of $\mathbf{w}$ can be arbitrarily selected, of course, subject to the controllability of the pair $(\mathbf{A}, \mathbf{B})$. However, if the rank of $\mathbf{B}$ is greater than that of $\mathbf{D}$, we cannot arbitrarily assign all the elements of $\mathbf{w}$ if we wish to arbitrarily assign a maximum number of eigenvalues of the closed-loop system. For the design of systems with state feedback, we do not have this problem, since the rank of $\mathbf{D}$, where $\mathbf{D}$ is a unity matrix of dimension n, is always n, and the rank of $\mathbf{B}$ cannot exceed n. The following example will illustrate the design of output feedback and the constraints mentioned above.

Example 10-26

Given the digital control system that is described by the following dynamic equations,

$$\mathbf{x}[(k+1)T] = \mathbf{A}\mathbf{x}(kT) + \mathbf{B}\mathbf{u}(kT) \tag{10-415}$$

$$\mathbf{c}(kT) = \mathbf{D}\mathbf{x}(kT) \tag{10-416}$$

where

$$\mathbf{A} = \begin{bmatrix} 0 & 1 & 0 \\ 0 & 0 & 1 \\ -1 & 0 & 0 \end{bmatrix} \qquad \mathbf{B} = \begin{bmatrix} 0 & 1 \\ 1 & 0 \\ 0 & 0 \end{bmatrix}$$

$$\mathbf{D} = \begin{bmatrix} 1 & 0 & 0 \\ 1 & 1 & 0 \end{bmatrix}$$

The problem is to find the feedback gain matrix $\mathbf{G}$ such that the output feedback

$$\mathbf{u}(kT) = -\mathbf{G}\mathbf{c}(kT) \tag{10-417}$$

places the eigenvalues of $\mathbf{A} - \mathbf{BGD}$ at certain desired values. Since $\mathbf{D}$ has a rank of two and the rank of $\mathbf{B}$ is also two, a maximum of two of the eigenvalues can be arbitrarily assigned. Let these two eigenvalues be at $z_1 = 0.1$ and $z_2 = 0.2$.

The characteristic equation of $\mathbf{A}$ is

$$|z\mathbf{I} - \mathbf{A}| = z^3 + 1 \tag{10-418}$$

Thus,

$$\mathbf{M} = \begin{bmatrix} a_4 & 0 & 0 \\ a_3 & a_4 & 0 \\ a_2 & a_3 & a_4 \end{bmatrix} = \begin{bmatrix} 1 & 0 & 0 \\ 0 & 1 & 0 \\ 0 & 0 & 1 \end{bmatrix} \tag{10-419}$$

From Eq. (10-410), $\mathbf{B}^*$ is written

$$\mathbf{B}^* = \begin{bmatrix} 0 & 1 \\ 1 & 0 \\ 0 & 0 \end{bmatrix} \begin{bmatrix} w_1 \\ w_2 \end{bmatrix} = \begin{bmatrix} w_2 \\ w_1 \\ 0 \end{bmatrix} \tag{10-420}$$

which has two independent parameters in w_1 and w_2, due to the fact that $\mathbf{B}$ has a rank of two. The controllability matrix for the pair $(\mathbf{A}, \mathbf{B}^*)$ is

$$\mathbf{S} = [\mathbf{B}^* \quad \mathbf{AB}^* \quad \mathbf{A}^2\mathbf{B}^*] = \begin{bmatrix} w_2 & w_1 & 0 \\ w_1 & 0 & -w_2 \\ 0 & -w_2 & -w_1 \end{bmatrix} \tag{10-421}$$

The matrix $\mathbf{S}$ is nonsingular if $w_1^3 - w_2^3 \neq 0$.

Let

$$\mathbf{G}^* = [g_1^* \quad g_2^*] \tag{10-422}$$

Then

$$\mathbf{G}^*\mathbf{D} = [g_1^* + g_2^* \quad g_2^* \quad 0] \tag{10-423}$$

Since $\mathbf{D}$ is of rank two, $\mathbf{G}^*\mathbf{D}$ has two independent parameters in g_1^* and g_2^*. Now using Eq. (10-408), with $\mathbf{G}$ replaced by $\mathbf{G}^*$, we have

$$\mathbf{G}^*\mathbf{D} = [(\mathbf{MS}')^{-1}(\boldsymbol{\alpha} - \mathbf{a})]' \tag{10-424}$$

or

$$\begin{bmatrix} g_1^* + g_2^* \\ g_2^* \\ 0 \end{bmatrix}' = \frac{1}{w_1^3 - w_2^3} \begin{bmatrix} -w_2^2\alpha_3 + \alpha_2 w_1^2 - w_1 w_2(\alpha_1 - 1) \\ \alpha_3 w_1^2 - w_1 w_2 \alpha_2 + w_2^2(\alpha_1 - 1) \\ -w_1 w_2 \alpha_3 + \alpha_2 w_2^2 - w_1^2(\alpha_1 - 1) \end{bmatrix}' \tag{10-425}$$

The last row in Eq. (10-425) corresponds to the constraint equation

$$-w_1 w_2 \alpha_3 + \alpha_2 w_2^2 - w_1^2(\alpha_1 - 1) = 0 \tag{10-426}$$

Since only two of the three eigenvalues, or, correspondingly, two of the three coefficients of the closed-loop characteristic equation can be arbitrarily assigned, although w_1 and w_2 can be arbitrary ($w_1^3 \neq w_2^3$), they should be selected so that the third eigenvalue is stable. This requirement on stability does impose additional constraints on the values of w_1 and w_2. For example, the necessary condition for the closed-loop system to be stable is that $|\alpha_1| < 1$. From Eq. (10-426), we see that w_2 cannot be zero.

For $z = 0.1$ and 0.2 to be roots of the characteristic equation

$$z^3 + \alpha_3 z^2 + \alpha_2 z + \alpha_1 = 0 \tag{10-427}$$

the following two equations must be satisfied:

$$\alpha_2 + 0.3\alpha_3 + 0.07 = 0 \tag{10-428}$$

$$\alpha_1 - 0.02\alpha_3 - 0.006 = 0 \tag{10-429}$$

Solving the last two equations together with Eq. (10-426) gives

$$\alpha_1 = \frac{0.02w_1^2 + 0.0004w_2^2 + 0.006w_1 w_2}{0.3w_2^2 + w_1 w_2 + 0.02w_1^2} \tag{10-430}$$

$$\alpha_2 = \frac{-0.2996w_1^2 - 0.07w_1 w_2}{0.3w_2^2 + w_1 w_2 + 0.02w_1^2} \tag{10-431}$$

$$\alpha_3 = \frac{0.994w_1^2 - 0.07w_2^2}{0.3w_2^2 + w_1 w_2 + 0.02w_1^2} \tag{10-432}$$

Setting $w_1 = 0$ and $w_2 = 1$, we have $\alpha_1 = 0.001333$, $\alpha_2 = 0$, and $\alpha_3 = -0.23333$.

With these coefficients, Eq. (10-427) gives the following three roots: $z_1 = 0.1$, $z_2 = 0.2$, and $z_3 = -0.06667$. Thus, with the choice of w_1 and w_2, the third root which we cannot place arbitrarily is at $z = -0.06667$, and the closed-loop system is stable. Apparently, there are other combinations of w_1 and w_2 that will correspond to a stable system with $z = 0.1$ and 0.2.

Substituting the values of $w_1 = 0$, $w_2 = 1$, and the corresponding values of α_1, α_2, and α_3 into Eq. (10-425), we get

$$\begin{bmatrix} g_1^* + g_2^* \\ g_2^* \\ 0 \end{bmatrix}' = \begin{bmatrix} \alpha_3 \\ -\alpha_1 + 1 \\ \alpha_2 \end{bmatrix}' = \begin{bmatrix} -0.23333 \\ 0.99866 \\ 0 \end{bmatrix}' \tag{10-433}$$

The feedback matrix $\mathbf{G}^*$ is written

$$\mathbf{G}^* = [-1.232 \quad 0.99866] \tag{10-434}$$

The feedback matrix $\mathbf{G}$ is

$$\mathbf{G} = \mathbf{w}\mathbf{G}^* = \begin{bmatrix} 0 \\ 1 \end{bmatrix} \quad \mathbf{G}^* = \begin{bmatrix} 0 & 0 \\ -1.232 & 0.99866 \end{bmatrix} \tag{10-435}$$

Example 10-27

Consider the same system as given in Example 10-26, except that

$$\mathbf{D} = \begin{bmatrix} 1 & 0 & 0 \\ 1 & 0 & 0 \end{bmatrix} \tag{10-436}$$

which has rank 1. This means that

$$\mathbf{G}^*\mathbf{D} = [g_1^* + g_2^* \quad 0 \quad 0] \tag{10-437}$$

which has only one independent parameter in $g_1^* + g_2^*$. Now Eq. (10-433) reads

$$\begin{bmatrix} g_1^* + g_2^* \\ 0 \\ 0 \end{bmatrix}' = \frac{1}{w_1^3 - w_2^3} \begin{bmatrix} -w_2^2\alpha_3 + \alpha_2 w_1^2 - w_1 w_2(\alpha_1 - 1) \\ \alpha_3 w_1^2 - w_1 w_2 \alpha_2 + w_2^2(\alpha_1 - 1) \\ -w_1 w_2 \alpha_3 + \alpha_2 w_2^2 - w_1^2(\alpha_1 - 1) \end{bmatrix}' \tag{10-438}$$

Since the last two rows in the last equation are constrained to be zero, we can only assign either w_1 or w_2 arbitrarily, but not both. Thus, from Eq. (10-438),

$$\alpha_3 w_1^2 - w_1 w_2 \alpha_2 + w_2^2(\alpha_1 - 1) = 0 \tag{10-439}$$

$$-w_1 w_2 \alpha_3 + \alpha_2 w_2^2 - w_1^2(\alpha_1 - 1) = 0 \tag{10-440}$$

For two of the closed-loop eigenvalues to be at $z_1 = 0.1$ and $z_2 = 0.2$, Eqs. (10-428) and (10-429) must also be satisfied. These two equations together with Eqs. (10-439) and (10-440) form a collection of four equations with five unknowns in $\alpha_1, \alpha_2, \alpha_3, w_1$, and w_2. Thus, only one of these unknowns can be arbitrarily assigned. Unfortunately, these four equations are nonlinear in w_1 and w_2 so that they are difficult to solve. In this case, it would be simpler to use the brute force method by writing

$$|z\mathbf{I} - \mathbf{A} + \mathbf{BGD}| = z^3 + (g_{21} + g_{22})z^2 + (g_{11} + g_{12})z + 1 = 0 \tag{10-441}$$

Hence, it is clear that only two of the three coefficients of the last equation can be arbitrarily assigned. Since the constant term is equal to one, the system cannot be stabilized by output feedback with the **D** matrix given.

10-14 DESIGN OF DIGITAL CONTROL SYSTEMS WITH STATE FEEDBACK AND DYNAMIC OUTPUT FEEDBACK

The state-feedback and output-feedback designs discussed in the preceding sections are suitable for the design of digital regulators. By properly selecting the eigenvalues of the closed-loop system, the natural responses of the state variables of the system can be properly controlled. When the design objective is to have the system track a certain input, the state-feedback and the output-feedback schemes should be properly modified. Since state feedback and output feedback do not increase the order of a system, they do not guarantee in general that the output or states of the system will follow the input in the steady state. In the conventional design technique, the PID controller has been used extensively in the industry for the compensation of a control process so that the transient response and the steady-state response behave in a prescribed manner. In fact, in the preceding sections the design of a control system with a digital PID controller was investigated. Since the PID controller always increases the order of the overall system (assuming that no cancellation of poles and zeros occurs), whereas the state or output feedback through constant gains does not affect the system order, the control effects of the two schemes are not equivalent. It may be stated that the state or output feedback through constant gains cannot achieve the same control objectives as the PID controller or any dynamic controllers in general.

In this section we present a method that will lead to a constant-gain feedback from the state variables in order to control the dynamics of the system and simultaneously feed the system output back through a dynamic controller for output regulation. In particular, the dynamic controller turns out to be a digital approximation of an integrator.

The digital control system under consideration is described by the following set of dynamic equations:

$$\mathbf{x}(k+1) = \mathbf{A}\mathbf{x}(k) + \mathbf{B}\mathbf{u}(k) + \mathbf{F}\mathbf{w} \tag{10-442}$$

$$\mathbf{c}(k) = \mathbf{D}\mathbf{x}(k) + \mathbf{E}\mathbf{u}(k) + \mathbf{H}\mathbf{w} \tag{10-443}$$

where

$\mathbf{x}(k)$ n-vector (state)
$\mathbf{u}(k)$ r-vector (input)
$\mathbf{c}(k)$ p-vector (output)
$\mathbf{w}$ q-vector (disturbance).

The dimensions of the matrices **A**, **B**, **D**, **E**, **F**, and **H** are defined according to that of the variables. The disturbance vector $\mathbf{w}$ is assumed to be constant. The elements of $\mathbf{w}$ may also include set points or reference inputs for the states or outputs of the system to follow. The true disturbance components of $\mathbf{w}$ are generally unknown, although their magnitudes are assumed to be constant.

The design objectives of the digital control system described by Eqs. (10-442) and (10-443) may be stated as follows. Find the control $\mathbf{u}(k)$ such that

$$\lim_{k\to\infty} \mathbf{x}(k+1) = \lim_{k\to\infty} \mathbf{x}(k) \tag{10-444}$$

and

$$\lim_{k\to\infty} \mathbf{c}(k) = \mathbf{0} \tag{10-445}$$

The condition in Eq. (10-444) is equivalent to requiring that the system be asymptotically stable, and Eq. (10-445) implies output regulation. The output vector $\mathbf{c}(k)$ need not be the true output of the system. In fact, by constructing

$\mathbf{c}(k)$ appropriately, a great variety of regulating and tracking problems can be defined.

Let us define the augmented state and control vectors as

$$\mathbf{y}(k) = \begin{bmatrix} \mathbf{x}(k+1) - \mathbf{x}(k) \\ \mathbf{c}(k) \end{bmatrix} \qquad (n+p) \times 1 \tag{10-446}$$

$$\mathbf{v}(k) = \mathbf{u}(k+1) - \mathbf{u}(k) \tag{10-447}$$

Then, from Eq. (10-446),

$$\mathbf{y}(k+1) = \begin{bmatrix} \mathbf{x}(k+2) - \mathbf{x}(k+1) \\ \mathbf{c}(k+1) \end{bmatrix} \tag{10-448}$$

From Eqs. (10-442) and (10-443), we have

$$\mathbf{x}(k+2) = \mathbf{A}\mathbf{x}(k+1) + \mathbf{B}\mathbf{u}(k+1) + \mathbf{F}\mathbf{w} \tag{10-449}$$

$$\mathbf{c}(k+1) = \mathbf{D}\mathbf{x}(k+1) + \mathbf{E}\mathbf{u}(k+1) + \mathbf{H}\mathbf{w} \tag{10-450}$$

Forming the difference vector between $\mathbf{y}(k+1)$ and $\mathbf{y}(k)$, we have

$$\begin{aligned} \mathbf{y}(k+1) - \mathbf{y}(k) &= \begin{bmatrix} \mathbf{A}[\mathbf{x}(k+1) - \mathbf{x}(k)] + \mathbf{B}[\mathbf{u}(k+1) - \mathbf{u}(k)] - [\mathbf{x}(k+1) - \mathbf{x}(k)] \\ \mathbf{D}[\mathbf{x}(k+1) - \mathbf{x}(k)] + \mathbf{E}[\mathbf{u}(k+1) - \mathbf{u}(k)] \end{bmatrix} \\ &= \begin{bmatrix} \mathbf{A} - \mathbf{I}_n & \mathbf{0} \\ \mathbf{D} & \mathbf{0} \end{bmatrix} \mathbf{y}(k) + \begin{bmatrix} \mathbf{B} \\ \mathbf{E} \end{bmatrix} \mathbf{v}(k) \end{aligned} \tag{10-451}$$

where $\mathbf{I}_n$ is an $n \times n$ identity matrix.

After rearranging, Eq. (10-451) is written as

$$\mathbf{y}(k+1) = \begin{bmatrix} \mathbf{A} & \mathbf{0} \\ \mathbf{D} & \mathbf{I}_p \end{bmatrix} \mathbf{y}(k) + \begin{bmatrix} \mathbf{B} \\ \mathbf{E} \end{bmatrix} \mathbf{v}(k) \tag{10-452}$$

Thus, the design objectives stated in Eqs. (10-444) and (10-445) are equivalent to driving the system in Eq. (10-452) from any initial state $\mathbf{y}(0)$ to $\mathbf{y}(k) \rightarrow 0$ as $k \rightarrow \infty$.

Condition on Controllability

To achieve the design objectives stated above, we must first investigate the controllability of the system described by Eq. (10-452).

Let

$$\hat{\mathbf{A}} = \begin{bmatrix} \mathbf{A} & \mathbf{0} \\ \mathbf{D} & \mathbf{I}_p \end{bmatrix} \qquad \hat{\mathbf{B}} = \begin{bmatrix} \mathbf{B} \\ \mathbf{E} \end{bmatrix}$$

Equation (10-452) is written

$$\mathbf{y}(k+1) = \hat{\mathbf{A}}\mathbf{y}(k) + \hat{\mathbf{B}}\mathbf{v}(k) \tag{10-453}$$

The necessary and sufficient conditions for the pair $(\hat{\mathbf{A}}, \hat{\mathbf{B}})$ to be completely controllable are that the $(n+p) \times (n+p+r)$ matrix

$$[\lambda\mathbf{I} - \hat{\mathbf{A}} \vdots \hat{\mathbf{B}}]$$

has rank $(n+p)$ for λ equals every eigenvalue of $\hat{\mathbf{A}}$. We have

$$[\lambda\mathbf{I} - \hat{\mathbf{A}} \vdots \hat{\mathbf{B}}] = \begin{bmatrix} \lambda\mathbf{I}_n - \mathbf{A} & \mathbf{0} & \mathbf{B} \\ -\mathbf{D} & (\lambda - 1)\mathbf{I}_m & \mathbf{E} \end{bmatrix} \tag{10-454}$$

where $\mathbf{I}_m$ is an $m \times m$ identity matrix. From the last equation we know that $\hat{\mathbf{A}}$ has at least p eigenvalues at $\lambda = 1$. When $\lambda = 1$, Eq. (10-454) becomes

$$[\lambda\mathbf{I} - \hat{\mathbf{A}} \vdots \hat{\mathbf{B}}] = \begin{bmatrix} \mathbf{I}_n - \mathbf{A} & \mathbf{0} & \mathbf{B} \\ -\mathbf{D} & \mathbf{0} & \mathbf{E} \end{bmatrix} \tag{10-455}$$

When $\lambda \neq 1$, $(\lambda - 1)\mathbf{I}_m$ is of rank p; in order that $[\lambda\mathbf{I}_n - \hat{\mathbf{A}} \vdots \hat{\mathbf{B}}]$ has a rank of $n + p$, the matrix $[\lambda\mathbf{I}_n - \hat{\mathbf{A}} \vdots \hat{\mathbf{B}}]$ must be of rank n, which implies that the pair $(\mathbf{A}, \mathbf{B})$ is controllable. Thus, the controllability of $(\hat{\mathbf{A}}, \hat{\mathbf{B}})$ requires that

1. $(\hat{\mathbf{A}}, \hat{\mathbf{B}})$ is controllable, and
2. $\begin{bmatrix} \mathbf{A} - \mathbf{I}_n & \mathbf{B} \\ \mathbf{D} & \mathbf{E} \end{bmatrix}$ is of rank $n + p$ (10-456)

Thus, we have established the controllability condition of the augmented system in Eq. (10-453) in terms of the coefficient matrices of the original system.

Now let us assume that the control vector $\mathbf{v}(k)$ is affected by state feedback, that is,

$$\mathbf{v}(k) = -\mathbf{G}\mathbf{y}(k) \tag{10-457}$$

where $\mathbf{G}$ is the $r \times (n+p)$ feedback matrix with constant-gain elements.

In view of the definitions of $\mathbf{v}(k)$ and $\mathbf{y}(k)$, Eq. (10-457) is written

$$\mathbf{v}(k) = \mathbf{u}(k+1) - \mathbf{u}(k) = -\mathbf{G}_1[\mathbf{x}(k+1) - \mathbf{x}(k)] - \mathbf{G}_2\mathbf{c}(k) \tag{10-458}$$

where $\mathbf{G}_1$ is an $r \times n$ matrix and $\mathbf{G}_2$ is an $r \times p$ matrix.

Taking the z-transform on both sides of Eq. (10-458) and rearranging, we get

$$\mathbf{U}(z) = -\mathbf{G}_1\mathbf{X}(z) - \frac{1}{z-1}\mathbf{G}_2\mathbf{C}(z) \tag{10-459}$$

The significance of the last equation is that the control $\mathbf{u}(k)$ is obtained as a combination of constant-gain state feedback and dynamic output feedback. The transfer function $1/(z-1)$ may be interpreted as a digital approximation of integration. Figure 10-74 shows the block diagram of the closed-loop system.

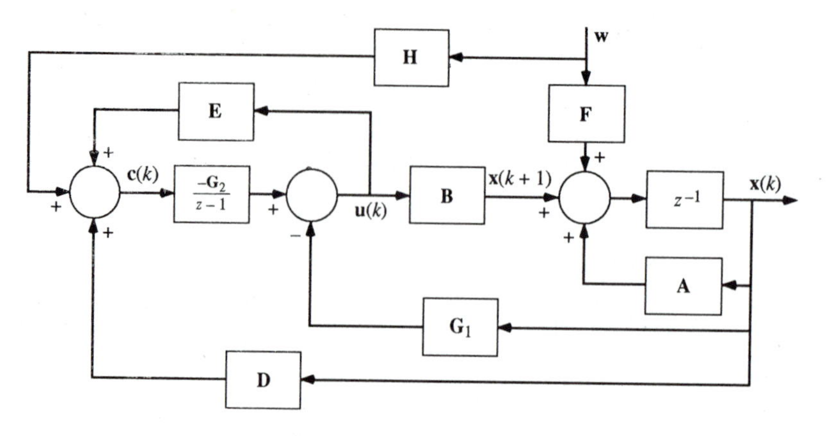

Figure 10-74. Block diagram of closed-loop digital control system with dynamic controller and state feedback.

Example 10-28

Consider the digital control system

$$\mathbf{x}(k+1) = \mathbf{A}\mathbf{x}(k) + \mathbf{B}u(k) + \mathbf{F}\mathbf{w} \tag{10-460}$$

where

$$\mathbf{A} = \begin{bmatrix} 0 & 1 \\ -1 & 0 \end{bmatrix} \quad \mathbf{B} = \begin{bmatrix} 0 \\ 1 \end{bmatrix} \quad \mathbf{w} = \begin{bmatrix} w_1 \\ w_2 \end{bmatrix} \quad \mathbf{F} = \begin{bmatrix} 0 & 1 \\ 0 & 0 \end{bmatrix}$$

w_2 denotes a constant disturbance whose amplitude is unknown.

The design objective is to find the control scheme such that

1. x_1 reaches the reference input $r = w_1$ as $k \to \infty$, and
2. the eigenvalues of the closed-loop system should be at certain specified values.

Let us define the output variable as

$$c(k) = w_1 - x_1(k) \tag{10-461}$$

Then, we want $c(k)$ to go to zero as k goes to infinity. Using the output equation model given in Eq. (10-443), we have

$$\mathbf{D} = [-1 \quad 0] \qquad \mathbf{E} = 0 \qquad \mathbf{H} = [1 \quad 0]$$

We can see that the pair (**A, B**) is completely controllable, and

$$\begin{bmatrix} \mathbf{A} - \mathbf{I}_n & \mathbf{B} \\ \mathbf{D} & \mathbf{E} \end{bmatrix} = \begin{bmatrix} -1 & 1 & 0 \\ -1 & -1 & 1 \\ -1 & 0 & 0 \end{bmatrix} \tag{10-462}$$

has rank three. Thus, the pair $(\hat{\mathbf{A}}, \hat{\mathbf{B}})$ is completely controllable, where

$$\hat{\mathbf{A}} = \begin{bmatrix} 0 & 1 & 0 \\ -1 & 0 & 0 \\ -1 & 0 & 1 \end{bmatrix} \qquad \hat{\mathbf{B}} = \begin{bmatrix} 0 \\ 1 \\ 0 \end{bmatrix}$$

The control in the z-domain is given by Eq. (10-459) and is written

$$U(z) = -g_1 X_1(z) - g_2 X_2(z) - \frac{g_3}{z-1} C(z) \tag{10-463}$$

where g_1, g_2, and g_3 are feedback gain constants. The characteristic equation of the closed-loop system is

$$|z\mathbf{I} - \hat{\mathbf{A}} + \hat{\mathbf{B}}\mathbf{G}| = \begin{bmatrix} z & -1 & 0 \\ 1 + g_1 & z + g_2 & g_3 \\ 1 & 0 & z - 1 \end{bmatrix}$$
$$= z^3 + (g_2 - 1)z^2 + (1 + g_1 - g_2)z - (1 + g_1 + g_3) = 0 \tag{10-464}$$

Let the closed-loop eigenvalues be at $z = 0.5, 0.3 + j0.3$, and $0.3 - j0.3$. Then the characteristic equation becomes

$$z^3 - 1.1z^2 + 0.39z - 0.045 = 0 \tag{10-465}$$

Equating the coefficients of Eqs. (10-464) and (10-465), we get $g_1 = -0.71$, $g_2 = -0.1$, and $g_3 = -0.245$. The block diagram of the closed-loop system is shown in Fig. 10-75, from which the z-transforms of $x_1(k)$ and $x_2(k)$ are written as

$$\begin{bmatrix} X_1(z) \\ X_2(z) \end{bmatrix} = \frac{1}{\Delta} \begin{bmatrix} \dfrac{-z^{-2}}{z-1} g_3 & z^{-1}(1 + g_2 z^{-1}) \\ \dfrac{-z^{-1}}{z-1} g_3 & -z^{-2} - g_1 z^{-2} + \dfrac{g_3}{z-1} z^{-2} \end{bmatrix} \begin{bmatrix} w_1 \\ w_2 \end{bmatrix} \frac{z}{z-1} \tag{10-466}$$

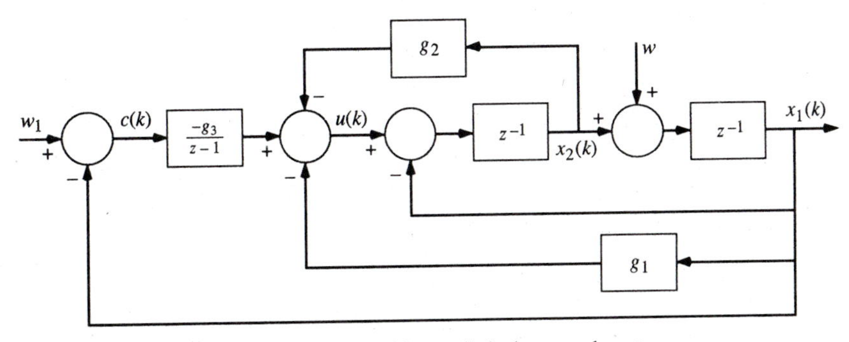

Figure 10-75. Block diagram of closed-loop digital control system.

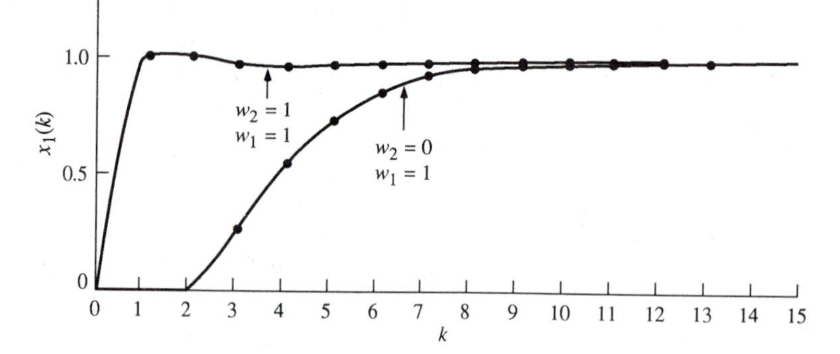

Figure 10-76. Responses of $x_1(k)$ for the system designed in Example 10-27.

where

$$\Delta = 1 + g_2 z^{-1} + z^{-2} + g_1 z^{-2} - \frac{g_3 z^{-2}}{z - 1} \tag{10-467}$$

Applying the final-value theorem of the z-transform to both sides of Eq. (10-466), we get

$$\lim_{k \to \infty} \begin{bmatrix} x_1(k) \\ x_2(k) \end{bmatrix} = \begin{bmatrix} 1 & 0 \\ 0 & -1 \end{bmatrix} \begin{bmatrix} w_1 \\ w_2 \end{bmatrix} = \begin{bmatrix} w_1 \\ w_1 - w_2 \end{bmatrix} \tag{10-468}$$

Therefore, the final value of $x_1(k)$ is equal to the reference input w_1 as desired.

Figure 10-76 illustrates the responses of $x_1(k)$ for $w_1 = 1$ and two values of w_2, the true disturbance. The actual responses are defined only at discrete intervals but are represented by smooth curves passing through the data points. When the true disturbance is zero, the response of $x_1(k)$ has a delay time of two sampling periods and approaches the reference input $w_1 = r = 1$ without overshoot. When $w_2 = 1$, which was chosen arbitrarily, the response of $x_1(k)$ actually approaches its final value a great deal faster. However, this result is better than the response for $w_2 = 0$ only because when $w_2 = 1$, the condition is most favorable. Apparently, when w_2 is of any other values, the response would not be as impressive.

10-15 REALIZATION OF STATE FEEDBACK BY DYNAMIC CONTROLLERS

The designer of control systems using state feedback often encounters one practical problem in that not all the state variables are accessible. Furthermore, for high-order systems, the implementation of state feedback would require a large number of transducers for the measurements of the state variables, a costly proposition. In Chapter 11 we shall see that if the process to be controlled is observable, then an "observer" can be designed that utilizes the input and the output signal measurements to generate a set of "observed" states. These observed states, which are estimates of the true state variables, are then used to control the process through state feedback.

In the conventional design principles, feedback from the output signal is usually used. This is natural, because the output is always measurable. The conventional design is also characterized in such a way that the controller configuration is selected *a priori* at the outset of the design. In this section we shall present a method which allows the approximation of state feedback by a cascade controller or a feedback controller. In a general sense the equivalent controller discussed here can be regarded as a *dynamic observer-controller* of the system.

Consider the digital system described by

$$\mathbf{x}(k+1) = \mathbf{A}\mathbf{x}(k) + \mathbf{B}\mathbf{u}(k) \tag{10-469}$$

$$\mathbf{c}(k) = \mathbf{D}\mathbf{x}(k) + \mathbf{E}\mathbf{u}(k) \tag{10-470}$$

where $\mathbf{x}(k)$ is an n-vector, $\mathbf{c}(k)$ is a q-vector, and $\mathbf{u}(k)$ is a p-vector; $\mathbf{A}$, $\mathbf{B}$, $\mathbf{D}$, and $\mathbf{E}$ are coefficients of the appropriate dimensions.

Let the control be realized by state feedback such that

$$\mathbf{u}(k) = -\mathbf{G}\mathbf{x}(k) \tag{10-471}$$

where $\mathbf{G}$ is the $p \times q$ feedback gain matrix.

The objective is to approximate the digital control system with state feedback shown in Fig. 10-77 by the system of Fig. 10-78 which has a feedback controller with input coming from the output $\mathbf{c}(k)$. Let the transfer function relation of the feedback controller be represented by

$$\mathbf{U}(z) = -\mathbf{H}(z)\mathbf{C}(z) \tag{10-472}$$

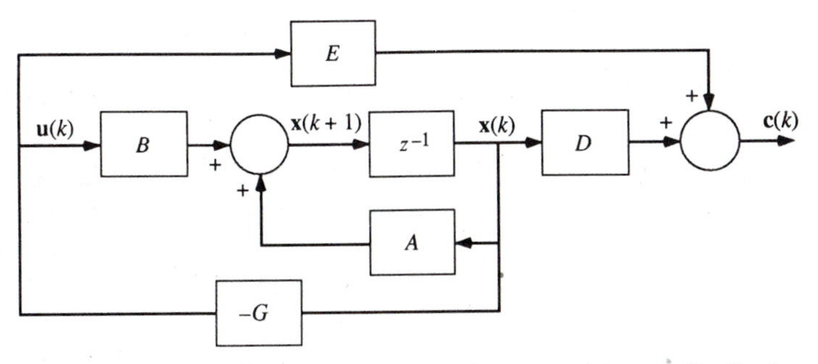

Figure 10-77. Digital control system with state feedback.

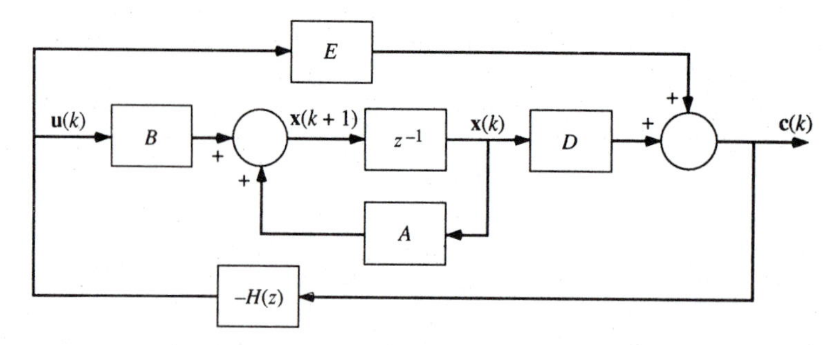

Figure 10-78. Digital control system with dynamic feedback controller from output.

where $\mathbf{H}(z)$ is the transfer function matrix. Let us express $\mathbf{H}(z)$ as

$$\mathbf{H}(z) = \begin{bmatrix} H_{11}(z) & H_{12}(z) & \cdots & H_{1q}(z) \\ H_{21}(z) & H_{22}(z) & \cdots & H_{2q}(z) \\ \vdots & \vdots & & \vdots \\ H_{p1}(z) & H_{p2}(z) & \cdots & H_{pq}(z) \end{bmatrix} \quad (p \times q) \tag{10-473}$$

Let $H_{ij}(z)$ $i = 1, 2, \ldots, p$, $j = 1, 2, \ldots, q$, be an mth-order transfer function.

$$H_{ij}(z) = \frac{K_{ij}(z^m + \alpha_{ij1} z^{m-1} + \alpha_{ij2} z^{m-2} + \cdots + \alpha_{ijm})}{z^m + \beta_{ij1} z^{m-1} + \beta_{ij2} z^{m-2} + \cdots + \beta_{ijm}} \tag{10-474}$$

Expanding $H_{ij}(z)$ into a Laurent's series about $z = 0$, we get

$$H_{ij}(z) = K_{ij} \sum_{k=0}^{\infty} d_{ijk} z^{-k} \tag{10-475}$$

where

$$\begin{aligned} d_{ijo} &= 1 \\ d_{ij1} &= \alpha_{ij1} - \beta_{ij1} \\ d_{ijk} &= \alpha_{ijk} - \beta_{ijk} - \sum_{v=1}^{k-1} \beta_{k-v} d_{ijv} \qquad k > 1 \end{aligned} \tag{10-476}$$

Truncating $H_{ij}(z)$ at m terms, we have

$$H_{ij}(z) \cong K_{ij} \sum_{k=0}^{m-1} d_{ijk} z^{-k} \tag{10-477}$$

where we have assumed that the infinite series converges. The significance of Eq. (10-472) and the truncated series representation of $H_{ij}(z)$ is that the control $\mathbf{u}(k)$ is implemented by feeding back the output $\mathbf{c}(k)$ and all the delayed outputs up to $\mathbf{c}(k - m + 1)$, where m is not yet specified. Let

$$H_{ijm}(z) = K_{ij} \sum_{k=0}^{m-1} d_{ijk} z^{-k} \tag{10-478}$$

Substituting the last equation into Eq. (10-472) as an approximation for $H_{ij}(z)$, we have

$$\mathbf{U}(z) \cong -\begin{bmatrix} K_{11}[d_{110} \cdots d_{11(m-1)}] & K_{12}[d_{120} \cdots d_{12(m-1)}] & \cdots & K_{1q}[d_{1q0} \cdots d_{1q(m-1)}] \\ K_{21}[d_{210} \cdots d_{21(m-1)}] & K_{22}[d_{220} \cdots d_{22(m-1)}] & \cdots & K_{2q}[d_{2q0} \cdots d_{2q(m-1)}] \\ \vdots & \vdots & & \vdots \\ K_{p1}[d_{p10} \cdots d_{p1(m-1)}] & K_{p2}[d_{p20} \cdots d_{p2(m-1)}] & \cdots & K_{pq}[d_{pq1} \cdots d_{pq(m-1)}] \end{bmatrix} \times \begin{bmatrix} C_1(z) \\ z^{-1}C_1(z) \\ \vdots \\ z^{-m+1}C_1(z) \\ C_2(z) \\ z^{-1}C_2(z) \\ \vdots \\ z^{-m+1}C_2(z) \\ \vdots \\ C_q(z) \\ z^{-1}C_q(z) \\ \vdots \\ z^{-m+1}C_q(z) \end{bmatrix} \tag{10-479}$$

The elements of the last equation are rearranged to give

$$\mathbf{U}(z) \cong -\begin{bmatrix} [K_{11}d_{110} \cdots K_{1q}d_{1q0}] & [K_{11}d_{111} \cdots K_{1q}d_{1q1}] & \cdots & [K_{11}d_{11(m-1)} \cdots K_{1q}d_{1q(m-1)}] \\ [K_{21}d_{210} \cdots K_{2q}d_{2q0}] & [K_{21}d_{211} \cdots K_{2q}d_{2q1}] & \cdots & [K_{21}d_{21(m-1)} \cdots K_{2q}d_{2q(m-1)}] \\ \vdots & \vdots & & \vdots \\ [K_{p1}d_{p10} \cdots K_{pq}d_{pq0}] & [K_{p1}d_{p11} \cdots K_{pq}d_{pq1}] & \cdots & [K_{p1}d_{p1(m-1)} \cdots K_{pq}d_{pq(m-1)}] \end{bmatrix} \times \begin{bmatrix} \mathbf{C}(z) \\ z^{-1}\mathbf{C}(z) \\ z^{-2}\mathbf{C}(z) \\ \vdots \\ z^{-m+1}\mathbf{C}(z) \end{bmatrix} \tag{10-480}$$

The time-domain equivalent of the last equation is

$$\mathbf{u}(k) = -\mathbf{P}\begin{bmatrix} \mathbf{c}(k) \\ \mathbf{c}(k-1) \\ \vdots \\ \mathbf{c}(k-m+1) \end{bmatrix} \tag{10-481}$$

where $\mathbf{P}$ denotes the $p \times qm$ coefficient matrix in Eq. (10-480).

Substituting Eq. (10-471) into Eq. (10-470), we have

$$\mathbf{c}(k) = (\mathbf{D} - \mathbf{EG})\mathbf{x}(k) \tag{10-482}$$

Thus,

$$\mathbf{c}(k-1) = (\mathbf{D} - \mathbf{EG})\mathbf{x}(k-1) \tag{10-483}$$

Also, from Eq. (10-469), we write

$$\begin{aligned}\mathbf{Ax}(k-1) &= \mathbf{x}(k) - \mathbf{Bu}(k-1) \\ &= \mathbf{x}(k) + \mathbf{BGx}(k-1)\end{aligned} \tag{10-484}$$

Solving $\mathbf{x}(k-1)$ from the last equation, we get

$$\mathbf{x}(k-1) = (\mathbf{A} - \mathbf{BG})^{-1}\mathbf{x}(k) \tag{10-485}$$

Now substituting the last equation in Eq. (10-483), we have

$$\mathbf{c}(k-1) = (\mathbf{D} - \mathbf{EG})(\mathbf{A} - \mathbf{BG})^{-1}\mathbf{x}(k) \tag{10-486}$$

Recursively, we can write the following relationships:

$$\mathbf{c}(k-2) = (\mathbf{D} - \mathbf{EG})(\mathbf{A} - \mathbf{BG})^{-2}\mathbf{x}(k) \tag{10-487}$$

$$\vdots$$

$$\mathbf{c}(k-m+1) = (\mathbf{D} - \mathbf{EG})(\mathbf{A} - \mathbf{BG})^{-m+1}\mathbf{x}(k) \tag{10-488}$$

Substituting Eqs. (10-482) and (10-486) through (10-488) in Eq. (10-481), the state-feedback control is written

$$\mathbf{u}(k) \cong -\underset{(p \times qm)}{\mathbf{P}} \underset{(qm \times n)}{\begin{bmatrix} \mathbf{D} - \mathbf{EG} \\ (\mathbf{D} - \mathbf{EG})(\mathbf{A} - \mathbf{BG})^{-1} \\ (\mathbf{D} - \mathbf{EG})(\mathbf{A} - \mathbf{BG})^{-2} \\ \vdots \\ (\mathbf{D} - \mathbf{EG})(\mathbf{A} - \mathbf{BG})^{-m+1} \end{bmatrix}} \underset{(n \times 1)}{\mathbf{x}(k)} \tag{10-489}$$

Now comparing Eq. (10-489) with Eq. (10-471), we have

$$\underset{(p \times qm)}{\mathbf{P}} \underset{(qm \times n)}{\begin{bmatrix} \mathbf{D} - \mathbf{EG} \\ (\mathbf{D} - \mathbf{EG})(\mathbf{A} - \mathbf{BG})^{-1} \\ (\mathbf{D} - \mathbf{EG})(\mathbf{A} - \mathbf{BG})^{-2} \\ \vdots \\ (\mathbf{D} - \mathbf{EG})(\mathbf{A} - \mathbf{BG})^{-m+1} \end{bmatrix}} = \underset{(p \times n)}{\mathbf{G}} \tag{10-490}$$

To solve for $\mathbf{P}$ from the last equation, qm must equal n, or $m = n/q$. This means that n/q must be an integer, and the series expansion of $H_{ij}(z)$ should be truncated at $m = n/q$ terms. Solving for $\mathbf{P}$ from Eq. (10-490), we get

$$\mathbf{P} = \mathbf{G}\begin{bmatrix} \mathbf{D} - \mathbf{EG} \\ (\mathbf{D} - \mathbf{EG})(\mathbf{A} - \mathbf{BG})^{-1} \\ (\mathbf{D} - \mathbf{EG})(\mathbf{A} - \mathbf{BG})^{-2} \\ \vdots \\ (\mathbf{D} - \mathbf{EG})(\mathbf{A} - \mathbf{BG})^{-m+1} \end{bmatrix}^{-1} \tag{10-491}$$

where the indicated matrix inverse must exist.

Since $\mathbf{P}$ is $p \times qm$, there are (pqm) unknowns in $\mathbf{P}$. However, there are only pn equations in Eq. (10-491). Thus, $p(qm - n)$ of the elements of $\mathbf{P}$ may be assigned arbitrarily. Also note that solution of the elements of the matrix $\mathbf{P}$ from Eq. (10-491) gives only the values of the coefficients K_{ij} and d_{ijk} in Eq. (10-477). The coefficients of the transfer function of Eq. (10-474) still have to be determined using Eq. (10-476). In general, there are more unknowns than equations in Eq. (10-476). This means that ideally we can simply set

$$d_{ijo} = 1 \tag{10-492}$$

and

$$d_{ijk} = \alpha_{ijk} \qquad (k \geq 1) \tag{10-493}$$

and set all β_{ijk} to 0, for $k = 1, 2, \ldots, q$. However, for a physically realizable transfer function, $H_{ijm}(z)$ must not have more zeros than poles. Therefore, the values of β_{ijk} should be assigned such that the dynamic behavior of the overall system is not appreciably affected by the presence of β_{ijk}, $k = 1, 2, \ldots, q$. This is similar to the classical design practice of designing the zeros of the transfer function of the controller $H_{ijm}(z)$ to control the dynamics of the system, while placing the poles of $H_{ijm}(z)$ so that they do not have appreciable effects on the system performance. In the case of digital control systems, the poles of $H_{ijm}(z)$ should be placed near the origin in the z-plane.

In the following we shall investigate the single-variable case of the dynamic feedback controller realization of state-feedback problems.

Single-Variable Systems

Consider that the digital control system described by Eqs. (10-469) through (10-471) has one input and one output; i.e., $p = q = 1$. Then, the feedback gain matrix is written as

$$\mathbf{G} = [g_1 \quad g_2 \quad \cdots \quad g_n] \qquad (1 \times n) \tag{10-494}$$

The dynamic feedback controller is described by the scalar transfer function

$$H(z) = \frac{K(1 + \alpha_1 z + \alpha_2 z^2 + \cdots + \alpha_n z^n)}{(1 + \beta_1 z + \beta_2 z^2 + \cdots + \beta_n z^n)} \tag{10-495}$$

The n-term Laurent's series expansion of $H(z)$ is

$$H_n(z) = K(1 + d_1 z^{-1} + d_2 z^{-2} + \cdots + d_{n-1} z^{-n+1}) \tag{10-496}$$

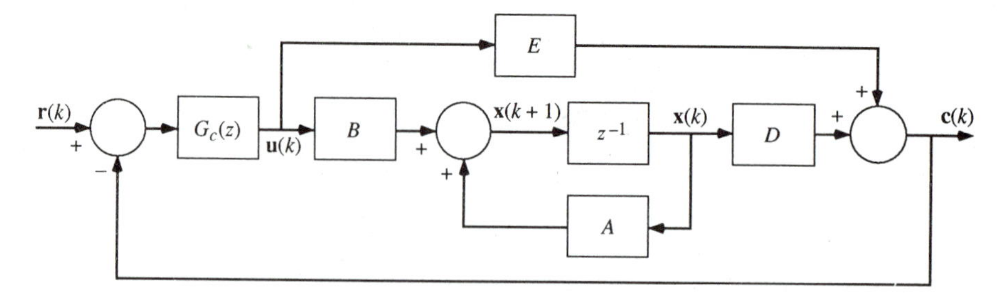

Figure 10-79. Digital control system with dynamic cascade controller.

where

$$d_k = \alpha_k - \beta_k - \sum_{v=1}^{k-1} \beta_{k-v} d_v \tag{10-497}$$

for $k = 1, 2, \ldots, n-1$. Equation (10-491) gives

$$\mathbf{P} = \mathbf{K}[1 \quad d_1 \quad d_2 \quad \cdots \quad d_{n-1}] = \mathbf{G}\begin{bmatrix} \mathbf{D} - \mathbf{EG} \\ (\mathbf{D} - \mathbf{EG})(\mathbf{A} - \mathbf{BG})^{-1} \\ (\mathbf{D} - \mathbf{EG})(\mathbf{A} - \mathbf{BG})^{-2} \\ \vdots \\ (\mathbf{D} - \mathbf{EG})(\mathbf{A} - \mathbf{BG})^{-n+1} \end{bmatrix}^{-1} \tag{10-498}$$

Equivalent Forward Cascade Controller

The development carried out in the preceding section is based on a dynamic controller $\mathbf{H}(z)$ being located in the feedback path of the system as shown in Fig. 10-78. When the reference input $\mathbf{r}(k)$ is zero, that is, when the system is designed as a regulator, it does not matter whether the dynamic controller is placed in the forward path or the feedback path. However, when the system is designed to track an input $\mathbf{r}(k)$, it may be desirable to place the dynamic controller in the forward path, as shown in Fig. 10-79. For a single-variable system, we can show that the equivalent forward cascade controller $\mathbf{G}_c(z)$ can be determined from $H(z)$.

$$\mathbf{G}_c(z) = \frac{1}{1 + \mathbf{D}(z\mathbf{I} - \mathbf{A})^{-1}\mathbf{B}[H(z) - 1]} \tag{10-499}$$

Example 10-29

Figure 10-80 shows the state diagram of a digital control process that is subject to state feedback. The feedback gains are chosen so that the eigenvalues of the closed-loop system are at $0.5 + j0.5$ and $0.5 - j0.5$.

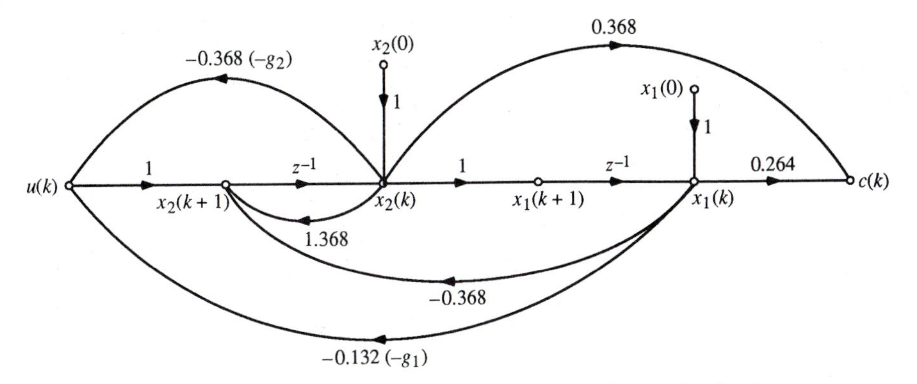

Figure 10-80. State diagram of a digital control system with state feedback.

The coefficient matrices of the process are

$$\mathbf{A} = \begin{bmatrix} 0 & 1 \\ -0.368 & 1.368 \end{bmatrix} \qquad \mathbf{B} = \begin{bmatrix} 0 \\ 1 \end{bmatrix}$$

$$\mathbf{D} = [0.264 \quad 0.368] \qquad E = 0$$

The feedback gain matrix is

$$\mathbf{G} = [0.132 \quad 0.368]$$

The equivalent feedback dynamic controller which approximates the state feedback is described by

$$H(z) = K\,\frac{z + \alpha_1}{z + \beta_1} \cong K(1 + d_1 z^{-1}) \tag{10-500}$$

Equation (10-498) gives

$$\mathbf{P} = \mathbf{K}[1 \quad d_1] = \mathbf{G}\begin{bmatrix} \mathbf{D} \\ \mathbf{D}(\mathbf{A} - \mathbf{BG})^{-1} \end{bmatrix}^{-1}$$

$$= [0.132 \quad 0.368]\begin{bmatrix} 0.264 & 0.368 \\ 0.896 & -0.528 \end{bmatrix}^{-1}$$

$$= 0.8514[1 \quad -0.1216] \tag{10-501}$$

Thus, $K = 0.8514$ and $d_1 = -0.1216$. For the dynamic controller to be physically realizable, we arbitrarily set β_1 to be a very small number compared with d_1, say, $\beta_1 = 0.0005$. Then

$$\alpha_1 = d_1 + \beta_1 = -0.1211 \tag{10-502}$$

The transfer function of the feedback dynamic controller is

$$H(z) = 0.8514\,\frac{z - 0.1211}{z - 0.0005} \tag{10-503}$$

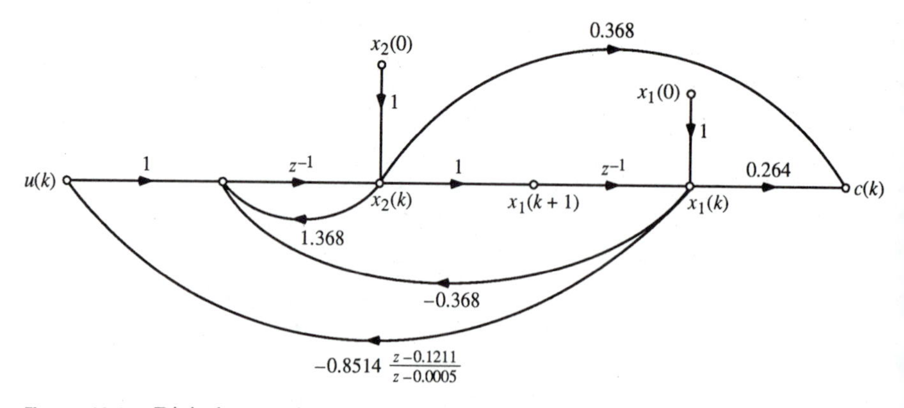

Figure 10-81. Digital control system with dynamic controller equivalent to that in Fig. 10-80.

The signal flow graph of the closed-loop system with the dynamic controller is shown in Fig. 10-81.

Now let us compare the characteristics and the performance of the two systems shown in Figs. 10-80 and 10-81. The characteristic equation of the system with the state feedback is

$$|z\mathbf{I} - \mathbf{A} + \mathbf{BG}| = z^2 - z + 0.5 = 0 \tag{10-504}$$

The roots of the last equation are $z = 0.5 + j0.5$ and $z = 0.5 - j0.5$, as specified. The characteristic equation of the system with the dynamic controller feeding back from the output is

$$z^3 - 1.05202z^2 + 0.557398z - 0.027678 = 0 \tag{10-505}$$

The roots of the last equation are $z = 0.0548$, $z = 0.4986 + j0.5043$, and $z = 0.4986 - j0.5043$. Thus we see that, although the use of the dynamic controller in place of the state feedback has made the overall system into a third-order one, the dominant characteristic roots are still very close to $0.5 \pm j0.5$. The root at $z = 0.0548$, which is very close to the origin of the z-plane, should not affect the dynamics of the system significantly.

Let us consider the initial state, $x_1(0) = 1$ and $x_2(0) = 0$. From Fig. 10-80 we have

$$C(z) = \frac{0.264z^2 - 0.448z}{z^2 - z + 0.5} x_1(0) \tag{10-506}$$

Figure 10-82 shows the response of $\mathbf{c}(k)$. From Fig. 10-81, the z-transform of the output of the system with the dynamic controller, when the above initial states are applied, is obtained as

$$C(z) = \frac{0.264z^3 - 0.496712z^2 + 0.000248z}{z^3 - 1.05202z^2 + 0.557398z - 0.027678} x_1(0) \tag{10-507}$$

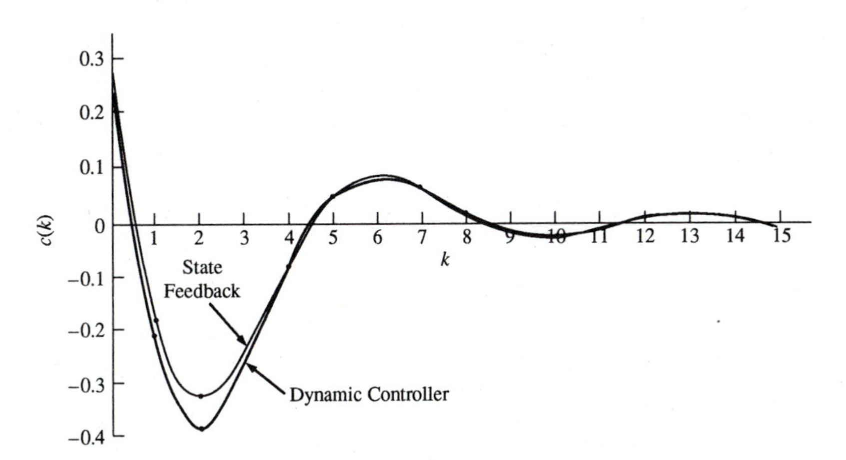

Figure 10-82. Time responses of the systems in Figs. 10-80 and 10-81.

The time response of $c(k)$ that corresponds to the last equation is shown in Fig. 10-82. From these responses we see that the system with the dynamic controller gives a higher overshoot, but the two responses are generally very close.

PROBLEMS

Digital Controllers

10-1 Find the block diagrams of direct digital programs for the digital controllers given.

a. $$D(z) = \frac{(z - 0.9543)(z - 0.386)}{(z - 1)(z + 0.222)(z + 0.5)}$$

b. $$D(z) = \frac{1.5 + 0.5z^{-1} + z^{-2}}{1 - z^{-1}}$$

c. $$D(z) = \frac{0.2 - 0.3z^{-1} + 0.2z^{-2}}{1 - 0.5z^{-1} - 0.4z^{-2}}$$

10-2 Find the block diagrams of a cascade digital program for the digital controller given in Problem 10-1-**a** using the following factoring.

$$D(z) = \left(\frac{1}{z - 1}\right)\left(\frac{z - 0.9543}{z + 0.222}\right)\left(\frac{z - 0.386}{z + 0.5}\right)$$

10-3 Given the transfer function,

$$D(z) = \frac{E_2(z)}{E_1(z)} = \frac{z^{-5}}{(1 + z^{-1} + z^{-2})^2}$$

a. Draw the block diagram of a direct digital program of $D(z)$ using a minimum number of data storage units.
b. Draw the block diagram of a cascade digital program of $D(z)$.
c. Draw the block diagram of a parallel digital program of $D(z)$ using a minimum number of data storage units.

Design of Continuous-Data Controller

10-4 The controlled process of the system described in Example 10-2 is

$$G_p(s) = \frac{887.3}{s(s + 8.871)}$$

The sampling period T is 0.01 s.

Design a cascade phase-lead continuous-data controller $G_c(s)$ so that the phase margin of the system is 50 degrees, where $G_c(s)$ should be of the form

$$G_c(s) = K_c \frac{s + z_1}{s + p_1} \qquad p_1 > z_1$$

where K_c, z_1, and p_1 are real constants. What is the maximum phase margin that can be obtained with a single-stage phase-lead controller subject to the condition that the pole of $G_c'(r)$ satisfies $-1 \leq r \leq 0$? Plot the Bode diagrams of $\mathscr{Z}[G_{h0}G_p(s)]$ and $\mathscr{Z}[G_c(s)G_{h0}(s)G_p(s)]$. Plot the unit-step response of the compensated system.

z-Domain Design

10-5 When traveling through the atmosphere, a guided missile usually encounters aerodynamic forces that tend to cause instability in the attitude of the missile. As shown in Fig. P10-5(a), the side force F applied at the center of

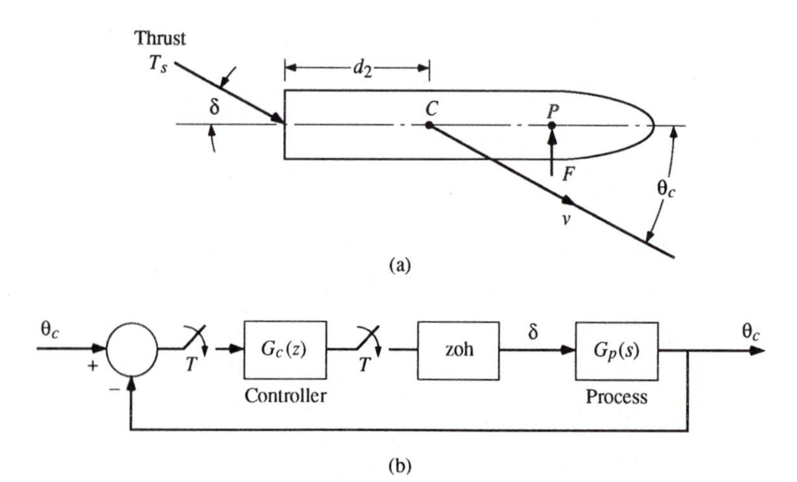

Figure P10-5.

pressure P tends to cause the missile to tumble. Let the angular acceleration of the missile about the center of gravity C be denoted by a. Normally, a is directly proportional to the angle of attack θ_c and is given by

$$\alpha = K\theta_c$$

where K is a constant described by

$$K = K_F d_1/J$$

and where K_F is a constant which depends on such parameters as dynamic pressure, velocity of the missile, air density, and so on, and J is the moment of inertia of the missile about C. The main objective of the flight control system is to provide the stabilization action to counter the effect of the side force by use of gas injection at the tail of the missile to deflect the direction of the rocket engine thrust T_s.

For small angle δ, $T_s \sin \delta \cong T_s\delta$. The torque due to T_s about the point C is $T_s\delta d_2$. Thus, the torque equation is written as

$$T_s d_2 \delta + JK\theta_c = J\frac{d^2\theta_c}{dt^2}$$

The block diagram of the closed-loop flight control system is shown in Fig. P10-5(b). The sampling period of the digital control system is 0.1 s. Let $T_s d_2/J = 10$ and $K = 1$. The transfer function of the digital controller is of the form

$$D(z) = K_c \frac{(z - 0.9048)(z - z_1)}{(z - 1)(z - p_1)}$$

Find the values of K_c, z_1, and p_1 so that the closed-loop system is stable (actually conditionally stable), and the maximum overshoot is less than 40 percent. Plot the unit-step response.

10-6 The digital controlled process of a unity-feedback control system is described by the transfer function

$$G_{h0}G_p(z) = \frac{K(z + 0.5)}{(z - 1)(z - 0.5)}$$

Design a cascade phase-lag controller with the transfer function

$$D(z) = K_c \frac{z - z_1}{z - p_1}$$

so that the following design specifications are satisfied.

a. The ramp-error constant $K_v = 6$.
b. The dominant roots of the closed-loop characteristic equation are approximately at $z = 0.71 + j0.19$ and $0.71 - j0.19$.
c. The maximum overshoot is ≤ 15 percent.

Determine the stability of the uncompensated system when the value of K is set for $K_v = 6$. Find the closed-loop transfer function of the compensated system and the roots of the characteristic equation. Plot the unit-step response of the compensated system.

Root-Locus Design

10-7 The controlled process of a discrete-data control system is described by the transfer function

$$G_{h0}G_p(z) = \frac{K(z+0.5)}{(z-1)(z-0.5)}$$

The sampling period is 0.1 s. Determine the value of K and design a cascade phase-lag digital controller with the transfer function

$$D(z) = K_c \frac{z - z_1}{z - p_1}$$

where $D(1) = 1$, so that the following design specifications are satisfied.

a. The ramp-error constant $K_v = 100$.

b. The dominant roots of the characteristic equation are approximately equal to $0.7 + j0.245$ and $0.7 - j0.245$.

Find the actual roots of the characteristic equation with the designed controller. Plot the unit-step response of the system.

10-8 Repeat Problem 10-7 with the following design specifications.

a. The ramp-error constant $K_v = 100$.

b. The dominant roots of the characteristic equation are approximately equal to $0.65 + j0.42$ and $0.65 - j0.42$.

10-9 Repeat Problem 10-7 with the following design specifications.

a. The ramp-error constant $K_v \geq 100$.

b. The dominant roots of the characteristic equation are approximately equal to $0.72 + j0.108$ and $0.72 - j0.108$.

PID Controller Design

10-10 The transfer function of the controlled process, including the zero-order hold, of the automatic forage and concentrate-proportioning system described in Problem 6-19 is

$$G_{h0}G_p(z) = \frac{1.338(z+0.7531)}{z^2 - 0.9398z + 0.4347}$$

The sampling period is 0.01 s.

a. Let the cascade digital PID controller be described by

$$D(z) = K_P + \frac{K_I T(z+1)}{2(z-1)} + \frac{K_D(z-1)}{Tz}$$

Find the values of K_P, K_I, and K_D so that the zeros of $D(z)$ cancel the poles of $G_{h0}G_p(z)$, and the ramp-error constant $K_v = 1$. Plot the unit-step response, and find the maximum overshoot.

b. Repeat part **a** with $K_v = 10$.

10-11 For the dc-motor speed control system described in Problem 7-11, the process transfer function is

$$G_p(s) = \frac{244680.85}{(s + 297.456)(s + 879.844)}$$

The sampling period T is 0.001 s. The digital controller has the transfer function

$$D(z) = K_P + \frac{K_R T}{2}\frac{z + 1}{z - 1}$$

Find the values of K_P and K_R so that the following specifications are satisfied.

a. Unit-step response reaches unity in less than 0.02 s.

b. Maximum overshoot is ≤ 0.5 percent.

Design $D(z)$ so that its zero cancels one of the poles of $G_{h0}G_p(z)$. Plot the unit-step response of the compensated system.

Frequency-Domain Design

10-12 The controlled process of a discrete-data control system is described by the transfer function

$$G_{h0}G_p(z) = \frac{K(z + 0.5)}{(z - 1)(z - 0.5)}$$

The sampling period is 0.1 s. Determine the value of K, and design a cascade phase-lag digital controller with the transfer function

$$D(z) = K_c \frac{z - z_1}{z - p_1}$$

where $D(1) = 1$, so that the following design specifications are satisfied.

a. The ramp-error constant $K_v = 100$.

b. The phase margin is 60 degrees.

Plot the Bode diagrams of the open-loop transfer functions of the uncompensated and the compensated systems. Find the phase margin, gain margin, M_p, and BW of the compensated system. Plot the unit-step response of the compensated system.

10-13 The controlled process of a discrete-data control system is

$$G_p(s) = \frac{887.3}{s(s + 8.871)}$$

The sampling period T is 0.01 s.

a. Design a cascade phase-lead digital controller $D(z)$ so that the phase margin of the system is 50 degrees. Use the r-transformation so that $D(r)$ is of the form of Eq. (10-78). Plot the Bode diagrams of $G_{h0}G_p(z)$ and $D(z)G_{h0}G_p(z)$. Plot the open-loop transfer functions on the Nichols chart and find the values of M_p and BW.

b. Plot the unit-step response of the compensated system.

10-14 Repeat Problem 10-13 with the cascade phase-lag digital controller $D(z)$ so that the phase margin is 50 degrees.

10-15 Repeat Problem 10-13 with a cascade lag-lead digital controller $D(z)$ so that the phase-lag portion contributes 20 degrees to the phase margin, and the remaining 30 degrees is from the phase-lead portion of the controller.

10-16 A discrete-data control system with parallel digital controllers is shown in Fig. P10-16. The transfer function of the controlled process is

$$G_p(s) = \frac{887.1}{s(s + 8.871)}$$

The sampling period T is 0.01 s.

a. Determine the transfer function $D(z)$ of the phase-lag digital controller so that the phase margin is 50 degrees. Plot the unit-step response.

b. Plot the unit-step response if one of the digital controllers fails.

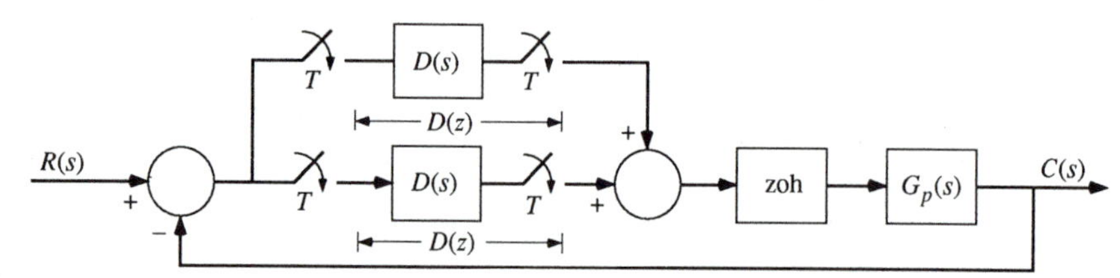

Figure P10-16.

10-17 Figure P10-17 shows the block diagram of the automobile idle-speed control system with a digital controller. The sampling period is 0.1 s.

a. Select the value of K so that the ramp-error constant is 8.

b. Design a phase-lag controller for $D(z)$ of the form

$$D(z) = K_c \frac{z - z_1}{z - p_1} \qquad p_1 > z_1$$

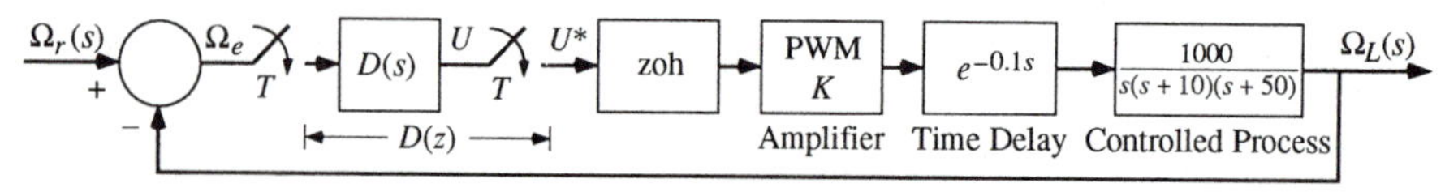

Figure P10-17.

so that the phase margin of the system is 60 degrees. Plot the Bode diagrams of the uncompensated and the compensated systems. Plot the unit-step response of the compensated system.

10-18 The z-transform of the zero-order-hold process of the dc-motor velocity control system described in Problem 7-12 is

$$G_{h0}G_p(z) = \frac{0.87517(z^2 - 0.8425z + 0.3753)}{(z - 0.9998)(z^2 - 1.80187z + 0.953)}$$

The sampling period is 0.0001 s.

a. Determine the stability condition of the uncompensated system.

b. Let the cascade digital controller be of the form

$$D(z) = \frac{z^2 - 1.80187z + 0.953}{(z - p_1)(z - p_2)}$$

where $D(1) = 1$. Determine the values of p_1 and p_2 so that the phase margin of the compensated system is at least 60 degrees. Plot the Bode plots of $G_{h0}G_p(z)$ and $D(z)G_{h0}G_p(z)$, and the unit-step response of the compensated system.

c. The system designed in part **b** may still have a high overshoot. Use the $D(z)G_{h0}G_p(z)$ as a basis, and design another stage of phase-lag control with the transfer function $D_1(z)$ so that the total phase margin is 80 degrees. Plot the Bode diagram of the compensated system and the unit-step response.

Pole-Zero Cancellation Design

10-19 Consider the liquid-level control system described in Problem 7-17. When the number of inlets N is equal to 5, the open-loop transfer function of the system without the digital controller (microprocessor) is

$$G_{h0}G_p(z) = \frac{H(z)}{E(z)} = \frac{0.001476(z + 0.2246)(z + 3.178)}{(z - 1)(z - 0.5353)(z - 0.9512)}$$

It is found in Problem 7-17 that when $N = 5$, the uncompensated system is stable, but the step response is quite oscillatory with a large overshoot. The objective of this problem is to design a digital controller $D(z)$ using the root loci with pole-zero cancellation. Let the transfer function of the zero-order hold be

$$D(z) = K_c \frac{z - z_1}{z - p_1}$$

with the condition that $D(1) = 1$. Consider that it is appropriate to cancel a pole and a zero of $G_{h0}G_p(z)$ by the zero and pole of $D(z)$, except for the pole at $z = 1$. The maximum overshoot of the compensated system is to be less than 10 percent. Display the unit-step response of the compensated system.

Deadbeat-Response Design

10-4 The digital controlled process of a unity-feedback control system is described by the transfer function

$$G_p(z) = \frac{10(z + 0.5)}{(z - 1)(z - 0.5)}$$

Design a cascade phase-lead controller with the transfer function

$$D(z) = K_c \frac{z - z_1}{z - p_1}$$

so that the output $c(k)$ follows a unit-step input in minimum k without overshoot.

10-21 Given the transfer function of a digital controlled process as

$$G_p(z) = \frac{z + 1}{(z - 0.5)(z - 0.8)}$$

a. Find the cascade digital controller $D(z)$ so that the system has a deadbeat response to a unit-step input.
b. With the digital controller designed in part **a**, find the output response when the input is a unit ramp, $R(z) = z/(z - 1)^2$.

10-22 The transfer function of a digital controlled process is

$$G_p(z) = \frac{z + 1}{(z - 1)(z - 0.5)}$$

Can you design a stable cascade digital controller so that the system output is deadbeat for a step function input?

10-23 For the dc-motor speed control system described in Problem 7-11, the process transfer function is

$$G_p(s) = \frac{244680.85}{(s + 297.456)(s + 879.844)}$$

The sampling period T is 0.001 s. Design a cascade digital controller $D(z)$ so that the system output is deadbeat when the input is a step function. Verify that no intersampling ripples occur in the output.

10-24 For the system in Example 10-19, design a digital controller so that the output response is deadbeat without intersampling ripples to a ramp input. Plot the output unit-ramp response.

10-25 Consider the liquid-level control system described in Problem 7-17. When the number of inlets N is equal to 5, the open-loop transfer function of the process excluding the microprocessor controller is

$$G_{h0}G_p(z) = \frac{H(z)}{E(z)} = \frac{0.001476(z^2 + 3.4031z + 0.7139)}{(z-1)(z-0.5353)(z-0.9512)}$$

where $E(z)$ is the input to the zero-order hold of the D/A converter. The sampling period is 0.05 s. Find the transfer function of the digital controller $D(z)$ so that the system exhibits a ripple-free response $h(t)$ to a step input. Plot the unit-step response $h(t)$.

10-26 The discrete-data control of the velocity loop of a printwheel control system of a word processor is shown in Fig. P10-26. The power amplifier is a pulsewidth modulator, so that the motor current can be considered to be a constant. The system parameters are

Amplifier gain	$K = 150$
Motor torque constant	$K_a = 3.5$ oz-in/A
Motor back EMF constant	$K_b = 0.0247$ V/rad/s
Motor inertia	$J_m = 2.7 \times 10^{-4}$ oz-in-s^2
Load inertia	$J_L = 12 \times 10^{-4}$ oz-in-s^2
Load frictional coefficient	$B_L = 0.1045$ oz-in/rad/s
Shaft compliance	$K_L = 3454$ oz-in/rad
Sampling period	$T = 0.0001$ s.

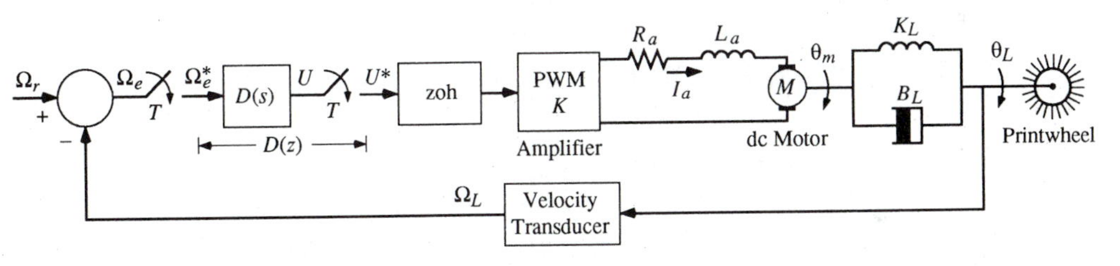

Figure P10-26.

a. Show that the transfer function between the motor current and the load velocity is

$$G_p(s) = \frac{\Omega_L(s)}{I_a(s)} = \frac{K_a(B_L s + K_L)}{s[J_m J_L s^2 + B_L(J_m + J_L)s + K_L(J_m + J_L)]}$$

b. Show that

$$G_{h0}G_p(z) = \frac{\Omega_L(z)}{U(z)} = \frac{0.01137(z^2 + 2.073z + 0.0454)}{(z-1)(z^2 - 1.80187z + 0.953)}$$

c. Design a deadbeat-response digital controller $D(z)$ so that the load velocity ω_L will follow a step input in finite time and without intersampling ripples. Show the unit-step response of the designed system.

10-27 The discrete-data control of the velocity loop of a printwheel control system of a word processor is shown in Fig. P10-27. The system parameters are identical to those given in Problem 10-26, and in addition,

Feedback resistance	$R_s = 0.18\ \Omega$
Motor resistance	$R_a = 4\ \Omega$
Motor inductance	$L_a = 0.002$ H.

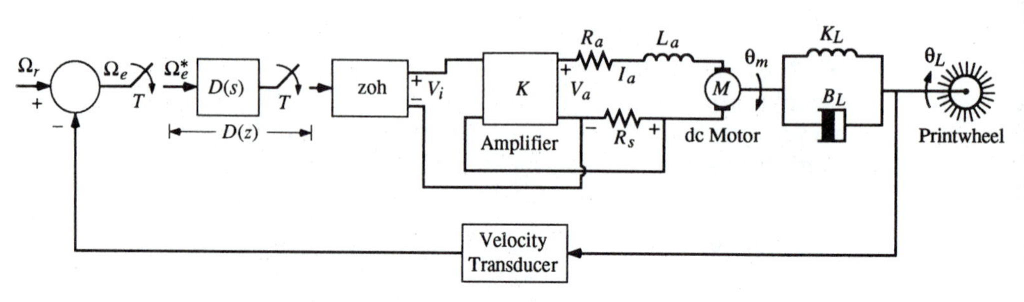

Figure P10-27.

a. Show that the transfer function between ω_L and v_i is

$$G_p(s) = \frac{\Omega_L(s)}{V_i(s)} = \frac{8.4664 \times 10^{10}(s + 33052.63)}{(s + 1.8864)(s + 15580)(s^2 + 482s + 1.568 \times 10^7)}$$

b. Show that the transfer function between $\omega_L^*(t)$ and $u^*(t)$ is

$$G_{h0}G_p(z) = \frac{\Omega_L(z)}{U(z)} = (1 - z^{-1})\mathscr{Z}[G_p(s)/s]$$

$$= \frac{0.87517(z^3 - 1.0527z^2 + 0.5524z - 0.0789)}{(z - 0.9998)(z - 0.2106)(z^2 - 1.80187z + 0.953)}$$

$$\cong \frac{0.87517(z^2 - 0.8425z + 0.3753)}{(z - 0.9998)(z^2 - 1.80187z + 0.953)}$$

c. Determine the roots of the characteristic equation and the system stability when $K = 150$.

d. Design a deadbeat-response controller $D(z)$ so that the output velocity ω_L will follow a step input with zero steady-state error in finite time, and with no intersampling ripples. Can you carry out the design so that the output has no overshoot? Plot the root loci of the characteristic equation of the deadbeat-response system as K varies. Show the location of the roots when $K = 150$.

e. Design a digital controller $D(z)$ so that the maximum overshoot is less than 10 percent.

z-Plane Forward Controller Design

10-28 The guided missile system described in Problem 10-5 is now subject to cascade and forward controls as shown in Fig. P10-28. Let the cascade controller be of the form:

$$D(z) = \frac{K_c(z - 0.98)(z - 0.905)}{(z - 1)(z + 0.5)}$$

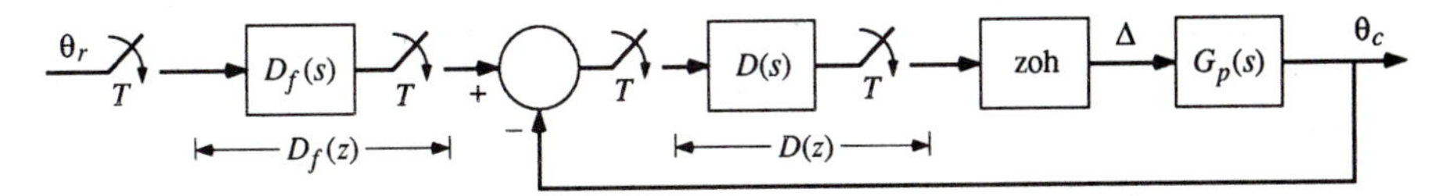

Figure P10-28.

Select the value of K_c so that the maximum overshoot of $c(t)$ is a minimum. Let the transfer function of the forward controller $D_f(z)$ be

$$D_f(z) = K_f \frac{z - z_1}{z - p_1}$$

Select the values of K_f, z_1, and p_1 so that the maximum overshoot of $c(t)$ is between 1 and 2 percent. [*Suggestion*: The pole of $D_f(z)$ cancels one of the zeros of the closed-loop transfer function.]

Robust Controller Design

10-29 A discrete-data control system is described by the transfer function

$$G_p(s) = \frac{887.3}{s(s + 8.871)}$$

The process is preceded by a zero-order hold, and the sampling period is 0.01 s.

a. Let the cascade controller be of the form

$$D(z) = K_c \frac{z^2 - 1.4z + 0.5576}{z(z - 0.5)}$$

and the forward controller is

$$D_f(z) = \frac{K_f z^2}{z^2 - 1.4z + 0.5576}$$

where K_f is set so that the steady-state unit-step response is unity. Determine the value of K_c so that the maximum overshoot of the output is a minimum. Plot the unit-step response of the compensated system.

b. Compute the sensitivity function S_G^M of the compensated system with the digital controller designed in part **a**. Repeat with the following cascade controllers and with the forward controller $D_f(z) = 1$.

$$D(z) = \frac{0.0688(z - 0.99424)}{z - 0.9996}$$

$$D(z) = \frac{60327(z - 0.8454)}{z - 0.06708}$$

$$D(z) = \frac{2.4023(z - 0.9834)(z - 0.9065)}{(z - 0.99423)(z - 0.35607)}$$

Pole-Zero Cancellation Design

10-30 The controlled process of a discrete-data control system is

$$G_{h0}G_p(z) = \frac{K(z + 0.8)}{z^2 - z + 0.34} \qquad K = 0.5$$

a. Design a cascade digital controller with zeros that cancel the complex poles of $G_{h0}G_p(z)$, and simultaneously achieve the following performance criteria:
(i) zero steady-state error to a step input.
(ii) unit-step response reaching 90 percent of the final value in less than five sampling periods, with the maximum overshoot less than 6 percent. Plot the unit-step response of the compensated system.

b. Plot the unit-step responses when $K = 0.3$ and 0.7 (0.5 ± 20 percent variation) with the controller designed in **a**. Are the performance specifications satisfied for these values of K?

Robust Controller Design

10-31 The controlled process of the discrete-data control system in Fig. P10-31 is

$$G_{h0}G_p(z) = \frac{K(z + 0.8)}{z^2 - z + 0.34} \qquad K = 0.5$$

Design the controller $D(z)$ and the forward controller $D_f(z)$ so that the system meets the following performance criteria:
(i) zero steady-state error to a step input.
(ii) unit-step response reaching 95 percent of its final value in less than 10 sampling periods, with the maximum overshoot less than 6 percent.
The system should be robust so that the performance specifications given above are satisfied for $0.3 \le K \le 0.7$.

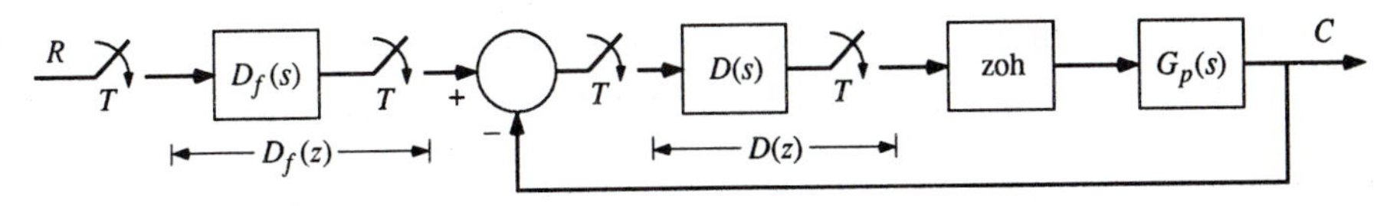

Figure P10-31.

Plot the unit-step responses of the compensated system for $K = 0.3$, 0.5, and 0.7. [*Suggestion*: Let the numerator of $D(z)$ contain the term $z^2 - z + 0.3$ or something similar.]

State-Feedback Control

10-32 Given a single-input digital control system

$$\mathbf{x}(k+1) = \mathbf{A}\mathbf{x}(k) + \mathbf{B}u(k)$$

where $\mathbf{x}(k)$ is an n-vector and the pair $[\mathbf{A}, \mathbf{B}]$ is completely controllable.

a. Show that given the closed-loop characteristic equation

$$\Delta_c(z) = z^n + a_1 z^{n-1} + a_2 z^{n-2} + \cdots + a_{n-1} z + a_n = 0$$

the state-feedback control is $u(k) = -\mathbf{G}\mathbf{x}(k)$, where

$$\mathbf{G} = \mathbf{F}\Delta_c(\mathbf{A})$$

$$\mathbf{F} = [0 \quad 0 \quad \cdots \quad 0 \quad 1][\mathbf{B} \quad \mathbf{AB} \quad \mathbf{A}^2\mathbf{B} \quad \cdots \quad \mathbf{A}^{n-1}\mathbf{B}]^{-1}$$

b. Show that the feedback gain matrix can be written as

$$\mathbf{G} = [1 \quad a_1 \quad a_2 \quad \cdots \quad a_n] \begin{bmatrix} \mathbf{FA}^n \\ \mathbf{FA}^{n-1} \\ \vdots \\ \mathbf{FA} \\ \mathbf{F} \end{bmatrix}$$

Note that the results presented in this problem are alternatives to the state-feedback algorithms for pole placement given in Eqs. (10-323) and (10-333).

c. Let

$$\mathbf{A} = \begin{bmatrix} 0 & 1 \\ -1 & 2 \end{bmatrix} \qquad \mathbf{B} = \begin{bmatrix} 0 \\ 1 \end{bmatrix}$$

Find the state-feedback matrix $\mathbf{G}$ such that the eigenvalues of $\mathbf{A} - \mathbf{BG}$ are at 0 and 0.3.

10-33

a. Given the discrete-data system

$$\mathbf{x}(k+1) = \mathbf{A}\mathbf{x}(k) + \mathbf{B}u(k)$$

where $\mathbf{A}$ is $n \times n$, $\mathbf{B}$ is $n \times 1$; $\mathbf{A}$ and $\mathbf{B}$ are in the phase-variable canonical form,

$$\mathbf{A} = \begin{bmatrix} 0 & 1 & 0 & 0 & 0 & \cdots & 0 \\ 0 & 0 & 1 & 0 & 0 & \cdots & 0 \\ 0 & 0 & 0 & 1 & 0 & \cdots & 0 \\ \cdots & \cdots & \cdots & \cdots & \cdots & \cdots & \cdots \\ 0 & 0 & 0 & 0 & 0 & \cdots & 1 \\ -a_1 & -a_2 & -a_3 & -a_4 & -a_5 & \cdots & -a_n \end{bmatrix}$$

$$\mathbf{B} = \begin{bmatrix} 0 \\ 0 \\ 0 \\ \vdots \\ 0 \\ 1 \end{bmatrix}$$

Let

$$u(k) = -\mathbf{G}\mathbf{x}(k)$$

It is desired to place the eigenvalues of the closed-loop system $\mathbf{x}(k+1) = (\mathbf{A} - \mathbf{B}\mathbf{G})\mathbf{x}(k)$ at $z_1, z_2, \ldots, z_n$, all distinct. Show that the feedback matrix $\mathbf{G}$ is given by

$$\mathbf{G} = -[\Delta_{01} \quad \Delta_{02} \quad \cdots \quad \Delta_{0n}]\mathbf{K}^{-1}$$

where

$$\Delta_{0i} = \Delta_0(z_i) = |z_i\mathbf{I} - \mathbf{A}|$$

and

$$\mathbf{K} = \begin{bmatrix} 1 & 1 & \cdots & 1 \\ z_1 & z_2 & \cdots & z_n \\ z_1^2 & z_2^2 & \cdots & z_n^2 \\ \cdots & \cdots & \cdots & \cdots \\ \cdots & \cdots & \cdots & \cdots \\ z_1^{n-1} & z_2^{n-1} & \cdots & z_n^{n-1} \end{bmatrix}$$

b. Let

$$\mathbf{A} = \begin{bmatrix} 0 & 1 \\ -1 & 2 \end{bmatrix} \qquad \mathbf{B} = \begin{bmatrix} 0 \\ 1 \end{bmatrix}$$

Find the state-feedback matrix **G** such that the eigenvalues of $\mathbf{A} - \mathbf{BG}$ are at 0 and 0.3.

10-34 An inventory control system is modeled by the following differential equations:

$$\frac{dx_1}{dt} = -x_2 + u$$

$$\frac{dx_2}{dt} = -bu$$

where

x_1 level of inventory
x_2 rate of sales of product
u production rate
b constant.

Let the production rate be controlled by discrete-time control so that

$$u(t) = u(kT) \qquad kT \le t < (k+1)T$$

where T is the sampling period. Furthermore, let

$$u(kT) = r(kT) - g_1 x_1(kT) - g_2 x_2(kT)$$

where $r(kT)$ is a reference set point used to control the inventory level; g_1 and g_2 are feedback gains.

a. Find the values of g_1 and g_2 so that the characteristic equation roots of the closed-loop system are all at the origin of the z-plane.
b. Find $x_1(kT)$ and $x_2(kT)$ for $k = 0, 1, 2, \ldots$, where $r(kT)$ is unity for all k. Assume that $x_1(0) = x_{10}$ and $x_2(0) = 0$.
c. Find the value of x_{10} so that, following from part **b**, $x_1(kT)$ is constant for all k; $x_2(0) = 0$. Find the corresponding $x_2(kT)$.

10-35 Given the digital control system,

$$\mathbf{x}(k+1) = \mathbf{A}\mathbf{x}(k) + \mathbf{B}\mathbf{u}(k)$$

where

$$\mathbf{A} = \begin{bmatrix} 0 & 1 \\ -1 & -2 \end{bmatrix} \qquad \mathbf{B} = \begin{bmatrix} 1 & 0 \\ 0 & 1 \end{bmatrix}$$

a. Find the state feedback $\mathbf{u}(k) = -\mathbf{G}\mathbf{x}(k)$ such that the eigenvalues of $\mathbf{A} - \mathbf{BG}$ are at $z = 0, 0$. It is also required that feedback is brought only to $u_1(k)$ and not to $u_2(k)$. Can this be achieved?
b. Repeat part **a** by requiring that feedback be brought only to $u_2(k)$.
c. Repeat part **a** by requiring that the feedback to u_1 is weighed twice that to u_2.

State-Feedback Design

10-36 The open-loop space shuttle control system described in Problem 7-12 is represented by the block diagram shown in Fig. P10-36. Find the gain matrix **G** such that the state feedback $u(kT) = -\mathbf{G}\mathbf{x}(kT)$ places the closed-loop eigenvalues at $z = 0.2$ and 0.5. The control u is subject to a sample-and-hold operation. The sampling period is 1 s. $M = 600$ and $K_s = 10$.

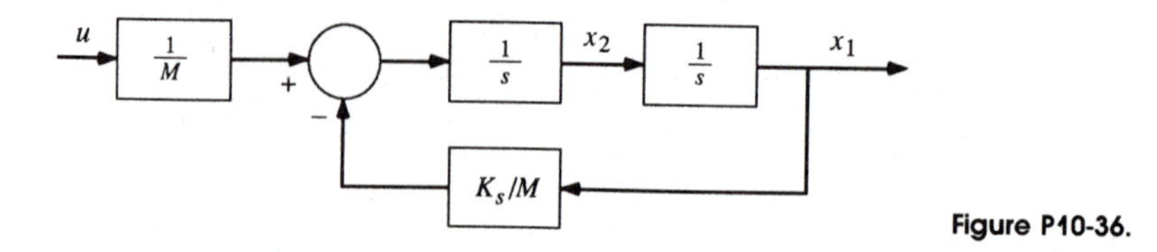

Figure P10-36.

Deadbeat-Response and PD Control

10-37 Consider the inventory control system described in Problem 10-34. The discrete-time control is given by

$$u(t) = u(kT) \qquad kT \le t < (k+1)T$$

where $u(t)$ is the output of the zero-order hold. The input of the zero-order hold is generated by a digital controller whose transfer function is given as

$$D(z) = \frac{U(z)}{E(z)}$$

Let $e(kT) = r(kT) - x_1(kT)$, where $r(kT)$ is the reference set point used to control the inventory level.

a. Given that $x_1(0) = 0$ and $x_2(0) = 0$, find the transfer function $D(z)$ so that the response of $x_1(kT)$ is deadbeat for $r(kT) = 1$ for all $k \ge 0$. Let $T = 1$ and $b = 1$. Find $x_2(kT)$ for $k \ge 0$.

b. Let the transfer function of the controller be

$$D(z) = K_P + \frac{K_D(z+1)}{Tz}$$

Find the values of K_P and K_D so that two of the roots of the characteristic equation are at $z = 0.5$ and 0.5. Where is the other root?

State and Dynamic Feedback Control

10-38 The temperature $x(t)$ in the electric furnace shown in Fig. P10-38 is governed by the differential equation

$$\frac{dx(t)}{dt} = -x(t) + u(t) + w_2(t)$$

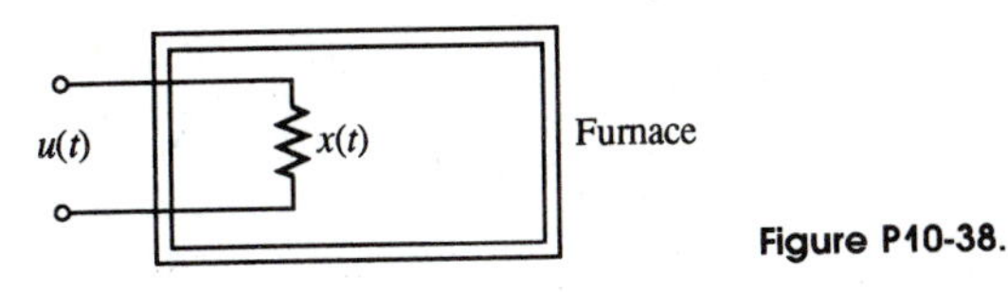

Figure P10-38.

where $u(t)$ is the control and $w_2(t)$ is an unknown constant disturbance due to heat losses. It is desired that the equilibrium temperature be at a set point w_1 = constant. Design a digital control with state and dynamic feedback by sampling the temperature $x(t)$ once every 0.2 s, and $u(t) = u(kT)$, $kT \leq t < (k+1)T$, such that the eigenvalues of the digital closed-loop system are at $z = 0$, and $x(kT)$ approaches w_1 as k approaches infinity. Assume zero initial conditions. Sketch $x(kT)$ for $k = 0, 1, 2, \ldots$ for positive w_1 and w_2.

10-39 Figure P10-39(a) shows the schematic diagram of a dc motor used for the speed control of a load. The characteristics of the permanent-magnet dc motor are described by the following parameters:

Armature resistance	$R = 2\ \Omega$
Armature inductance	L = negligible
Torque constant	$K_a = 10$ oz-in/A
Back EMF constant	$K_b = 0.052$ V/rad/s
Load and motor inertia at motor shaft	$J = 0.01$ oz-in-s^2
Load and motor friction	negligible.

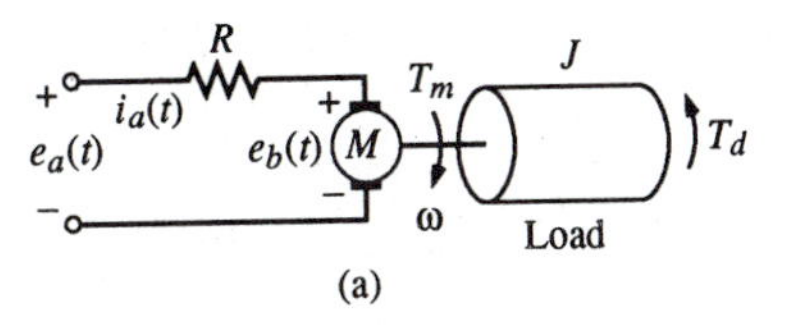

(a)

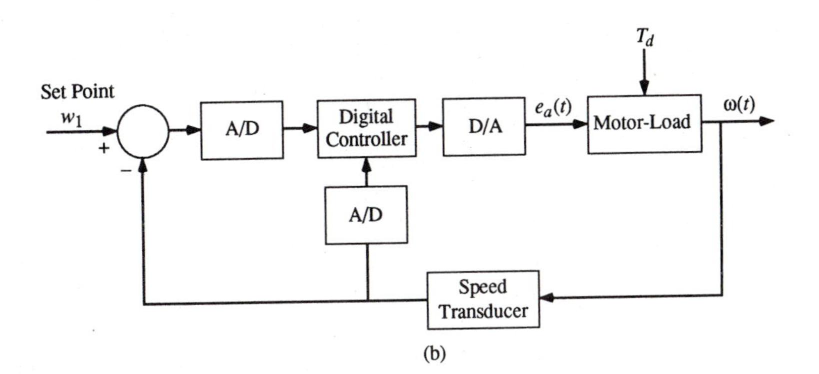

(b)

Figure P10-39.

The dynamics of the motor control system are described by the following equations:

$$e_a(t) = Ri_a(t) + e_b(t)$$

$$e_b = K_b \omega(t)$$

$$T_m(t) = K_a i_a(t)$$

$$T_m(t) = J \frac{d\omega(t)}{dt} + T_d$$

where

$e_a(t)$ motor applied voltage
$i_a(t)$ armature current
$e_b(t)$ back EMF
$T_m(t)$ motor torque
T_d constant load torque disturbance
$\omega(t)$ motor speed.

It is desired that the equilibrium speed of the motor-load shaft be at a set point w_1 = constant. The block diagram of the closed-loop system is shown in Fig. P10-39(b). Let the sampling period be 1/26 s. Design the configuration of the digital controller, and find the values of its parameters so that $\omega(kT)$ approaches w_1 as k approaches infinity. The eigenvalues of the closed-loop system should all be at $z = 0$.

10-40 For the controlled process shown in Fig. P10-40, design state and dynamic feedback control so that the state variable $x_1(k)$ will follow the set point w_1 = constant as $k \to \infty$. The noise signals w_2 and w_3 are unknown

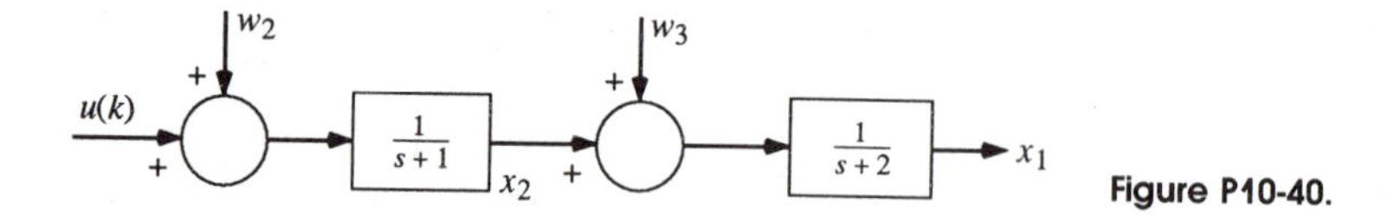

Figure P10-40.

constants. The roots of the characteristic equation of the closed-loop system should all be at $z = 0$. The control is subject to a sample-and-hold operation; that is, $u(t) = u(kT)$ for $k \le t < (k + 1)T$. The sampling period is 1 s.

References

1. Kuo, B. C., *Automatic Control Systems*, 6th ed., Prentice-Hall, Englewood Cliffs, N.J., 1991.
2. Whitbeck, R. F., and L. G. Hofmann, "Digital Control Law Synthesis in the w′ Domain," *Journal Guidance and Control*, vol. 1, no. 5, pp. 319–326, September–October 1978.

3. Sirisena, H. R., "Ripple-Free Deadbeat Control of SISO Discrete Systems," *IEEE Trans. Automatic Control*, vol. AC-30, pp. 168–170, February 1985.
4. Sebakhy, O. A., "Comments on 'Ripple-Free Deadbeat Control of SISO Discrete Systems'," *IEEE Trans. Automatic Control*, vol. AC-35, no. 6, pp. 765–766, June 1990.
5. Urikura, O., and A. Nagata, "Ripple-Free Deadbeat Control for Sampled-Data Systems," *IEEE Trans. Automatic Control*, vol. AC-32, pp. 474–483, June 1987.
6. Sebakhy, O. A., and J. A. Assiri, "Optimal Response in SISO Ripple-Free Deadbeat Systems," *Optimal Control Application Methods*, vol. 11, pp. 103–109, 1990.
7. Davison, E. J., "On Pole Assignment in Linear Systems with Incomplete State Feedback," *IEEE Trans. Automatic Control*, vol. AC-15, pp. 348–351, June 1970.
8. Wonham, W. M., "On Pole Assignment in Multi-Input Controllable Linear Systems," *IEEE Trans. Automatic Control*, vol. AC-12, pp. 660–665, December 1976.
9. Smith, H. W., and E. J. Davison, "Design of Industrial Regulators," *Proc. IEE (London)*, vol. 119, no. 8, pp. 1210–1216, August 1972.
10. Kuo, B. C., "Design of Digital Control Systems with State Feedback and Dynamic Output Feedback," *Journal Astronautical Sciences*, vol. XXVII, no. 2, pp. 207–214, April–June 1979.

Optimal Control

KEYWORDS AND TOPICS

discrete maximum principle • time-optimal control • linear regulator design • dynamic programming • Riccati equation • sampling period sensitivity • digital state observer

11-1 INTRODUCTION

The classical design methods discussed in the preceding chapters make use of such performance criteria as the phase margin, gain margin, bandwidth, maximum overshoot, and damping ratio, which lead to systems that would satisfy the prescribed performance requirements but are not optimized. The modern optimal control theory, on the other hand, relies on design techniques that maximize or minimize a performance index, resulting in a designed system that is optimal in the prescribed sense.

In this chapter we shall discuss the application of the optimal design techniques of the maximum principle and linear regulators to discrete-data control systems. The design of digital state observers is also covered.

11-2 THE DISCRETE EULER–LAGRANGE EQUATION

A large class of optimal designs for digital control systems has the objective of minimizing or maximizing a performance index of the form

$$J = \sum_{k=0}^{N-1} F[\mathbf{x}(k), \mathbf{x}(k+1), \mathbf{u}(k), k] \tag{11-1}$$

where $F[\mathbf{x}(k), \mathbf{x}(k+1), \mathbf{u}(k), k]$ is a differentiable scalar function.

The minimization or maximization of J is subject to the equality constraint of

$$\mathbf{x}(k+1) = \mathbf{f}[\mathbf{x}(k), \mathbf{u}(k), k] \tag{11-2}$$

which is recognized as the state equation of the system, as well as other equality and inequality constraints. For Eq. (11-2), $\mathbf{x}(k)$ is n-dimensional and $\mathbf{u}(k)$ is p-dimensional.

A majority of the design techniques for optimal control rely on the calculus principle of variation. According to the principle of variation, the problem of minimizing one function while it is subject to equality constraints is solved by adjoining the constraint to the function to be minimized or maximized.

Let $\boldsymbol{\lambda}(k+1)$, an $n \times 1$ vector, be defined as the *Lagrange multiplier*. We adjoin the performance index J of Eq. (11-1) with the equality constraint in Eq. (11-2) to form the adjoined performance index

$$J_c = \sum_{k=0}^{N-1} F[\mathbf{x}(k), \mathbf{x}(k+1), \mathbf{u}(k), k] + \langle \boldsymbol{\lambda}(k+1), [\mathbf{x}(k+1) - \mathbf{f}(\mathbf{x}, \mathbf{u}, k)] \rangle \tag{11-3}$$

where $\langle \cdot \rangle$ denotes the inner product of matrix vectors.

The calculus of variation asserts that the minimization or maximization of J subject to Eq. (11-2) is equivalent to minimizing or maximizing J_c without any constraints.

Let $\mathbf{x}(k)$, $\mathbf{x}(k+1)$, $\mathbf{u}(k)$, and $\boldsymbol{\lambda}(k+1)$ take on various variations as follows:

$$\mathbf{x}(k) = \mathbf{x}^\circ(k) + \varepsilon \boldsymbol{\eta}(k) \tag{11-4}$$

$$\mathbf{x}(k+1) = \mathbf{x}^\circ(k+1) + \varepsilon \boldsymbol{\eta}(k+1) \tag{11-5}$$

$$\mathbf{u}(k) = \mathbf{u}^\circ(k) + \delta \boldsymbol{\mu}(k) \tag{11-6}$$

$$\boldsymbol{\lambda}(k+1) = \boldsymbol{\lambda}^\circ(k+1) + \gamma \boldsymbol{\omega}(k+1) \tag{11-7}$$

where $\mathbf{x}^\circ(k)$, $\mathbf{x}^\circ(k+1)$, $\mathbf{u}^\circ(k)$, and $\boldsymbol{\lambda}^\circ(k+1)$ represent the vectors that correspond to the optimal trajectories; $\boldsymbol{\eta}(k)$, $\boldsymbol{\mu}(k)$, and $\boldsymbol{\omega}(k)$ are arbitrary vector variables.

Substitution of Eqs. (11-4) through (11-7) into Eq. (11-3) yields

$$\begin{aligned} J_c = \sum_{k=0}^{N-1} & F[\mathbf{x}^\circ(k) + \varepsilon\boldsymbol{\eta}(k), \mathbf{x}^\circ(k+1) + \varepsilon\boldsymbol{\eta}(k+1), \mathbf{u}^\circ(k) + \delta\boldsymbol{\mu}(k), k] \\ & + \langle \boldsymbol{\lambda}^\circ(k+1) + \gamma\boldsymbol{\omega}(k+1), \mathbf{x}^\circ(k+1) + \varepsilon\boldsymbol{\eta}(k+1) \\ & - \mathbf{f}[\mathbf{x}^\circ(k) + \varepsilon\boldsymbol{\eta}(k), \mathbf{u}^\circ(k) + \delta\boldsymbol{\mu}(k), k] \rangle \end{aligned} \tag{11-8}$$

To simplify the notation, we express J_c as

$$J_c = \sum_{k=0}^{N-1} F_c[\mathbf{x}(k), \mathbf{x}(k+1), \boldsymbol{\lambda}(k+1), \mathbf{u}(k), k] \tag{11-9}$$

Expanding F_c into a Taylor series about $\mathbf{x}^\circ(k)$, $\mathbf{x}^\circ(k+1)$, $\boldsymbol{\lambda}^\circ(k+1)$, $\mathbf{u}^\circ(k)$, we get

$$F_c[\mathbf{x}(k), \mathbf{x}(k+1), \boldsymbol{\lambda}(k+1), \mathbf{u}(k), k] = F_c[\mathbf{x}^\circ(k), \mathbf{x}^\circ(k+1), \boldsymbol{\lambda}^\circ(k+1), \mathbf{u}^\circ(k), k]$$
$$+ \left\langle \varepsilon\boldsymbol{\eta}(k), \frac{\partial F_c^\circ(k)}{\partial \mathbf{x}^\circ(k)} \right\rangle + \left\langle \varepsilon\boldsymbol{\eta}(k+1), \frac{\partial F_c^\circ(k)}{\partial \mathbf{x}^\circ(k+1)} \right\rangle + \left\langle \gamma\boldsymbol{\omega}(k+1), \frac{\partial F_c^\circ(k)}{\partial \boldsymbol{\lambda}^\circ(k+1)} \right\rangle$$
$$+ \left\langle \delta\boldsymbol{\mu}(k), \frac{\partial F_c^\circ(k)}{\partial \mathbf{u}^\circ(k)} \right\rangle + \text{higher order terms} \tag{11-10}$$

where

$$F_c^\circ(k) = F_c[\mathbf{x}^\circ(k), \mathbf{x}^\circ(k+1), \boldsymbol{\lambda}^\circ(k+1), \mathbf{u}^\circ(k), k] \tag{11-11}$$

The necessary condition for J_c to be a minimum is

$$\left.\frac{\partial J_c}{\partial \varepsilon}\right|_{\varepsilon=\delta=\gamma=0} = 0 \tag{11-12}$$

$$\left.\frac{\partial J_c}{\partial \gamma}\right|_{\varepsilon=\delta=\gamma=0} = 0 \tag{11-13}$$

$$\left.\frac{\partial J_c}{\partial \delta}\right|_{\varepsilon=\delta=\gamma=0} = 0 \tag{11-14}$$

Substituting the Taylor series expansion of F_c into Eq. (11-9) and applying the necessary condition of minimum J_c in Eqs. (11-12) through (11-14), we get

$$\sum_{k=0}^{N-1} \left[\left\langle \boldsymbol{\eta}(k), \frac{\partial F_c^\circ(k)}{\partial \mathbf{x}^\circ(k)} \right\rangle + \left\langle \boldsymbol{\eta}(k+1), \frac{\partial F_c^\circ(k)}{\partial \mathbf{x}^\circ(k+1)} \right\rangle \right] = 0 \tag{11-15}$$

$$\sum_{k=0}^{N-1} \left\langle \boldsymbol{\omega}(k+1), \frac{\partial F_c^\circ(k)}{\partial \boldsymbol{\lambda}^\circ(k+1)} \right\rangle = 0 \tag{11-16}$$

$$\sum_{k=0}^{N-1} \left\langle \boldsymbol{\mu}(k), \frac{\partial F_c^\circ(k)}{\partial \mathbf{u}^\circ(k)} \right\rangle = 0 \tag{11-17}$$

Equation (11-15) can be written as

$$\sum_{k=0}^{N-1} \left\langle \boldsymbol{\eta}(k), \frac{\partial F_c^\circ(k)}{\partial \mathbf{x}^\circ(k)} \right\rangle = -\sum_{k=1}^{N} \left\langle \boldsymbol{\eta}(k), \frac{\partial F_c^\circ(k-1)}{\partial \mathbf{x}^\circ(k)} \right\rangle$$
$$= -\sum_{k=0}^{N-1} \left\langle \boldsymbol{\eta}(k), \frac{\partial F_c^\circ(k-1)}{\partial \mathbf{x}^\circ(k)} \right\rangle$$
$$+ \left.\left\langle \boldsymbol{\eta}(k), \frac{\partial F_c^\circ(k-1)}{\partial \mathbf{x}^\circ(k)} \right\rangle\right|_{k=0} - \left.\left\langle \boldsymbol{\eta}(k), \frac{\partial F_c^\circ(k-1)}{\partial \mathbf{x}^\circ(k)} \right\rangle\right|_{k=N} \tag{11-18}$$

where

$$F_c^\circ(k-1) = F_c[\mathbf{x}^\circ(k-1), \mathbf{x}^\circ(k), \boldsymbol{\lambda}^\circ(k), \mathbf{u}^\circ(k-1), k-1] \tag{11-19}$$

Rearranging the terms in the last equation, we have

$$\sum_{k=0}^{N-1} \left\langle \boldsymbol{\eta}(k), \frac{\partial F_c^\circ(k)}{\partial \mathbf{x}^\circ(k)} + \frac{\partial F_c^\circ(k-1)}{\partial \mathbf{x}^\circ(k)} \right\rangle + \left\langle \boldsymbol{\eta}(k), \frac{\partial F_c^\circ(k-1)}{\partial \mathbf{x}^\circ(k)} \right\rangle \Bigg|_{k=0}^{k=N} = 0 \tag{11-20}$$

According to the fundamental lemma of the calculus of variation, Eq. (11-20) is satisfied for any $\boldsymbol{\eta}(k)$ when the following equations are satisfied:

$$\frac{\partial F_c^\circ(k)}{\partial \mathbf{x}^\circ(k)} + \frac{\partial F_c^\circ(k-1)}{\partial \mathbf{x}^\circ(k)} = 0 \tag{11-21}$$

$$\left\langle \boldsymbol{\eta}(k), \frac{\partial F_c^\circ(k-1)}{\partial \mathbf{x}^\circ(k)} \right\rangle \Bigg|_{k=0}^{k=N} = 0 \tag{11-22}$$

The reason behind the fundamental lemma is that for *arbitrary* $\boldsymbol{\eta}(k)$, the only way that Eq. (11-20) can be satisfied is that the two components of the equation be individually zero.

Equation (11-21) is called the *discrete Euler–Lagrange equation* which is the necessary condition that must be satisfied for J_c to be an extremum (maximum or minimum). The equation in Eq. (11-22) is known as the *transversality condition*, or simply the boundary condition needed to solve the partial differential equations in Eq. (11-21).

Now referring to the two additional conditions in Eqs. (11-16) and (11-17), we have for arbitrary $\boldsymbol{\mu}(k)$ and $\boldsymbol{\omega}(k+1)$,

$$\frac{\partial F_c^\circ(k)}{\partial \lambda_i^\circ(k+1)} = 0 \qquad i = 1, 2, \ldots, n \tag{11-23}$$

$$\frac{\partial F_c^\circ(k)}{\partial u_j^\circ(k)} = 0 \qquad j = 1, 2, \ldots, p \tag{11-24}$$

Equation (11-23) leads to

$$\mathbf{x}^\circ(k+1) = \mathbf{f}[\mathbf{x}^\circ(k), \mathbf{u}^\circ(k), k] \tag{11-25}$$

which implies that the state equation must satisfy the optimal trajectory. Equation (11-24), when applied to $F_c^\circ(k)$, gives the optimal control $\mathbf{u}^\circ(k)$ in terms of $\boldsymbol{\lambda}^\circ(k+1)$.

In a majority of the design problems the initial state $\mathbf{x}(0)$ is given at the outset. Then, the perturbation to $\mathbf{x}(k)$ at $k = 0$ is zero since $\mathbf{x}(0)$ is fixed; thus, $\boldsymbol{\eta}(0) = \mathbf{0}$. The transversality condition in Eq. (11-22) is reduced to

$$\left\langle \boldsymbol{\eta}(k), \frac{\partial F_c^\circ(k-1)}{\partial \mathbf{x}^\circ(k)} \right\rangle \Bigg|_{k=N} = 0 \tag{11-26}$$

Furthermore, most optimal control design problems are classified according to the end conditions. For instance, if $\mathbf{x}(N)$ is given and fixed, the problem is defined as a *fixed-endpoint* design problem. On the other hand, if $\mathbf{x}(N)$ is free or belongs to a certain target set, we have a free-endpoint problem. The transversality condition in Eq. (11-26) should be applied according to the end conditions, as follows.

Fixed-Endpoint

$$\mathbf{x}(N) = \text{fixed} \qquad \boldsymbol{\eta}(N) = \mathbf{0}$$

then nothing can be said about

$$\left.\frac{\partial F_c^\circ(k-1)}{\partial \mathbf{x}^\circ(k)}\right|_{k=N}$$

and no transversality condition is needed to solve Eq. (11-21).

Free-Endpoint

$$\mathbf{x}(N) = \text{free} \qquad \boldsymbol{\eta}(N) \neq 0$$

then

$$\left.\frac{\partial F_c^\circ(k-1)}{\partial \mathbf{x}^\circ(k)}\right|_{k=N} = 0 \tag{11-27}$$

which is the transversality condition required to solve Eq. (11-21).

In many cases some components of $\mathbf{x}(N)$ are fixed and the others are free; then the transversality conditions discussed above can be applied accordingly.

The following example illustrates the application of the calculus of variation and the discrete Euler–Lagrange equation method to the design of digital control systems.

Example 11-1

Find the optimal control $u^\circ(k)$, $k = 0, 1, 2, \ldots, 10$, such that the performance index

$$J = \frac{1}{2}\sum_{k=0}^{10} [x^2(k) + 2u^2(k)] \tag{11-28}$$

is minimized, subject to the equality constraint

$$x(k+1) = x(k) + 2u(k) \tag{11-29}$$

1. The initial state is $x(0) = 1$ and the final state is $x(11) = 0$.
2. The initial state is $x(0) = 1$ but the final state $x(11)$ is free.

Solution:

1. We first form the adjoined performance index

$$J_c = \sum_{k=0}^{10} F_c[x(k), u(k)] \tag{11-30}$$

where

$$F_c[x(k), u(k)] = \tfrac{1}{2}[x^2(k) + 2u^2(k)] + \lambda(k+1)[x(k+1) - x(k) - 2u(k)]$$
$$= F_c(k) \tag{11-31}$$

Using Eq. (11-21), the discrete Euler–Lagrange equation is determined,

$$\lambda^\circ(k+1) - \lambda^\circ(k) - x^\circ(k) = 0 \tag{11-32}$$

From Eq. (11-23),

$$\frac{\partial F_c^\circ(k)}{\partial \lambda^\circ(k+1)} = x^\circ(k+1) - x^\circ(k) - 2u^\circ(k) = 0 \tag{11-33}$$

which is the equality constraint or the state equation in Eq. (11-29) under the optimal condition. The optimal control is determined from Eq. (11-24),

$$\frac{\partial F_c^\circ(k)}{\partial u^\circ(k)} = 2u^\circ(k) - 2\lambda^\circ(k+1) = 0 \tag{11-34}$$

Thus,

$$u^\circ(k) = \lambda^\circ(k+1) \tag{11-35}$$

After substituting Eq. (11-35) into Eq. (11-33), Eqs. (11-32) and (11-33) form a set of two simultaneous first-order difference equations that must be solved to give $\lambda^\circ(k+1)$ for substitution in Eq. (11-35). In the present case, since $x(0)$ and $x(11)$ are fixed, these values provide the two necessary boundary conditions for solving Eqs. (11-32) and (11-33), and the transversality condition of Eq. (11-26) is not needed.

The two simultaneous difference equations to be solved are

$$\lambda^\circ(k+1) - \lambda^\circ(k) - x^\circ(k) = 0 \tag{11-36}$$

$$x^\circ(k+1) - 2\lambda^\circ(k+1) - x^\circ(k) = 0 \tag{11-37}$$

with $x(0) = 1$ and $x(11) = 0$. These equations may be solved by the z-transform method or the state transition method, and the solutions are

$$x^\circ(k) = 0.289[2.732 + 2\lambda^\circ(0)](3.732)^k + 0.289[0.732 - 2\lambda^\circ(0)](0.268)^k \tag{11-38}$$

$$\lambda^\circ(k) = [0.289 + 0.211\lambda^\circ(0)](3.732)^k + [-0.289 + 0.789\lambda^\circ(0)](0.268)^k \tag{11-39}$$

The initial value of $\lambda^\circ(k)$ is found by substituting $x^\circ(11) = 0$ into Eq. (11-38),

$$\lambda^\circ(0) = -1.366 \tag{11-40}$$

Then, the optimal trajectory of $x^\circ(k)$ is described by

$$x^\circ(k) = (0.268)^k \tag{11-41}$$

and the optimal control is

$$u^\circ(k) = -2.732(0.268)^{k+1}$$
$$= -0.732x^\circ(k) \tag{11-42}$$

Strictly, when $k = 11$, $x^\circ(11)$ should be equal to zero; however, due to the numerical nature of the problem, an error of $(0.268)^{11}$ is generated.

2. When $x(0) = 1$ and $x(11)$ is free, we need to use the transversality condition:

$$\left. \frac{\partial F_c^\circ(k-1)}{\partial x^\circ(k)} \right|_{k=11} = 0 \tag{11-43}$$

The last condition leads to

$$\lambda^\circ(11) = 0 \tag{11-44}$$

Now substituting Eq. (11-44) into Eq. (11-39), we get $\lambda^\circ(0) = -1.366$ and the same results as in Eqs. (11-41) and (11-42).

The reason that the same results are obtained when the endpoint $x^\circ(11)$ is fixed at 0 or free is that the performance index already includes a constraint on $x(k)$, so that the design naturally forces $x(k)$ to go to zero as quickly as possible. In general, of course, when $x^\circ(N)$ is not fixed it may reach any final value.

In summary, the calculus of variation method requires the solution of the discrete Euler–Lagrange equation. However, for an nth-order system the Euler–Lagrange equation is of the order $2n$. This means that solution of the Euler–Lagrange equation is usually tedious.

Although controllability was not mentioned in the formulation of the design using the discrete Euler–Lagrange equation, it is apparent that state controllability is required when $\mathbf{x}(N)$ is specified and when N is a finite integer.

11-3 THE DISCRETE MAXIMUM (MINIMUM) PRINCIPLE

The maximum principle (or minimum principle) introduced by Pontryagin is a very powerful method of solving a wide class of continuous-data control systems. The design by maximum principle is also based on the calculus of variation, but the mechanics are more refined and elegant than the use of the Euler–Lagrange equation. The discrete maximum principle may be considered as an extension to the design of digital control systems. Strictly speaking, the application of the discrete maximum principle requires the investigation of the condition of convexity of the system. The presentation of the subject in this section is intended only for a working knowledge of the method so that abstract mathematics are kept to a minimum.

The design problem can be stated as follows. Find the optimal control $\mathbf{u}^\circ(k)$ over $[0, N]$ such that the performance index

$$J = G[\mathbf{x}(N), N] + \sum_{k=0}^{N-1} F[\mathbf{x}(k), \mathbf{u}(k), k] \tag{11-45}$$

is minimized, subject to the equality constraint,

$$\mathbf{x}(k+1) = \mathbf{f}[\mathbf{x}(k), \mathbf{u}(k), k] \tag{11-46}$$

The term $G[\mathbf{x}(N), (N)]$ in Eq. (11-45) is the terminal cost of the performance index. It is required as a terminal constraint on the end condition only if $\mathbf{x}(N)$ is not fixed.

As with the Lagrange multiplier, we define a *costate* vector $\mathbf{p}(k)$ $(n \times 1)$. Then, the optimization problem is equivalent to minimizing

$$J_c = G[\mathbf{x}(N), N] + \sum_{k=0}^{N-1} [F[\mathbf{x}(k), \mathbf{u}(k), k] - \langle \mathbf{p}(k+1), [\mathbf{x}(k+1) - \mathbf{f}(\mathbf{x}, \mathbf{u}, k)] \rangle] \tag{11-47}$$

Let us define the scalar function $H[\mathbf{x}(k), \mathbf{u}(k), \mathbf{p}(k+1), k]$ as the Hamiltonian, such that

$$H[\mathbf{x}(k), \mathbf{u}(k), \mathbf{p}(k+1), k] = F[\mathbf{x}(k), \mathbf{u}(k), k] - \langle \mathbf{p}(k+1), \mathbf{f}[\mathbf{x}(k), \mathbf{u}(k), k] \rangle \tag{11-48}$$

When the Hamiltonian is defined as in the last equation, it forms the basis of the discrete maximum principle. For the discrete minimum principle, the Hamiltonian is defined as

$$H[\mathbf{x}(k), \mathbf{u}(k), \mathbf{p}(k+1), k] = F[\mathbf{x}(k), \mathbf{u}(k), k] + \langle \mathbf{p}(k+1), \mathbf{f}[\mathbf{x}(k), \mathbf{u}(k), k] \rangle \tag{11-49}$$

As we shall see later in the discussions, the maximum principle refers to the property that the Hamiltonian is a maximum along the optimal trajectory, whereas for the minimum principle the Hamiltonian is a minimum.

Substituting the definition of the Hamiltonian of Eq. (11-49) into Eq. (11-47), we have

$$J_c = G[\mathbf{x}(N), N] + \sum_{k=0}^{N-1} [H[\mathbf{x}(k), \mathbf{u}(k), \mathbf{p}(k+1), k] - \langle \mathbf{p}(k+1), \mathbf{x}(k+1) \rangle] \tag{11-50}$$

which is for the minimum principle.

Let $\mathbf{x}(k)$, $\mathbf{x}(k+1)$, and $\mathbf{u}(k)$ take on variations as follows:

$$\mathbf{x}(k) = \mathbf{x}^\circ(k) + \varepsilon\boldsymbol{\eta}(k) \qquad (n \times 1) \tag{11-51}$$

$$\mathbf{x}(k+1) = \mathbf{x}^\circ(k+1) + \varepsilon\boldsymbol{\eta}(k+1) \qquad (n \times 1) \tag{11-52}$$

$$\mathbf{u}(k) = \mathbf{u}^\circ(k) + \delta\boldsymbol{\mu}(k) \qquad (p \times 1) \tag{11-53}$$

$$\mathbf{p}(k+1) = \mathbf{p}^\circ(k+1) + \gamma\boldsymbol{\omega}(k+1) \qquad (n \times 1) \tag{11-54}$$

Equation (11-50) is now written

$$\begin{aligned} J_c = {} & G[\mathbf{x}^\circ(N) + \varepsilon\boldsymbol{\eta}(N), N] \\ & + \sum_{k=0}^{N-1} H[\mathbf{x}^\circ(k) + \varepsilon\boldsymbol{\eta}(k), \mathbf{u}^\circ(k) + \delta\boldsymbol{\mu}(k), \mathbf{p}^\circ(k+1) + \gamma\boldsymbol{\omega}(k+1), k] \\ & - \langle \mathbf{p}^\circ(k+1) + \gamma\boldsymbol{\omega}(k+1), \mathbf{x}^\circ(k+1) + \varepsilon\boldsymbol{\eta}(k+1) \rangle \end{aligned} \tag{11-55}$$

Expanding $G[\mathbf{x}(N), N]$ into a Taylor series about $G[\mathbf{x}^\circ(N), N]$, we get

$$G[\mathbf{x}(N), N] = G[\mathbf{x}^\circ(N), N] + \varepsilon\left\langle \boldsymbol{\eta}(N), \frac{\partial G^\circ(N)}{\partial \mathbf{x}^\circ(N)} \right\rangle + \cdots \tag{11-56}$$

Similarly, we expand $H[\mathbf{x}(k), \mathbf{u}(k), \mathbf{p}(k+1), k]$ into a Taylor series about $\mathbf{x}^\circ(k)$, $\mathbf{u}^\circ(k)$, $\mathbf{p}^\circ(k+1)$, and $\mathbf{x}^\circ(k+1)$,

$$\begin{aligned} H[\mathbf{x}(k), \mathbf{u}(k), \mathbf{p}(k+1), k] = {} & H[\mathbf{x}^\circ(k), \mathbf{u}^\circ(k), \mathbf{p}^\circ(k+1), k] + \varepsilon\left\langle \boldsymbol{\eta}(k), \frac{\partial H^\circ(k)}{\partial \mathbf{x}^\circ(k)} \right\rangle \\ & + \delta\left\langle \boldsymbol{\mu}(k), \frac{\partial H^\circ(k)}{\partial \mathbf{u}^\circ(k)} \right\rangle \\ & + \gamma\left\langle \boldsymbol{\omega}(k+1), \frac{\partial H^\circ(k)}{\partial \mathbf{p}^\circ(k+1)} \right\rangle + \cdots \end{aligned} \tag{11-57}$$

where

$$H^\circ(k) = H[\mathbf{x}^\circ(k), \mathbf{u}^\circ(k), \mathbf{p}^\circ(k+1), k] \tag{11-58}$$

Substituting into Eq. (11-55) the Taylor series expansions of $G[\mathbf{x}(N), N]$ and $H[\mathbf{x}(k), \mathbf{u}(k), \mathbf{p}(k+1), k]$ and carrying out the following necessary conditions for minimum J_c:

$$\left.\frac{\partial J_c}{\partial \varepsilon}\right|_{\varepsilon=\delta=\gamma=0} = 0 \tag{11-59}$$

$$\left.\frac{\partial J_c}{\partial \delta}\right|_{\varepsilon=\delta=\gamma=0} = 0 \tag{11-60}$$

$$\left.\frac{\partial J_c}{\partial \gamma}\right|_{\varepsilon=\delta=\gamma=0} = 0 \tag{11-61}$$

we have

$$\left\langle \boldsymbol{\eta}(N), \frac{\partial G^\circ(N)}{\partial \mathbf{x}^\circ(N)} \right\rangle + \sum_{k=0}^{N-1} \left\langle \boldsymbol{\eta}(k), \frac{\partial H^\circ(k)}{\partial \mathbf{x}^\circ(k)} \right\rangle - \sum_{k=0}^{N-1} \langle \mathbf{p}^\circ(k+1), \boldsymbol{\eta}(k+1) \rangle = 0 \tag{11-62}$$

$$\left\langle \boldsymbol{\mu}(k), \frac{\partial H^\circ(k)}{\partial \boldsymbol{\mu}^\circ(k)} \right\rangle = 0 \tag{11-63}$$

$$\left\langle \boldsymbol{\omega}(k+1), \frac{\partial H^\circ(k)}{\partial \mathbf{p}^\circ(k+1)} - \mathbf{x}^\circ(k+1) \right\rangle = 0 \tag{11-64}$$

Equation (11-64) leads to

$$\frac{\partial H^\circ(k)}{\partial \mathbf{p}^\circ(k+1)} = \mathbf{x}^\circ(k+1) \tag{11-65}$$

which is recognized as the original state equation, Eq. (11-46). Equation (11-63) gives

$$\frac{\partial H^\circ(k)}{\partial \mathbf{u}^\circ(k)} = 0 \tag{11-66}$$

which has the significance that the Hamiltonian has an extremum along the optimal trajectory with respect to the optimal control.

The last term on the left-hand side of Eq. (11-62) is written as

$$\begin{aligned}\sum_{k=0}^{N-1} \langle \mathbf{p}^\circ(k+1), \boldsymbol{\eta}(k+1) \rangle &= \sum_{k=1}^{N} \langle \mathbf{p}^\circ(k), \boldsymbol{\eta}(k) \rangle \\ &= \sum_{k=0}^{N-1} \langle \mathbf{p}^\circ(k), \boldsymbol{\eta}(k) \rangle + \langle \mathbf{p}^\circ(N), \boldsymbol{\eta}(N) \rangle - \langle \mathbf{p}^\circ(0), \boldsymbol{\eta}(0) \rangle\end{aligned} \tag{11-67}$$

Since $\mathbf{x}(0)$ is given, $\boldsymbol{\eta}(0) = \mathbf{0}$; the last equation becomes

$$\sum_{k=0}^{N-1} \langle \mathbf{p}^\circ(k+1), \boldsymbol{\eta}(k+1) \rangle = \sum_{k=0}^{N-1} \langle \mathbf{p}^\circ(k), \boldsymbol{\eta}(k) \rangle + \langle \mathbf{p}^\circ(N), \boldsymbol{\eta}(N) \rangle \tag{11-68}$$

After substituting the last expression into Eq. (11-62) and rearranging, we have

$$\left\langle \frac{\partial G^\circ(N)}{\partial \mathbf{x}^\circ(N)} - \mathbf{p}^\circ(N), \boldsymbol{\eta}(N) \right\rangle + \sum_{k=0}^{N-1} \left\langle \frac{\partial H^\circ(k)}{\partial \mathbf{x}^\circ(k)} - \mathbf{p}^\circ(k), \boldsymbol{\eta}(k) \right\rangle = 0 \tag{11-69}$$

Since the variations are mutually independent, the only way to satisfy the last equation is

$$\frac{\partial G^\circ(N)}{\partial \mathbf{x}^\circ(N)} = \mathbf{p}^\circ(N) \tag{11-70}$$

$$\frac{\partial H^\circ(k)}{\partial \mathbf{x}^\circ(k)} = \mathbf{p}^\circ(k) \tag{11-71}$$

In summarizing the results, the necessary condition for J_c to be an extremum is that

$$\frac{\partial H^\circ(k)}{\partial \mathbf{x}^\circ(k)} = \mathbf{p}^\circ(k) \tag{11-72}$$

$$\frac{\partial H^\circ(k)}{\partial \mathbf{p}^\circ(k+1)} = \mathbf{x}^\circ(k+1) \tag{11-73}$$

$$\frac{\partial H^\circ(k)}{\partial \mathbf{u}^\circ(k)} = 0 \tag{11-74}$$

$$\frac{\partial G^\circ(N)}{\partial \mathbf{x}^\circ(N)} = \mathbf{p}^\circ(N) \tag{11-75}$$

Equations (11-72) and (11-73) represent $2n$ first-order difference equations, which are defined as the *canonical state equations.* Equation (11-74) gives the relation for the optimal control $\mathbf{u}^\circ(k)$, and Eq. (11-75) gives the transversality condition, which should be used when $\mathbf{x}(N)$ is not fixed. If any component of $\mathbf{x}(N)$ is fixed, the corresponding transversality condition of $\mathbf{p}^\circ(N)$ does not apply.

The following example illustrates the application of the discrete minimum principle to the design of a digital control system.

Example 11-2

Find the optimal control $u^\circ(k)$, $k = 0, 1, 2, \ldots, 10$, such that the performance index

$$J = \frac{1}{2} \sum_{k=0}^{10} [x^2(k) + 2u^2(k)] \tag{11-76}$$

is minimized, subject to the equality constraint

$$x(k+1) = x(k) + 2u(k) \tag{11-77}$$

The initial state is $x(0) = 1$ and the final state is $x(11) = 0$. Notice that this is the same design problem stated in part 1 of Example 11-1.

The first step in the discrete minimum principle design is to define the Hamiltonian, according to Eq. (11-49),

$$H[\mathbf{x}(k), \mathbf{u}(k), \mathbf{p}(k+1)] = \tfrac{1}{2}[x^2(k) + 2u^2(k)] + p(k+1)[x(k) + 2u(k)] \tag{11-78}$$

The canonical state equations are obtained from Eqs. (11-72) and (11-73). These are

$$p^\circ(k+1) - p^\circ(k) = -x^\circ(k) \tag{11-79}$$

$$x^\circ(k+1) - x^\circ(k) = 2u^\circ(k) \tag{11-80}$$

The optimal control is found from Eq. (11-74),

$$u^\circ(k) = -p^\circ(k+1) \tag{11-81}$$

Since the endpoint $x(11)$ is fixed, the transversality condition of Eq. (11-75) is not needed.

The solutions of Eqs. (11-79), (11-80), and (11-81) are

$$x^\circ(k) = 0.289[2.732 - 2p^\circ(0)](3.732)^k + 0.289[0.732 + 2p^\circ(0)](0.268)^k \tag{11-82}$$

$$p^\circ(k) = [-0.289 + 0.211p^\circ(0)](3.732)^k + [0.289 + 0.789p^\circ(0)](0.268)^k \tag{11-83}$$

These equations are similar to the solutions in Eqs. (11-38) and (11-39) except for a few differences in signs. Substitution of $x^\circ(11) = 0$ in Eq. (11-82) yields the result $p^\circ(0) = 1.366$. Thus,

$$x^\circ(k) = (0.268)^k$$

and

$$u^\circ(k) = -2.732(0.268)^{k+1}$$

which are the same results as in Example 11-1, part 1.

Now let us make $x(11)$ a free-end condition and add a terminal cost to the performance index of Eq. (11-76),

$$J = \frac{1}{2}x^2(11) + \frac{1}{2}\sum_{k=0}^{10}[x^2(k) + 2u^2(k)] \tag{11-84}$$

The transversality condition of Eq. (11-75) gives

$$x^\circ(11) = p^\circ(11) \tag{11-85}$$

Now equating Eqs. (11-82) and (11-83) at $k = 11$ again results in

$$p^\circ(0) = 1.366$$

and the same results are obtained for $x^\circ(k)$ and $u^\circ(k)$.

It should be pointed out that the use of the terminal cost of the performance index of the minimum principle design of free-endpoint problems may yield different results than the calculus of variation method.

11-4 TIME-OPTIMAL CONTROL WITH ENERGY CONSTRAINT

In general, the time-optimal control design can be described as the problem of bringing $\mathbf{x}(0)$ to $\mathbf{x}(N)$ in minimum time. The control is subject to the amplitude constraint $|\mathbf{u}(kT)| \le U$. In this section we shall present a time-optimal control design with a quadratic constraint on the control. The design is carried out by use of the discrete minimum principle. The development will also pave the way for the discussions on the linear digital regulator design presented in the next chapter. The design problem is stated as follows.

The digital system

$$\mathbf{x}(k+1) = \mathbf{A}\mathbf{x}(k) + \mathbf{B}\mathbf{u}(k) \tag{11-86}$$

where $\mathbf{x}(k)$ is $n \times 1$, $\mathbf{u}(k)$ is $p \times 1$, and $\mathbf{A}$ is nonsingular, is given. It is assumed that the pair $[\mathbf{A}, \mathbf{B}]$ is completely controllable. Find the optimal control $\mathbf{u}^\circ(k)$, $k = 0, 1, 2, \ldots, N-1$, such that the initial state $\mathbf{x}(0)$ is driven to the final state $\mathbf{x}(N) = \mathbf{0}$, subject to the control constraint

$$J = \frac{1}{2} \sum_{k=0}^{N-1} \mathbf{u}'(k)\mathbf{R}\mathbf{u}(k) = \text{minimum} \tag{11-87}$$

where $\mathbf{R}$ is symmetric and positive definite.

The performance index in Eq. (11-87) is generally known as the quadratic form, and in this case it represents an energy constraint on the designed system. Therefore, the performance index J in Eq. (11-87) represents an alternate, although not equivalent, constraint on the control $\mathbf{u}(k)$.

The solution of the problem begins by defining the Hamiltonian function

$$H[\mathbf{x}(k), \mathbf{p}(k+1), \mathbf{u}(k)] = \tfrac{1}{2}\langle \mathbf{u}(k), \mathbf{R}\mathbf{u}(k)\rangle + \langle \mathbf{p}(k+1), \mathbf{x}(k+1)\rangle \tag{11-88}$$

The necessary conditions for J to be a minimum are

$$\frac{\partial H^\circ(k)}{\partial \mathbf{x}^\circ(k)} = \mathbf{p}^\circ(k) = \mathbf{A}'\mathbf{p}^\circ(k+1) \tag{11-89}$$

$$\frac{\partial H^\circ(k)}{\partial \mathbf{p}^\circ(k+1)} = \mathbf{x}^\circ(k+1) = \mathbf{A}\mathbf{x}^\circ(k) + \mathbf{B}\mathbf{u}^\circ(k) \tag{11-90}$$

$$\frac{\partial H^\circ(k)}{\partial \mathbf{u}^\circ(k)} = \mathbf{R}\mathbf{u}^\circ(k) + \mathbf{B}'\mathbf{p}^\circ(k+1) = 0 \tag{11-91}$$

where

$$H^\circ(k) = \tfrac{1}{2}\langle \mathbf{u}^\circ(k), \mathbf{R}\mathbf{u}^\circ(k)\rangle + \langle \mathbf{p}^\circ(k+1), \mathbf{A}\mathbf{x}^\circ(k) + \mathbf{B}\mathbf{u}^\circ(k)\rangle \tag{11-92}$$

Since $\mathbf{x}(0)$ and $\mathbf{x}(N)$ are all fixed, the transversality condition is not needed.

The optimal control is obtained from Eq. (11-91),

$$\mathbf{u}^\circ(k) = -\mathbf{R}^{-1}\mathbf{B}'\mathbf{p}^\circ(k+1) \tag{11-93}$$

where the inverse of $\mathbf{R}$ exists since it is positive definite.

In the present problem the costate equation, Eq. (11-89), is not coupled to the state variable $\mathbf{x}(k)$. Thus, Eq. (11-89) can be solved by itself to give

$$\mathbf{p}^\circ(k) = (\mathbf{A}^{-k})'\mathbf{p}^\circ(0) \tag{11-94}$$

where we have assumed that $\mathbf{A}$ has an inverse.

Substitution of Eq. (11-93) into Eq. (11-90) gives

$$\mathbf{x}^\circ(k+1) = \mathbf{A}\mathbf{x}^\circ(k) - \mathbf{B}\mathbf{R}^{-1}\mathbf{B}'\mathbf{p}^\circ(k+1) \tag{11-95}$$

The solution of the last equation is

$$\mathbf{x}^\circ(N) = \mathbf{A}^N\mathbf{x}(0) - \sum_{k=0}^{N-1} \mathbf{A}^{N-k-1}\mathbf{B}\mathbf{R}^{-1}\mathbf{B}'(\mathbf{A}^{-1})'\mathbf{p}^\circ(k) \tag{11-96}$$

Setting $\mathbf{x}^\circ(N) = \mathbf{0}$ and solving for $\mathbf{x}(0)$ in the last equation, we get

$$\mathbf{x}(0) = \sum_{k=0}^{N-1} \mathbf{A}^{-k-1}\mathbf{B}\mathbf{R}^{-1}\mathbf{B}'(\mathbf{A}^{-k-1})'\mathbf{p}^\circ(0) \tag{11-97}$$

where Eq. (11-94) has been used.

It is interesting to note that the matrix

$$\mathbf{W} = \sum_{k=0}^{N-1} \mathbf{A}^{-k-1}\mathbf{B}\mathbf{R}^{-1}\mathbf{B}'(\mathbf{A}^{-k-1})' = \sum_{k=0}^{N-1} \mathbf{S}_k\mathbf{S}_k' \tag{11-98}$$

is the controllability matrix if $\mathbf{R}$ is an identity matrix [refer to Eq. (6-10)]. For any $\mathbf{R}$ that is not an identity matrix but still symmetric and positive definite, we let

$$\mathbf{R}^{-1} = \mathbf{K}\mathbf{K}' \tag{11-99}$$

where $\mathbf{K}$ is a $p \times p$ matrix. Then in Eq. (11-98),

$$\mathbf{S}_k = \mathbf{A}^{-k-1}\mathbf{B}\mathbf{K} \tag{11-100}$$

Since $\mathbf{R}$ is symmetric and positive definite, $\mathbf{R}^{-1}$ has the same properties. Equation (11-99) has a unique solution $\mathbf{K}$, which is symmetric and positive definite, since the equation is a simple form of the Riccati-type nonlinear matrix equation.

$$\mathbf{K}\mathbf{\Theta}\mathbf{Q}^{-1}\mathbf{\Theta}'\mathbf{K}' - \mathbf{K}\mathbf{\Phi} - \mathbf{\Phi}'\mathbf{K} - \mathbf{R}^{-1} = \mathbf{0} \tag{11-101}$$

with $\mathbf{\Theta} = \mathbf{I}$, $\mathbf{Q}^{-1} = \mathbf{I}$, and $\mathbf{\Phi} = \mathbf{0}$.

The $n \times n$ matrix $\mathbf{W}$ in Eq. (11-98) is nonsingular if the matrix

$$[\mathbf{A}^{-1}\mathbf{B}\mathbf{K} \quad \mathbf{A}^{-2}\mathbf{B}\mathbf{K} \quad \cdots \quad \mathbf{A}^{-N}\mathbf{B}\mathbf{K}] \qquad (n \times pN) \tag{11-102}$$

is of rank n. The last matrix can be written as

$$\underset{(n \times pN)}{[\mathbf{A}^{-1}\mathbf{B} \quad \mathbf{A}^{-2}\mathbf{B} \quad \cdots \quad \mathbf{A}^{-N}\mathbf{B}]} \underset{(pN \times pN)}{\begin{bmatrix} \mathbf{K} & & & \\ & \mathbf{K} & & \\ & & \ddots & \\ & & & \mathbf{K} \end{bmatrix}} \tag{11-103}$$

The rank of the matrix in Eq. (11-102) is the same as the rank of $[\mathbf{A}^{-1}\mathbf{B} \quad \mathbf{A}^{-2}\mathbf{B} \quad \cdots \quad \mathbf{A}^{-N}\mathbf{B}]$ or of $[\mathbf{B} \quad \mathbf{A}\mathbf{B} \quad \cdots \quad \mathbf{A}^{N-1}\mathbf{B}]$, since $\mathbf{K}$ is positive definite. Therefore, if the pair $[\mathbf{A}, \mathbf{B}]$ is completely controllable, so is $[\mathbf{A}, \mathbf{B}\mathbf{K}]$.

We have now established that if $[\mathbf{A}, \mathbf{B}]$ is completely controllable, $\mathbf{W}$ is nonsingular. From Eq. (11-97) we have

$$\mathbf{p}^\circ(0) = \mathbf{W}^{-1}\mathbf{x}(0) \tag{11-104}$$

The optimal control can be expressed in terms of the initial state $\mathbf{x}(0)$; from Eq. (11-93),

$$\begin{aligned} \mathbf{u}^\circ(k) &= -\mathbf{R}^{-1}\mathbf{B}'(\mathbf{A}^{-1})'\mathbf{p}(k) \\ &= -\mathbf{R}^{-1}\mathbf{B}'(\mathbf{A}^{-1})'(\mathbf{A}^{-k})'\mathbf{p}(0) \end{aligned} \tag{11-105}$$

Thus,

$$\mathbf{u}^\circ(k) = -\mathbf{R}^{-1}\mathbf{B}'(\mathbf{A}^{-k-1})'\mathbf{W}^{-1}\mathbf{x}(0) \tag{11-106}$$

Substituting the last equation into Eq. (11-87), we get, after simplification, the optimal performance index

$$J^\circ = \tfrac{1}{2}\mathbf{x}'(0)\mathbf{W}^{-1}\mathbf{x}(0) \tag{11-107}$$

This result is significant in the sense that the optimal performance index is dependent upon the initial state $\mathbf{x}(0)$. Since $\mathbf{W}$ depends only upon $\mathbf{A}$, $\mathbf{B}$, and $\mathbf{R}$, which are all given, Eq. (11-107) implies that once a maximum bound on J° is fixed, a region of controllable states can be established for $\mathbf{x}(0)$. In other words, Eq. (11-107) defines a domain in the state space for $\mathbf{x}(0)$ which can be brought to $\mathbf{x}(N) = \mathbf{0}$ for a given N and a given J°.

Example 11-3

Given the digital system

$$\mathbf{x}(k+1) = \mathbf{A}\mathbf{x}(k) + \mathbf{B}\mathbf{u}(k) \tag{11-108}$$

where

$$\mathbf{A} = \begin{bmatrix} 0 & 1 \\ -0.5 & -0.2 \end{bmatrix} \qquad \mathbf{B} = \begin{bmatrix} 0 \\ 1 \end{bmatrix}$$

the following parts are to be carried out in this problem.

1. Find the constant state feedback gain matrix $\mathbf{G}$, such that $u(k) = -\mathbf{G}\mathbf{x}(k)$ will bring any initial state $\mathbf{x}(0)$ to $\mathbf{x}(N) = \mathbf{0}$, for $N = 2$. Determine the optimal control $u^\circ(k)$ for $k = 0, 1$, and the optimal state trajectory $\mathbf{x}^\circ(k)$ when $\mathbf{x}(0) = [1 \quad 1]'$ and $\mathbf{x}^\circ(2) = \mathbf{0}$.
2. The control is subject to amplitude constraint $|u(k)| \le 1$. Determine the region of controllable states of $\mathbf{x}(0)$ in the state plane for $N \le 2$, such that $\mathbf{x}(N) = \mathbf{0}$.
3. Find the optimal control $u^\circ(k)$ which will bring the initial state $\mathbf{x}(0) = [1 \quad 1]'$ to $\mathbf{x}(2) = \mathbf{0}$, and simultaneously satisfy

$$J = \frac{1}{2}\sum_{k=0}^{1} u^2(k) = \text{minimum} \tag{11-109}$$

   Determine the optimal trajectory $\mathbf{x}^\circ(k)$ and the optimal value of J.
4. Determine the region of controllable states of $\mathbf{x}(0)$ in the state plane for $N = 2$ such that $\mathbf{x}(N) = 0$, and $J \le 1$, with the J given in Eq. (11-109). Repeat the problem for $J \le 0.25$.

Solution:

1. Setting $\mathbf{x}(2)$ to $\mathbf{0}$ in the solution of Eq. (11-108), it can be shown that the time-optimal control is

$$u^\circ(k) = -[1 \quad 0][\mathbf{A}^{-1}\mathbf{B} \quad \mathbf{A}^{-2}\mathbf{B}]^{-1}\mathbf{x}(k) \tag{11-110}$$

Thus,

$$u^\circ(k) = [0.5 \quad 0.2]\mathbf{x}(k) \tag{11-111}$$

The optimal state feedback gain matrix is

$$\mathbf{G} = [-0.5 \quad -0.2] \tag{11-112}$$

For the given initial state, the following results are obtained:

$$u^\circ(0) = 0.7 \qquad \mathbf{x}^\circ(1) = \begin{bmatrix} 1 \\ 0 \end{bmatrix}$$

$$u^\circ(1) = 0.5 \qquad \mathbf{x}^\circ(2) = \begin{bmatrix} 0 \\ 0 \end{bmatrix}$$

The optimal state trajectory is shown in Fig. 11-1.

2. The control is now subject to the amplitude constraint, $|u(k)| \leq 1$. For $N = 2$, the state transition equation is written

$$\mathbf{x}(2) = \mathbf{A}^2\mathbf{x}(0) + \mathbf{AB}u(0) + \mathbf{B}u(1) = 0 \tag{11-113}$$

Solving for $\mathbf{x}(0)$ from the last equation, we have

$$\mathbf{x}(0) = \begin{bmatrix} 2 & -0.8 \\ 0 & 2 \end{bmatrix}\begin{bmatrix} u(0) \\ u(1) \end{bmatrix} \tag{11-114}$$

The vertices of the region of controllable states are found by substituting the four possible combinations of $u(k) = +1$ and -1 into Eq. (11-114). The

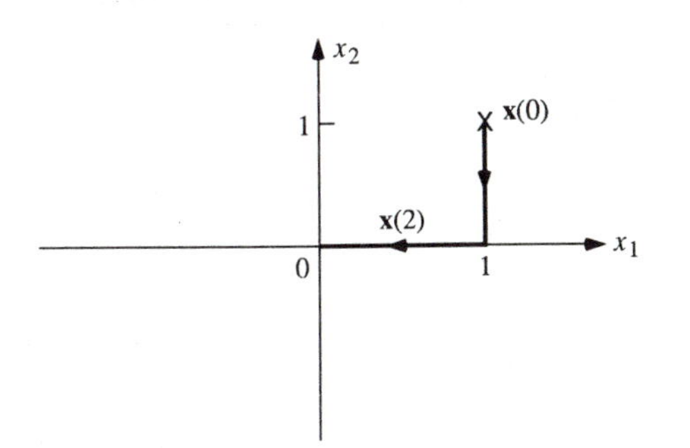

Figure 11-1. Optimal state trajectory.

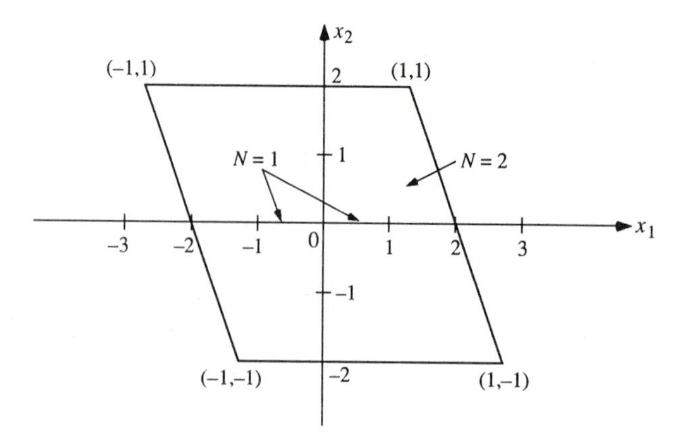

Figure 11-2. Region of controllable states.

results are

$$u(0) = u(1) = 1 \qquad \mathbf{x}(0) = \begin{bmatrix} 1.2 \\ 2 \end{bmatrix} \qquad u(0) = -u(1) = 1 \qquad \mathbf{x}(0) = \begin{bmatrix} 2.8 \\ -2 \end{bmatrix}$$

$$u(0) = u(1) = -1 \qquad \mathbf{x}(0) = \begin{bmatrix} -1.2 \\ -2 \end{bmatrix} \qquad -u(0) = u(1) = 1 \qquad \mathbf{x}(0) = \begin{bmatrix} -2.8 \\ 2 \end{bmatrix}$$

The convex polygon bound by these four vertices is shown in Fig. 11-2.

3. Referring to the performance index in Eq. (11-109), $R = 1$. The matrix $\mathbf{W}$ is determined using Eq. (11-98), for $N = 2$.

$$\mathbf{W} = \mathbf{A}^{-1}\mathbf{B}\mathbf{B}'(\mathbf{A}^{-1})' + \mathbf{A}^{-2}\mathbf{B}\mathbf{B}'(\mathbf{A}^{-2})' = \begin{bmatrix} 4.64 & -1.6 \\ -1.6 & 4 \end{bmatrix} \tag{11-115}$$

The optimal control is given by Eq. (11-106),

$$u^{\circ}(k) = -\mathbf{B}'(\mathbf{A}^{-k-1})'\mathbf{W}^{-1}\mathbf{x}(0) \tag{11-116}$$

Thus, $u^{\circ}(0) = 0.7$ and $u^{\circ}(1) = 0.5$. Notice that these solutions are identical to those obtained in part 1. This is because, as long as N is less than or equal to two for the second-order system, the solution for $\mathbf{x}(N) = \mathbf{0}$ is unique.

The optimal performance index is

$$J^{\circ} = \tfrac{1}{2}\mathbf{x}'(0)\mathbf{W}^{-1}\mathbf{x}(0) = 0.37 \tag{11-117}$$

4. For $J \leq 1$, we have

$$\tfrac{1}{2}\mathbf{x}'(0)\mathbf{W}^{-1}\mathbf{x}(0) \leq 1 \tag{11-118}$$

or

$$0.125x_1^2(0) + 0.1x_1(0)x_2(0) + 0.145x_2^2(0) \leq 1 \tag{11-119}$$

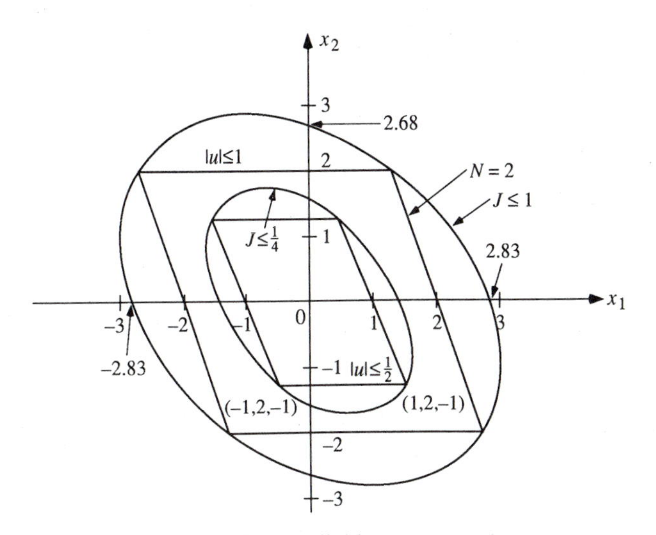

Figure 11-3. Region of controllable states.

The boundary described by the last expression is an ellipse in the state plane. The regions of controllable states for $J \leq 1$ and $J \leq 0.25$ are shown in Fig. 11-3.

11-5 OPTIMAL LINEAR DIGITAL REGULATOR DESIGN

One of the optimal control design techniques that has found general practical applications is the linear regulator design. A regulator problem is defined with reference to a system with zero reference inputs, and the design objective is to drive the states or outputs to the neighborhood of the equilibrium state. The condition of zero inputs is not a severe limitation to the design since the linear regulator design (infinite time) assures that the resultant system is stable and possesses certain damping characteristics, so that the performance of the system will be satisfactory in practice even if the inputs are nonzero.

In the following discussions we shall first formulate the performance index with reference to a linear continuous-data process with sampled data, and then with respect to a completely digital process.

The linear digital regulator problem may be stated as follows. Given the linear system

$$\dot{\mathbf{x}}(t) = \mathbf{A}\mathbf{x}(t) + \mathbf{B}\mathbf{u}(t) \tag{11-120}$$

where $\mathbf{x}(t)$ is the $n \times 1$ state vector and $\mathbf{u}(t)$ is the $p \times 1$ control vector given by

$$\mathbf{u}(t) = \mathbf{u}(kT) \qquad kT \leq t < (k+1)T \tag{11-121}$$

Find the optimal control $\mathbf{u}^\circ(kT)$ for $k = 0, 1, 2, \ldots, N-1$, such that the quadratic performance index

$$J = \frac{1}{2}\langle \mathbf{x}(t_f), \mathbf{S}\mathbf{x}(t_f)\rangle + \frac{1}{2}\int_0^{t_f} [\langle \mathbf{x}(t), \mathbf{Q}\mathbf{x}(t)\rangle + \langle \mathbf{u}(t), \mathbf{R}\mathbf{u}(t)\rangle]\, dt \tag{11-122}$$

is minimized. In the last equation, $t_f = NT$, and

S symmetric positive-semidefinite matrix ($n \times n$)
Q symmetric positive-semidefinite matrix ($n \times n$)
R symmetric positive-definite matrix ($p \times p$).

We first discretize the system of Eq. (11-120) by forming the difference equation,

$$\mathbf{x}[(k+1)T] = \boldsymbol{\phi}(T)\mathbf{x}(kT) + \boldsymbol{\theta}(T)\mathbf{u}(kT) \tag{11-123}$$

where

$$\boldsymbol{\phi}(T) = e^{\mathbf{A}T} \tag{11-124}$$

$$\boldsymbol{\theta}(T) = \int_0^T \boldsymbol{\phi}(T-\tau)\mathbf{B}\, d\tau \tag{11-125}$$

To discretize the performance index of Eq. (11-122), we write

$$J_N = \frac{1}{2}\langle \mathbf{x}(NT), \mathbf{S}\mathbf{x}(NT)\rangle + \frac{1}{2}\sum_{k=0}^{N-1}\int_{kT}^{(k+1)T} [\langle \mathbf{x}(t), \mathbf{Q}\mathbf{x}(t)\rangle + \langle \mathbf{u}(t), \mathbf{R}\mathbf{u}(t)\rangle]\, dt \tag{11-126}$$

The state transition equation of Eq. (11-120) is written for $t \geq kT$,

$$\mathbf{x}(t) = \boldsymbol{\phi}(t-kT)\mathbf{x}(kT) + \boldsymbol{\theta}(t-kT)\mathbf{u}(kT) \tag{11-127}$$

Then,

$$\mathbf{x}'(t) = \mathbf{x}'(kT)\boldsymbol{\phi}'(t-kT) + \mathbf{u}'(kT)\boldsymbol{\theta}'(t-kT) \tag{11-128}$$

Substitution of Eqs. (11-121), (11-127), and (11-128) into Eq. (11-126) yields

$$J_N = \frac{1}{2}\mathbf{x}'(NT)\mathbf{S}\mathbf{x}(NT) + \frac{1}{2}\sum_{k=0}^{N-1}[\mathbf{x}'(kT)\hat{\mathbf{Q}}(T)\mathbf{x}(kT) + 2\mathbf{x}'(kT)\mathbf{M}(T)\mathbf{u}(kT) + \mathbf{u}'(kT)\hat{\mathbf{R}}(T)\mathbf{u}(kT)] \tag{11-129}$$

where

$$\hat{\mathbf{Q}}(T) = \int_{kT}^{(k+1)T} \boldsymbol{\phi}'(t - kT)\mathbf{Q}\boldsymbol{\phi}(t - kT)\, dt \tag{11-130}$$

$$\mathbf{M}(T) = \int_{kT}^{(k+1)T} \boldsymbol{\phi}'(t - kT)\mathbf{Q}\boldsymbol{\theta}(t - kT)\, dt \tag{11-131}$$

$$\hat{\mathbf{R}}(T) = \int_{kT}^{(k+1)T} [\boldsymbol{\theta}'(t - kT)\mathbf{Q}\boldsymbol{\theta}(t - kT) + \mathbf{R}]\, dt \tag{11-132}$$

In view of the properties of $\mathbf{Q}$ and $\mathbf{R}$, we see that $\hat{\mathbf{Q}}(T)$ is symmetric and positive semidefinite and $\hat{\mathbf{R}}(T)$ is symmetric and positive definite. However, nothing can be said with regard to $\mathbf{M}(T)$.

The problem now is that, given the digital system of Eq. (11-123), we must find the optimal control so that the performance index in Eq. (11-129) is minimized.

In general, the digital system may be described by the state equation of Eq. (11-123) at the outset; there is no special reason why the performance index should be as complex as that in Eq. (11-129), since it is not clear, in general, how the weighting matrix $\mathbf{M}$ should be chosen. It is more natural to define the quadratic performance index as

$$J_N = \frac{1}{2}\langle \mathbf{x}(N), \mathbf{S}\mathbf{x}(N)\rangle + \frac{1}{2}\sum_{k=0}^{N-1} [\langle \mathbf{x}(k), \mathbf{Q}\mathbf{x}(k)\rangle + \langle \mathbf{u}(k), \mathbf{R}\mathbf{u}(k)\rangle] \tag{11-133}$$

where $\mathbf{Q}$ and $\mathbf{R}$ are weighting matrices with properties specified earlier, and the sampling period T has been dropped for convenience.

11-5-1 Linear Digital Regulator Design (Finite-Time Problem)

The linear digital regulator design problem stated in the last section is to be solved by use of the discrete minimum principle.

The digital control process is described by

$$\mathbf{x}(k + 1) = \boldsymbol{\phi}\mathbf{x}(k) + \boldsymbol{\theta}\mathbf{u}(k) \tag{11-134}$$

with $\mathbf{x}(0)$ given. The design objective is to find $\mathbf{u}^{\circ}(k)$ so that the performance index

$$J_N = \frac{1}{2}\langle \mathbf{x}(N), \mathbf{S}\mathbf{x}(N)\rangle$$
$$+ \frac{1}{2}\sum_{k=0}^{N-1} [\langle \mathbf{x}(k), \hat{\mathbf{Q}}\mathbf{x}(k)\rangle + \langle \mathbf{x}(k), 2\mathbf{M}\mathbf{u}(k)\rangle + \langle \mathbf{u}(k), \hat{\mathbf{R}}\mathbf{u}(k)\rangle] \tag{11-135}$$

is minimized.

We form the Hamiltonian,

$$H(k) = H[\mathbf{x}(k), \mathbf{p}(k+1), \mathbf{u}(k)]$$
$$= \tfrac{1}{2}\langle \mathbf{x}(k), \hat{\mathbf{Q}}\mathbf{x}(k)\rangle + \langle \mathbf{x}(k), \mathbf{M}\mathbf{u}(k)\rangle$$
$$+ \tfrac{1}{2}\langle \mathbf{u}(k), \hat{\mathbf{R}}\mathbf{u}(k)\rangle + \langle \mathbf{p}(k+1), \boldsymbol{\phi}\mathbf{x}(k) + \boldsymbol{\theta}\mathbf{u}(k)\rangle \tag{11-136}$$

The necessary conditions for J_N to be an extremum are

$$\frac{\partial H^\circ(k)}{\partial \mathbf{x}^\circ(k)} = \mathbf{p}^\circ(k) = \hat{\mathbf{Q}}\mathbf{x}^\circ(k) + \boldsymbol{\phi}'\mathbf{p}^\circ(k+1) + \mathbf{M}\mathbf{u}^\circ(k) \tag{11-137}$$

$$\frac{\partial H^\circ(k)}{\partial \mathbf{p}^\circ(k+1)} = \mathbf{x}^\circ(k+1) = \boldsymbol{\phi}\mathbf{x}^\circ(k) + \boldsymbol{\theta}\mathbf{u}^\circ(k) \tag{11-138}$$

$$\frac{\partial H^\circ(k)}{\partial \mathbf{u}^\circ(k)} = \mathbf{M}'\mathbf{x}^\circ(k) + \hat{\mathbf{R}}\mathbf{u}^\circ(k) + \boldsymbol{\theta}'\mathbf{p}^\circ(k+1) = \mathbf{0} \tag{11-139}$$

In this case, since $\mathbf{x}^\circ(N)$ is not specified, the transversality condition is

$$\frac{\partial G[\mathbf{x}(N), N]}{\partial \mathbf{x}(N)} = \frac{\partial}{\partial \mathbf{x}(N)}\left[\frac{1}{2}\langle \mathbf{x}(N), \mathbf{S}\mathbf{x}(N)\rangle\right] = \mathbf{S}\mathbf{x}(N) = \mathbf{p}(N) \tag{11-140}$$

The optimal control is obtained from Eq. (11-139),

$$\mathbf{u}^\circ(k) = -\hat{\mathbf{R}}^{-1}[\boldsymbol{\theta}'\mathbf{p}^\circ(k+1) + \mathbf{M}'\mathbf{x}^\circ(k)] \tag{11-141}$$

Substituting the last equation into Eqs. (11-137) and (11-138), the canonical state equations are written

$$\mathbf{x}^\circ(k+1) = (\boldsymbol{\phi} - \boldsymbol{\theta}\hat{\mathbf{R}}^{-1}\mathbf{M}')\mathbf{x}^\circ(k) - \boldsymbol{\theta}\hat{\mathbf{R}}^{-1}\boldsymbol{\theta}'\mathbf{p}^\circ(k+1) \tag{11-142}$$

$$(\boldsymbol{\phi}' - \mathbf{M}\hat{\mathbf{R}}^{-1}\boldsymbol{\theta}')\mathbf{p}^\circ(k+1) = \mathbf{p}^\circ(k) - (\hat{\mathbf{Q}} - \mathbf{M}\hat{\mathbf{R}}^{-1}\mathbf{M}')\mathbf{x}^\circ(k) \tag{11-143}$$

These represent $2n$ difference equations that are to be solved with the known boundary conditions of $\mathbf{x}(0)$ and $\mathbf{p}^\circ(N) = \mathbf{S}\mathbf{x}^\circ(N)$. Notice that the two equations in Eqs. (11-142) and (11-143) are coupled in $\mathbf{x}^\circ(k)$ and $\mathbf{p}^\circ(k)$. These are more general than Eqs. (11-89) and (11-90), which are derived for a much simpler performance index. Since Eq. (11-89) is completely decoupled from $\mathbf{x}^\circ(k)$, it can be solved directly.

The coupled canonical state equations in Eqs (11-142) and (11-143) are difficult to solve directly. However, it can be shown that the solution is of the form

$$\mathbf{p}(k) = \mathbf{K}(k)\mathbf{x}(k) \tag{11-144}$$

where $\mathbf{K}(k)$ is an $n \times n$ matrix with yet unknown properties, except that at $k = N$, from Eq. (11-140),

$$\mathbf{K}(N) = \mathbf{S} \tag{11-145}$$

Substituting Eq. (11-144) into Eq. (11-142) and rearranging terms, we get

$$\mathbf{x}^\circ(k+1) = [\mathbf{I} + \boldsymbol{\theta}\hat{\mathbf{R}}^{-1}\boldsymbol{\theta}'\mathbf{K}(k+1)]^{-1}(\boldsymbol{\phi} - \boldsymbol{\theta}\hat{\mathbf{R}}^{-1}\mathbf{M}')\mathbf{x}^\circ(k) \tag{11-146}$$

where it is assumed that the inverse of $[\mathbf{I} + \boldsymbol{\theta}\hat{\mathbf{R}}^{-1}\boldsymbol{\theta}'\mathbf{K}(k+1)]$ exists. Later it will be shown that $\mathbf{K}(k+1)$ is at least positive semidefinite so that the assertion is true. Similarly, substituting Eq. (11-144) into Eq. (11-143), we have

$$(\boldsymbol{\phi}' - \mathbf{M}\hat{\mathbf{R}}^{-1}\boldsymbol{\theta}')\mathbf{K}(k+1)\mathbf{x}^\circ(k+1) = [\mathbf{K}(k) - \hat{\mathbf{Q}} + \mathbf{M}\hat{\mathbf{R}}^{-1}\mathbf{M}']\mathbf{x}^\circ(k) \tag{11-147}$$

Now substituting Eq. (11-146) into Eq. (11-147), we get

$$\begin{aligned}(\boldsymbol{\phi}' - \mathbf{M}\hat{\mathbf{R}}^{-1}\boldsymbol{\theta}')\mathbf{K}(k+1)[\mathbf{I} + \boldsymbol{\theta}\hat{\mathbf{R}}^{-1}\boldsymbol{\theta}'\mathbf{K}(k+1)]^{-1}(\boldsymbol{\phi} - \boldsymbol{\theta}\hat{\mathbf{R}}^{-1}\mathbf{M}')\mathbf{x}^\circ(k) \\ = [\mathbf{K}(k) - \hat{\mathbf{Q}} + \mathbf{M}\hat{\mathbf{R}}^{-1}\mathbf{M}']\mathbf{x}^\circ(k)\end{aligned} \tag{11-148}$$

For any $\mathbf{x}^\circ(k)$, the following equation must hold:

$$\begin{aligned}(\boldsymbol{\phi}' - \mathbf{M}\hat{\mathbf{R}}^{-1}\boldsymbol{\theta}')\mathbf{K}(k+1)[\mathbf{I} + \boldsymbol{\theta}\hat{\mathbf{R}}^{-1}\boldsymbol{\theta}'\mathbf{K}(k+1)]^{-1}(\boldsymbol{\phi} - \boldsymbol{\theta}\hat{\mathbf{R}}^{-1}\mathbf{M}') \\ = \mathbf{K}(k) - \mathbf{Q} + \mathbf{M}\hat{\mathbf{R}}^{-1}\mathbf{M}'\end{aligned} \tag{11-149}$$

The last equation represents a nonlinear matrix difference equation in $\mathbf{K}(k)$ which is of the Riccati type and is generally called the *discrete Riccati equation.* The $n \times n$ matrix $\mathbf{K}(k)$ is often referred to as the *Riccati gain*. The boundary condition for the Riccati equation is given in Eq. (11-145). In general, Eq. (11-149) contains n^2 scalar equations with the same number of unknowns in the elements of $\mathbf{K}(k)$. However, we shall show later that $\mathbf{K}(k)$ is symmetrical so that there are only $n(n+1)$ unknowns.

The optimal control is determined by substituting Eq. (11-144) into Eq. (11-141). Thus,

$$\begin{aligned}\mathbf{u}^\circ(k) &= -[\mathbf{I} + \hat{\mathbf{R}}^{-1}\boldsymbol{\theta}'\mathbf{K}(k+1)\boldsymbol{\theta}]^{-1}\hat{\mathbf{R}}^{-1}[\boldsymbol{\theta}'\mathbf{K}(k+1)\boldsymbol{\phi} + \mathbf{M}']\mathbf{x}^\circ(k) \\ &= -[\hat{\mathbf{R}} + \boldsymbol{\theta}'\mathbf{K}(k+1)\boldsymbol{\theta}]^{-1}[\boldsymbol{\theta}'\mathbf{K}(k+1)\boldsymbol{\phi} + \mathbf{M}']\mathbf{x}^\circ(k)\end{aligned} \tag{11-150}$$

which is in the form of state feedback.

When $\mathbf{M} = \mathbf{0}$, $\hat{\mathbf{Q}} = \mathbf{Q}$, and $\hat{\mathbf{R}} = \mathbf{R}$, the Riccati equation in Eq. (11-149) becomes

$$\boldsymbol{\phi}'\mathbf{K}(k+1)[\mathbf{I} + \boldsymbol{\theta}\mathbf{R}^{-1}\boldsymbol{\theta}'\mathbf{K}(k+1)]^{-1}\boldsymbol{\phi} + \mathbf{Q} = \mathbf{K}(k) \tag{11-151}$$

The corresponding optimal control is given by

$$\mathbf{u}^\circ(k) = -[\mathbf{R} + \boldsymbol{\theta}'\mathbf{K}(k+1)\boldsymbol{\theta}]^{-1}\boldsymbol{\theta}'\mathbf{K}(k+1)\boldsymbol{\phi}\mathbf{x}^\circ(k) \tag{11-152}$$

Before embarking on the discussion of the various methods of solving the Riccati equation, let us investigate the important properties of the Riccati equation and the Riccati gain $\mathbf{K}(k)$.

The Riccati equation given in Eq. (11-149) is only one of many equivalent forms that satisfy the optimal linear discrete regulator design. In general, a particular form of the Riccati equation is best suited for a given purpose.

First we shall show that Eq. (11-149) is equivalent to the following form:

$$\begin{aligned}\mathbf{K}(k) = \boldsymbol{\phi}'\mathbf{K}(k+1)\boldsymbol{\phi} + \hat{\mathbf{Q}} \\ - [\boldsymbol{\theta}'\mathbf{K}(k+1)\boldsymbol{\phi} + \mathbf{M}']'[\hat{\mathbf{R}} + \boldsymbol{\theta}'\mathbf{K}(k+1)\boldsymbol{\theta}]^{-1}[\boldsymbol{\theta}'\mathbf{K}(k+1)\boldsymbol{\phi} + \mathbf{M}']\end{aligned} \tag{11-153}$$

The equivalence of the two Riccati equations in Eqs. (11-149) and (11-153) is not obvious. To show that these two equations are equivalent, we shall first prove the following identity:

$$\mathbf{K}(k+1)[\mathbf{I}+\boldsymbol{\theta}\hat{\mathbf{R}}^{-1}\boldsymbol{\theta}'\mathbf{K}(k+1)]^{-1} = \mathbf{K}(k+1) - \mathbf{K}(k+1)\boldsymbol{\theta}[\hat{\mathbf{R}}+\boldsymbol{\theta}'\mathbf{K}(k+1)\boldsymbol{\theta}]^{-1}\boldsymbol{\theta}'\mathbf{K}(k+1) \qquad (11\text{-}154)$$

We let

$$\mathbf{P} = \mathbf{K}(k+1)[\mathbf{I}+\boldsymbol{\theta}\hat{\mathbf{R}}^{-1}\boldsymbol{\theta}'\mathbf{K}(k+1)]^{-1} \qquad (11\text{-}155)$$

Post-multiplying both sides of the last equation by $[\mathbf{I}+\boldsymbol{\theta}\hat{\mathbf{R}}^{-1}\boldsymbol{\theta}'\mathbf{K}(k+1)]\boldsymbol{\theta}$, we have

$$\mathbf{P}\boldsymbol{\theta}\hat{\mathbf{R}}^{-1}[\hat{\mathbf{R}}+\boldsymbol{\theta}'\mathbf{K}(k+1)\boldsymbol{\theta}] = \mathbf{K}(k+1)\boldsymbol{\theta} \qquad (11\text{-}156)$$

We then post-multiply both sides of the last equation by $[\hat{\mathbf{R}}+\boldsymbol{\theta}'\mathbf{K}(k+1)\boldsymbol{\theta}]^{-1}\cdot\boldsymbol{\theta}'\mathbf{K}(k+1)$; the result is

$$\mathbf{P}\boldsymbol{\theta}\hat{\mathbf{R}}^{-1}\boldsymbol{\theta}'\mathbf{K}(k+1) = \mathbf{K}(k+1)\boldsymbol{\theta}[\hat{\mathbf{R}}+\boldsymbol{\theta}'\mathbf{K}(k+1)\boldsymbol{\theta}]^{-1}\boldsymbol{\theta}'\mathbf{K}(k+1) \qquad (11\text{-}157)$$

Equation (11-155) indicates that the left-hand side of the last equation is $\mathbf{K}(k+1)-\mathbf{P}$. Thus, Eq. (11-157) is written

$$\mathbf{P} = \mathbf{K}(k+1) - \mathbf{K}(k+1)\boldsymbol{\theta}[\hat{\mathbf{R}}+\boldsymbol{\theta}'\mathbf{K}(k+1)\boldsymbol{\theta}]^{-1}\boldsymbol{\theta}'\mathbf{K}(k+1) \qquad (11\text{-}158)$$

Comparing Eq. (11-158) with Eq. (11-155), we clearly have the identity in Eq. (11-154). Now substituting Eq. (11-154) into Eq. (11-149), we get

$$\begin{aligned}\mathbf{K}(k) - \hat{\mathbf{Q}} + \mathbf{M}\mathbf{R}^{-1}\mathbf{M}' &= (\boldsymbol{\phi}'-\mathbf{M}\hat{\mathbf{R}}^{-1}\boldsymbol{\theta}')[\mathbf{K}(k+1)-\mathbf{K}(k+1)\boldsymbol{\theta}[\hat{\mathbf{R}}+\boldsymbol{\theta}'\mathbf{K}(k+1)\boldsymbol{\theta}]^{-1}\boldsymbol{\theta}'\mathbf{K}(k+1)]\\ &\quad\cdot(\boldsymbol{\phi}-\boldsymbol{\theta}\hat{\mathbf{R}}^{-1}\mathbf{M}')\end{aligned} \qquad (11\text{-}159)$$

The last expression may be regarded as still another form of the discrete Riccati equation. Equation (11-159) is further manipulated to show that it is equivalent to Eq. (11-153).

Multiplying out the terms in Eq. (11-159) gives

$$\begin{aligned}\mathbf{K}(k) - \hat{\mathbf{Q}} + \mathbf{M}\mathbf{R}^{-1}\mathbf{M}' &= (\boldsymbol{\phi}'-\mathbf{M}\hat{\mathbf{R}}^{-1}\boldsymbol{\theta}')[\mathbf{K}(k+1)\boldsymbol{\phi}-\mathbf{K}(k+1)\boldsymbol{\theta}\hat{\mathbf{R}}^{-1}\mathbf{M}']\\ &\quad-(\boldsymbol{\phi}'-\mathbf{M}\hat{\mathbf{R}}^{-1}\boldsymbol{\theta}')\mathbf{K}(k+1)\boldsymbol{\theta}[\hat{\mathbf{R}}+\boldsymbol{\theta}'\mathbf{K}(k+1)\boldsymbol{\theta}]^{-1}\boldsymbol{\theta}'\mathbf{K}(k+1)(\boldsymbol{\phi}-\boldsymbol{\theta}\hat{\mathbf{R}}^{-1}\mathbf{M}')\end{aligned} \qquad (11\text{-}160)$$

$$\begin{aligned}\mathbf{K}(k) - \hat{\mathbf{Q}} + \mathbf{M}\hat{\mathbf{R}}^{-1}\mathbf{M}' &= \boldsymbol{\phi}'\mathbf{K}(k+1)\boldsymbol{\phi} - \boldsymbol{\phi}'\mathbf{K}(k+1)\boldsymbol{\theta}\hat{\mathbf{R}}^{-1}\mathbf{M}' - \mathbf{M}\hat{\mathbf{R}}^{-1}\boldsymbol{\theta}'\mathbf{K}(k+1)\boldsymbol{\phi}\\ &\quad+\mathbf{M}\hat{\mathbf{R}}^{-1}\boldsymbol{\theta}'\mathbf{K}(k+1)\boldsymbol{\theta}\hat{\mathbf{R}}^{-1}\mathbf{M}' - [\boldsymbol{\phi}'\mathbf{K}(k+1)\boldsymbol{\theta}-\mathbf{M}\hat{\mathbf{R}}^{-1}\boldsymbol{\theta}'\mathbf{K}(k+1)\boldsymbol{\theta}]\\ &\quad\cdot[\hat{\mathbf{R}}+\boldsymbol{\theta}'\mathbf{K}(k+1)\boldsymbol{\theta}]^{-1}[\boldsymbol{\theta}'\mathbf{K}(k+1)\boldsymbol{\phi}-\boldsymbol{\theta}'\mathbf{K}(k+1)\boldsymbol{\theta}\hat{\mathbf{R}}^{-1}\mathbf{M}']\end{aligned} \qquad (11\text{-}161)$$

The last term inside the brackets in Eq. (11-161) is written as

$$\begin{aligned}&\boldsymbol{\theta}'\mathbf{K}(k+1)\boldsymbol{\phi}-\boldsymbol{\theta}'\mathbf{K}(k+1)\boldsymbol{\theta}\hat{\mathbf{R}}^{-1}\mathbf{M}'\\ &\quad=\boldsymbol{\theta}'\mathbf{K}(k+1)\boldsymbol{\phi}-[\hat{\mathbf{R}}+\boldsymbol{\theta}'\mathbf{K}(k+1)\boldsymbol{\theta}]\hat{\mathbf{R}}^{-1}\mathbf{M}'+\mathbf{M}'\end{aligned} \qquad (11\text{-}162)$$

Substituting the right-hand side of the last equation into Eq. (11-161), we get

$$\begin{aligned}\mathbf{K}(k) - \hat{\mathbf{Q}} + \mathbf{M}\hat{\mathbf{R}}^{-1}\mathbf{M}' &= \boldsymbol{\phi}'\mathbf{K}(k+1)\boldsymbol{\phi} - \boldsymbol{\phi}'\mathbf{K}(k+1)\boldsymbol{\theta}\hat{\mathbf{R}}^{-1}\mathbf{M}' - \mathbf{M}\hat{\mathbf{R}}^{-1}\boldsymbol{\theta}'\mathbf{K}(k+1)\boldsymbol{\phi} \\ &\quad + \mathbf{M}\hat{\mathbf{R}}^{-1}\boldsymbol{\theta}'\mathbf{K}(k+1)\boldsymbol{\theta}\hat{\mathbf{R}}^{-1}\mathbf{M}' - \boldsymbol{\phi}'\mathbf{K}(k+1)\boldsymbol{\theta}[\hat{\mathbf{R}} + \boldsymbol{\theta}'\mathbf{K}(k+1)\boldsymbol{\theta}]^{-1} \\ &\quad \cdot [\mathbf{M}' + \boldsymbol{\theta}'\mathbf{K}(k+1)\boldsymbol{\phi}] + \boldsymbol{\phi}'\mathbf{K}(k+1)\boldsymbol{\theta}\hat{\mathbf{R}}^{-1}\mathbf{M}' - \mathbf{M}\mathbf{R}^{-1}\boldsymbol{\theta}'\mathbf{K}(k+1)\boldsymbol{\theta}\hat{\mathbf{R}}^{-1}\mathbf{M}' \\ &\quad + \mathbf{M}\hat{\mathbf{R}}^{-1}\boldsymbol{\theta}'\mathbf{K}(k+1)\boldsymbol{\theta}[\hat{\mathbf{R}} + \boldsymbol{\theta}'\mathbf{K}(k+1)\boldsymbol{\theta}]^{-1}[\mathbf{M}' + \boldsymbol{\theta}'\mathbf{K}(k+1)\boldsymbol{\phi}] \end{aligned} \tag{11-163}$$

After cancellation of equal and opposite terms, and conditioning the last term, the last equation becomes

$$\begin{aligned}\mathbf{K}(k) - \hat{\mathbf{Q}} + \mathbf{M}\hat{\mathbf{R}}^{-1}\mathbf{M}' &= \boldsymbol{\phi}'\mathbf{K}(k+1)\boldsymbol{\phi} - \mathbf{M}\hat{\mathbf{R}}^{-1}\boldsymbol{\theta}'\mathbf{K}(k+1)\boldsymbol{\phi} \\ &\quad - \boldsymbol{\phi}'\mathbf{K}(k+1)\boldsymbol{\theta}[\hat{\mathbf{R}} + \boldsymbol{\theta}'\mathbf{K}(k+1)\boldsymbol{\theta}]^{-1}[\mathbf{M}' + \boldsymbol{\theta}'\mathbf{K}(k+1)\boldsymbol{\phi}] \\ &\quad + \mathbf{M}\hat{\mathbf{R}}^{-1}[\boldsymbol{\theta}'\mathbf{K}(k+1)\boldsymbol{\theta} + \hat{\mathbf{R}} - \hat{\mathbf{R}}][\hat{\mathbf{R}} + \boldsymbol{\theta}'\mathbf{K}(k+1)\boldsymbol{\theta}]^{-1}[\mathbf{M}' + \boldsymbol{\theta}'\mathbf{K}(k+1)\boldsymbol{\phi}] \end{aligned} \tag{11-164}$$

The last equation is finally simplified to

$$\begin{aligned}\mathbf{K}(k) = \boldsymbol{\phi}'\mathbf{K}(k+1)\boldsymbol{\phi} + \hat{\mathbf{Q}} - [\boldsymbol{\theta}'\mathbf{K}'(k+1)\boldsymbol{\phi} + \mathbf{M}']'[\hat{\mathbf{R}} + \boldsymbol{\theta}'\mathbf{K}(k+1)\boldsymbol{\theta}]^{-1} \\ \cdot [\boldsymbol{\theta}'\mathbf{K}(k+1)\boldsymbol{\phi} + \mathbf{M}'] \end{aligned} \tag{11-165}$$

which is identical to Eq. (11-153).

The following theorems concern the properties of the Riccati gain $\mathbf{K}(k)$.

□ Theorem 11-1

$\mathbf{K}(k)$ is a symmetric matrix. ■

Proof: Taking the transpose on both sides of the Riccati equation in Eq. (11-165), we have

$$\begin{aligned}\mathbf{K}'(k) &= \boldsymbol{\phi}'\mathbf{K}'(k+1)\boldsymbol{\phi} + \hat{\mathbf{Q}} \\ &\quad - [\boldsymbol{\theta}'\mathbf{K}(k+1)\boldsymbol{\phi} + \mathbf{M}']'[\hat{\mathbf{R}} + \boldsymbol{\theta}'\mathbf{K}'(k+1)\boldsymbol{\theta}]^{-1}[\boldsymbol{\phi}'\mathbf{K}(k+1)\boldsymbol{\theta} + \mathbf{M}]' \\ &= \boldsymbol{\phi}'\mathbf{K}'(k+1)\boldsymbol{\phi} + \hat{\mathbf{Q}} \\ &\quad - [\boldsymbol{\theta}'\mathbf{K}(k+1)\boldsymbol{\phi} + \mathbf{M}']'[\hat{\mathbf{R}} + \boldsymbol{\theta}'\mathbf{K}'(k+1)\boldsymbol{\theta}]^{-1}[\boldsymbol{\theta}'\mathbf{K}'(k+1)\boldsymbol{\phi} + \mathbf{M}'] \end{aligned} \tag{11-166}$$

Since Eq. (11-166) is identical to Eq. (11-165) except that $\mathbf{K}(k)$ and $\mathbf{K}(k+1)$ are replaced by $\mathbf{K}'(k)$ and $\mathbf{K}'(k+1)$, respectively, these equations should have the same solution. Therefore,

$$\mathbf{K}(k) = \mathbf{K}'(k) \tag{11-167}$$

●

□ **Theorem 11-2**

The performance index J_N in Eq. (11-135) is for the time interval $[0, N]$, or a total of N stages if time is not involved as the independent variable. We define $J_{N-i}[\mathbf{x}(i)]$ as the performance index over the interval from i to N, or the last $N - i$ stages; that is,

$$J_{N-i}[\mathbf{x}(i)] = \frac{1}{2}\langle \mathbf{x}(N), \mathbf{S}\mathbf{x}(N)\rangle + \frac{1}{2}\sum_{k=i}^{N-1}[\langle \mathbf{x}(k), \hat{\mathbf{Q}}\mathbf{x}(k)\rangle + \langle \mathbf{x}(k), 2\mathbf{M}\mathbf{u}(k)\rangle + \langle \mathbf{u}(k), \hat{\mathbf{R}}\mathbf{u}(k)\rangle] \tag{11-168}$$

When $i = 0$,

$$J_N[\mathbf{x}(0)] = J_N \tag{11-169}$$

When $i = N$,

$$J_0[\mathbf{x}(N)] = \tfrac{1}{2}\langle \mathbf{x}(N), \mathbf{S}\mathbf{x}(N)\rangle \tag{11-170}$$

which is the performance index over the last stage only.

The theorem states that

$$\min_{\mathbf{u}(i)} J_{N-i}[\mathbf{x}(i)] = \tfrac{1}{2}\mathbf{x}'(i)\mathbf{K}(i)\mathbf{x}(i) \tag{11-171}$$

where $\mathbf{K}(i)$ is the Riccati gain matrix. ■

The proof of this theorem is given in the process of derivation of the Riccati equation using the principle of optimality which is found in Sec. 11-6.

□ **Theorem 11-3**

The Riccati gain $\mathbf{K}(k)$ is at least positive semidefinite for $k = 0, 1, 2, \ldots, N$. ■

Proof: Since the performance index J_N is in a quadratic form, it is nonnegative. In view of Eq. (11-171), $\mathbf{K}(k)$ must be nonnegative. ●

Remarks on Controllability, Observability, and Stability

For the finite-time (N = finite) linear digital regulator problem, we do not require that the process be controllable, observable, or even stable. The performance index J_N can be finite for finite N even if an uncontrollable state is unstable. The design objective using the quadratic performance index is to drive any initial state $\mathbf{x}(0)$ to the equilibrium state $\mathbf{0}$ as closely as possible, but $\mathbf{x}(N)$ is not specified. Therefore, for finite N, stability is not a problem.

11-5-2 Linear Digital Regulator Design (Infinite-Time Problem)

For infinite time or an infinite number of stages, $N = \infty$, the performance index in Eq. (11-135) is written

$$J = \frac{1}{2} \sum_{k=0}^{\infty} [\langle \mathbf{x}(k), \hat{\mathbf{Q}}\mathbf{x}(k) \rangle + \langle \mathbf{x}(k), 2\mathbf{M}\mathbf{u}(k) \rangle + \langle \mathbf{u}(k), \hat{\mathbf{R}}\mathbf{u}(k) \rangle] \tag{11-172}$$

In this case the terminal cost is eliminated, since as N approaches infinity, the final state $\mathbf{x}(N)$ should approach the equilibrium state $\mathbf{0}$, so that the terminal constraint is no longer necessary.

One important requirement of the infinite-time linear regulator design is that the closed-loop system should be asymptotically stable. The following conditions are required with regard to the system modeled by Eq. (11-134):

1. The pair $[\boldsymbol{\phi}, \boldsymbol{\theta}]$ must be either completely controllable or stabilizable by state feedback (necessary condition).
2. The pair $[\boldsymbol{\phi}, \mathbf{D}]$ must be completely observable, where $\mathbf{D}$ is any $n \times n$ matrix such that $\mathbf{D}\mathbf{D}' = \mathbf{Q}$ (sufficient condition).

Then, the solution to the infinite-time linear digital regulator problem can be obtained by setting $k \to -\infty$. As $N \to \infty$, the Riccati gain matrix $\mathbf{K}(k)$ becomes a constant matrix, that is,

$$\lim_{k \to -\infty} \mathbf{K}(k) = \mathbf{K} \tag{11-173}$$

Replacing $\mathbf{K}(k+1)$ and $\mathbf{K}(k)$ by $\mathbf{K}$ in Eq. (11-153), we have the steady state or infinite-time Riccati equation

$$\mathbf{K} = \boldsymbol{\phi}'\mathbf{K}\boldsymbol{\phi} + \hat{\mathbf{Q}} - (\boldsymbol{\phi}'\mathbf{K}\boldsymbol{\theta} + \mathbf{M})(\hat{\mathbf{R}} + \boldsymbol{\theta}'\mathbf{K}\boldsymbol{\theta})^{-1}(\boldsymbol{\theta}'\mathbf{K}\boldsymbol{\phi} + \mathbf{M}') \tag{11-174}$$

This equation is often referred to as the *algebraic Riccati equation.*

The optimal control is, from Eq. (11-150),

$$\mathbf{u}^\circ(k) = -(\hat{\mathbf{R}} + \boldsymbol{\theta}'\mathbf{K}\boldsymbol{\theta})^{-1}(\boldsymbol{\theta}'\mathbf{K}\boldsymbol{\phi} + \mathbf{M}')\mathbf{x}^\circ(k) \tag{11-175}$$

In this case the feedback matrix

$$\mathbf{G} = (\hat{\mathbf{R}} + \boldsymbol{\theta}'\mathbf{K}\boldsymbol{\theta})^{-1}(\boldsymbol{\theta}'\mathbf{K}\boldsymbol{\phi} + \mathbf{M}') \tag{11-176}$$

is a constant matrix.

The optimal performance index for $N = \infty$ follows directly from Eq. (11-171),

$$J_\infty^\circ = \tfrac{1}{2}\mathbf{x}'(0)\mathbf{K}\mathbf{x}(0) \tag{11-177}$$

The controllability, stabilizability, and observability conditions 1 and 2 given earlier require further discussion.

For the infinite-time regulator, it is necessary that the process in Eq. (11-134) be either controllable or stabilizable by state feedback. Controllability is a stronger requirement than stabilizability, since an uncontrollable system can still be stabilized if the uncontrollable states are stable. However, just because the system in Eq. (11-134) is controllable or stabilizable does not mean that the closed-loop system designed by the optimal linear regulator theory is asymptotically stable. As it turns out, the observability condition (condition 2, above) should also be satisfied. Therefore, controllability and stabilizability are necessary conditions, whereas observability of $[\boldsymbol{\phi}, \mathbf{D}]$ is a sufficient condition. The following example and theorems will further amplify the ideas behind the requirements on controllability, observability, and stabilizability.

Example 11-4

Consider the first-order digital process

$$x(k+1) = x(k) + u(k) \tag{11-178}$$

It is apparent that the process is controllable, unstable, but stabilizable. With the constant state feedback $u(k) = -Gx(k)$, the closed-loop system is asymptotically stable for $|1 - G| < 1$.

For the infinite-time linear regulator design, let us choose the performance index to be

$$J = \frac{1}{2} \sum_{k=0}^{\infty} u^2(k) \tag{11-179}$$

Then, $\phi = 1$, $\theta = 1$, $\hat{Q} = 0$, $M = 0$, and $\hat{R} = 1$. The optimal control which minimizes J is given by Eq. (11-175), reduced to

$$u^\circ(k) = -\frac{K}{1+K} x^\circ(k) \tag{11-180}$$

where K is the scalar constant Riccati gain which is the solution of the Riccati equation [Eq. (11-174)]:

$$K = K - \frac{K^2}{1+K} \tag{11-181}$$

The solution of the last equation is $K = 0$. Therefore, $u^\circ(k) = 0$ for all k, and the closed-loop system is not asymptotically stable.

The reason that the optimal linear regulator design does not provide an asymptotically stable system in this case is that $\hat{Q} = 0$, so that the state variable is not observed by the performance index. More important, it is the unstable state that is unobservable. To ensure that all the states are observed by J, it is *sufficient* that $\hat{Q}$ is positive definite. However, in general, we may require the pair $[\boldsymbol{\phi}, \mathbf{D}]$ to be completely observable, where $\mathbf{D}$ is any matrix such that $\mathbf{DD}' = \hat{\mathbf{Q}}$. Of course, if $\hat{\mathbf{Q}}$ is positive definite, we can always find a $\mathbf{D}$ that is a square matrix, then $[\boldsymbol{\phi}, \mathbf{D}]$ is always observable or of rank n for any $\boldsymbol{\phi}$.

To see why the observability of $[\boldsymbol{\phi}, \mathbf{D}]$ is important for the states to be reflected in J, let us consider only the homogeneous part of the state transition equation,

$$\mathbf{x}(k) = \boldsymbol{\phi}(k)\mathbf{x}(0) \tag{11-182}$$

and let the performance index be

$$J = \sum_{k=0}^{\infty} \mathbf{x}'(k)\hat{\mathbf{Q}}\mathbf{x}(k) \tag{11-183}$$

Substitution of Eq. (12-182) into Eq. (12-183) yields

$$J = \sum_{k=0}^{\infty} \mathbf{x}'(0)\boldsymbol{\phi}'(k)\hat{\mathbf{Q}}\boldsymbol{\phi}(k)\mathbf{x}(0) \tag{11-184}$$

From Chapter 6 we learned that $\mathbf{x}(0)$ will be observable by J if the matrix

$$\sum_{k=0}^{\infty} \boldsymbol{\phi}'(k)\hat{\mathbf{Q}}\boldsymbol{\phi}(k)$$

is nonsingular or positive definite. Therefore, letting $\mathbf{DD}' = \hat{\mathbf{Q}}$, we have the condition that $[\boldsymbol{\phi}, \mathbf{D}]$ must be observable.

□ Theorem 11-4

For the performance index

$$J = \frac{1}{2}\sum_{k=0}^{\infty} [\mathbf{x}'(k)\mathbf{Q}\mathbf{x}(k) + \mathbf{u}'(k)\mathbf{R}\mathbf{u}(k)] \tag{11-185}$$

if $\mathbf{Q}$ and $\mathbf{R}$ are both positive definite, $\mathbf{K}$ is positive definite. ■

Proof: Since J is of the quadratic form, for positive definite $\mathbf{Q}$ and $\mathbf{R}$, J is positive. Since J and $\mathbf{K}$ are related through Eq. (11-177) it follows that $\mathbf{K}$ is positive definite. ●

□ Theorem 11-5

For the digital system in Eq. (11-134), if the performance index is

$$J = \frac{1}{2}\sum_{k=0}^{\infty} [\mathbf{x}'(k)\mathbf{Q}\mathbf{x}(k) + \mathbf{u}'(k)\mathbf{R}\mathbf{u}(k)] \tag{11-186}$$

where $\mathbf{Q}$ and $\mathbf{R}$ are both positive definite, then the optimal control that minimizes J,

$$\mathbf{u}^{\circ}(k) = -(\mathbf{R} + \boldsymbol{\theta}'\mathbf{K}\boldsymbol{\theta})^{-1}\boldsymbol{\theta}'\mathbf{K}\boldsymbol{\phi}\mathbf{x}^{\circ}(k) \tag{11-187}$$

provides an asymptotically stable closed-loop system

$$\mathbf{x}^{\circ}(k+1) = [\boldsymbol{\phi} - \boldsymbol{\theta}(\mathbf{R} + \boldsymbol{\theta}'\mathbf{K}\boldsymbol{\theta})^{-1}\boldsymbol{\theta}'\mathbf{K}\boldsymbol{\phi}]\mathbf{x}^{\circ}(k) \tag{11-188}$$

■

Proof: From Theorem 11-4, since **Q** and **R** are both positive definite, **K** is also positive definite. Let the Liapunov function be defined as

$$V[\mathbf{x}(k)] = \tfrac{1}{2}\mathbf{x}(k)\mathbf{K}\mathbf{x}(k) \tag{11-189}$$

which is positive definite. Then,

$$\begin{aligned}\Delta V[\mathbf{x}(k)] &= V[\mathbf{x}(k+1)] - V[\mathbf{x}(k)] \\ &= \tfrac{1}{2}\mathbf{x}'(k+1)\mathbf{K}\mathbf{x}(k+1) - \tfrac{1}{2}\mathbf{x}'(k)\mathbf{K}\mathbf{x}(k)\end{aligned} \tag{11-190}$$

Substituting Eq. (11-188) into Eq. (11-190), we have

$$\begin{aligned}\Delta V[\mathbf{x}(k)] = \tfrac{1}{2}\mathbf{x}'(k)[&\boldsymbol{\phi}'\mathbf{K}\boldsymbol{\phi} - \boldsymbol{\phi}'\mathbf{K}\boldsymbol{\theta}(\mathbf{R} + \boldsymbol{\theta}'\mathbf{K}\boldsymbol{\theta})^{-1}\boldsymbol{\theta}'\mathbf{K}\boldsymbol{\phi} \\ &+ \boldsymbol{\phi}'\mathbf{K}\boldsymbol{\theta}(\mathbf{R} + \boldsymbol{\theta}'\mathbf{K}\boldsymbol{\theta})^{-1}\boldsymbol{\theta}'\mathbf{K}\boldsymbol{\theta}(\mathbf{R} + \boldsymbol{\theta}'\mathbf{K}\boldsymbol{\theta})^{-1}\boldsymbol{\theta}'\mathbf{K}\boldsymbol{\phi} \\ &- \boldsymbol{\phi}'\mathbf{K}\boldsymbol{\theta}(\mathbf{R} + \boldsymbol{\theta}'\mathbf{K}\boldsymbol{\theta})^{-1}\boldsymbol{\theta}'\mathbf{K}\boldsymbol{\phi} - \mathbf{K}]\mathbf{x}(k)\end{aligned} \tag{11-191}$$

The Riccati equation for this case is

$$\mathbf{K} = \mathbf{Q} + \boldsymbol{\phi}'\mathbf{K}\boldsymbol{\phi} - \boldsymbol{\phi}'\mathbf{K}\boldsymbol{\theta}(\mathbf{R} + \boldsymbol{\theta}'\mathbf{K}\boldsymbol{\theta})^{-1}\boldsymbol{\theta}'\mathbf{K}\boldsymbol{\phi} \tag{11-192}$$

Using the last equation, Eq. (11-191) is simplified to

$$\Delta V[\mathbf{x}(k)] = \tfrac{1}{2}\mathbf{x}'(k)[-\mathbf{Q} - \boldsymbol{\phi}'\mathbf{K}\boldsymbol{\theta}(\mathbf{R} + \boldsymbol{\theta}'\mathbf{K}\boldsymbol{\theta})^{-1}\boldsymbol{\theta}'\mathbf{K}[\boldsymbol{\phi} - \boldsymbol{\theta}(\mathbf{R} + \boldsymbol{\theta}'\mathbf{K}\boldsymbol{\theta})^{-1}\boldsymbol{\theta}'\mathbf{K}\boldsymbol{\phi}]]\mathbf{x}(k) \tag{11-193}$$

Since **Q** and **K** are both positive definite, the quantity inside the brackets on the right-hand side of Eq. (11-193) is negative definite. Therefore, $\Delta V[\mathbf{x}(k)]$ is negative, and according to the Liapunov stability theorem, the system in Eq. (11-188) is asymptotically stable. ●

11-6 PRINCIPLE OF OPTIMALITY AND DYNAMIC PROGRAMMING

The optimal linear digital regulator design carried out in the preceding sections can be affected by use of the principle of optimality. The design using the principle of optimality is also known as the method of dynamic programming. Let us first state the principle of optimality.

Principle of Optimality

An optimal control strategy has the property that, whatever the initial state and the control of the initial stages, the remaining control must form an optimal control with respect to the state resulting from the control of the initial stages. Stated in another way, any control strategy that is optimal over the interval $[i, N]$ is necessarily optimal over $[i+1, N]$ for $i = 0, 1, 2, \ldots, N-1$.

The finite-time optimal linear digital regulator design problem is repeated as follows.

Find the optimal control $\mathbf{u}^\circ(k)$, $k = 0, 1, 2, \ldots, N-1$, so that

$$J_N = G[\mathbf{x}(N)] + \sum_{k=0}^{N-1} F_k[\mathbf{x}(k), \mathbf{u}(k)] = \text{minimum} \tag{11-194}$$

where

$$G[\mathbf{x}(N)] = \tfrac{1}{2}\mathbf{x}'(N)\mathbf{S}\mathbf{x}(N) \tag{11-195}$$

$$F_k[\mathbf{x}(k), \mathbf{u}(k)] = \tfrac{1}{2}\mathbf{x}'(k)\hat{\mathbf{Q}}\mathbf{x}(k) + \mathbf{x}'(k)\mathbf{M}\mathbf{u}(k) + \tfrac{1}{2}\mathbf{u}'(k)\hat{\mathbf{R}}\mathbf{u}(k) \tag{11-196}$$

subject to the constraint

$$\mathbf{x}(k+1) = \boldsymbol{\phi}\mathbf{x}(k) + \boldsymbol{\theta}\mathbf{u}(k)$$

with $\mathbf{x}(0)$ given.

Let $J_{N-i}[\mathbf{x}(i)]$ be the performance index over the interval $[i, N]$, i.e., over the last $N - i$ intervals or stages. Then,

$$J_{N-i}[\mathbf{x}(i)] = G[\mathbf{x}(N)] + \sum_{k=i}^{N-1} F_k[\mathbf{x}(k), \mathbf{u}(k)] \qquad i = 0, 1, 2, \ldots, N \tag{11-197}$$

Let the minimum value of $J_{N-i}[\mathbf{x}(i)]$ be represented by

$$f_{N-i}[\mathbf{x}(i)] = \min_{\mathbf{u}(i)} J_{N-i}[\mathbf{x}(i)] \tag{11-198}$$

For $i = N$, the last equation represents the performance index or return over the last 0 stage, which is the terminal cost. Therefore,

$$f_0[\mathbf{x}(N)] = G[\mathbf{x}(N)] = \tfrac{1}{2}\mathbf{x}'(N)\mathbf{S}\mathbf{x}(N) \tag{11-199}$$

For $i = N - 1$, we have a one-stage or one-interval process which is the last stage. Then, the optimal performance index is

$$\begin{aligned} f_1[\mathbf{x}(N-1)] &= \min_{\mathbf{u}(N-1)} J_1[\mathbf{x}(N-1)] \\ &= \min_{\mathbf{u}(N-1)} [G[\mathbf{x}(N)] + F_{N-1}[\mathbf{x}(N-1), \mathbf{u}(N-1)]] \end{aligned} \tag{11-200}$$

Substituting Eq. (11-196) and

$$G[\mathbf{x}(N)] = \tfrac{1}{2}[\boldsymbol{\phi}\mathbf{x}(N-1) + \boldsymbol{\theta}\mathbf{u}(N-1)]'\mathbf{S}[\boldsymbol{\phi}\mathbf{x}(N-1) + \boldsymbol{\theta}\mathbf{u}(N-1)] \tag{11-201}$$

into Eq. (11-200) and rearranging, we have

$$\begin{aligned} f_1[\mathbf{x}(N-1)] &= \min_{\mathbf{u}(N-1)} [\tfrac{1}{2}\, \mathbf{x}'(N-1)(\hat{\mathbf{Q}} + \boldsymbol{\phi}'\mathbf{S}\boldsymbol{\phi})\mathbf{x}(N-1) \\ &\quad + \mathbf{x}'(N-1)(\mathbf{M} + \tfrac{1}{2}\, \boldsymbol{\phi}'\mathbf{S}\boldsymbol{\theta})\mathbf{u}(N-1) \\ &\quad + \tfrac{1}{2}\, \mathbf{u}'(N-1)\boldsymbol{\theta}'\mathbf{S}\boldsymbol{\phi}\mathbf{x}(N-1) \\ &\quad + \tfrac{1}{2}\, \mathbf{u}'(N-1)(\hat{\mathbf{R}} + \boldsymbol{\theta}'\mathbf{S}\boldsymbol{\theta})\mathbf{u}(N-1)] \\ &= \min_{\mathbf{u}(N-1)} J_1[\mathbf{x}(N-1)] \end{aligned} \tag{11-202}$$

For minimum $J_1[\mathbf{x}(N-1)]$ we set

$$\frac{\partial J_1[\mathbf{x}(N-1)]}{\partial \mathbf{u}(N-1)} = \mathbf{0} \tag{11-203}$$

The result is

$$[(\mathbf{M} + \tfrac{1}{2}\boldsymbol{\phi}'\mathbf{S}\boldsymbol{\theta})' + \tfrac{1}{2}\boldsymbol{\theta}'\mathbf{S}\boldsymbol{\phi}]\mathbf{x}^\circ(N-1) + (\hat{\mathbf{R}} + \boldsymbol{\theta}'\mathbf{S}\boldsymbol{\theta})\mathbf{u}^\circ(N-1) = 0 \tag{11-204}$$

Thus, the optimal control is

$$\mathbf{u}^\circ(N-1) = -(\hat{\mathbf{R}} + \boldsymbol{\theta}'\mathbf{S}\boldsymbol{\theta})^{-1}(\mathbf{M}' + \boldsymbol{\theta}'\mathbf{S}\boldsymbol{\phi})\mathbf{x}^\circ(N-1) \tag{11-205}$$

Notice that this is the same result as in Eq. (11-150) by setting $k = N - 1$. Now substituting Eq. (11-205) into Eq. (11-202) for $\mathbf{x}(N-1)$, we get, after simplification,

$$\begin{aligned} f_1[\mathbf{x}(N-1)] = \tfrac{1}{2}\mathbf{x}'(N-1)[\hat{\mathbf{Q}} + \boldsymbol{\phi}'\mathbf{S}\boldsymbol{\phi} \\ - (\mathbf{M}' + \boldsymbol{\theta}'\mathbf{S}\boldsymbol{\phi})'(\hat{\mathbf{R}} + \boldsymbol{\theta}'\mathbf{S}\boldsymbol{\theta})^{-1}(\mathbf{M}' + \boldsymbol{\theta}'\mathbf{S}\boldsymbol{\phi})]\mathbf{x}(N-1) \end{aligned} \tag{11-206}$$

We make the following definitions:

$$\mathbf{K}(N) = \mathbf{S} \tag{11-207}$$

$$\mathbf{K}(N-1) = \hat{\mathbf{Q}} + \boldsymbol{\phi}'\mathbf{S}\boldsymbol{\phi} - (\mathbf{M}' + \boldsymbol{\theta}'\mathbf{S}\boldsymbol{\phi})'(\hat{\mathbf{R}} + \boldsymbol{\theta}'\mathbf{S}\boldsymbol{\theta})^{-1}(\mathbf{M}' + \boldsymbol{\theta}'\mathbf{S}\boldsymbol{\phi}) \tag{11-208}$$

We see that Eq. (11-208) is the same as the Riccati equation of Eq. (11-153) with $k = N - 1$.

The optimal returns in Eqs. (11-199) and (11-200) are, respectively,

$$f_0[\mathbf{x}(N)] = \tfrac{1}{2}\mathbf{x}'(N)\mathbf{K}(N)\mathbf{x}(N) \tag{11-209}$$

$$f_1[\mathbf{x}(N-1)] = \tfrac{1}{2}\mathbf{x}'(N-1)\mathbf{K}(N-1)\mathbf{x}(N-1) \tag{11-210}$$

Continuing the process, we let $i = N - 2$; that is, an optimization problem consisting of two (last) stages. The optimal performance index for the two-stage process is written

$$\begin{aligned} f_2[\mathbf{x}(N-2)] &= \min_{\substack{\mathbf{u}(N-2)\\ \mathbf{u}(N-1)}} J_2[\mathbf{x}(N-2)] \\ &= \min_{\substack{\mathbf{u}(N-2)\\ \mathbf{u}(N-1)}} [F_{N-2}[\mathbf{x}(N-2), \mathbf{u}(N-2)] + F_{N-1}[\mathbf{x}(N-1), \mathbf{u}(N-1)] \\ &\quad + G[\mathbf{x}(N)]] \end{aligned} \tag{11-211}$$

Applying the principle of optimality which asserts that for the two-stage process to be optimal, regardless of the control strategy for the first stage, the last stage must be optimal by itself. Therefore, Eq. (11-211) is written

$$f_2[\mathbf{x}(N-2)] = \min_{\mathbf{u}(N-2)} [F_{N-2}[\mathbf{x}(N-2), \mathbf{u}(N-2)] + f_1[\mathbf{x}(N-1)]] \tag{11-212}$$

where $f_1[\mathbf{x}(N-1)]$ is the optimal return from the last stage and is given by Eq. (11-210). Substituting

$$F_{N-2}[\mathbf{x}(N-2), \mathbf{u}(N-2)] = \tfrac{1}{2}\mathbf{x}'(N-2)\hat{\mathbf{Q}}\mathbf{x}(N-2) + \mathbf{x}'(N-2)\mathbf{M}\mathbf{u}(N-2) + \tfrac{1}{2}\mathbf{u}'(N-2)\hat{\mathbf{R}}\mathbf{u}(N-2) \tag{11-213}$$

and

$$f_1[\mathbf{x}(N-1)] = \tfrac{1}{2}[\boldsymbol{\phi}\mathbf{x}(N-2) + \boldsymbol{\theta}\mathbf{u}(N-2)]'\mathbf{K}(N-1)[\boldsymbol{\phi}\mathbf{x}(N-2) + \boldsymbol{\theta}\mathbf{u}(N-2)] \tag{11-214}$$

into Eq. (11-212), rearranging terms, and setting

$$\frac{\partial J_2[\mathbf{x}(N-2)]}{\partial \mathbf{u}(N-2)} = 0 \tag{11-215}$$

we can show that the optimal control is

$$\mathbf{u}^\circ(N-2) = -[\hat{\mathbf{R}} + \boldsymbol{\theta}'\mathbf{K}(N-1)\boldsymbol{\theta}]^{-1}[\mathbf{M}' + \boldsymbol{\theta}'\mathbf{K}(N-1)\boldsymbol{\phi}]\mathbf{x}^\circ(N-2) \tag{11-216}$$

and

$$\begin{aligned} f_2[\mathbf{x}(N-2)] = \tfrac{1}{2}\mathbf{x}'(N-2)[&\hat{\mathbf{Q}} + \boldsymbol{\phi}'\mathbf{K}(N-1)\boldsymbol{\phi} \\ &- [\mathbf{M}' + \boldsymbol{\theta}'\mathbf{K}(N-1)\boldsymbol{\phi}]'[\hat{\mathbf{R}} + \boldsymbol{\theta}'\mathbf{K}(N-1)\boldsymbol{\theta}]^{-1} \\ &\cdot [\mathbf{M}' + \boldsymbol{\theta}'\mathbf{K}(N-1)\boldsymbol{\phi}]]\mathbf{x}(N-2) \end{aligned} \tag{11-217}$$

Letting

$$\begin{aligned} \mathbf{K}(N-2) = &\hat{\mathbf{Q}} + \boldsymbol{\phi}'\mathbf{K}(N-1)\boldsymbol{\phi} \\ &- [\mathbf{M}' + \boldsymbol{\theta}'\mathbf{K}(N-1)\boldsymbol{\phi}]'[\hat{\mathbf{R}} + \boldsymbol{\theta}'\mathbf{K}(N-1)\boldsymbol{\theta}]^{-1}[\mathbf{M}' + \boldsymbol{\theta}'\mathbf{K}(N-1)\boldsymbol{\phi}] \end{aligned} \tag{11-218}$$

Eq. (11-217) is simplified to

$$f_2[\mathbf{x}(N-2)] = \tfrac{1}{2}\mathbf{x}'(N-2)\mathbf{K}(N-2)\mathbf{x}(N-2) \tag{11-219}$$

Continuing the induction process, we can show that, in general,

$$f_{N-i}[\mathbf{x}(i)] = \tfrac{1}{2}\mathbf{x}'(i)\mathbf{K}(i)\mathbf{x}(i) \tag{11-220}$$

where

$$\begin{aligned} \mathbf{K}(i) = &\hat{\mathbf{Q}} + \boldsymbol{\phi}'\mathbf{K}(i+1)\boldsymbol{\phi} \\ &- [\mathbf{M}' + \boldsymbol{\theta}'\mathbf{K}(i+1)\boldsymbol{\phi}]'[\hat{\mathbf{R}} + \boldsymbol{\theta}'\mathbf{K}(i+1)\boldsymbol{\theta}]^{-1}[\mathbf{M}' + \boldsymbol{\theta}'\mathbf{K}(i+1)\boldsymbol{\phi}] \end{aligned} \tag{11-221}$$

The optimal control is

$$\mathbf{u}^\circ(i) = -[\hat{\mathbf{R}} + \boldsymbol{\theta}'\mathbf{K}(i+1)\boldsymbol{\theta}]^{-1}[\mathbf{M}' + \boldsymbol{\theta}'\mathbf{K}(i+1)\boldsymbol{\phi}]\mathbf{x}^\circ(i) \tag{11-222}$$

Thus, we have derived the Riccati equation using the principle of optimality. The method of solution is also known as dynamic programming.

It is interesting to note that the dynamic programming method does not require that $\hat{\mathbf{R}}$ be positive definite, since $\hat{\mathbf{R}}^{-1}$ was not encountered. Furthermore, $\hat{\mathbf{R}}$ can be a null matrix. However, the matrix $\hat{\mathbf{R}} + \boldsymbol{\theta}'\mathbf{K}(i+1)\boldsymbol{\theta}$ must have an inverse.

For infinite time or an infinite number of stages, $N = \infty$, similar to the results obtained in the preceding sections, $\mathbf{K}(i)$ becomes $\mathbf{K}$, and Eqs. (11-220), (11-221), and (11-222) are reduced to Eqs. (11-177), (11-174), and (11-175), respectively.

11-7 SOLUTION OF THE DISCRETE RICCATI EQUATION

The amount of work that has been published on the solution and properties of the Riccati equation can probably fill a book by itself. In general, the algebraic Riccati equation of Eq. (11-174) is more difficult to solve than the difference equation of Eq. (11-211).

The difference Riccati equation is generally solved by one of the following methods:

1. numerical computation method
2. recursive method
3. eigenvalue-eigenvector method.

The numerical computation method involves iterative solution of the nonlinear difference Riccati equation. The recursive and the eigenvalue-eigenvector methods are discussed in the following.

11-7-1 Recursive Method of Solving the Riccati Equation

The dynamic programming method presented in the last section is generally known as a recursive solution of optimal control problems.

Let

$$\mathbf{G}(i) = [\hat{\mathbf{R}} + \boldsymbol{\theta}'\mathbf{K}(i+1)\boldsymbol{\theta}]^{-1}[\mathbf{M}' + \boldsymbol{\theta}'\mathbf{K}(i+1)\boldsymbol{\phi}] \tag{11-223}$$

which is the feedback gain matrix of the optimal linear digital regulator; the optimal control in Eq. (11-222) is written

$$\mathbf{u}^{\circ}(i) = -\mathbf{G}(i)\mathbf{x}^{\circ}(i) \tag{11-224}$$

Similarly, the Riccati equation in Eq. (11-221) is simplified to

$$\mathbf{K}(i) = \hat{\mathbf{Q}} + \boldsymbol{\phi}'\mathbf{K}(i+1)\boldsymbol{\phi} - [\mathbf{M}' + \boldsymbol{\theta}'\mathbf{K}(i+1)\boldsymbol{\phi}]'\mathbf{G}(i) \tag{11-225}$$

Starting with the boundary condition $\mathbf{K}(N) = \mathbf{S}$, Eqs. (11-223) and (11-225) are solved by recursion in the backward direction.

As an alternative, we can use the Riccati equation in Eq. (11-149). Letting

$$\mathbf{H}(k+1) = \mathbf{K}(k+1)[\mathbf{I} + \boldsymbol{\theta}\hat{\mathbf{R}}^{-1}\boldsymbol{\theta}'\mathbf{K}(k+1)]^{-1} \tag{11-226}$$

Eq. (11-149) becomes

$$\mathbf{K}(k) = (\boldsymbol{\phi}' - \mathbf{M}\hat{\mathbf{R}}^{-1}\boldsymbol{\theta}')\mathbf{H}(k+1)(\boldsymbol{\phi} - \boldsymbol{\theta}\hat{\mathbf{R}}^{-1}\mathbf{M}') + \hat{\mathbf{Q}} - \mathbf{M}\hat{\mathbf{R}}^{-1}\mathbf{M}' \tag{11-227}$$

Again, these last two equations can be solved recursively backward in time starting from the boundary condition $\mathbf{K}(N) = \mathbf{S}$.

Equations (11-226) and (11-227) have the nice property that $\mathbf{K}(k+1)$ appears only in $\mathbf{H}(k+1)$. However, the recursive method using Eqs. (11-223) and (11-225) has the advantage that the feedback matrix is given directly by Eq. (11-223).

Example 11-5

Consider that a first-order digital process is described by

$$x(k+1) = x(k) + u(k) \tag{11-228}$$

with $x(0) = x_0$. The design objective is to find the optimal control $u^\circ(k)$, $k = 0, 1, 2, \ldots, 9$, such that the following performance index is minimized:

$$J_{10} = \frac{1}{2}[10x^2(10)] + \frac{1}{2}\sum_{k=0}^{9}[x^2(k) + u^2(k)] \tag{11-229}$$

For the problem we can identify that $S = 10$, $\hat{Q} = 1$, $\hat{R} = 1$, $M = 0$, and $\phi = \theta = 1$. The Riccati equation is obtained by substituting these parameters into Eq. (11-225),

$$K(i) = 1 + K(i+1) - K(i+1)G(i) \tag{11-230}$$

where

$$G(i) = \frac{K(i+1)}{1 + K(i+1)} \tag{11-231}$$

The optimal control is

$$u^\circ(i) = -G(i)\mathbf{x}^\circ(i) \tag{11-232}$$

Equations (11-230) and (11-231) are solved recursively starting with the boundary condition $K(10) = S = 10$. The results are tabulated as follows.

i	$K(i)$	$G(i)$
0	1.6180	0.6180
1	1.6180	0.6180
2	1.6180	0.6180
3	1.6180	0.6180
4	1.6180	0.6180
5	1.6182	0.6182
6	1.6188	0.6188
7	1.6236	0.6236
8	1.6562	0.6562
9	1.9091	0.9091
10	10	0

Substituting the optimal control into Eq. (11-228) we can show that the optimal trajectory of $x(k)$ is given by

$$x^\circ(k) = \prod_{i=0}^{k-1} [1 - G(i)]x_0 \tag{11-233}$$

for $k = 0, 1, \ldots, 10$. For any nonzero x_0, we see that $\mathbf{x}^\circ(k)$ approaches zero rapidly as k increases.

Example 11-6

A second-order digital process is described by the state equation

$$\mathbf{x}(k+1) = \boldsymbol{\phi}\mathbf{x}(k) + \boldsymbol{\theta}u(k) \tag{11-234}$$

where

$$\boldsymbol{\phi} = \begin{bmatrix} 0 & 1 \\ -1 & 1 \end{bmatrix} \qquad \boldsymbol{\theta} = \begin{bmatrix} 0 \\ 1 \end{bmatrix} \tag{11-235}$$

Given that $\mathbf{x}(0) = [1 \quad 1]'$, find the optimal control $u(k)$, $k = 0, 1, 2, \ldots, 7$, such that the performance index

$$J_8 = \sum_{k=0}^{7} [x_1^2(k) + u^2(k)] \tag{11-236}$$

is minimized.

For this problem we identify that $M = 0$, $\hat{R} = 2$,

$$\hat{\mathbf{Q}} = \begin{bmatrix} 2 & 0 \\ 0 & 0 \end{bmatrix} \quad \text{and} \quad \mathbf{S} = \begin{bmatrix} 0 & 0 \\ 0 & 0.5 \end{bmatrix} \tag{11-237}$$

The Riccati equation is obtained by substituting these parameters into Eq. (11-225); we have

$$\mathbf{K}(i) = \begin{bmatrix} 2 & 0 \\ 0 & 0 \end{bmatrix} + \begin{bmatrix} 0 & -1 \\ 1 & 1 \end{bmatrix} \mathbf{K}(i+1) \left[\begin{bmatrix} 0 & 1 \\ -1 & 1 \end{bmatrix} - \begin{bmatrix} 0 \\ 1 \end{bmatrix} G(i) \right] \tag{11-238}$$

The optimal control is

$$u^\circ(i) = -\mathbf{G}(i)\mathbf{x}^\circ(i) \tag{11-239}$$

where from Eq. (11-223),

$$\mathbf{G}(i) = \left[2 + [0 \quad 1]\mathbf{K}(i+1) \begin{bmatrix} 0 \\ 1 \end{bmatrix} \right]^{-1} [0 \quad 1]\mathbf{K}(i+1) \begin{bmatrix} 0 & 1 \\ -1 & 1 \end{bmatrix} \tag{11-240}$$

Starting with the boundary condition,

$$\mathbf{K}(8) = \mathbf{S} = \begin{bmatrix} 0 & 0 \\ 0 & 0 \end{bmatrix} \tag{11-241}$$

Eqs. (11-238) and (11-240) are solved recursively to give

$$\mathbf{K}(7) = \begin{bmatrix} 2 & 0 \\ 0 & 0 \end{bmatrix} \qquad \mathbf{G}(7) = [0 \quad 0]$$

$$\mathbf{K}(6) = \begin{bmatrix} 2 & 0 \\ 0 & 2 \end{bmatrix} \qquad \mathbf{G}(6) = [0 \quad 0]$$

$$\mathbf{K}(5) = \begin{bmatrix} 3 & -1 \\ -1 & 3 \end{bmatrix} \qquad \mathbf{G}(5) = [-0.5 \quad 0.5]$$

$$\mathbf{K}(4) = \begin{bmatrix} 3.2 & -0.8 \\ -0.8 & 3.2 \end{bmatrix} \qquad \mathbf{G}(4) = [-0.6 \quad 0.4]$$

$$\mathbf{K}(3) = \begin{bmatrix} 3.23 & -0.922 \\ -0.922 & 3.69 \end{bmatrix} \qquad \mathbf{G}(3) = [-0.615 \quad 0.462]$$

$$\mathbf{K}(2) = \begin{bmatrix} 3.297 & -0.973 \\ -0.973 & 3.729 \end{bmatrix} \qquad \mathbf{G}(2) = [-0.651 \quad 0.481]$$

$$\mathbf{K}(1) = \begin{bmatrix} 3.301 & -0.962 \\ -0.962 & 3.75 \end{bmatrix} \qquad \mathbf{G}(1) = [-0.652 \quad 0.485]$$

$$\mathbf{K}(0) = \begin{bmatrix} 3.305 & -0.97 \\ -0.97 & 3.777 \end{bmatrix} \qquad \mathbf{G}(0) = [-0.6538 \quad 0.486]$$

The optimal control and the optimal trajectories are computed by substituting the feedback gains into Eqs. (11-239) and (11-234). The results are tabulated as follows:

i	$u^\circ(i)$	$x_1^\circ(i)$	$x_2^\circ(i)$
0	0.1678	1	1
1	0.5708	1	0.1678
2	0.235	0.1678	−0.2614
3	−0.071	−0.2614	−0.1942
4	−0.115	−0.1942	−0.0038
5	0.0358	−0.0038	0.0754
6	0	0.0754	0.115
7	0	0.115	0.0396
8	0	0.0396	−0.0754

For large values of N we can show that the Riccati gain approaches the steady-state solution of

$$\mathbf{K} = \begin{bmatrix} 3.308 & -0.972 \\ -0.972 & 3.780 \end{bmatrix} \tag{11-242}$$

and the constant optimal control is

$$\mathbf{G} = [-0.654 \quad 0.486]$$

For $N = 8$, the finite-time problem has solutions already rapidly approaching these values. In general, however, the recursive method could only lead to the steady-state solutions by using as large a value of N as necessary. In this case, since the pair $[\boldsymbol{\phi}, \boldsymbol{\theta}]$ is completely controllable, and we can find a 2×2 matrix $\mathbf{D}$ such that $\mathbf{DD}' = \mathbf{Q}$ and $[\boldsymbol{\phi}, \mathbf{D}]$ is observable, the closed-loop system will be asymptotically stable for $N = \infty$.

11-7-2 The Eigenvalue-Eigenvector Method

The nonlinear difference Riccati equation given in Eq. (11-221) can be solved by the eigenvalue-eigenvector method. The result of the method is a closed-form solution for the Riccati gain, and the solution of the algebraic Riccati equation can be obtained by applying a limiting process.

The canonical state equations in Eqs. (11-142) and (11-143) are rewritten as

$$\mathbf{x}^\circ(k+1) = \boldsymbol{\Omega}\mathbf{x}^\circ(k) - \boldsymbol{\theta}\hat{\mathbf{R}}^{-1}\boldsymbol{\theta}'\mathbf{p}^\circ(k+1) \tag{11-243}$$

$$\mathbf{p}^\circ(k) = \boldsymbol{\Gamma}\mathbf{x}^\circ(k) + \boldsymbol{\Omega}'\mathbf{p}^\circ(k+1) \tag{11-244}$$

where

$$\boldsymbol{\Omega} = \boldsymbol{\phi} - \boldsymbol{\theta}\hat{\mathbf{R}}^{-1}\mathbf{M}' \tag{11-245}$$

$$\boldsymbol{\Gamma} = \hat{\mathbf{Q}} - \mathbf{M}\hat{\mathbf{R}}^{-1}\mathbf{M}' \tag{11-246}$$

Solving for $\mathbf{x}^\circ(k)$ from Eq. (11-243) and writing the canonical state equations in vector-matrix form, we have

$$\begin{bmatrix} \mathbf{x}^\circ(k) \\ \mathbf{p}^\circ(k) \end{bmatrix} = \mathbf{V}\begin{bmatrix} \mathbf{x}^\circ(k+1) \\ \mathbf{p}^\circ(k+1) \end{bmatrix} \tag{11-247}$$

where

$$\mathbf{V} = \begin{bmatrix} \boldsymbol{\Omega}^{-1} & \boldsymbol{\Omega}^{-1}\boldsymbol{\theta}\hat{\mathbf{R}}^{-1}\boldsymbol{\theta}' \\ \boldsymbol{\Gamma}\boldsymbol{\Omega}^{-1} & \boldsymbol{\Omega}' + \boldsymbol{\Gamma}\boldsymbol{\Omega}^{-1}\boldsymbol{\theta}\hat{\mathbf{R}}^{-1}\boldsymbol{\theta}' \end{bmatrix} \tag{11-248}$$

Equation (11-247) represents $2n$ difference equations in backward time with boundary conditions, $\mathbf{x}(0) = \mathbf{x}_0$ and $\mathbf{p}(N) = \mathbf{S}\mathbf{x}(N)$.

An important property of the matrix $\mathbf{V}$ is that the reciprocal of every eigenvalue is also an eigenvalue. We can show this by letting λ be an eigenvalue of $\mathbf{V}$ and $\mathbf{h}$ be the corresponding eigenvector. Then, by the definition of eigenvectors,

$$\mathbf{V}\mathbf{h} = \lambda\mathbf{h} \tag{11-249}$$

Let us partition $\mathbf{h}$ so that the last equation is written

$$\begin{bmatrix} \boldsymbol{\Omega}^{-1} & \boldsymbol{\Omega}^{-1}\boldsymbol{\theta}\hat{\mathbf{R}}^{-1}\boldsymbol{\theta}' \\ \boldsymbol{\Gamma}\boldsymbol{\Omega}^{-1} & \boldsymbol{\Omega}' + \boldsymbol{\Gamma}\boldsymbol{\Omega}^{-1}\boldsymbol{\theta}\hat{\mathbf{R}}^{-1}\boldsymbol{\theta}' \end{bmatrix}\begin{bmatrix} \mathbf{f} \\ \mathbf{g} \end{bmatrix} = \lambda\begin{bmatrix} \mathbf{f} \\ \mathbf{g} \end{bmatrix} \tag{11-250}$$

The determinant of $\mathbf{V}$ is written as

$$\begin{aligned} \Delta &= |\boldsymbol{\Omega}^{-1}(\boldsymbol{\Omega}' + \boldsymbol{\Gamma}\boldsymbol{\Omega}^{-1}\boldsymbol{\theta}\hat{\mathbf{R}}^{-1}\boldsymbol{\theta}') - \boldsymbol{\Omega}^{-1}\boldsymbol{\Gamma}\boldsymbol{\Omega}^{-1}\boldsymbol{\Omega}\boldsymbol{\Omega}^{-1}\boldsymbol{\theta}\hat{\mathbf{R}}^{-1}\boldsymbol{\theta}'| \\ &= |\boldsymbol{\Omega}^{-1}\boldsymbol{\Omega}' + \boldsymbol{\Omega}^{-1}\boldsymbol{\Gamma}\boldsymbol{\Omega}^{-1}\boldsymbol{\theta}\hat{\mathbf{R}}^{-1}\boldsymbol{\theta}' - \boldsymbol{\Omega}^{-1}\boldsymbol{\Gamma}\boldsymbol{\Omega}^{-1}\boldsymbol{\theta}\hat{\mathbf{R}}^{-1}\boldsymbol{\theta}'| \\ &= |\boldsymbol{\Omega}^{-1}\boldsymbol{\Omega}'| = |\boldsymbol{\Omega}^{-1}||\boldsymbol{\Omega}'| = 1 \end{aligned} \tag{11-251}$$

Therefore, the determinant of $\mathbf{V}$ is unity.

Now taking the inverse of $\mathbf{V}$ and then the transpose, the following identity results:

$$(\mathbf{V}^{-1})'\begin{bmatrix} \mathbf{g} \\ -\mathbf{f} \end{bmatrix} = \lambda\begin{bmatrix} \mathbf{g} \\ -\mathbf{f} \end{bmatrix} \tag{11-252}$$

The significance of the last equation is that λ is an eigenvalue of $(\mathbf{V}^{-1})'$ and also of $\mathbf{V}^{-1}$. Thus, $1/\lambda$ is an eigenvalue of $\mathbf{V}$. This also means that n eigenvalues of $\mathbf{V}$ are inside the unit circle and n are outside. Let us introduce a nonsingular transformation

$$\begin{bmatrix} \mathbf{x}^\circ(k) \\ \mathbf{p}^\circ(k) \end{bmatrix} = \mathbf{W}\begin{bmatrix} \mathbf{q}(k) \\ \mathbf{r}(k) \end{bmatrix} \tag{11-253}$$

where $\mathbf{W}$ is of the form

$$\mathbf{W} = \begin{bmatrix} \mathbf{W}_{11} & \mathbf{W}_{12} \\ \mathbf{W}_{21} & \mathbf{W}_{22} \end{bmatrix} \tag{11-254}$$

and $\mathbf{W}$ has the property that

$$\mathbf{W}^{-1}\mathbf{V}\mathbf{W} = \begin{bmatrix} \boldsymbol{\Lambda} & 0 \\ 0 & \boldsymbol{\Lambda}^{-1} \end{bmatrix} \tag{11-255}$$

For distinct eigenvalues, the elements of $\boldsymbol{\Lambda}$ form either a diagonal matrix with λ_i on the main diagonal, where λ_i represents the eigenvalues of $\mathbf{V}$ that are outside the unit circle, or a modal form matrix for complex conjugate eigenvalues. For example, let λ_1 and λ_2 be the real eigenvalues of $\mathbf{V}$ that are outside the unit circle. Then Eq. (11-255) is written

$$\mathbf{W}^{-1}\mathbf{V}\mathbf{W} = \left[\begin{array}{cc|cc} \lambda_1 & 0 & 0 & 0 \\ 0 & \lambda_2 & 0 & 0 \\ \hline 0 & 0 & \dfrac{1}{\lambda_1} & 0 \\ 0 & 0 & 0 & \dfrac{0}{\lambda_2} \end{array}\right] = \left[\begin{array}{c|c} \boldsymbol{\Lambda} & 0 \\ \hline 0 & \boldsymbol{\Lambda}^{-1} \end{array}\right] \tag{11-256}$$

If $\mathbf{V}$ has complex conjugate eigenvalues, $\sigma_1 + j\omega_1$ and $\sigma_1 - j\omega_1$, all outside the unit circle, then Eq. (11-255) becomes

$$\mathbf{W}^{-1}\mathbf{V}\mathbf{W} = \left[\begin{array}{cc:cc} \sigma_1 & \omega_1 & 0 & 0 \\ -\omega_1 & \sigma_1 & 0 & 0 \\ \hdashline 0 & 0 & \sigma_2 & -\omega_2 \\ 0 & 0 & \omega_2 & \sigma_2 \end{array}\right] = \left[\begin{array}{c:c} \mathbf{\Lambda} & \mathbf{0} \\ \hdashline \mathbf{0} & \mathbf{\Lambda}^{-1} \end{array}\right] \tag{11-257}$$

where

$$\sigma_2 - j\omega_2 = \frac{1}{\sigma_1 + j\omega_1} \tag{11-258}$$

Of course, when $\mathbf{V}$ has both real and complex eigenvalues, a combination of Eqs. (11-256) and (11-257) should be used.

Now writing Eq. (11-253) as

$$\begin{bmatrix} \mathbf{q}(k) \\ \mathbf{r}(k) \end{bmatrix} = \mathbf{W}^{-1}\begin{bmatrix} \mathbf{x}^\circ(k) \\ \mathbf{p}^\circ(k) \end{bmatrix}$$

and using Eq. (11-247), we have

$$\begin{bmatrix} \mathbf{q}(k) \\ \mathbf{r}(k) \end{bmatrix} = \mathbf{W}^{-1}\mathbf{V}\mathbf{W}\begin{bmatrix} \mathbf{q}(k+1) \\ \mathbf{r}(k+1) \end{bmatrix} = \begin{bmatrix} \mathbf{\Lambda} & \mathbf{0} \\ \mathbf{0} & \mathbf{\Lambda}^{-1} \end{bmatrix}\begin{bmatrix} \mathbf{q}(k+1) \\ \mathbf{r}(k+1) \end{bmatrix} \tag{11-259}$$

The last equation is solved by recursion in the backward direction with the boundary condition $[\mathbf{q}(N) \quad \mathbf{r}(N)]'$ to give

$$\begin{bmatrix} \mathbf{q}(N-k) \\ \mathbf{r}(N-k) \end{bmatrix} = \begin{bmatrix} \mathbf{\Lambda}^{k} & \mathbf{0} \\ \mathbf{0} & \mathbf{\Lambda}^{-k} \end{bmatrix}\begin{bmatrix} \mathbf{q}(N) \\ \mathbf{r}(N) \end{bmatrix} \tag{11-260}$$

which is rearranged to the form

$$\begin{bmatrix} \mathbf{q}(N) \\ \mathbf{r}(N-k) \end{bmatrix} = \begin{bmatrix} \mathbf{\Lambda}^{-k} & \mathbf{0} \\ \mathbf{0} & \mathbf{\Lambda}^{-k} \end{bmatrix}\begin{bmatrix} \mathbf{q}(N-k) \\ \mathbf{r}(N) \end{bmatrix} \tag{11-261}$$

Let $\mathbf{r}(N)$ and $\mathbf{q}(N)$ be related through

$$\mathbf{r}(N) = \mathbf{U}\mathbf{q}(N) \tag{11-262}$$

where $\mathbf{U}$ is an $n \times n$ matrix. Substituting Eq. (11-262) into Eqs. (11-261) and (11-253) for $k = N$, we have

$$\mathbf{U} = -(\mathbf{W}_{22} - \mathbf{S}\mathbf{W}_{12})^{-1}(\mathbf{W}_{21} - \mathbf{S}\mathbf{W}_{11}) \tag{11-263}$$

and

$$\mathbf{r}(N-k) = \mathbf{\Lambda}^{-k}\mathbf{U}\mathbf{\Lambda}^{-k}\mathbf{q}(N-k) \tag{11-264}$$

Letting

$$\mathbf{H}(k) = \mathbf{\Lambda}^{-k}\mathbf{U}\mathbf{\Lambda}^{-k} \tag{11-265}$$

Eq. (11-264) is written

$$\mathbf{r}(N-k) = \mathbf{H}(k)\mathbf{q}(N-k) \tag{11-266}$$

Now substituting the last equation into Eq. (11-253), we can show that the following relations are true:

$$\mathbf{x}(N-k) = [\mathbf{W}_{11} + \mathbf{W}_{12}\mathbf{H}(k)]\mathbf{q}(N-k) \tag{11-267}$$

$$\mathbf{p}(N-k) = \mathbf{K}(N-k)\mathbf{x}(N-k) = [\mathbf{W}_{21} + \mathbf{W}_{22}\mathbf{H}(k)]\mathbf{q}(N-k) \tag{11-268}$$

Comparing the last two equations, the Riccati gain is written as

$$\mathbf{K}(N-k) = [\mathbf{W}_{21} + \mathbf{W}_{22}\mathbf{H}(k)][\mathbf{W}_{11} + \mathbf{W}_{12}\mathbf{H}(k)]^{-1} \tag{11-269}$$

Although the last equation gives a closed-form solution of the difference Riccati equation, in general, because of the computations involved in finding the matrix $\mathbf{W}$ and the eigenvalues of $\mathbf{V}$, the eigenvalue-eigenvector method does not have a clear-cut advantage over the recursive method. However, the eigenvalue-eigenvector method leads directly to the solution of the algebraic Riccati equation for infinite-time problems.

Letting $i = N - k$, $\mathbf{K}(N-k)$ becomes $\mathbf{K}(i)$, which is the Riccati gain used in the preceding sections, $i = 0, 1, 2, \ldots, N-1$. The constant Riccati gain is written

$$\mathbf{K} = \lim_{i=-\infty} \mathbf{K}(i) = \lim_{k\to\infty} \mathbf{K}(N-k) \tag{11-270}$$

Since

$$\lim_{k\to\infty} \mathbf{H}(k) = \lim_{k\to\infty} \mathbf{\Lambda}^{-k}\mathbf{U}\mathbf{\Lambda}^{-k} = 0 \tag{11-271}$$

Eq. (11-269) leads to

$$\mathbf{K} = \lim_{k\to\infty} \mathbf{K}(N-k) = \mathbf{W}_{21}\mathbf{W}_{11}^{-1} \tag{11-272}$$

Example 11-7

The first-order system treated in Example 11-5 is considered here again. The finite-time problem is to be solved by the eigenvalue-eigenvector method, and in addition, the problem is to be solved for $N = \infty$.

Substitution of the system parameters into Eq. (11-248) yields

$$\mathbf{V} = \begin{bmatrix} 1 & 1 \\ 1 & 2 \end{bmatrix} \tag{11-273}$$

The eigenvalues of $\mathbf{V}$ are $\lambda_1 = 2.618$ and $\lambda_2 = 1/\lambda_1 = 0.382$. Therefore, we form the matrix

$$\mathbf{W}^{-1}\mathbf{V}\mathbf{W} = \begin{bmatrix} \mathbf{\Lambda} & 0 \\ 0 & \mathbf{\Lambda}^{-1} \end{bmatrix} = \begin{bmatrix} 2.618 & 0 \\ 0 & 0.382 \end{bmatrix} \tag{11-274}$$

Since $\mathbf{W}$ is a similarity transformation, the columns of $\mathbf{W}$ are the eigenvectors of $\mathbf{V}$. Therefore,

$$\mathbf{W} = \begin{bmatrix} \mathbf{W}_{11} & \mathbf{W}_{12} \\ \mathbf{W}_{21} & \mathbf{W}_{22} \end{bmatrix} = \begin{bmatrix} 1 & 1 \\ 1.618 & -0.618 \end{bmatrix} \tag{11-275}$$

From Eq. (11-263) we have

$$U = -(-0.618 - 10)^{-1}(1.618 - 10) = -0.789 \tag{11-276}$$

and Eq. (11-265) gives

$$\mathbf{H}(k) = \mathbf{\Lambda}^{-k} U \mathbf{\Lambda}^{-k} = -0.789(0.382)^{2k} \tag{11-277}$$

The Riccati gain of the finite-time problem is

$$\mathbf{K}(N-k) = [1.618 - 0.618(-0.789)(0.382)^{2k}][1 - 0.789(0.382)^{2k}]^{-1} \tag{11-278}$$

or

$$\mathbf{K}(i) = [1.618 + 0.488(0.382)^{2(N-i)}][1 + 0.789(0.382)^{2(N-i)}]^{-1} \tag{11-279}$$

It is simple to show that this will give the same results as those obtained in Example 11-5 for $N = 10$.

For $N = \infty$, the constant Riccati gain is given by Eq. (11-272); we have

$$\mathbf{K} = \mathbf{W}_{21}\mathbf{W}_{11}^{-1} = 1.618 \tag{11-280}$$

Example 11-8

It is informative to apply the eigenvalue-eigenvector method to a system higher than the first order. The second-order system in Example 11-6 is considered.

The matrix $\mathbf{V}$ is written

$$\mathbf{V} = \left[\begin{array}{cc:cc} 1 & -1 & 0 & -0.5 \\ 1 & 0 & 0 & 0 \\ \hdashline 2 & -2 & 0 & -2 \\ 0 & 0 & 1 & 1 \end{array}\right] \tag{11-281}$$

The eigenvalues of $\mathbf{V}$ are $\lambda_1 = \sigma_1 \pm j\omega_1 = 0.743 \pm j1.529$ and $\lambda_2 = 1/\lambda_1 = \sigma_2 \pm j\omega_2 = 0.257 \pm j0.529$. These eigenvalues are complex conjugate pairs with two outside and two inside the unit circle. Then,

$$\mathbf{W}^{-1}\mathbf{V}\mathbf{W} = \left|\begin{array}{c:c} \mathbf{\Lambda} & 0 \\ \hdashline 0 & \mathbf{\Lambda}^{-1} \end{array}\right| = \left[\begin{array}{cc:cc} \sigma_1 & \omega_1 & 0 & 0 \\ -\omega_1 & \sigma_1 & 0 & 0 \\ \hdashline 0 & 0 & \sigma_2 & -\omega_2 \\ 0 & 0 & \omega_2 & \sigma_2 \end{array}\right]$$

$$= \left[\begin{array}{cc:cc} 0.743 & 1.529 & 0 & 0 \\ -1.529 & 0.743 & 0 & 0 \\ \hdashline 0 & 0 & 0.257 & -0.529 \\ 0 & 0 & 0.529 & 0.257 \end{array}\right] \tag{11-282}$$

It is known that, given the coefficient matrix

$$\mathbf{A} = \begin{bmatrix} \sigma_1 + j\omega_1 & 0 \\ 0 & \sigma_1 - j\omega_1 \end{bmatrix} \tag{11-283}$$

with its eigenvectors designated as $\boldsymbol{\alpha}_1 + j\boldsymbol{\beta}_1$ and $\boldsymbol{\alpha}_1 - j\boldsymbol{\beta}_1$, then the eigenvectors of the modal form

$$\mathbf{\Lambda} = \begin{bmatrix} \sigma_1 & \omega_1 \\ -\omega_1 & \sigma_1 \end{bmatrix} \tag{11-284}$$

are $\boldsymbol{\alpha}_1$ and $\boldsymbol{\beta}_1$. For the modal form matrix in Eq. (11-257) the matrix $\mathbf{W}$ is found by using the real and imaginary parts of the eigenvectors of the eigenvalues of $\mathbf{W}^{-1}\mathbf{V}\mathbf{W}$. Thus,

$$\mathbf{W} = [\boldsymbol{\alpha}_1 \,\vdots\, \boldsymbol{\beta}_1 \,\vdots\, \boldsymbol{\alpha}_2 \,\vdots\, \boldsymbol{\beta}_2] = \left[\begin{array}{cc:cc} 0.318 & -0.053 & 0 & -0.5 \\ 0.053 & -0.182 & -0.765 & -0.371 \\ \hdashline 1 & 0 & -0.743 & 0.529 \\ -0.107 & -0.636 & 1 & 0 \end{array}\right]^{-1} \tag{11-285}$$

where $\boldsymbol{\alpha}_1 + j\boldsymbol{\beta}_1$ is the eigenvector of $\sigma_1 + j\omega_1$ and $\boldsymbol{\alpha}_2 + j\boldsymbol{\beta}_2$ is the eigenvector of $\sigma_2 + j\omega_2$.

From Eq. (11-263), we get

$$\mathbf{U} = \begin{bmatrix} 0.108 & 0.635 \\ -0.174 & 0.894 \end{bmatrix} \tag{11-286}$$

and $\mathbf{H}(k)$ is given by Eq. (11-265). Then, the time-varying Riccati gain $\mathbf{K}(N-k)$ is determined by using Eq. (11-269). It is apparent that the solution process requires lengthy matrix multiplications, and even for a second-order system a digital computer is needed.

The infinite-time solution is obtained easily from Eq. (11-272). Therefore, for $N = \infty$,

$$\mathbf{K} = \mathbf{W}_{21}\mathbf{W}_{11}^{-1} = \begin{bmatrix} 1 & 0 \\ -0.107 & -0.636 \end{bmatrix} \begin{bmatrix} 0.318 & -0.053 \\ 0.053 & -0.182 \end{bmatrix}^{-1}$$
$$= \begin{bmatrix} 3.308 & -0.972 \\ -0.972 & 3.780 \end{bmatrix} \tag{11-287}$$

which is the same result predicted by the recursive method in Eq. (11-242). It has been pointed out that the closed-loop system with $N = \infty$ will be asymptotically stable.

11-8 SAMPLING PERIOD SENSITIVITY

The unique properties of the discrete Riccati equation lead to a sensitivity study of the optimal digital regulator system with respect to the sampling period T, when the system model is described by Eqs. (11-120) and (11-121). The sampling period sensitivity also provides a method of approximating the constant Riccati gain **K** for the infinite-time linear regulator problem, based on the knowledge of the Riccati gain of the continuous-data system, that is, when $T = 0$. In the end it is possible to approximate the feedback gain matrix **G** of the optimal linear digital regulator in terms of the feedback gain matrix of the optimal continuous-data linear regulator, that is when $T = 0$. It is interesting to note that the development of the sampling period sensitivity and its application to the optimal linear digital regulator design originally motivated the solution to the digital redesign problem presented in Chapter 9. Therefore, in this section we shall look at the digital redesign problem again, but from a different viewpoint.

We begin by referring to the linear digital regulator design problem stated in Eqs. (11-120), (11-121), and (11-122). It has been shown in Sec. 11-5 that the optimal digital control is [Eq. (11-150)]

$$\begin{aligned}\mathbf{u}^\circ(k) &= -[\hat{\mathbf{R}} + \boldsymbol{\theta}'\mathbf{K}(k+1)\boldsymbol{\theta}]^{-1}[\boldsymbol{\theta}'\mathbf{K}(k+1)\boldsymbol{\phi} + \mathbf{M}']\mathbf{x}^\circ(k) \\ &= -\mathbf{G}(k)\mathbf{x}^\circ(k)\end{aligned} \tag{11-288}$$

where $\mathbf{K}(k)$ is the positive-semidefinite solution of the Riccati equation [Eq. (11-153)]

$$\begin{aligned}\mathbf{K}(k) &= \boldsymbol{\phi}'\mathbf{K}(k+1)\boldsymbol{\phi} + \hat{\mathbf{Q}} \\ &\quad - [\boldsymbol{\theta}'\mathbf{K}(k+1)\boldsymbol{\phi} + \mathbf{M}']'[\hat{\mathbf{R}} + \boldsymbol{\theta}'\mathbf{K}(k+1)\boldsymbol{\theta}]^{-1}[\boldsymbol{\theta}'\mathbf{K}(k+1)\boldsymbol{\phi} + \mathbf{M}']\end{aligned} \tag{11-289}$$

with the boundary condition $\mathbf{K}(t_f) = \mathbf{S}$.

The last equation is conditioned by rearranging the terms, adding and subtracting $\mathbf{K}(k+1)$ on the same side, and dividing both sides by T; we have

$$\frac{\mathbf{K}(k+1)-\mathbf{K}(k)}{T}+\frac{\boldsymbol{\phi}'\mathbf{K}(k+1)\boldsymbol{\phi}-\mathbf{K}(k+1)+\hat{\mathbf{Q}}}{T}$$
$$=\left[\frac{\mathbf{M}+\boldsymbol{\phi}'\mathbf{K}(k+1)\boldsymbol{\theta}}{T}\right]\left[\frac{\hat{\mathbf{R}}+\boldsymbol{\theta}'\mathbf{K}(k+1)\boldsymbol{\theta}}{T}\right]^{-1}\left[\frac{\boldsymbol{\theta}'\mathbf{K}(k+1)\boldsymbol{\phi}+\mathbf{M}'}{T}\right] \tag{11-290}$$

It can be shown that as the sampling period T approaches zero, the last equation becomes

$$\dot{\mathbf{K}}(t)+\mathbf{A}'\mathbf{K}(t)+\mathbf{K}(t)\mathbf{A}+\mathbf{Q}=\mathbf{K}(t)\mathbf{B}\mathbf{R}^{-1}\mathbf{B}'\mathbf{K}(t) \tag{11-291}$$

which is known as the differential Riccati equation of the optimal continuous-data linear regulator, and $\mathbf{K}(t)$ is the Riccati gain, that is, when the system in Eq. (11-120) is subject to continuous-data control $\mathbf{u}(t)$.

For infinite time $t_f=\infty$,

$$\mathbf{K}(k)=\mathbf{K}(k+1)=\mathbf{K}(T) \tag{11-292}$$

where, instead of using the notation $\mathbf{K}$ as the steady-state discrete Riccati gain, we have adopted the notation $\mathbf{K}(T)$ to indicate the dependence on T. For infinite time, Eq. (11-290) becomes the algebraic Riccati equation,

$$\frac{\boldsymbol{\phi}'\mathbf{K}(T)\boldsymbol{\phi}-\mathbf{K}(T)+\hat{\mathbf{Q}}}{T}=\left[\frac{\mathbf{M}+\boldsymbol{\phi}'\mathbf{K}(T)\boldsymbol{\theta}}{T}\right]\left[\frac{\hat{\mathbf{R}}+\boldsymbol{\theta}'\mathbf{K}(T)\boldsymbol{\theta}}{T}\right]^{-1}\left[\frac{\boldsymbol{\theta}'\mathbf{K}(T)\boldsymbol{\phi}+\mathbf{M}'}{T}\right] \tag{11-293}$$

As T approaches zero, the last equation becomes

$$\mathbf{K}(0)\mathbf{A}+\mathbf{A}'\mathbf{K}(0)+\mathbf{Q}=\mathbf{K}(0)\mathbf{B}\mathbf{R}^{-1}\mathbf{B}'\mathbf{K}(0) \tag{11-294}$$

where

$$\mathbf{K}(0)=\lim_{T\to 0}\mathbf{K}(T) \tag{11-295}$$

or $\mathbf{K}(0)$ is also the steady-state solution of the differential Riccati equation in Eq. (11-291).

It follows that the optimal feedback gain matrix of the infinite-time digital regulator can be written as

$$\mathbf{G}(T)=\left[\frac{\hat{\mathbf{R}}+\boldsymbol{\theta}'\mathbf{K}(T)\boldsymbol{\theta}}{T}\right]^{-1}\left[\frac{\boldsymbol{\theta}'\mathbf{K}(T)\boldsymbol{\phi}+\mathbf{M}'}{T}\right] \tag{11-296}$$

For $T=0$, the optimal feedback gain matrix of the infinite-time continuous-data regulator is

$$\mathbf{G}(0)=\lim_{T\to 0}\mathbf{G}(T)=\mathbf{R}^{-1}\mathbf{B}'\mathbf{K}(0) \tag{11-297}$$

Sampling Period Sensitivity Study of the Riccati Gain Matrix

Let us expand the steady-state discrete Riccati gain $\mathbf{K}(T)$ into a Taylor series about $T = 0$:

$$\mathbf{K}(T) = \mathbf{K}(0) + \sum_{i=0}^{\infty} \mathbf{S}_i^K \frac{T^i}{i!} \tag{11-298}$$

where

$$\mathbf{S}_i^K = \left. \frac{\partial^i \mathbf{K}(T)}{\partial T^i} \right|_{T=0} \tag{11-299}$$

is defined as the ith-order *sampling period sensitivity* of $\mathbf{K}(T)$. It is assumed that the infinite series in Eq. (11-298) converges over the sampling period $[0, T_c]$. For most practical problems it is adequate to use only a few terms of the series to approximate $\mathbf{K}(T)$. In the following, the first-order and the second-order sensitivities of $\mathbf{K}(T)$ are derived. For simplicity of notation, it is understood that all the partial derivatives are evaluated at $T = 0$.

For convenience, we define

$$\mathbf{X}(T) = \frac{\boldsymbol{\phi}'\mathbf{K}(T)\boldsymbol{\phi} - \mathbf{K}(T) + \hat{\mathbf{Q}}}{T} \tag{11-300}$$

$$\mathbf{X}(0) = \lim_{T \to 0} \mathbf{X}(T) = \mathbf{K}(0)\mathbf{A} + \mathbf{A}'\mathbf{K}(0) + \mathbf{Q} \tag{11-301}$$

$$\mathbf{Y}(T) = \frac{\mathbf{M} + \boldsymbol{\phi}'\mathbf{K}(T)\boldsymbol{\theta}}{T} \tag{11-302}$$

$$\mathbf{Y}(0) = \lim_{T \to 0} \mathbf{Y}(T) = \mathbf{K}(0)\mathbf{B} \tag{11-303}$$

$$\mathbf{Z}(T) = \frac{\hat{\mathbf{R}} + \boldsymbol{\theta}'\mathbf{K}(T)\boldsymbol{\theta}}{T} \tag{11-304}$$

$$\mathbf{Z}(0) = \mathbf{R} \tag{11-305}$$

Then, the optimal feedback gain in Eq. (11-296) is simply

$$\mathbf{G}(T) = \mathbf{Z}^{-1}(T)\mathbf{Y}'(T) \tag{11-306}$$

and the discrete Riccati equation in Eq. (11-293) becomes

$$\begin{aligned} \mathbf{X}(T) &= \mathbf{Y}(T)\mathbf{Z}^{-1}(T)\mathbf{Y}'(T) \\ &= \mathbf{Y}(T)\mathbf{G}(T) \end{aligned} \tag{11-307}$$

For first-order sensitivities, Eqs. (11-306) and (11-307) are differentiated with respect to T and then evaluated at $T = 0$. We have

$$\frac{\partial \mathbf{G}(T)}{\partial T} = \mathbf{R}^{-1} \frac{\partial \mathbf{Y}'(T)}{\partial T} - \mathbf{R}^{-1} \frac{\partial \mathbf{Z}(T)}{\partial T} \mathbf{R}^{-1}\mathbf{B}'\mathbf{K}(0) \tag{11-308}$$

$$\frac{\partial \mathbf{X}(T)}{\partial T} = \mathbf{K}(0)\mathbf{B} \frac{\partial \mathbf{G}(T)}{\partial T} + \frac{\partial \mathbf{Y}(T)}{\partial T} \mathbf{R}^{-1}\mathbf{B}'\mathbf{K}(0) \tag{11-309}$$

where $\partial^i \mathbf{G}(T)/\partial T^i$ denotes the ith-order derivative of $\mathbf{G}(T)$ with respect to T, evaluated at $T = 0$, and is called the ith-order sampling period sensitivity of $\mathbf{G}(T)$.

The partial derivatives of $\mathbf{X}(T)$, $\mathbf{Y}(T)$, and $\mathbf{Z}(T)$ with respect to T can be found from Eqs. (11-300), (11-302), and (11-304), respectively, by using the infinite series expansion of $\boldsymbol{\phi} = e^{\mathbf{A}T}$ in Eqs. (11-125), (11-130), (11-131), and (11-132), retaining terms only up to the third order in T. The following results are obtained:

$$\frac{\partial \mathbf{X}(T)}{\partial T} = \mathbf{A}' \frac{\partial \mathbf{K}(T)}{\partial T} + \frac{\partial \mathbf{K}(T)}{\partial T} \mathbf{A} + \frac{1}{2} \mathbf{A}'\mathbf{K}(0)\mathbf{B}\mathbf{R}^{-1}\mathbf{B}'\mathbf{K}(0) + \frac{1}{2} \mathbf{K}(0)\mathbf{B}\mathbf{R}^{-1}\mathbf{B}'\mathbf{K}(0)\mathbf{A} \tag{11-310}$$

$$\frac{\partial \mathbf{Y}(T)}{\partial T} = \frac{\partial \mathbf{K}(T)}{\partial T} \mathbf{B} + \frac{1}{2} [\mathbf{A}'\mathbf{K}(0) + \mathbf{K}(0)\mathbf{B}\mathbf{R}^{-1}\mathbf{B}'\mathbf{K}(0)]\mathbf{B} \tag{11-311}$$

$$\frac{\partial \mathbf{Z}(T)}{\partial T} = \mathbf{B}'\mathbf{K}(0)\mathbf{B} \tag{11-312}$$

Substituting Eqs. (11-311) and (11-312) into Eq. (11-308), we have, after simplification,

$$\frac{\partial \mathbf{G}(T)}{\partial T} = \mathbf{R}^{-1}\mathbf{B}' \frac{\partial \mathbf{K}(T)}{\partial T} + \frac{1}{2} \mathbf{R}^{-1}\mathbf{B}'\mathbf{K}(0)[\mathbf{A} - \mathbf{B}\mathbf{R}^{-1}\mathbf{B}'\mathbf{K}(0)] \tag{11-313}$$

Substituting Eqs. (11-310), (11-311), and (11-313) into Eq. (11-309), we get, after simplification,

$$[\mathbf{A}' - \mathbf{K}(0)\mathbf{B}\mathbf{R}^{-1}\mathbf{B}'\mathbf{K}(0)] \frac{\partial \mathbf{K}(T)}{\partial T} + \frac{\partial \mathbf{K}(T)}{\partial T} [\mathbf{A} - \mathbf{B}\mathbf{R}^{-1}\mathbf{B}'\mathbf{K}(0)] = \mathbf{0} \tag{11-314}$$

which is known as the Liapunov equation. Since $\mathbf{A} - \mathbf{B}\mathbf{R}^{-1}\mathbf{B}'\mathbf{K}(0)$ is the coefficient matrix of the optimal closed-loop continuous-data linear regulator system that is asymptotically stable, it follows that the only possible solution to Eq. (11-314) is

$$\frac{\partial \mathbf{K}(T)}{\partial T} = \mathbf{0} \tag{11-315}$$

Therefore, Eq. (11-313) gives

$$\frac{\partial \mathbf{G}(T)}{\partial T} = \tfrac{1}{2} \mathbf{R}^{-1}\mathbf{B}'\mathbf{K}(0)[\mathbf{A} - \mathbf{B}\mathbf{R}^{-1}\mathbf{B}'\mathbf{K}(0)] \tag{11-316}$$

Equation (11-315) also leads to the interesting result on the sensitivity of the performance index with respect to T for small T. Equation (11-177) gives the optimal performance index for the infinite-time digital regulator and is rewritten as follows:

$$J_\infty^\circ = \tfrac{1}{2}\mathbf{x}'(0)\mathbf{K}(T)\mathbf{x}(0) \tag{11-317}$$

Therefore,

$$\left.\frac{\partial J^\circ_\infty}{\partial T}\right|_{T=0} = \frac{1}{2}\,\mathbf{x}'(0)\,\frac{\partial \mathbf{K}(T)}{\partial T}\,\mathbf{x}(0) = 0 \tag{11-318}$$

which means that the optimal performance index J°_∞ is insensitive to variations in T for small T.

For second-order sensitivities we must evaluate the second-order derivatives of $\mathbf{G}(T)$, $\mathbf{X}(T)$, $\mathbf{Y}(T)$, and $\mathbf{Z}(T)$. These are given as follows:

$$\frac{\partial^2 \mathbf{X}(T)}{\partial T^2} = \mathbf{A}'\,\frac{\partial^2 \mathbf{K}(T)}{\partial T^2} + \frac{\partial^2 \mathbf{K}(T)}{\partial T^2}\,\mathbf{A} + \frac{1}{3}(\mathbf{A}')^2\left[\mathbf{K}(0)\mathbf{B}\mathbf{R}^{-1}\mathbf{B}'\mathbf{K}(0)\right. \\ \left. + \frac{1}{3}\,\mathbf{K}(0)\mathbf{B}\mathbf{R}^{-1}\mathbf{B}'\mathbf{K}(0)\right]\mathbf{A}^2 + \frac{2}{3}\,\mathbf{A}'\mathbf{K}(0)\mathbf{B}\mathbf{R}^{-1}\mathbf{B}'\mathbf{K}(0)\mathbf{A} \tag{11-319}$$

$$\frac{\partial^2 \mathbf{Y}(T)}{\partial T^2} = \frac{\partial^2 \mathbf{K}(T)}{\partial T^2}\,\mathbf{B} + \frac{1}{3}(\mathbf{A}')^2\mathbf{K}(0)\mathbf{B} + \frac{2}{3}\,\mathbf{A}'\mathbf{K}(0)\mathbf{B}\mathbf{R}^{-1}\mathbf{B}'\mathbf{K}(0)\mathbf{B} \\ + \frac{1}{3}\,\mathbf{K}(0)\mathbf{B}\mathbf{R}^{-1}\mathbf{B}'\mathbf{K}(0)\mathbf{A}\mathbf{B} \tag{11-320}$$

$$\frac{\partial^2 \mathbf{Z}(T)}{\partial T^2} = \frac{1}{3}\,\mathbf{B}'\mathbf{K}(0)\mathbf{A}\mathbf{B} + \frac{1}{3}\,\mathbf{B}'\mathbf{A}'\mathbf{K}(0)\mathbf{B} + \frac{2}{3}\,\mathbf{B}'\mathbf{K}(0)\mathbf{B}\mathbf{R}^{-1}\mathbf{B}'\mathbf{K}(0)\mathbf{B} \tag{11-321}$$

Differentiating Eqs. (11-308) and (11-309) once with respect to T gives

$$\frac{\partial^2 \mathbf{G}(T)}{\partial T^2} = \mathbf{R}^{-1}\left[\frac{\partial^2 \mathbf{Y}'(T)}{\partial T^2} - 2\,\frac{\partial \mathbf{Z}(T)}{\partial T}\,\frac{\partial \mathbf{G}(T)}{\partial T} - \frac{\partial^2 \mathbf{Z}(T)}{\partial T^2}\,\mathbf{R}^{-1}\mathbf{B}'\mathbf{K}(0)\right] \tag{11-322}$$

$$\frac{\partial^2 \mathbf{X}(T)}{\partial T^2} = 2\,\frac{\partial \mathbf{Y}(T)}{\partial T}\,\frac{\partial \mathbf{G}(T)}{\partial T} + \mathbf{K}(0)\mathbf{B}\,\frac{\partial^2 \mathbf{G}(T)}{\partial T^2} + \frac{\partial^2 \mathbf{Y}(T)}{\partial T^2}\,\mathbf{R}^{-1}\mathbf{B}'\mathbf{K}(0) \tag{11-323}$$

It should be reiterated that all the derivatives are evaluated at $T = 0$; that is,

$$\frac{\partial \mathbf{G}(T)}{\partial T} \quad \text{implies} \quad \left.\frac{\partial \mathbf{G}(T)}{\partial T}\right|_{T=0} \tag{11-324}$$

Substituting Eqs. (11-312), (11-316), and (11-321) into Eq. (11-322) and simplifying, we have

$$\frac{\partial^2 \mathbf{G}(T)}{\partial T^2} = \mathbf{R}^{-1}\mathbf{B}'\,\frac{\partial^2 \mathbf{K}(T)}{\partial T^2} + \frac{1}{3}\,\mathbf{R}^{-1}\mathbf{B}'\mathbf{K}(0)[\mathbf{A} - \mathbf{B}\mathbf{R}^{-1}\mathbf{B}'\mathbf{K}(0)]^2 \tag{11-325}$$

Similarly, it can be shown that $\partial^2\mathbf{K}(T)/\partial T^2$ is the solution of the Liapunov equation

$$[\mathbf{A}' - \mathbf{K}(0)\mathbf{B}\mathbf{R}^{-1}\mathbf{B}']\,\frac{\partial^2 \mathbf{K}(T)}{\partial T^2} + \frac{\partial^2 \mathbf{K}(T)}{\partial T^2}\,[\mathbf{A} - \mathbf{B}\mathbf{R}^{-1}\mathbf{B}'\mathbf{K}(0)] \\ + \tfrac{1}{6}\,[\mathbf{A}' - \mathbf{K}(0)\mathbf{B}\mathbf{R}^{-1}\mathbf{B}'][\mathbf{K}(0)\mathbf{B}\mathbf{R}^{-1}\mathbf{B}'\mathbf{K}(0)][\mathbf{A} - \mathbf{B}\mathbf{R}^{-1}\mathbf{B}'\mathbf{K}(0)] = \mathbf{0} \tag{11-326}$$

Since the third term of the last equation is always positive semidefinite, it follows that $\partial^2\mathbf{K}(T)/\partial T^2$ is unique and positive semidefinite.

The amount of work involved in generating third and higher order sensitivities is apparently excessive, and the results are not available. However, in general, the second-order approximation usually provides adequate accuracy.

Equation (11-326) can be solved using other equivalent forms. Assume that $\partial^2\mathbf{K}(T)/\partial T^2$ is of the following form:

$$\frac{\partial^2\mathbf{K}(T)}{\partial T^2} = [\mathbf{A}' - \mathbf{K}(0)\mathbf{B}\mathbf{R}^{-1}\mathbf{B}'] \frac{\mathbf{\Pi}}{6} [\mathbf{A} - \mathbf{B}\mathbf{R}^{-1}\mathbf{B}'\mathbf{K}(0)] \tag{11-327}$$

Then $\mathbf{\Pi}$ must be the unique positive-semidefinite solution to

$$[\mathbf{A}' - \mathbf{K}(0)\mathbf{B}\mathbf{R}^{-1}\mathbf{B}']\mathbf{\Pi} + \mathbf{\Pi}[\mathbf{A} - \mathbf{B}\mathbf{R}^{-1}\mathbf{B}'\mathbf{K}(0)] + \mathbf{K}(0)\mathbf{B}\mathbf{R}^{-1}\mathbf{B}'\mathbf{K}(0) = \mathbf{0} \tag{11-328}$$

Equation (11-294) is reconditioned to read

$$[\mathbf{A}' - \mathbf{K}(0)\mathbf{B}\mathbf{R}^{-1}\mathbf{B}']\mathbf{K}(0) + \mathbf{K}(0)[\mathbf{A} - \mathbf{B}\mathbf{R}^{-1}\mathbf{B}'\mathbf{K}(0)] + \mathbf{K}(0)\mathbf{B}\mathbf{R}^{-1}\mathbf{B}'\mathbf{K}(0) + \mathbf{Q} = \mathbf{0} \tag{11-329}$$

Subtraction of Eq. (11-328) from Eq. (11-329) yields

$$[\mathbf{A}' - \mathbf{K}(0)\mathbf{B}\mathbf{R}^{-1}\mathbf{B}'][\mathbf{K}(0) - \mathbf{\Pi}] + [\mathbf{K}(0) - \mathbf{\Pi}][\mathbf{A} - \mathbf{B}\mathbf{R}^{-1}\mathbf{B}'\mathbf{K}(0)] + \mathbf{Q} = \mathbf{0} \tag{11-330}$$

The unique positive-semidefinite solution of Eq. (11-330), $\mathbf{K}(0) - \mathbf{\Pi}$, is enough to determine $\partial^2\mathbf{K}(T)/\partial T^2$ from Eq. (11-327).

Sampling Period Sensitivity Study of the Feedback Gain Matrix

The best approximation of $\mathbf{K}(T)$ given by a finite number (two) of terms of Eq. (11-298) can be substituted in Eq. (11-296) to give an approximation to the optimal feedback matrix $\mathbf{G}(T)$. For the second-order approximation, $\mathbf{G}(T)$ is given by

$$\mathbf{G}(T) = \left[\frac{\hat{\mathbf{R}} + \mathbf{\theta}'\left[\mathbf{K}(0) + \dfrac{\partial^2\mathbf{K}(T)}{\partial T^2}\dfrac{T^2}{2}\right]\mathbf{\theta}}{T}\right]^{-1} \left[\frac{\mathbf{M}' + \mathbf{\theta}'\left[\mathbf{K}(0) + \dfrac{\partial^2\mathbf{K}(T)}{\partial T^2}\dfrac{T^2}{2}\right]\mathbf{\phi}}{T}\right] \tag{11-331}$$

As an alternative, we can expand $\mathbf{G}(T)$ into a Taylor series about $T = 0$; that is

$$\mathbf{G}(T) = \mathbf{G}(0) + \sum_{i=1}^{\infty} \frac{\partial^i\mathbf{G}(T)}{\partial T^i}\frac{T^i}{i!} \tag{11-332}$$

where $\mathbf{G}(0)$ is given by Eq. (11-297), and the first and second derivatives of $\mathbf{G}(T)$ are given by Eqs. (11-316) and (11-325), respectively.

Example 11-9

Consider that the state equations of a second-order process are given by

$$\dot{\mathbf{x}}(t) = \mathbf{A}\mathbf{x}(t) + \mathbf{B}\mathbf{u}(t) \tag{11-333}$$

where

$$\mathbf{A} = \begin{bmatrix} 0 & 0 \\ 0 & 0 \end{bmatrix} \qquad \mathbf{B} = \begin{bmatrix} 1 & 0 \\ 0 & 1 \end{bmatrix} \tag{11-334}$$

The problem is to find the optimal sampled-data control $\mathbf{u}(t) = \mathbf{u}(kT)$, $k = 0, 1, 2, \ldots$, such that the performance index

$$J = \frac{1}{2}\int_0^\infty [[x_1(t) - x_2(t)]^2 + x_1^2(t) + x_2^2(t) + u_1^2(t) + u_2^2(t)]\, dt \tag{11-335}$$

is minimized.

According to the notation used in Eq. (11-122), the following weighting matrices are defined:

$$\mathbf{Q} = \begin{bmatrix} 2 & -1 \\ -1 & 2 \end{bmatrix} \qquad \mathbf{R} = \begin{bmatrix} 1 & 0 \\ 0 & 1 \end{bmatrix} \tag{11-336}$$

The sampling period of the system is not specified at the outset since its effect on the optimal linear digital regulator is to be investigated.

Since the process in Eq. (11-333) is controllable (as well as stabilizable), and the pair $[\mathbf{A}, \mathbf{D}]$ is observable, where $\mathbf{D}\mathbf{D}' = \mathbf{Q}$, the linear continuous-data regulator problem when $T = 0$ has an asymptotically stable solution. The coefficient matrix of the sampled-data process is $\boldsymbol{\phi}(T) = e^{\mathbf{A}T} = \mathbf{I}$, and

$$\boldsymbol{\theta}(T) = \int_0^T \boldsymbol{\phi}(\lambda)\mathbf{B}\, d\lambda = \begin{bmatrix} T & 0 \\ 0 & T \end{bmatrix} \tag{11-337}$$

Therefore, the sampled-data system

$$\mathbf{x}[(k+1)T] = \boldsymbol{\phi}(T)\mathbf{x}(kT) + \boldsymbol{\theta}(T)\mathbf{u}(kT) \tag{11-338}$$

is completely controllable, and since $\hat{\mathbf{R}}$ as given by Eq. (11-132) is nonsingular, the pair $[\boldsymbol{\phi}, \mathbf{D}]$ is observable for any $\mathbf{D}\mathbf{D}' = \mathbf{Q}$. Thus, the system in Eq. (11-338) has an asymptotically stable solution for the optimal linear digital regulator problem.

Ordinarily, the optimal linear digital regulator problem can be solved by formulating the discrete Riccati equation of Eq. (11-149) or Eq. (11-153) and solving for the Riccati gain $\mathbf{K}(T)$; then the optimal control is given by Eq. (11-175). In this case since the sampling period T is not specified, it would be difficult to solve the Riccati equation whether the recursive method or the eigenvalue-eigenvector method is used.

Let us solve the problem by the sampling sensitivity method. Since the matrices $\mathbf{K}(0)$ and $\mathbf{G}(0)$ are required, we must first solve the optimal linear continuous-data regulator problem. The Riccati equation for the continuous-data regulator is given in Eq. (11-294), and upon substitution of the system parameters, it is simplified to

$$\mathbf{K}^2(0) = \mathbf{Q} = \begin{bmatrix} 2 & -1 \\ -1 & 2 \end{bmatrix} \tag{11-339}$$

Thus,

$$\mathbf{K}(0) = \frac{1}{2}\begin{bmatrix} 1+\sqrt{3} & 1-\sqrt{3} \\ 1-\sqrt{3} & 1+\sqrt{3} \end{bmatrix} \tag{11-340}$$

The optimal feedback gain is determined by using Eq. (11-297), which leads to

$$\mathbf{G}(0) = \mathbf{R}^{-1}\mathbf{B}'\mathbf{K}(0) = \mathbf{K}(0) \tag{11-341}$$

The first-order sampling period sensitivity of $\mathbf{K}(T)$ is always zero. The second-order sensitivity $\partial^2\mathbf{K}(T)/\partial T^2$ is obtained by first solving for $\mathbf{\Pi}$ from Eq. (11-330). Equation (11-330) becomes

$$-2\mathbf{K}^2(0) + 2\mathbf{K}(0)\mathbf{\Pi} + \mathbf{Q} = \mathbf{0} \tag{11-342}$$

from which we have

$$\mathbf{\Pi} = \frac{1}{4}\begin{bmatrix} 1+\sqrt{3} & 1-\sqrt{3} \\ 1-\sqrt{3} & 1+\sqrt{3} \end{bmatrix} \tag{11-343}$$

Substitution of $\mathbf{\Pi}$ into Eq. (11-327) gives

$$\frac{\partial^2\mathbf{K}(T)}{\partial T^2} = \frac{1}{6}\mathbf{K}(0)\mathbf{\Pi}\mathbf{K}(0) = \frac{1}{24}\begin{bmatrix} 1+3\sqrt{3} & 1-3\sqrt{3} \\ 1-3\sqrt{3} & 1+3\sqrt{3} \end{bmatrix} \tag{11-344}$$

Thus, the second-order approximation of $\mathbf{K}(T)$ is

$$\mathbf{K}(T) \cong \mathbf{K}(0) + \frac{T^2}{2}\frac{\partial^2\mathbf{K}(T)}{\partial T^2} = \begin{bmatrix} e_1+e_2 & e_1-e_2 \\ e_1-e_2 & e_1+e_2 \end{bmatrix} \tag{11-345}$$

where

$$e_1 = \frac{1}{2} + \frac{T^2}{48} \tag{11-346}$$

$$e_2 = \frac{\sqrt{3}}{2} + \frac{\sqrt{3}\,T^2}{16} \tag{11-347}$$

We can show that, by solving the Riccati equation in Eq. (11-293), the exact Riccati gain is found to be

$$\mathbf{K}(T) = \begin{bmatrix} d_1 + d_2 & d_1 - d_2 \\ d_1 - d_2 & d_1 + d_2 \end{bmatrix} \tag{11-348}$$

where

$$d_1 = \frac{1}{2}\left[1 + \frac{T^2}{12}\right]^{1/2} \tag{11-349}$$

$$d_2 = \frac{\sqrt{3}}{2} + \frac{\sqrt{3}T^2}{8} \tag{11-350}$$

Substituting the approximation for $\mathbf{K}(T)$ in Eq. (11-345) into Eq. (11-296), $\mathbf{G}(T)$ is approximated by (after lengthy matrix manipulation)

$$\mathbf{G}(T) \cong \frac{1}{2}\begin{bmatrix} f_1 + f_2 & f_1 - f_2 \\ f_1 - f_2 & f_1 + f_2 \end{bmatrix} \tag{11-351}$$

where

$$f_1 = \frac{24 + 12T + T^2}{24 + 24T + 8T^2 + T^3} \tag{11-352}$$

$$f_2 = \frac{8\sqrt{3} + 12T + \sqrt{3}T^2}{8 + 8\sqrt{3}T + 8T^2 + \sqrt{3}T^3} \tag{11-353}$$

As an alternative we can approximate $\mathbf{G}(T)$ by the first three terms of the series expansion in Eq. (11-332). From Eq. (11-341), we have found that $\mathbf{G}(0) = \mathbf{K}(0)$. The first-order sensitivity of $\mathbf{G}(T)$ is given by Eq. (11-313); we have

$$\frac{\partial \mathbf{G}(T)}{\partial T} = \frac{1}{2}\begin{bmatrix} -2 & 1 \\ 1 & -2 \end{bmatrix} \tag{11-354}$$

The second-order sensitivity of $\mathbf{G}(T)$ is determined from Eq. (11-325),

$$\frac{\partial^2 \mathbf{G}(T)}{\partial T^2} = \frac{5}{24}\begin{bmatrix} 1 + 3\sqrt{3} & 1 - 3\sqrt{3} \\ 1 - 3\sqrt{3} & 1 + 3\sqrt{3} \end{bmatrix} \tag{11-355}$$

Thus, the series approximation of $\mathbf{G}(T)$ gives

$$\mathbf{G}(T) \cong \mathbf{G}(0) + T\frac{\partial \mathbf{G}(T)}{\partial T} + \frac{T^2}{2!}\frac{\partial^2 \mathbf{G}(T)}{\partial T^2}$$

$$= \begin{bmatrix} \frac{1+\sqrt{3}}{2} + T + \frac{5T^2}{48}(1 + 3\sqrt{3}) & \frac{1-\sqrt{3}}{2} + \frac{T}{2} + \frac{5T^2}{48}(1 - 3\sqrt{3}) \\ \frac{1-\sqrt{3}}{2} + \frac{T}{2} + \frac{5T^2}{48}(1 - 3\sqrt{3}) & \frac{1+\sqrt{3}}{2} + T + \frac{5T^2}{48}(1 + 3\sqrt{3}) \end{bmatrix} \tag{11-356}$$

The exact feedback gain matrix is obtained by substituting the exact expression of $\mathbf{K}(T)$ in Eq. (11-348) into Eq. (11-296), and the result is

$$\mathbf{G}(T) = \frac{1}{2}\begin{bmatrix} g_1 + g_2 & g_1 - g_2 \\ g_1 - g_2 & g_1 + g_2 \end{bmatrix} \tag{11-357}$$

where

$$g_1 = \frac{\left[1 + \dfrac{T^2}{12}\right]^{1/2} + \dfrac{T}{2}}{1 + T\left[1 + \dfrac{T^2}{12}\right]^{1/2} + \dfrac{T^2}{3}} \tag{11-358}$$

$$g_2 = \frac{\sqrt{3}\left[1 + \dfrac{T^2}{4}\right]^{1/2} + \dfrac{3T}{2}}{1 + \sqrt{3}T\left[1 + \dfrac{T^2}{4}\right]^{1/2} + T^2} \tag{11-359}$$

Sampling Period Sensitivity and Digital Redesign

The digital redesign problem studied in Chapter 9 was motivated by the development of the sampling period sensitivity of the Riccati gain and the optimal feedback gain. In Chapter 9 the digital redesign problem is solved by means of a point-by-point comparison of the state responses of the continuous-data system and the digital system at the sampling instants. The development given in the preceding section leads directly to another solution to the digital redesign problem using the sampling period sensitivity.

With reference to the notation used in this chapter, the digital redesign problem may be stated as follows. Given the continuous-data system of Eq. (11-120) which has the optimal control

$$\mathbf{u}(t) = -\mathbf{G}(0)\mathbf{x}(t) \tag{11-360}$$

that minimizes the performance index in Eq. (11-122) for $t_f = \infty$. The optimal feedback gain $\mathbf{G}(0)$ is given by Eq. (11-297); that is,

$$\mathbf{G}(0) = \mathbf{R}^{-1}\mathbf{B}'\mathbf{K}(0) \tag{11-361}$$

where $\mathbf{K}(0)$ is the positive-semidefinite solution of the Riccati equation in Eq. (11-294).

Let the continuous-data system in Eq. (11-120) be approximated by a digital system in such a way that the control $\mathbf{u}(t)$ is derived through sample-and-hold units. Thus,

$$\mathbf{u}(t) = \mathbf{u}(kT) \qquad kT \le t < (k+1)T \tag{11-362}$$

The problem is to find the optimal state-feedback control

$$\mathbf{u}(kT) = -\mathbf{G}(T)\mathbf{x}(kT) \tag{11-363}$$

such that the discretized performance index of Eq. (11-126) is minimized for $N = \infty$.

Summarizing the results obtained in the preceding section, the feedback gain of the digital system is expanded into a Taylor series about $T = 0$, and by taking only the first three terms, we have

$$\mathbf{G}(T) = \mathbf{G}(0) + T\frac{\partial \mathbf{G}(T)}{\partial T} + \frac{T^2}{2!}\frac{\partial^2 \mathbf{G}(T)}{\partial T^2} \tag{11-364}$$

where $\mathbf{G}(0)$ is given by Eq. (11-361) and from Eq. (11-313),

$$\frac{\partial \mathbf{G}(T)}{\partial T} = \frac{1}{2}\mathbf{G}(0)[\mathbf{A} - \mathbf{B}\mathbf{G}(0)] \tag{11-365}$$

Equation (11-325) gives

$$\frac{\partial^2 \mathbf{G}(T)}{\partial T^2} = \mathbf{R}^{-1}\mathbf{B}'\frac{\partial^2 \mathbf{K}(T)}{\partial T^2} + \frac{1}{3}\mathbf{G}(0)[\mathbf{A} - \mathbf{B}\mathbf{G}(0)]^2 \tag{11-366}$$

where $\partial^2\mathbf{K}(T)/\partial T^2$ is solved from Eq. (11-326).

Comparing these results with those in Chapter 9, it is interesting to note that the first two terms of the Taylor series expansion of $\mathbf{G}(T)$ by the two methods are identical.

11-9 DIGITAL STATE OBSERVER

A significant portion of the optimal control theory of digital control systems relies on the feeding back of the state variables to form the control. Unfortunately, in practice, not all the state variables are accessible, and, in general, only the outputs of the system are measurable. Therefore, when feedback from all the state variables is required in a given design, and not all the state variables are accessible, it is necessary to "observe" the states from information contained in the output as well as the input variables. The subsystem that performs the observation of the state variables based on information received from the measurements of the input and the output is called a *state observer*, or simply an *observer*. Figure 11-4 shows the block diagram of a digital control system which has a state observer. The observed state vector $\mathbf{x}_e(k)$ is fed to the feedback gain $\mathbf{G}$ which in turn generates the control signal $\mathbf{u}(k)$. From Fig. 11-4 the control signal is expressed as

$$\mathbf{u}(k) = \mathbf{E}\mathbf{r}(k) - \mathbf{G}\mathbf{x}_e(k) \tag{11-367}$$

We must first establish the condition under which an observer exists. The following theorem indicates that the design of the digital state observer is closely related to the observability criterion.

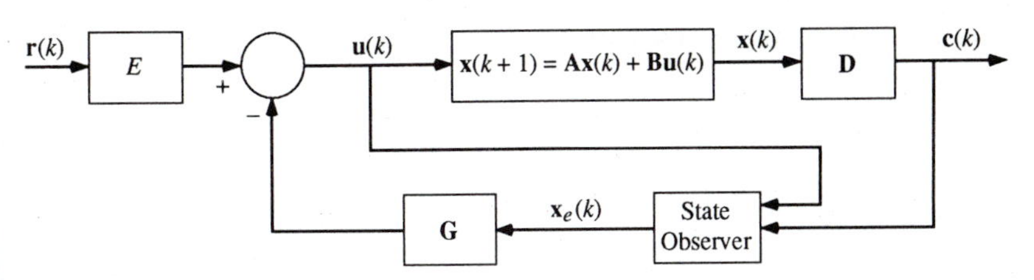

Figure 11-4. Digital control system with state observer.

□ Theorem 11-6

Given a linear digital system that is described by the dynamic equations

$$\mathbf{x}(k+1) = \mathbf{A}\mathbf{x}(k) + \mathbf{B}\mathbf{u}(k) \tag{11-368}$$

$$\mathbf{c}(k) = \mathbf{D}\mathbf{x}(k) \tag{11-369}$$

where $\mathbf{x}(k)$ is an n-vector, $\mathbf{u}(k)$ is a p-vector, and $\mathbf{c}(k)$ is a q-vector. It is assumed that the matrix $\mathbf{A}$ is nonsingular. The state vector $\mathbf{x}(k)$ may be constructed from linear combinations of the output $\mathbf{c}(k)$, input $\mathbf{u}(k)$, and the past values of these variables, if the digital system is completely observable. ■

Proof: Equation (11-368) leads to

$$\mathbf{x}(k-1) = \mathbf{A}^{-1}\mathbf{x}(k) - \mathbf{A}^{-1}\mathbf{B}\mathbf{u}(k-1) \qquad k \geq 1 \tag{11-370}$$

In general,

$$\mathbf{x}(k-n) = \mathbf{A}^{-1}\mathbf{x}(k-n+1) - \mathbf{A}^{-1}\mathbf{B}\mathbf{u}(k-n) \qquad k \geq n \tag{11-371}$$

From Eq. (11-369) we write

$$\mathbf{c}(k-1) = \mathbf{D}\mathbf{x}(k-1) \qquad k \geq 1 \tag{11-372}$$

Substitution of Eq. (11-370) into Eq. (11-372) yields

$$\mathbf{c}(k-1) = \mathbf{D}\mathbf{A}^{-1}\mathbf{x}(k) - \mathbf{D}\mathbf{A}^{-1}\mathbf{B}\mathbf{u}(k-1) \qquad k \geq 1 \tag{11-373}$$

Similarly,

$$\begin{aligned}\mathbf{c}(k-2) &= \mathbf{D}\mathbf{x}(k-2)\\ &= \mathbf{D}\mathbf{A}^{-2}\mathbf{x}(k) - \mathbf{D}\mathbf{A}^{-2}\mathbf{B}\mathbf{u}(k-1) - \mathbf{D}\mathbf{A}^{-1}\mathbf{B}\mathbf{u}(k-2)\end{aligned} \tag{11-374}$$

Continuing the process, we have

$$\mathbf{c}(k-N) = \mathbf{D}\mathbf{A}^{-N}\mathbf{x}(k) - \sum_{i=1}^{N} \mathbf{D}\mathbf{A}^{-N+i-1}\mathbf{B}\mathbf{u}(k-i) \qquad k \geq N \tag{11-375}$$

The preceding equations are written in matrix form as follows:

$$\begin{bmatrix} \mathbf{c}(k-1) \\ \mathbf{c}(k-2) \\ \vdots \\ \mathbf{c}(k-N) \end{bmatrix} = \begin{bmatrix} \mathbf{DA}^{-1} \\ \mathbf{DA}^{-2} \\ \vdots \\ \mathbf{DA}^{-N} \end{bmatrix} \mathbf{x}(k)$$

$$- \begin{bmatrix} \mathbf{DA}^{-1}\mathbf{B} & 0 & 0 & \cdots & 0 \\ \mathbf{DA}^{-2}\mathbf{B} & \mathbf{DA}^{-1}\mathbf{B} & 0 & \cdots & 0 \\ \mathbf{DA}^{-3}\mathbf{B} & \mathbf{DA}^{-2}\mathbf{B} & \mathbf{DA}^{-1}\mathbf{B} & \cdots & 0 \\ \vdots & \vdots & \vdots & & \vdots \\ \mathbf{DA}^{-N}\mathbf{B} & \mathbf{DA}^{-N+1}\mathbf{B} & \mathbf{DA}^{-N+2}\mathbf{B} & \cdots & \mathbf{DA}^{-1}\mathbf{B} \end{bmatrix} \begin{bmatrix} \mathbf{u}(k-1) \\ \mathbf{u}(k-2) \\ \mathbf{u}(k-3) \\ \vdots \\ \mathbf{u}(k-N) \end{bmatrix} \quad \text{(11-376)}$$

This matrix equation represents Nq equations with n unknowns in the state vector $\mathbf{x}(k)$. With $Nq \geq n$, given the control inputs and the outputs in Eq. (11-376), $\mathbf{x}(k)$ can be determined if the matrix

$$[\mathbf{DA}^{-1} \quad \mathbf{DA}^{-2} \quad \cdots \quad \mathbf{DA}^{-N}]' \qquad (Nq \times n) \quad \text{(11-377)}$$

is of rank n. It is apparent that the rank condition on the matrix in Eq. (11-377) is the observability criterion on the pair $[\mathbf{A}, \mathbf{D}]$ for $N = n$.

Under certain conditions of the state vector $\mathbf{x}(k)$ can be uniquely solved from Eq. (11-377). If $Nq = n$, the matrix in Eq. (11-377) is square, and if it is nonsingular, $\mathbf{x}(k)$ can be determined for $k \geq N$ by obtaining N past output measurements and N past input measurements. ●

11-9-1 Design of the Full-Order State Observer

In general, the state observer shown in Fig. 11-4 is to be designed so that the observed state $\mathbf{x}_e(k)$ will be as close as possible to the actual state $\mathbf{x}(k)$. There are many ways of designing a digital state observer, and generally there is more than one way of judging the closeness of $\mathbf{x}_e(k)$ to $\mathbf{x}(k)$. Intuitively, the state observer should have the same state equations as the original system. However, the observer should have a configuration with $\mathbf{u}(k)$ and $\mathbf{c}(k)$ as inputs and should have the capability of minimizing the error between $\mathbf{x}(k)$ and $\mathbf{x}_e(k)$ automatically.

The digital state observer design presented in the following is a parallel of the observer design for continuous-data systems.

Since $\mathbf{x}(k)$ cannot be measured directly, we cannot compare $\mathbf{x}_e(k)$ with $\mathbf{x}(k)$. As an alternative, $\mathbf{c}_e(k)$ is compared with $\mathbf{c}(k)$, where

$$\mathbf{c}_e(k) = \mathbf{D}\mathbf{x}_e(k) \quad \text{(11-378)}$$

Based on the above discussion, a logical configuration for a digital state observer is shown in Fig. 11-5. The observer is formulated as a feedback control system with $\mathbf{G}_e$ as the feedback gain matrix. The design objective is to select $\mathbf{G}_e$ such that $\mathbf{c}_e(k)$ will approach $\mathbf{c}(k)$ as fast as possible.

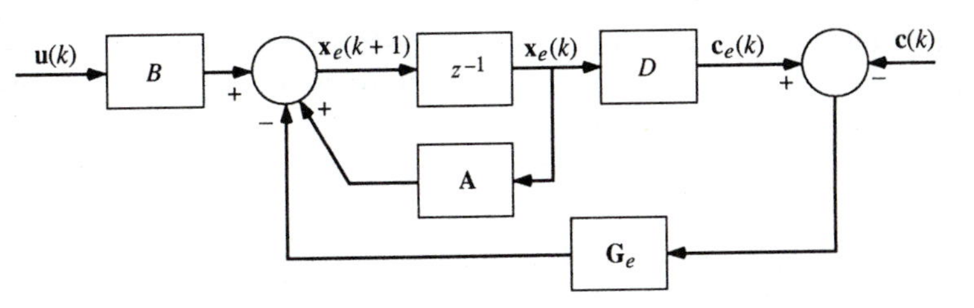

Figure 11-5. A digital state observer with feedback.

The state equation of the closed-loop observer is

$$\mathbf{x}_e(k+1) = (\mathbf{A} - \mathbf{G}_e\mathbf{D})\mathbf{x}_e(k) + \mathbf{B}\mathbf{u}(k) + \mathbf{G}_e\mathbf{c}(k) \tag{11-379}$$

where the matrices **A**, **B**, and **D** are the same as those in Eqs. (11-368) and (11-369) and $\mathbf{G}_e$ is an $n \times q$ feedback matrix. When $\mathbf{c}_e(k)$ is equal to $\mathbf{c}(k)$, Eq. (11-379) becomes

$$\mathbf{x}_e(k+1) = \mathbf{A}\mathbf{x}_e(k) + \mathbf{B}\mathbf{u}(k) \tag{11-380}$$

which is identical to the state equation of the original system.

Substituting the block diagram in Fig. 11-5 in the state observer block of Fig. 11-4, we have the combined system shown in Fig. 11-6.

Since $\mathbf{c}(k)$ and $\mathbf{x}(k)$ are related through Eq. (11-369), Eq. (11-379) is written

$$\mathbf{x}_e(k+1) = \mathbf{A}\mathbf{x}_e(k) + \mathbf{B}\mathbf{u}(k) + \mathbf{G}_e\mathbf{D}[\mathbf{x}(k) - \mathbf{x}_e(k)] \tag{11-381}$$

The significance of this equation is that if the initial states $\mathbf{x}(0)$ and $\mathbf{x}_e(0)$ are identical, the equation is identical to Eq. (11-380), and the response of the

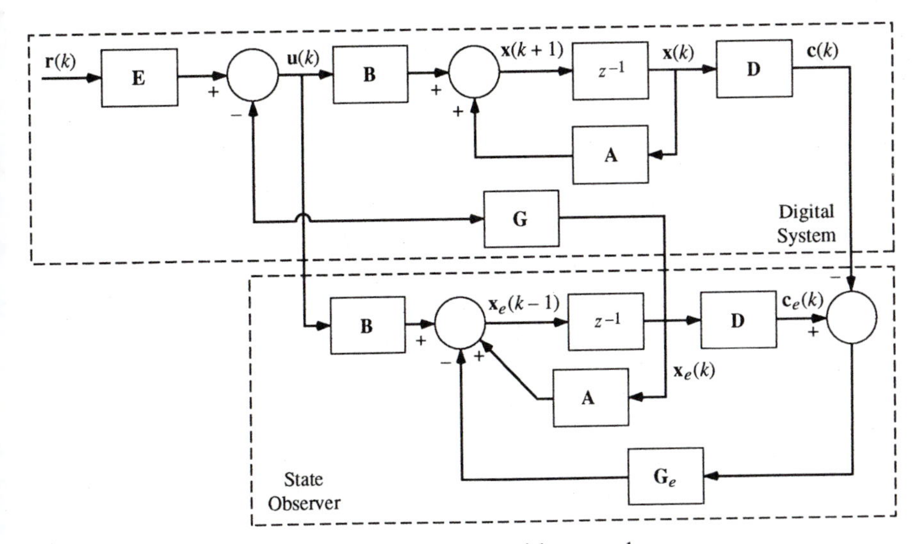

Figure 11-6. Digital control system with state observer.

observer will be identical to that of the original system. Therefore, the design of the observer is significant only if the initial conditions of $\mathbf{x}(k)$ and $\mathbf{x}_e(k)$ are different.

Subtracting Eq. (11-381) from Eq. (11-368) leads to the following equation:

$$\mathbf{x}(k+1) - \mathbf{x}_e(k+1) = (\mathbf{A} - \mathbf{G}_e\mathbf{D})[\mathbf{x}(k) - \mathbf{x}_e(k)] \tag{11-382}$$

which may be regarded as the homogeneous difference state equation of a linear digital system with coefficient matrix $\mathbf{A} - \mathbf{G}_e\mathbf{D}$. One way of achieving a rapid convergence of $\mathbf{x}_e(k)$ to $\mathbf{x}(k)$ would be to design $\mathbf{G}_e$ in such a way that the eigenvalues of $\mathbf{A} - \mathbf{G}_e\mathbf{D}$ are appropriately placed in the z-plane. Using the eigenvalue-assignment technique discussed in Chapter 5, the elements of $\mathbf{G}_e$ can be so selected that the natural response of the system in Eq. (11-382) decays to zero as quickly as possible.

Since the eigenvalues of $\mathbf{A} - \mathbf{G}_e\mathbf{D}$ and $(\mathbf{A} - \mathbf{G}_e\mathbf{D})' = \mathbf{A}' - \mathbf{D}'\mathbf{G}_e'$ are identical, it follows from the discussions in Sec. 5-13 that the requirement for arbitrary assignment of the eigenvalues of $\mathbf{A} - \mathbf{G}_e\mathbf{D}$ is that the pair $[\mathbf{A}', \mathbf{D}']$ be completely controllable. Since the controllability of $[\mathbf{A}', \mathbf{D}']$ is equivalent to the complete observability of $[\mathbf{A}, \mathbf{D}]$, satisfying this requirement would assure not only the existence of a state observer but also the solution of arbitrary eigenvalue assignment.

The following example will illustrate the design of a state observer by eigenvalue assignment.

Example 11-10

Consider the digital process in Example 11-6, with the state equations described by

$$\mathbf{x}(k+1) = \mathbf{A}\mathbf{x}(k) + \mathbf{B}u(k) \tag{11-383}$$

where

$$\mathbf{A} = \begin{bmatrix} 0 & 1 \\ -1 & 1 \end{bmatrix} \qquad \mathbf{B} = \begin{bmatrix} 0 \\ 1 \end{bmatrix}$$

Let the output equation be

$$c(k) = \mathbf{D}\mathbf{x}(k) \tag{11-384}$$

where

$$\mathbf{D} = [2 \quad 0] \tag{11-385}$$

A digital observer is to be designed which will observe the states $x_1(k)$ and $x_2(k)$ from the output $c(k)$.

The digital observer has the block diagram shown in Fig. 11-5. The characteristic equation of the observer is

$$|\lambda\mathbf{I} - \mathbf{A} + \mathbf{G}_e\mathbf{D}| = 0 \tag{11-386}$$

or

$$\lambda^2 + (2g_{e1} - 1)\lambda + 1 + 2g_{e2} - 2g_{e1} = 0 \tag{11-387}$$

where g_{e1} and g_{e2} are the elements of the 2×1 feedback matrix $\mathbf{G}_e$. We can attempt to design the observer to have a deadbeat response so that $\mathbf{x}_e(k)$ will reach $\mathbf{x}(k)$ in one sampling period. For a deadbeat response the characteristic equation should be of the form

$$\lambda^2 = 0 \tag{11-388}$$

Thus, from Eq. (11-387), $g_{e1} = 0.5$ and $g_{e2} = 0$. The corresponding coefficient matrix for the closed-loop observer is now

$$\mathbf{A} - \mathbf{G}_e\mathbf{D} = \begin{bmatrix} -1 & 1 \\ -1 & 1 \end{bmatrix} \tag{11-389}$$

For $u(k) = 0$, the state equations of the observer are written

$$\mathbf{x}_e(k+1) = \begin{bmatrix} -1 & 1 \\ -1 & 1 \end{bmatrix}\mathbf{x}_e(k) + \begin{bmatrix} 1 & 0 \\ 0 & 0 \end{bmatrix}\mathbf{x}(k) \tag{11-390}$$

Let us arbitrarily assume that the initial states of the system and the observer are

$$\mathbf{x}(0) = \begin{bmatrix} 1 \\ 0 \end{bmatrix} \quad \text{and} \quad \mathbf{x}_e(0) = \begin{bmatrix} 0.5 \\ 0 \end{bmatrix} \tag{11-391}$$

respectively. Equations (11-383) and (11-390) are now solved, starting with the given initial states. The true states, $x_1(k)$ and $x_2(k)$, and the observed states, $x_{e1}(k)$ and $x_{e2}(k)$, are sketched as shown in Fig. 11-7. The observed states reach the actual states in at most two sampling periods.

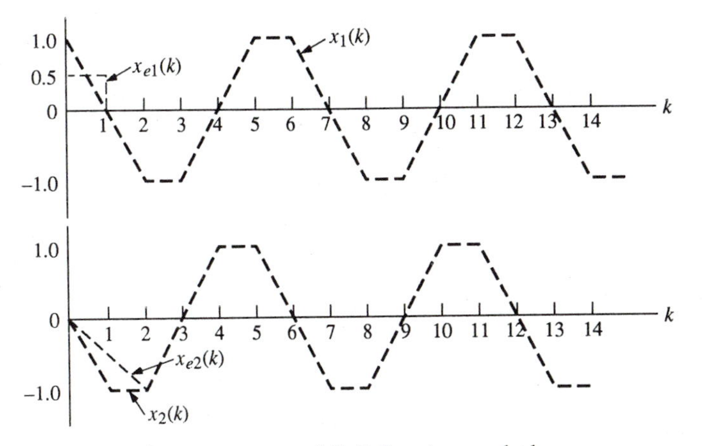

Figure 11-7. State responses of digital system and observer.

Closed-Loop Control with Observer

The digital observer design carried out thus far is for an open-loop control system; that is, $\mathbf{G} = 0$ in Fig. 11-6. For the general case shown in Fig. 11-6 the state equations are written as follows:

$$\mathbf{x}(k+1) = \mathbf{A}\mathbf{x}(k) - \mathbf{B}\mathbf{G}\mathbf{x}_e(k) + \mathbf{B}\mathbf{E}\mathbf{r}(k) \tag{11-392}$$

$$\mathbf{x}_e(k+1) = (\mathbf{A} - \mathbf{B}\mathbf{G} - \mathbf{G}_e\mathbf{D})\mathbf{x}_e(k) + \mathbf{G}_e\mathbf{D}\mathbf{x}(k) + \mathbf{B}\mathbf{E}\mathbf{r}(k) \tag{11-393}$$

We can investigate the effects of the initial states of the system and the observer and the mutual effects of these two systems by taking the z-transforms on both sides of Eqs. (11-392) and (11-393). After rearranging the terms, the z-transform equations are

$$(z\mathbf{I} - \mathbf{A})\mathbf{X}(z) = z\mathbf{x}(0) + \mathbf{B}\mathbf{E}\mathbf{R}(z) - \mathbf{B}\mathbf{G}\mathbf{X}_e(z) \tag{11-394}$$

$$(z\mathbf{I} - \mathbf{A} + \mathbf{B}\mathbf{G} + \mathbf{G}_e\mathbf{D})\mathbf{X}_e(z) = z\mathbf{x}_e(0) + \mathbf{G}_e\mathbf{D}\mathbf{X}(z) + \mathbf{B}\mathbf{E}\mathbf{R}(z) \tag{11-395}$$

When $\mathbf{x}(0) = \mathbf{x}_e(0)$, subtracting Eq. (11-395) from Eq. (11-394), we get

$$(z\mathbf{I} - \mathbf{A} + \mathbf{G}_e\mathbf{D})\mathbf{X}(z) = (z\mathbf{I} - \mathbf{A} + \mathbf{G}_e\mathbf{D})\mathbf{X}_e(z) \tag{11-396}$$

Thus,

$$\mathbf{x}(k) = \mathbf{x}_e(k) \tag{11-397}$$

and Eq. (11-394) becomes

$$(z\mathbf{I} - \mathbf{A} + \mathbf{B}\mathbf{G})\mathbf{X}(z) = z\mathbf{x}(0) + \mathbf{B}\mathbf{E}\mathbf{R}(z) \tag{11-398}$$

This result shows that when $\mathbf{x}(0) = \mathbf{x}_e(0)$, the dynamics of the original system are independent of that of the observer. However, in general, the transient response of the original system will be affected by the observer when $\mathbf{x}(0) \neq \mathbf{x}_e(0)$.

Subtracting Eq. (11-393) from Eq. (11-392), we have

$$\mathbf{x}(k+1) - \mathbf{x}_e(k+1) = (\mathbf{A} - \mathbf{G}_e\mathbf{D})[\mathbf{x}(k) - \mathbf{x}_e(k)] \tag{11-399}$$

which shows that given $\mathbf{A}$ and $\mathbf{D}$, the natural response of $\mathbf{x}(k) - \mathbf{x}_e(k)$ depends only upon $\mathbf{G}_e$. Therefore, the design of the observer for a system with a state feedback $\mathbf{G}$ can still be accomplished by placing the eigenvalues of $\mathbf{A} - \mathbf{G}_e\mathbf{D}$. The only difference is that, in general, when $\mathbf{x}(0) \neq \mathbf{x}_e(0)$, the response of the system $\mathbf{x}(k)$ will be affected by the dynamics of the observer, due to the feedback of $\mathbf{x}_e(k)$ through $\mathbf{G}$. The following example will illustrate the performance of the system and its observer in Example 11-10 when there is a state-feedback matrix $\mathbf{G}$.

Example 11-11

For the digital process in Example 11-10, let us adopt the optimal feedback gain designed in Example 11-6,

$$\mathbf{G} = [-0.654 \quad 0.486] \tag{11-400}$$

As designed in Example 11-10, the feedback gain matrix of the digital observer for the response $\mathbf{x}(k) - \mathbf{x}_e(k)$ to be deadbeat is

$$\mathbf{G}_e = \begin{bmatrix} 0.5 \\ 0 \end{bmatrix} \tag{11-401}$$

Thus, for $\mathbf{r}(k) = 0$, the state equations of the digital system with feedback and the observer in Eqs. (11-392) and (11-393) are written

$$\begin{bmatrix} \mathbf{x}(k+1) \\ \hline \mathbf{x}_e(k+1) \end{bmatrix} = \left[\begin{array}{c:c} \mathbf{A} & -\mathbf{BG} \\ \hdashline \mathbf{G}_e\mathbf{D} & \mathbf{A} - \mathbf{BG} - \mathbf{G}_e\mathbf{D} \end{array}\right] \begin{bmatrix} \mathbf{x}(k) \\ \hline \mathbf{x}_e(k) \end{bmatrix}$$

$$= \left[\begin{array}{cc:cc} 0 & 1 & 0 & 0 \\ -1 & 1 & 0.654 & -0.486 \\ \hdashline 1 & 0 & -1 & 1 \\ 0 & 0 & -0.346 & 0.514 \end{array}\right] \begin{bmatrix} \mathbf{x}(k) \\ \hline \mathbf{x}_e(k) \end{bmatrix} \tag{11-402}$$

Figure 11-8 shows the response of $x_1(k)$ and $x_2(k)$ when $\mathbf{x}(0) = \mathbf{x}_e(0)$, which is equivalent to the condition that the observer is absent. The responses of the original system with the observer are also shown in Fig. 11-8. As explained earlier, the observer dynamics *do* affect the transient responses of the system. For the initial states chosen, $\mathbf{x}_e(k)$ reaches $\mathbf{x}(k)$ after two sampling periods.

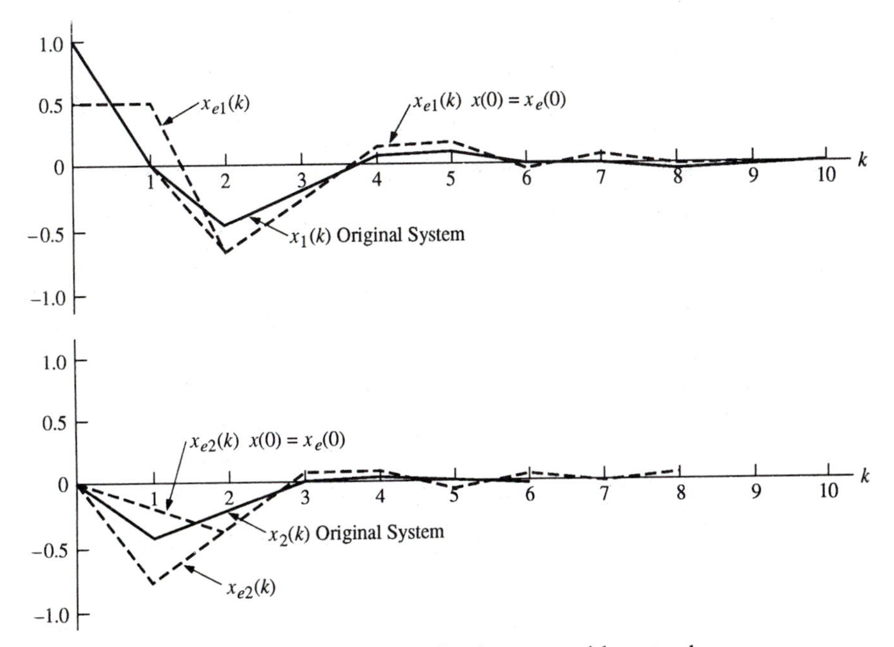

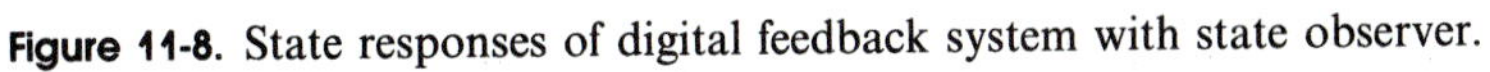
Figure 11-8. State responses of digital feedback system with state observer.

Design by Adjoint Phase-Variable Canonical Form

Since the observer design presented in the preceding section is based on the eigenvalue assignment of the closed-loop observer, the phase-variable canonical form transformation discussed in Sec. 5-13 may be utilized. We shall consider only systems with single input and output.

Let us restate the observer design problem. The states to be observed are described by the state equation

$$\mathbf{x}(k+1) = \mathbf{A}\mathbf{x}(k) + \mathbf{B}u(k) \tag{11-403}$$

where $\mathbf{x}(k)$ is an n-vector and $\mathbf{u}(k)$ is the scalar input. The control is obtained through state feedback, $u(k) = -\mathbf{G}\mathbf{x}(k)$, and the output equation is

$$c(k) = \mathbf{D}\mathbf{x}(k) \tag{11-404}$$

where $c(k)$ is the scalar output, and $\mathbf{D}$ is a $1 \times n$ matrix. The observer dynamics are described by [Eq. (11-379)]

$$\mathbf{x}_e(k+1) = (\mathbf{A} - \mathbf{G}_e\mathbf{D})\mathbf{x}_e(k) + \mathbf{B}u(k) + \mathbf{G}_e c(k) \tag{11-405}$$

The original closed-loop system is represented by

$$\mathbf{x}(k+1) = (\mathbf{A} - \mathbf{B}\mathbf{G})\mathbf{x}(k) \tag{11-406}$$

It is known that if the pair $[\mathbf{A}, \mathbf{B}]$ is completely controllable, then the eigenvalues of $\mathbf{A} - \mathbf{B}\mathbf{G}$ can be arbitrarily assigned by properly choosing the elements of the feedback matrix $\mathbf{G}$. In Sec. 5-13 we have shown that if $[\mathbf{A}, \mathbf{B}]$ is controllable, $\mathbf{A}$ and $\mathbf{B}$ can be transformed into the phase-variable canonical form of Eqs. (5-240) and (5-241). Then each of the feedback gain elements is associated only with one of the coefficients of the characteristic equation. In view of Eq. (11-405), the design of the observer by eigenvalue assignment creates a similar problem. In this case the eigenvalues of $\mathbf{A} - \mathbf{G}_e\mathbf{D}$ are to be located specifically by proper selection of the elements of the feedback matrix $\mathbf{G}_e$. Since the eigenvalues of $\mathbf{A} - \mathbf{G}_e\mathbf{D}$ are the same as those of $(\mathbf{A} - \mathbf{G}_e\mathbf{D})' = \mathbf{A}' - \mathbf{D}'\mathbf{G}_e'$, we can identify a parallel between the $\mathbf{A}$, $\mathbf{B}$, $\mathbf{G}$ system and the $\mathbf{A}'$, $\mathbf{D}'$, $\mathbf{G}_e'$ system. Notice that the prerequisite of arbitrary eigenvalue assignment for the observer is the complete controllability of the pair $[\mathbf{A}', \mathbf{D}']$ which is equivalent to the observability of $[\mathbf{A}, \mathbf{D}]$, a requirement that was established earlier in this section.

We must first transform the observer system into a so-called *adjoint phase-variable canonical form*. The reason for this modification is that the coefficient matrix in the form of $\mathbf{A}' - \mathbf{D}'\mathbf{G}_e'$ is required for isolation of the feedback matrix elements in the coefficients of the characteristic equation, and the matrix corresponds to a system with the state equation

$$\mathbf{y}(k+1) = \mathbf{A}'\mathbf{y}(k) + \mathbf{D}'\mathbf{v}(k) \tag{11-407}$$

where

$$\mathbf{v}(k) = -\mathbf{G}_e'\mathbf{y}(k) \tag{11-408}$$

$$\mathbf{y}(k+1) = (\mathbf{A}' - \mathbf{D}'\mathbf{G}_e')\mathbf{y}(k) \tag{11-409}$$

We need to transform the observer system in Eq. (11-405) into the following form:

$$\mathbf{y}_e(k+1) = (\mathbf{A}_1 - \mathbf{K}_e\mathbf{D}_1)\mathbf{y}_e(k) + \mathbf{B}_1 u(k) + \mathbf{K}_e c(k) \tag{11-410}$$

where $\mathbf{A}_1$ is $n \times n$, $\mathbf{K}_e$ is $n \times 1$, $\mathbf{D}_1$ is $1 \times n$, and $\mathbf{B}_1$ is $n \times 1$. In general, we do not have to restrict $u(k)$ to be a scalar input as $\mathbf{B}_1$ will not enter the design equations. The transformation is

$$\mathbf{y}_e(k) = \mathbf{P}\mathbf{x}_e(k) \tag{11-411}$$

where $\mathbf{P}$ is nonsingular. Then,

$$\mathbf{K}_e = \mathbf{P}\mathbf{G}_e = [k_{e1} \quad k_{e2} \quad \cdots \quad k_{en}]' \tag{11-412}$$

$$\mathbf{D}_1 = \mathbf{D}\mathbf{P}^{-1} = [0 \quad 0 \quad \cdots \quad 1] \tag{11-413}$$

$$\mathbf{B}_1 = \mathbf{P}\mathbf{B} \tag{11-414}$$

and

$$\mathbf{A}_1 = \mathbf{P}\mathbf{A}\mathbf{P}^{-1} = \begin{bmatrix} 0 & 0 & \cdots & 0 & -a_n \\ 1 & 0 & \cdots & 0 & -a_{n-1} \\ 0 & 1 & \cdots & 0 & -a_{n-2} \\ \cdots & \cdots & \cdots & \cdots & \cdots \\ 0 & 0 & \cdots & 1 & -a_1 \end{bmatrix} \tag{11-415}$$

Thus, the system with $\mathbf{A}_1$ and $\mathbf{B}_1$ is known as the adjoint phase-variable canonical form since these matrices are the transposes of the phase-variable canonical form matrices $\mathbf{A}$ and $\mathbf{B}$, respectively. Then,

$$\mathbf{A}_1 - \mathbf{K}_e\mathbf{D}_1 = \begin{bmatrix} 0 & 0 & \cdots & 0 & -a_n - k_{e1} \\ 1 & 0 & \cdots & 0 & -a_{n-1} - k_{e2} \\ 0 & 1 & \cdots & 0 & -a_{n-2} - k_{e3} \\ \cdots & \cdots & \cdots & \cdots & \cdots \\ 0 & 0 & \cdots & 1 & -a_1 - k_{en} \end{bmatrix} \tag{11-416}$$

The characteristic equation of the transformed observer in Eq. (11-410) is

$$|\lambda\mathbf{I} - \mathbf{A}_1 + \mathbf{K}_e\mathbf{D}_1| = \lambda^n + (a_1 + k_{en})\lambda^{n-1} + \cdots + (a_{n-1} + k_{e2})\lambda + (a_n + k_{e1}) = 0 \tag{11-417}$$

Therefore, the coefficients of the feedback matrix $\mathbf{K}_e$ are all isolated in the coefficients of the characteristic equation.

To determine the matrix $\mathbf{P}$, we let

$$\mathbf{P} = \begin{bmatrix} \mathbf{P}_1 \\ \mathbf{P}_2 \\ \vdots \\ \mathbf{P}_n \end{bmatrix} \tag{11-418}$$

where $\mathbf{P}_i$ is a $1 \times n$ row matrix, $i = 1, 2, \ldots, n$. Substituting Eq. (11-418) into Eq. (11-415), we have

$$\mathbf{A}_1\mathbf{P} = \mathbf{P}\mathbf{A} = \begin{bmatrix} -a_n\mathbf{P}_n \\ \mathbf{P}_1 - a_{n-1}\mathbf{P}_n \\ \mathbf{P}_2 - a_{n-2}\mathbf{P}_n \\ \vdots \\ \mathbf{P}_{n-1} - a_1\mathbf{P}_n \end{bmatrix} \tag{11-419}$$

From Eq. (11-413),

$$\begin{aligned} \mathbf{D}_1\mathbf{P} = \mathbf{D} &= [0 \quad 0 \quad \cdots \quad 1]\mathbf{P} \\ &= \mathbf{P}_n \end{aligned} \tag{11-420}$$

Equation (11-419) becomes

$$\begin{bmatrix} -a_n\mathbf{D} \\ \mathbf{P}_1 - a_{n-1}\mathbf{D} \\ \mathbf{P}_2 - a_{n-2}\mathbf{D} \\ \vdots \\ \mathbf{P}_{n-1} - a_1\mathbf{D} \end{bmatrix} = \begin{bmatrix} \mathbf{P}_1\mathbf{A} \\ \mathbf{P}_2\mathbf{A} \\ \vdots \\ \mathbf{P}_n\mathbf{A} \end{bmatrix} \tag{11-421}$$

The last equation leads to

$$\mathbf{P}_{n-1} = a_1\mathbf{D} + \mathbf{D}\mathbf{A} \tag{11-422}$$

$$\begin{aligned} \mathbf{P}_{n-2} &= a_2\mathbf{D} + \mathbf{P}_{n-1}\mathbf{A} \\ &= a_2\mathbf{D} + a_1\mathbf{D}\mathbf{A} + \mathbf{D}\mathbf{A}^2 \end{aligned} \tag{11-423}$$

$$\vdots$$

$$\mathbf{P}_2 = a_{n-2}\mathbf{D} + a_{n-3}\mathbf{D}\mathbf{A} + \cdots + a_1\mathbf{D}\mathbf{A}^{n-3} + \mathbf{D}\mathbf{A}^{n-2} \tag{11-424}$$

$$\mathbf{P}_1 = a_{n-1}\mathbf{D} + a_{n-2}\mathbf{D}\mathbf{A} + \cdots + a_1\mathbf{D}\mathbf{A}^{n-2} + \mathbf{D}\mathbf{A}^{n-1} \tag{11-425}$$

Combining Eqs. (11-420) and (11-422) through (11-425), the matrix $\mathbf{P}$ is given by

$$\mathbf{P} = \begin{bmatrix} a_{n-1} & a_{n-2} & \cdots & a_1 & 1 \\ a_{n-2} & a_{n-3} & \cdots & 1 & 0 \\ a_{n-3} & a_{n-4} & \cdots & 0 & 0 \\ \hline a_1 & 1 & & 0 & 0 \\ 1 & 0 & & 0 & 0 \end{bmatrix} \begin{bmatrix} \mathbf{D} \\ \mathbf{D}\mathbf{A} \\ \mathbf{D}\mathbf{A}^2 \\ \vdots \\ \mathbf{D}\mathbf{A}^{n-2} \\ \mathbf{D}\mathbf{A}^{n-1} \end{bmatrix} \tag{11-426}$$

where it should be emphasized that the coefficients $a_1, a_2, \ldots, a_{n-1}$ are identified from the last column of $\mathbf{A}_1$ in Eq. (11-415), but more important, these coefficients are found from the characteristic equation of $\mathbf{A}$ which has the form

$$\lambda^n + a_1\lambda^{n-1} + a_2\lambda^{n-2} + \cdots + a_{n-1}\lambda + a_n = 0 \tag{11-427}$$

The adjoint phase-variable canonical form method just described can be applied to systems with multiple outputs. Consider that in Eq. (11-404), $c(k)$ is now a q-vector, and $\mathbf{D}$ is $q \times n$. Since changing the state variables of an observable system does not destroy the observability of the system, we may transform the state equations to the following form:

$$\mathbf{y}(k+1) = \begin{bmatrix} \mathbf{A}_1 & & & 0 \\ & \mathbf{A}_2 & & \\ & & \ddots & \\ 0 & & & \mathbf{A}_q \end{bmatrix} \mathbf{y}(k) + \begin{bmatrix} \mathbf{B}_1 \\ \mathbf{B}_2 \\ \vdots \\ \mathbf{B}_q \end{bmatrix} \mathbf{u}(k) \tag{11-428}$$

where $\mathbf{A}_i$, $i = 1, 2, \ldots, q$, are all in the adjoint phase-variable canonical form of Eq. (11-415). The output equation is transformed into

$$\mathbf{c}(k) = \begin{bmatrix} \mathbf{D}_1 & & & 0 \\ & \mathbf{D}_2 & & \\ & & \ddots & \\ 0 & & & \mathbf{D}_q \end{bmatrix} \mathbf{y}(k) \tag{11-429}$$

where $\mathbf{D}_i$, $i = 1, 2, \ldots, q$, are of the form of the row matrix in Eq. (11-413).

We shall use the same observer design problem in Example 11-11 to illustrate the adjoint phase-variable canonical form method.

Example 11-12

For the digital process described in Examples 11-10 and 11-11, the characteristic equation of $\mathbf{A}$ is

$$|\lambda \mathbf{I} - \mathbf{A}| = \begin{vmatrix} \lambda & -1 \\ 1 & \lambda - 1 \end{vmatrix} = \lambda^2 - \lambda + 1 = 0 \tag{11-430}$$

Comparing Eqs. (11-427) and (11-430), we see that $a_1 = -1$ and $a_2 = 1$. The adjoint phase-variable canonical form of $\mathbf{A}$ is

$$\mathbf{A}_1 = \begin{bmatrix} 0 & -a_2 \\ 1 & -a_1 \end{bmatrix} = \begin{bmatrix} 0 & -1 \\ 1 & 1 \end{bmatrix} \tag{11-431}$$

For a deadbeat response for $\mathbf{x}_e(k) - \mathbf{x}(k)$, Eq. (11-417) gives

$$k_{e2} = -a_1 = 1$$
$$k_{e1} = -a_2 = -1$$

Thus,

$$\mathbf{K}_e = \begin{bmatrix} -1 \\ 1 \end{bmatrix} \tag{11-432}$$

The transformation matrix **P** is determined from Eq. (11-426).

$$\mathbf{P} = \begin{bmatrix} a_1 & 1 \\ 1 & 0 \end{bmatrix} \begin{bmatrix} \mathbf{D} \\ \mathbf{DA} \end{bmatrix} = \begin{bmatrix} -1 & 1 \\ 1 & 0 \end{bmatrix} \begin{bmatrix} 2 & 0 \\ 0 & 2 \end{bmatrix} = \begin{bmatrix} -2 & 2 \\ 2 & 0 \end{bmatrix} \quad \text{(11-433)}$$

The feedback matrix of the observer is found by use of Eq. (11-426). We have

$$\mathbf{G}_e = \mathbf{P}^{-1}\mathbf{K}_e = \begin{bmatrix} \frac{1}{2} \\ 0 \end{bmatrix} \quad \text{(11-434)}$$

which is the same result as in Eq. (11-401).

11-9-2 Design of the Reduced-Order State Observer

The observer design discussed in the preceding sections is for the full-order observer. In other words, the order of the observer is the same as that of the system. In general, since the q outputs are linear combinations of the n state variables, only a maximum of $n - q$ states need be observed. This results in a reduced-order observer. In fact, in the system described in Example 11-10, since $c(k) = 2x_1(k)$, we can get $x_1(k)$ directly from $c(k)$ without any observer dynamics. In this case only a first-order observer is necessary. However, it should be pointed out that while q of the states are obtained directly from the output, we may have less flexibility in the design of the dynamics of the reduced-order observer.

The principle of the reduced-order observer is illustrated by the block diagram of Fig. 11-9. The reduced-order observer is of the $(n - q)$th order. The n observed states $\mathbf{x}_e(k)$ are obtained from the $(n - q)$ observed states $\bar{\mathbf{w}}_e(k)$ and the $q \times 1$ output vector $\mathbf{c}(k)$.

The reduced-order observer can be designed using the principle of the adjoint phase-variable canonical form transformation. We start with the system

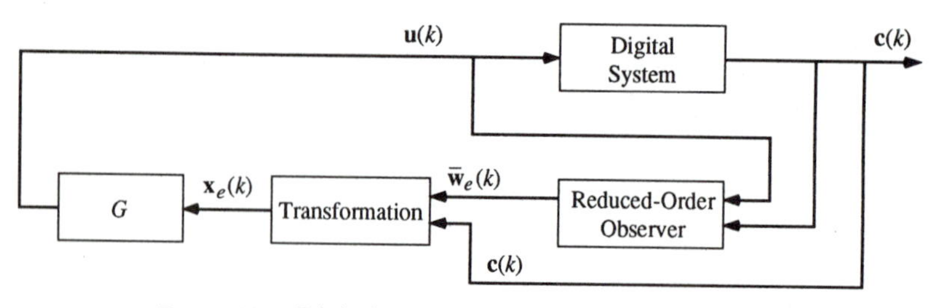

Figure 11-9. Digital system with a reduced-order observer.

with single output and input,

$$\mathbf{x}(k+1) = \mathbf{A}\mathbf{x}(k) + \mathbf{B}u(k) \tag{11-435}$$

$$c(k) = \mathbf{D}\mathbf{x}(k) \tag{11-436}$$

which is controllable and observable. The system is transformed into the adjoint phase-variable canonical form

$$\mathbf{y}(k+1) = \mathbf{A}_1\mathbf{y}(k) + \mathbf{B}_1u(k) \tag{11-437}$$

where $\mathbf{A}_1 = \mathbf{PAP}^{-1}$ is given by Eq. (11-415), and $\mathbf{B}_1 = \mathbf{PB}$. The output equation is transformed into

$$c(k) = \mathbf{D}_1\mathbf{y}(k) \tag{11-438}$$

where

$$\mathbf{y}(k) = \mathbf{P}\mathbf{x}(k)$$

$$\mathbf{D}_1 = \mathbf{DP}^{-1} = [0 \quad 0 \quad \cdots \quad 1] \qquad (1 \times n) \tag{11-439}$$

The full-order observer of the system described by Eqs. (11-435) and (11-436) is

$$\mathbf{x}_e(k+1) = (\mathbf{A} - \mathbf{G}_e\mathbf{D})\mathbf{x}_e(k) + \mathbf{B}u(k) + \mathbf{G}_ec(k) \tag{11-440}$$

where $\mathbf{G}_e$ is the $n \times 1$ feedback gain matrix of the observer. The full-order observer for the system described by the adjoint phase-variable canonical form system in Eqs. (11-437) and (11-438) is represented by

$$\mathbf{y}_e(k+1) = (\mathbf{A}_1 - \mathbf{K}_e\mathbf{D}_1)\mathbf{y}_e(k) + \mathbf{B}_1u(k) + \mathbf{K}_ec(k) \tag{11-441}$$

where $\mathbf{K}_e$ is the $n \times 1$ feedback gain matrix of the observer. Since $\mathbf{D}_1$ is of the form given in Eq. (11-439), it implies that $c(k) = y_n(k)$, so that $y_n(k)$ does not need to be observed. To observe the remaining $n - 1$ states of $\mathbf{y}(k)$ and be able to place the eigenvalues arbitrarily, we let

$$\mathbf{Q} = \begin{bmatrix} 1 & 0 & \cdots & 0 & -\alpha_{n-1} \\ 0 & 1 & \cdots & 0 & -\alpha_{n-2} \\ \hdotsfor{5} \\ 0 & 0 & \cdots & 1 & -\alpha_1 \\ 0 & 0 & \cdots & 0 & 1 \end{bmatrix} \qquad (n \times n) \tag{11-442}$$

Then

$$\mathbf{Q}^{-1} = \begin{bmatrix} 1 & 0 & \cdots & 0 & \alpha_{n-1} \\ 0 & 1 & \cdots & 0 & \alpha_{n-2} \\ \hdotsfor{5} \\ 0 & 0 & \cdots & 1 & \alpha_1 \\ 0 & 0 & \cdots & 0 & 1 \end{bmatrix} \tag{11-443}$$

Note that $\mathbf{QQ}^{-1} = \mathbf{I}$ (identity matrix).

Let

$$\mathbf{A}_2 = \mathbf{Q}\mathbf{A}_1\mathbf{Q}^{-1} = \begin{bmatrix} 0 & 0 & \cdots & -\alpha_{n-1} & -\alpha_1\alpha_{n-1} - a_n + a_1\alpha_{n-1} \\ 1 & 0 & \cdots & -\alpha_{n-2} & -\alpha_1\alpha_{n-2} - a_{n-1} + a_1\alpha_{n-2} \\ 0 & 1 & \cdots & -\alpha_{n-3} & -\alpha_1\alpha_{n-3} - a_{n-2} + a_1\alpha_{n-3} \\ \cdots & & & & \cdots \\ 0 & 0 & \cdots & -\alpha_1 & -\alpha_1^2 - a_2 + a_1\alpha_1 \\ 0 & 0 & \cdots & 1 & -a_1 + \alpha_1 \end{bmatrix} \tag{11-444}$$

Now we transform the system in $\mathbf{y}(k)$ into the following system:

$$\mathbf{w}(k+1) = \mathbf{A}_2\mathbf{w}(k) + \mathbf{B}_2 u(k) \tag{11-445}$$

$$c(k) = \mathbf{D}_2\mathbf{w}(k) \tag{11-446}$$

where

$$\mathbf{w}(k) = \mathbf{Q}\mathbf{y}(k)$$

$\mathbf{A}_2$ is defined in Eq. (11-444), and $\mathbf{B}_2 = \mathbf{Q}\mathbf{B}_1$,

$$\mathbf{D}_2 = \mathbf{D}_1\mathbf{Q}^{-1} = [0 \quad 0 \quad \cdots \quad 1] \tag{11-447}$$

The observer which observes $\mathbf{w}(k)$ is of the form

$$\mathbf{w}_e(k+1) = \mathbf{A}_2\mathbf{w}_e(k) + \mathbf{B}_2 u(k) + \mathbf{L}_e\mathbf{D}_2[\mathbf{w}(k) - \mathbf{w}_e(k)] \tag{11-448}$$

In view of the form of $\mathbf{D}_2$ in Eq. (11-447), we have $c(k) = w_n(k)$; that is, the nth state of $\mathbf{w}(k)$ is the output so that it does not have to be observed. Therefore, $w_{en}(k) = w_n(k)$, and

$$\mathbf{L}_e\mathbf{D}_2[\mathbf{w}(k) - \mathbf{w}_e(k)] = \mathbf{0} \tag{11-449}$$

Equation (11-446) becomes

$$\mathbf{w}_e(k+1) = \mathbf{A}_2\mathbf{w}_e(k) + \mathbf{B}_2 u(k) \tag{11-450}$$

The main objective now is to reduce the observer in $\mathbf{w}_e(\mathbf{k})$ to an $(n-1)$th order one and at the same time to be able to place its eigenvalues. Equation (11-450) is written

$$\begin{bmatrix} \bar{\mathbf{w}}_e(k+1) \\ \hline w_{en}(k+1) \end{bmatrix} = \begin{bmatrix} \bar{\mathbf{A}}_2 & \vdots & \mathbf{E}_2 \\ \hline 0 \quad 0 \quad \cdots \quad 1 & \vdots & -a_1 + \alpha_1 \end{bmatrix} \begin{bmatrix} \bar{\mathbf{w}}_e(k) \\ \hline c(k) \end{bmatrix} + \begin{bmatrix} \bar{\mathbf{B}}_2 \\ \hline b_2 \end{bmatrix} u(k) \tag{11-451}$$

Thus, the $(n-1)$th-order observer is described by

$$\bar{\mathbf{w}}_e(k+1) = \bar{\mathbf{A}}_2\bar{\mathbf{w}}_e(k) + \mathbf{E}_2 c(k) + \bar{\mathbf{B}}_2 u(k) \tag{11-452}$$

where

$$\bar{\mathbf{A}}_2 = \begin{bmatrix} 0 & 0 & \cdots & 0 & -\alpha_{n-1} \\ 1 & 0 & \cdots & 0 & -\alpha_{n-2} \\ 0 & 1 & \cdots & 0 & -\alpha_{n-3} \\ \cdots & \cdots & \cdots & \cdots & \cdots \\ 0 & 0 & \cdots & 1 & -\alpha_1 \end{bmatrix} \tag{11-453}$$

$$\mathbf{E}_2 = \begin{bmatrix} -\alpha_1\alpha_{n-1} - a_n + a_1\alpha_{n-1} \\ -\alpha_1\alpha_{n-2} - a_{n-1} + a_1\alpha_{n-2} + \alpha_{n-1} \\ \cdots \\ -\alpha_1^2 - a_2 + a_1\alpha_1 \end{bmatrix} \tag{11-454}$$

and $\bar{\mathbf{B}}_2$ is an $(n-1) \times 1$ matrix.

Since $\bar{\mathbf{A}}_2$ is in the adjoint phase-variable canonical form, the last column of the matrix contains the coefficients of the characteristic equation

$$|\lambda\mathbf{I} - \bar{\mathbf{A}}_2| = \lambda^{n-1} + \alpha_1\lambda^{n-2} + \alpha_2\lambda^{n-3} + \cdots + \alpha_{n-2}\lambda + \alpha_{n-1} = 0 \tag{11-455}$$

Once the reduced-order observed state vector $\bar{\mathbf{w}}_e(k)$ is determined, $\mathbf{w}_e(k)$ is given by

$$\mathbf{w}_e(k) = \begin{bmatrix} \bar{\mathbf{w}}_e(k) \\ c(k) \end{bmatrix} \tag{11-456}$$

Then,

$$\mathbf{x}_e(k) = (\mathbf{QP})^{-1}\mathbf{w}_e(k) \tag{11-457}$$

which is the transformation indicated in Fig. 11-9.

The effectiveness of the reduced-order observer is studied by first comparing the observed state $\mathbf{w}_e(k)$ with the state $\mathbf{w}(k)$. Subtracting Eq. (11-450) from Eq. (11-445), we get

$$\mathbf{w}(k+1) - \mathbf{w}_e(k+1) = \mathbf{A}_2[\mathbf{w}(k) - \mathbf{w}_e(k)] \tag{11-458}$$

Since $w_n(k) = c(k) = w_{en}(k)$, the last equation leads to

$$\bar{\mathbf{w}}(k+1) - \bar{\mathbf{w}}_e(k+1) = \bar{\mathbf{A}}_2[\bar{\mathbf{w}}(k) - \bar{\mathbf{w}}_e(k)] \tag{11-459}$$

Thus, $\bar{\mathbf{w}}_e(k)$ approaches $\bar{\mathbf{w}}(k)$ according to the dynamics determined by the eigenvalues of $\bar{\mathbf{A}}_2$. Since the transformation in Eq. (11-457) is nondynamic and $w_{en}(k) = c(k)$, the $(n-1)$ states of $\mathbf{x}(k)$ that have to be observed are also controlled by the eigenvalues of $\bar{\mathbf{A}}_2$.

Example 11-13

The digital process described in Example 11-10 is again considered in this example. We shall first consider that the system is open loop. The state equation is

$$\mathbf{x}(k+1) = \mathbf{A}\mathbf{x}(k) + \mathbf{B}u(k) \tag{11-460}$$

where

$$\mathbf{A} = \begin{bmatrix} 0 & 1 \\ -1 & 1 \end{bmatrix} \qquad \mathbf{B} = \begin{bmatrix} 0 \\ 1 \end{bmatrix}$$

The output equation is

$$c(k) = \mathbf{D}\mathbf{x}(k) = [2 \quad 0]\mathbf{x}(k) \tag{11-461}$$

The objective is to design a first-order observer for the system. The matrix **P** for the transformation of **A** into the adjoint phase-variable canonical form is given in Eq. (11-433). The matrix $\mathbf{A}_2$ in Eq. (11-444) is written for the second-order case as

$$\mathbf{A}_2 = \begin{bmatrix} -\alpha_1 & -\alpha_1^2 - a_2 + a_1\alpha_1 \\ 1 & -a_1 + \alpha_1 \end{bmatrix} \tag{11-462}$$

where a_1 and a_2 are the coefficients of the characteristic equation of **A**, that is,

$$|\lambda\mathbf{I} - \mathbf{A}| = \lambda^2 - \lambda + 1 = 0 \tag{11-463}$$

Thus, according to Eq. (11-427), $a_1 = -1$ and $a_2 = 1$. The coefficient α_1 is identified with the characteristic equation of $\bar{\mathbf{A}}_2$, which is

$$\lambda + \alpha_1 = 0 \tag{11-464}$$

The reduced-order observer in the transformed domain is described by

$$\bar{w}_e(k+1) = \bar{A}_2\bar{w}_e(k) + E_2c(k) + \bar{B}_2u(k) \tag{11-465}$$

and

$$\mathbf{w}_e(k) = [\bar{w}_e(k) \quad c(k)]' \tag{11-466}$$

$$\bar{A}_2 = -\alpha_1$$

$$E_2 = -\alpha_1^2 - a_2 + a_1\alpha_1 = -\alpha_1^2 - \alpha_1 - 1$$

$$\bar{B}_2 = 2$$

For a deadbeat response for the error of the observer we set α_1 to zero. Equation (11-465) becomes

$$\bar{w}_e(k+1) = -c(k) + 2u(k) \tag{11-467}$$

A state diagram for the system and the first-order observer is shown in Fig. 11-10. For arbitrary initial states $\mathbf{x}(0)$ and $\bar{w}_e(0)$ we can show that $\mathbf{x}_e(k)$

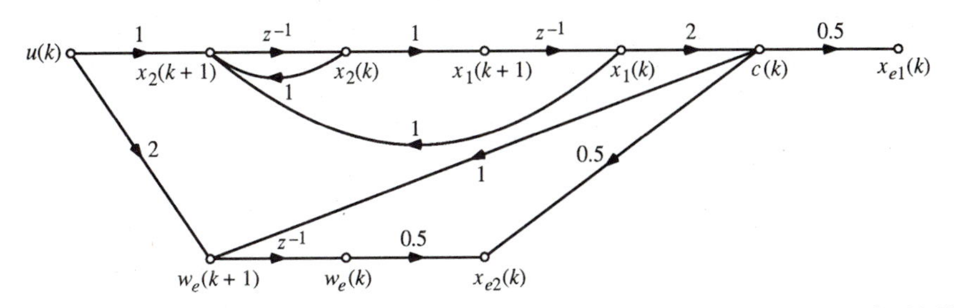

Figure 11-10. Reduced-order observer and open-loop digital system in Example 11-13.

reaches $\mathbf{x}(k)$ for $k \geq 1$. For $\mathbf{x}(0) = [1 \quad 0]'$, $\bar{w}_e(0) = 0.5$, and $u(k) = 0$ for all k, the responses of the system are tabulated below for $k \leq 5$.

k	0	1	2	3	4	5
$x_1(k)$	1	0	-1	-1	0	1
$x_{e1}(k)$	1	0	-1	-1	0	1
$x_2(k)$	0	-1	-1	0	1	1
$x_{e2}(k)$	1.25	-1	-1	0	1	1
$\bar{w}_e(k)$	0.50	-2	0	2	2	0
$c(k)$	2	0	-2	-2	0	2

For the closed-loop system $u(k) = -\mathbf{G}\mathbf{x}_e(k)$, where the feedback gain matrix $\mathbf{G}$ is given in Eq. (11-400). Substituting Eqs. (11-466), (11-457), and the control into Eq. (11-467), we have

$$\bar{w}_e(k+1) = -c(k) - 2\mathbf{G}\mathbf{x}_e(k)$$

$$= -2x_1(k) + [1.308 \quad -0.972]\begin{bmatrix} x_1(k) \\ 0.5\bar{w}_e(k) + x_1(k) \end{bmatrix} \tag{11-468}$$

or

$$\bar{w}_e(k+1) = -1.664x_1(k) - 0.486\bar{w}_e(k) \tag{11-469}$$

The state equations of the digital system in Eq. (11-460) are conditioned with the state feedback, and the resulting state equations are

$$x_1(k+1) = x_2(k) \tag{11-470}$$

$$x_2(k+1) = -0.832x_1(k) + x_2(k) - 0.243\bar{w}_e(k) \tag{11-471}$$

Figure 11-11 shows the state diagram of the feedback system and its first-order observer.

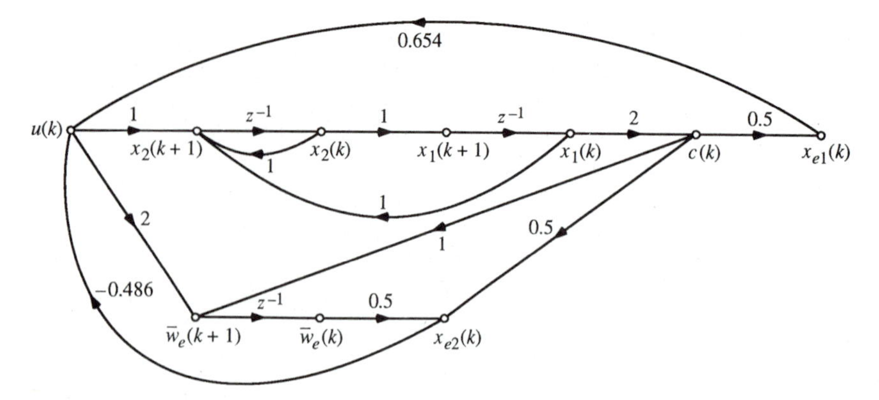

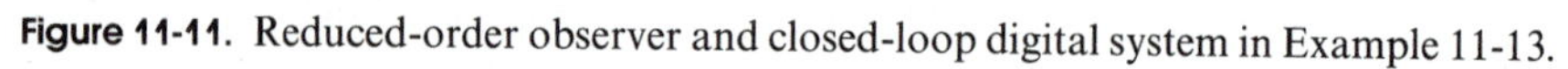
Figure 11-11. Reduced-order observer and closed-loop digital system in Example 11-13.

For the initial states $\mathbf{x}(0) = [1 \quad 0]'$ and $\bar{w}_e(0) = 0.5$, the last three equations are solved to give the responses of $\mathbf{x}(k)$ and $\bar{w}_e(k)$, $k = 1, 2, \ldots$. The observed states $\mathbf{x}_e(k)$ are determined from Eq. (11-457); that is,

$$\mathbf{x}_e(k) = (\mathbf{QP})^{-1} w_e(k) = \begin{bmatrix} x_1(k) \\ 0.5\bar{w}_e(k) + x_1(k) \end{bmatrix} \tag{11-472}$$

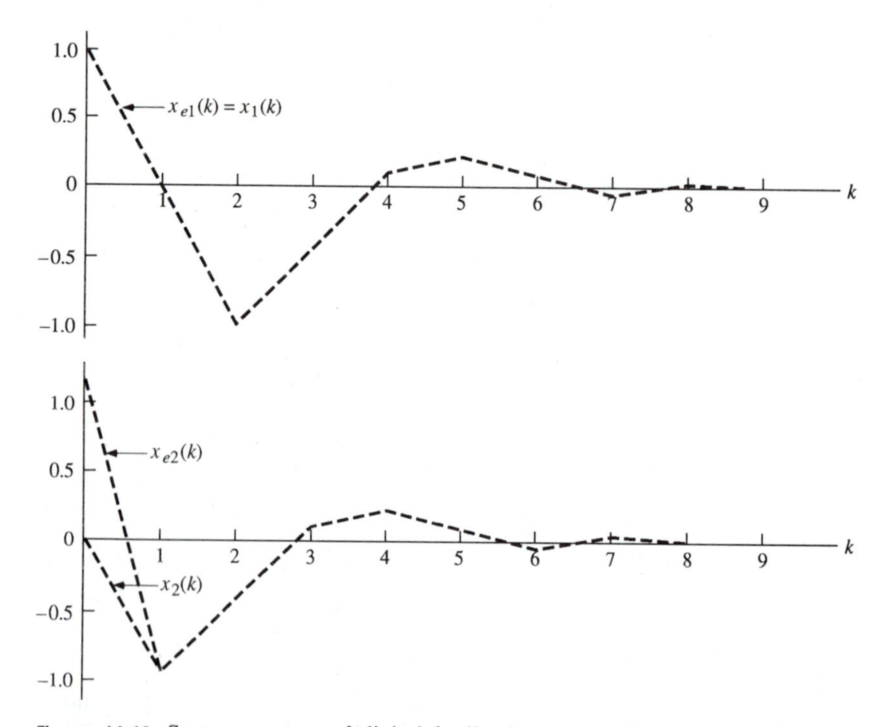

Figure 11-12. State responses of digital feedback system with reduced-order state observer.

Thus, in this system, $x_{e1}(k)$ is directly $x_1(k)$, and $x_2(k)$ is the only state that is observed through the first-order observer.

Figure 11-12 shows the time responses of $\mathbf{x}(k)$ and $\mathbf{x}_e(k)$ for the closed-loop system. It is interesting to note that since the reduced-order observer is of the first order, $\mathbf{x}_e(k)$ reaches $\mathbf{x}(k)$ in only one sampling period for the deadbeat observer design. However, comparing the responses in Fig. 11-12 with those of the system with a full observer (Fig. 11-8), we see that the reduced-order observer causes the closed-loop system to have a higher overshoot.

PROBLEMS

11-1 For a linear digital control process

$$\mathbf{x}(k+1) = \mathbf{A}\mathbf{x}(k) + \mathbf{B}\mathbf{u}(k)$$

where $\mathbf{A}$ is $n \times n$ and is nonsingular, show that the necessary conditions for minimizing the performance index

$$J = \frac{1}{2}\mathbf{x}'(N)\mathbf{S}\mathbf{x}(N) + \frac{1}{2}\sum_{k=0}^{N-1}[\mathbf{x}'(k)\mathbf{Q}\mathbf{x}(k) + \mathbf{u}'(k)\mathbf{R}\mathbf{u}(k)]$$

where $\mathbf{S}$ and $\mathbf{Q}$ are symmetric positive-semidefinite matrices and $\mathbf{R}$ is a symmetric positive-definite matrix, are

$$\mathbf{p}(k+1) = (\mathbf{A}')^{-1}\mathbf{p}(k) - (\mathbf{A}')^{-1}\mathbf{Q}\mathbf{x}(k)$$

$$\mathbf{x}(k+1) = \mathbf{A}\mathbf{x}(k) - \mathbf{B}\mathbf{R}^{-1}\mathbf{B}'\mathbf{p}(k+1)$$

11-2 Given the digital control system

$$\mathbf{x}(k+1) = \mathbf{A}\mathbf{x}(k) + \mathbf{B}u(k)$$

where

$$\mathbf{A} = \begin{bmatrix} 0 & 1 \\ -1 & 1 \end{bmatrix} \qquad \mathbf{B} = \begin{bmatrix} 0 \\ 1 \end{bmatrix} \qquad \mathbf{x}(0) = \begin{bmatrix} 1 \\ 1 \end{bmatrix}$$

Find the optimal control sequence $u(0)$, $u(1)$, and $u(2)$ so that the performance index

$$J_3 = \frac{1}{2}x_1^2(3) + \frac{1}{2}\sum_{k=0}^{2}[x_1^2(k) + u^2(k)] \qquad x_1(0) = x_2(0) = 0$$

is minimized. The end condition is $x_1(3) = $ free and $x_2(3) = 0$.

11-3 The dynamic equations of a permanent-magnet step motor are written as

$$\frac{d\lambda_a(\theta)}{d\theta} = \frac{1}{2}\frac{v_a}{\omega(\theta)} - \frac{R}{2L\omega(\theta)}\lambda_a(\theta) + \frac{RK_1\cos\theta}{2L\omega(\theta)}$$

$$\frac{d\lambda_b(\theta)}{d\theta} = \frac{1}{2}\frac{v_b}{\omega(\theta)} - \frac{R}{2L\omega(\theta)}\lambda_b(\theta) + \frac{RK_1\sin\theta}{2L\omega(\theta)}$$

$$\frac{d\omega(\theta)}{d\theta} = -\frac{N_rK_T}{JL}\frac{\lambda_a(\theta)\sin\theta - \lambda_b(\theta)\cos\theta}{\omega(\theta)} - \frac{T_DN_r}{J\omega(\theta)} - \frac{BN_r}{J}$$

where

v_a, v_b	constant applied voltages
λ_a, λ_b	flux linkages of motor phases
L	inductance of each phase
R	resistance of each phase
N_r	number of rotor teeth
K_T	torque constant
K_1	constant linking the flux between the permanent-magnet rotor and the stator
ω	angular velocity of rotor shaft
θ	angular displacement of rotor shaft
J	inertia of rotor and load
B	viscous friction coefficient
T_D	Coulomb friction torque

The initial state is given as $\lambda_a(\theta_0) = \lambda_{a0}$, $\lambda_b(\theta_0) = \lambda_{b0}$, and $\omega(\theta_0) = \omega_0$. The performance index to be minimized is

$$J = \int_{\theta_0}^{\theta_f} \frac{1}{\omega(\theta)}\, d\theta$$

The controls v_a and v_b are subject to the constraints $-V \le v_a \le V$, $-V \le v_b \le V$.

a. Discretize the state equations and the performance index by using

$$\frac{dx(\theta)}{d\theta} \cong \frac{1}{\Delta\theta}\,[x(k+1) - x(k)]$$

where $x(k)$ is a state variable evaluated at $\theta = k$.

b. Set up the Hamiltonian, the necessary conditions for optimal control, and the transversality condition.

11-4 Consider the optimal capital allocation problem:

c	initial amount of capital to be invested in investment A or investment B over a total of N periods
x	amount of capital to be invested in the first period in investment choice A
$g(x) = 0.4\sqrt{x}$,	return from investment A for one period
$h(c - x) = 0.2\sqrt{c - x}$,	return from investment B for one period
ax	capital left at the end of the first stage from investment A
$b(c - x)$	capital left at the end of the first stage from investment B

$a = 0.6$
$b = 0.8.$

a. Set up the capital allocation problem as an optimal control problem. Find the maximum return for a three-stage process using the principle of optimality.

b. Determine the optimum allocation of capital at the beginning of each period.

11-5 A linear digital system is characterized by the state equation

$$\frac{d\mathbf{x}(t)}{dt} = \mathbf{A}\mathbf{x}(t) + \mathbf{B}u(t)$$

where

$$\mathbf{A} = \begin{bmatrix} 0 & 1 \\ -4 & -5 \end{bmatrix} \qquad \mathbf{B} = \begin{bmatrix} 0 \\ 1 \end{bmatrix}$$

The control is described by $u(t) = u(kT)$ for $kT \le t < (k+1)T$, where $T = 0.5$ s. Determine the optimal control $u(kT)$ as a function of $\mathbf{x}(kT)$ for $k = 0, 1, 2, 3, 4$, which minimizes the performance index

$$J = \sum_{k=0}^{4} x_1^2[(k+1)T]$$

Determine the optimal trajectories for $x_1(kT)$ and $x_2(kT)$ as functions of $\mathbf{x}(0)$. Repeat the problem for $N = \infty$.

11-6 A linear digital process is represented by the state equation

$$\mathbf{x}(k+1) = \mathbf{A}\mathbf{x}(k) + \mathbf{B}u(k)$$

where

$$\mathbf{A} = \begin{bmatrix} 0 & 1 \\ -1 & -2 \end{bmatrix} \qquad \mathbf{B} = \begin{bmatrix} 0 \\ 1 \end{bmatrix}$$

a. Determine the optimal control $u(0)$, $u(1)$, ..., $u(4)$ in the form of state feedback so that the performance index

$$J = \frac{1}{2} \sum_{k=0}^{5} [x_1^2(k) + x_2^2(k)]$$

is minimized.

b. Repeat part **a** when $N = \infty$, infinite time.

11-7 Repeat Problem 11-6 when the performance index is

$$J = \frac{1}{2} \sum_{k=0}^{5} [x_1^2(k) + x_2^2(k) + u^2(k)]$$

11-8

a. Find the optimal control sequence $u(0)$, $u(1)$, $u(2)$, such that the performance index

$$J = \frac{1}{2} \sum_{k=0}^{2} [x_1^2(k) + u^2(k)]$$

is minimized, subject to the constraint

$$\mathbf{x}(k+1) = \mathbf{A}\mathbf{x}(k) + \mathbf{B}u(k)$$

$$\mathbf{A} = \begin{bmatrix} 0 & 1 \\ -1 & 1 \end{bmatrix} \qquad \mathbf{B} = \begin{bmatrix} 0 \\ 1 \end{bmatrix}$$

$$\mathbf{x}(0) = \begin{bmatrix} 1 \\ 1 \end{bmatrix} \qquad \mathbf{x}(3) = \text{free}$$

Use the Riccati equation method, and solve for the Riccati gain matrices $\mathbf{K}(0)$, $\mathbf{K}(1)$, $\mathbf{K}(2)$, and $\mathbf{K}(3)$.

b. Repeat part **a** for the performance index

$$J = \frac{1}{2} \mathbf{x}'(3)\mathbf{S}\mathbf{x}(3) + \sum_{k=0}^{2} [x_1^2(k) + u^2(k)]$$

where $\mathbf{S} = \mathbf{I}$ (identity matrix).

c. Repeat part **a** for the infinite-time problem.

11-9 A linear digital process with input delays is described by the state equation

$$\mathbf{x}[(k+1)T] = \mathbf{A}\mathbf{x}(kT) + \mathbf{B}u[(k-p)T]$$

where $\mathbf{x}(kT)$ is an $n \times 1$ state vector and $u[(k-p)T]$ is the delayed scalar control. The delay time pT is assumed to be an integral multiple of T. Given the initial state $\mathbf{x}(0)$ and the past inputs, $u(-pT)$, $u[(-p+1)T]$, ..., $u(-T)$, determine the optimal control $u(0)$, $u(T)$, ..., $u[(N-1)T]$ in terms of state feedback and the past inputs so that the performance index

$$J = \frac{1}{2} \sum_{k=0}^{N+p-1} [\mathbf{x}'(kT)\mathbf{Q}(kT)\mathbf{x}(kT) + u^2[(k-p-1)T]]$$

is minimized for $N > p$.

11-10 Consider that a linear digital process with input delay is described by the first-order difference equation

$$x(k+1) = Ax(k) + Bu(k-1)$$

where $A = 0.368$ and $B = 0.632$. Determine the optimal control $u(0)$ through $u(3)$ in terms of state feedback so that the performance index

$$J = \frac{1}{2} \sum_{k=0}^{4} [x^2(k) + u^2(k-1)]$$

is minimized. For $x(0) = -1$ and $u(-1) = 0.178$, determine the optimal trajectory for $x(k)$, $k = 0, 1, 2, 3, 4$.

11-11 A first-order digital control process with state delay is described by the state equation

$$x(k+1) = x(k) + 2x(k-1) + u(k)$$

The performance index is

$$J = \frac{1}{2} \sum_{k=0}^{N-1} [x^2(k) + u^2(k)]$$

Find the optimal control $u(0)$, $u(1)$ so that J is minimized for $N = 2$. The initial states are given as $x(0)$ and $x(-1)$.

11-12 A tracking problem of the linear regulator design is described as one which attempts to match the system output $\mathbf{c}(k)$ with a desired output $\mathbf{c}_d(k)$. The tracking problem can be stated as follows. Given the linear process

$$\mathbf{x}(k+1) = \mathbf{A}\mathbf{x}(k) + \mathbf{B}\mathbf{u}(k)$$

$$\mathbf{c}(k) = \mathbf{D}\mathbf{x}(k)$$

where $\mathbf{x}(k)$ is the $n \times 1$ state vector, $\mathbf{u}(k)$ is the $r \times 1$ input vector, and $\mathbf{c}(k)$ is the $p \times 1$ output vector. It is assumed that $\mathbf{A}$ is nonsingular. Find the optimal control $\mathbf{u}(k)$, $k = 0, 1, \ldots, N-1$, so that the performance index

$$J = \frac{1}{2} \mathbf{e}'(N)\mathbf{S}\mathbf{e}(N) + \frac{1}{2} \sum_{k=0}^{N-1} [\mathbf{e}'(k)\mathbf{Q}\mathbf{e}(k) + \mathbf{u}'(k)\mathbf{R}\mathbf{u}(k)]$$

is a minimum. $\mathbf{S}$ and $\mathbf{Q}$ are positive semidefinite, and $\mathbf{R}$ is positive definite, and all are symmetric matrices.

$$\mathbf{e}(k) = \mathbf{c}_d(k) - \mathbf{c}(k)$$

where $\mathbf{c}_d(k)$ is a constant vector for a given k.

11-13 Given the digital control system

$$\mathbf{x}(k+1) = \mathbf{A}\mathbf{x}(k) + \mathbf{B}u(k)$$

where

$$\mathbf{A} = \begin{bmatrix} 0 & 1 \\ -0.5 & -1 \end{bmatrix} \qquad \mathbf{B} = \begin{bmatrix} 0 \\ 1 \end{bmatrix}$$

The control is

$$u(k) = -\mathbf{G}\mathbf{x}(k)$$

where

$$\mathbf{G} = [-0.4 \quad -0.2]$$

Determine for what $\mathbf{Q}$ is the system optimal in the sense that

$$J = \frac{1}{2} \sum_{k=0}^{\infty} [\mathbf{x}'(k)\mathbf{Q}\mathbf{x}(k) + u^2(k)]$$

is a minimum.

11-14 Given the digital process

$$\mathbf{x}(k+1) = \mathbf{A}\mathbf{x}(k) + \mathbf{B}u(k)$$

where

$$\mathbf{A} = \begin{bmatrix} 0 & 1 \\ -0.4 & -0.6 \end{bmatrix} \qquad \mathbf{B} = \begin{bmatrix} 0 \\ 1 \end{bmatrix}$$

The control $u(k)$ is obtained through partial state feedback,

$$u(k) = -\mathbf{G}\mathbf{x}(k)$$

where

$$\mathbf{G} = [-0.3 \quad 0]$$

Determine the admissible $\mathbf{Q}$ so that the system is optimal in the sense that

$$J = \frac{1}{2} \sum_{k=0}^{\infty} [\mathbf{x}'(k)\mathbf{Q}\mathbf{x}(k) + u^2(k)]$$

is a minimum.

11-15 Given the digital control process

$$\mathbf{x}(k+1) = \mathbf{A}\mathbf{x}(k) + \mathbf{B}u(k)$$

where

$$\mathbf{A} = \begin{bmatrix} 0 & 1 \\ -0.4 & -0.6 \end{bmatrix} \qquad \mathbf{B} = \begin{bmatrix} 0 \\ 1 \end{bmatrix}$$

$$u(k) = -\mathbf{G}\mathbf{x}(k)$$

The characteristic equation of the closed-loop system is specified as

$$|\lambda\mathbf{I} - \mathbf{A} + \mathbf{B}\mathbf{G}| = \lambda^2 + 6\lambda = 0$$

Find the admissible $\mathbf{Q}$ so that the system is optimal in the sense that

$$J = \frac{1}{2} \sum_{k=0}^{\infty} [\mathbf{x}'(k)\mathbf{Q}\mathbf{x}(k) + u^2(k)] = \text{minimum}$$

11-16 The dynamic equations of a digital control system are described by

$$\mathbf{x}(k+1) = \mathbf{A}\mathbf{x}(k) + \mathbf{B}u(k)$$

$$c(k) = \mathbf{D}\mathbf{x}(k)$$

where

$$\mathbf{A} = \begin{bmatrix} 0 & 1 \\ 0.5 & 0 \end{bmatrix} \qquad \mathbf{B} = \begin{bmatrix} 0 \\ 1 \end{bmatrix}$$

$$\mathbf{D} = [1 \quad 1]$$

The state vector $\mathbf{x}(k)$ at $k = N$ is to be determined from the measurements of $u(k)$ and $\mathbf{c}(k)$ at instants previous to $k = N$. Determine how many samples of $c(k)$ and $u(k)$ are needed for exact measurements for $\mathbf{x}(k)$. Express $\mathbf{x}(k)$ in terms of these measurements of $c(k)$ and $u(k)$.

11-17 The dynamic equations of a digital process are given as

$$\mathbf{x}(k+1) = \mathbf{A}\mathbf{x}(k) + \mathbf{B}u(k)$$

$$c(k) = \mathbf{D}\mathbf{x}(k)$$

where

$$\mathbf{A} = \begin{bmatrix} 0 & 1 \\ 0 & 0 \end{bmatrix} \qquad \mathbf{B} = \begin{bmatrix} 0 \\ 1 \end{bmatrix} \qquad \mathbf{D} = [1 \quad 1]$$

The state-feedback control is $u(k) = -\mathbf{G}\mathbf{x}(k)$, where

$$\mathbf{G} = [g_1 \quad g_2]$$

a. Assuming that the state variables are unaccessible, design a full-order state observer so that $\mathbf{x}(k)$ is observed from $c(k)$. Find the elements of $\mathbf{G}_e$ in terms of g_1 and g_2 so that the dynamics of the observer are the same as that of the closed-loop digital process.

b. Design a first-order observer for the system with the eigenvalue at 0.5. Write the state equations of the system with feedback realized from the first-order observer.

11-18 For the stick-balancing control system described in Problem 6-5, design a full-order observer that will observe the states $x_1(kT)$, $x_2(kT)$, $x_3(kT)$, and $x_4(kT)$ from the output $c(kT) = [0 \quad 0 \quad 1 \quad 0]\mathbf{x}(kT)$. The eigenvalues of the closed-loop observer should be zero. Solve the problem first by working with the characteristic equation and then by means of the adjoint phase-variable canonical form.

11-19 Design a reduced-order observer for the system described in Problem 11-18.

References

1. Rozonoer, L. I., "The Maximum Principle of L. S. Pontryagin in Optimal System Theory," *Automation and Remote Control*, parts I, II, III, vol. 20, 1959.
2. Chang, S. S. L., "Digitized Maximum Principle," *Proc. IRE*, vol. 48, pp. 2030–2031, December 1960.

3. Butkovskii, A. G., "The Necessary and Sufficient Conditions for Optimality of Discrete Control Systems," *Automation and Remote Control*, vol. 24, pp. 1056–1064, August 1963.
4. Katz, S., "A Discrete Version of Pontryagin's Maximum Principle," *J. Electron. Control*, vol. 13, no. 2, pp. 179–184, 1962.
5. Halkin, H., "Optimal Control for Systems Described by Difference Equations," *Advances in Control Systems*, chap. 4, C. T. Leondes, Ed., Academic Press, New York, 1964.
6. Sage, A. P., *Optimum System Control*, Prentice-Hall, Englewood Cliffs, N.J., 1968.
7. Dorato, P., and A. H. Levis, "Optimal Linear Regulator: The Discrete-Time Case," *IEEE Trans. Automatic Control*, vol. AC-16, pp. 613–620, December 1971.
8. Kleinman, D. L., "Stabilizing a Discrete, Constant, Linear System with Application to Iterative Methods for Solving the Riccati Equation," *IEEE Trans. Automatic Control*, vol. AC-19, pp. 252–254, June 1974.
9. Vaughan, D. R., "A Nonrecursive Algebraic Solution for the Discrete Riccati Equation," *IEEE Trans. Automatic Control*, vol. AC-15, pp. 597–599, October 1970.
10. Howerton, R. D., "A New Solution of the Discrete Algebraic Riccati Equation," *IEEE Trans. Automatic Control*, vol. AC-19, pp. 90–92, February 1974.
11. Lainiotis, D. G., "Discrete Riccati Equation Solutions: Partitioned Algorithms," *IEEE Trans. Automatic Control*, vol. AC-20, pp. 555–556, August 1975.
12. Molinari, B. P., "The Stabilizing Solution of the Discrete Algebraic Riccati Equation," *IEEE Trans. Automatic Control*, vol. AC-20, pp. 396–399, June 1975.
13. Molinari, B. P., "The Stabilizing Solution of the Algebraic Riccati Equation," *SIAM Journal Control*, vol. 11, pp. 262–271, May 1973.
14. Payne, H. J., and L. M. Silverman, "On the Discrete Time Algebraic Riccati Equation," *IEEE Trans. Automatic Control*, vol. AC-18, pp. 226–234, June 1973.
15. Caines, P. E., and D. Q. Mayne, "On the Discrete-Time Matrix Riccati Equation of Optimal Control," *International J. Control*, vol. 12, pp. 785–794, November 1970.
16. Hewer, G. A., "An Iterative Technique for the Computation of the Steady-State Gains for the Discrete Optimal Regulator," *IEEE Trans. Automatic Control*, vol. AC-16, pp. 382–384, August 1971.
17. Hewer, G. A., "Analysis of a Discrete Matrix Riccati Equation of Linear Control and Kalman Filtering," *J. Math. Analysis and Applications*, vol. 42, pp. 226–236, April 1973.
18. Cadzow, J. A., "Nilpotency Property of the Discrete Regulator," *IEEE Trans. Automatic Control*, vol. AC-13, pp. 734–735, December 1968.
19. Tse, E., and M. Athans, "Optimal Minimal-Order Observer-Estimators

for Discrete Linear Time-Varying Systems," *IEEE Trans. Automatic Control*, vol. AC-15, pp. 416–426, August 1970.

20. Luenberger, D. G., "An Introduction to Observers," *IEEE Trans. Automatic Control*, vol. AC-16, pp. 596–602, December 1971.

21. Leondes, C. T., and L. M. Novak, "Reduced-Order Observers for Linear Discrete-Time Systems," *IEEE Trans. Automatic Control*, vol. AC-19, pp. 42–46, February 1974.

Microprocessor and DSP Control

KEYWORDS AND TOPICS

microprocessor control • single-board controllers • digital signal processor (DSP)

12-1 INTRODUCTION

In the preceding chapters we presented the analysis and design of discrete-data control systems mainly from the analytical standpoint. We have covered the z-transform and the state-variable methods of analysis and design. In all instances, the sampling periods were chosen based on the sampling theorem, the system bandwidth, or the stability and overall performance of the discrete-data system. The digital controller parameters were selected purely from the standpoint of performance and physical realizability in the analytical and theoretical sense. However, we must realize that discrete-data control systems have physical limitations due to the inherent discrete nature of the system components which are not found in analog or continuous-data control systems. For example, the sampling period of a discrete-data control system is governed by the clock rate and how fast the numerical operations and instructions are executed by the digital processor. In the case of a digital signal processor (DSP) the execution speed can be extremely fast. However, in some microprocessors, especially those that are time shared, the speed of execution can be relatively slow. Thus, the sampling rate has inherent limits based on the hardware used.

Another limitation of the discrete-data control system design is the finite-wordlength characteristic of digital computers or processors. This means that not all numbers can be represented precisely by a digital processor. Many microprocessors have a wordlength of only 8 bits. From Chapter 2 we learned that an 8-bit word would allow only $2^8 = 256$ levels of resolution. This is also known as *quantization.* In the preceding sections, the design examples often yielded results for digital controller parameters such as 0.995 or 1.316, etc. Clearly, these numbers cannot be realized precisely on an 8-bit microprocessor; hence the effects of quantization must be studied. It is important to be aware of these limitations and consider them as the design is being carried out. While the pencil-and-paper design of discrete-data control systems can easily yield deadbeat responses, in practice, we have to face the reality of what the hardware can provide.

The objectives of this chapter are outlined as follows:

1. Introduce the applications of microprocessors and microcomputers for controls.
2. Introduce single-board controllers with custom-designed chips.
3. Introduce single-chip digital signal processor basics.
4. Discuss the effects of finite wordlength, quantization, and time delay due to the finite computation time of digital processors.

12-2 BASIC COMPUTER ARCHITECTURE

The main purpose of this section is to familiarize the reader with the basic components of a digital computer. Although each machine may have its own special features and hardware, certain components are common to most existing computers. No details on the construction and programming of the microprocessor are given, since these are beyond the scope of this text.

A basic digital computer consists of the following components:

central processing unit (CPU)
memory
input and output (I/O) devices

A microprocessor is essentially the central processing unit of a digital computer. When a microprocessor is equipped with memory and I/O devices, the microcomputer is complete. Figure 12-1 illustrates the interconnections of the basic building blocks of a microcomputer.

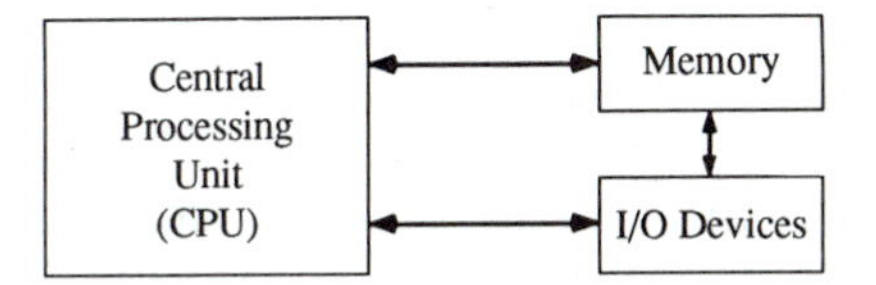

Figure 12-1. Basic components of a microcomputer.

In general, the CPU also contains the arithmetic logic unit (ALU), registers, instruction register and decoder (IRD), and timing and control circuits. These basic elements of the microcomputers are described separately in the following.

12-2-1 Input-Output Devices

Information is transferred in and out of the computer by means of input-output devices. The computer is generally provided with one or more *input ports*. The CPU can address these input ports and issue appropriate control signals to receive data at these ports. The addition of input ports enables the computer to receive information from external input devices such as a manual keyboard, disk drives, or switches.

For the CPU to communicate the results of its processing to the outside world, a computer also needs one or more *output ports*. The CPU can then address these output ports and send the data to the output devices. Typical output devices are printers, plotter, disk drives, or CRT displays. The outputs of the computer may also constitute process control signals that control the operation of another part of a system, or another system such as an automated assembly line.

12-2-2 Memory

One device that all digital computers have is the memory. The memory serves as a place to store *instructions*, the operation codes that direct the activities of the CPU, and *data*, the binary-coded information that is processed by the CPU.

The memory of a computer contains a large number of memory cells or locations. The smallest unit of the memory is a *bit*, which can be put into one of two states, 0 or 1. Each location can be used to store n bits of binary information.

A basic memory cell which is identifiable by a distinct memory location is called a *word*. In general, the words of a computer memory may be of almost any bit length, from 4 bits in the simple hand-held calculators to 128 bits or more for large-scale scientific computers. Microprocessors generally have a wordlength of 4 to 32 bits. For instance, the wordlength of the Intel 8086 microprocessor is 16 bits, and the Motorola MC68020 has a 32-bit wordlength.

Each of the words in the memory has a unique and permanent identifying number called the *address*. The individual numbering of the memory locations is called *addressing*. Access to the particular word or location is accomplished by specifying the address of the word location.

It is important to distinguish between the address and the numerical content of that address in the memory. The memory address is normally occupied by one word, but the numerical address itself may occupy more than one word. For instance, an 8-bit microprocessor could use two consecutive words for addressing. Thus, each address is 16 bits, or the largest number the microprocessor can address is 2^{16}, or 65,536. In computer jargon we say that a maximum of 64K bytes of memory can be attached to the microprocessor, whereas in reality it is 65,536 bytes. Actually, the address number runs from 0 through 65,535. Similarly, a memory capacity of 1K bytes is actually $2^{10} = 1024$ bytes, and 2K bytes is $2^{11} = 2048$ bytes, etc.

Random Access Memory

The memory of a microprocessor is available in several forms. The random access memory (RAM) can be accessed by the computer at any time and at any location in a random fashion. The RAM can be used for both read and write operations. The important feature of the RAM is that when the power to the computer is turned off, the contents of the memory are entirely lost. Thus, the RAM can be used for data storage, temporary instructions, and operation codes.

Read-Only Memory

The read-only memory (ROM) is used for the storage of programs that are executed repeatedly and must be stored permanently. Typically, the contents of the ROM are written initially and then stored permanently. The microprocessor does not write any new information into the program, but merely reads the contents of the ROM and executes them.

Certain types of ROM can be reprogrammed by a special process and are known as *programmable* read-only memories (PROMs).

12-2-3 The Central Processing Unit

The CPU of a microcomputer is the heart of the machine. It is a piece of hardware, in this case, the microprocessor, in which data processing and control take place. For example, inside the CPU two operands may be added together, data may be transferred into the CPU from some input device through an input port, or the CPU may transmit information to the outside through an output port.

The CPU is responsible for centrally controlling almost the entire computer. It generates most of the key timing pulses and control signals which the external devices will need to interact properly with the computer. Most of the computer control functions are derived from a finite-state machine (FSM) that is a part of the CPU. The FSM will direct the processor to follow the sequential processing of the stored program. When the computer is first turned on, the FSM knows where to retrieve the first instruction from the memory. It retrieves this instruction and places it in some decoder so that the CPU can understand what to do with the instruction.

A typical CPU consists of the following interconnected functional blocks:

registers
arithmetic logic unit
control circuitry

These units and the operations of a microprocessor are briefly described in the following.

Arithmetic Logic Unit

The ALU is a device that performs all the arithmetic computation and logical operations of the computer. The ALU can receive two numbers and perform arithmetic operations such as addition, subtraction, or logic operations such as AND, OR, NOT, and EXCLUSIVE-OR. The ALUs of most microprocessors are fairly simple so that they cannot perform hardware operations such as multiplication and division. However, through the use of software, these arithmetic operations can be performed, for instance, by repeated addition.

The ALU has some registers for storing sequences of digits. The main register is called the accumulator. At the beginning of an operation, data are sent to the ALU by the CPU control circuitry. The ALU performs the arithmetic or logic operation requested by the CPU controller and then stores the answer in some register such as the *accumulator* in the CPU.

One final function performed by the ALU is to report interpretive information on the results. For example, if the ALU subtracts two numbers, it is often required to report if the result is greater than, equal to, or less than zero. Similarly, the ALU may have to report if the sum of two numbers or the result of any operation overflows the 8 bits of a byte and is carried over to the ninth bit. Therefore, one of the functions of the ALU is to wave a flag when the result of an operation satisfies a certain condition.

Registers

Registers are special memory locations located inside the CPU. Because registers are physically a part of the CPU, they are directly wired to the control logic, the ALU, and other internal CPU devices. On some machines, data must be first transferred from addressable memory locations to a register before they can be operated on. Even on machines that can operate directly on data stored in the core memory, executing instructions in which all the operands are already in the registers is always faster.

As mentioned earlier, many machines have a special register called the accumulator (AC) or the *result register*. The AC is special in two respects. On some computers, we cannot process the contents of any two registers directly. On these machines, one of the operands to be processed in any operation must be in the AC. Furthermore, the result of the operation always stays in the AC. For example, if we add the content of the AC to that of register B, the result will be placed in the AC; the original number that was in the AC is automatically replaced by the sum. Therefore, in general, we can regard the accumulator as both a source (operand) and the destination (result) register.

Another important function of the AC is that it forms a link: to travel between the core memory and the CPU, data must first be transferred to the AC.

In addition to the AC, a typical microprocessor has other general-purpose registers and special-purpose registers such as program counters, flags, stack pointer, index pointers, etc.

Buses

One important aspect of microprocessor operation is that a rapid and accurate information link between the memory, input/output devices, and the CPU can be maintained. For this purpose, these microprocessor components are interconnected by various buses. A bus is a set of wires, grouped together because of the similarity of their functions, that runs between all the functional blocks of a microcomputer system. Figure 12-2 shows the simplified block diagram of a microcomputer system with interconnecting buses. The three basic types of bus in a microprocessor are described in the following.

Data bus — This is a set of wires in which data can flow between the CPU, the memory, and the I/O devices.

Address bus — This is a set of wires that connects the memory and the I/O devices to the CPU. The address bus is used to identify a particular memory location or I/O device. Information flows only in the direction from the CPU to the memory or the I/O.

Control bus — This is a set of wires that connects the memory and the I/O devices to the CPU, in order to indicate the type of activity in the current process. Information flows only in the direction from the CPU to the memory or the I/O devices. The activity may be *memory read*, *memory write*, *I/O read*, *I/O write*, or *interrupt acknowledge*.

The basic advantage of using the bus structure is that it simplifies the communication between the main hardware components of a microcomputer. Instead of having separate wires connecting each pair of devices, we have one

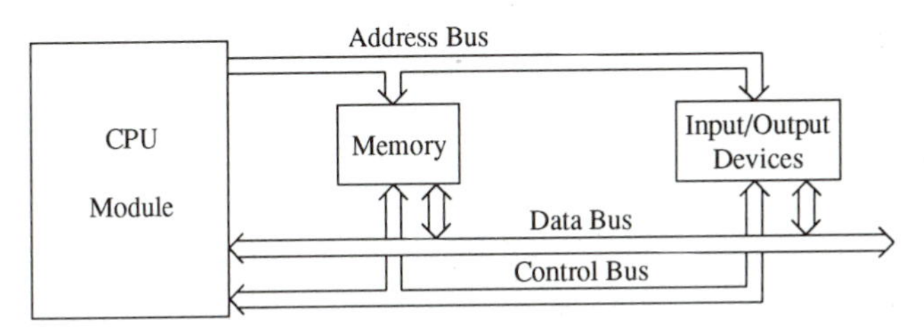

Figure 12-2. Block diagram of a microcomputer system with buses.

set of wires running throughout the system. The bus effectively connects all the devices together.

12-3 MICROPROCESSOR CONTROL OF CONTROL SYSTEMS

In this section we shall describe some of the applications of microprocessors in control systems. One way of using a microprocessor as a controller is shown in Fig. 12-3. In this case, the controlled process consists of a dc motor, load, and the power amplifier. The analog process and the microprocessor are interfaced through analog-to-digital and digital-to-analog (A/D and D/A) converters. Thus, the overall system is analytically considered to be a digital control system, with a sampling period of T s.

Let us consider that the purpose of the dc-motor control system is to drive the load speed $\omega(t)$ to follow the constant command speed ω_d. The error between the command speed and the load speed is

$$e(t) = \omega_d - \omega(t) \tag{12-1}$$

Thus, the input to the microprocessor is the digitized error signal $e(kT)$, $k = 0, 1, 2, \ldots$, and we let the output of the microprocessor be $u(kT)$.

Consider that the microprocessor is to perform the digital computation to implement a proportional-integral (PI) controller, so that the continuous-data form of the microprocessor output is

$$u(t) = K_P e(t) + K_I \int e(t)\, dt \tag{12-2}$$

The integral in Eq. (12-2) is written as

$$x(t) = \int_{t_0}^{t} [\omega_d - \omega(\tau)]\, d\tau + x(t_0) \tag{12-3}$$

where t_0 is the initial time and $x(t_0)$ is the initial value of $x(t)$. To approximate the integral by a digital model, several schemes may be used. Let us use the

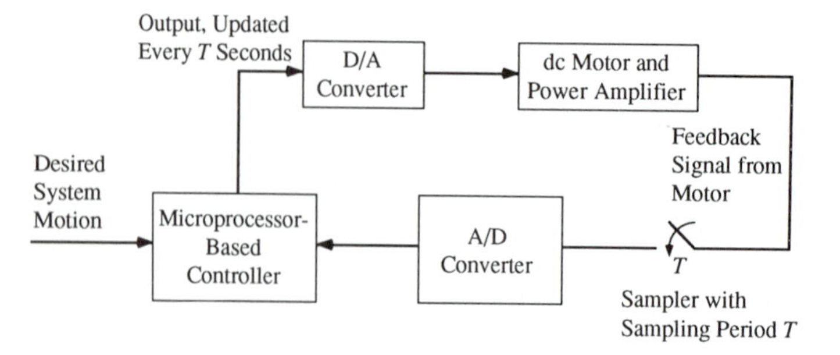

Figure 12-3. Block diagram of a microcomputer-controlled dc motor system.

trapezoidal rule here, and let $t = kT, t_0 = (k - 1)T$. Then, the definite integral in Eq. (12-3) is approximated as

$$\int_{(k-1)T}^{kT} [\omega_d - \omega(t)]\, dt \cong \omega_d T - \frac{T}{2}[\omega(kT) + \omega(k-1)T] \tag{12-4}$$

for $k = 1, 2, \ldots$. Therefore, the value of the integral in the last equation at $t = kT$ can be computed based on the given input ω_d and the data for $\omega(kT)$ and $\omega[(k-1)T]$. However, in reality, the microprocessor takes a finite amount of time to compute the integral in Eq. (12-4), so that given the input data $\omega[(k-1)T]$ and $\omega(kT)$, the result of the integral computation is not available at $t = kT$. In general, one has to add up all the time intervals required to execute the operation codes or *op-codes* of the integration subroutine on the microprocessor to find out what this time delay is. For convenience, we shall assume that this computation time delay is equal to one sampling period T. This means that the right-hand side of Eq. (12-4) gives the computational result of the integral at $t = (k+1)T$. Thus, Eq. (12-3) is discretized to

$$x[(k+1)T] = \omega_d T - \frac{T}{2}[\omega(kT) + \omega[(k-1)T]] + x(kT) \tag{12-5}$$

Notice that we are using $x(kT)$ rather than $x[(k-1)T]$ as the initial state of $x(t)$. Substituting $x[(k+1)T]$ for the integral in Eq. (12-2), we can write the discretized version of $u(t)$ as

$$u[(k+1)T] = K_P[\omega_d - \omega(kT)] + K_I x[(k+1)T] \tag{12-6}$$

This control is applied to the dc motor system at $t = (k+1)T$, $k = 0, 1, 2, \ldots$. The control is updated every T s and is held constant between the sampling instants.

The block diagram of the typical microprocessor system used to implement the digital PI controller is shown in Fig. 12-4. The system shown in this figure uses an analog timer to determine the start of the next sampling period. Alternatively, the microprocessor can use a software timing loop to keep track of when T s have elapsed. The timer outputs a pulse every T s. This pulse is used in two ways. First, the pulse is applied to the interrupt line of the microprocessor. This will cause the processor to stop what it is doing and execute the interrupt routine, which in this case would be to output the next value of the control, $u[(k+1)T]$. This control is sent to the D/A converter whose output in turn controls the power amplifier. The timing pulse from the timer is also sent to the sample command line of the accumulator. A pulse on the sample line triggers the sample-and-hold circuitry within the accumulator, at which instant the motor velocity $\omega(t)$ is sampled and held constant for one sampling period. The value of $\omega(kT)$ is then converted to an N-bit binary number by the rest of the A/D circuitry. As described in Chapter 2, a finite conversion time is associated with the conversion process.

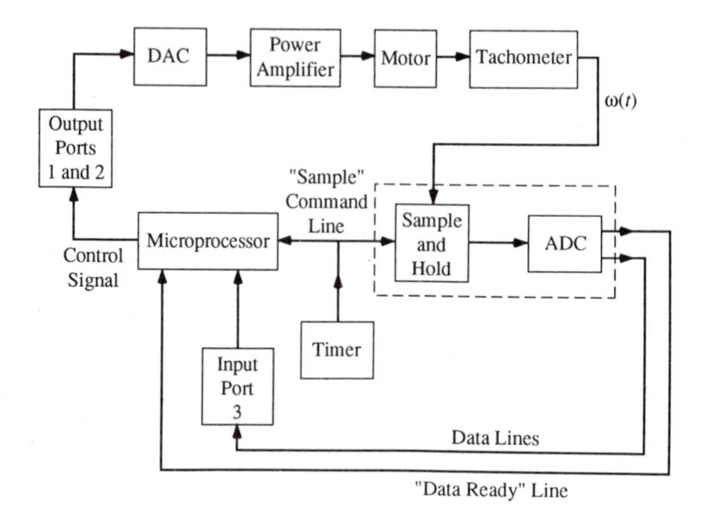

Figure 12-4. Block diagram of implementation of discretized PI controller.

The implementation of the PI controller is normally done by writing the assembly-language program. The operation code of the assembly language depends on the microprocessor used. In general, the PD or the PID controller can be implemented in a similar manner, once the derivative operation is approximated numerically.

12-4 SINGLE-BOARD CONTROLLERS WITH CUSTOM-DESIGNED CHIPS

In general, the implementation of digital controllers can be achieved in at least two levels. For high-volume cost-conscious original-equipment manufacturer (OEM) users, individual integrated circuits (ICs) and microprocessors, such as that discussed in the last section, are preferred. The advantages of this approach are low cost and more flexibility. The disadvantages are longer design and development lead time. To simplify the design effort and shorten the turnaround time, many single-board controllers with custom-designed microprocessor chips are available. These commercially available boards provide certain control features that simplify the control engineer's task considerably. All the control engineer has to do is to set the parameters of the predesigned chips with a personal computer or smart terminal, depending on the performance requirements of the control system. The applications of one of such boards are described in the following.

12-4-1 The Galil DMC-105 Board

The DMC-105 manufactured by Galil Motion Control, Inc., is a general-purpose motion controller for small dc and brushless motors. It is programmed

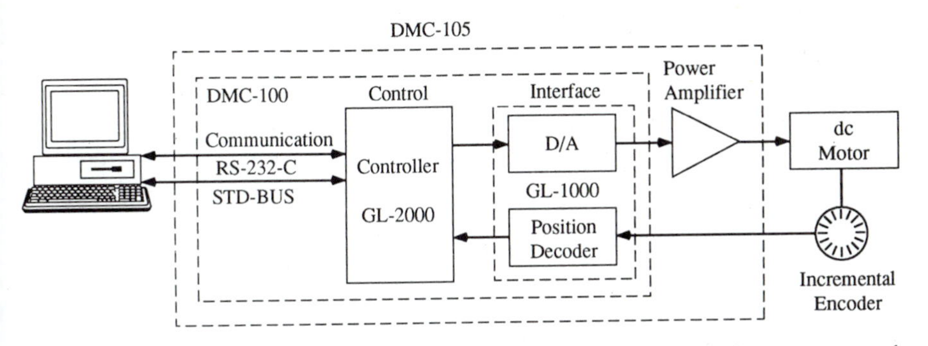

Figure 12-5. Functional blocks of the Galil DMC-105 in a dc-motor control system.

and monitored by an IBM PC by using an RS-232-C serial interface or STD bus. Figure 12-5 shows the application of the DMC-105 board, which contains the DMC-100 controller and the linear power amplifier, to the closed-loop control of a dc motor. A photograph and more detailed information on the DMC-100 controller board are given in Appendix F. Position information is fed back to the DMC-100 controller by an incremental encoder coupled to the motor shaft. The controller can control motor position, velocity, or torque, depending on the mode selected by the user [4].

Two ICs, the MC1488 and the MC1489, interface between the controller and the RS-232-C cable. The heart of the controller is the GL-2000 programmable microprocessor, a 40-pin custom IC. The GL-2000 microprocessor communicates with the host computer via the RS-232-C ICs. The microprocessor interprets and executes commands sent to it by the host computer. There are 38 permissible commands, each consisting of a 2-character operation code (sent in upper-case ASCII), followed by an integer-valued operand when appropriate. The commands can be either directions for motor control or interrogations of motor status and internal controller parameters.

In addition, the GL-2000 microprocessor implements the controller's first-order compensator. The user can set the compensator's gain, pole, and zero locations individually.

The dc motor is coupled to a Hewlett-Packard incremental encoder containing a fixed light source and sensor, separated by a rotating disk. The disk contains finely spaced radial slits which allow the sensor to be alternately illuminated by and shielded from the light source. The output of the sensor is a sinusoidally varying voltage when the motor runs at a constant speed. The incremental encoder also includes a second sensor positioned so that its sinusoidal signal is in quadrature (90 degrees out of phase) with that of the first sensor.

The GL-2000 microprocessor communicates with the encoder interface IC, the GS-1000, via an 8-bit bus. The encoder interface sends the microprocessor motor position information over this bus. The microprocessor, in turn, sends the encoder interface the control signal over the same bus. The GL-1000 encoder interface is a custom IC that contains the decoder for the motor's incremental encoder and the digital-to-analog converter for the motor driver. The GL-1000 decodes the quadrature incremental encoder signals. By comparing the phase sign of the second sensor channel with that of the first, the GL-1000 can determine the direction of rotation. By counting the number of cycles of one of the channels, the GL-1000 can accurately determine the relative change in position of the motor shaft. Note that the decoding of the analog quadrature signals results in an inherent quantization of the position information. Position information is sent over the 8-bit bus to the GL-2000 microprocessor where the motor velocity can be computed for velocity control.

The sampling period T of the DMC-105 is fixed at 0.0005 s with the clock rate of 2 kHz.

A sampled-data model of the dc-motor position-control system utilizing the DMC-105 board is shown in Fig. 12-6.

The Digital Controller

In Fig. 12-6, the transfer function of the digital controller is given as

$$D_c(z) = \text{GN}\frac{z - (\text{ZR}/256)}{z - (\text{PL}/256)} \tag{12-7}$$

where GN is the gain factor which can be set to any integer from 1 through 255. The values of ZR and PL can be set to any integer from 0 to 255. Thus, the zero and the pole of the controller can assume values from 0 to 255/256 (=0.9961).

The D/A Model

The 8-bit D/A converter is modeled by a gain and a zero-order hold (zoh). The value of the gain is

$$K_h = 0.078 \qquad \text{V/counts} \tag{12-8}$$

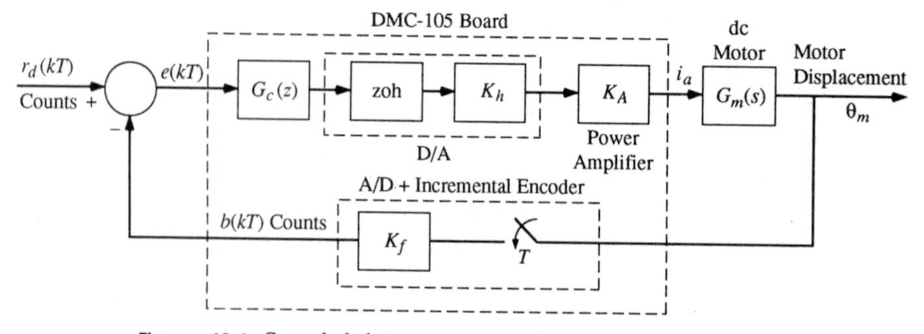

Figure 12-6. Sampled-data system model of the dc-motor control system with the DMC-105 controller board.

The Power Amplifier

The current-source chopper power amplifier is modeled by a gain with the value given as

$$K_a = 0.2 \qquad \text{A/V} \tag{12-9}$$

The Encoder-Decoder

The DMC-105 contains position decoding circuitry in the GL-1000 IC that decodes the output of the incremental encoder. The encoder-decoder combination performs the A/D conversion of the motor position from an analog signal in radians to a discrete-time signal in counts. As shown in Fig. 12-6, the encoder-decoder operation is modeled by an ideal sampler connected in cascade with an A/D gain of K_f. The value of K_f is

$$K_f = 343.77 \qquad \text{counts/rad} \tag{12-10}$$

The DC Motor

Since the output of the current-source power amplifier is current $i_a(t)$, the torque developed by the motor is written as

$$T_m(t) = K_m i_a(t) \tag{12-11}$$

where K_m is the torque constant in oz-in/A.

The differential equations that model the dynamics between the motor torque and the motor displacement are

$$\frac{d\omega_m(t)}{dt} = T_m(t) - \frac{B_m}{J_m}\,\omega_m(t) \tag{12-12}$$

$$\frac{d\theta_m(t)}{dt} = \omega_m(t) \tag{12-13}$$

where $\omega_m(t)$ and $\theta_m(t)$ are the motor velocity and displacement of the motor, respectively. The dc motor is a Pittman model 14203 with the following parameters:

Rated voltage	$V = 24$ V
Stall current	$I_s = 17.4$ A
No-load current	$I = 0.195$ A
Torque constant	$K_m = 9.26$ oz-in/A
Back EMF constant	$K_b = 0.065$ V/rad/s
Armature resistance	$R_a = 1.38\ \Omega$
Armature inductance	$L_a = 0.00226$ H
Rotor inertia	$J_m = 0.0047$ oz-in-s^2
Rotor viscous friction coefficient	$B_m \cong 0$ oz-in-s
Peak torque	160 oz-in
No-load speed	358 rad/s

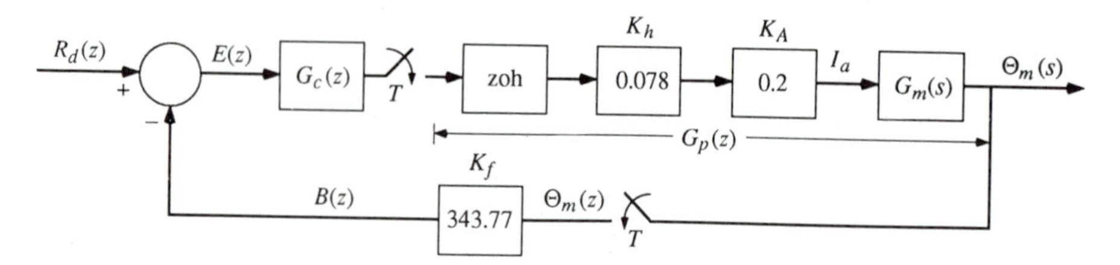

Figure 12-7. Analytical model of the sampled-data system in Fig. 12-6.

Since the viscous friction of the motor is negligible, the transfer function between the motor current and the displacement is simply

$$G_m(s) = \frac{\Theta_m(s)}{I_a(s)} = \frac{K_m/J_m}{s^2} = \frac{1970.21}{s^2} \tag{12-14}$$

An analytical model of the system is shown in Fig. 12-7. The open-loop transfer function of the system is written

$$G(z) = \frac{\Theta_m(z)}{E(z)} = G_c(z)(1 - z^{-1})(0.078)(0.2)\mathscr{Z}\left(\frac{1970.21}{s^3}\right)$$

$$= \text{GN}\left(\frac{z - \text{ZR}/256}{z - \text{PL}/256}\right)\left(\frac{15.37T^2(z+1)}{(z-1)^2}\right) \tag{12-15}$$

The closed-loop transfer function of the system is

$$\frac{\Theta_m(z)}{R_d(z)} = \frac{G(z)}{1 + 343.77G(z)} \tag{12-16}$$

The characteristic equation of the uncompensated closed-loop system, with GN = 1, ZR = 0, and PL = 0, is obtained from the numerator polynomial of $1 + 343.77G(z)$, or

$$z^2 - 1.99868z + 1.001321 = 0 \tag{12-17}$$

Since the coefficient of the z^0 term in the characteristic equation is greater than unity, from the stability criterion discussed in Chapter 6, the uncompensated system is unstable. The roots of the characteristic equation are $z = 0.99934 + j0.051386$ and $z = 0.99934 - j0.051386$, which are just outside the unit circle.

Since the system is of type 2, the steady-state error $e(kT)$, as $k \to \infty$ due to a step input, is zero. Since the feedback path gain is 343.77, if a unit-step input of one count is applied as $r_d(kT)$, the steady-state value of the motor displacement $\theta_m(t)$ would be 1/343.77 rad.

The GN, ZR, and PL parameters of the digital controller are designed by trial-and-error. When GN = 127, ZR = 255, and PL = 0, the output response

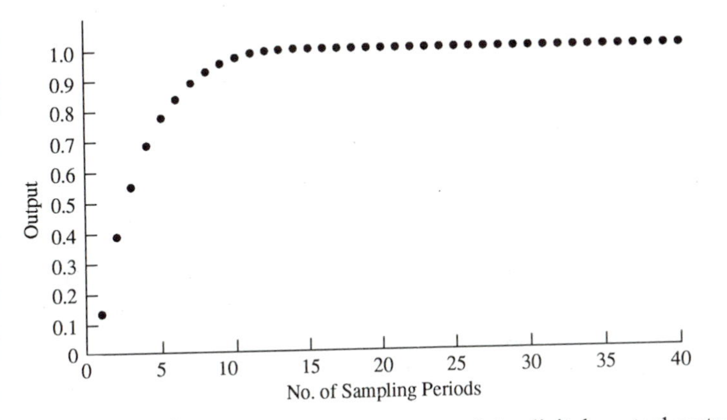

Figure 12-8. Unit-step response of the digital control system in Fig. 12-7.

reaches the reference input in approximately 14 sampling instants, or 0.007 s. The maximum overshoot is approximately 1 percent. Figure 12-8 shows the unit-step response of the compensated system. The transfer function of the digital controller is

$$G_c(z) = 127(z - 0.9961) \tag{12-18}$$

The closed-loop transfer function of the compensated system is

$$\frac{\Theta_m(z)}{R_d(z)} = \frac{0.000488(z^2 + 0.0039z - 0.9961)}{1.16764975z^2 - 1.99934617z + 0.83300408} \tag{12-19}$$

Apparently, as with any conventional design, the controller parameters are not unique.

The single-board controller with custom-designed IC chips does have the disadvantage of being restricted to the configuration and parameters designed into the system, such as the fixed controller type and the fixed sampling period. However, the advantages are that the board consists of all the hardware interfaces and interconnections required to accomplish the control function. The ICs have already been programmed, so that the control system engineer is relieved of the chores of writing the assembly or machine language programs.

Just to list another example, the National Semiconductor LM628 controller is a custom chip that is assembled on a circuit board that is compatible with the IBM PC. The board, which is installed in an IBM PC, also contains the D/A converter that converts the digital PID controller output to a -10-V to $+10$-V analog signal. This analog signal is then amplified by a power amplifier to produce the needed control signal for the dc motor. The user need only enter the parameters of the PID controller via the PC keyboard to implement the control algorithm.

12-5 DIGITAL SIGNAL PROCESSORS

While microprocessors have many advantages as digital controllers in control systems, they have many limitations. Since a microprocessor is designed to perform a multitude of functions, including managing input/output, buses, and data manipulation, the speed at which these functions can be executed is relatively low. In addition, the arithmetic logic unit of a microprocessor is not designed for high-speed number crunching, and the speed for real-time computation is limited. For example, a general-purpose microprocessor takes at least 5 μs to multiply two 16-bit numbers, since multiplication is done by repetitive addition. This is why the Galil GL-2000 chip is limited to a 2-kHz sampling frequency. Another reason for speed limitation is that a standard microprocessor has only a single bus which has to be shared by program commands and data. As a result, many microprocessor applications in control systems employ the microprocessor essentially as a look-up table that contains precomputed control parameters. This practice severely limits the flexibility of the controller, as well as consuming valuable memory space.

In recent years, digital signal processors have been designed specifically for high-speed computing, and they are suitable for high-speed control applications. These single-chip processors are commercially available in VLSI circuits with competitive prices. The main difference between the DSP and the microprocessor is that the ALU of the DSP contains hardware multipliers that can handle the multiplication and accumulate operations rapidly. The DSP also features a dual-bus architecture that allows simultaneous processing of program instructions and data. Thus, the typical cycle time for a DSP to multiply two 16-bit numbers is between 50 and 200 ns, which is about 100 times faster than that of the standard microprocessor. These improvements in execution speed allow the sampling frequency in a DSP to be over 20 kHz. Other hardware enhancements in the DSP also improve the precision of the calculations. The improved frequency response of the DSP allows the design of notch controllers with sharp cutoff characteristics that cannot be achieved by analog and standard microprocessor controllers. The high-speed capability of the DSP also allows the device to be applied to adaptive control, in which case, the processor can simultaneously perform monitor and control functions.

The applications of the DSPs in control systems are very similar to those of the microprocessor. The block diagram of a DSP-controlled motor system is illustrated in Fig. 12-9.

Many DSPs are available commercially that can be applied to a wide range of applications. The DSPs of the TMS320 family manufactured by Texas Instruments are among the most popular. We shall describe briefly the TMS320 DSPs in the following section.

12-5-1 The Texas Instruments TMS320 DSPs

In 1982 Texas Instruments released the TMS320 family of DSPs, starting with the TMS32010. New generations of the TMS320 DSPs support floating-

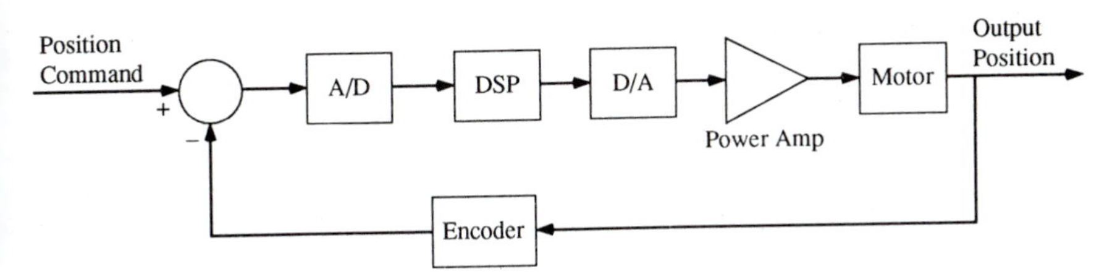

Figure 12-9. Block diagram of a DSP-controlled motor system.

point as well as fixed-point arithmetics. We shall describe in this section the architecture of the TMS320C30, a third-generation member of the TMS320 family. The reader may refer to the literature [6]–[9] for more detailed information and application data.

The functional block diagram of the TMS320C30 is shown in Fig. 12-10. The diagram gives the key feature of the device.

The CPU

As shown in Fig. 12-10, the CPU consists of the following key elements: *integer/floating-point multiplier, the ALU for performing floating-point, integer, and logical operations.*

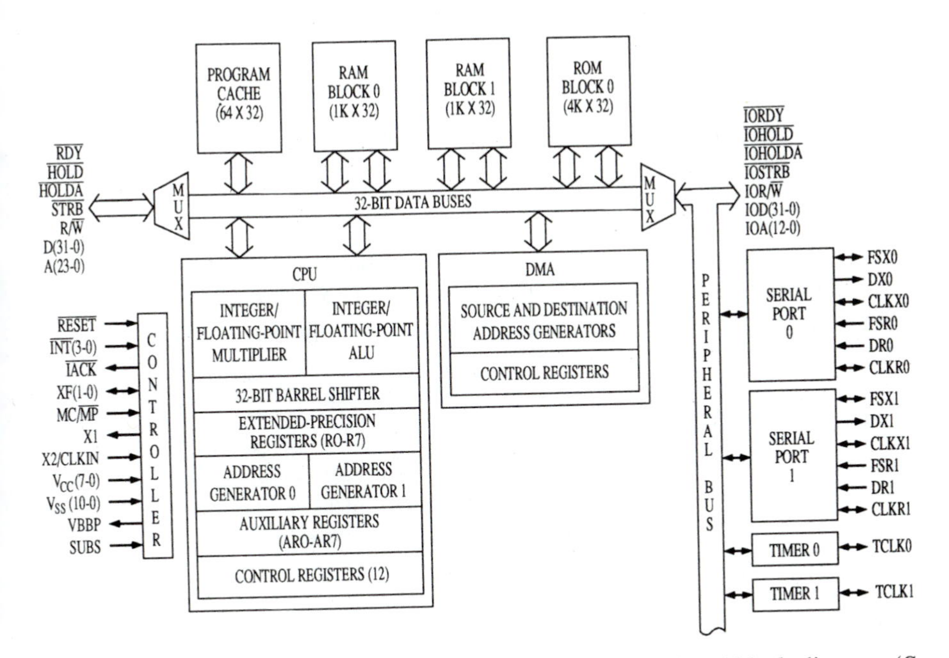

Figure 12-10. Texas Instruments TMS320C30 functional block diagram. (*Source*: K. S. Lin, G. A. Frantz, and R. Simar, "The TMS320 Family of Digital Signal Processors," *Proc. IEEE*, vol. 75, no. 9. © 1987, IEEE.)

The multiplier of the CPU performs floating-point and integer multiplications. When performing floating-point multiplication, the inputs are 32-bit floating-point numbers, and the result is a 40-bit floating-point number. When performing integer multiplication, the input data are 24-bit numbers, and the result is a 32-bit integer. The ALU performs 32-bit integer, 32-bit logical, and 40-bit floating-point operations.

The register file has a total of 28 registers. The first eight registers are the extended-precision registers which support operations on the 32-bit integers and the 40-bit floating-point numbers. The next eight registers are the auxiliary registers whose primary purpose is related to the generation of addresses. The remaining 12 registers are the control registers which support a variety of system functions such as addressing, stack management, processor status, block repeat, and interrupts.

Memory

The TMS320C30 features on-chip memory that minimizes system cost. The total memory of the TMS320C30 is 16 million × 32 bits which contain program, data, and I/O space. A machine word is 32 bits, and all addressing is performed by word. The RAM and ROM blocks are shown in Fig. 12-10.

Direct Memory Access

The on-chip direct memory access (DMA) controller of the TMS320C30 is capable of reading from and writing to any location in the memory without interfering with the operation of the CPU. Dedicated DMA address and data buses allow data transfer without conflicts between the CPU and the DMA controller.

Peripherals and External Interfaces

Figure 12-10 shows that the TMS320C30 has two independent serial ports and two timers, all connected to the peripheral bus. In addition, there are two external interfaces: the parallel interface and the I/O interface, both consisting of a data bus, an address bus, and a set of control signals.

12-5-2 Development System and Support Tools

Although DSPs offer many benefits, as with microprocessors, designing them into control systems applications is difficult. Unless a compiler is available for a specific DSP, the operation codes must be written in assembly language. This means that it is essential to have development tools in the form of evaluation modules for application evaluation, assembler/linkages, software simulators, and full-capacity hardware emulators. Some of these development systems have emerged commercially which will make the tasks of applying DSPs much simpler. The Power-14 board [10] offered by Teknic, Inc., of

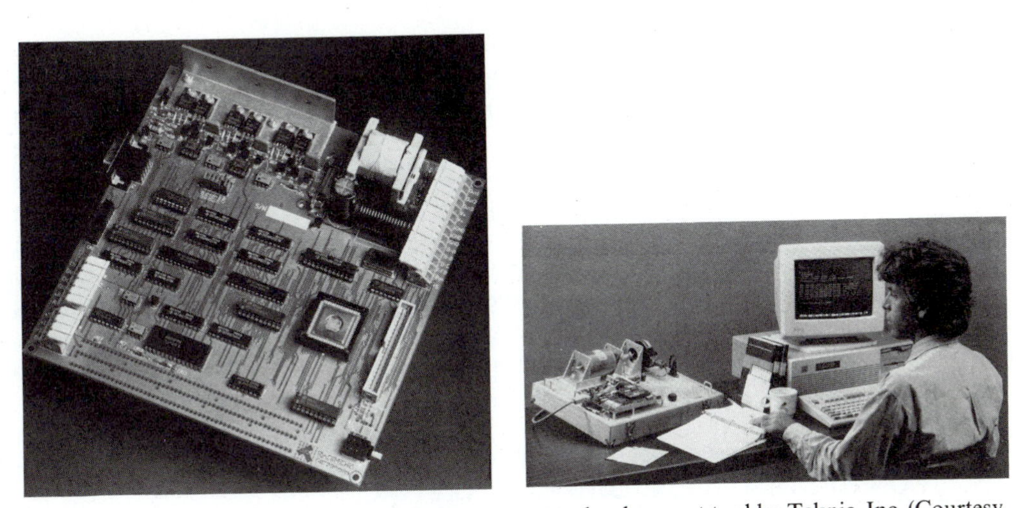

Figure 12-11. Power-14 DSP microprocessor development tool by Teknic, Inc. (Courtesy of Teknic, Inc.)

Rochester, New York, is designed for the TMS320C14 DSP. The Power-14 development system allows the engineer to prototype DSP control systems on PC platforms, using methods similar to those for microprocessors. Figure 12-11 shows the circuit board of the Power-14, which consists of a Texas Instruments TMS320C14 processor, emulation hardware providing downloadable program memory, RS-232 communications hardware, a monitor/debugger program, and

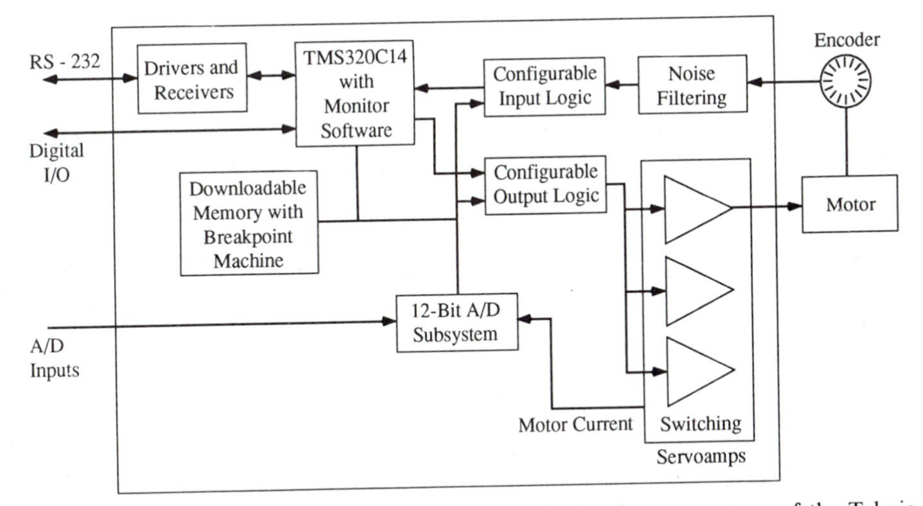

Figure 12-12. Block diagram showing the hardware structure of the Teknic Power-14 DSP microcontroller development board. (Courtesy of Teknic, Inc.)

three channels of switching servo-amplifiers. An A/D subsystem is also provided. The block diagram showing the hardware structure of the Power-14 is shown in Fig. 12-12. A dc motor and an incremental encoder are connected at the output to demonstrate the control application of the Power-14. A software package is loaded on the PC to provide full control of the system through the RS-232 link. The user may vary the coefficients of the PID control algorithm preprogrammed in the software package. The user may also generate custom control codes by writing program modules in the TMS320C14 assembly language on the monitor/debugger. Programs may be tested with various emulator and simulator software.

12-6 EFFECTS OF FINITE WORDLENGTH AND QUANTIZATION ON CONTROLLABILITY AND CLOSED-LOOP POLE PLACEMENT

Because microprocessors and DSPs have finite wordlengths, in general, the signals in and output of these processors will be truncated or quantized, and the parameters used in the control law will be truncated when the latter are being realized by the op-codes of the processor. Most microprocessors and the early-generation DSPs use fixed-point arithmetic processors in which only a finite amount of wordlength is available to represent the magnitude of the signal or a gain constant. Thus, signals and coefficients must be scaled to fit the wordlength of the processor. Therefore, it is necessary to investigate the effects on the system performance when the system parameters can be realized only by a finite set of numbers.

In the preceding chapters the design of digital control systems through pole placement has been discussed. The design parameters are the state-feedback gains or the output-feedback gains. It was pointed out in the pole-placement design that if the system is completely controllable, then the poles of the closed-loop system can be arbitrarily placed in the z-plane. However, if the feedback gains cannot assume an arbitrary set of numbers, the closed-loop poles do not have a domain with an infinite resolution. Put another way, if the feedback gains can be realized by only a finite set of numbers, any initial state can be driven to only a finite number of final states in finite time. Therefore, it seems appropriate to state that in the rigorous sense of the definition of controllability, when a digital control system is subject to finite wordlength and quantization, it cannot be a controllable system. However, from a practical standpoint, if the quantization levels are small, we should not be overly concerned with the controllability problems. Rather, the limitation on the resolution of the system parameters is translated to system errors and stability problems which are direct results of amplitude quantization. The same is true when the gain and coefficients of a digital controller such as a PID controller must be set according to the analytical design.

The effect of quantization on the control input $u(k)$ can be investigated by referring to the system

$$\mathbf{x}(k+1) = \mathbf{A}\mathbf{x}(k) + \mathbf{B}u(k) \tag{12-20}$$

where $u(k)$ is a scalar input that is restricted to a quantized set of values, that is,

$$u(k) = n_k q \tag{12-21}$$

where $n_k = 0, \pm 1, \pm 2, \ldots$, and q is the quantization level.

The solution of Eq. (12-20) is

$$\mathbf{x}(N) = \mathbf{A}^N\mathbf{x}(0) + \sum_{k=0}^{N-1} \mathbf{A}^{N-k-1}\mathbf{B}u(k) \tag{12-22}$$

Substituting Eq. (12-21) into Eq. (12-22), we have

$$\mathbf{x}(N) = \mathbf{A}^N\mathbf{x}(0) + q\sum_{k=0}^{N-1} \mathbf{A}^{N-k-1}\mathbf{B}n_k \tag{12-23}$$

Since the magnitude of $u(k)$ is constrained to distinct levels of $n_k q$, the final state $\mathbf{x}(N)$ is parameterized by the set of integers n_k, $k = 0, 1, 2, \ldots, N-1$. If the pair $[\mathbf{A}, \mathbf{B}]$ is controllable, $\mathbf{x}(0)$ can be driven to any state $\mathbf{x}(N)$ in the state space only if $u(k)$ can be selected from a continuum of values. However, since $u(k)$ is limited to quantized values, $\mathbf{x}(N)$ is also restricted. The following example illustrates the controllability problem due to quantization in digital control systems.

Example 12-1

Consider the digital control system that is described by the state equation in Eq. (12-20), where

$$\mathbf{A} = \begin{bmatrix} 0 & 1 \\ -1 & -2 \end{bmatrix} \qquad \mathbf{B} = \begin{bmatrix} 0 \\ 1 \end{bmatrix} \tag{12-24}$$

and $u(k)$ is subject to amplitude quantization with quantization level q. The state transition equation of Eq. (12-20) is given in Eq. (12-22), where

$$\mathbf{A}^N = \begin{bmatrix} (-1)^{N+1}(N-1) & (-1)^{N+1}N \\ (-1)^N N & (-1)^N(N+1) \end{bmatrix} \tag{12-25}$$

Since the pair $[\mathbf{A}, \mathbf{B}]$ is completely controllable, the vector $\mathbf{A}^{N-k-1}\mathbf{B}$ can be expressed as a linear combination of the two vectors $\mathbf{B}$ and $\mathbf{AB}$. In the present case, we can show that

$$\mathbf{A}^{N-k-1}\mathbf{B} = (-1)^{N-k-2}[(N-k-2)\mathbf{B} + (N-k-1)\mathbf{AB}] \tag{12-26}$$

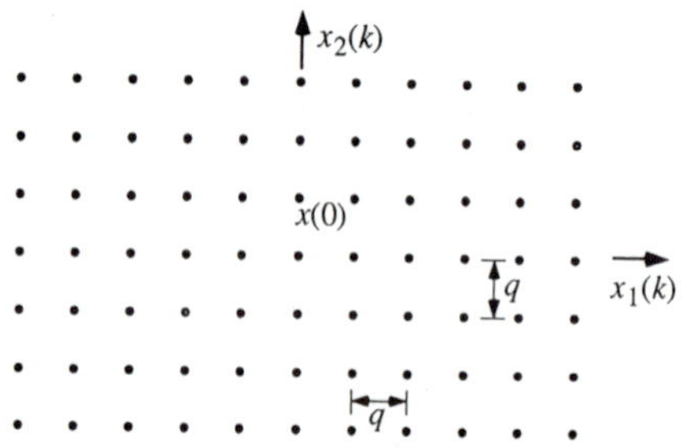

Figure 12-13. Realizable states of $x_1(k)$ and $x_2(k)$ due to amplitude quantization in states and controls.

Since $\mathbf{A}^{N-k-1}$ is linearly and integrally dependent of $\mathbf{B}$ and $\mathbf{AB}$, we can write Eq. (12-22) as

$$\mathbf{x}(N) = \mathbf{A}^N\mathbf{x}(0) + n_1 q\mathbf{B} + n_2 q\mathbf{AB} \tag{12-27}$$

$$= \mathbf{A}^N\mathbf{x}(0) + n_1 q\begin{bmatrix} 0 \\ 1 \end{bmatrix} + n_2 q\begin{bmatrix} 1 \\ -1 \end{bmatrix} \tag{12-28}$$

where n_1 and n_2 are integers and q is the quantization level.

Equation (12-28) shows that at any particular sampling instant, $k = N$, $x_1(N)$ and $x_2(N)$ can assume a discrete set of values only. For any given initial state $\mathbf{x}(0)$, the sets of reachable states for $x_1(k)$ and $x_2(k)$ are spaced at intervals of q in the x_1-versus-x_2 plane, as shown in Fig. 12-13.

When the feedback gains of state feedback or output feedback are implemented by a microprocessor or DSP, these gains are again subject to amplitude quantization. This simply means that if the design is for the objective of pole placement, the poles of the closed-loop system cannot be placed with arbitrarily fine resolution in the z-plane. The following example illustrates the effect of quantization on the pole-placement design.

Example 12-2

Consider the digital system modeled in Eq. (12-20), where the coefficient matrices are given as

$$\mathbf{A} = \begin{bmatrix} 0 & 1 \\ 0 & 0 \end{bmatrix} \qquad \mathbf{B} = \begin{bmatrix} 0 \\ 1 \end{bmatrix} \tag{12-29}$$

The state-feedback control is described by

$$u(k) = -\mathbf{G}\mathbf{x}(k) \tag{12-30}$$

$$\mathbf{G} = [g_1 \quad g_2] \tag{12-31}$$

The values of g_1 and g_2 are quantized with quantization level q. The characteristic equation of the closed-loop system is written

$$|z\mathbf{I} - \mathbf{A} + \mathbf{BG}| = z^2 + g_2 z + g_1$$
$$= (z - r_1)(z - r_2) = 0 \tag{12-32}$$

where r_1 and r_2 are the two closed-loop eigenvalues. Then, the feedback gain matrix is

$$\mathbf{G} = [g_1 \quad g_2] = [r_1 r_2 \quad -(r_1 + r_2)] \tag{12-33}$$

Figure 12-14 shows the region of stable operations in the g_1-versus-g_2 parameter plane when the feedback gains g_1 and g_2 are subject to fixed-point data representation with a 3-bit wordlength. In this case, the quantization level is

$$q = 2^{-3} = 1/8 \tag{12-34}$$

The realizable values of g_1 and g_2 are indicated by dots in the parameter plane, and only the values that correspond to a stable closed-loop system are shown.

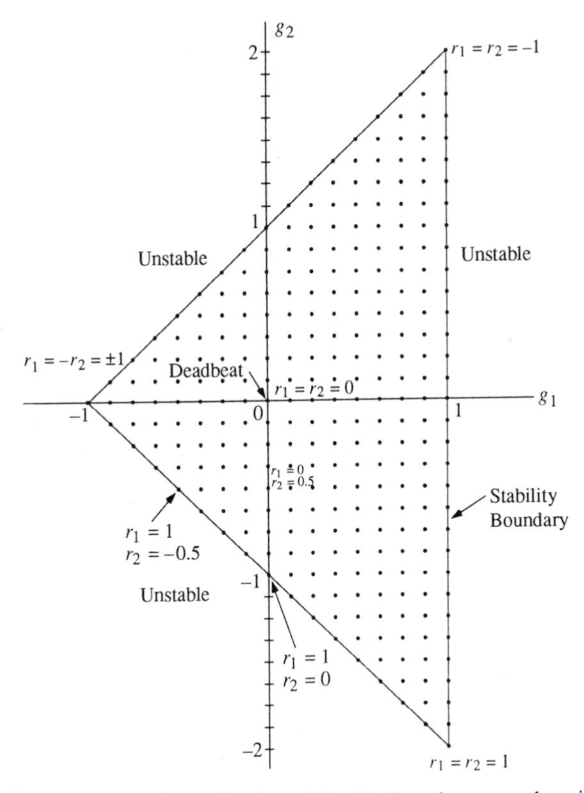

Figure 12-14. Quantization of feedback gains g_1 and g_2 in the parameter plane.

It is useful to investigate the realizable closed-loop poles due to the quantization of g_1 and g_2. Let the roots of Eq. (12-32) be represented in polar form,

$$z_1 = re^{j\theta} \tag{12-35}$$

and

$$z_2 = re^{-j\theta} \tag{12-36}$$

The characteristic equation in Eq. (12-32) is written

$$z^2 - 2r\cos\theta z + r^2 = 0 \tag{12-37}$$

Then,

$$r = \pm\sqrt{g_1} \tag{12-38}$$

Thus, quantizing g_1 with a quantization level q forces the roots to lie on concentric circles in the z-plane. These circles are all centered at the origin, with radii equal to $0, \sqrt{q}, \sqrt{2q}, \ldots$. These circles are shown in Fig. 12-15 for the present case with $q = 1/8$.

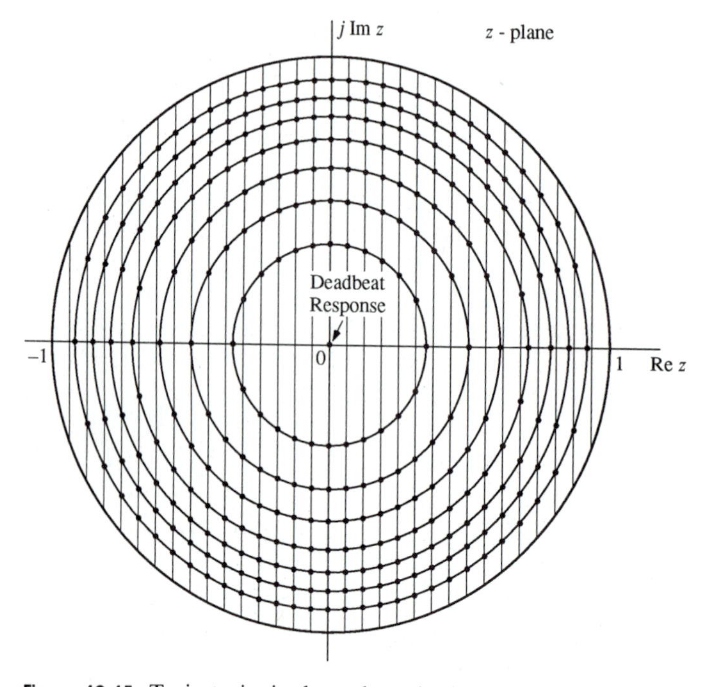

Figure 12-15. Trajectories in the z-plane showing the realizable poles due to the quantization of feedback gains g_1 and g_2.

Since the real parts of the roots are $r \cos \theta$, quantizing g_2 $(= -2r \cos \theta)$ is equivalent to limiting the real parts of the roots to a finite set of numbers. Let the real parts of the roots be denoted by σ, then

$$\sigma = r \cos \theta = -g_2/2 \tag{12-39}$$

The intersections of the circles and the vertical lines represent the realizable poles due to the prescribed quantization of g_1 and g_2.

Another problem with finite wordlength and quantization in closed-loop digital control systems is limit cycles. That is, the system may exhibit sustained oscillations in the steady state. The amplitude of oscillation is usually an integral multiple of the quantization level q. For small values of q, the limit cycles may be tolerable in noncritical systems. For high-precision control systems, limit cycles are not permissible.

The analytical approach to detecting limit cycles in digital control systems due to quantization and ways to avoid the phenomena from a design standpoint are beyond the scope of this text.

12-7 EFFECTS OF QUANTIZATION—LEAST UPPER BOUND ON QUANTIZATION ERROR

In the preceding section the effect of magnitude quantization and finite wordlength was demonstrated in terms of pole-placement design and controllability. In this section the effect of quantization on steady-state error is investigated.

The block diagram of an amplitude quantizer, together with its input-output characteristics, is shown in Fig. 12-16. The dotted line represents the desired transfer characteristic, and the staircase function is the actual characteristic of the quantizer. Notice that the input to the quantizer $r(t)$ could be of any form of amplitude, but the output $y(t)$ can take on only those discrete quantized levels that are nearest to the value of $r(t)$. The transfer characteristic shown in Fig. 12-16 has uniformly spaced quantized levels, so that when the input lies between $-q/2$ and $q/2$, the output is zero; when the amplitude of the input lies between $q/2$ and $3q/2$, the output amplitude is q, and so on. As explained in Chapter 2, the relation between the quantization level q and the wordlength is

$$q = 2^{-N}\text{FS} \tag{12-40}$$

where N is the number of significant binary bits or wordlength and FS represents full scale. The quantization error is $q/2$ and is given by

$$\frac{q}{2} = 2^{-N-1}\text{FS} \tag{12-41}$$

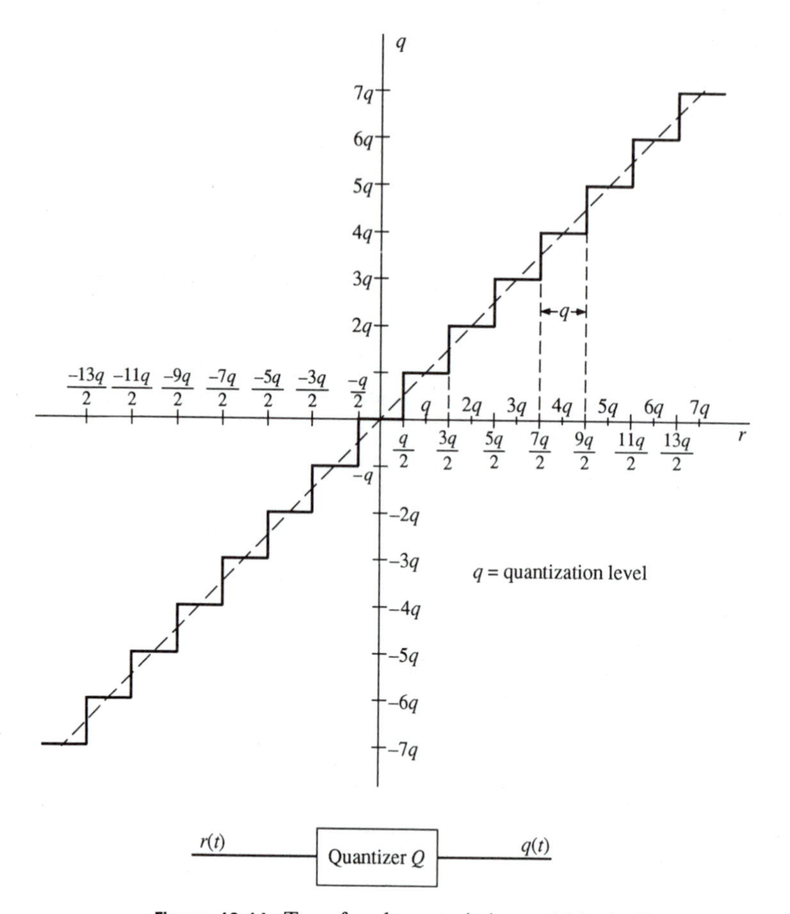

Figure 12-16. Transfer characteristics and block diagram representation of a quantizer.

In view of the transfer characteristic of the quantizer shown in Fig. 12-16, it is apparent that the device is nonlinear. Therefore, the analysis of the effects of quantization in a digital control system can be complex analytically.

The two most common effects of amplitude quantization in closed-loop systems are steady-state error and sustained oscillations. We shall illustrate these effects by means of a simple example.

Example 12-3

Let us refer to the digital control systems shown in Fig. 12-17. The quantization effects have been neglected in these systems. The only difference between the two systems in Fig. 12-17(a) and (b) is that one has negative feedback and the other has positive feedback. It is simple to show that both systems are

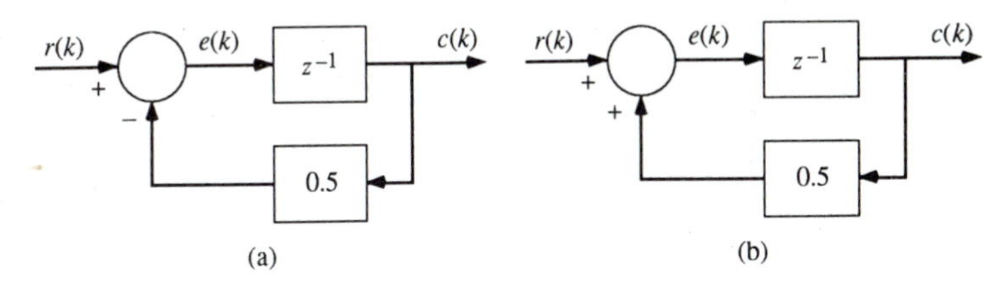

Figure 12-17. Two digital control systems.

asymptotically stable. For $r(k) = 0$, the z-transform of the output of the system in Fig. 12-17(a) is written

$$C(z) = \frac{z}{z + 0.5} c(0) \tag{12-42}$$

and that of the system in Fig. 12-17(b) is

$$C(z) = \frac{z}{z - 0.5} c(0) \tag{12-43}$$

where $c(0)$ is the initial value of $c(k)$.

The inverse z-transform of Eq. (12-42) is

$$c(k) = (0.5)^k \cos k\pi c(0) \tag{12-44}$$

and the inverse z-transform of Eq. (12-43) is

$$c(k) = (0.5)^k c(0) \tag{12-45}$$

Thus, in both cases the output response due to any arbitrary initial condition $c(0)$ goes to zero as k goes to infinity.

Figure 12-18 illustrates the same two digital control systems but with an amplitude quantizer in each loop. Assuming that a 4-bit wordlength is used in

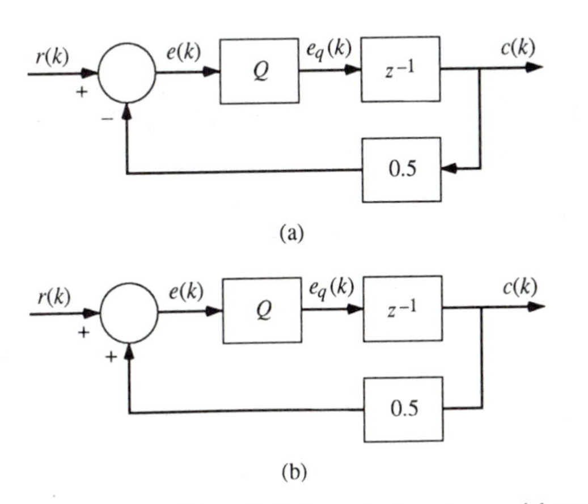

Figure 12-18. Two digital control systems with quantizers.

Table 12-1 Response of $c(k)$ with Negative Feedback; $r(k) = 0$ and $c(0) = 0.58$

k	$c(k)$	$e(k)$	$e_q(k)$
0	0.58000	−0.29000	−0.31250
1	−0.31250	0.15625	0.18750
2	0.18750	−0.09375	−0.12500
3	−0.12500	0.06250	0.06250
4	0.06250	−0.03125	−0.06250
5	−0.06250	0.03125	0.06250
6	0.06250	−0.03125	−0.06250
7	−0.06250	0.03125	0.06250

the digital process, so that the quantization level is $q = 2^{-4} = 0.0625$, the characteristics of the quantizer are defined as

$$e_q(k) = 0 \qquad -q/2 < e(k) < q/2 \tag{12-46}$$

$$e_q(k) = nq \qquad nq - q/2 \le e(k) < nq + q/2 \qquad n \ge 1 \tag{12-47}$$

$$e_q(k) = -nq \qquad -nq - q/2 < e(k) \le -nq + q/2 \qquad n \le -1 \tag{12-48}$$

For the system with negative feedback, shown in Fig. 12-17(a), the response of $c(k)$ with $r(k) = 0$ and $c(0) = 0.58$ is tabulated in Table 12-1. In this case, the output $c(k)$ exhibits a sustained oscillation with an amplitude of $\pm q$, and a period of two. For the system with positive feedback shown in Fig. 12-17(b), the response $c(k)$ for the same input and initial condition is given in Table 12-2. Notice that, in this case, the system has a constant steady-state error of q in $c(k)$ as k becomes large.

Table 12-2 Response of $c(k)$ with Positive Feedback; $r(k) = 0$ and $c(0) = 0.58$

k	$c(k)$	$e(k)$	$e_q(k)$
0	0.58000	0.29000	0.31250
1	0.31250	0.15625	0.18750
2	0.18750	0.09375	0.12500
3	0.12500	0.06250	0.06250
4	0.06250	0.03125	0.06250
5	0.06250	0.03125	0.06250
6	0.06250	0.03125	0.06250
7	0.06250	0.03125	0.06250

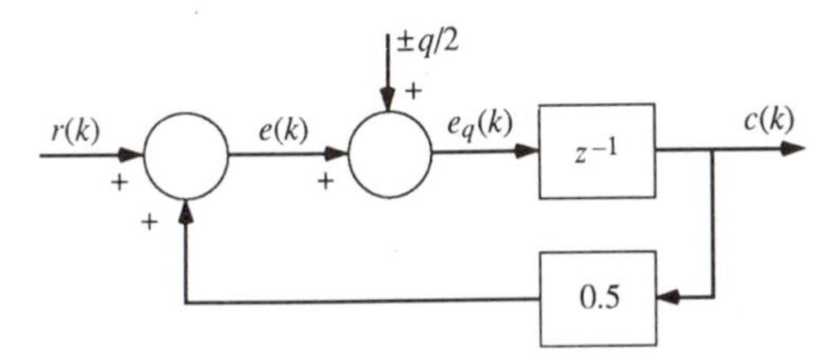

Figure 12-19. Block diagram of digital control systems with quantizer for least upper bound error analysis.

Although the two systems with quantizers shown in Fig. 12-17 both show imperfection in the responses that are directly dependent upon the quantization level q, the steady-state error and the sustained oscillation phenomena are grossly different in character. In practice, both the steady-state error and the sustained oscillations are undesirable in control systems, and their amplitudes should be kept as small as possible.

The least upper bound on the steady-state error due to quantization can be estimated by replacing the quantizers by equivalent noise sources. The prediction of sustained oscillations due to quantization is a nonlinear problem, and the analysis may be carried out by such techniques as the discrete describing function [14].

Since the error of the quantized signal has a least upper bound of $\pm q/2$, the "worst" error due to quantization in a digital control system can be studied by replacing the quantizer by an external noise source with a signal magnitude of $\pm q/2$. As an illustration, the block diagram of the digital control system with quantization shown in Fig. 12-18(b) is represented by the equivalent system in Fig. 12-19, where it is understood that the latter system model is used solely for the purpose of analyzing the least upper bound on the steady-state errors due to quantization.

From Fig. 12-19, with $r(k) = 0$, the steady-state value of $c(k)$ is obtained as

$$\lim_{k\to\infty} c(k) = \lim_{z\to 1}(1 - z^{-1})C(z) = \lim_{z\to 1}(1 - z^{-1})\frac{z^{-1}}{1 - 0.5^{-1}}\frac{(\pm q/2)z}{z - 1} \tag{12-49}$$

Therefore, in this case, the least upper bound predicted on the quantization error is identical to the value actually calculated. However, when we apply the same technique to the system in Fig. 12-18(a), the predicted least upper bound on the quantization error in $c(k)$ is $\pm q/3$. However, we have shown that the system actually has a sustained oscillation with an amplitude that varies between $-q$ and $+q$. Therefore, we can only say that the least upper bound error analysis using the equivalent noise sources does not predict the sustained-oscillation behavior in quantized systems. In general, we have to investigate both the steady-state error and the sustained-oscillation characteristics of a given digital system with quantization.

12-7-1 State-Variable Analysis

The least upper bound error analysis of quantized systems may be analyzed by the state-variable method or the z-transform method. Let the dynamic equations of a digital control system without quantization be

$$\mathbf{x}(k+1) = \mathbf{A}\mathbf{x}(k) + \mathbf{B}\mathbf{u}(k) \tag{12-50}$$

$$\mathbf{c}(k) = \mathbf{D}\mathbf{x}(k) + \mathbf{E}\mathbf{u}(k) \tag{12-51}$$

where $\mathbf{x}(k)$ is an n-vector, $\mathbf{u}(k)$ is an r-vector, and $\mathbf{c}(k)$ is a p-vector. Now let us consider that the same system described above has m quantizers which quantize the signals of the system. The quantization levels of these m quantizers are denoted by q_i, $i = 1, 2, \ldots, m$. As described earlier, we may represent the m quantizers by inputs with signals of magnitude $\pm q_i/2, i = 1, 2, \ldots, m$. The digital control system with quantizers is now described by the following dynamic equations:

$$\mathbf{x}_q(k+1) = \mathbf{A}\mathbf{x}_q(k) + \mathbf{B}\mathbf{u}(k) + \mathbf{F}\mathbf{q} \tag{12-52}$$

$$\mathbf{c}_q(k) = \mathbf{D}\mathbf{x}_q(k) + \mathbf{E}\mathbf{u}(k) + \mathbf{G}\mathbf{q} \tag{12-53}$$

where $\mathbf{x}_q(k)$ is the $n \times 1$ state vector of the quantized system and $\mathbf{c}_q(k)$ is the $p \times 1$ output vector; $\mathbf{F}$ is an $n \times m$ matrix which denotes the dependence of $\mathbf{x}_q(k+1)$ upon the equivalent inputs due to the quantizers, and $\mathbf{G}$ is a $p \times m$ matrix representing the dependence of $\mathbf{c}_q(k)$ on $\mathbf{q}$, where $\mathbf{q}$ is the vector

$$\mathbf{q} = \begin{bmatrix} \pm q_1/2 \\ \pm q_2/2 \\ \vdots \\ \pm q_m/2 \end{bmatrix} \tag{12-54}$$

Let the quantization error in the state vector at the kth sampling instant due to quantization be represented by $\mathbf{e}_x(k)$; then

$$\mathbf{e}_x(k) = \mathbf{x}(k) - \mathbf{x}_q(k) \tag{12-55}$$

Subtracting Eq. (12-52) from Eq. (12-50), we have

$$\mathbf{e}_x(k+1) = \mathbf{A}\mathbf{e}_x(k) - \mathbf{F}\mathbf{q} \tag{12-56}$$

Similarly, the difference between Eqs. (12-51) and (12-53) is

$$\mathbf{e}_c(k) = \mathbf{c}(k) - \mathbf{c}_q(k) = \mathbf{D}\mathbf{e}_x(k) - \mathbf{G}\mathbf{q} \tag{12-57}$$

where $\mathbf{e}_c(k)$ is the quantization error in the output $\mathbf{c}_q(k)$ at the kth sampling instant.

The solution of Eq. (12-56) at $k = N$ is

$$\mathbf{e}_x(N) = \mathbf{A}^N\mathbf{e}_x(N) - \sum_{k=0}^{N-1} \mathbf{A}^{N-k-1}\mathbf{F}\mathbf{q} \tag{12-58}$$

The ith element of $\mathbf{e}_x(N)$ is written

$$\mathbf{e}_{xi}(N) = \mathbf{P}_i\mathbf{A}^N\mathbf{e}_x(0) - \sum_{j=1}^{m}\sum_{k=0}^{N-1}\mathbf{P}_i\mathbf{A}^{N-k-1}\mathbf{F}_j\left(\pm\frac{q_j}{2}\right) \tag{12-59}$$

$i = 1, 2, \ldots, n$, where $\mathbf{P}_i\mathbf{A}^N$ denotes a row matrix containing the ith row of $\mathbf{A}^N$; $\mathbf{F}_j$ represents the jth column of $\mathbf{F}$.

For an asymptotically stable system,

$$\lim_{N\to\infty} \mathbf{A}^N = \mathbf{0} \tag{12-60}$$

The least upper bound of the steady-state quantization error of the ith state is

$$\left|\lim_{N\to\infty} \mathbf{e}_{xi}(N)\right| = \left|\lim_{N\to\infty}\sum_{j=1}^{m}\sum_{k=0}^{N-1}\mathbf{P}_i\mathbf{A}^{N-k-1}\mathbf{F}_j\frac{q_j}{2}\right| \qquad i = 1, 2, \ldots, n \tag{12-61}$$

In a similar manner, the least upper bound of the steady-state quantization error of the ith output is obtained by using Eq. (12-57),

$$\left|\lim_{N\to\infty} \mathbf{e}_{ci}(N)\right| = \left|\lim_{N\to\infty}\sum_{j=1}^{m}\left(\sum_{k=0}^{N-1}\mathbf{D}_i\mathbf{A}^{N-k-1}\mathbf{F}_j - g_{ij}\right)\frac{q_j}{2}\right| \tag{12-62}$$

where $\mathbf{D}_i$ is a $1 \times n$ matrix which is formed by the ith row of $\mathbf{D}$, and g_{ij} is the ijth element of $\mathbf{G}$; $i = 1, 2, \ldots, p$, and $j = 1, 2, \ldots, m$.

12-7-2 z-Transform Analysis

The z-transform analysis of the least upper bound error analysis is conceptually simpler than the state-variable method described above. Taking the z-transform on both sides of Eq. (12-56) and solving for $\mathbf{E}_x(z)$, we have

$$\mathbf{E}_x(z) = (z\mathbf{I} - \mathbf{A})^{-1}\mathbf{e}_x(0) - (z\mathbf{I} - \mathbf{A})^{-1}\mathbf{F}\mathbf{q}\frac{z}{z-1} \tag{12-63}$$

The ith element of $\mathbf{E}_x(z)$ is written

$$E_{xi}(z) = \mathbf{P}_i(z\mathbf{I} - \mathbf{A})^{-1}\mathbf{e}_x(0) - \sum_{j=1}^{m}\mathbf{P}_i(z\mathbf{I} - \mathbf{A})^{-1}\mathbf{F}_j\frac{\pm q_j}{2}\frac{z}{z-1} \tag{12-64}$$

The least upper bound of the quantization error in the ith state variable is

$$\left|\lim_{N\to\infty} e_{xi}(N)\right|\lim_{z\to 1}(1 - z^{-1})E_{xi}(z)\Bigg| = \left|\lim_{z\to 1}\sum_{j=1}^{m}\mathbf{P}_i(z\mathbf{I} - \mathbf{A})^{-1}\mathbf{F}_j\frac{q_i}{2}\right| \tag{12-65}$$

$i = 1, 2, \ldots, p$. Similarly, for the output,

$$\left|\lim_{N\to\infty} e_{ci}(N)\right| = \left|\lim_{z\to 1}\sum_{j=1}^{m}(\mathbf{D}_i(z\mathbf{I} - \mathbf{A})^{-1}\mathbf{F}_j - g_{ij})\frac{q_j}{2}\right| \tag{12-66}$$

The following example serves to illustrate the least upper bound quantization error analysis.

Example 12-4

A digital controller in a control system is usually implemented digitally so that the round-off of digital data is modeled by amplitude quantization. Let us consider a typical first-order digital controller with the transfer function

$$D(z) = \frac{C(z)}{U(z)} = \frac{1 + bz^{-1}}{1 + az^{-1}} \qquad a < 1 \tag{12-67}$$

A state diagram of the controller is shown in Fig. 12-20(a), and a model with quantizers positioned at appropriate locations is shown in Fig. 12-20(b). The state diagram with the quantizers replaced by a branch with unity gain and an external source with signal magnitude of $\pm q/2$ is shown in Fig. 12-20(c). In the present case, it is assumed that all four quantizers have the same quantization levels.

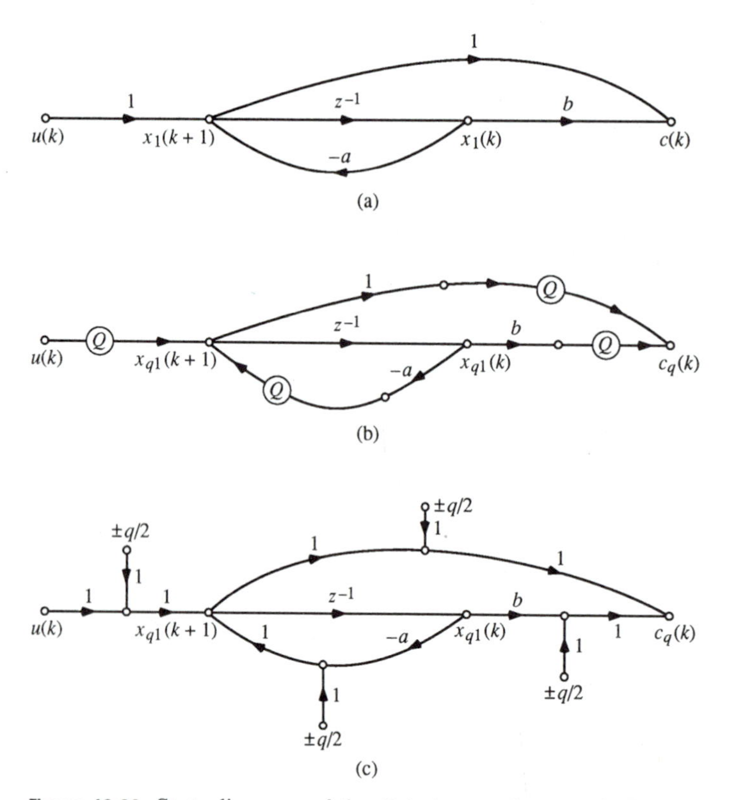

Figure 12-20. State diagrams of the digital control system in Example 12-4.

Without quantization, the dynamic equations of the controller are written directly from Fig. 12-20(a):

$$x_1(k+1) = -ax_1(k) + u(k) \tag{12-68}$$

$$c(k) = (b-a)x_1(k) + u(k) \tag{12-69}$$

For the system with quantizers, the dynamic equations are

$$x_{q1}(k+1) = -ax_{q1}(k) + u(k) \pm q \tag{12-70}$$

$$c_q(k) = (b-a)x_{q1}(k) + u(k) \pm 2q \tag{12-71}$$

The least upper bound of the quantization error in the state variable $x_1(k)$ is defined as

$$e_x(k) = x_1(k) - x_{q1}(k) \tag{12-72}$$

From Eqs. (12-68) and (12-69), we have

$$e_x(k+1) = -ae_x(k) \pm q \tag{12-73}$$

The solution of the last equation is

$$e_x(N) = (-a)^N e_x(0) + \sum_{k=0}^{N-1} (-a)^{N-k-1}(\pm q) \tag{12-74}$$

The magnitude of the steady-state value of $e_x(N)$ is

$$\left| \lim_{N\to\infty} e_x(N) \right| = \left| \lim_{N\to\infty} \sum_{N\to\infty}^{N-1} (-a)^{N-k-1}(\pm q) \right| = \frac{q}{1+a} \tag{12-75}$$

The least upper bound of the quantization error in the output is obtained from Eqs. (12-69) and (12-71). We have

$$e_c(N) = c(N) - c_q(N) = (b-a)e_x(N) \pm 2q \tag{12-76}$$

Thus, the magnitude of the steady-state error bound is

$$\left| \lim_{N\to\infty} e_c(N) \right| = \frac{(b-a)q}{1+a} + 2q = \frac{2+a+b}{1+a} \tag{12-77}$$

The z-transform analysis is carried out by evaluating the z-transforms of $x_1(k)$ and $x_{q1}(k)$ as functions of the equivalent signal sources from Fig. 12-20. Thus,

$$E_x(z) = X_1(z) - X_{q1}(z) = \frac{-z^{-1}}{1+az^{-1}} \frac{\pm qz}{z-1} \tag{12-78}$$

where the initial states have been neglected, since they do not affect the steady-state error. Then,

$$\left| \lim_{N\to\infty} e_x(N) \right| = \left| \lim_{z\to 1} (1-z^{-1})E_x(z) \right| = \frac{q}{1+a} \tag{12-79}$$

which agrees with the result obtained in Eq. (12-75).

Similarly, the least upper bound of the quantization error in the output may be obtained by evaluating the z-transform of the output.

$$E_c(z) = C(z) - C_q(z) = \left(\frac{1 + bz^{-1}}{1 + az^{-1}} + 1\right)\frac{\pm qz}{z - 1} \tag{12-80}$$

Therefore,

$$\left|\lim_{N\to\infty} e_c(N)\right| = \left|\lim_{z\to 1} (1 - z^{-1})E_c(z)\right| = \left|\lim_{z\to 1}\left(\frac{1 + bz^{-1}}{1 + az^{-1}} + 1\right)q\right| \tag{12-81}$$

$$= \frac{2 + a + b}{1 + a} q \tag{12-82}$$

12-8 TIME DELAYS IN MICROPROCESSOR-BASED CONTROL SYSTEMS

Besides having a finite wordlength, microprocessors are relatively slow digital computing machines. The DSPs are much faster, but the computing time is still nonzero. In many digital computing applications, which may not be concerned with the real-time handling of data, the slow computing speed may not be important. However, in control systems applications, real-time computation is often necessary, and the time delays encountered in handling the data may have a significant effect on the system performance. In general, it is important to know how large these delays are in order to deal with the problem analytically. Two immediate problems may be attributed to time delays in control systems. One is that if the time delay is too large, there would not be enough time to carry out all the necessary computations required to execute the control algorithms, and the other is the adverse effect that time delay has on the stability of closed-loop control systems.

The time delays due to microprocessor and DSP computation may be identified by analyzing the program used for the control law along with the subroutines that may be called from the available utility package. Each program is made up of a set of instructions, and each instruction requires a particular number of machine cycles to execute. Each machine cycle in turn requires a certain number of machine states. The time required for the microprocessor to execute a particular instruction is directly proportional to the total number of machine states that the microprocessor must go through to complete that particular instruction. For the Intel 8080 microprocessor, for example, each state requires 500 ns. This allows the machine to go through two million machine states per second. This may seem extremely fast, but even a simple program may require thousands of machine states to execute.

In general, the information on the number of machine states necessary for a particular instruction is found in the user's manual of the processor. With this information, it is possible to go through any program, instruction by instruction, adding up the number of machine states required, to come up with both the total time necessary to complete the program and the time required to reach a particular point in the computations. For example, the time delay in executing the PID control by the Intel 8080 microprocessor is approximately 2 ms, whereas the same for the Texas Instruments TMC320C14 DSP is only around 2 μs.

In summary, the main objective of this section is to point out the time delays encountered in the execution of a digital program. For slow microprocessors, the time delays cannot always be neglected in the modeling of the digital controller. Since it is well known that time delays often cause instability in closed-loop systems, digital or analog, it is essential that these imperfect conditions be considered when designing a digital control system.

PROBLEMS

12-1 The state diagram of a digital control system is shown in Fig. P12-1. The digital controller is represented by the dynamics between $e(k)$ and $u(k)$, and the controlled process is modeled between the nodes $u(k)$ and $c(k)$. Consider that amplitude quantization exists in the digital controller in the branches with gains 2.72, -1, and -0.72. Also, quantization appears in the overall feedback path due to the use of a digital encoder, and at the input to the digital controller. Insert the five quantizers in Fig. P12-1, and conduct a least upper bound error analysis. Determine the magnitudes of the least upper bound errors of the three state variables and the output. The quantization level is q.

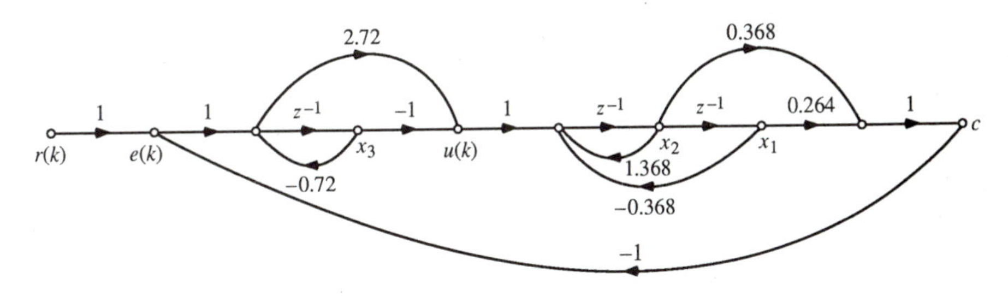

Figure P12-1.

12-2 The block diagram of a digital positioning control system with quantization is shown in Fig. P12-2. Determine the maximum value of the quantization level q so that the error in the output due to quantization either in the form of steady-state error or sustained oscillation will not exceed 0.01. The input is a unit-step function, and the initial conditions are zero.

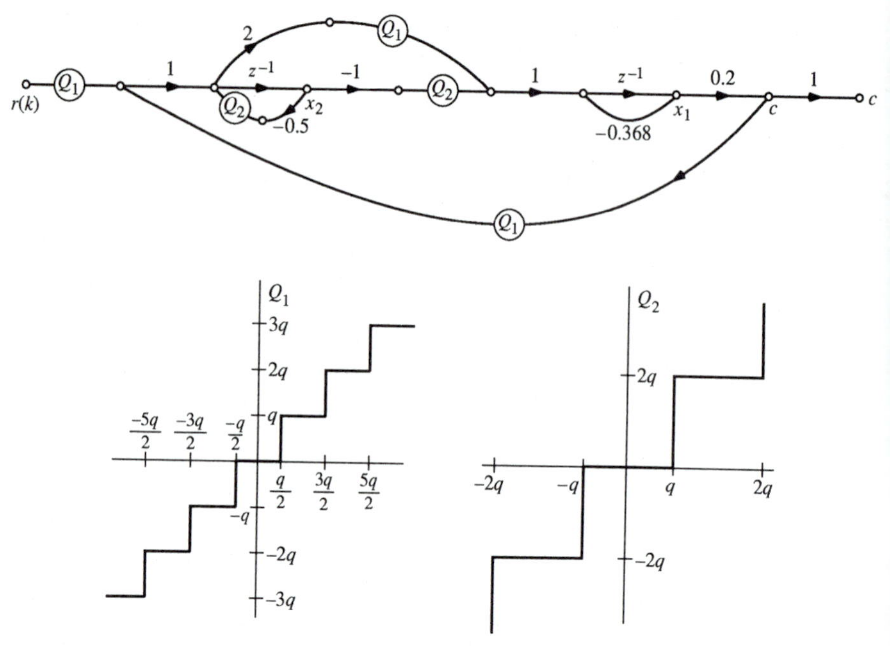

Figure P12-2.

12-3 The state diagram of a digital control system is shown in Fig. P12-3. The characteristics of the quantizers located at the various positions in the system are given as shown. Determine the magnitudes of the least upper bound errors in the variables $x_1(k)$, $x_2(k)$, and $c(k)$.

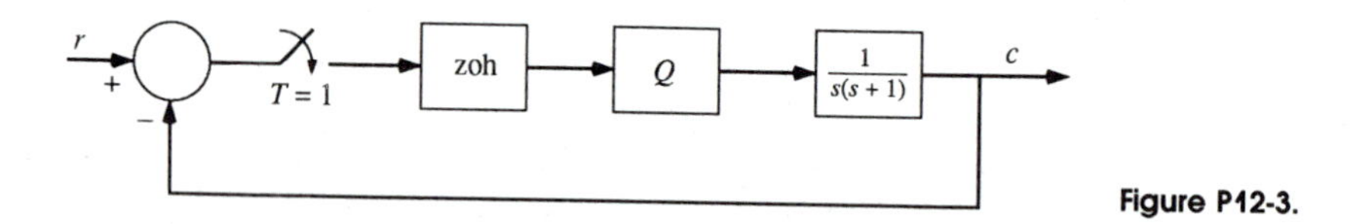

Figure P12-3.

12-4 Figure P12-4 shows the block diagram of the LST (large space telescope) system with amplitude quantization. The transfer functions of $G_a(z)$, $G_b(z)$, and $G_c(z)$ are given as

$$G_a(z) = \frac{K_P z + K_I T - K_P}{z - 1}$$

$$G_b(z) = \frac{T^2(z + 1)}{2J_v(z - 1)^2}$$

$$G_c(z) = \frac{K_R T}{J_v(z - 1)}$$

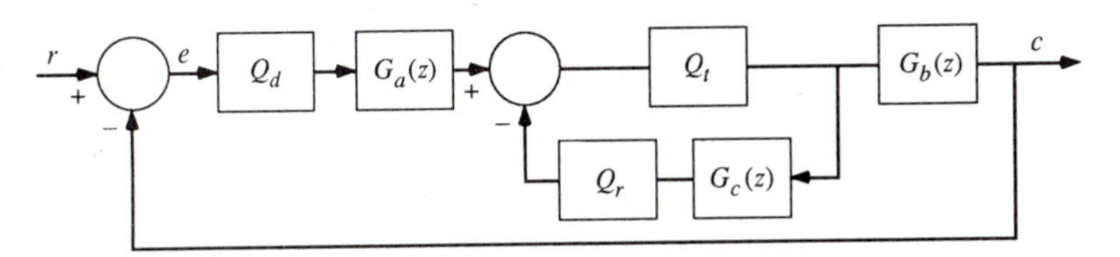

Figure P12-4.

The quantization levels of Q_d, Q_t, and Q_r are q_d, q_t, and q_r, respectively. Find the least upper bound error in $c(k)$ due to the quantization. Discuss the effect of the integral control constant K_I on the quantization error.

References

1. Cassell, D. A., *Microcomputers and Modern Control Engineering*, Reston Publishing Co., Reston, Va., 1983.
2. Andrews, M., *Programming Microprocessor Interfaces for Control and Instrumentation*, Prentice-Hall, Englewood Cliffs, N.J., 1982.
3. Gibson, G. A., and Y. C. Liu, *Microcomputers for Engineers and Scientists*, Prentice-Hall, Englewood Cliffs, N.J., 1980.
4. Tal, J., "The Galil Chip Set—Building Blocks for Motion Control Systems," *Proc. 13th Annual Symp. Incremental Motion Control Systems and Devices*, pp. 49–53, 1984.
5. Brey, B. B., *Microprocessors and Peripherals*, 2nd ed., Merrill, Columbus, Ohio, 1988.
6. Lin, K. S., Ed., *Digital Signal Processing Applications with the TMS320 Family*, vol. 1, Texas Instruments, 1986.
7. Papamichalis, P., Ed., *Digital Signal Processing Applications with the TMS320 Family*, vol. 2, Texas Instruments, 1990.
8. Papamichalis, P., Ed., *Digital Signal Processing Applications with the TMS320 Family*, vol. 3, Texas Instruments, 1990.
9. Lin, K. S., G. A. Frantz, and R. Simar, Jr., "The TMS320 Family of Digital Signal Processors," *Proc. IEEE*, vol. 75, no. 9, pp. 1143–1159, September 1989.
10. Bucella, T., and I. Ahmed, "Taking Control with DSPs," *Machine Design*, pp. 73–80, October 12, 1989.
11. Jain Y., "Tapping the Power of DSPs," *Machine Design*, pp. 73–80, June 7, 1990.
12. Slaughter, J. B., "Quantization Errors in Digital Control Systems," *IEEE Trans. Automatic Control*, vol. AC-9, pp. 70–74, January 1964.
13. Curry, E. E., "The Analysis of Round-Off and Truncation Errors in a Hybrid Control System," *IEEE Trans. Automatic Control*, vol. AC-12, pp. 601–604, October 1967.
14. Kuo, B. C., "The *z*-Transform Describing Function for Nonlinear Sampled-Data Control Systems," *Proc. IRE*, vol. 43, no. 5, pp. 941–942, May 1960.

Fixed-Point and Floating-Point Numbers

A-1 FIXED-POINT NUMBER REPRESENTATION

In general, an n-bit binary word representing a fixed-point integer number N is written as

$$N = a_{n-1}2^{n-1} + \cdots + a_2 2^2 + a_1 2^1 + a_0 2^0 \tag{A-1}$$

where the coefficients a_i, $i = 0, 1, 2, \ldots, n-1$, are either zero or one. The digits of the number in Eq. (A-1) are ordered from left to right with the most significant bit (MSB) being a_{n-1} on the left and the least significant bit (LSB), a_0, on the right. Figure A-1 illustrates a fixed-point number that is represented by a 16-bit word where each bit can be a zero or a one.

As a simple example, consider a 3-bit binary word,

$$N = a_2 2^2 + a_1 2^1 + a_0 2^0 \tag{A-2}$$

By assigning various combinations of zeros and ones to the coefficients a_0, a_1, and a_2, the word N can represent eight (2^3) distinct integral states or numbers. The conversion relationship between binary and decimal integers for a 3-bit word is shown in Table A-1.

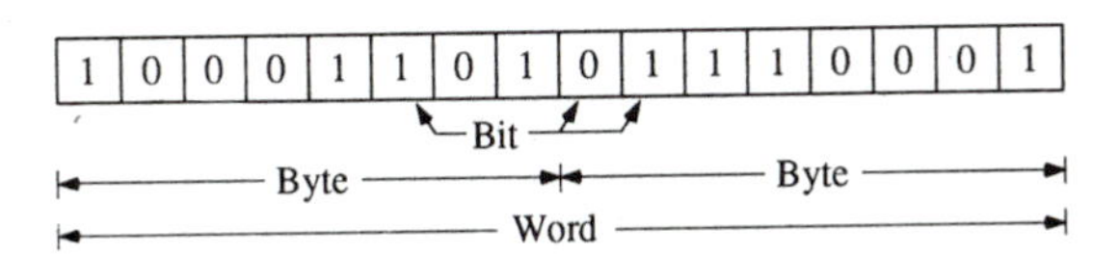

Figure A-1. Relations between word, byte, and bit.

Table A-1 Conversion Relationship Between Natural Binary and Decimal Codes for Integer Coding

Decimal Integer	Binary Integer	Binary Integer Code MSB (×4)	(×2)	LSB (×1)
0	000	0	0	0
1	001	0	0	1
2	010	0	1	0
3	011	0	1	1
4	100	1	0	0
5	101	1	0	1
6	110	1	1	0
7	111	1	1	1

We can also use a fixed-point notation to represent nonintegers or fractions. Using a fictitious *binary point* in a data word, a part of the word can be used to represent integers, and the other part fractions. For example, the 8-bit word shown in Fig. A-2 has its first five digits representing the integral part, and the last three digits representing the fractional part of the number. Note that the binary point is only symbolic so that it does not occupy any bit.

With reference to the 8-bit word shown in Fig. A-2, the number N is written

$$N = a_4 2^4 + a_3 2^3 + a_2 2^2 + a_1 2^1 + a_0 2^0 + a_{-1} 2^{-1} + a_{-2} 2^{-2} + a_{-3} 2^{-3} \tag{A-3}$$

Thus, for instance, the binary number 01011.101 is equivalent to the decimal number

$$\begin{aligned} N &= 0 \times 2^4 + 1 \times 2^3 + 0 \times 2^2 + 1 \times 2^1 + 1 \times 2^0 \\ &\quad + 1 \times 2^{-1} + 0 \times 2^{-2} + 1 \times 2^{-3} \\ &= 8 + 2 + 1 + 1/2 + 1/8 \\ &= 11.625 \end{aligned} \tag{A-4}$$

In general, an n-bit fraction can be represented as

$$N = a_{-1} 2^{-1} + a_{-2} 2^{-2} + \cdots + a_{-n} 2^{-n} \tag{A-5}$$

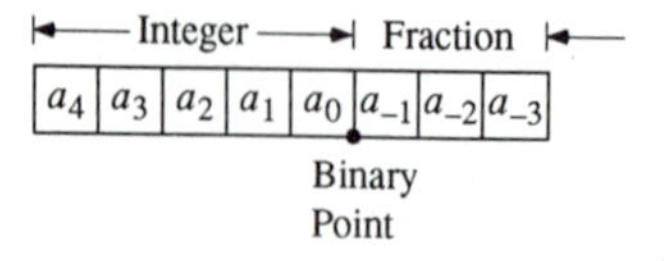

Figure A-2. Fixed-point representation of an 8-bit noninteger.

Table A-2 Conversion Relationship Between Natural Binary and Decimal Codes for Fractional Coding

		Binary Fraction Code		
Decimal Fraction	Binary Fraction	MSB $(\times\frac{1}{2})$	$(\times\frac{1}{4})$	LSB $(\times\frac{1}{8})$
0	.000	0	0	0
1/8	.001	0	0	1
1/4	.010	0	1	0
3/8	.011	0	1	1
1/2	.100	1	0	0
5/8	.101	1	0	1
3/4	.110	1	1	0
7/8	.111	1	1	1

where the coefficients a_i, $i = -1, -2, \ldots, -n$, assume the value of zero or one. The first coefficient a_{-1} represents the MSB, and the LSB is a_{-n}. As a simple illustration, the eight distinct states or fractional numbers that are representable by a 3-bit word are shown in Table A-2.

Thus far we have only considered the binary representation of positive numbers. Negative numbers may be represented by assigning the first bit of the binary word as a *sign bit*; i.e., zero for positive and one for negative. An alternative method of representing negative numbers is to use the "two's-complement" algorithm (refer to any elementary text on digital computers).

If the first bit of a 3-bit word is used as a sign bit, then the largest integer that the word can represent is $2^2 - 1 = 3$, and the smallest integer that can be represented is $-(2^2 - 1) = -3$. In general, the integer representable by an n-bit word with a sign bit lies between $(2^{n-1} - 1)$ and $-(2^{n-1} - 1)$, inclusive, zeros included. For example, for a 3-bit word, with the first bit used for sign, the eight states or integers representable are tabulated in Table A-3. Notice that in this case 0 (zero) is represented twice, $+0$ and -0. However, the 3-bit word still defines eight distinct states, as in the case shown in Table A-1.

In a similar fashion, a sign bit can be used for the representation of a noninteger or fraction. The number that is representable by an n-bit word with a sign bit and m fractional bits (m bits after the "binary point") lies between $-(2^{n-1} - 1)2^{-m}$ and $(2^{n-1} - 1)2^{-m}$, inclusively.

The fixed-point representation of real numbers has a serious disadvantage due to the limited range in which the numbers can be represented by a given wordlength once the fictitious binary point has been assigned. When multiplying two large numbers, frequently, the result will exceed the capacity of the wordlength of the fixed-point representation, and overflow will take place, which causes inaccuracy in the computation.

Table A-3 Conversion Relationship Between Binary and Decimal Codes for a 3-Bit Integer with a Sign Bit

Decimal Integer	Binary Integer	Sign Bit	Binary Integer Code MSB (×2)	LSB (×1)
−3	111	1	1	1
−2	110	1	1	0
−1	101	1	0	1
−0	100	1	0	0
+0	000	0	0	0
+1	001	0	0	1
+2	010	0	1	0
+3	011	0	1	1

A-2 FLOATING-POINT NUMBER REPRESENTATION

A representation system of numbers which is easier to work with and has more range is the *floating-point* representation. This method is also known as the scientific notation where the first part of the data word is used to store a number called the *mantissa*, and the second part is the *exponent*. For instance, in the decimal system the number 5 can be written as 0.5×10^1, 50×10^{-1}, or 0.05×10^2, etc. In digital computers and systems, binary floating-point numbers are usually represented as

$$N = M \times 2^E \tag{A-6}$$

where M is the mantissa and E is the exponent of the number N. Furthermore, M is usually scaled to be a fraction whose decimal value lies in the range of $0.5 \leq M < 1$.

Figure A-3 shows a floating-point representation of an 8-bit word with a 5-bit mantissa and a 3-bit exponent. Since the mantissa and the exponent can both be positive or negative, the first bits of the mantissa and the exponent are the sign bits. (We can also use the first two bits of the entire word as the sign bits of the mantissa and the exponents, respectively.) For microcomputers that

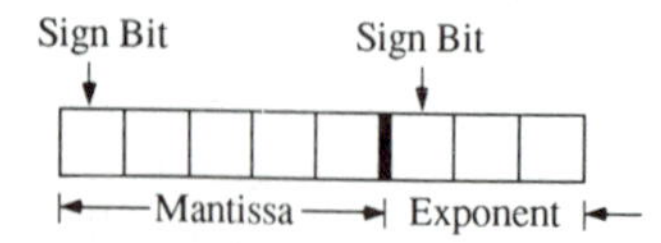

Figure A-3. Floating-point representations of an 8-bit number.

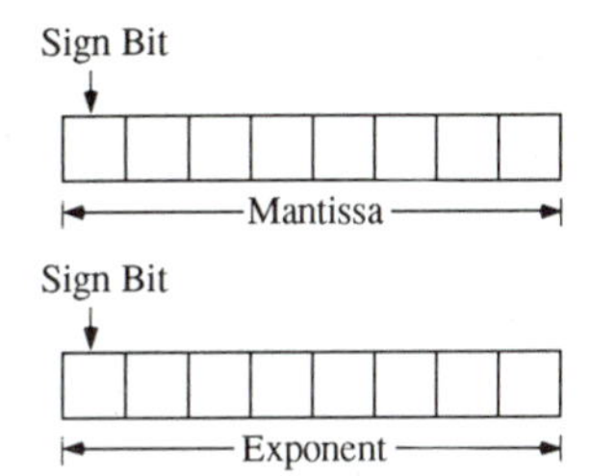

Figure A-4. Floating-point representation using two computer words.

have small wordlengths, two consecutive data words can be strung together to form a floating-point number, such as that shown in Fig. A-4.

Since the mantissa is normalized to be a fraction between one-half and one, *the first bit after the sign bit should always be a one, and there is always a fictitious binary point just after the sign bit.*

The exponent E represents how many places the binary point should be shifted to the right (for $E > 0$) or to the left (for $E < 0$). As an illustrative example, the decimal number 6.5 is represented by the 8-bit floating-point binary word shown in Fig. A-5. In this case, the mantissa consists of 4 bits for the fraction .1101, and the exponent is the 2-bit binary integer 11 which represents the decimal number 3. Thus, the number represented is

$$N = .1101 \times 2^{11} \qquad \text{(binary)}$$

$$= [\tfrac{1}{2} + \tfrac{1}{4} + \tfrac{1}{16}]2^3 = 6.5 \qquad \text{(decimal)}$$

Also, we see that if the binary point is moved to the right by 3 bits, where 3 is the exponent in decimal, we have the fixed-point binary number 110.1 which is 6.5 decimal.

For an n-bit word with an m-bit mantissa and an e-bit exponent, both including the sign bits, the *largest* number $N_{\max}$ that the word can represent is shown in Fig. A-6(a). In this case, all the nonsign bits are filled with ones, and $N_{\max}$ is expressed as

$$N_{\max} = (1 - 2^{-m+1})2^{(2^{e-1}-1)} \tag{A-7}$$

For example, for the 8-bit floating-point word shown in Fig. A-5, $m = 5$ and $e = 3$. The largest number that can be represented is

$$N_{\max} = (1 - 2^{-4})2^{(2^2-1)} = (1 - \tfrac{1}{16})2^3 = 7.5 \tag{A-8}$$

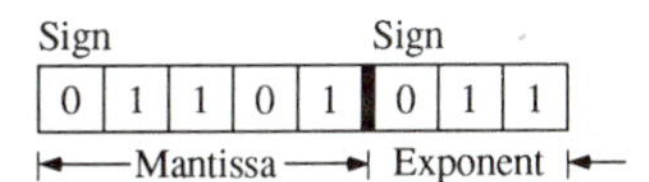

Figure A-5. Floating-point representation of the decimal number 6.5 by an 8-bit binary word.

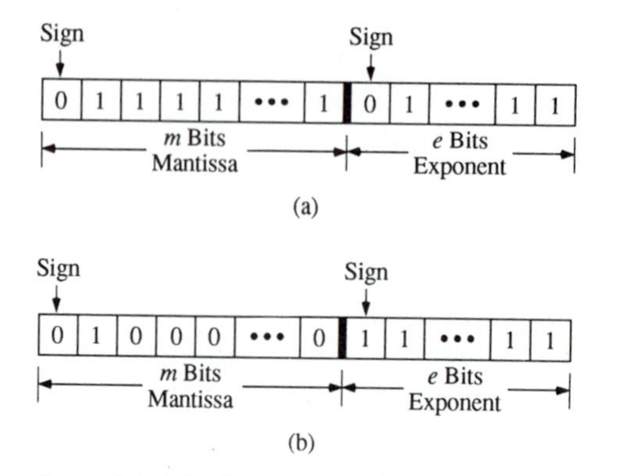

Figure A-6. Maximum and minimum magnitudes of floating-point numbers.

For the same 8-bit word, if we allocate 4 bits for the mantissa and the same for the exponent, then

$$N_{max} = (1 - 2^{-3})2^{(2^3 - 1)} = (1 - \tfrac{1}{8})2^7 = 112 \tag{A-9}$$

Figure A-6(b) shows the contents of the *smallest* positive number N_{min} that can be represented by an n-bit word with an m-bit mantissa and an e-bit exponent. In this case, the first bit after the sign bit in the mantissa is a one, and the rest of the bits are all zeros. The bits of the exponent are all ones. Thus, N_{min} is written

$$N_{min} = 0.5 \times 2^{-(2^{e-1} - 1)} \tag{A-10}$$

Mathematical Modeling of Sampling by Convolution Integral

B-1 EXPRESSIONS FOR THE FINITE-PULSEWIDTH SAMPLER OUTPUT $F_p^*(s)$

Alternate descriptions of the sampled signal $f_p^*(t)$ in the transform domain are derived by means of the complex convolution method of the Laplace transform. With reference to Eq. (2-17), the Laplace transform of $f_p^*(t)$ is written

$$F_p^*(s) = \mathscr{L}[f_p^*(t)] = \mathscr{L}[f(t)p(t)] \tag{B-1}$$

where $\mathscr{L}$ denotes the Laplace transform operation.

Equation (B-1) is written as

$$F_p^*(s) = F(s) * P(s) \tag{B-2}$$

where the asterisk represents *complex convolution* in the Laplace domain and $F(s)$ and $P(s)$ are the Laplace transforms of $f(t)$ and $p(t)$, respectively.

The Laplace transform of $p(t)$ is determined from Eq. (2-16),

$$P(s) = \sum_{k=-\infty}^{\infty} \frac{1 - e^{-ps}}{s} e^{-kTs} \tag{B-3}$$

The summation on the right-hand side of Eq. (B-3) actually begins at $k = 0$, since the one-sided Laplace transform is defined for $0 \leq t < \infty$. The infinite series in Eq. (B-3) is written in closed form as

$$P(s) = \frac{1 - e^{-ps}}{s(1 - e^{-Ts})} \tag{B-4}$$

Substituting $P(s)$ from the last equation into Eq. (B-2), we get

$$F_p^*(s) = F(s) * \frac{1 - e^{-ps}}{s(1 - e^{-Ts})} \tag{B-5}$$

Applying the complex convolution integral of the Laplace transform to the last equation, we have

$$F_p^*(s) = \frac{1}{2\pi j} \int_{c-j\infty}^{c+j\infty} F(\xi)P(s-\xi)\, d\xi \tag{B-6}$$

Thus, using Eq. (B-4), the last equation becomes

$$F_p^*(s) = \frac{1}{2\pi j} \int_{c-j\infty}^{c+j\infty} F(\xi) \frac{1 - e^{-p(s-\xi)}}{(s-\xi)[1 - e^{-T(s-\xi)}]}\, d\xi \tag{B-7}$$

where ξ is the variable of integration. The constant c in the integral limits must satisfy the following conditions:

$$\sigma_1 < c < \sigma - \sigma_2$$

$$\sigma > \max(\sigma_1, \sigma_2, \sigma_1 + \sigma_2)$$

where σ is the real part of s, and σ_1 and σ_2 are the abscissas of convergence of $F(\xi)$ and $P(\xi)$, respectively. The path of integration of the integral of Eq. (B-7) is along the straight line from $\xi = c - j\infty$ to $c + j\infty$ in the complex ξ-plane as shown in Fig. B-1.

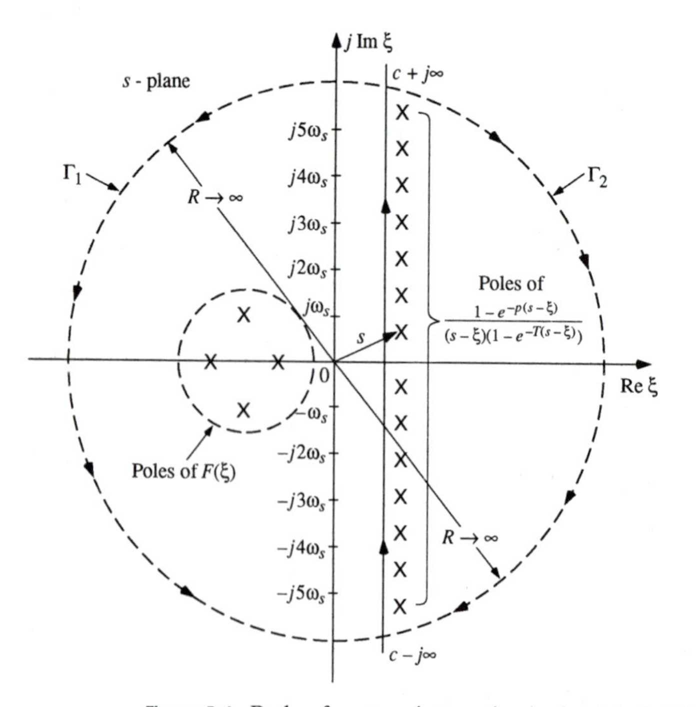

Figure B-1. Paths of contour integration in the right half ξ-plane and the left half ξ-plane.

The convolution integral of Eq. (B-7) may be carried out by taking the contour integration along a closed path Γ_1 formed by the $\xi = c - j\infty$ to $c + j\infty$ line and a semicircle of infinitely large radius going around the left half of the ξ-plane, as shown in Fig. B-1. As an alternative, the convolution integral can be evaluated along the closed path Γ_2 formed by the $\xi = c - j\infty$ to $c + j\infty$ line and the semicircle of infinite radius in the right half ξ-plane. Either contour integral can then be evaluated by means of the residues theorem of complex variables.

Contour Integration Around the Left Half ξ-Plane

$$F_p^*(s) = \frac{1}{2\pi j} \oint_{\Gamma_1} F(\xi) \frac{1 - e^{-p(s-\xi)}}{(s-\xi)[1 - e^{-T(s-\xi)}]} d\xi - \frac{1}{2\pi j} \oint F(\xi) \frac{1 - e^{-p(s-\xi)}}{(s-\xi)[1 - e^{-T(s-\xi)}]} d\xi \tag{B-8}$$

where the second integral is taken along the semicircle with infinite radius on Γ_1 which encloses the entire left half of the ξ-plane.

Contour Integration Around the Right Half ξ-Plane

$$F_p^*(s) = \frac{1}{2\pi j} \oint_{\Gamma_2} F(\xi) \frac{1 - e^{-p(s-\xi)}}{(s-\xi)[1 - e^{-T(s-\xi)}]} d\xi - \frac{1}{2\pi j} \oint F(\xi) \frac{1 - e^{-p(s-\xi)}}{(s-\xi)[1 - e^{-T(s-\xi)}]} d\xi \tag{B-9}$$

where the second integral is taken along the semicircle with infinite radius on Γ_2 which encloses the entire right half of the ξ-plane.

In order that the line integral of Eq. (B-7) can be properly evaluated by the contour integration of Eq. (B-8) or Eq. (B-9), we must investigate the poles of $F(\xi)$ and $P(s - \xi)$ which form the integrand of Eq. (B-7). Normally, the poles of $F(\xi)$ are located in the left half, and perhaps on the imaginary axis of the ξ-plane, and are finite in number. In general, $F(s)$ depends on the function $f(t)$, and the specific values of the poles are unknown. The poles of $P(s - \xi)$ are the values of ξ that make the following function go to infinity:

$$P(s - \xi) = \frac{1 - e^{-p(s-\xi)}}{(s-\xi)[1 - e^{-T(s-\xi)}]} \tag{B-10}$$

The function $P(s - \xi)$ has simple poles at

$$\xi_n = s + \frac{2\pi nj}{T} = s + jn\omega_s \tag{B-11}$$

where n is an integer that varies from $-\infty$ to ∞, T is the sampling period in seconds, and ω_s is the sampling frequency in rad/s. Equation (B-11) shows that the poles of $P(s-\xi)$ are infinite in number and are located at frequency intervals of $n\omega_s$, $n = 0, \pm 1, \pm 2, \ldots$, along the path of Re $(\xi) =$ Re (s) in the ξ-plane. The poles of $P(s-\xi)$ and the assumed poles of $F(\xi)$ are shown in Fig. B-1.

If the function $F(\xi)$ has *at least one more pole than zero*, that is,

$$\lim_{\xi \to \infty} F(\xi) = 0 \tag{B-12}$$

the integrals along the infinite-radius semicircular paths Γ_1 and Γ_2 are zero in Eqs. (B-8) and (B-9), respectively. Then $F_p^*(s)$ can be evaluated by the residues theorem.

Evaluation of the Contour Integral Around the Left Half Plane

If we use the closed path Γ_1 around the left half ξ-plane, the path encloses all the poles of $F(\xi)$, and $F_p^*(s)$ is written

$$F_p^*(s) = \sum \text{residues of} \left[F(\xi) \frac{1 - e^{-p(s-\xi)}}{(s-\xi)[1 - e^{-T(s-\xi)}]} \right]$$
$$\text{at the poles of } F(\xi) \tag{B-13}$$

If $F(\xi)$ is a rational function with k *simple* poles, then Eq. (B-13) can be written as

$$F_p^*(s) = \sum_{n=1}^{k} \frac{N(\xi_n)}{D'(\xi_n)} \frac{1 - e^{-p(s-\xi_n)}}{(s-\xi_n)[1 - e^{-T(s-\xi_n)}]} \tag{B-14}$$

where

$$F(\xi) = \frac{N(\xi)}{D(\xi)} \tag{B-15}$$

$$D'(\xi_n) = \left. \frac{dD(\xi)}{d\xi} \right|_{\xi=\xi_n} \tag{B-16}$$

and ξ_n denotes the nth pole of $F(\xi)$, $n = 1, 2, \ldots, k$.

Note that the coefficient $N(\xi_n)/D'(\xi_n)$ is equal to the coefficient of the partial-fraction expansion of $F(\xi)$ associated with the pole at $\xi = \xi_n$. Thus, as an alternative, Eq. (B-14) can also be carried out by partial-fraction expansion.

In general, if the function $F(s)$ has poles at $s_1, s_2, \ldots, s_k$, with multiplicity, $m_1, m_2, \ldots, m_k$, respectively, $F_p^*(s)$ is given by

$$F_p^*(s) = \sum_{n=1}^{k} \sum_{i=1}^{m_n} \frac{(-1)^{m_n - i} K_{ni}}{(m_n - 1)!} \frac{\partial^{m_n - i}}{\partial s^{m_n - i}} \left[\frac{1 - e^{-ps}}{s(1 - e^{-Ts})} \right] \Bigg|_{s = s - s_n} \tag{B-17}$$

where

$$K_{ni} = \frac{1}{(i-1)!} \frac{\partial^{i-1}}{\partial s^{i-1}} [(s - s_n)^{m_n} F(s)] |_{s = s_n} \tag{B-18}$$

Note that the solutions in Eqs. (B-17) and (B-18) also include the case of the simple poles. For simple poles, the multiplicity $m_n = 1$ for all n, and we can show that these results are identical to those of Eqs. (B-14) through (B-16).

If we can sort out the simple poles and the multiple-order poles of $F(s)$, then the residues that correspond to the simple poles can be obtained from Eq. (B-14), and the residue at the multiple-order pole ξ_i of multiplicity m_i, $m_i > 1$, is given by

$$\frac{1}{(m_i - 1)!} \frac{\partial^{m_i - 1}}{\partial \xi^{m_i - 1}} \left[(\xi - \xi_i)^{m_i} F(\xi) \frac{1 - e^{-p(s-\xi)}}{(s - \xi)[1 - e^{-T(s-\xi)}]} \right]_{\xi = \xi_i} \quad \text{(B-19)}$$

Evaluation of the Contour Integral Around the Right Half Plane

Now let us use the contour integration of Fig. B-1 around the right half plane. If the condition in Eq. (B-12) is satisfied, the integral along the infinite-radius semicircular path Γ_2 vanishes, and we can use the residues method to find $F_p^*(s)$ in Eq. (B-9). We have

$$F_p^*(s) = -\sum \text{residues of } F(\xi) \frac{1 - e^{-p(s-\xi)}}{(s - \xi)[1 - e^{-T(s-\xi)}]}$$

$$\text{at the poles of } \frac{1 - e^{-p(s-\xi)}}{(s - \xi)[1 - e^{-T(s-\xi)}]} \quad \text{(B-20)}$$

where the minus sign is due to the contour integration in Eq. (B-9) being taken in the clockwise direction. Since $P(s - \xi)$ has only simple poles that lie at periodic intervals along Re (ξ) = Re (s) in the ξ-plane, Eq. (B-20) is written by using Eq. (B-14),

$$F_p^*(s) = -\sum_{n=-\infty}^{\infty} \frac{N(\xi_n)}{D'(\xi_n)} F(\xi_n) \quad \text{(B-21)}$$

where $\xi_n = s + jn\omega_s$ denotes the poles of $P(s - \xi)$, $n = 0, \pm 1, \pm 2 \ldots$. Also, in this case,

$$N(\xi_n) = 1 - e^{-p(s-\xi)}|_{\xi = \xi_n = s + jn\omega_s} = 1 - e^{jn\omega_s p} \quad \text{(B-22)}$$

and

$$D'(\xi_n) = \frac{d}{d\xi} [(s - \xi)[1 - e^{-T(s-\xi)}]]|_{\xi = s + jn\omega_n} = jn\omega_s T \quad \text{(B-23)}$$

After carrying out the steps in the last equation and substituting the results in Eq. (B-21), we have

$$F_p^*(s) = \sum_{n=-\infty}^{\infty} \frac{1 - e^{jn\omega_s p}}{-jn\omega_s T} F(s + jn\omega_s) \quad \text{(B-24)}$$

Notice that when s is replaced by $j\omega$, we can show that Eq. (B-24) is identical to Eq. (2-29) which was obtained by a different method.

B-2 EXPRESSIONS FOR THE IDEAL SAMPLER OUTPUT $F^*(s)$

We can derive alternate expressions for $F^*(s)$ that are comparable to those for $F_p^*(s)$ in Eqs. (B-14), (B-17), and (B-24).

Since $\delta_T(t)$ and $p(t)$ are related through

$$\delta_T(t) = \lim_{p \to 0} \frac{1}{p} p(t) \tag{B-25}$$

from Eq. (B-24), $F^*(s)$ can be expressed as

$$F^*(s) = \lim_{p \to 0} \frac{1}{p} F_p^*(s) = \lim_{p \to 0} \frac{1}{p} \sum_{n=-\infty}^{\infty} \frac{1 - e^{-jn\omega_s p}}{jn\omega_s T} F(s + jn\omega_s)$$

$$= \frac{1}{T} \sum_{n=-\infty}^{\infty} F(s + jn\omega_s) \tag{B-26}$$

Note that Eq. (B-24) is obtained from the contour integration around the path Γ_2 which is in the right half ζ-plane. It is important to examine the convergence properties of the contour integration taken along the semicircle with infinite radius along Γ_2, as a result of approximating the pulses by impulses. Applying the limit as $p \to 0$ to $F_p^*(s)/p$ in the first integral on the right-hand side of Eq. (B-9), we have

$$F^*(s) = \lim_{p \to 0} \frac{1}{2\pi jp} \oint_{\Gamma_2} F(\xi) \frac{1 - e^{-p(s-\xi)}}{(s - \xi)[1 - e^{-T(s-\xi)}]} d\xi$$

$$= \frac{1}{2\pi j} \oint_{\Gamma_2} F(\xi) \frac{1}{1 - e^{-T(s-\xi)}} d\xi \tag{B-27}$$

where, in taking the limit, we have applied L'Hopital's rule of calculus. In the present case since $1/[1 - e^{-T(s-\xi)}]$ has a simple pole at infinity in the ξ-plane, the part of the contour integral of Eq. (B-27) along the infinite-radius semicircular path in the right half ξ-plane may not vanish. In fact, if the degree of the denominator of $F(\xi)$ is not *higher than the degree of the numerator by at least two*, the integral along the semicircle may have a finite value or may not even converge. Therefore, Eq. (B-26) is valid for the ideal sampling only if $F(s)$ *has a pole-zero excess greater than or equal to two*. In other words, the signal $f(t)$ *must not have a jump discontinuity at* $t = 0$. We can show that if $F(\xi)$ has two more poles than zeros, the integral along the infinite-radius semicircle in the right half plane yields a constant value of $f(0+)/2$, and Eq. (B-26) should appear as

$$F^*(s) = \frac{f(0+)}{2} + \frac{1}{T} \sum_{n=-\infty}^{\infty} F(s + jn\omega_s) \tag{B-28}$$

The last equation can also be obtained using the Fourier series approach. Let us define the unit-impulse train $\delta_T(t)$ as an even function so that the unit impulse at $t = 0$ is a pulse with amplitude $1/p$, width p, extending from $-p/2$ to $p/2$, and in the limit $p \to 0$. The Fourier series representation of $\delta_T(t)$ is

$$\delta_T(t) = \sum_{n=-\infty}^{\infty} C_n e^{jn\omega_s t} \tag{B-29}$$

where it can be easily shown that the Fourier coefficient C_n is equal to $1/T$. For the ideal sampler that starts the sampling at $t = 0$, the Fourier series for $\delta_T(t)$ for $t > 0$ is written as

$$\delta_T(t) = \frac{1}{2}\delta(t) + \sum_{n=-\infty}^{\infty} \frac{1}{T} e^{jn\omega_s t} \tag{B-30}$$

The term $\delta(t)/2$ in the last equation is included because, when the sampling starts at $t = 0$, only one half of the unit impulse $\delta(t)$ at $t = 0$ is included. Now substituting Eq. (B-30) into the following equation and taking the Laplace transform, we have

$$f^*(t) = f(t)\delta_T(t) = \sum_{k=0}^{\infty} f(kT)\delta(t - kT) \tag{B-31}$$

$$F^*(s) = \frac{f(0+)}{2} + \frac{1}{T}\sum_{n=-\infty}^{\infty} F(s + jn\omega_s) \tag{B-32}$$

which is the same as Eq. (B-28).

In a similar manner, we can show that for the ideal sampler the expression for $F^*(s)$ can be obtained from Eq. (2-72) by taking the limit of $F_p^*(s)/p$ as p approaches zero. The result is

$$F^*(s) = \sum_{n=1}^{k} \frac{N(\xi_n)}{D'(\xi_n)} \frac{1}{1 - e^{-T(s-\xi_n)}} \tag{B-33}$$

where $N(\xi_n)$ and $D'(\xi_n)$ are defined in Eqs. (B-15) and (B-16), respectively. The same limiting process can be applied to Eq. (B-17) to obtain the $F^*(s)$ expression when $F(s)$ has multiple-order poles, and the results are given in Eqs. (2-58), (2-59), and (2-60).

B-3 ALTERNATE EXPRESSIONS FOR THE MODIFIED z-TRANSFORM

The expressions of the modified z-transform in Eqs. (3-199) and (3-200) are derived in the following.

The sampled signal $c^*(t - \Delta T)$ in Fig. 3-14 can be written as the product of $c(t - \Delta T)$ and the impulse train $\delta_T(t)$; i.e.,

$$c^*(t - \Delta T) = c(t - \Delta T) \sum_{k=0}^{\infty} \delta(t - kT) \tag{B-34}$$

The Laplace transform of $c^*(t - \Delta T)$ is written as the complex convolution of the Laplace transform of $c(t - \Delta T)$ and $\delta_T(t)$.

$$\mathscr{L}[c^*(t-\Delta T)] = \mathscr{L}[c(t-\Delta T)] * \mathscr{L}\left(\sum_{k=0}^{\infty} \delta(t-kT)\right) \tag{B-35}$$

where the asterisk indicates complex convolution in the Laplace domain. Taking the z-transform on both sides of Eq. (B-35), we get

$$\begin{aligned} \mathscr{Z}[c^*(t-\Delta T)] &= \left(C(s)e^{-\Delta Ts} * \frac{1}{1-e^{-Ts}}\right)\Bigg|_{z=e^{Ts}} \\ &= C(z, \Delta) \end{aligned} \tag{B-36}$$

Since $c^*(t - \Delta T)$ is already a sampled function, taking its z-transform simply involves the substitution of $z = e^{Ts}$ in the Laplace transform of $c^*(t - \Delta T)$. Using the convolution integral of the Laplace transform, the last equation is written

$$C(z, \Delta) = \frac{1}{2\pi j}\left(\int_{c-j\infty}^{c+j\infty} C(\xi)e^{-\Delta T\xi} \frac{1}{1-e^{-T(s-\xi)}}\, d\xi\right)\Bigg|_{z=e^{Ts}} \tag{B-37}$$

The line integral in the last equation can be evaluated along the line from $c - j\infty$ to $c + j\infty$ and the semicircle with infinite radius in either the left half or the right half of the complex ξ-plane. These contours of integration are similar to those of Fig. B-1. The contour-integration method will yield a correct result for the line integral of Eq. (B-37) if the integral evaluated along the semicircles is zero. The results generated using the two closed paths will generally be of different forms.

Let us first consider the contour integration along the infinite semicircle in the left-half ξ-plane. Since the term $e^{-\Delta T\xi}$ in Eq. (B-37) has a pole at $\xi = -\infty$, there is a pole on the infinite semicircle in the left half ξ-plane. Thus, the integral along this semicircle will be infinite. This problem is overcome by the change of parameters from Δ to m in the modified z-transform. Substituting $\Delta = 1 - m$ into Eq. (B-37), we get

$$C(z, m) = \left(\frac{1}{2\pi j} \oint C(\xi)e^{-T\xi}e^{mT\xi} \frac{1}{1-e^{-T(s-\xi)}}\, d\xi\right)\Bigg|_{z=e^{Ts}} \tag{B-38}$$

or

$$C(z, m) = z^{-1}\left(\frac{1}{2\pi j} \oint C(\xi)e^{mT\xi} \frac{1}{1-e^{-T(s-\xi)}}\, d\xi\right)\Bigg|_{z=e^{Ts}} \tag{B-39}$$

Now the last integral along the infinite-radius semicircle in the left half ξ-plane is zero, and the contour integral is equivalent to the line integral of Eq. (B-37). Using the residues theorem of complex-variable theory, Eq. (B-39) is written

$$C(z, m) = z^{-1} \sum \text{residues of } C(\xi)\frac{e^{mT\xi}}{1-e^{T\xi}z^{-1}} \text{ at the poles of } C(\xi) \tag{B-40}$$

Equation (B-40) gives an alternative expression of the modified z-transform to that in Eq. (3-193). In this case, the modified z-transform is determined using the Laplace transform function $C(s)$.

By taking the contour integral along the infinite-radius semicircular path in the right-half ξ-plane, the line integral of Eq. (B-36) is written as

$$C(z, \Delta) = \frac{1}{2\pi j} \oint C(\xi)e^{-\Delta T\xi} \frac{1}{1 - e^{-T(s-\xi)}} d\xi \bigg|_{z=e^{Ts}} \tag{B-41}$$

In this case, the integral along the semicircle is zero if $C(s)$ approaches zero as s approaches infinity, or $c(0) = 0$. Then, Eq. (B-41) can be expressed as

$$C(z, \Delta) = -\sum \text{residues of } C(\xi)e^{-\Delta T\xi} \frac{1}{1 - e^{-T(s-\xi)}} \bigg|_{z=e^{Ts}}$$

$$\text{at the poles of } 1/[1 - e^{-T(s-\xi)}] \tag{B-42}$$

Since the poles of $1/[1 - e^{-T(s-\xi)}]$ are simple and are at $\xi = s \pm jn\omega_s$, $n = 0, 1, 2, \ldots$, Eq. (B-42) gives

$$C(z, \Delta) = \frac{1}{T} \sum_{n=-\infty}^{\infty} C(s + jn\omega_s)e^{-\Delta T(s+jn\omega_s)}\big|_{z=e^{Ts}} \tag{B-43}$$

To get the modified z-transform, we set $\Delta = 1 - m$, and the last equation gives

$$C(z, m) = \frac{1}{T} \sum_{n=-\infty}^{\infty} C(s + jn\omega_s)e^{-(1-m)(s+jn\omega_s)T}\big|_{z=e^{Ts}} \tag{B-44}$$

Table of Laplace Transforms, z-Transforms, and Modified z-Transforms

Laplace Transform $F(s)$	Time Function $f(t), t > 0$	z-Transform $F(z)$	Modified z-Transform $F(z, m)$
1	$\delta(t)$	1	0
e^{-kTs}	$\delta(t - kT)$	z^{-k}	z^{-k-1+m}
$\frac{1}{s}$	$u_s(t)$	$\frac{z}{z-1}$	$\frac{1}{z-1}$
$\frac{1}{s^2}$	t	$\frac{Tz}{(z-1)^2}$	$\frac{mT}{z-1} + \frac{T}{(z-1)^2}$
$\frac{2}{s^3}$	t^2	$\frac{T^2 z(z+1)}{(z-1)^3}$	$T^2 \frac{m^2 z^2 + (2m - 2m^2 + 1)z + (m-1)^2}{(z-1)^3}$
$\frac{(k-1)!}{s^k}$	t^{k-1}	$\lim_{a \to 0} (-1)^{k-1} \frac{\partial^{k-1}}{\partial a^{k-1}} \left[\frac{z}{z - e^{-aT}} \right]$	$\lim_{a \to 0} (-1)^{k-1} \frac{\partial^{k-1}}{\partial a^{k-1}} \left(\frac{e^{-amT}}{z - e^{-aT}} \right)$
$\frac{1}{s+a}$	e^{-at}	$\frac{z}{z - e^{-aT}}$	$\frac{e^{-amT}}{z - e^{-aT}}$
$\frac{1}{(s+a)^2}$	te^{-at}	$\frac{Tze^{-aT}}{(z - e^{-aT})^2}$	$\frac{Te^{-amT}[e^{-aT} + m(z - e^{-aT})]}{(z - e^{-aT})^2}$
$\frac{(k-1)!}{(s+a)^k}$	$t^k e^{-at}$	$(-1)^k \frac{\partial^k}{\partial a^k} \left[\frac{z}{z - e^{-aT}} \right]$	$(-1)^k \frac{\partial^k}{\partial a^k} \left[\frac{e^{-amT}}{z - e^{-aT}} \right]$
$\frac{a}{s(s+a)}$	$1 - e^{-at}$	$\frac{z(1 - e^{-aT})}{(z-1)(z - e^{-aT})}$	$\frac{(1 - e^{-amT})z + (e^{-amT} - e^{-aT})}{(z-1)(z - e^{-aT})}$

Laplace Transform $F(s)$	Time Function $f(t),\ t > 0$	z-Transform $F(z)$	Modified z-Transform $F(z, m)$
$\frac{1}{(s+a)(s+b)}$	$\frac{1}{(b-a)}(e^{-at} - e^{-bt})$	$\frac{1}{(b-a)}\left[\frac{z}{z-e^{-aT}} - \frac{z}{z-e^{-bT}}\right]$	$\frac{1}{(b-a)}\left[\frac{e^{-amT}}{z-e^{-aT}} - \frac{e^{-bmT}}{z-e^{-bT}}\right]$
$\frac{a}{s^2(s+a)}$	$t - \frac{1}{a}(1 - e^{-at})$	$\frac{Tz}{(z-1)^2} - \frac{(1-e^{-aT})z}{a(z-1)(z-e^{-aT})}$	$\frac{T}{(z-1)^2} + \frac{amT-1}{a(z-1)} + \frac{e^{-amT}}{a(z-e^{-aT})}$
$\frac{1}{(s+a)^2}$	te^{-at}	$\frac{Tze^{-aT}}{(z-e^{-aT})^2}$	$\frac{Te^{-amT}[e^{-aT} + m(z-e^{-aT})]}{(z-e^{-aT})^2}$
$\frac{a}{s^3(s+a)}$	$\frac{1}{2}\left(t^2 - \frac{2}{a}t + \frac{2}{a^2} - \frac{2}{a^2}e^{-at}\right)$	$\frac{T^2z}{(z-1)^3} + \frac{(aT-2)Tz}{2a(z-1)^2} + \frac{z}{a^2(z-1)} - \frac{z}{a^2(z-e^{-aT})}$	$\frac{T^2}{(z-1)^3} + \frac{T^2(m+\frac{1}{2})a - T}{a(z-1)^2} + \frac{(amT)^2/2 - amT + 1}{a^2(z-1)} - \frac{e^{-amT}}{a^2(z-e^{-aT})}$
$\frac{a^2}{s(s+a)^2}$	$u_s(t) - (1+at)e^{-at}$	$\frac{z}{z-1} - \frac{z}{z-e^{-aT}} - \frac{aTe^{-aT}z}{(z-e^{-aT})^2}$	$\frac{1}{z-1} - \left[\frac{1+amT}{z-e^{-aT}} + \frac{aTe^{-aT}}{(z-e^{-aT})^2}\right]e^{-amT}$
$\frac{a^2}{s^2(s+a)^2}$	$t - \frac{2}{a} + \left(t + \frac{2}{a}\right)e^{-at}$	$\frac{1}{a}\left[\frac{(aT+2)z - 2z^2}{(z-1)^2} + \frac{2z}{z-e^{-aT}} + \frac{aTe^{-aT}z}{(z-e^{-aT})^2}\right]$	$\frac{1}{a}\left\{\frac{aT}{(z-1)^2} + \frac{amT-2}{z-1} + \left[\frac{aTe^{-aT}}{(z-e^{-aT})^2} - \frac{amT-2}{z-e^{-aT}}\right]e^{-amT}\right\}$
$\frac{\omega}{s^2+\omega^2}$	$\sin \omega t$	$\frac{z \sin \omega T}{z^2 - 2z\cos \omega T + 1}$	$\frac{\sin m\omega T + \sin (1-m)\omega T}{z^2 - 2z\cos \omega T + 1}$

Laplace Transform $F(s)$	Time Function $f(t), t > 0$	z-Transform $F(z)$	Modified z-Transform $F(z, m)$
$\frac{s}{s^2 + \omega^2}$	$\cos \omega t$	$\frac{z(z - \cos \omega T)}{z^2 - 2z \cos \omega T + 1}$	$\frac{\cos m\omega T - \cos (1 - m)\omega T}{z^2 - 2z \cos \omega T + 1}$
$\frac{\omega}{s^2 - \omega^2}$	$\sinh \omega t$	$\frac{z \sinh \omega T}{z^2 - 2z \cosh \omega T + 1}$	$\frac{\sinh m\omega T + \sinh (1 - m)\omega T}{z^2 - 2z \cosh \omega T + 1}$
$\frac{s}{s^2 - \omega^2}$	$\cosh \omega t$	$\frac{z(z - \cosh \omega T)}{z^2 - 2z \cosh \omega T + 1}$	$\frac{\cosh m\omega Tz - \cosh (1 - m)\omega T}{z^2 - 2z \cosh \omega T + 1}$
$\frac{\omega}{(s + a)^2 + \omega^2}$	$e^{-at} \sin \omega t$	$\frac{ze^{-aT} \sin \omega T}{z^2 - 2ze^{-aT} \cos \omega T + e^{-2aT}}$	$\frac{e^{-amT}[z \sin m\omega T + e^{-aT} \sin (1 - m)\omega T]}{z^2 - 2ze^{-aT} \cos \omega T + e^{-2aT}}$
$\frac{a^2 + \omega^2}{s[(s + a)^2 + \omega^2]}$	$1 - e^{-at} \sec \phi \cos (\omega t + \phi)$ $\phi = \tan^{-1} (-a/\omega)$	$\frac{z}{z - 1} - \frac{z^2 - ze^{-aT} \sec \phi \cos (\omega T - \phi)}{z^2 - 2ze^{-aT} \cos \omega T + e^{-2aT}}$	$\frac{1}{z - 1} - \frac{e^{-maT} \sec \phi(Az - B)}{z^2 - 2ze^{-aT} \cos \omega T + e^{-2aT}}$ $A = \cos (m\omega T + \phi)$ $B = e^{-aT} \cos [(1 - m)\omega T - \phi]$
$\frac{s + a}{(s + a)^2 + \omega^2}$	$e^{-at} \cos \omega t$	$\frac{z^2 - ze^{-aT} \cos \omega T}{z^2 - 2ze^{-aT} \cos \omega T + e^{-2aT}}$	$\frac{e^{-maT}[z \cos m\omega T + e^{-aT} \sin (1 - m)\omega T]}{z^2 - 2ze^{-aT} \cos \omega T + e^{-2aT}}$

General Gain Formula for Signal Flow Graphs

The gain G between an input variable y_{in} and an output variable y_{out} is

$$G = \frac{y_{\text{out}}}{y_{\text{in}}} = \sum_{k=1}^{N} \frac{G_k \Delta_k}{\Delta} \tag{D-1}$$

where

- y_{in} input variable
- y_{out} output variable
- G gain between y_{in} and y_{out}
- N total number of forward paths between y_{in} and y_{out}
- G_k gain of the kth forward path between y_{in} and y_{out}
- Δ $1 - \sum P_{m1} + \sum P_{m2} - \sum P_{m3} + \cdots$ (D-2)
- P_{mi} product of gains of the mth possible combination of i nontouching loops $(1 \le i \le N)$. Two parts of a signal flow graph are nontouching if they do not share a common node.
- Δ_k the Δ for that part of the signal flow graph that is not touching the kth forward path.

An *input node* is a node that has only outgoing branches. An *output node* is a node that has only incoming branches. A *forward path* is any collection of branches with the same signal direction that starts from an input node and ends at an output node. A *loop* is a collection of branches that starts and ends at the same node, and along which no other node is encountered more than once.

Routh's Tabulation for Stability Analysis

Given the nth-order characteristic equation of a linear time-invariant system:

$$a_0 s^n + a_1 s^{n-1} + a_2 s^{n-2} + \cdots + a_{n-1} s + a_n = 0 \tag{E-1}$$

The necessary conditions for all the roots of the equation to be in the left half s-plane are as follows.

1. All the coefficients have the same sign.
2. None of the coefficients is zero.

The Routh's tabulation using the coefficients of the characteristic equation is given below for $n = 4$.

$$\begin{array}{llll}
s^4 & a_0 & a_2 & a_4 \\
s^3 & a_1 & a_3 & 0 \\
s^2 & \dfrac{a_1 a_2 - a_0 a_3}{a_1} = A & \dfrac{a_1 a_4 - a_0 \times 0}{a_1} = B = a_4 & 0 \\
s^1 & \dfrac{A a_3 - a_1 B}{A} = C & 0 & \\
s^0 & \dfrac{BC - A \times 0}{C} = B = a_4 & &
\end{array}$$

The necessary and sufficient condition for the fourth-order equation to have no roots in the right half plane and on the $j\omega$ axis is that all the coefficients in the first column of the Routh's tabulation are of the same sign.

Galil DMC-100 Motion Controller Board

(Courtesy of GALIL Motion Control, Inc.)

Features

- STD Bus compatible; RS232 Port
- For servo motors with incremental encoder feedback
- Controls motion of one axis
- Position and velocity control
- Programmable velocity profiling
- Change position, velocity, acceleration "on-the-fly"
- Incremental position mode for continuous path
- Position "learn mode"
- 2 KHz sample and update
- 250,000 counts/sec maximum speed
- Programmable digital filter with gain, damping and integration—eliminates Tach
- ± Overtravel limits, home inputs, emergency stop
- Programmable torque and error limits
- PWM or Analog output
- Optional on-board 30 watt amplifier

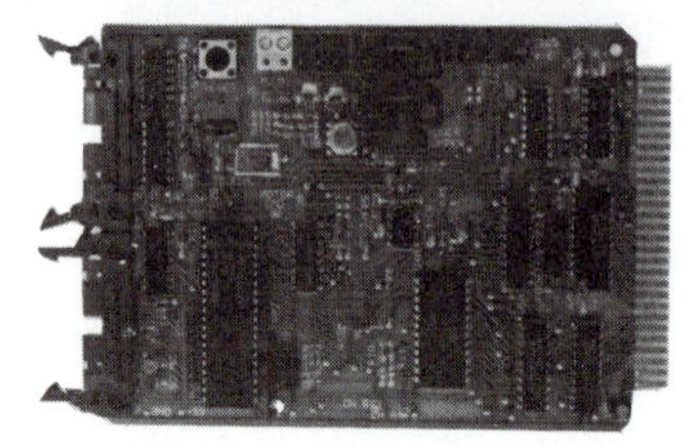

General Description

The DMC-100-10 Series are general purpose motion control cards for servo motors with encoder feedback. The DMC-100-10 controls one motor of any size and requires an external power driver. The DMC-105-10 contains an on-board 30 watt linear amplifier suitable for driving small motors. Each controller plugs into the STD Bus.

The controller contains a microprocessor dedicated to the time-intensive motion control tasks. The controller functions include quadrature decoding of the encoder, generating the velocity profile and position trajectory, digital filtering of the control signal and generation of a ± 10 volt analog motor command signal. In addition, the controller provides overtravel, homing, emergency stop, and error handling functions.

The DMC-100-10 is programmable, accepting ASCII commands from the host computer via RS232 or the STD Bus. The controller responds to over 40 instructions for specifying system parameters and motion profiles. Controller status and motor position can also be interrogated at any time.

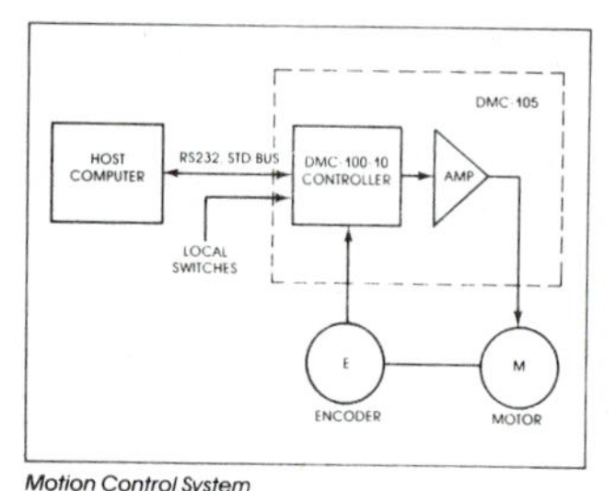

Motion Control System

Index